Lecture Notes in Mathematics

Edited by A. Dold, F. Takens and B. Teissier

Editorial Policy
for the publication of monographs

1. Lecture Notes aim to report new developments in all areas of mathematics – quickly, informally and at a high level. Monograph manuscripts should be reasonably-self-contained and rounded off. Thus they may, and often will, presentnot only results of the author but also related work by other people. They may bebased on specialized lecture courses. Furthermore, the manuscripts should provide sufficient motivation, examples and applications. This clearly distinguishes Lecture Notes from journal articles or technical reports which normally are very concise. Articles intended for a journal but too long to be accepted by most journals, usually do not have this "lecture notes" character. For similar reasons it is unusual for doctoral theses to be accepted for the Lecture Notes series.

2. Manuscripts should be submitted (preferably in duplicate) either to one of the series editors or to Springer-Verlag, Heidelberg. In general, manuscripts will be sent out to 2 external referees for evaluation. If a decision cannot yet be reached on the basis of the first 2 reports, further referees may be contacted: The author will be informed of this. A final decision to publish can be made only on the basis of the complete manuscript, however a refereeing process leading to a preliminary decision can be based on a pre-final or incomplete manuscript. The strict minimum amount of material that will be considered should include a detailed outline describing the planned contents of each chapter, a bibliography and several sample chapters.
Authors should be aware that incomplete or insufficiently close to final manuscripts almost always result in longer refereeing times and nevertheless unclear referees' recommendations, making further refereeing of a final draft necessary.
Authors should also be aware that parallel submission of their manuscript to another publisher while under consideration for LNM will in general lead to immediate rejection.

3. Manuscripts should in general be submitted in English.
Final manuscripts should contain at least 100 pages of mathematical text and should include
– a table of contents;
– an informative introduction, with adequate motivation and perhaps some
   historical remarks: it should be accessible to a reader not intimately familiar
   with the topic treated;
– a subject index: as a rule this is genuinely helpful for the reader.

Lecture Notes in Mathematics　　　　1704

Editors:
A. Dold, Heidelberg
F. Takens, Groningen
B. Teissier, Paris

Springer
Berlin
Heidelberg
New York
Barcelona
Hong Kong
London
Milan
Paris
Singapore
Tokyo

Hirotaka Tamanoi

Elliptic Genera
and Vertex Operator
Super-Algebras

Springer

Author

Hirotaka Tamanoi
Department of Mathematics
University of California
Santa Cruz, CA 95064, USA
E-mail: tamanoi@math.ucsc.edu

Cataloging-in-Publication Data applied for

Die Deutsche Bibliothek - CIP-Einheitsaufnahme

Tamanoi, Hirotaka:
Elliptic genera and vertex operator super algebras / Hirotaka
Tamanoi. - Berlin ; Heidelberg ; New York ; Barcelona ; Budapest ;
Hong Kong ; London ; Milan ; Paris ; Santa Clara ; Singapore ;
Tokyo : Springer, 1999
(Lecture notes in mathematics ; 1704)
ISBN 3-540-66006-2

Mathematics Subject Classification (1991): Primary: 17B65, 55N22, 81R10
Secondary: 17A70, 17B68, 53C55, 81T40

ISSN 0075-8434
ISBN 3-540-66006-2 Springer-Verlag Berlin Heidelberg New York

© Springer-Verlag Berlin Heidelberg 1999
Printed in Germany

The use of general descriptive names, registered names, trademarks, etc. in this
publication does not imply, even in the absence of a specific statement, that such
names are exempt from the relevant protective laws and regulations and therefore
free for general use.

Typesetting: Camera-ready TEX output by the author
SPIN: 10650205   41/3143-543210 - Printed on acid-free paper

CONTENTS

INTRODUCTION AND SUMMARY OF RESULTS

§0.1 Towards the signature of loop spaces: Elliptic genus

We briefly review the signature of closed oriented manifolds [BGV, H1, Wa] Let $M^{4N'}$ be a closed oriented smooth manifold of dimension $4N'$. On the middle dimensional real cohomology $H^{2N'}(M;\mathbb{R})$, we can consider a symmetric bilinear pairing Q defined by

$$Q([\alpha],[\beta]) = \int_M \alpha \wedge \beta = ([\alpha] \cup [\beta])[M], \quad \text{for } [\alpha],[\beta] \in H^{2N'}(M;\mathbb{R}),$$

where α, β are closed differential $2N'$-forms on M, and their cohomology classes are denoted by $[\alpha],[\beta]$. Recall that the Hodge $*$-operator is an involution on $H^{2N'}$, and let $H_+^{2N'}$, $H_-^{2N'}$ be the ± 1 eigenspaces of the $*$-operator. Since the usual inner product (α, β) on $H^*(M;\mathbb{R})$ is defined to be the integral of $\alpha \wedge *\beta$ over M, the signature $\text{Sig}(M) \in \mathbb{Z}$ of the quadratic form Q on $H^{2N'}(M;\mathbb{R})$ is given by

$$\text{Sig}(M) = \dim_{\mathbb{R}}(H_+^{2N'}) - \dim_{\mathbb{R}}(H_-^{2N'}).$$

Let M^{2N} be a closed oriented Riemannian manifold. [Throughout this book, an even dimensional manifold has dimension $2N$. If dimension of a manifold is a multiple of 4, then its dimension is denoted by $4N'$]. Let $\Delta = \Delta^+ \oplus \Delta^-$ be the decomposition of the total Spin representation into the half Spin representations of $\text{Spin}(2N)$ [BD]. On $\Delta^\pm$, the element $-1 \in \text{Spin}(2N)$ acts as ± 1. Although the associated Spin bundles $\underline{\Delta}^\pm$, $\underline{\Delta}$ are defined only for Spin manifolds, their tensor products $\underline{\Delta}^\pm \otimes \underline{\Delta}$ can be defined for oriented manifolds since $-1 \in \text{Spin}(2N)$ acts as 1 on the tensor product. (A good reference of Spin manifolds is [LM].) We can consider the following twisted Dirac operator, the signature operator, on the oriented manifold M:

$$d_\Delta : \Gamma(\underline{\Delta}^+ \otimes \underline{\Delta}) \to \Gamma(\underline{\Delta}^- \otimes \underline{\Delta}).$$

Since the above d_Δ is defined on a closed manifold, the kernel and the cokernel of the operator d_Δ are finite dimensional because d_Δ is elliptic. It is not difficult to see that the index of the operator d_Δ calculates the signature $\text{Sig}(M)$ of the manifold M when the manifold dimension $2N$ is a multiple of 4. By index theory, this integer is also known as the L-genus $L(M)$ of $M^{4N'}$ which can be expressed in terms of Pontrjagin numbers. The L-genus is a cobordism invariant and is multiplicative on closed oriented manifolds, defining a ring map, or genus $L : \Omega_*^{SO} \to \mathbb{Z}$. Since twisted Dirac operators are formally self-adjoint [§5.3 Proposition 7], the cokernel space coincides with the kernel space of d_Δ applied to $\Gamma(\underline{\Delta}^- \otimes \underline{\Delta})$. Furthermore, d_Δ is nothing but $d + d^*$ restricted to $\underline{\Delta}^\pm \otimes \underline{\Delta}$ through the isomorphism $\underline{\Delta} \otimes \underline{\Delta} \cong \bigwedge^* T_{\mathbb{C}}^* M$. Hence, the vector space $\operatorname{Ker} d_\Delta \oplus \operatorname{Coker} d_\Delta$ is the entire space of *harmonic* forms on M. To preserve this very nice geometric meaning of the kernel space and the cokernel space of d_Δ, we define $\mathcal{L}(M)$ as the super-pair of vector spaces of harmonic forms as follows:

$$\mathcal{L}(M) = [\operatorname{Ker} d_\Delta \; ; \; \operatorname{Coker} d_\Delta].$$

Here the vector space $\operatorname{Ker} d_\Delta$ is regarded as the "even" part of $\mathcal{L}(M)$ and $\operatorname{Coker} d_\Delta$ is regarded as the "odd" part of $\mathcal{L}(M)$. Its super-dimension is the L-genus:

$$L(M) = \text{s-dim}_{\mathbb{C}} \, \mathcal{L}(M) = \dim_{\mathbb{C}}(\operatorname{Ker} d_\Delta) - \dim_{\mathbb{C}}(\operatorname{Coker} d_\Delta).$$

Let $LM = \operatorname{Map}(S^1, M)$ be the free loop space of smooth loops in M. On LM, the circle group S^1 acts by rotating loops. The free loop space even admits the action of the infinite dimensional group $\operatorname{Diff}^+ S^1$ of diffeomorphisms of the circle by reparametrizing loops in M. If M is Riemannian, then there is an obvious metric on LM with respect to which the group S^1 acts by isometries. We want to consider an analogue of the signature for infinite dimensional free loop spaces. The structure group of LM^{2N} is the loop group $LSO(2N) = \operatorname{Map}(S^1, SO(2N))$ with two connected components. The free loop space LM is called *orientable* when the structure group can be reduced to the identity component $L_0 SO(2N)$. One can show that the free loop space LM is orientable if the original manifold M has a Spin structure [Mc]. From now on we assume that our manifold M^{2N} is a closed Riemannian Spin manifold. Since S^1 acts on LM, we even want to consider the *character valued signature* of LM. However, since LM is infinite dimensional, there is no "middle dimensional cohomology groups" for LM on which we can consider quadratic forms. So, the approach using quadratic forms doesn't generalize. But the operator formalism does generalize via a formal application of Localization Theorem for G-signatures [W3, W4]. The resulting character valued signature on LM is the *elliptic genus*. It is a super-pair of graded vector spaces denoted by

$$\Phi_{\text{ell}}(M) = [\Phi_{\text{ell}}(M)_\ell \; ; \; \Phi_{\text{ell}}(M)_r].$$

For the notation and the precise definition, see §0.2. The grading will also be called the weight. It turns out that there are at least two essentially different signatures on the loop space LM, due to the fact that there are two different total Spin representations for the loop group $\tilde{L}\operatorname{Spin}(2N)$ with the central extension [FFR, PS]. Each vector of finite weight in the elliptic genus $\Phi_{\text{ell}}(M)$ is a twisted harmonic spinor with respect to a certain twisting bundle on M. The graded super-dimension of the elliptic genus is denoted by $\varphi_{\text{ell}}(M)$:

$$\varphi_{\text{ell}}(M) = \text{s-dim}_* \, \Phi_{\text{ell}}(M) = \dim_* \left(\Phi_{\text{ell}}(M)_\ell \right) - \dim_* \left(\Phi_{\text{ell}}(M)_r \right).$$

It turns out that $\varphi_{ell}(M)$ can be interpreted as a modular function of weight 0 on the upper half plane with respect to an index 6 subgroup Γ of $SL_2(\mathbb{Z})$. By general index theory for elliptic operators, one can see that the numerical elliptic genus $\varphi_{ell}(M)$ is completely determined by the Pontrjagin numbers of M, so $\varphi_{ell}(M)$ is a Spin cobordism invariant. Furthermore, it is multiplicative on closed Spin manifolds, giving rise to a genus $\varphi_{ell} : \Omega_*^{Spin} \to Mod_0(\Gamma)$, where $Mod_0(\Gamma)$ is the ring of modular functions with respect to Γ. The word "elliptic" for the elliptic genus comes from the fact that the logarithm series for this genus can be expressed in terms of the elliptic integral of the Jacobi quartic. It is known that the limiting value of the modular function $\varphi_{ell}(M)$ at one cusp is essentially the L-genus and the limit at the other cusp is the $\hat{A}$-genus up to an inessential factor. Thus, the numerical elliptic genus $\varphi_{ell}(M)$ gives rise to an analytic function connecting the two important integer valued cobordism invariants for Spin manifolds.

Note that the elliptic genus $\varphi_{ell}(M)$ we will be dealing with is different from the "usual" one which has values in the ring of modular forms of non-negative weight with respect to index 3 subgroups of $SL_2(\mathbb{Z})$ [L2]. The difference is merely a multiplicative factor which depends only on the dimension of the manifold. However, our version is more natural geometrically. For a discussion on this issue, see Chapter 1.

The elliptic genus $\Phi_{ell}(M)$ as a super-pair of graded vector spaces is not just a pair of graded vector spaces. We will show that various infinite dimensional (Lie) algebras act on both even and odd parts of the elliptic genus. Depending on certain differential geometric properties of M, the elliptic genus of M^{2N} admits actions of an infinite dimensional Clifford algebra, an affine Lie algebra, a Virasoro algebra of rank N, or its variants, etc. The unifying notion of these algebras is the notion of *vertex operator super algebras* [FFR, FHL, FLM]. For the definition and their properties, see §0.3 below and Chapter 2. Our main result says that to each closed Riemannian Spin manifold, we can associate a vertex operator super algebra denoted by $\mathcal{P}_M(\underline{V})$ which acts on the elliptic genus $\Phi_{ell}(M)$. $\mathcal{P}_M(\underline{V})$ has non-negative half-integral grading. Its structure is determined by the holonomy group of the Riemannian manifold M. Thus, the elliptic genus is a geometric device which produces a super-pair of modules $\Phi_{ell}(M) = [\Phi_{ell}(M)_\ell \; ; \; \Phi_{ell}(M)_r]$ over a vertex operator super algebra $\mathcal{P}_M(\underline{V})$ for each closed Riemannian Spin manifold M^{2N}:

$$\left\{ \begin{array}{c} \text{closed Riemannian} \\ \text{Spin manifolds} \end{array} \right\} \xrightarrow{\text{elliptic genus}} \left\{ \begin{array}{c} \text{pairs of vertex operator super} \\ \text{algebras and their modules} \end{array} \right\}$$

$$M \longrightarrow (\mathcal{P}_M(\underline{V}), \Phi_{ell}(M))$$

In this correspondence, $\mathcal{P}_M(\underline{V})$ and $\Phi_{ell}(M)$ are determined by the geometry of M, whereas the modular function $\varphi_{ell}(M)$ is determined by the topology of M, more precisely by Pontrjagin numbers of M.

The purpose of this paper is two-fold:

(i) to describe properties of the modular function valued elliptic genus $\varphi_{ell}(M)$;

(ii) to investigate infinite dimensional symmetries of the elliptic genus $\Phi_{\text{ell}}(M)$ for various closed Riemannian Spin manifolds.

For (i), we will calculate the logarithm series for the two versions of elliptic genera and study their modular properties. We will also make some comments on so-called elliptic cohomology.

$\mathcal{P}_M(\underline{V})$ always contains a canonical weight 2 vector corresponding to the Riemannian metric tensor on M^{2N}. The corresponding vertex operator generates a Virasoro algebra of rank N. A Virasoro algebra is, roughly speaking, the Lie algebra of the infinite dimensional Lie group $\widetilde{\text{Diff}}^+S^1$ with a central extension [PS]. So, an elliptic genus is always a representation of the Virasoro algebra. Thus the elliptic genus is not only an S^1-character valued signature, but it is a $\widetilde{\text{Diff}}^+S^1$-character valued genus. We remark that the reparametrization group Diff^+S^1 does not act on the loop space LM isometrically with respect to the obvious metric induced from M. For isometric action, we need to introduce the notion of half densities on the circle.

For other symmetries in the elliptic genus $\Phi_{\text{ell}}(M)$, an important role is played by differential forms on M which are parallel with respect to the Levi-Civita connection. It turns out that through vertex operators parallel differential 1-forms on M generate an infinite dimensional Clifford algebra acting on the even and odd parts of the elliptic genus $\Phi_{\text{ell}}(M)$ separately, and parallel differential 2-forms on M generate an affine Lie algebra over which both parts of $\Phi_{\text{ell}}(M)$ are representations. Many interesting geometric structures on manifolds have associated parallel 2-forms [§0.5]. For example, Kähler forms on Kähler manifolds are parallel differential 2-forms and they generate Heisenberg algebras. Furthermore, certain parallel differential 4-forms, the quaternion-Kähler forms [Bes], [§5.1 Proposition 18], which are closely related to Casimir operators for Lie algebras of the holonomy groups play an interesting role. They give rise to splittings of the above rank N Virasoro algebra into various mutually commuting pairs of Virasoro algebras [§0.4 Theorem 5].

For the rest of this introduction, we describe and summarize our results.

§0.2 Elliptic genera

The Fixed Point Index Formula for the S^1-equivariant signature applied formally to loop spaces gives rise to elliptic genera [W1, W2]. We describe this resulting object. Let $T_{\mathbb{C}} = T^*M \otimes \mathbb{C}$ be the complexification of the *cotangent* bundle of a closed Riemannian Spin manifold. For a non-negative integer k, let $\bigwedge^k(T_{\mathbb{C}})$, $S^k(T_{\mathbb{C}})$ be the k-th exterior power and the k-th symmetric power bundle of $T_{\mathbb{C}}$. Let $\Lambda_t(T_{\mathbb{C}}) = \sum_{k \geq 0} \bigwedge^k(T_{\mathbb{C}})t^k$ and $S_t(T_{\mathbb{C}}) = \sum_{k \geq 0} S^k(T_{\mathbb{C}})t^k$ be formal series in t with bundle coefficients. We define bundles Q_ℓ, R_ℓ for $\ell \in \frac{1}{2}\mathbb{Z}_+$ by

$$Q(T_{\mathbb{C}}) = \sum_{0 \leq \ell \in \frac{1}{2}\mathbb{Z}} Q_\ell q^\ell = \bigotimes_{0 \leq m \in \mathbb{Z}} S_{q^m}(T_{\mathbb{C}}) \otimes \bigotimes_{1 \leq m \in \mathbb{Z}} \Lambda_{q^{m-\frac{1}{2}}}(T_{\mathbb{C}}),$$

$$R(T_{\mathbb{C}}) = \sum_{0 \leq \ell \in \mathbb{Z}} R_\ell q^\ell = \bigotimes_{1 \leq m \in \mathbb{Z}} S_{q^m}(T_{\mathbb{C}}) \otimes \underline{\Delta} \otimes \bigotimes_{1 \leq m \in \mathbb{Z}} \Lambda_{q^m}(T_{\mathbb{C}})$$

where $\Delta = \Delta^+ \oplus \Delta^-$ is the complex 2^N-dimensional total Spin representation which is a sum of half Spin representations of complex dimension 2^{N-1} for the group $\mathrm{Spin}(2N)$. The bundles $\underline{\Delta}^{\pm} \otimes Q_\ell$ make sense only for Spin manifolds, whereas $\underline{\Delta}^{\pm} \otimes R_\ell$ is well defined on oriented manifolds. Here, q is a formal variable which originates in the S^1-character in the formal S^1-equivariant signature on the loop space LM [W3]. The use of cotangent bundles is not only convenient to define twisted Dirac operators but it also allows us to identify the total exterior bundle $\bigwedge^* T_{\mathbb{C}}^* M$ as a subbundle of $\underline{V}$ (defined in §0.6). We consider Dirac operators twisted by the above bundles R_ℓ and Q_ℓ for $\ell \in \frac{1}{2}\mathbb{Z}$, $\ell > 0$:

$$d_{Q_\ell} : \Gamma(\Delta^+ \otimes Q_\ell) \longrightarrow \Gamma(\Delta^- \otimes Q_\ell),$$

$$d_{R_\ell} : \Gamma(\Delta^+ \otimes R_\ell) \longrightarrow \Gamma(\Delta^- \otimes R_\ell).$$

Since the twisted Dirac operator d_* for $* = Q_\ell, R_\ell$ with $\ell \in \frac{1}{2}\mathbb{Z}$ is elliptic, the kernel space $(\mathrm{Ker}\, d_*)$ and the cokernel space $(\mathrm{Coker}\, d_*)$ are finite dimensional. Since twisted Dirac operators are formally self-adjoint (§5.3), we have $d_{Q_\ell}^* = d_{Q_\ell}$ and $d_{R_\ell}^* = d_{R_\ell}$ for $\ell \in \frac{1}{2}\mathbb{Z}$. Thus, the cokernel space $\mathrm{Coker}\, d_{Q_\ell}$ can be identified with a kernel space of the adjoint operator $d_{Q_\ell}^* = d_{Q_\ell}$:

$$\mathrm{Coker}\, d_{Q_\ell} = \mathrm{Ker}\{d_{Q_\ell} : \Gamma(\Delta^- \otimes Q_\ell) \longrightarrow \Gamma(\Delta^+ \otimes Q_\ell)\},$$

and $(\mathrm{Ker}\, d_{Q_\ell}) \oplus (\mathrm{Coker}\, d_{Q_\ell})$ is the entire space of twisted harmonic spinors in $\Gamma(\underline{\Delta} \otimes Q_\ell)$. Similarly for d_{R_ℓ}.

DEFINITION 1 [§1.1 DEFINITION 1]. For a closed Riemannian Spin manifold M^{2N}, the *elliptic genus* $\Phi_{\mathrm{ell}}^*(M)$ for $* = Q, R$ is a super-pair of graded vector spaces defined by

$$\Phi_{\mathrm{ell}}^Q(M) = \Big[\bigoplus_{0 \le \ell \in \frac{1}{2}\mathbb{Z}} (\mathrm{Ker}\, d_{Q_\ell})q^{\ell - \frac{N}{8}} ; \bigoplus_{0 \le \ell \in \frac{1}{2}\mathbb{Z}} (\mathrm{Coker}\, d_{Q_\ell})q^{\ell - \frac{N}{8}} \Big],$$

$$\Phi_{\mathrm{ell}}^R(M) = \Big[\bigoplus_{0 \le \ell \in \mathbb{Z}} (\mathrm{Ker}\, d_{R_\ell})q^{\ell} ; \bigoplus_{0 \le \ell \in \mathbb{Z}} (\mathrm{Coker}\, d_{R_\ell})q^{\ell} \Big].$$

Note that $\Phi_{\mathrm{ell}}^R(M)$ is well defined for closed oriented manifolds. Being a super-pair means that when we calculate the graded super-dimension of $\Phi_{\mathrm{ell}}^Q(M)$, we take the difference of the graded dimensions of "even" part and "odd" part:

$$\varphi_{\mathrm{ell}}^Q(M) \equiv \text{s-dim}_* \Phi_{\mathrm{ell}}^Q(M) = \sum_{0 \le \ell \in \frac{1}{2}\mathbb{Z}} \mathrm{Index}\,(d_{Q_\ell})q^{\ell - \frac{N}{8}}.$$

It has its value in the power series ring $\mathbb{Z}[q^{-\frac{1}{4}}][[q^{\frac{1}{2}}]]$, and every coefficient is a linear combination of finitely many Pontrjagin numbers of M in view of index theory. So, it is a Spin cobordism invariant. Similarly, the graded super-dimension φ_{ell}^R for $\Phi_{\mathrm{ell}}^R(M)$ has its value in $\mathbb{Z}[[q]]$, and it is an oriented cobordism invariant.

Since the power series $Q(\cdot)$ and $R(\cdot)$ in q with bundle coefficients behaves exponentially (by this we mean $Q(E \oplus F) = Q(E) \otimes Q(F)$ for any complex vector bundles E, F), these genera $\varphi_{\text{ell}}^Q(M)$, $\varphi_{\text{ell}}^R(M)$ are multiplicative on closed Spin manifolds or on oriented manifolds, respectively. Hence they give rise to genera

$$\varphi_{\text{ell}}^Q : \Omega_*^{\text{Spin}} \to \mathbb{Z}[q^{-\frac{1}{4}}][[q^{\frac{1}{2}}]], \qquad \varphi_{\text{ell}}^R : \Omega_*^{\text{SO}} \to \mathbb{Z}[[q]].$$

Of course, our graded vector space valued elliptic genera $\Phi_{\text{ell}}^*(M)$ are not cobordism invariants: they depend on Riemannian metrics.

For a multiplicative genus, the logarithm series is of particular interest. Let Ω_*^U be the complex cobordism ring. Recall that for a multiplicative genus $\varphi : \Omega_*^U \to \Lambda$ with values in a commutative $\mathbb{Q}$-algebra Λ, its logarithm is a power series $\log_\varphi(X) \in \Lambda[[X]]$ given by

$$\log_\varphi(X) = \sum_{n \geq 0} \frac{\varphi([\mathbb{C}P^n])}{n+1} X^{n+1}.$$

Recall that the cobordism classes of complex projective spaces $\mathbb{C}P^n$ are rational generators of Ω_*^U.

Another feature of elliptic genera is their modular invariance. When we let $q = e^{2\pi i \tau}$ for $\tau \in \mathbb{C}$ in the upper half plane, values of elliptic genera as q-series are modular invariant for suitable subgroups of $\text{SL}_2(\mathbb{Z})$. Let

$$\Gamma_0(2) = \left\{ \begin{pmatrix} a & b \\ c & d \end{pmatrix} \in \text{SL}_2(\mathbb{Z}) \,\middle|\, c \equiv 0 \mod 2 \right\},$$

$$\Gamma_\theta = \left\{ \begin{pmatrix} a & b \\ c & d \end{pmatrix} \in \text{SL}_2(\mathbb{Z}) \,\middle|\, \begin{pmatrix} a & b \\ c & d \end{pmatrix} \equiv \begin{pmatrix} 1 & 0 \\ 0 & 1 \end{pmatrix}, \begin{pmatrix} 0 & 1 \\ 1 & 0 \end{pmatrix} \mod 2 \right\}.$$

These subgroups have index 3 in $\text{SL}_2(\mathbb{Z})$.

We recall that a holomorphic function $f(\tau)$ on the upper half plane $\text{Im}\, \tau > 0$ is said to be *modular invariant* with respect to a subgroup $\Gamma \subset \text{SL}_2(\mathbb{Z})$ if

$$f\left(\frac{a\tau + b}{c\tau + d}\right) = f(\tau) \qquad \text{for all } \begin{pmatrix} a & b \\ c & d \end{pmatrix} \in \Gamma.$$

THEOREM 2 [§1.1 THEOREM 4, §1.4 THEOREM 2]. (i) *For the elliptic genus* $\varphi_{\text{ell}}^Q : \Omega_*^{\text{Spin}} \to \mathbb{Z}[q^{-\frac{1}{4}}][[q^{\frac{1}{2}}]]$, *its logarithm series* $\log_{\text{ell}}^Q(X)$ *is given by an elliptic integral of the form*

$$\log_{\text{ell}}^Q(X) = \int_0^X \frac{dt}{\sqrt{1 - 2\delta^Q t^2 - t^4}}, \qquad \text{where}$$

$$\delta^Q(q) = \frac{1}{8} q^{-\frac{1}{4}} \left\{ \prod_{1 \leq \ell \in \mathbb{Z}} \left(\frac{1 - q^{\ell - \frac{1}{2}}}{1 + q^\ell}\right)^4 - 16 q^{\frac{1}{2}} \prod_{1 \leq \ell \in \mathbb{Z}} \left(\frac{1 + q^\ell}{1 - q^{\ell - \frac{1}{2}}}\right)^4 \right\}.$$

Here, $\delta^Q(q)$ is modular invariant for an index 2 subgroup of Γ_θ.

(ii) For the elliptic genus $\varphi_{\text{ell}}^R : \Omega_*^{SO} \to \mathbb{Z}[[q]]$, its logarithm $\log_{\text{ell}}^R(X)$ is given by an elliptic integral of the form

$$\log_{\text{ell}}^R(X) = \int_0^X \frac{dt}{\sqrt{1 - 2\delta^R t^2 + t^4}}, \qquad \text{where}$$

$$\delta^R(q) = \frac{1}{2} \left\{ \prod_{1 \leq \ell \in \mathbb{Z}} \left(\frac{1 + q^{\ell - \frac{1}{2}}}{1 - q^{\ell - \frac{1}{2}}} \right)^4 + \prod_{1 \leq \ell \in \mathbb{Z}} \left(\frac{1 - q^{\ell - \frac{1}{2}}}{1 + q^{\ell - \frac{1}{2}}} \right)^4 \right\}.$$

$\delta^R(q)$ is modular invariant for an index 2 subgroup of $\Gamma_0(2)$.

For the precise definition of these index 2 subgroups of $\Gamma_0(2)$ and of Γ_θ appearing in the above statements, see Proposition 9 in §1.4. It is interesting to note that all of the coefficients of $\delta^R = \varphi_{\text{ell}}^R([\mathbb{C}P^2])$ are nonzero and in fact they are positive integers. In Theorem 2, the expression of $\delta^R(q)$ is proved by deducing a differential equation satisfied by the elliptic function which is inverse to the above elliptic integral [§1.2 Proposition 2]. Here, an expression of elliptic functions as quotients of certain theta functions is used. The modular invariance of δ^R follows from the invariance of the solution (which is an elliptic function) under the change of basis for the lattices of zeros and poles of the solution function, together with the modular properties of certain theta constants [§1.4 Lemma 5, Lemma 7].

Since the coefficient of t^4 in the above elliptic integral is 1, the Landweber Exact Functor Theorem [L1] applies and we get a cohomology theory whose coefficient ring is $\mathbb{Z}[\frac{1}{2}][\delta^R(q)]$. There is no obvious grading in this ring which corresponds to the dimension of manifolds, for δ^R is a weight 0 modular function. But in view of [HS] in which elliptic genera of manifolds of different dimension are compared, the manifold dimension may not have significant meaning in the setting of elliptic cohomology.

In §1.5, we discuss some divisibility properties of coefficients of the elliptic genera $\varphi_{\text{ell}}^*(M)$ for Spin manifolds and oriented manifolds.

In §1.6, we define the Witten genus $\Phi_\theta(M)$ and show that its super-dimension $\varphi_\theta(M)$ is a "theta" genus rather than an elliptic genus.

§0.3 Vertex operator super algebras

Let $(E^{2N}; \langle \, , \, \rangle)$ be a Euclidean vector space of even dimension. Let $A = A^+ \oplus A^- = E \otimes \mathbb{C}$ be a decomposition of its complexification into maximal isotropic subspaces. For $m \in \frac{1}{2}\mathbb{Z}$, let $A(-m)$ be a copy of A carrying the weight m. The $\mathbb{C}$-linear pairing on A can be extended to $A_{\mathbb{Z}+\frac{1}{2}} = \bigoplus_{m \in \mathbb{Z}} A(-m - \frac{1}{2})$ by letting $\langle a(m), b(n) \rangle = \langle a, b \rangle \delta_{m+n,0}$ for $a, b \in A$. So, the dual vector space of $A(m)$ is $A(-m)$. The Clifford algebra $\text{Cliff}(\mathbb{Z} + \frac{1}{2})$ associated to $(A_{\mathbb{Z}+\frac{1}{2}}, \langle \, , \, \rangle)$ acts on the Clifford module given by

$$V = V_q = \bigoplus_{0 \leq \ell \in \frac{1}{2}\mathbb{Z}} (V)_\ell q^\ell = \bigotimes_{0 \leq m \in \mathbb{Z}} \bigwedge_{q^{m+\frac{1}{2}}}^* A(-m - \frac{1}{2}).$$

Note that the group $\mathrm{Spin}(2N)$ acts on V_q preserving the weight via the action of $\mathrm{SO}(2N)$. Similarly, we can introduce a nondegenerate $\mathbb{C}$-linear pairing $\langle\,,\,\rangle$ on the graded vector space $A_{\mathbb{Z}} = \bigoplus_{m\in\mathbb{Z}} A(m)$, and we can construct the Clifford algebra $\mathrm{Cliff}(\mathbb{Z})$. The associated Clifford module is given by

$$W_q = (\textstyle\bigwedge^* A^-) \otimes \bigotimes_{0\leq m\in\mathbb{Z}} \textstyle\bigwedge^*_{q^m} A(-m).$$

Let even and odd parity parts of V_q and W_q be V_0, V_1 and W_0, W_1, respectively. It is well known that these graded vector spaces are irreducible level 1 Spin representations for the orthogonal affine Lie algebra $\hat{o}(2N)$ [Fr].

It is known that V_q has a structure of a vertex operator super algebra and has W_q as its ($\mathbb{Z}_2$-twisted) module [FFR].

The main structure of a general vertex operator super algebra $(V, Y(\,,\zeta), 1, \omega)$ is an injective vertex operator map

$$Y(\,,\zeta) : V \to \mathrm{End}(V)[[\zeta, \zeta^{-1}]], \quad \text{where } Y(v,\zeta) = \sum_{n\in\mathbb{Z}} \{v\}_n \zeta^{-n-1}.$$

The operators $\{v\}_n$ lower the weight by $n + 1 - \mathrm{wt}(v)$ for any $n \in \mathbb{Z}$ and satisfy the following *Jacobi identity*:

$$\sum_{0\leq i\in\mathbb{Z}} (-1)^i \binom{r}{i}\left(\{v_1\}_{m+r-i}\{v_2\}_{n+i} - (-1)^{|v_1||v_2|+r}\{v_2\}_{n+r-i}\{v_1\}_{m+i} \right)$$
$$= \sum_{0\leq k\in\mathbb{Z}} \binom{m}{k} \{\{v_1\}_{r+k} v_2\}_{m+n-k}$$

for any $v_1, v_2 \in V$ and $m, n, r \in \mathbb{Z}$. We let $|v| = 2\,\mathrm{wt}(v) \mod 2$. As a special but very useful case, we obtain the super commutator formula by letting $r = 0$ in the above Jacobi identity:

$$\{v_1\}_m\{v_2\}_n - (-1)^{|v_1||v_2|}\{v_2\}_n\{v_1\}_m = \sum_{0\leq k\in\mathbb{Z}} \binom{m}{k} \{\{v_1\}_k v_2\}_{m+n-k},$$

for $v_1, v_2 \in V$ and $m, n \in \mathbb{Z}$. There exists a distinguished element $\omega \in (V)_2$ of weight 2, the canonical Virasoro element, whose associated vertex operator $Y(\omega,\zeta) = \sum_{n\in\mathbb{Z}} D(n)\zeta^{-n-2}$ generates a Virasoro algebra $\underline{Vir} = \bigoplus_{n\in\mathbb{Z}} \mathbb{C}D(n) \oplus \mathbb{C}\mathrm{Id}_V$ satisfying the following commutation relation:

$$[D(m), D(n)] = (n-m)D(m+n) + \frac{m(m^2-1)}{12}\delta_{m,-n} c \cdot \mathrm{Id}_V,$$

for $m, n \in \mathbb{Z}$. Here $c \in \mathbb{C}$ is a constant called the rank of V. It is also called the central charge. For the "vacuum" vector $1 \in (V)_0$, we require that the associated vertex operator is the identity map: $Y(1,\zeta) = \mathrm{Id}_V$. We require a few more conditions for a vertex operator super algebra. For details, see §2.1. The notion of a module over a vertex operator super algebra can be defined in an obvious way.

Vertex operator (super) algebras often contain affine Lie algebras and Heisenberg algebras. We briefly recall definitions of these algebras. Let $\mathfrak{g}$ be a Lie algebra with a $\mathfrak{g}$-invariant bilinear pairing $\langle \, , \, \rangle$ on $\mathfrak{g}$. Let t be an indeterminate and let $\mathbb{C}[t, t^{-1}]$ be the ring of Laurent polynomials. Let $\hat{\mathfrak{g}} = \mathfrak{g} \otimes \mathbb{C}[t, t^{-1}] \oplus \mathbb{C}c$. We define a bracket product in this vector space by

$$[x \otimes t^m, y \otimes t^n] = [x, y] \otimes t^{m+n} + m\langle x, y \rangle \delta_{m+n,0} c, \qquad [\hat{\mathfrak{g}}, c] = 0.$$

The $\mathfrak{g}$-invariance and the bilinearity of $\langle \, , \, \rangle$ make $\hat{\mathfrak{g}}$ into a Lie algebra with the Jacobi identity with respect to the above bracket. When $\mathfrak{g} = \mathfrak{sl}_2(\mathbb{C})$, $\hat{\mathfrak{g}}$ is also denoted by $A_1^{(1)}$.

Now, let $\mathfrak{h}_{(N)}$ be a complex N-dimensional abelian Lie algebra. The Heisenberg algebra $\hat{\mathfrak{h}}'_{(N)}$ is the subalgebra of $\hat{\mathfrak{h}}_{(N)}$ given by

$$\hat{\mathfrak{h}}'_{(N)} = \bigoplus_{n \in \mathbb{Z}, n \neq 0} \mathfrak{h}_{(N)} \otimes t^n \oplus \mathbb{C}c,$$

with respect to the above commutation relations. Note that the center $\mathbb{C}c$ coincides with the commutator ideal $[\hat{\mathfrak{h}}'_{(N)}, \hat{\mathfrak{h}}'_{(N)}]$.

One can show that for a general vertex operator (super) algebra V, the vectors in $(V)_{1/2}$ generate an infinite dimensional Clifford algebra, and the vectors in $(V)_1$ generate an affine Lie algebra with an additional mild assumption on V. See §2.1 for details.

In §2.2, we summarize constructions of vertex operators $Y(v, \zeta)$ acting on V_q and on W_q above. In §2.3, we give a detailed proof of well known commutator relations among the infinite dimensional Clifford algebra, the orthogonal affine Lie algebra and Virasoro algebra in V. The contents of §2.1, §2.2, §2.3 are basically well known, although we supply more details than usual accounts for some of the topics covered there. See also [T3].

In §2.4, we discuss adjoint vertex operators with respect to the naturally defined Hermitian pairing on V. On a general connected vertex operator (super) algebra (that is, those whose weight 0 space is spanned by the vacuum) with a real structure, we can define a sesquilinear pairing (u, v) for any homogeneous vectors u, v by $(u, v) = Y_{\text{wt}(u^*)}(v)$. Some properties of this pairing are discussed. It is interesting that Gramians [§2.4 Definition 11] show up in our calculations.

§0.4   *G*-invariant vertex operator super subalgebras and their vertex operators

Among various vertex operator super algebras, the one which is most important for elliptic genera is one which arises as one of the two total Spin representations of the orthogonal affine Lie algebra $\hat{\mathfrak{o}}(2N)$. Recall that this vertex operator super algebra is an infinite tensor product of the form $V = \bigotimes_{0 \leq n \in \mathbb{Z}} \bigwedge^* A(-n - \frac{1}{2})$. We identify A with $A(-\frac{1}{2})$, so that $\bigwedge^* A$ can be regarded as a subspace of V. This subspace is special among all the exterior algebras $\bigwedge^* A(-n - \frac{1}{2})$, $n \geq 0$, in the following sense. We say a vector $v \in V$ is called a *conformal highest weight vector* if $D(k)v = 0$ for $k \geq 1$, where $D(k)$'s are Virasoro operators.

PROPOSITION 3 [§3.5 PROPOSITION 8]. *The subspace $\bigwedge^* A(-\frac{1}{2})$ of V consists of conformal highest weight vectors.*

The action of the group $\mathrm{Spin}(2N)$ on V preserves the weight of vectors. In the context of elliptic genera, not only V but also invariant subspaces V^G for various Lie subgroups $G \subset \mathrm{Spin}(2N)$ or $\mathrm{SO}(2N)$ play crucial roles.

THEOREM 4 [§3.1 THEOREM 2]. *For any Lie subgroup $G \subset \mathrm{Spin}(2N)$, the invariant subspace V^G of the vertex operator super algebra V also has a structure of a vertex operator super algebra.*

Given general vertex operator algebras, various methods can be used to construct new vertex operator algebras. Taking tensor products of vertex operator algebras and considering their subalgebras or quotient algebras are the standard methods. Theorem 4 gives another method to construct new vertex operator (super) algebras. When G is a finite group, the representation theory of V^G has been discussed in [DM] from the point of view of twisted sectors.

Since we have an application to elliptic genera of Riemannian manifolds in mind, we only consider those groups G which appear as holonomy groups of Riemannian manifolds. In view of the Classification Theorem of holonomy groups [§5.1 Theorem 2], we consider those cases in which $G = \mathrm{SO}(2N)$, $\mathrm{U}(N)$, $\mathrm{SU}(N)$, $\mathrm{Sp}(N')$, and $\mathrm{Sp}(N') \cdot \mathrm{Sp}(1)$ acting on $\mathbb{R}^{2N}$ or $\mathbb{R}^{4N'}$. We do not consider exceptional cases $\mathrm{Spin}(9)$, $\mathrm{Spin}(7)$, and G_2 acting on $\mathbb{R}^{16}$, $\mathbb{R}^8$, and $\mathbb{R}^7$, respectively.

For the groups G in the above list, we identify many basic vectors in V^G [§3.4]. But the complete description of V^G is beyond our scope. We do, however, completely determine V^G up to weight 2 in §4.12. For our present purpose, this is sufficient.

Any Lie subalgebra $\mathfrak{g} \subset \mathfrak{o}(2N)$ can be identified with a subspace of the weight 1 space of V, $(V)_1 = \bigwedge^2 A(-\frac{1}{2}) \cong \mathfrak{o}(2N)$. For various Lie algebras $\mathfrak{g}$, we give a very explicit description of these subspaces in §3.2. The wight 1 space $(V)_1 \cong \bigwedge^2 A(-\frac{1}{2})$ has a natural $\mathfrak{o}(2N)$-invariant pairing induced from that of A. With respect to this pairing, we can consider the Casimir element $\phi_{\mathfrak{g}}$ for $\mathfrak{g} \subset \mathfrak{o}(2N)$ in $S^2(\mathfrak{g}) \subset \bigwedge^4 A(-\frac{1}{2}) \subset (V)_2$. For any Lie algebra $\mathfrak{g}$, the corresponding Casimir element $\phi_{\mathfrak{g}}$ in $(V)_2$ is a $\mathfrak{g}$-invariant conformal highest weight vector (§3.5 Proposition 8). We calculate the Casimir elements for those Lie algebras in the above list in §3.3. Using the unitary Casimir element $\phi_{\mathfrak{u}(N)}$ and the symplectic Casimir element $\phi_{\mathfrak{sp}(N')}$, we let

$$
\begin{cases}
\theta = \dfrac{1}{2N}(2\omega + \phi_{\mathfrak{u}(N)}), \\[2mm]
\lambda = \dfrac{1}{2N}\left(2(N-1)\omega - \phi_{\mathfrak{u}(N)}\right),
\end{cases}
\qquad
\begin{cases}
\sigma = \dfrac{1}{N+4}(3\omega + \phi_{\mathfrak{sp}(N')}), \\[2mm]
\tau = \dfrac{1}{N+4}\left((N+1)\omega - \phi_{\mathfrak{sp}(N')}\right).
\end{cases}
$$

Since $\theta + \lambda = \omega$ and $\sigma + \tau = \omega$, the above elements give two different splittings of the canonical Virasoro element ω. Let the corresponding vertex operators be

$$
\begin{cases}
Y(\theta,\zeta) = \sum_{n \in \mathbb{Z}} \Theta(n)\zeta^{-n-2}, \\
Y(\lambda,\zeta) = \sum_{n \in \mathbb{Z}} \Lambda(n)\zeta^{-n-2},
\end{cases}
\qquad
\begin{cases}
Y(\sigma,\zeta) = \sum_{n \in \mathbb{Z}} S(n)\zeta^{-n-2}, \\
Y(\tau,\zeta) = \sum_{n \in \mathbb{Z}} T(n)\zeta^{-n-2}.
\end{cases}
$$

The above splittings of ω give rise to splittings of the Virasoro algebra into pairs of mutually commuting Virasoro algebras.

THEOREM 5. (i) (Unitary Virasoro algebra [§3.6 THEOREM 2]) *The two families of vertex operators* $\{\Theta(n)\}_{n \in \mathbb{Z}}$, $\{\Lambda(n)\}_{n \in \mathbb{Z}}$ *generate mutually commuting Virasoro algebras of rank 1 and* $N-1$, *respectively. That is,*

$$
\begin{cases}
[\Theta(m), \Theta(n)] = (n-m)\Theta(m+n) + \dfrac{m(m^2-1)}{12}\delta_{m+n,0}\,\mathrm{Id} \\[2mm]
[\Lambda(m), \Lambda(n)] = (n-m)\Lambda(m+n) + \dfrac{m(m^2-1)}{12}\delta_{m+n,0}(N-1)\,\mathrm{Id} \\[2mm]
[\Theta(m), \Lambda(n)] = 0.
\end{cases}
$$

(ii) (Symplectic Virasoro algebra [§3.8 THEOREM 2]) *The two families of vertex operators* $\{S(n)\}_{n \in \mathbb{Z}}$ *and* $\{T(n)\}_{n \in \mathbb{Z}}$ *generate mutually commuting Virasoro algebras of rank* $3N/(N+4)$ *and* $N(N+1)/(N+4)$. *Namely,*

$$
\begin{cases}
[S(m), S(n)] = (n-m)S(m+n) + \dfrac{m(m^2-1)}{12}\left\{\dfrac{3N}{(N+4)}\right\}\delta_{m+n,0}\cdot\mathrm{Id} \\[3mm]
[T(m), T(n)] = (n-m)T(m+n) + \dfrac{m(m^2-1)}{12}\left\{\dfrac{N(N+1)}{(N+4)}\right\}\delta_{m+n,0}\cdot\mathrm{Id} \\[3mm]
[T(m), S(n)] = 0.
\end{cases}
$$

When the vector space $(E^{2N}; \langle\ ,\ \rangle)$ has an isometric complex structure I, let $A = E \otimes \mathbb{C} = A^+ \oplus A^-$ be its decomposition into $\pm i$ eigenspaces of I. Note that $\dim_{\mathbb{C}} A^\pm = N$. Let $\rho_\pm \in \bigwedge^N A^\pm(-\frac{1}{2}) \subset (V)_{\frac{N}{2}}$ be the complex volume forms of unit length. Let the corresponding vertex operators be $Y(\rho_\pm,\zeta) = \sum_{n \in \mathbb{Z}}\{\rho_\pm\}_n \zeta^{-n-1}$.

PROPOSITION 6 [§3.7 PROPOSITION 3]. *The vertex operators* $\{\rho_+\}_n$, $n \in \mathbb{Z}$, *generate an infinite dimensional abelian Lie algebra when* N *is even and an infinite dimensional exterior algebra when* N *is odd. Similar statement holds for the vertex operators* $\{\rho_-\}_n$, $n \in \mathbb{Z}$.

Some calculations of mixed super commutators $[\{\rho_+\}_m, \{\rho_-\}_n]_\pm$ are also given in §3.7.

An isometric complex structure I in $(E^{2N}; \langle\ ,\ \rangle)$ defines a Kähler form $\omega_I \in \bigwedge^2(A^*)$ by $\omega_I(a_1, a_2) = \langle I(a_1), a_2 \rangle$ for $a_1, a_2 \in A$. Through the nondegenerate $\mathbb{C}$-bilinear pairing $\langle\ ,\ \rangle$ on $\bigwedge^* A$, we can regard $\omega_I \in \bigwedge^2 A$. For a quaternionic vector space $(E^{4N'}; \langle\ ,\ \rangle, I, J, K)$, we can consider three Kähler forms ω_I, ω_J, ω_K associated to three complex structures I, J, K satisfying $I^2 = J^2 = -1$, $IJ = -JI = K$. By letting $I(v) = -v \cdot i$, $J(v) = -v \cdot j$, $K(v) = -v \cdot k$ for $v \in E^{4N'}$, we can define a right $\mathbb{H}$-module structure on $E^{4N'}$, where $i^2 = j^2 = -1$, $ij = -ji = k$. The reason of the minus sign in the definition of the action by i, j, k is that we switched from the left action by I, J, K to the right action by i, j, k so that we ought to set $I(v) = v \cdot i^{-1}$, etc. Any endomorphism $\mathcal{J}$ of E of the form $\mathcal{J} = aI + bJ + cK$ for $a^2 + b^2 + c^2 = 1$, $a, b, c \in \mathbb{R}$, defines an isometric complex structure in $(E; \langle\ ,\ \rangle)$. Thus, a quaternionic vector space E has real 2-dimensional family of distinguished isometric complex structures parametrized by S^2. For any such complex structure $\mathcal{J}$, let $\mathfrak{u}(N; \mathcal{J})$ be the associated unitary Lie algebra.

THEOREM 7. *Let* $(E^{4N'}; \langle\ ,\ \rangle, I, J, K)$ *be a quaternionic vector space.*

(i) [§3.9 THEOREM 15] *The Kähler form* $\omega_{\mathcal{J}}$ *associated to a complex structure* $\mathcal{J}$ *as above is given by* $\omega_{\mathcal{J}} = a\omega_I + b\omega_J + c\omega_K$.

(ii) [§3.10] *The unitary Casimir element* $\phi_{\mathfrak{u}(N;\mathcal{J})} \in \bigwedge^4 A$ *associated to the unitary Lie algebra* $\mathfrak{u}(N; \mathcal{J}) \subset \mathfrak{o}(2N)$ *with respect to the complex structure* $\mathcal{J}$ *as above is given by* $\phi_{\mathfrak{u}(N;\mathcal{J})} = \omega_{\mathcal{J}}^2$.

(iii) [§3.3 PROPOSITION 3, §3.9 COROLLARY 17] *The symplectic Casimir element* $\phi_{\mathfrak{sp}(N')} \in \bigwedge^4 A$ *associated to the symplectic algebra* $\mathfrak{sp}(N') \subset \mathfrak{o}(2N)$ *is given by* $2\phi_{\mathfrak{sp}(N')} = \omega_I^2 + \omega_J^2 + \omega_K^2$. •

The unitary Casimir elements $\omega_{\mathcal{J}}^2$ together with the canonical Virasoro element ω generate a unitary Virasoro algebra. Since the commutation relations for these algebras are the same for any complex structure $\mathcal{J}$ in the above real 2-dimensional family, we can deduce some relations among vertex operators [§3.10].

In the total exterior algebra $\bigwedge^* A$ arising from a quaternionic vector space E, we consider a commutative subalgebra generated by three Kähler forms ω_I, ω_J, $\omega_K \in \bigwedge^2 A$. Their images in V through the embedding $\bigwedge^2 A \xrightarrow{\cong} \bigwedge^2 A(-\frac{1}{2}) \subset \bigwedge^* A(-\frac{1}{2}) \subset V$ are denoted by $[\omega_I]$, $[\omega_J]$, $[\omega_K]$. By (ii) of Theorem 7, squares of any Kähler form $\omega_{\mathcal{J}}$ are unitary Casimir elements and they have nice properties. The vertex operators for general quadratic elements in Kähler forms ω_I, ω_J, ω_K still have the following interesting properties:

THEOREM 8 [§3.11 THEOREM 24]. *Let* $[v] \in S^2(\mathbb{C}[\omega_I] \oplus \mathbb{C}[\omega_J] \oplus \mathbb{C}[\omega_K])$ *be a quadratic Kähler form. Then*

$$\begin{cases} \{v\}_0[v] = -\frac{1}{2}D(-1)\left(\{v\}_1[v]\right), \\ \{v\}_1[v] \in S^2\left(\mathbb{C}[\omega_I] \oplus \mathbb{C}[\omega_J] \oplus \mathbb{C}[\omega_K]\right) \oplus \mathbb{C}[\omega], \\ \{v\}_2[v] = 0, \\ \{v\}_3[v] \in \mathbb{C}\Omega, \\ \{v\}_k[v] = 0, \quad for\ k \geq 4. \end{cases}$$

A corollary of this Theorem is that the commutators $[\{v\}_{m+1}, \{v\}_{n+1}]$ for quadratic Kähler forms close in $S^2(\mathbb{C}[\omega_I] \oplus \mathbb{C}[\omega_J] \oplus \mathbb{C}[\omega_K]) \oplus \mathbb{C}[\omega]$ [§3.11 Corollary

25, §0.7 Theorem 16]. The first identity of Theorem 8 doesn't hold for $\{v_1\}[v_2]$ when different quadratic Kähler elements v_1, v_2 are used. In general, the vector $\{v_1\}_0[v_2]$ is not even in the image of the Virasoro operator $D(-1)$. In §3.11, we calculate all possible commutators among quadratic Kähler operators. Since $\phi_{u(N)} = \omega_I^2$ and $2\phi_{sp(N')} = \omega_I^2 + \omega_J^2 + \omega_K^2$, using the results in §3.11, we can calculate the commutation relations of the unitary Virasoro algebra and the symplectic Virasoro algebra in a much simpler way compared with the previous calculations of these algebras done in §3.6 (for unitary Virasoros) and in §3.8 (for symplectic Virasoros).

§0.5   Classifying spaces of geometric structures on manifolds

Since various geometric structures on Riemannian manifolds give rise to various symmetries in elliptic genera, in this section (and in Chapter 4) we give a systematic treatment of general geometric structures on manifolds from the point of view of "classifying spaces". We start with a detailed account of various geometric structures in a real n-dimensional vector space E^n. In §4.1, we show how interesting geometric structures in E arise from a frame in the vector space E. A *frame* is a linear isomorphism e : $\mathbb{R}^n \xrightarrow{\cong} E$. This provides us with a surjective map from the set of all frames $\mathcal{F}(E)$ in E to the moduli space of geometric structures of a given type Σ which is of interest to us. See the list in §4.1 for a list of geometric types we consider. Note that $\mathcal{F}(E)$ is both a left $GL(E)$- and right $GL_n(\mathbb{R})$-space at the same time, and that these two actions commute. In §4.2, we realize the moduli spaces of various geometric structures in the standard vector space $\mathbb{R}^n$ as single $GL_n(\mathbb{R})$-orbits in (the projectivizations of) various representation spaces of $GL_n(\mathbb{R})$.

From these examples, we abstract the notion of *classifying spaces* for geometric structures and we show that the usual formalism of classifying spaces in algebraic topology applies here. The virtue of this abstraction is that we can define the notion of compatible geometric structures for *any* two types of geometric structures [Definition 9 below].

To be more precise, a *classifying space* $B(\Sigma)$ of geometric structures of type Σ in real n-dimensional vector spaces is a single $GL_n(\mathbb{R})$-orbit in some $GL_n(\mathbb{R})$-space equipped with a base point Σ_0 such that for any real n-dimensional vector space E, we have

$$\mathfrak{M}^E(\Sigma) = \mathrm{Hom}_{GL_n(\mathbb{R})}\left(\mathcal{F}(E), B(\Sigma)\right).$$

Here, $\sigma : \mathcal{F}(E) \to B(\Sigma)$ is $GL_n(\mathbb{R})$-equivariant if it satisfies $\sigma(eg) = g^{-1}\sigma(e)$ for any frame e $\in \mathcal{F}(E)$ and any element $g \in GL_n(\mathbb{R})$. The moduli space $\mathfrak{M}^E(\Sigma)$ of geometric structures of type Σ has a structure of left $GL(E)$-space using its left action on $\mathcal{F}(E)$. Let $G(\Sigma)$ be the isotropy subgroup of $GL_n(\mathbb{R})$ at $\Sigma_0 \in B(\Sigma)$. Then, $B(\Sigma) \cong GL_n(\mathbb{R})/G(\Sigma)$ [§4.5 Lemma 8]. Thus our classifying spaces are always finite dimensional homogeneous spaces. Note that the usual classifying spaces of vector bundles of various types are also homogeneous spaces, although they are infinite dimensional. Such classifying spaces always exist. We can take

$B(\Sigma) = \mathfrak{M}^{\mathbb{R}^n}(\Sigma)$, the moduli space of geometric structures of type Σ in the standard vector space $\mathbb{R}^n$. For any geometric structure $\sigma \in \mathfrak{M}^E(\Sigma)$ of type Σ in a vector space E, we let $\mathcal{F}_\sigma(E) = \sigma^{-1}(\Sigma_0) \subset \mathcal{F}(E)$. This is the set of "$\sigma$-frames", i.e., those frames adapted to the geometric structure σ.

For example, if Σ is the Euclidean structure, then σ is a particular choice of an inner product in E, $\mathcal{F}_\sigma(E)$ is the set of all the orthonormal frames in E, $\mathfrak{M}^E(\Sigma)$ is the totality of inner products in E, and $G(\Sigma)$ is the orthogonal group $O(n)$, and the classifying space of the Euclidean structure is $B(\Sigma) = \mathrm{GL}_n(\mathbb{R})/O(n)$.

Using the base point $\Sigma_0 \in B(\Sigma)$, we can construct a $\mathrm{GL}_n(E)$-equivariant surjective map $\Sigma : \mathcal{F}(E) \to \mathfrak{M}^E(\Sigma)$ by letting $\Sigma(e)$, $e \in \mathcal{F}(E)$, be the $\mathrm{GL}_n(\mathbb{R})$-equivariant map sending e to $\Sigma_0 \in B(\Sigma)$. This induces a $\mathrm{GL}_n(\mathbb{R})$-equivariant homeomorphism

$$\Sigma : \mathcal{F}(E)/G(\Sigma) \overset{\cong}{\to} \mathfrak{M}^E(\Sigma),$$

where $G(\Sigma)$ is the isotropy group at the base point $\Sigma_0 \in B(\Sigma)$ as above (§4.3 Corollary 13). The above homeomorphism says that $G(\Sigma)$ is the right automorphism group of any geometric structure σ of type Σ, independent of the specific choice of the geometric structure of type Σ. What is discussed in §4.1 is the detailed proof of the above general $\mathrm{GL}(E)$-equivariant homeomorphisms for some concrete geometric structures of interest to us.

Since we are dealing with Riemannian manifolds, we are interested in those geometric structures which are "compatible" with the given Riemannian structure on M. But it is not always clear what is meant when we say a given geometric structure is compatible with another arbitrarily given geometric structure. For example, a symplectic form ω in a vector space E^{2N} is a nondegenerate 2-form in $\bigwedge^2 E^*$. If E has a Euclidean structure $\langle \, , \, \rangle$, what can it possibly mean when we say that a symplectic structure ω is compatible with a Euclidean structure $\langle \, , \, \rangle$? Our formalism of (based) classifying spaces for geometric structures gives rise to a systematic definition of compatibilities for an arbitrary collection of geometric types.

DEFINITION 9 [§4.3 DEFINITION 14]. A pair of geometric structures (σ_1, σ_2) of types Σ_1, Σ_2, respectively, in a vector space E are *compatible* if the pair (σ_1, σ_2) is in the image of the diagonal map $(\Sigma_1, \Sigma_2) : \mathcal{F}(E) \to \mathfrak{M}^E(\Sigma_1) \times \mathfrak{M}^E(\Sigma_2)$.

Although this definition depends on the choice of base points of the classifying spaces $B(\Sigma_1)$ and $B(\Sigma_2)$, in all types of geometric structures we deal with, there are natural choices of base points in the corresponding classifying spaces. With these obvious choices of base points, the above definition agrees with our usual notion of compatible geometric structures whenever the meaning of compatibility is clear (§4.3 Proposition 15). Let $\mathfrak{M}^E(\Sigma_1, \Sigma_2)$ be the set of compatible pairs of geometric structures (σ_1, σ_2) of types Σ_1, Σ_2. Let $B(\Sigma_1, \Sigma_2)$ be the $\mathrm{GL}_n(\mathbb{R})$-orbit of the pair of base points $((\Sigma_1)_0, (\Sigma_2)_0) \in B(\Sigma_1) \times B(\Sigma_2)$. Then,

$$\mathcal{F}(E)/G(\Sigma_1) \cap G(\Sigma_2) \overset{\cong}{\to} \mathfrak{M}^E(\Sigma_1, \Sigma_2) \cong \mathrm{Hom}_{\mathrm{GL}_n(\mathbb{R})}\left(\mathcal{F}(E), B(\Sigma_1, \Sigma_2)\right).$$

That is, the classifying space of pairs of compatible geometric structures of types Σ_1, Σ_2 is given by $B(\Sigma_1, \Sigma_2) \cong \mathrm{GL}_n(\mathbb{R})/G(\Sigma_1) \cap G(\Sigma_2)$.

Let $\sigma_1 \in \mathfrak{M}^E(\Sigma_1)$ be a geometric structure of type Σ_1. The moduli space of geometric structures σ_2 of type Σ_2 which are compatible with σ_1 is denoted by $\mathfrak{M}^E_{\sigma_1}(\Sigma_2)$. We are most interested in the case in which σ_1 is a Euclidean structure in E. Let $B_{\Sigma_1}(\Sigma_2)$ be the $G(\Sigma_1)$-orbit of the base point $(\Sigma_2)_0$ in $B(\Sigma_2)$. Then, $B_{\Sigma_1}(\Sigma_2) \cong G(\Sigma_1)/G(\Sigma_1) \cap G(\Sigma_2)$ is the classifying space of geometric structures of type Σ_2 compatible with σ_1 [§4.3 Proposition 19]:

$$\mathcal{F}_{\sigma_1}(E)/G(\Sigma_1) \cap G(\Sigma_2) \xrightarrow{\cong} \mathfrak{M}^E_{\sigma_1}(\Sigma_2) \cong \operatorname{Hom}_{\operatorname{GL}_n(\mathbb{R})}(\mathcal{F}_{\sigma_1}(E), B_{\Sigma_1}(\Sigma_2)).$$

Examples of the classifying spaces $B_{\Sigma_1}(\Sigma_2)$ of various types of geometric structures are given in §4.5 Lemma 11.

Let the (right) automorphism groups of three types of geometric structures $\Sigma_1, \Sigma_2, \Sigma_3$ be $G(\Sigma_1)$, $G(\Sigma_2)$, $G(\Sigma_3)$. We have the following square diagram of inclusion homomorphisms:

$$
\begin{array}{ccc}
G(\Sigma_1) \cap G(\Sigma_2) & \longrightarrow & G(\Sigma_1) \\
\uparrow & & \uparrow \\
G(\Sigma_1) \cap G(\Sigma_2) \cap G(\Sigma_3) & \longrightarrow & G(\Sigma_1) \cap G(\Sigma_3).
\end{array}
$$

To each inclusion map of groups, we can associate a moduli space of suitable compatible geometric structures. For example, given any geometric structures σ_1, σ_3 of types Σ_1, Σ_3, we can associate moduli spaces $\mathfrak{M}^E_{\sigma_1}(\Sigma_2)$ and $\mathfrak{M}^E_{(\sigma_1,\sigma_3)}(\Sigma_2)$ to the upper and lower horizontal arrows in the above diagram. The pair of vertical inclusion arrows between inclusion pairs (upper and lower inclusion pairs in the above diagram) induces an inclusion map of the corresponding moduli spaces $\mathfrak{M}^E_{(\sigma_1,\sigma_3)}(\Sigma_2) \hookrightarrow \mathfrak{M}^E_{\sigma_1}(\Sigma_2)$. At the end of §4.3, we discuss these inclusion maps of moduli spaces associated to square diagrams of inclusions of Lie groups which arise as right automorphism groups of geometric structures of interest to us (Diagram (39) of §4.3).

For geometric structures on manifolds, statements similar to those for vector spaces can be made using the same classifying spaces $B(\Sigma)$, $B_{\Sigma_1}(\Sigma_2)$, etc. Namely, the moduli space of geometric structures of type Σ on a manifold M^n is given by

$$\mathfrak{M}^M(\Sigma) \cong \operatorname{Hom}_{\operatorname{GL}_n(\mathbb{R})}(\mathcal{F}(M), B(\Sigma)),$$

where $\mathcal{F}(M)$ is the principal $\operatorname{GL}_n(\mathbb{R})$-bundle of frames on M. For any geometric structure $\sigma \in \mathfrak{M}^M(\Sigma)$, $\mathcal{F}_\sigma(M) = \sigma^{-1}(\Sigma_0) \subset \mathcal{F}(M)$ is a principal $G(\Sigma)$ bundle of σ-frames. Similarly, given two types of geometric structures Σ_1, Σ_2, the moduli space of geometric structures of type Σ_2 compatible with a geometric structure σ_1 of type Σ_1 is given by

$$\mathfrak{M}^M_{\sigma_1}(\Sigma_2) \cong \operatorname{Hom}_{G(\Sigma_1)}(\mathcal{F}_{\sigma_1}(M), B_{\Sigma_1}(\Sigma_2)).$$

See §4.5 for various concrete examples. In the context of Riemannian geometry, we are particularly interested in the case in which σ_1 in the above formula is the Riemannian metric g. Let $\mathcal{F}_g(M) = \mathcal{Q}(M)$ be the $O(n)$-principal bundle of orthonormal frames with respect to g.

Many important geometric structures on Riemannian manifolds are parallel with respect to the Levi-Civita connection on the orthonormal frame bundle $\mathcal{Q}(M)$. In §4.4, we give a brief account of three types of connections on manifolds and their relationships. These connections are Ehresmann connections, principal connections, and covariant derivatives. We also discuss the relationships of curvatures of these connections. Since we cannot differentiate geometric structures $\sigma : \mathcal{Q}(M) \to B_\eta(\Sigma)$ by a tangent vector because the classifying space $B_\eta(\Sigma)$ of Σ-structure compatible with the Euclidean structure is not a vector space in general, we cannot use the covariant derivatives to define the notion of parallel geometric structures in this general framework. Here η denotes general Euclidean structures on M. We call a geometric structure σ *parallel* if the $O(2N)$-equivariant map σ is constant along any horizontal paths in $\mathcal{Q}(M)$. Let

$$\text{P-Hom}_{O(n)}\left(\mathcal{Q}(M), B_\eta(\Sigma)\right) = \left\{ \begin{array}{l} \text{parallel geometric structures of type } \Sigma \text{ on } M \\ \text{compatible with the Riemannian structure} \end{array} \right\}.$$

Interesting and important geometric structures such as Kählerian structures, special Kähler structures, hyperkähler structures and quaternion-Kähler structures can all be formulated in this framework of equivariant parallel maps from the orthonormal frame bundles to suitable classifying spaces. See Example 19 in §4.5. In Proposition 20, we show that such a parallel geometric structure exists if the holonomy group of M is contained in the right automorphism group $G(\Sigma)$ for the geometric structures of type Σ.

§0.6   Elliptic genera as modules over vertex operator super algebras

Let (M^{2N}, g) be a closed Riemannian Spin manifold. Let $T_{\mathbb{C}}M(-m)$ be a copy of the complexified tangent bundle $T_{\mathbb{C}}M$ carrying weight m for $m \in \frac{1}{2}\mathbb{Z}$. At each point $x \in M$, we can construct a vertex operator super algebra V_x from the Euclidean vector space $(T_x M, g_x)$. These graded vector spaces form a vector bundle on M of vertex operator super algebras given by

$$\underline{V} = \underline{V}_q = \bigoplus_{0 \le \ell \in \frac{1}{2}\mathbb{Z}} (\underline{V})_\ell q^\ell = \bigotimes_{0 \le m \in \mathbb{Z}} \bigwedge\nolimits^*_{q^{m+\frac{1}{2}}} T_{\mathbb{C}}M(-m - \tfrac{1}{2}).$$

$T_{\mathbb{C}}M$ and $T_{\mathbb{C}}^*M$ are identified with $T_{\mathbb{C}}M(\frac{1}{2})$, $T_{\mathbb{C}}M(-\frac{1}{2})$, respectively. So the total exterior bundle $\bigwedge^* T_{\mathbb{C}}^*M$ is identified with $\bigwedge^* T_{\mathbb{C}}M(-\frac{1}{2}) \subset \underline{V}$. This subbundle consists of conformal highest weight vectors (essentially due to §3.5 Proposition 8). We can also consider a graded vector bundle

$$\underline{W} = \underline{W}_q = \Delta \otimes \bigotimes_{0 \le m \in \mathbb{Z}} \bigwedge\nolimits^*_{q^m} T_{\mathbb{C}}M(-m)$$

of $\mathbb{Z}_2$-twisted modules over the bundle of vertex operator super algebras $\underline{V}$. The Levi-Civita connection on M induces a covariant derivative ∇ on the graded vector spaces of smooth sections of $\underline{V}$ and of $\underline{W}$. Let $\mathcal{P}_M(\underline{V})$ be the graded vector space of parallel sections in $\underline{V}$. For connected manifold M, we have $\mathcal{P}_M(\underline{V})_0 \cong \mathbb{C} \cdot 1_M$, where 1_M is the constant function on M with value 1 everywhere.

THEOREM 10 [§5.4 PROPOSITION 5, THEOREM 7, THEOREM 9]. *The graded vector space* $\mathcal{P}_M(\underline{V})$ *has the following properties*:

(i) $\mathcal{P}_M(\underline{V})$ *has a structure of a vertex operator super algebra.*

(ii) *For any parallel section* $\sigma \in \mathcal{P}_M(\underline{V})$, *the components of the associated vertex operator acting on* $\underline{V}$ *and on* $\underline{W}$ *commute with covariant derivative, that is,* $[\nabla, \{\sigma\}_n] = 0$ *for* $n \in \frac{1}{2}\mathbb{Z}$.

(iii) *If* $G \subset \mathrm{Spin}(2N)$ *is the structure group of a holonomy bundle on* M, $\mathcal{P}_M(\underline{V})$ *is canonically isomorphic to* V^G *as vertex operator super algebras.*

In (iii) above, note that to have a holonomy bundle we need to choose a frame in the tangent space of a point of M. In §5.2, we identify many parallel sections in $\underline{V}$ for various subclasses of Riemannian manifolds such as Kählerian, special Kähler, hyperkähler and quaternion-Kähler manifolds.

THEOREM 11 (MODULE STRUCTURE ON ELLIPTIC GENERA) [§5.4]. *For a closed Spin Riemannian manifold* M, *the elliptic genus* $\Phi^*_{\mathrm{ell}}(M)$ *for* $* = Q, R$ *has a structure of a super-pair of modules over the vertex operator super algebra* $\mathcal{P}_M(\underline{V})$. *The diagonal action of* $\mathcal{P}_M(\underline{V})$ *on* $\Phi^*_{\mathrm{ell}}(M)$ *is effective if the Spin index* $\hat{A}(M)$ *doesn't vanish.*

Theorem 11 can be reformulated to a statement [§5.4 Proposition 15] that the elliptic genus $\Phi^*_{\mathrm{ell}}(M)$ is a super-pair of modules over V^G using the G-equivariance of the vertex operators $\{v\}_n$ for $v \in V^G$ as operators acting on both graded vector spaces V and W.

REMARK. Even if $\hat{A}(M) = 0$, if the Witten genus $\varphi_\theta(M)$ [§1.6] doesn't vanish, then the vertex operator map $Y(\ , \zeta) : \mathcal{P}_M(\underline{V}) \to \mathrm{End}\,(\Phi^*_{\mathrm{ell}}(M))\,[[\zeta^{\frac{1}{2}}, \zeta^{-\frac{1}{2}}]]$ is still injective. In fact, the elliptic genus $\Phi^*_{\mathrm{ell}}(M)$ contains a super-pair of free $\mathcal{P}_M(\underline{V})$-modules $\Phi_\theta(M) \otimes \mathcal{P}_M(\underline{V})$.

The collection of component operators $\{\sigma\}_0$ for $\sigma \in \mathcal{P}_M(\underline{V})$ is closed under the super bracket and forms a Lie super algebra. Similarly, the collection of operators $Y_0(\sigma)$ for $\sigma \in \mathcal{P}_M(\underline{V})$ is also closed under the super commutator, although it doesn't satisfy the usual Jacobi identity. As a corollary of the Main Theorem, the elliptic genus is a super-pair of modules over these algebras [§5.4 Corollaries 11, 12].

Some interesting conclusions follow when the tangent bundle splits into a sum of subbundles. For example, when the tangent bundle TM^{2N} splits into a sum of plane bundles, then the corresponding elliptic genus is a super-pair of representations of the Heisenberg algebra $\mathfrak{h}_{(N)}$ [§5.4 Corollary 13]. When there are r linearly independent vector fields on M, then the elliptic genus is a super-pair of repesentations of an (infinite dimensional) Clifford algebra [§5.4 Proposition 15].

§0.7 Infinite dimensional symmetries in
elliptic genera for Kählerian manifolds

We describe some consequences of Module Theorem (Theorem 11) for various types of closed Riemannian manifolds [Bes]. In §5.1, we give a brief account of various properties of Kählerian manifolds relevant to our purpose. It is very interesting to note that the vertex operators associated to parallel geometric tensors have very nice properties.

In what follows, we use the notation $[v]$ to denote a vector in $\mathcal{P}_M(\underline{V})$, and use $\{v\}_*$ to denote the corresponding vertex operators.

Riemannian manifolds. The Riemannian metric $g : TM \otimes TM \to \mathbb{R}$ is parallel on M. A generalized Riemannian metric tensor $\hat{g} : T_\mathbb{C}M(3/2) \otimes T_\mathbb{C}M(1/2) \to \mathbb{C}$ defined by $\hat{g}(v_1(3/2), v_2(1/2)) = g(v_1, v_2)$ for $v_1, v_2 \in T_\mathbb{C}M$ is a weight 2 parallel section in $\Gamma\left(T_\mathbb{C}M(-3/2) \otimes T_\mathbb{C}M(-1/2)\right)$ [§5.2 Proposition 3]. So, $[\hat{g}] \in \mathcal{P}_M(\underline{V})_2$.

THEOREM 12 [§5.5 THEOREM 2]. *Let (M^{2N}, g) be a closed Riemannian Spin manifold. The vertex operator $Y(-[\hat{g}]/2, \zeta) = \sum_{m \in \mathbb{Z}} D(m) \zeta^{-m-1}$ for the generalized Riemannian metric tensor $\hat{g} \in \mathcal{P}_M(\underline{V})_2$ generates a Virasoro algebra of rank N. Hence, the elliptic genus $\Phi_{\mathrm{ell}}^*(M^{2N})$, $* = Q, R$, is a super-pair of rank N representations of the Virasoro algebra.*

Kählerian manifolds. A Riemannian manifold (M^{2N}, g) with an isometric almost complex structure I is called Kählerian if I is parallel. In this case, I is integrable and M is actually a complex manifold and the holonomy group of M is contained in $U(N)$ [§5.1 Proposition 8]. The Kähler form $\kappa \in \bigwedge^2 T^*M$, defined by

$$\kappa(X, Y) = \kappa_I(X, Y) = g(I(X), Y)$$

for $X, Y \in TM$, is parallel and hence closed, because g, I are parallel [§5.2 Proposition 9]. Through the embedding $\bigwedge^* TM \hookrightarrow \underline{V}$, we obtain $[\kappa] \in \mathcal{P}_M(\underline{V})_1$ and $[\kappa^2] \in \mathcal{P}_M(\underline{V})_2$. Let

$$[\theta] = \frac{1}{2N}([\kappa^2] - [\hat{g}]), \qquad [\lambda] = \frac{-1}{2N}([\kappa^2] + (N-1)[\hat{g}]).$$

be vectors in $\mathcal{P}_M(\underline{V})_2$. Note that $[\theta] + [\lambda] = -[\hat{g}]/2$ generates the rank N Virasoro algebra above. Note also that $[\kappa^2]$ is actually equal to the unitary Casimir form $[\phi_{\mathrm{u}(N)}]$ for the unitary Lie algebra $\mathrm{u}(N)$.

THEOREM 13 [§5.5 THEOREM 5]. *Let M^{2N} be a closed Kählerian Spin manifold. The vertex operator super algebra $\mathcal{P}_M(\underline{V})$ contains the following Lie algebras.*

(i) (Heisenberg algebra) *The vertex operator $Y([\kappa], \zeta) = \sum_{n \in \mathbb{Z}} K(n) \zeta^{-n-1}$ generates a Heisenberg algebra $\mathfrak{h} = \bigoplus_{m \neq 0} \mathbb{C}K(m) \oplus \mathbb{C}\mathrm{Id}_V$ with commutation relations $[K(m), K(n)] = nN\delta_{m+n,0}\mathrm{Id}_V$, for $m, n \in \mathbb{Z}$.*

(ii) (Unitary Virasoro algebra) *The vertex operators $Y([\theta], \zeta)$ and $Y([\lambda], \zeta)$ generate two commuting Virasoro algebras of rank 1, $N-1$, respectively.*

The elliptic genus $\Phi_{\mathrm{ell}}^*(M)$ is a super-pair of representations of the above algebras.

REMARK. One can show that the direct sum of the above algebras

$$\mathfrak{k}_{(N)} = \bigoplus_{n \in \mathbb{Z}} \mathbb{C}K(n) \oplus \bigoplus_{n \in \mathbb{Z}} \mathbb{C}\Theta(n) \oplus \bigoplus_{n \in \mathbb{Z}} \mathbb{C}\Lambda(n) \oplus \mathbb{C}\mathrm{Id},$$

has a very interesting property that the bracket product closes. Thus, $\mathfrak{k}_{(N)}$ has a structure of a Lie algebra. In $\mathfrak{k}_{(N)}$, the subalgebra $\bigoplus_{n\in\mathbb{Z}}\mathbb{C}K(n)\oplus\mathbb{C}\mathrm{Id}$ is actually an ideal. We call this Lie algebra $\mathfrak{k}_{(N)}$ the Heisenberg Unitary-Virasoro algebra.

Special Kähler manifolds with SU(N)-holonomy. On a Riemannian manifold, contraction of the curvature tensor R yields the Ricci tensor r defined by $r(X,Y) = \mathrm{Tr}(Z \mapsto R(X,Z)Y)$ for $X,Y \in TM$. A Ricci-flat ($r \equiv 0$) Kählerian manifold is called a special Kähler manifold. The restricted holonomy group (the connected component of the holonomy group) of a special Kähler manifold is contained in SU(N) [§5.1 Proposition 11]. However, in our present context, we require that the *entire* holonomy group be contained in SU(N). In this case, the algebra $\mathcal{P}_M(\underline{V})$ contains not only the Kähler form $[\kappa] \in \mathcal{P}_M(\underline{V})_1$ and its square and metric tensor $[\kappa^2], [\hat{g}] \in \mathcal{P}_M(\underline{V})_2$, but also the complex volume form and its conjugate $[\xi^\pm] \in \mathcal{P}_M(\underline{V})_{\frac{N}{2}}$ [§5.2 Proposition 13].

THEOREM 14 [§5.5 THEOREM 9]. *Let M^{2N} be a special Kähler manifold with SU(N) holonomy. The vertex operator super algebra $\mathcal{P}_M(\underline{V})$ contains the following subalgebras.*

(i) (Heisenberg Unitary-Virasoro algebra) *The Kähler form $[\kappa]$, the generalized Riemannian tensor $[\hat{g}]$, and the unitary Casimir form $[\phi_{\mathrm{u}(N)}] = [\kappa^2]$ generate the Heisenberg Unitary-Virasoro algebra $\mathfrak{k}_{(N)}$.*

(ii) (Complex volume form operators) *The vertex operator $Y([\xi^+], \zeta)$ generates an infinite dimensional abelian Lie algebra when N is even, and an infinite dimensional exterior algebra when N is odd. The same statement holds for the vertex operator $Y([\xi^-], \zeta)$.*

The elliptic genus $\Phi^*_{\mathrm{ell}}(M)$ is a super-pair of representations of the above algebras.

REMARK. Let $\mathfrak{s}^+ = \mathfrak{k}_{(N)} \oplus \bigoplus_{n\in\mathbb{Z}}\mathbb{C}\Xi^+(n)$, $\mathfrak{s}^- = \mathfrak{k}_{(N)} \oplus \bigoplus_{n\in\mathbb{Z}}\mathbb{C}\Xi^-(n)$ be the direct sum of the Heisenberg Unitary-Virasoro algebra $\mathfrak{k}_{(N)}$ and the vector spaces of complex volume form operators where $Y([\xi^\pm], \zeta) = \sum_{n\in\mathbb{Z}}\Xi^\pm(n)\zeta^{-n-1}$. It can be shown that each of these vector spaces has a structure of a Lie super algebra (i.e., the super bracket product closes) and $\bigoplus_{n\in\mathbb{Z}}\mathbb{C}\Xi^\pm(n)$ forms an ideal in $\mathfrak{s}^\pm$.

Quaternion-Kähler manifolds. A Riemannian manifold $(M^{4N'}, g)$, $N = 2N'$, is called quaternion-Kähler if there exists a real 3-dimensional parallel subbundle $L \subset \mathrm{End}(TM)$ of the endomorphism bundle of the tangent bundle TM which is locally spanned by isometric almost complex structures I, J, K satisfying the quaternion relations. Here, the locally defined tensor fields I, J, K are not necessarily parallel. If such parallel tensors exist globally, then we have hyperkähler manifolds below. The holonomy group of M is then contained in Sp(N') · Sp(1) [§5.1 Proposition 17]. Quaternion-Kähler manifolds are not necessarily Kähler manifolds. Corresponding to locally defined almost complex structures I, J, K which may not be parallel, we have locally defined 2-forms $\kappa_I, \kappa_J, \kappa_K$ which may be neither closed nor parallel. However, the sum of squares $\kappa_{Q-K} = \kappa_I^2 + \kappa_J^2 + \kappa_K^2$ is a globally defined parallel closed differential 4-form on M, the quaternion-Kähler form [§5.1 Proposition 18]. We have $[\kappa_{Q-K}], [\hat{g}] \in$

$\mathcal{P}_M(\underline{V})_2$, and we let

$$[\sigma] = \frac{1}{2(N+4)}([\kappa_{Q-K}] - 3[\hat{g}]), \qquad [\tau] = \frac{-1}{2(N+4)}([\kappa_{Q-K}] + (N+1)[\hat{g}]).$$

Again note that $[\sigma] + [\tau] = -[\hat{g}]/2$ generates the rank N Virasoro algebra. Also note that half of the quaternion-Kähler form $\frac{1}{2}[\kappa_{Q-K}]$ is equal to the symplectic Casimir form $[\phi_{\mathfrak{sp}(N')}]$ for the symplectic Lie algebra $\mathfrak{sp}(N')$ [§3.9 Corollary 17].

THEOREM 15 [§5.5 THEOREM 14]. *Let $(M^{4N'}; g, L)$ be a closed quaternion-Kähler Spin manifold. The elliptic genus $\Phi^*_{\text{ell}}(M)$ is a super-pair of representations of a symplectic Virasoro algebra which is a direct sum of two commuting Virasoro algebras of rank $3N/(N+4)$, $N(N+1)/(N+4)$, respectively, generated by vertex operators $Y([\sigma], \zeta)$ and $Y([\tau], \zeta)$.*

Hyperkähler manifolds. A hyperkähler manifold $M^{4N'}$ possesses three isometric parallel almost complex structures I, J, K satisfying the quaternion relations. The holonomy group of M is contained in $\text{Sp}(N')$ [§5.1 Proposition 13]. The corresponding Kähler forms $\kappa_I, \kappa_J, \kappa_K$ are parallel and closed, as is the associated quaternion-Kähler form $\kappa_{Q-K} = \kappa_I^2 + \kappa_J^2 + \kappa_K^2$. For any integrable almost complex structure $\mathcal{J} = aI + bJ + cK$ with $a^2 + b^2 + c^2 = 1$, $a, b, c \in \mathbb{R}$, the corresponding Kähler form is given by $\kappa_{\mathcal{J}} = a\kappa_I + b\kappa_J + c\kappa_K$ [§0.4 Theorem 7 (i)]. Thus, we have $[\kappa_I], [\kappa_J], [\kappa_K], [\kappa_{\mathcal{J}}] \in \mathcal{P}_M(\underline{V})_1$ and $[\kappa_{Q-K}], [\kappa_{\mathcal{J}}^2], [\hat{g}] \in \mathcal{P}_M(\underline{V})_2$. The finite dimensional commutative subalgebra of $\mathcal{P}_M(\underline{V})$ generated by Kähler forms $[\kappa_I], [\kappa_J], [\kappa_K]$ plays an important role for hyperkähler manifolds. Recall that for a connected Riemannian manifold M, $\mathcal{P}_M(\underline{V})_0 = \mathbb{C}1_M$. We let $N = 2N'$.

THEOREM 16 [§5.5 THEOREM 16]. *Let $(M^{4N'}; g, I, J, K)$ be a connected closed hyperkähler manifold. The vertex operator super algebra $\mathcal{P}_M(\underline{V})$ contains the following algebras.*

(1) *(Affine Lie algebra $A_1^{(1)}$) The vertex operators corresponding to Kähler forms $[\kappa_I], [\kappa_J], [\kappa_K] \in \mathcal{P}_M(\underline{V})_1$ generate a level N' affine Lie algebra $A_1^{(1)}$.*

(2) *(Unitary Virasoro algebra) For any parallel almost complex structure $\mathcal{J} = aI + bJ + cK$ as above, the subspace $\mathbb{C}[\kappa_{\mathcal{J}}^2] \oplus \mathbb{C}[\hat{g}] \subset \mathcal{P}_M(\underline{V})_2$ generates a Unitary Virasoro algebra of rank $(1, N-1)$.*

(3) *(Symplectic Virasoro algebra) The subspace $\mathbb{C}[\kappa_{Q-K}] \oplus \mathbb{C}[\hat{g}] \subset \mathcal{P}_M(\underline{V})_2$ generates a symplectic Virasoro algebra (sum of commuting Virasoro algebras) of rank $(3N/(N+4), N(N+1)/(N+4))$.*

(4) *Let $N' \geq 2$. For any vector $[\vartheta] \in S^2(\mathbb{C}[\kappa_I] \oplus \mathbb{C}[\kappa_J] \oplus \mathbb{C}[\kappa_K]) \subset \mathcal{P}_M(\underline{V})_2$, the corresponding vertex operators $\{\vartheta\}_*$ satisfy the commutation relations of the form*

$$[\{\vartheta\}_{m+1}, \{\vartheta\}_{n+1}] = \binom{m-n}{2}\{\{\vartheta\}_1[\vartheta]\}_{m+n+1} + \binom{m+1}{3}(\{\vartheta\}_3[\vartheta], 1)\delta_{m,-n} \cdot \text{Id}_V,$$

where $\{\vartheta\}_1[\vartheta] \in S^2(\mathbb{C}[\kappa_I] \oplus \mathbb{C}[\kappa_J] \oplus \mathbb{C}[\kappa_K]) \oplus \mathbb{C}[\hat{g}] \subset \mathcal{P}_M(\underline{V})_2$ and the pairing $(\ ,\ )$ on $\mathcal{P}_M(\underline{V})_0 = \mathbb{C} \cdot 1_M$ is such that $(1_M, 1_M) = 1$.

The elliptic genus $\Phi_{\text{ell}}^*(M)$ *is a super-pair of representations of the above algebras.*

Note that in (4) above, $[\vartheta]$ can be $[\kappa_{Q-K}]$ or $[\kappa_{\mathcal{J}}^2]$ or any other quadratic expression of Kähler forms, and the formula there is very general.

ACKNOWLEDGEMENT. The author is grateful for the hospitality of the Max-Planck-Institut für Mathematik at Bonn, where the essential part of this work started. The author also wishes to thank Peter Landweber for his careful reading of the manuscript and his numerous helpful suggestions which improved the exposition of this paper.

CHAPTER I

ELLIPTIC GENERA

§1.1   Elliptic Genera

The word "elliptic" in elliptic genus comes from the fact that its logarithm $\log X$ is described in terms of an elliptic integral of the form

$$(1) \qquad \int_0^X \frac{dt}{\sqrt{1 - 2\delta t^2 + \varepsilon t^4}}, \qquad \delta, \varepsilon \in \Lambda,$$

where the integral should be expanded as a power series in X, and Λ is a commutative ring in which 2 is invertible. The origin of the elliptic genus is a problem in cobordism theory. One was interested in characterizing the ideal in the localized cobordism ring $\Omega_*^{\text{Spin}}[\frac{1}{2}] = \Omega_*^{\text{SO}}[\frac{1}{2}]$ generated by the cobordism classes of Spin manifolds admitting semi-free S^1-actions of odd type [Bor, LS]. It turned out that this ideal was precisely the kernel of the genus defined by the above elliptic integral [O]. For basic materials on elliptic genera, see [L2].

Later Witten [W3, W4] shed a considerable light on the subject by showing that the elliptic genus of a Spin manifold M can be interpreted as the S^1-equivariant signature of the free loop space LM via Fixed Point Index Formula. Our formulation of elliptic genera closely follows his approach, but with some refinement.

Let $M^{4N'}$ be an oriented Riemannian Spin manifold of real dimension $4N'$. [Throughout this paper, if the dimension of the manifold M^n is even, then $n = 2N$. If it is a multiple of 4, then $n = 4N'$.] Using the Riemannian metric on M, we can construct the principal $\text{SO}(n)$ bundle $P_{\text{SO}(n)} \to M^n$ of orthonormal frames in the cotangent bundle. On $P_{\text{SO}(n)}$ we can introduce a canonical torsion free connection, the Levi-Civita connection with respect to which the metric tensor is parallel. Since M^n has a Spin structure, we can double cover the principal $\text{SO}(n)$-bundle by a principal $\text{Spin}(n)$ bundle, $P_{\text{Spin}(n)} \to M^n$. We can lift the Levi-Civita connection on $P_{\text{SO}(n)}$ to $P_{\text{Spin}(n)}$. With respect to this Riemannian connection on the Spin bundle, we can construct twisted Dirac operators on M. For details of Dirac operators, see §5.3, where certain generalizations are discussed. The twisting bundles we are interested in are constructed in the following way.

Let $T_{\mathbb{C}} = T \otimes_{\mathbb{R}} \mathbb{C}$ be the complexification of the cotangent bundle $T = T^*M$. Although we can canonically identify the cotangent bundle and the tangent bundle of M by the Riemannian metric, our use of cotangent bundles is partly motivated by the use of Dirac operators and partly motivated by later applications in Chapter 5.

Let the total exterior powers and the total symmetric powers of $T_{\mathbb{C}}$ be

$$(2) \qquad \Lambda_t(T_{\mathbb{C}}) = \sum_{k \geq 0} \Lambda^k(T_{\mathbb{C}}) t^k, \qquad S_t(T_{\mathbb{C}}) = \sum_{k \geq 0} S^k(T_{\mathbb{C}}) t^k.$$

We define bundles R_ℓ, Q_ℓ, $\ell \geq 0$, $\ell \in \mathbb{Z}$ by the formulae

$$(3) \qquad \begin{aligned} \sum_{0 \leq \ell \in \mathbb{Z}} R_\ell q^\ell &= \bigotimes_{\ell \geq 1} S_{q^\ell}(T_{\mathbb{C}}) \otimes \underline{\Delta} \otimes \bigotimes_{\ell \geq 1} \Lambda_{q^\ell}(T_{\mathbb{C}}) \\ \sum_{0 \leq \ell \in \frac{1}{2}\mathbb{Z}} Q_\ell q^\ell &= \bigotimes_{\ell \geq 1} S_{q^\ell}(T_{\mathbb{C}}) \otimes \bigotimes_{\ell \geq 1} \Lambda_{q^{\ell - \frac{1}{2}}}(T_{\mathbb{C}}). \end{aligned}$$

Here $\underline{\Delta} = \underline{\Delta}^+ \oplus \underline{\Delta}^-$ is the complex $2^{2N'}$ dimensional total spin bundle for M.

We consider Dirac operators twisted by bundles R_ℓ, Q_ℓ, $\ell \geq 0$:

$$(4) \qquad \begin{aligned} d_{R_\ell} &: \Gamma(\Delta^+ \otimes R_\ell) \to \Gamma(\Delta^- \otimes R_\ell), \\ d_{Q_\ell} &: \Gamma(\Delta^+ \otimes Q_\ell) \to \Gamma(\Delta^- \otimes Q_\ell). \end{aligned}$$

A detailed account of twisted Dirac operators are given in §5.3. The usual elliptic genus [Definition 2 below] is a modular form valued cobordism invariant. Our definition is a refinement of this usual one.

DEFINITION 1 (GEOMETRIC ELLIPTIC GENERA). Let $M^{4N'}$ be a closed Spin manifold whose Spin principal bundle is equipped with a connection. We define two types of elliptic genus as super-pairs of graded vector spaces given by the following formulae:

$$(5) \qquad \Phi_{\text{ell}}^Q(M^{4N'}) = \left[\bigoplus_{0 \leq \ell \in \mathbb{Z}} [\text{Ker } d_{Q_\ell}] q^{\ell - \frac{N'}{4}} \ ; \ \bigoplus_{0 \leq \ell \in \mathbb{Z}} [\text{Coker } d_{Q_\ell}] q^{\ell - \frac{N'}{4}} \right]$$

$$(6) \qquad \Phi_{\text{ell}}^R(M^{4N'}) = \left[\bigoplus_{0 \leq \ell \in \mathbb{Z}} [\text{Ker } d_{R_\ell}] q^\ell \ ; \ \bigoplus_{0 \leq \ell \in \mathbb{Z}} [\text{Coker } d_{R_\ell}] q^\ell \right]$$

Note that in (6), only integral powers of q appear. We refer to these vector spaces as "left" and "right" elliptic genus and we use the notation

$$\Phi_{\text{ell}}^*(M) = [\Phi_{\text{ell}}^*(M)_\ell \ ; \ \Phi_{\text{ell}}^*(M)_r], \qquad \text{for } * = Q, R.$$

There is a physical reason for the fractional power of q in (5), but mathematically this adjustment gives better modular properties [§1.4 Theorem 2].

The usual elliptic genera are obtained by taking the graded super-dimension of the geometric elliptic genera above. We assume familiarity with index theory of elliptic operators on closed Spin manifolds. References are [BGV, H1, LM].

LEMMA AND DEFINITION 2 (ELLIPTIC GENERA). The super-dimensions of the above geometric elliptic genera are given as follows:

$$(7) \qquad \varphi_{\text{ell}}^Q(M^{4N'}) = q^{-\frac{N'}{4}} \sum_{0 \le \ell \in \frac{1}{2}\mathbb{Z}} (\text{Index } d_{Q_\ell}) \, q^\ell$$

$$= q^{-\frac{N'}{4}} \hat{A}(M) \text{ch} \Big[\bigotimes_{\ell \ge 1} S_{q^\ell}(T_{\mathbb{C}}) \otimes \bigotimes_{\ell \ge 1} \Lambda_{q^{\ell-\frac{1}{2}}}(T_{\mathbb{C}}) \Big][M^{4N'}],$$

$$(8) \qquad \varphi_{\text{ell}}^R(M^{4N'}) = \sum_{0 \le \ell \in \mathbb{Z}} (\text{Index } d_{R_\ell}) \, q^\ell$$

$$= \hat{A}(M) \text{ch} \Big[\bigotimes_{\ell \ge 1} S_{q^\ell}(T_{\mathbb{C}}) \otimes \underline{\Delta} \otimes \bigotimes_{\ell \ge 1} \Lambda_{q^\ell}(T_{\mathbb{C}}) \Big][M^{4N'}].$$

Index theory justifies the second and the fourth equalities in the formulae above. Actually, there is the third version of elliptic genus given by

$$(9) \qquad \varphi_{\text{ell}}^{Q'}(M^{4N'}) = q^{-\frac{N'}{4}} \hat{A}(M) \text{ch} \Big[\bigotimes_{\ell \ge 1} S_{q^\ell}(T_{\mathbb{C}}) \otimes \bigotimes_{\ell \ge 1} \Lambda_{(-q^{\ell-\frac{1}{2}})}(T_{\mathbb{C}}) \Big][M^{4N'}].$$

This is obtained from $\varphi_{\text{ell}}^Q(M^{4N'})$ by replacing $q^{\frac{1}{2}}$ by $-q^{\frac{1}{2}}$. The corresponding geometric elliptic genus is essentially the same as Φ_{ell}^Q. But when we let $q = e^{2\pi i \tau}$, as a function of τ in the upper half plane the above three genera exhibit different behaviours, and these three modular functions span a complex 3-dimensional representation of $\text{SL}_2(\mathbb{C})$ for each manifold M.

Note that φ_{ell}'s can be defined on any oriented manifold and we don't need to choose a metric nor a connection. From index theory, all coefficients of φ_{ell}^Q and $\varphi_{\text{ell}}^{Q'}$ are integers for closed Spin manifolds. On the other hand, all coefficients of φ_{ell}^R are integral even for oriented closed manifolds. Thus the three elliptic genera define the following ring maps:

$$(10) \qquad \begin{aligned} \varphi_{\text{ell}}^Q, \; \varphi_{\text{ell}}^{Q'} &: \Omega_*^{\text{Spin}} \to \mathbb{Z}[q^{(-\frac{1}{4})}][[q^{\frac{1}{2}}]], \\ \varphi_{\text{ell}}^R &: \Omega_*^{\text{SO}} \to \mathbb{Z}[[q]]. \end{aligned}$$

To describe these genera in more detail, we recall the definition of the logarithm of a genus [A2, H1, HS].

DEFINITION 3 (LOGARITHM OF A MULTIPLICATIVE GENUS). Let $\varphi : \Omega_*^{\text{U}} \to R \subset R \otimes \mathbb{Q}$ be a multiplicative genus with values in a torsion free commutative ring R. Then its logarithm $\log_\varphi(X) \in R \otimes \mathbb{Q}[[X]]$ is defined by

$$(11) \qquad \log_\varphi(X) = \sum_{n \ge 0} \frac{\varphi([\mathbb{C}P^n])}{n+1} X^{n+1}.$$

Given a genus, its logarithms completely recovers the original genus if the value ring is torsion free. We now describe the above genera (7), (8), (9) in terms of their logarithms.

THEOREM 4 (LOGARITHMS OF ELLIPTIC GENERA). *The logarithms of elliptic genera* φ_{ell}^R, φ_{ell}^Q, $\varphi_{\text{ell}}^{Q'}$ *are given by elliptic integrals of the following forms:*

(12)
$$\begin{cases} \log_{\varphi_{\text{ell}}^R}(X) = \displaystyle\int_0^X \frac{dt}{\sqrt{1 - 2\delta^R t^2 + t^4}}, & where \\[2ex] \delta^R = \delta^R(q) = \dfrac{1}{2}\left\{ \displaystyle\prod_{\ell \geq 1}\left(\frac{1 + q^{\ell - \frac{1}{2}}}{1 - q^{\ell - \frac{1}{2}}}\right)^4 + \prod_{\ell \geq 1}\left(\frac{1 - q^{\ell - \frac{1}{2}}}{1 + q^{\ell - \frac{1}{2}}}\right)^4 \right\}. \end{cases}$$

(13)
$$\begin{cases} \log_{\varphi_{\text{ell}}^Q}(X) = \displaystyle\int_0^X \frac{dt}{\sqrt{1 + 2\delta^Q t^2 - t^4}}, & where \\[2ex] \delta^Q = \delta^Q(q) = \dfrac{1}{8}q^{(-\frac{1}{4})}\left\{ \displaystyle\prod_{\ell \geq 1}\left(\frac{1 - q^{\ell - \frac{1}{2}}}{1 + q^{\ell}}\right)^4 - 16q^{\frac{1}{2}}\prod_{\ell \geq 1}\left(\frac{1 + q^{\ell}}{1 - q^{\ell - \frac{1}{2}}}\right)^4 \right\}. \end{cases}$$

(14)
$$\begin{cases} \log_{\varphi_{\text{ell}}^{Q'}}(X) = \displaystyle\int_0^X \frac{dt}{\sqrt{1 + 2\delta^{Q'} t^2 + t^4}}, & where \\[2ex] \delta^{Q'} = \delta^{Q'}(q) = \dfrac{1}{8}q^{(-\frac{1}{4})}\left\{ \displaystyle\prod_{\ell \geq 1}\left(\frac{1 + q^{\ell - \frac{1}{2}}}{1 + q^{\ell}}\right)^4 + 16q^{\frac{1}{2}}\prod_{\ell \geq 1}\left(\frac{1 + q^{\ell}}{1 + q^{\ell - \frac{1}{2}}}\right)^4 \right\}. \end{cases}$$

The above elliptic integrals should be expanded into non-negative powers of X. We prove Theorem 4 later in §1.3 using differential equations satisfied by certain elliptic functions. These equations are deduced by manipulating theta functions of half integral characteristics.

But before proving Theorem 4, we list its corollaries. First note that the elliptic genera above have values in a polynomial ring in one variable δ^*.

COROLLARY 5. *The three types of elliptic genera* φ_{ell}^R, φ_{ell}^Q, $\varphi_{\text{ell}}^{Q'}$, *have values in the following rings:*

(15)
$$\begin{aligned} \varphi_{\text{ell}}^R &: \Omega_*^{SO} \to \mathbb{Z}[\tfrac{1}{2}][\delta^R] \cap \mathbb{Z}[[q]], \\ \varphi_{\text{ell}}^Q &: \Omega_*^{Spin} \to \mathbb{Z}[\tfrac{1}{2}][\delta^Q] \cap \mathbb{Z}[q^{-\frac{1}{4}}][[q^{\frac{1}{2}}]], \\ \varphi_{\text{ell}}^{Q'} &: \Omega_*^{Spin} \to \mathbb{Z}[\tfrac{1}{2}][\delta^{Q'}] \cap \mathbb{Z}[q^{-\frac{1}{4}}][[q^{\frac{1}{2}}]]. \end{aligned}$$

We can say something slightly more precise on the values of elliptic genera on complex projective spaces. We recall that the n-th Legendre polynomial $P_n(\delta)$ is defined by

$$\frac{1}{\sqrt{1 - 2\delta t^2 + t^4}} = \sum_{n=0}^{\infty} P_n(\delta)t^{2n},$$

where $P_0(\delta) = 1$ and

(16)
$$P_n(\delta) = \frac{1}{2^n n!} \cdot \frac{d^n}{d\delta^n} (\delta^2 - 1)^n = \sum_{r=0}^{[\frac{n}{2}]} (-1)^r \frac{(2n-2r)!}{2^n r!(n-r)!(n-2r)!} \delta^{n-2r}$$

$$= \sum_{r=0}^{[\frac{n}{2}]} \binom{n}{2r} \binom{2r}{r} \delta^{n-2r} \left(\frac{\delta^2 - 1}{4} \right)^r, \qquad n > 0.$$

A simple calculation shows the following:

COROLLARY 6. *For any of the above elliptic genera in* (15), *their values on complex projective spaces are given by* $\varphi_{\text{ell}}^*([\mathbb{C}P^{2n+1}]) = 0$ *for* $n \geq 0$, *and on* $[\mathbb{C}P^{2n}]$, $n \geq 0$, *we have*

(17)
$$\varphi_{\text{ell}}^R([\mathbb{C}P^{2n}]) = P_n(\delta^R),$$
$$\varphi_{\text{ell}}^Q([\mathbb{C}P^{2n}]) = i^n P_n(i\delta^Q), \qquad i^2 = -1,$$
$$\varphi_{\text{ell}}^{Q'}([\mathbb{C}P^{2n}]) = P_n(-\delta^{Q'}),$$

where δ^R, δ^Q, $\delta^{Q'}$ *are as in* (12), (13), (14).

REMARK. In the usual formulation of elliptic genera [L2], one uses virtual vector bundles $T_{\mathbb{C}} - 4N'1_{\mathbb{C}}$ of virtual dimension 0 instead of an honest bundle $T_{\mathbb{C}}$. The use of virtual bundles has the effect of obtaining elliptic genera whose values are modular forms of *positive weight*. However, we prefer to use non-virtual $T_{\mathbb{C}}$ for several reasons. First of all, when we calculate the signature of a finite dimensional manifold, we use signature complexes and these are not complexes of virtual bundles. So, in the infinite dimensional case it would be more natural not to use virtual bundles. Secondly, when one formally applies Fixed Point Index Formula to free loop spaces with the standard S^1-action, one is naturally lead to expressions which involve only "honest" vector bundles instead of virtual bundles. Thirdly, the use of the "honest" bundle $T_{\mathbb{C}}$ results in elliptic genera with values in a polynomial ring in *one variable* as shown in Corollary 5, whereas the usual version takes values in a polynomial ring in two variables, usually denoted by δ, ε, which are modular forms of weight 2 and 4, respectively, for some index 3 subgroup of $SL_2(\mathbb{Z})$ [L2]. This extra variable is essentially the contribution of trivial bundles $4N'1_{\mathbb{C}}$. Their effect is multiplication by suitable powers of the eta function $\eta(q)$ so that the weight of modular forms is half the dimension of manifolds, and that the multiplicative sequence becomes stable. So, having two variables δ, ε instead of just one is essentially to keep track of dimension of manifolds. As the fourth reason, we mention the paper [HS] in which they compared modular function valued elliptic genera of homogeneous Spin manifolds of *different* dimension to show that the elliptic genera of such manifolds are constants. Being able to directly compare elliptic genera of manifolds of different dimension would prove to be useful. However, our elliptic genera don't have as nice modular properties as the usual one; ours are modular invariant for some index 6 subgroups of $SL_2(\mathbb{Z})$, instead of index 3 subgroups [§1.4 Theorem 2]. But this difference comes from the subtle modular property of the eta function $\eta(q)$ and it is not important for elliptic genera from topological and geometric point of view.

One of the most interesting things about elliptic genera is that they have a very close relationship with theta functions and elliptic functions, allowing us to use analytic properties of these functions to prove topological results. This will be evident when we look at the characteristic power series for these genera.

PROPOSITION 7 (CHARACTERISTIC POWER SERIES). *For each of the elliptic genus* φ_{ell}^*, $* = R, Q, Q'$, *let its characteristic power series be given by* $x/E_*(x)$. *Then* $E_*(x)$ *are given as follows*:

$$(19) \qquad E_R(x) = \tanh\frac{x}{2} \cdot \prod_{\ell \geq 1} \frac{(1 - q^\ell e^x)(1 - q^\ell e^{-x})}{(1 + q^\ell e^x)(1 + q^\ell e^{-x})}$$

$$(20) \qquad E_Q(x) = 2q^{\frac{1}{8}}\sinh\frac{x}{2} \cdot \prod_{\ell \geq 1} \frac{(1 - q^\ell e^x)(1 - q^\ell e^{-x})}{(1 + q^{\ell-\frac{1}{2}}e^x)(1 + q^{\ell-\frac{1}{2}}e^{-x})}$$

$$(21) \qquad E_{Q'}(x) = 2q^{\frac{1}{8}}\sinh\frac{x}{2} \cdot \prod_{\ell \geq 1} \frac{(1 - q^\ell e^x)(1 - q^\ell e^{-x})}{(1 - q^{\ell-\frac{1}{2}}e^x)(1 - q^{\ell-\frac{1}{2}}e^{-x})}.$$

PROOF. Let $P(M^{4n}) = \prod_{j=1}^{2n}(1 + x_j^2)$ be a formal decomposition of the total Pontrjagin class of M^{2n}. In terms of these x_j's, we can easily calculate Chern characters of tensor products of bundles:

$$\text{ch}\left[\bigotimes_{\ell \geq 1} S_{q^\ell}(T_\mathbb{C})\right] = \prod_{j=1}^{2n}\left\{\prod_{\ell \geq 1}(1 - q^\ell e^{x_j})(1 - q^\ell e^{-x_j})\right\}^{-1}$$

$$\text{ch}\left[\bigotimes_{\ell \geq 1} \Lambda_{q^\ell}(T_\mathbb{C})\right] = \prod_{j=1}^{2n}\left\{\prod_{\ell \geq 1}(1 + q^\ell e^{x_j})(1 + q^\ell e^{-x_j})\right\}$$

$$\text{ch}\left[\bigotimes_{\ell \geq 1} \Lambda_{q^{\ell-\frac{1}{2}}}(T_\mathbb{C})\right] = \prod_{j=1}^{2n}\left\{\prod_{\ell \geq 1}(1 + q^{\ell-\frac{1}{2}}e^{x_j})(1 + q^{\ell-\frac{1}{2}}e^{-x_j})\right\}$$

$$\text{ch}\left[\bigotimes_{\ell \geq 1} \Lambda_{(-q^{\ell-\frac{1}{2}})}(T_\mathbb{C})\right] = \prod_{j=1}^{2n}\left\{\prod_{\ell \geq 1}(1 - q^{\ell-\frac{1}{2}}e^{x_j})(1 - q^{\ell-\frac{1}{2}}e^{-x_j})\right\}$$

From these calculations and the fact that

$$\hat{A}(M) = \prod_{j=1}^{2n} \frac{x_j}{2\sinh\frac{x_j}{2}}, \qquad \text{ch}\,\Delta = \prod_{j=1}^{2n} 2\cosh\frac{x_j}{2},$$

the above results follow. $\square$

Note that these power series in x don't have the leading coefficient 1. We need to extend the usual formalism of characteristic power series and logarithm to accommodate the present context.

PROPOSITION 8 (LOGARITHM FOR GENERAL CHARACTERISTIC POWER SERIES). *Let Λ be a torsion free commutative ring. Let a genus $\varphi : \Omega_*^{SO} \to \Lambda$ have the characteristic power series of the form $X/E(X)$ such that $E(X) \equiv \alpha X$ mod (X^2) where $\alpha \in \Lambda$ is a unit in the ring Λ. Then, the logarithm $\log_\varphi(X)$ of the genus φ is given by*

$$(22) \qquad \log_\varphi(X) = \alpha E^{-1}(X) \equiv X \quad \text{mod } (X^2).$$

PROOF. For $\mathbb{C}P^n$, we have $\tau_{\mathbb{C}P^n} \oplus 1_{\mathbb{C}} \cong (2n+1)\xi$, where ξ is the line bundle dual to the tautological line bundle. So, if $P(\mathbb{C}P^n) = \prod_{j=1}^{n}(1 + x_j^2)$ is a formal decomposition of the total Pontrjagin class, then, letting $x = c_1(\xi)$, we have

$$\prod_{j=1}^{n} \frac{x_j}{E(x_j)} = \left(\frac{x}{E(x)}\right)^{2n+1} \cdot \left(\frac{E(x)}{x}\right)\bigg|_{x=0} = \alpha \cdot \left(\frac{x}{E(x)}\right)^{2n+1}.$$

Here, α is the contribution from the trivial line bundle $1_{\mathbb{C}}$. Thus the value of the genus φ on $[\mathbb{C}P^n]$ is given by

$$\varphi([\mathbb{C}P^n]) = \left(\prod_{j=1}^{n} \frac{x_j}{E(x_j)}\right)[\mathbb{C}P^n] = \frac{1}{2\pi i}\int_C \alpha \left(\frac{x}{E(x)}\right)^{2n+1} \cdot \frac{dx}{x^{2n+1}}.$$

Letting $E(x) = y$, we have

$$\varphi([\mathbb{C}P^n]) = \alpha \frac{1}{2\pi i}\int_{C'} \frac{(E^{-1})'(y)}{y^{2n+1}} dy = \alpha \cdot (\text{coefficient of } y^{2n} \text{ in } (E^{-1})'(y)).$$

Since the genus φ is defined on Ω_*^{SO}, $E(X)$ is an odd power series, as is $E^{-1}(X)$. So the above formula means that $\sum_{n \geq 0} \varphi([\mathbb{C}P^{2n}])y^{2n} = \alpha \cdot (E^{-1})'(y)$. In other words, in the integrated version,

$$\sum_{n \geq 0} \frac{\varphi([\mathbb{C}P^{2n}])}{2n+1} y^{2n+1} = \alpha \cdot E^{-1}(y).$$

Thus the logarithm of the genus φ is given by $\alpha \cdot E^{-1}(X)$. $\square$

Note that the logarithm series of a genus is not simply the inverse power series of its exponential series. By definition (11), the logarithm series always starts with the constant term 1.

In view of Proposition 8, to prove Theorem 4 we must find the inverse power series of (19), (20), (21) above. At first, this may look almost an impossible task, since these power series are so huge. But it turns out that as a function in a complex variable x, these series represent elliptic functions. Thus, to find the inverse function, we need the theory of elliptic functions and elliptic integrals. In our present context, the theory of theta functions with half integral characteristics is very convenient. We now turn to these theta functions.

§1.2 Theta functions and elliptic functions

The basic reference for this section is Mumford [Mum]. See also [WW]. Theta functions we use are those with half integral characteristics. There are four of them. These are holomorphic functions in two complex variables $(z, \tau) \in \mathbb{C} \times H$, where H is the upper half plane. For each $\tau \in H$, these theta functions are quasi-periodic holomorphic functions on $\mathbb{C}$ with respect to the lattice $\mathbb{Z} + \mathbb{Z}\tau$. We list various formulae for these theta functions for future reference, starting with their definition.

Theta functions with half integral characteristics. Let $q = e^{2\pi i \tau}$.

(1)
$$\vartheta_{00}(z; \tau) = \sum_{n \in \mathbb{Z}} q^{\frac{1}{2}n^2} e^{2\pi i n z}.$$

$$\vartheta_{01}(z; \tau) = \sum_{n \in \mathbb{Z}} q^{\frac{1}{2}n^2} e^{2\pi i n (z + \frac{1}{2})}.$$

$$\vartheta_{10}(z; \tau) = \sum_{n \in \mathbb{Z}} q^{\frac{1}{2}(n+\frac{1}{2})^2} e^{2\pi i (n+\frac{1}{2}) z}.$$

$$\vartheta_{11}(z; \tau) = \sum_{n \in \mathbb{Z}} q^{\frac{1}{2}(n+\frac{1}{2})^2} e^{2\pi i (n+\frac{1}{2})(z+\frac{1}{2})}.$$

The general scheme is, for $a, b = 0, 1$,

(2)
$$\vartheta_{ab}(z; \tau) = \sum_{n \in \mathbb{Z}} q^{\frac{1}{2}(n+\frac{a}{2})^2} e^{2\pi i (n+\frac{a}{2})(z+\frac{b}{2})}.$$

Transformation properties of theta functions at half quasi-periods.
(I) The quasi periodicity property on the lattice $\mathbb{Z} + \mathbb{Z}\tau$:

(3)
$$\begin{cases} \vartheta_{ab}(z + 1; \tau) = (-1)^a \vartheta_{ab}(z; \tau) \\ \vartheta_{ab}(z + \tau; \tau) = (-1)^b q^{-\frac{1}{2}} e^{-2\pi i z} \vartheta_{ab}(z; \tau) \end{cases}$$

(II) The transformation property of theta functions for $z \mapsto z + \frac{1}{2}$:

(4)
$$\vartheta_{00}(z + \tfrac{1}{2}; \tau) = \vartheta_{01}(z; \tau), \qquad \vartheta_{01}(z + \tfrac{1}{2}; \tau) = \vartheta_{00}(z; \tau)$$
$$\vartheta_{10}(z + \tfrac{1}{2}; \tau) = \vartheta_{11}(z; \tau), \qquad \vartheta_{11}(z + \tfrac{1}{2}; \tau) = -\vartheta_{10}(z; \tau)$$

(III) The transformation property of theta functions for $z \mapsto z + \frac{1}{2}\tau$:

(5)
$$\vartheta_{00}(z + \tfrac{1}{2}\tau; \tau) = q^{-\frac{1}{8}} e^{-\pi i z} \vartheta_{10}(z; \tau),$$
$$\vartheta_{01}(z + \tfrac{1}{2}\tau; \tau) = -i q^{-\frac{1}{8}} e^{-\pi i z} \vartheta_{11}(z; \tau)$$
$$\vartheta_{10}(z + \tfrac{1}{2}\tau; \tau) = q^{-\frac{1}{8}} e^{-\pi i z} \vartheta_{00}(z; \tau),$$
$$\vartheta_{11}(z + \tfrac{1}{2}\tau; \tau) = -i q^{-\frac{1}{8}} e^{-\pi i z} \vartheta_{01}(z; \tau)$$

Although theta functions have many beautiful properties, one of them relevant for our purpose is their product expansion formulae.

Product formulae for theta functions. Let $q = e^{2\pi i \tau}$ and let $q^{\frac{1}{8}} = e^{\frac{\pi i \tau}{4}}$.

(6)

$$\vartheta_{00}(z; \tau) = \prod_{n \geq 1}(1 - q^n) \cdot \prod_{n \geq 1}\left\{(1 + q^{n-\frac{1}{2}}e^{2\pi iz})(1 + q^{n-\frac{1}{2}}e^{-2\pi iz})\right\},$$

$$\vartheta_{01}(z; \tau) = \prod_{n \geq 1}(1 - q^n) \cdot \prod_{n \geq 1}\left\{(1 - q^{n-\frac{1}{2}}e^{2\pi iz})(1 - q^{n-\frac{1}{2}}e^{-2\pi iz})\right\},$$

$$\vartheta_{10}(z; \tau) = 2q^{\frac{1}{8}}\cos \pi z \cdot \prod_{n \geq 1}(1 - q^n) \cdot \prod_{n \geq 1}\left\{(1 + q^n e^{2\pi iz})(1 + q^n e^{-2\pi iz})\right\},$$

$$\vartheta_{11}(z; \tau) = -2q^{\frac{1}{8}}\sin \pi z \cdot \prod_{n \geq 1}(1 - q^n) \cdot \prod_{n \geq 1}\left\{(1 - q^n e^{2\pi iz})(1 - q^n e^{-2\pi iz})\right\}.$$

For $a, b = 0, 1$, $\vartheta_{ab} = \vartheta_{ab}(0; \tau)$ are called *Theta Nullwert* or *theta constants*. Since $\vartheta_{11}(0; \tau) = 0$, we let $\vartheta'_{11} = \vartheta'_{11}(0; \tau)$. These are given as follows.

Theta Nullwerts. Values of $\vartheta_{ab}(z; \tau)$ or their derivatives at $z = 0$:

$$\vartheta_{00} = \prod_{n \geq 1}(1 - q^n) \cdot \prod_{n \geq 1}(1 + q^{n-\frac{1}{2}})^2,$$

$$\vartheta_{01} = \prod_{n \geq 1}(1 - q^n) \cdot \prod_{n \geq 1}(1 - q^{n-\frac{1}{2}})^2,$$

(7)

$$\vartheta_{10} = 2q^{\frac{1}{8}}\prod_{n \geq 1}(1 - q^n) \cdot \prod_{n \geq 1}(1 + q^n)^2,$$

$$\vartheta'_{11} = -2\pi q^{\frac{1}{8}}\prod_{n \geq 1}(1 - q^n) \cdot \prod_{n \geq 1}(1 - q^n)^2.$$

Among these theta constants, there is an important and very useful formula.

Jacobi's derivative formula.

(8) $$\vartheta'_{11} = -\pi\vartheta_{00}\vartheta_{01}\vartheta_{10}.$$

There are several relations among the theta functions with theta constants coefficients.

Equations for ϑ_{ab}'s. In the following formulae, we suppress τ's from our notation for convenience.

$$\vartheta_{00}(z)^2\vartheta_{00}^2 = \vartheta_{01}(z)^2\vartheta_{01}^2 + \vartheta_{10}(z)^2\vartheta_{10}^2$$

(9)

$$\vartheta_{11}(z)^2\vartheta_{00}^2 = \vartheta_{01}(z)^2\vartheta_{10}^2 - \vartheta_{10}(z)^2\vartheta_{01}^2$$

$$\vartheta_{01}(z)^2\vartheta_{00}^2 = \vartheta_{00}(z)^2\vartheta_{01}^2 + \vartheta_{11}(z)^2\vartheta_{10}^2$$

$$\vartheta_{10}(z)^2\vartheta_{00}^2 = \vartheta_{00}(z)^2\vartheta_{10}^2 - \vartheta_{11}(z)^2\vartheta_{01}^2$$

The last two identities in (9) are obtained from the first two identities by the transformation caused by $z \mapsto z + \frac{1}{2}$. Another transformation $z \mapsto z + \frac{1}{2}\tau$ doesn't give us new formulae.

From the above product expansion formulae, one can easily see that each of the above theta functions has one simple zero in each quasi-period lattice. The location of zeros is described below. This information is necessary for Proposition 1 below.

Zeroes of theta functions.

(10)
$$
\begin{aligned}
&\text{(i)} \quad \vartheta_{00}(z;\tau) = 0, \quad \text{if and only if} \quad z \in \tfrac{1}{2}(1+\tau) + \mathbb{Z} + \mathbb{Z}\tau, \\
&\text{(ii)} \quad \vartheta_{01}(z;\tau) = 0, \quad \text{if and only if} \quad z \in \tfrac{\tau}{2} + \mathbb{Z} + \mathbb{Z}\tau, \\
&\text{(iii)} \quad \vartheta_{10}(z;\tau) = 0, \quad \text{if and only if} \quad z \in \tfrac{1}{2} + \mathbb{Z} + \mathbb{Z}\tau, \\
&\text{(iv)} \quad \vartheta_{11}(z;\tau) = 0, \quad \text{if and only if} \quad z \in \mathbb{Z} + \mathbb{Z}\tau.
\end{aligned}
$$

The quotient of any two theta functions above is an elliptic function on a suitable lattice. We are interested in their derivatives and differential equations they satisfy.

PROPOSITION 1 (DERIVATIVES OF ELLIPTIC FUNCTIONS).

(11)
$$
\begin{aligned}
\frac{d}{dz}\left(\frac{\vartheta_{11}(z;\tau)}{\vartheta_{10}(z;\tau)}\right) &= -\pi\vartheta_{10}^2 \, \frac{\vartheta_{00}(z;\tau)\vartheta_{01}(z;\tau)}{\vartheta_{10}(z;\tau)^2}. \\
\frac{d}{dz}\left(\frac{\vartheta_{11}(z;\tau)}{\vartheta_{00}(z;\tau)}\right) &= -\pi\vartheta_{00} \, \frac{\vartheta_{01}(z;\tau)\vartheta_{10}(z;\tau)}{\vartheta_{00}(z;\tau)^2}. \\
\frac{d}{dz}\left(\frac{\vartheta_{11}(z;\tau)}{\vartheta_{01}(z;\tau)}\right) &= -\pi\vartheta_{01}^2 \, \frac{\vartheta_{00}(z;\tau)\vartheta_{10}(z;\tau)}{\vartheta_{01}(z;\tau)^2}.
\end{aligned}
$$

PROOF. We prove the one in the middle. The other formulae can be proved in a similar way, but easier.

First we check the quasi-periodicity property of functions on both sides on the lattice $\mathbb{Z} + \mathbb{Z}\tau$. $\vartheta_{11}(z)/\vartheta_{00}(z)$ changes sign under $z \mapsto z + 1$ and $z \mapsto z + \tau$. So, the same is true for its derivative. The same property can be checked for the right hand side using (3). Thus, their quotient

$$
\phi(z) = \frac{\vartheta'_{11}(z;\tau)\vartheta_{00}(z;\tau) - \vartheta_{11}(z;\tau)\vartheta'_{00}(z;\tau)}{-\pi\vartheta_{00}^2\vartheta_{01}(z;\tau)\vartheta_{10}(z;\tau)}
$$

is periodic, hence elliptic on the lattice $\mathbb{Z} + \mathbb{Z}\tau$. In fact, this function $\phi(z)$ is even periodic on a smaller lattice $\mathbb{Z} + \frac{1+\tau}{2}\mathbb{Z}$. To see this, we calculate $\phi(z + \frac{1+\tau}{2})$ using (4), (5). This can be seen to be equal to

$$
\frac{iq^{-\frac{1}{8}}\left(e^{-\pi iz}\vartheta_{00}(z)\right)' q^{-\frac{1}{8}} e^{-\pi iz}\vartheta_{11}(z) - iq^{-\frac{1}{8}} e^{-\pi iz}\vartheta_{00}(z) q^{-\frac{1}{8}} \left(e^{-\pi iz}\vartheta_{11}(z)\right)'}{-\pi\vartheta_{00}^2 \cdot q^{-\frac{1}{8}} e^{-\pi iz}\vartheta_{10}(z) \cdot (-i)q^{-\frac{1}{8}} e^{-\pi iz}\vartheta_{01}(z)}
$$

After some calculations, this can be seen to be equal to $\phi(z)$.

The only possible poles of $\phi(z)$ come from simple zeroes of $\vartheta_{01}(z;\tau)$ and $\vartheta_{10}(z;\tau)$ and from (10), we see that there is only one possible simple pole for $\phi(z)$ in each period parallelogram for the lattice $\mathbb{Z} + \frac{1+\tau}{2}\mathbb{Z}$. Thus, this elliptic function must be a constant, because the sum of the residues of poles in each period parallelogram must be zero for elliptic functions, and elliptic functions with no poles are constant functions. This constant can be calculated by evaluating $\phi(z)$ at $z = 0$. Using Jacobi's derivative formula (8), we have

$$\phi(0) = \frac{\vartheta'_{11}\vartheta_{00}}{-\pi\vartheta^2_{00}\vartheta_{01}\vartheta_{10}} = 1.$$

This proves the second identity in (11).

The two other identities can be proved similarly. In each case, the quotient is an elliptic function on the lattice $\frac{1}{2}\mathbb{Z} + \mathbb{Z}\tau$, or on $\mathbb{Z} + \frac{\tau}{2}\mathbb{Z}$, respectively. One can easily check that there is only one possible simple pole in each period lattice, hence, they are constant functions.   □

PROPOSITION 2 (DIFFERENTIAL EQUATIONS SATISFIED BY ELLIPTIC FUNCTIONS).

(I) *The elliptic function* $w(z) = \dfrac{\vartheta_{11}(z;\tau)}{\vartheta_{10}(z;\tau)}$ *satisfies the differential equation*

$$(12) \qquad \left(\frac{dw}{dz}\right)^2 = \pi^2\vartheta^2_{00}\vartheta^2_{01}\left(1 + \frac{\vartheta^2_{01}}{\vartheta^2_{00}}w(z)^2\right)\left(1 + \frac{\vartheta^2_{00}}{\vartheta^2_{01}}w(z)^2\right).$$

(II) *The elliptic function* $w(z) = \dfrac{\vartheta_{11}(z;\tau)}{\vartheta_{00}(z;\tau)}$ *satisfies the differential equation*

$$(13) \qquad \left(\frac{dw}{dz}\right)^2 = \pi^2\vartheta^2_{01}\vartheta^2_{10}\left(1 + \frac{\vartheta^2_{10}}{\vartheta^2_{01}}w(z)^2\right)\left(1 - \frac{\vartheta^2_{01}}{\vartheta^2_{10}}w(z)^2\right).$$

(III) *The elliptic function* $w(z) = \dfrac{\vartheta_{11}(z;\tau)}{\vartheta_{01}(z;\tau)}$ *satisfies the differential equation*

$$(14) \qquad \left(\frac{dw}{dz}\right)^2 = \pi^2\vartheta^2_{00}\vartheta^2_{10}\left(1 - \frac{\vartheta^2_{10}}{\vartheta^2_{00}}w(z)^2\right)\left(1 - \frac{\vartheta^2_{00}}{\vartheta^2_{10}}w(z)^2\right).$$

PROOF. We only have to rewrite the square of the derivatives in Proposition 1 using (9). For (I), we use the second and fourth of (9). For (II), we use the third and fourth of (9). For (III), we use the second and third of (9).   □

§1.3   Proof of Theorem 4 of §1.1

We give a proof of Theorem 4 of §1.1. In view of Proposition 8 of §1.1, we must calculate the inverse power series of the power series in Proposition 7 of §1.1. The main point here is that we regard the formal variable x in Proposition 7 of §1.1 as a complex variable and use the analytic theory of theta functions and elliptic functions to find the inverse power series.

LEMMA 1 ($E_*(z)$ IN TERMS OF THETA FUNCTIONS). *The power series $E_*(z)$ in Proposition 7 of §1.1 for $* = R, Q, Q'$ can be written in terms of theta functions as follows:*

(1)
$$E_R(z) = \frac{1}{i} \frac{\vartheta_{11}\left(\frac{z}{2\pi i}; \tau\right)}{\vartheta_{10}\left(\frac{z}{2\pi i}; \tau\right)}, \qquad E_R'(0) = \frac{1}{2}\vartheta_{00}\vartheta_{01}$$

$$E_Q(z) = \frac{1}{i} \frac{\vartheta_{11}\left(\frac{z}{2\pi i}; \tau\right)}{\vartheta_{00}\left(\frac{z}{2\pi i}; \tau\right)}, \qquad E_Q'(0) = \frac{1}{2}\vartheta_{01}\vartheta_{10}$$

$$E_{Q'}(z) = \frac{1}{i} \frac{\vartheta_{11}\left(\frac{z}{2\pi i}; \tau\right)}{\vartheta_{01}\left(\frac{z}{2\pi i}; \tau\right)}, \qquad E_{Q'}'(0) = \frac{1}{2}\vartheta_{00}\vartheta_{10}.$$

PROOF. This is a direct consequence of the product expansion formulae (6) in §1.2. For example, from the third and fourth formulae there, we have

$$\vartheta_{11}\left(\frac{z}{2\pi i}; \tau\right) = 2iq^{\frac{1}{8}}\sinh\frac{z}{2} \cdot \prod_{n\geq 1}(1 - q^n) \cdot \prod_{n\geq 1}\left\{(1 - q^n e^z)(1 - q^n e^{-z})\right\},$$

$$\vartheta_{10}\left(\frac{z}{2\pi i}; \tau\right) = 2q^{\frac{1}{8}}\cosh\frac{z}{2} \cdot \prod_{n\geq 1}(1 - q^n) \cdot \prod_{n\geq 1}\left\{(1 + q^n e^z)(1 + q^n e^{-z})\right\},$$

so, their quotient is $E_R(z)$ given in (19) of §1.1 except for the constant multiple. Also, by direct calculations, we see that $E_R'(0) = -\dfrac{\vartheta_{11}'}{2\pi\vartheta_{10}}$, since $\vartheta_{11}(0; \tau) = 0$. Using Jacobi's derivative formula (8) of §1.2, this is equal to $\frac{1}{2}\vartheta_{00}\vartheta_{01}$. This proves the first set of identities above.

Other two formulae can be shown similarly. □

Since the above functions are quotients of theta functions, they are elliptic functions on some period lattices. We can now use the theory of elliptic functions to calculate their inverse functions.

If we let $w(z) = \dfrac{\vartheta_{11}(z; \tau)}{\vartheta_{10}(z; \tau)}$ as in (12) of §1.2, then $E_R(z) = -i \cdot w\left(\frac{z}{2\pi i}\right)$. From the differential equation in (12) of §1.2 satisfied by $w(z)$, we can easily deduce the equation satisfied by $E_R(z)$, which turns out to be

(2)
$$\left(\frac{dE_R}{dz}\right)^2 = \left(\frac{\vartheta_{00}\vartheta_{01}}{2}\right)^2 \left(1 - \frac{\vartheta_{01}^2}{\vartheta_{00}^2}E_R(z)^2\right)\left(1 - \frac{\vartheta_{00}^2}{\vartheta_{01}^2}E_R(z)^2\right).$$

Let $x = E_R(z)$, so $z = E_R^{-1}(x)$. Then $\dfrac{dE_R^{-1}}{dx} = \dfrac{dz}{dx} = \left(\dfrac{dx}{dz}\right)^{-1} = \left(\dfrac{dE_R}{dz}\right)^{-1}.$

Using the above equation, we have

$$\frac{dE_R^{-1}}{dx} = \frac{2}{\vartheta_{00}\vartheta_{01}} \frac{1}{\sqrt{\left(1 - \frac{\vartheta_{01}^2}{\vartheta_{00}^2}x^2\right)\left(1 - \frac{\vartheta_{00}^2}{\vartheta_{01}^2}x^2\right)}}.$$

Here we should take the positive sign for the square root because (1) above gives $E_R'(0) = \frac{1}{2}\vartheta_{00}\vartheta_{01}$. Since $E_R^{-1}(0) = 0$, $E_R^{-1}(x)$ is given by

$$(3) \qquad E_R^{-1}(x) = \frac{2}{\vartheta_{00}\vartheta_{01}} \int_0^x \frac{dt}{\sqrt{\left(1 - \frac{\vartheta_{01}^2}{\vartheta_{00}^2}x^2\right)\left(1 - \frac{\vartheta_{00}^2}{\vartheta_{01}^2}x^2\right)}}.$$

In view of Proposition 8 of §1.1, we finally have

$$\log_{\varphi_{\mathrm{ell}}^R}(X) = \int_0^x \frac{dt}{\sqrt{1 - \left(\frac{\vartheta_{01}^2}{\vartheta_{00}^2} + \frac{\vartheta_{00}^2}{\vartheta_{01}^2}\right)t^2 + t^4}}.$$

Expressing the theta constants in terms of $q = e^{2\pi i \tau}$ using (7) of §1.2, we have

$$\frac{\vartheta_{01}^2}{\vartheta_{00}^2} + \frac{\vartheta_{00}^2}{\vartheta_{01}^2} = \prod_{n \geq 1} \frac{(1 - q^{n-\frac{1}{2}})^4}{(1 + q^{n-\frac{1}{2}})^4} + \prod_{n \geq 1} \frac{(1 + q^{n-\frac{1}{2}})^4}{(1 - q^{n-\frac{1}{2}})^4}.$$

This finishes the proof of (12) of Theorem 4 in §1.1.

Similarly, from (13) and (14) of §1.2, we can see that $E_Q(z)$ and $E_{Q'}(z)$ satisfy the following differential equations:

$$(4) \qquad \left(\frac{dE_Q}{dz}\right)^2 = \left(\frac{\vartheta_{01}\vartheta_{10}}{2}\right)^2 \left(1 - \frac{\vartheta_{10}^2}{\vartheta_{01}^2}E_Q(z)^2\right)\left(1 + \frac{\vartheta_{01}^2}{\vartheta_{10}^2}E_Q(z)^2\right),$$

$$(5) \qquad \left(\frac{dE_{Q'}}{dz}\right)^2 = \left(\frac{\vartheta_{00}\vartheta_{10}}{2}\right)^2 \left(1 + \frac{\vartheta_{10}^2}{\vartheta_{00}^2}E_{Q'}(z)^2\right)\left(1 + \frac{\vartheta_{00}^2}{\vartheta_{10}^2}E_{Q'}(z)^2\right).$$

Using (1) as before, we can write down their inverse functions:

$$(6) \qquad E_Q^{-1}(x) = \frac{2}{\vartheta_{01}\vartheta_{10}} \int_0^x \frac{dt}{\sqrt{1 + \left(\frac{\vartheta_{01}^2}{\vartheta_{10}^2} - \frac{\vartheta_{10}^2}{\vartheta_{01}^2}\right)t^2 - t^4}},$$

$$(7) \qquad E_{Q'}^{-1}(x) = \frac{2}{\vartheta_{00}\vartheta_{10}} \int_0^x \frac{dt}{\sqrt{1 + \left(\frac{\vartheta_{00}^2}{\vartheta_{10}^2} + \frac{\vartheta_{10}^2}{\vartheta_{00}^2}\right)t^2 + t^4}}.$$

Rewriting these theta constants in terms of q using (7) of §1.2, we get

$$\frac{\vartheta_{01}^2}{\vartheta_{10}^2} - \frac{\vartheta_{10}^2}{\vartheta_{01}^2} = \frac{1}{4}q^{-\frac{1}{4}} \prod_{n \geq 1} \frac{(1 - q^{n-\frac{1}{2}})^4}{(1 + q^n)^4} - 4q^{\frac{1}{4}} \prod_{n \geq 1} \frac{(1 + q^n)^4}{(1 - q^{n-\frac{1}{2}})^4},$$

$$\frac{\vartheta_{00}^2}{\vartheta_{10}^2} + \frac{\vartheta_{10}^2}{\vartheta_{00}^2} = \frac{1}{4}q^{-\frac{1}{4}} \prod_{n \geq 1} \frac{(1 + q^{n-\frac{1}{2}})^4}{(1 + q^n)^4} + 4q^{\frac{1}{4}} \prod_{n \geq 1} \frac{(1 + q^n)^4}{(1 + q^{n-\frac{1}{2}})^4}.$$

Now Proposition 8 of §1.1 gives (13) and (14) of Theorem 4 of §1.1. This completes the proof of Theorem 4 of §1.1.

§1.4   The modular properties of elliptic genera

In this section, we study modular properties of elliptic genera φ^*_{ell} for $* = R, Q, Q'$. This amounts to the study of modular properties of $\delta^*(q)$ appearing in the expressions of their logarithms [§1.1 Theorem 4] as elliptic integrals when $q = e^{2\pi i \tau}$ with $\operatorname{Im} \tau > 0$.

First, we briefly review the concept of modular forms and modular functions of level N, where N is a positive integer. Let

$$(1) \qquad\qquad \gamma_N : \mathrm{SL}_2(\mathbb{Z}) \to \mathrm{SL}_2(\mathbb{Z}/N\mathbb{Z})$$

be the natural reduction map mod N. Its kernel, denoted by $\Gamma(N)$, is called the level N principal congruence subgroup. Explicitly,

$$(2) \qquad \Gamma(N) = \left\{ \begin{pmatrix} a & b \\ c & d \end{pmatrix} \in \mathrm{SL}_2(\mathbb{Z}) \,\middle|\, \begin{pmatrix} a & b \\ c & d \end{pmatrix} \equiv \begin{pmatrix} 1 & 0 \\ 0 & 1 \end{pmatrix} \mod N \right\}.$$

We are primarily interested in three groups containing the level 2 congruence subgroup $\Gamma(2)$. These groups are given by

$$(3)$$
$$\Gamma_\theta = \left\{ \begin{pmatrix} a & b \\ c & d \end{pmatrix} \in \mathrm{SL}_2(\mathbb{Z}) \,\middle|\, \begin{pmatrix} a & b \\ c & d \end{pmatrix} \equiv \begin{pmatrix} 1 & 0 \\ 0 & 1 \end{pmatrix}, \begin{pmatrix} 0 & 1 \\ 1 & 0 \end{pmatrix} \mod 2 \right\}$$
$$= \gamma_2^{-1}\left(\begin{pmatrix} 1 & 0 \\ 0 & 1 \end{pmatrix}, \begin{pmatrix} 0 & 1 \\ 1 & 0 \end{pmatrix} \right)$$

$$(4)$$
$$\Gamma^0(2) = \left\{ \begin{pmatrix} a & b \\ c & d \end{pmatrix} \in \mathrm{SL}_2(\mathbb{Z}) \,\middle|\, b \equiv 0 \mod 2 \right\} = \gamma_2^{-1}\left(\begin{pmatrix} 1 & 0 \\ 0 & 1 \end{pmatrix}, \begin{pmatrix} 1 & 0 \\ 1 & 1 \end{pmatrix} \right)$$

$$(5)$$
$$\Gamma_0(2) = \left\{ \begin{pmatrix} a & b \\ c & d \end{pmatrix} \in \mathrm{SL}_2(\mathbb{Z}) \,\middle|\, c \equiv 0 \mod 2 \right\} = \gamma_2^{-1}\left(\begin{pmatrix} 1 & 0 \\ 0 & 1 \end{pmatrix}, \begin{pmatrix} 1 & 1 \\ 0 & 1 \end{pmatrix} \right)$$

Here, we note that the group $\mathrm{SL}_2(\mathbb{Z}/2\mathbb{Z})$ consists of the following six matrices

$$(6) \qquad \begin{pmatrix} 1 & 0 \\ 0 & 1 \end{pmatrix}, \begin{pmatrix} 0 & 1 \\ 1 & 0 \end{pmatrix}, \begin{pmatrix} 1 & 1 \\ 0 & 1 \end{pmatrix}, \begin{pmatrix} 1 & 1 \\ 1 & 0 \end{pmatrix}, \begin{pmatrix} 1 & 0 \\ 1 & 1 \end{pmatrix}, \begin{pmatrix} 0 & 1 \\ 1 & 1 \end{pmatrix}.$$

We see that all the above three groups have index 3 in $\mathrm{SL}_2(\mathbb{Z})$. As is well known, the group $\mathrm{SL}_2(\mathbb{Z})$ is generated by two elements $T = \begin{pmatrix} 1 & 1 \\ 0 & 1 \end{pmatrix}$ and $S = \begin{pmatrix} 0 & -1 \\ 1 & 0 \end{pmatrix}$. The above three groups are obtained from $\Gamma(2)$ by adding a single generator:

$$(7) \qquad \Gamma_0(2) = \langle \Gamma(2), T \rangle, \quad \Gamma_\theta = \langle \Gamma(2), S \rangle, \quad \Gamma^0(2) = \langle \Gamma(2), TST \rangle.$$

Now we give the definition of modular forms.

DEFINITION 1 (MODULAR FORMS). Let $k \in \mathbb{Z}^+$, $N \in \mathbb{N}$. A holomorphic function $f(\tau)$ on the upper half plane $H = \{\tau \in \mathbb{C} \mid \operatorname{Im} \tau > 0\}$ is called a *modular form of weight k, of level N*, if the following conditions hold:

(1) For any element $\begin{pmatrix} a & b \\ c & d \end{pmatrix} \in \Gamma(N)$, we have

$$f\left(\frac{a\tau + b}{c\tau + d}\right) = (c\tau + d)^k f(\tau), \qquad \tau \in H.$$

(2) There exists constants $M, d > 0$, such that $|f(\tau)| \leq M$ if $\operatorname{Im} \tau > d$.

(3) For any $p/q \in \mathbb{Q}$, there exists positive real numbers $M_{p,q}$ and $d_{p,q}$ such that

$$|f(\tau)| \leq M_{p,q}/(\tau - p/q)^k, \qquad \text{if} \quad |\tau - (p/q + i \cdot d_{p,q})| < d_{p,q}.$$

A modular form of weight 0 is called a *modular function*.

The set of modular forms of weight k and of level N is a vector space denoted by $\operatorname{Mod}_k^{(N)}$. The graded vector space $\operatorname{Mod}^{(N)} = \bigoplus_{k \geq 0} \operatorname{Mod}_k^{(N)}$ has the structure of a ring and the group $\operatorname{SL}_2(\mathbb{Z})/\Gamma(N) = \operatorname{SL}_2(\mathbb{Z}/N\mathbb{Z})$ acts on $\operatorname{Mod}_k^{(N)}$ for any $k \in \mathbb{Z}^+$. More precisely, for $\gamma = \begin{pmatrix} a & b \\ c & d \end{pmatrix} \in \operatorname{SL}_2(\mathbb{Z})$, γ acts on $f \in \operatorname{Mod}_k^{(N)}$ by

$$(8) \qquad (f^\gamma)(\tau) = (c\tau + d)^{-k} \cdot f\left(\frac{a\tau + b}{c\tau + d}\right).$$

This action has the property $f^{(\gamma_1 \gamma_2)} = (f^{\gamma_1})^{\gamma_2}$.

A well known fact states that the ring of level 4 modular forms are generated by squares of theta constants ϑ_{00}, ϑ_{01}, ϑ_{10} [Mum, p52]:

$$(9) \qquad \operatorname{Mod}^{(4)} = \mathbb{C}[\vartheta_{00}^2, \vartheta_{01}^2, \vartheta_{10}^2]/(\vartheta_{00}^4 - \vartheta_{01}^4 - \vartheta_{10}^4).$$

In particular, any powers of these squared theta constants satisfy appropriate growth conditions near $\mathbb{Q} \cup \{\infty\}$. The action of $\operatorname{SL}_2(\mathbb{Z})/\Gamma(4)$ on this ring is discussed later. See (16).

Our goal is to prove modular properties of elliptic genera. In the expressions of $\delta^*(q)$ in Theorem 4 of §1.1, we let $q = e^{2\pi i \tau}$ and let $\delta^*(\tau)$ be the resulting functions on the upper half plane. The next Theorem is our main result in this section, whose proof occupies the rest of this section. For the definition of the groups $\Gamma(\delta^R)$, $\Gamma(\delta^Q)$, $\Gamma(\delta^{Q'})$, see Proposition 9 below.

THEOREM 2 (MODULARITY OF ELLIPTIC GENERA).

(I) (The modularity of δ^*)

(1) $\vartheta_{00}^4 + \vartheta_{01}^4$ and $\vartheta_{00}^4 \vartheta_{01}^4$ are modular forms of weight 2, 4, respectively, for the group $\Gamma_0(2)$. The modular function

$$\delta^R(\tau) = \frac{\vartheta_{00}^4 + \vartheta_{01}^4}{\vartheta_{00}^2 \vartheta_{01}^2}$$

is invariant with respect to an index 2 subgroup $\Gamma(\delta^R)$ of $\Gamma_0(2)$.

(2) $\vartheta_{01}^4 - \vartheta_{10}^4$ and $\vartheta_{01}^4 \vartheta_{10}^4$ are modular forms of weight 2, 4, respectively, for the group Γ_θ. The modular function

$$\delta^Q(\tau) = \frac{\vartheta_{01}^4 - \vartheta_{10}^4}{\vartheta_{01}^2 \vartheta_{10}^2}$$

is invariant with respect to an index 2 subgroup $\Gamma(\delta^Q)$ of Γ_θ.

(3) $\vartheta_{00}^4 + \vartheta_{10}^4$ and $\vartheta_{00}^4 \vartheta_{10}^4$ are modular forms of weight 2, 4, respectively, for the group $\Gamma^0(2)$. The modular function

$$\delta^{Q'}(\tau) = \frac{\vartheta_{00}^4 + \vartheta_{10}^4}{\vartheta_{00}^2 \vartheta_{10}^2}$$

is invariant with respect to an index 2 subgroup $\Gamma(\delta^{Q'})$ of $\Gamma^0(2)$.

The above groups are conjugate to each other. In fact,

$$\Gamma(\delta^{Q'}) = T \cdot \Gamma(\delta^Q) \cdot T^{-1}, \qquad \Gamma(\delta^R) = S \cdot \Gamma(\delta^{Q'}) \cdot S^{-1}.$$

The same conjugation formulae hold for Γ_θ, $\Gamma_0(2)$ and $\Gamma^0(2)$. Namely,

$$\Gamma^0(2) = T \cdot \Gamma_\theta \cdot T^{-1}, \qquad \Gamma_0(2) = S \cdot \Gamma^0(2) \cdot S^{-1}.$$

These groups have the following structure:

$$\Gamma_0(2) = \langle \Gamma(\delta^R), STST^{-1}S^{-1} \rangle, \quad \Gamma_\theta = \langle \Gamma(\delta^Q), S \rangle, \quad \Gamma^0(2) = \langle \Gamma(\delta^{Q'}), TST^{-1} \rangle.$$

(II) (The modularity of elliptic genera) The elliptic genera φ_{ell}^* take values in the following rings of modular functions:

$$\varphi_{\text{ell}}^R : \Omega_*^{SO} \longrightarrow \text{Mod}_0\left(\Gamma(\delta^R)\right) \cap \mathbb{Z}[\tfrac{1}{2}][\delta^R] \cap \mathbb{Z}[[q]],$$

$$\varphi_{\text{ell}}^Q : \Omega_*^{\text{Spin}} \longrightarrow \text{Mod}_0\left(\Gamma(\delta^Q)\right) \cap \mathbb{Z}[\tfrac{1}{2}][\delta^Q] \cap \mathbb{Z}[q^{(-\frac{1}{4})}][[q^{\frac{1}{2}}]],$$

$$\varphi_{\text{ell}}^{Q'} : \Omega_*^{\text{Spin}} \longrightarrow \text{Mod}_0\left(\Gamma(\delta^{Q'})\right) \cap \mathbb{Z}[\tfrac{1}{2}][\delta^{Q'}] \cap \mathbb{Z}[q^{(-\frac{1}{4})}][[q^{\frac{1}{2}}]].$$

Note that the part (II) follows from (I) and Corollary 5 of §1.1. Our method to deduce the modular property is to use the fact that two elliptic functions differ by a constant multiple if and only if they have the same lattices of zeros and poles. But there are many choices of basis for these lattices. Different choices of basis lead to the modular properties above.

We start from the elliptic functions $E_*(z)$ in Lemma 1 of §1.3. We introduce a lattice parameter $\omega \in \mathbb{C}$ which controls the size of lattices and we consider the normalized elliptic function $\hat{E}_*(z)$ given by

(10) $$\hat{E}_*(z) = \frac{\omega}{2\pi i} \cdot \frac{E_*\left(\frac{2\pi i}{\omega} z\right)}{E_*'(0)}, \qquad * = R, Q, Q'.$$

This is normalized in the sense that $\hat{E}_*(0) = 0$, $\hat{E}_*'(0) = 1$. More explicitly in terms of theta functions, we have

$$\hat{E}_R(z) = -\left(\frac{\omega}{2\pi}\right)\left(\frac{2}{\vartheta_{00}\vartheta_{01}}\right)\frac{\vartheta_{11}(\frac{z}{\omega};\tau)}{\vartheta_{10}(\frac{z}{\omega};\tau)},$$

(11)
$$\hat{E}_Q(z) = -\left(\frac{\omega}{2\pi}\right)\left(\frac{2}{\vartheta_{01}\vartheta_{10}}\right)\frac{\vartheta_{11}(\frac{z}{\omega};\tau)}{\vartheta_{00}(\frac{z}{\omega};\tau)},$$

$$\hat{E}_{Q'}(z) = -\left(\frac{\omega}{2\pi}\right)\left(\frac{2}{\vartheta_{00}\vartheta_{10}}\right)\frac{\vartheta_{11}(\frac{z}{\omega};\tau)}{\vartheta_{01}(\frac{z}{\omega};\tau)}.$$

From the differential equations satisfied by $E_*(z)$'s given in (2), (4), (5) of §1.3, we can easily deduce differential equations satisfied by $\hat{E}_*(z)$'s.

LEMMA 3.

(1) *The elliptic function* $u = \hat{E}_R(z)$ *is the unique solution to the differential equation*

(12)
$$\left(\frac{du}{dz}\right)^2 = 1 + \left(\frac{\pi}{\omega}\right)^2 (\vartheta_{00}^4 + \vartheta_{01}^4)\, u^2 + \left(\frac{\pi}{\omega}\right)^4 \vartheta_{00}^4 \vartheta_{01}^4 u^4$$

with the initial condition $u(0) = 0$ *and* $u'(0) = 1$.

(2) *The elliptic function* $u = \hat{E}_Q(z)$ *is the unique solution to the differential equation*

(13)
$$\left(\frac{du}{dz}\right)^2 = 1 - \left(\frac{\pi}{\omega}\right)^2 (\vartheta_{01}^4 - \vartheta_{10}^4)\, u^2 - \left(\frac{\pi}{\omega}\right)^4 \vartheta_{01}^4 \vartheta_{10}^4 u^4$$

with the initial condition $u(0) = 0$ *and* $u'(0) = 1$.

(3) *The elliptic function* $u = \hat{E}_{Q'}(z)$ *is the unique solution to the differential equation*

(14)
$$\left(\frac{du}{dz}\right)^2 = 1 - \left(\frac{\pi}{\omega}\right)^2 (\vartheta_{00}^4 + \vartheta_{10}^4)\, u^2 + \left(\frac{\pi}{\omega}\right)^4 \vartheta_{00}^4 \vartheta_{10}^4 u^4$$

with the initial condition $u(0) = 0$ *and* $u'(0) = 1$.

Next, we locate zeros and poles of the above three elliptic functions. In view of (10) of §1.2, we immediately have the following result using the explicit expressions of these elliptic functions given in (11) above.

LEMMA 4 (LATTICES).

(1) *The elliptic function* $\hat{E}_R(z)$ *has the period lattice* $\omega(\mathbb{Z} + 2\tau\mathbb{Z})$, *simple zeros at* $\omega(\mathbb{Z} + \tau\mathbb{Z})$, *and simple poles at* $\omega(\frac{1}{2} + \mathbb{Z} + \tau\mathbb{Z})$.

(2) *The elliptic function* $\hat{E}_Q(z)$ *has the period lattice* $\omega((1-\tau)\mathbb{Z} + (1+\tau)\mathbb{Z})$, *simple zeros at* $\omega(\mathbb{Z} + \tau\mathbb{Z})$, *and simple poles at* $\omega(\frac{1+\tau}{2} + \mathbb{Z} + \tau\mathbb{Z})$.

(3) *The elliptic function* $\hat{E}_{Q'}(z)$ *has the period lattice* $\omega(2\mathbb{Z} + \tau\mathbb{Z})$, *simple zeros at* $\omega(\mathbb{Z} + \tau\mathbb{Z})$, *and simple poles at* $\omega(\frac{\tau}{2} + \mathbb{Z} + \tau\mathbb{Z})$.

In Lemma 4, the lattices are described in terms of (ω, τ)'s, which are equivalent to choosing bases of lattices. Two bases of the same lattice are related by an element in $SL_2(\mathbb{Z})$, but not all elements in this group preserve both lattices of zeros and poles. We examine subgroups of $SL_2(\mathbb{Z})$ which preserve period lattices and lattices of zeroes and poles for the three cases in Lemma 4.

LEMMA 5 (SUBGROUPS PRESERVING LATTICES). *Suppose (ω, τ) and (ω', τ') are related by $\tau' = \dfrac{a\tau + b}{c\tau + d}$ and $\omega' = (c\tau + d)\omega$ for some $\begin{pmatrix} a & b \\ c & d \end{pmatrix} \in SL_2(\mathbb{Z})$. Then the following statements hold:*
(i) *We have $\gamma \in \Gamma_0(2)$ if and only if*

$$\omega(\mathbb{Z} + 2\tau\mathbb{Z}) = \omega'(\mathbb{Z} + 2\tau'\mathbb{Z}),$$
$$\omega(\mathbb{Z} + \tau\mathbb{Z}) = \omega'(\mathbb{Z} + \tau'\mathbb{Z}), \quad \omega(\tfrac{1}{2} + \mathbb{Z} + \tau\mathbb{Z}) = \omega'(\tfrac{1}{2} + \mathbb{Z} + \tau'\mathbb{Z}).$$

(ii) *We have $\gamma \in \Gamma_\theta$ if and only if*

$$\omega((1 - \tau)\mathbb{Z} + (1 + \tau)\mathbb{Z}) = \omega'((1 - \tau')\mathbb{Z} + (1 + \tau')\mathbb{Z}),$$
$$\omega(\mathbb{Z} + \tau\mathbb{Z}) = \omega'(\mathbb{Z} + \tau'\mathbb{Z}), \quad \omega(\tfrac{1+\tau}{2} + \mathbb{Z} + \tau\mathbb{Z}) = \omega'(\tfrac{1+\tau'}{2} + \mathbb{Z} + \tau'\mathbb{Z}).$$

(iii) *We have $\gamma \in \Gamma^0(2)$ if and only if*

$$\omega(2\mathbb{Z} + \tau\mathbb{Z}) = \omega'(2\mathbb{Z} + \tau'\mathbb{Z}),$$
$$\omega(\mathbb{Z} + \tau\mathbb{Z}) = \omega'(\mathbb{Z} + \tau'\mathbb{Z}), \quad \omega(\tfrac{\tau}{2} + \mathbb{Z} + \tau\mathbb{Z}) = \omega'(\tfrac{\tau'}{2} + \mathbb{Z} + \tau'\mathbb{Z}).$$

PROOF. (i) From the second identity of the lattices, we have that

$$\begin{pmatrix} \omega'\tau' \\ \omega' \end{pmatrix} = \begin{pmatrix} a & b \\ c & d \end{pmatrix} \begin{pmatrix} \omega\tau \\ \omega \end{pmatrix}, \quad \text{for some } \begin{pmatrix} a & b \\ c & d \end{pmatrix} \in SL_2(\mathbb{Z}).$$

This implies that $\tau' = \dfrac{a\tau + b}{c\tau + d}$, $\omega' = (c\tau + d)\omega$. Assuming this, the third identity of lattices is equivalent to the statement that $(\omega' - \omega)/2$ belongs to the lattice $\omega(\mathbb{Z} + \tau\mathbb{Z})$. Since $\omega' - \omega = (c\tau + (d - 1))\omega$, c must be even and d must be odd. This shows that the above matrix actually belongs to $\Gamma_0(2)$ defined in (5). In this case, it is easy to see that $\omega'(\mathbb{Z} + 2\tau'\mathbb{Z}) \subset \omega(\mathbb{Z} + 2\tau\mathbb{Z})$. Since $\Gamma_0(2)$ is a group, the inverse of the above matrix also belongs to the same group and we have the opposite inclusion relation. Thus, the first lattice identity is automatically satisfied.

Conversely, it is immediate to check that elements in $\Gamma_0(2)$ preserve the lattice in part (i).

(ii) As before, the second lattice identity implies the existence of the relations $\tau' = \dfrac{a\tau + b}{c\tau + d}$, $\omega' = (c\tau + d)\omega$, for some $\begin{pmatrix} a & b \\ c & d \end{pmatrix} \in SL_2(\mathbb{Z})$. The third identity is

equivalent to $\omega\mathbb{Z}+\omega\tau\mathbb{Z} \ni (1+\tau')\omega'/2 - (1+\tau)\omega/2 = \{(a+c-1)\tau + (b+d-1)\}\omega/2$.
This implies that $a + c$ is odd, $b + d$ is odd, and $\begin{pmatrix} a & b \\ c & d \end{pmatrix} \equiv \begin{pmatrix} 1 & 0 \\ 0 & 1 \end{pmatrix}, \begin{pmatrix} 0 & 1 \\ 1 & 0 \end{pmatrix}$
mod 2. Thus $\begin{pmatrix} a & b \\ c & d \end{pmatrix} \in \Gamma_\theta$. The first identity is automatically satisfied. The converse is immediate.

(iii) can be shown similarly. $\square$

Now we are ready to prove a part of Theorem 2.

PROOF OF THEOREM 2; PART I. Since our elliptic functions depend on (ω, τ), we denote them by $\hat{E}_*(z; \omega, \tau)$ for $* = R, Q, Q'$. For different (ω', τ') the elliptic functions $\hat{E}_*(z; \omega', \tau')$ satisfy similar differential equations. Namely, $(12), (13), (14)$ with (ω, τ) replaced by (ω', τ').

For the proof of a part of (1) of Theorem 2, we let $\tau' = \dfrac{a\tau + b}{c\tau + d}$, $\omega' = (c\tau + d)\omega$,
where $\begin{pmatrix} a & b \\ c & d \end{pmatrix} \in \Gamma_0(2)$. Then, by Lemma 4 and Lemma 5, both of the elliptic functions $\hat{E}_R(z; \omega, \tau)$ and $\hat{E}_R(z; \omega'\tau')$ have the same zeros and poles and the same period lattices. Thus they differ by a multiplicative constant. Since both functions have the same first derivative at $z = 0$, they must be equal. Thus the differential equations (12) in Lemma 3 satisfied by these functions must be the same. Equating the coefficients, we get

$$\frac{\vartheta_{00}^4(\tau) + \vartheta_{01}^4(\tau)}{\omega^2} = \frac{\vartheta_{00}^4(\tau') + \vartheta_{01}^4(\tau')}{\omega'^2}, \quad \frac{\vartheta_{00}^4(\tau) \cdot \vartheta_{01}^4(\tau)}{\omega^4} = \frac{\vartheta_{00}^4(\tau') \cdot \vartheta_{01}^4(\tau')}{\omega'^4}.$$

Since $\tau' = \dfrac{a\tau + b}{c\tau + d}$, $\omega' = (c\tau + d)\omega$, we have

$$\vartheta_{00}^4\left(\frac{a\tau + b}{c\tau + d}\right) + \vartheta_{01}^4\left(\frac{a\tau + b}{c\tau + d}\right) = (c\tau + d)^2 \left\{\vartheta_{00}^4(\tau) + \vartheta_{01}^4(\tau)\right\}$$

$$\vartheta_{00}^4\left(\frac{a\tau + b}{c\tau + d}\right) \cdot \vartheta_{01}^4\left(\frac{a\tau + b}{c\tau + d}\right) = (c\tau + d)^4 \left\{\vartheta_{00}^4(\tau) \cdot \vartheta_{01}^4(\tau)\right\}.$$

This shows that $\vartheta_{00}^4(\tau) + \vartheta_{01}^4(\tau)$, $\vartheta_{00}^4(\tau)\vartheta_{01}^4(\tau)$ are modular forms of weight 2,4, respectively, for the group $\Gamma_0(2)$. The first half of (2) and (3) of Theorem 2 can be proved in a similar way. $\square$

To prove the rest of Theorem 2, we need more preparation.

LEMMA 6. *The groups* Γ_θ, $\Gamma^0(2)$ *and* $\Gamma_0(2)$ *are conjugate to each other in* $SL_2(\mathbb{Z})$. *In fact, we have* $\Gamma^0(2) = T \cdot \Gamma_\theta \cdot T^{-1}$ *and* $\Gamma_0(2) = S \cdot \Gamma^0(2) \cdot S$.

PROOF. Since the above three groups contain the normal subgroup $\Gamma(2) \subset SL_2(\mathbb{Z})$, we only have to verify the above relations modulo $\Gamma(2)$, that is, in $SL_2(\mathbb{Z}/2\mathbb{Z})$. With respect to the map γ_2 of (1) with $N = 2$, we have

$$\gamma_2(\Gamma_\theta) = \left\{\begin{pmatrix} 1 & 0 \\ 0 & 1 \end{pmatrix}, \begin{pmatrix} 0 & 1 \\ 1 & 0 \end{pmatrix}\right\}, \quad \gamma_2(\Gamma^0(2)) = \left\{\begin{pmatrix} 1 & 0 \\ 0 & 1 \end{pmatrix}, \begin{pmatrix} 1 & 0 \\ 1 & 1 \end{pmatrix}\right\},$$

$$\gamma_2(\Gamma_0(2)) = \left\{\begin{pmatrix} 1 & 0 \\ 0 & 1 \end{pmatrix}, \begin{pmatrix} 1 & 1 \\ 0 & 1 \end{pmatrix}\right\}.$$

Since $\gamma_2(T) = \begin{pmatrix} 1 & 1 \\ 0 & 1 \end{pmatrix}$ and $\gamma_2(S) = \begin{pmatrix} 0 & 1 \\ 1 & 0 \end{pmatrix}$, we have

$$\gamma_2(T \cdot \Gamma_\theta \cdot T) = \begin{pmatrix} 1 & 1 \\ 0 & 1 \end{pmatrix} \gamma_2(\Gamma_\theta) \begin{pmatrix} 1 & 1 \\ 0 & 1 \end{pmatrix} = \gamma_2(\Gamma^0(2)).$$

Thus, $\Gamma^0(2) = T \cdot \Gamma_\theta \cdot T$. The other statement can be checked similarly. $\square$

To complete the proof of Theorem 2, we need to examine the action of the group $SL_2(\mathbb{Z})/\Gamma(4)$ on the ring of level 4 modular forms $\text{Mod}^{(4)}$ generated by ϑ_{00}^2, ϑ_{01}^2, ϑ_{10}^2 described in (9). We have the following table of transformations of theta functions under $S, T \in SL_2(\mathbb{Z})$. (See [Mum, p36].)

Transformations of theta functions.

(15)
$$\vartheta_{00}(z; \tau + 1) = \vartheta_{01}(z; \tau) \qquad \vartheta_{00}(z/\tau; -1/\tau) = (-i\tau)^{\frac{1}{2}} e^{\pi i z^2/\tau} \vartheta_{00}(z; \tau)$$
$$\vartheta_{01}(z; \tau + 1) = \vartheta_{00}(z; \tau) \qquad \vartheta_{01}(z/\tau; -1/\tau) = (-i\tau)^{\frac{1}{2}} e^{\pi i z^2/\tau} \vartheta_{10}(z; \tau)$$
$$\vartheta_{10}(z; \tau + 1) = e^{\pi i/4} \vartheta_{10}(z; \tau) \qquad \vartheta_{10}(z/\tau; -1/\tau) = (-i\tau)^{\frac{1}{2}} e^{\pi i z^2/\tau} \vartheta_{01}(z; \tau)$$
$$\vartheta_{11}(z; \tau + 1) = e^{\pi i/4} \vartheta_{11}(z; \tau) \qquad \vartheta_{11}(z/\tau; -1/\tau) = -(-i\tau)^{\frac{1}{2}} e^{\pi i z^2/\tau} \vartheta_{11}(z; \tau)$$

We can easily calculate the action (8) of $SL_2(\mathbb{Z})$ on the ring $\text{Mod}^{(4)}$ as follows:

(16)
$$[\vartheta_{00}^2(\tau)]^T = \vartheta_{01}^2(\tau), \qquad [\vartheta_{00}^2(\tau)]^S = -i\tau \vartheta_{00}^2(\tau)$$
$$[\vartheta_{01}^2(\tau)]^T = \vartheta_{00}^2(\tau), \qquad [\vartheta_{01}^2(\tau)]^S = -i\tau \vartheta_{10}^2(\tau)$$
$$[\vartheta_{10}^2(\tau)]^T = i\vartheta_{10}^2(\tau), \qquad [\vartheta_{10}^2(\tau)]^S = -i\tau \vartheta_{01}^2(\tau).$$

The next lemma is an immediate consequence of this.

LEMMA 7. *The following transformation properties hold:*

$$\left(\vartheta_{00}^2 \vartheta_{01}^2\right)^S = -\tau^2 \vartheta_{00}^2 \vartheta_{10}^2, \qquad \left(\vartheta_{00}^2 \vartheta_{10}^2\right)^T = i\vartheta_{01}^2 \vartheta_{10}^2, \qquad \left(\vartheta_{01}^2 \vartheta_{10}^2\right)^S = -\tau^2 \vartheta_{01}^2 \vartheta_{10}^2.$$

Lemma 7 and (7), (9) imply that the theta constants $\vartheta_{00}^4 \vartheta_{01}^4$, $\vartheta_{01}^4 \vartheta_{10}^4$, $\vartheta_{00}^4 \vartheta_{10}^4$ are invariant under the action (8) of the groups $\Gamma_0(2)$, Γ_θ, $\Gamma^0(2)$, respectively. However, their square roots may not be invariant and may acquire minus sign when these groups act. We let

(17)
$$\left(\vartheta_{00}^2 \vartheta_{01}^2\right)^\gamma = \varepsilon_R(\gamma) \cdot \vartheta_{00}^2 \vartheta_{01}^2, \quad \text{for } \gamma \in \Gamma_0(2),$$
$$\left(\vartheta_{01}^2 \vartheta_{10}^2\right)^\gamma = \varepsilon_Q(\gamma) \cdot \vartheta_{01}^2 \vartheta_{10}^2, \quad \text{for } \gamma \in \Gamma_\theta,$$
$$\left(\vartheta_{00}^2 \vartheta_{10}^2\right)^\gamma = \varepsilon_{Q'}(\gamma) \cdot \vartheta_{00}^2 \vartheta_{10}^2, \quad \text{for } \gamma \in \Gamma^0(2),$$

Then, ε_R, ε_Q, $\varepsilon_{Q'}$ define homomorphisms,

$$\varepsilon_R : \Gamma_0(2) \to \{\pm 1\}, \quad \varepsilon_Q : \Gamma_\theta \to \{\pm 1\}, \quad \varepsilon_{Q'} : \Gamma^0(2) \to \{\pm 1\}.$$

These maps are nontrivial.

LEMMA 8. *The homomorphisms ε_R, ε_Q, $\varepsilon_{Q'}$ are surjective. In fact,*

$$\varepsilon_R(STST^{-1}S^{-1}) = -1, \quad \varepsilon_Q(S) = -1, \quad \varepsilon_{Q'}(TST^{-1}) = -1.$$

PROOF. We calculate using Lemma 7. First $\left(\vartheta_{01}^2\vartheta_{10}^2\right)^S = -\vartheta_{01}^2\vartheta_{10}^2$. This proves the second statement. Next, the third statement follows from

$$\left(\vartheta_{00}^2\vartheta_{10}^2\right)^{TST^{-1}} = \left(i\vartheta_{01}^2\vartheta_{10}^2\right)^{ST^{-1}} = -i\left(\vartheta_{01}^2\vartheta_{10}^2\right)^{T^{-1}} = -\vartheta_{00}^2\vartheta_{10}^2.$$

Using this calculation, the first statement can be checked as follows :

$$\left(\vartheta_{00}^2\vartheta_{01}^2\right)^{STST^{-1}S^{-1}} = -\left(\vartheta_{00}^2\vartheta_{10}^2\right)^{(TST^{-1})S^{-1}} = \left(\vartheta_{00}\vartheta_{10}\right)^{S^{-1}} = -\vartheta_{00}^2\vartheta_{10}^2.$$

This completes the proof. □

Next we look at the kernels of these homomorphisms.

PROPOSITION 9. *Let ε_R, ε_Q, $\varepsilon_{Q'}$ be as above. Let $\Gamma(\delta^Q) = \ker \varepsilon_Q \subset \Gamma_\theta$, $\Gamma(\delta^R) = \ker \varepsilon_R \subset \Gamma_0(2)$, and $\Gamma(\delta^{Q'}) = \ker \varepsilon_{Q'} \subset \Gamma^0(2)$. All these inclusions have index 2. The theta constants $\vartheta_{00}^2\vartheta_{01}^2$, $\vartheta_{01}^2\vartheta_{10}^2$, $\vartheta_{00}^2\vartheta_{10}^2$ are invariant under the above groups, $\Gamma(\delta^R)$, $\Gamma(\delta^Q)$, $\Gamma(\delta^{Q'})$, respectively. Furthermore, these groups are conjugate to each other in $SL_2(\mathbb{Z})$. Namely,*

$$\Gamma(\delta^{Q'}) = T \cdot \Gamma(\delta^Q) \cdot T^{-1}, \qquad \Gamma(\delta^R) = S \cdot \Gamma(\delta^{Q'}) \cdot S^{-1}.$$

PROOF. Lemma 8 implies that the above inclusions of groups have index 2. So, we only have to prove the conjugation relations. The first one is proved as follows. From Lemma 7 and (17), we have

$$\left(\vartheta_{00}^2\vartheta_{10}^2\right)^{T\cdot\Gamma(\delta^Q)\cdot T^{-1}} = i\left(\vartheta_{01}^2\vartheta_{10}^2\right)^{\Gamma(\delta^Q)\cdot T^{-1}} = i\left(\vartheta_{01}^2\vartheta_{10}^2\right)^{T^{-1}} = \vartheta_{00}^2\vartheta_{10}^2.$$

Thus, $T \cdot \Gamma(\delta^Q) \cdot T^{-1} \subset \Gamma(\delta^{Q'})$. Since both of these groups have index 6 in $SL_2(\mathbb{Z})$, they must be equal. The other identity can be proved similarly. □

(PROOF OF THEOREM 2; PART II). The rest of Theorem 2 is now clear from Lemma 8 and Proposition 9. □

§1.5   On the divisibility of coefficients of the elliptic genus φ_{ell}^R

We consider the elliptic genus $\varphi_{ell}^R : \Omega_*^{SO} \to \mathbb{Z}[[q]]$ in Definition 2 of §1.1. The coefficients of $\varphi_{ell}^R(M^{4n})$ for an oriented manifold M^{4n} is the signature twisted by bundles R_ℓ's given by (3) of §1.1. We study divisibilities of these integers for oriented and Spin manifolds.

The logarithm of φ_{ell}^R is described in Theorem 4 of §1.1:

(*)

$$\log_{\varphi_{\text{ell}}^R}(X) = \int_0^X \frac{dt}{\sqrt{1 - 2\delta t^2 + t^4}},$$

$$\delta = \frac{1}{2}\left\{ \prod_{\ell \geq 1}\left(\frac{1 + q^{\ell - \frac{1}{2}}}{1 - q^{\ell - \frac{1}{2}}}\right)^4 + \prod_{\ell \geq 1}\left(\frac{1 - q^{\ell - \frac{1}{2}}}{1 + q^{\ell - \frac{1}{2}}}\right)^4 \right\}.$$

Here $q = e^{2\pi i \tau}$, $\operatorname{Im} \tau > 0$. So $|q| < 1$.

NOTATION. Let $x_1, x_2, \ldots, x_k, \ldots$ be any collection of elements in $\mathbb{Z}[[q]]$. Let $\mathbb{Z}\{x_1, x_2, \ldots, x_k, \ldots\}$ be the subalgebra generated by the x_k's. This may not be a polynomial algebra since there may be some algebraic relations among the x_k's.

The following theorem is due to Chudnovskys, Landweber, Ochanine, Stong.

THEOREM 1 (IMAGE OF ELLIPTIC GENERA).

Let $\Omega_*^{\text{Spin}} \to \Omega_*^{\text{SO}} \xrightarrow{\varphi_{\text{ell}}^R} \mathbb{Z}[\frac{1}{2}][\delta] \cap \mathbb{Z}[[q]]$ be the elliptic genus.

(1) The image of the Spin bordism ring under φ_{ell}^R is given by

$$\varphi_{\text{ell}}^R(\Omega_*^{\text{Spin}}) = \mathbb{Z}\{16\delta, (8\delta)^2\}.$$

(2) The image of the oriented bordism ring under φ_{ell}^R is given by

$$\varphi_{\text{ell}}^R(\Omega_*^{\text{SO}}) = \mathbb{Z}\{\delta, 2\gamma, 2\gamma^2, \ldots, 2\gamma^{2^s}, \ldots\}, \quad \text{where } \gamma = \frac{\delta^2 - 1}{4}.$$

To study divisibility of the coefficients of $\varphi_{\text{ell}}^R(M)$ for oriented or Spin manifold M, we first examine divisibility of the coefficients of δ and γ.

LEMMA 2. We have the following additive expansions of δ and γ:

$$\delta = 1 + \sum_{m \geq 1} \frac{2^{6m} q^m}{(2m)!} \left\{ \sum_{\ell \geq 0} \frac{1}{2\ell + 1}\left(\frac{q^\ell}{1 - q^{2\ell+1}}\right) \right\}^{2m},$$

$$\gamma = \frac{\delta^2 - 1}{4} = \sum_{m \geq 1} \frac{2^{8m-3} q^m}{(2m)!} \left\{ \sum_{\ell \geq 0} \frac{1}{2\ell + 1}\left(\frac{q^\ell}{1 - q^{2\ell+1}}\right) \right\}^{2m}.$$

All the coefficients of the above powers of q are integers.

PROOF. For $k = 1, 2$, observe that

$$\log\left\{ \prod_{n \geq 1} \frac{(1 + q^{n - \frac{1}{2}})^{4k}}{(1 - q^{n - \frac{1}{2}})^{4k}} \right\} = 4k \sum_{n \geq 1}\left\{ \log(1 + q^{n - \frac{1}{2}}) - \log(1 - q^{n - \frac{1}{2}}) \right\}$$

$$= 8k \sum_{n \geq 1}\sum_{\ell \geq 0} \frac{q^{(2\ell+1)(n - \frac{1}{2})}}{2\ell + 1} = 8k q^{\frac{1}{2}} \sum_{\ell \geq 0} \frac{1}{2\ell + 1}\left(\frac{q^\ell}{1 - q^{2\ell+1}}\right).$$

Here the expansion of log in the second identity, and the summation over n in the last identity make sense since $|q| < 1$.

Taking the exponential and expanding it, we get

$$\prod_{n \geq 1} \frac{(1 + q^{n-\frac{1}{2}})^{4k}}{(1 - q^{n-\frac{1}{2}})^{4k}} = 1 + \sum_{m \geq 1} 2^{3m} q^{\frac{m}{2}} k^m \frac{1}{m!} \left\{ \sum_{\ell \geq 0} \frac{1}{2\ell + 1} \left(\frac{q^\ell}{1 - q^{2\ell+1}} \right) \right\}^m.$$

By taking the part of integral powers of q in this expansion with $k = 1$, we get the formula for δ in view of (∗). As for γ, since

$$\gamma = \frac{\delta^2 - 1}{4} = \frac{1}{2^4} \left\{ \prod_{n \geq 1} \frac{(1 + q^{n-\frac{1}{2}})^8}{(1 - q^{n-\frac{1}{2}})^8} + \prod_{n \geq 1} \frac{(1 - q^{n-\frac{1}{2}})^8}{(1 + q^{n-\frac{1}{2}})^8} - 2 \right\},$$

we take $k = 2$ in the above expansion and take nonconstant integral power part, then divide by 2^3. This gives the formula for γ. □

To extract the information on 2-divisibility from our calculation in Lemma 2, we first note the following fact.

LEMMA 3. *Let $m \in \mathbb{N}$ and let $k \in \mathbb{Z}$ be such that $2^k \leq m < 2^{k+1}$. Let $\nu_2(n)$ be the number of times 2 divides n. Then*

$$\nu_2(m!) = \left[\frac{m}{2}\right] + \left[\frac{m}{4}\right] + \cdots + \left[\frac{m}{2^k}\right] \leq m \left(1 - \frac{1}{2^k}\right).$$

The equality holds when m is a power of 2.

PROOF. We only have to note that $[m/2^r] \leq m/2^r$ is the number of positive integers $\leq m$ which are divisible by 2^r. □

As a simple consequence of these calculations, we have

PROPOSITION 4. *As power series in q, the modular functions δ and γ have the following form of q-expansions:*

$$\delta \in 1 + 2^5 q + 2^5 q^2 \mathbb{Z}[[q]], \qquad \gamma \in 2^4 q + 2^4 q^2 \mathbb{Z}[[q]].$$

PROOF. We calculate $\nu_2(\ )$ of the coefficient of q^m, $m \geq 1$. Let k be an integer such that $2^k \leq m < 2^{k+1}$. Now,

$$\nu_2 \left(\frac{2^{6m}}{(2m)!} \right) = 6m - \nu_2((2m)!) \geq 6m - 2m \left(1 - \frac{1}{2^{k+1}}\right) \geq 4m + 1 \geq 5,$$

for any $m \geq 1$. From this, the statement for δ follows. As for γ, we have

$$\nu_2 \left(\frac{2^{8m-3}}{(2m)!} \right) = 8m - 3 - \nu_2((2m)!) \geq 8m - 3 - 2m \left(1 - \frac{1}{2^{k+1}}\right) \geq 6m - 2 \geq 4.$$

for any $m \geq 1$. This shows that all the coefficients of γ are divisible by 2^4.   □

Let R_ℓ, $\ell \geq 0$ be the bundles defined in (3) of §1.1. Note that $R_\ell = 0$ if ℓ is not integral. For an oriented manifold M of dimension $8n + 4\varepsilon$, $\varepsilon = 0, 1$, we let

$$\varphi_{\text{ell}}^R(M^{8n+4\varepsilon}) = \sum_{0 \leq \ell \in \mathbb{Z}} a_\ell q^\ell,$$

where $a_\ell = \text{Index } d_{R_\ell}$. The divisibility property of coefficients are in a way at random. However, the following theorem shows that for manifolds $M^{8n+4\varepsilon}$ of dimension $8n + 4\varepsilon$ the first n coefficients $a_0, a_1, a_2, \ldots, a_{n-1}$ are obstructions to higher divisibility of the rest of coefficients.

THEOREM 5. *The following statements hold for a closed manifold M:*
(I) *Let M be an oriented manifold of $\dim_{\mathbb{R}} M = 8n + 4\varepsilon$, $\varepsilon = 0, 1$. Suppose for some $k \leq n$, we have $2^{4k+\alpha(k)} | a_\ell$, for $\ell = 0, 1, \ldots, k-1$. Then $2^{4k+\alpha(k)} | \varphi_{\text{ell}}^R(M)$.*
(II) *Let M be a Spin manifold of $\dim_{\mathbb{R}} M = 8n + 4\varepsilon$, $\varepsilon = 0, 1$. Suppose for some $k \leq n$, we have $2^{12k+4\varepsilon} | a_\ell$ for $\ell = 0, 1, \ldots, k-1$. Then $2^{12k+4\varepsilon} | \varphi_{\text{ell}}^R(M)$.*

PROOF. From (2) of Theorem 1, we see that $\varphi_{\text{ell}}^R(\Omega_{8n+4\varepsilon}^{SO})$ is spanned by the following elements:

$$\varphi_{\text{ell}}^R(\Omega_{8n+4\varepsilon}^{SO}) = \bigoplus_{\ell=0}^{n} \mathbb{Z} 2^{\alpha(\ell)} \delta^{2n-2\ell+\varepsilon} \gamma^\ell.$$

Here $\alpha(\ell)$ denotes the number of 1's in its dyadic expansion. Note that in view of Proposition 4, $2^{\alpha(\ell)} \delta^{2n-2\ell+\varepsilon} \gamma^\ell \in 2^{4\ell+\alpha(\ell)} q^\ell (1 + q\mathbb{Z}[[q]])$.

Let $\varphi_{\text{ell}}^R(M) = \sum_{\ell=0}^{n} b_\ell 2^{\alpha(\ell)} \delta^{2n-2\ell+\varepsilon} \gamma^\ell$. Observe that the coefficient a_ℓ of q^ℓ is such that

$$a_\ell = 2^{4\ell+\alpha(\ell)} b_\ell + \{\text{integer linear combination of } 2^{4r+\alpha(r)} b_r, \, 0 \leq r \leq \ell - 1\}.$$

We claim that our hypothesis $2^{4k+\alpha(k)} | a_\ell$ for $0 \leq \ell \leq k - 1$ implies that we have $2^{4k+\alpha(k)} | 2^{4\ell+\alpha(\ell)} b_\ell$ for $\ell = 0, 1, \ldots, k-1$. This can be seen by induction on ℓ. When $\ell = 0$, this is obvious since $a_0 = b_0$. The inductive step goes through because of the above expression of a_ℓ in terms of $b_0, b_1, \ldots, b_\ell$. Therefore, under our hypothesis, discarding the first k terms in $\varphi_{\text{ell}}^R(M)$, we have

$$\varphi_{\text{ell}}^R(M) \equiv \sum_{\ell=k}^{n} b_\ell 2^{\alpha(\ell)} \delta^{2n-2\ell+\varepsilon} \gamma^\ell \quad \text{mod } 2^{4k+\alpha(k)}.$$

But from Proposition 4, we know that $2^{\alpha(\ell)} \delta^{2n-2\ell+\varepsilon} \gamma^\ell \equiv 0 \mod 2^{4k+\alpha(k)}$ when $\ell \geq k$. Hence $\varphi_{\text{ell}}^R(M) \equiv 0 \mod 2^{4k+\alpha(k)}$. This completes the proof of (I).
For (II), first note that from Theorem 1, we have

$$\varphi_{\text{ell}}^R(\Omega_{8n+4}^{\text{Spin}}) = \bigoplus_{\ell=0}^{n} \mathbb{Z} 2^{4+6\ell} \delta^{2\ell+1} = \bigoplus_{\ell=0}^{n} \mathbb{Z} 2^{4+6\ell} \delta(\delta^2 - 1)^\ell,$$

$$\varphi_{\text{ell}}^R(\Omega_{8n}^{\text{Spin}}) = \bigoplus_{\ell=0}^{n} \mathbb{Z} 2^{6\ell} \delta^{2\ell} = \bigoplus_{\ell=0}^{n} \mathbb{Z} 2^{6\ell} (\delta^2 - 1)^\ell.$$

The second equality in each of the above formulae is easy to see. From Lemma 2, we have

$$2^{4\varepsilon+6\ell}\delta^{\varepsilon}(\delta^2 - 1)^{\ell} \equiv 2^{4\varepsilon+12\ell}q^{\ell}(1 + q\mathbb{Z}[[q]])$$

The rest of the argument is the same as (I). This completes the proof. $\square$

COROLLARY 6. *Let M be a closed manifold of dimension* $\dim_{\mathbb{R}} M = 8n + 4\varepsilon$ *with $\varepsilon = 0, 1$.*

(1) *Suppose M is an oriented manifold. If the first n coefficients $a_0, a_1, \ldots,$ a_{n-1} of q-expansion of $\varphi^R_{\mathrm{ell}}(M)$ are divisible by $2^{4n+\alpha(n)}$, then we have* $2^{4n+\alpha(n)}|\varphi^R_{\mathrm{ell}}(M^{8n+4\varepsilon})$.

(2) *Suppose M is a Spin manifold. If the first n coefficients $a_0, a_1, \ldots, a_{n-1}$ of the q-expansion of $\varphi^R_{\mathrm{ell}}(M)$ are divisible by $2^{12n+4\varepsilon}$, then we have* $2^{12n+4\varepsilon}|\varphi^R_{\mathrm{ell}}(M^{8n+4\varepsilon})$.

§1.6   Spinors on loop spaces: Witten genus

Let M^{2N} be a closed Riemannian Spin manifold. Let d denote the Dirac operator on M, $d: \Gamma(\underline{\Delta}^+) \to \Gamma(\underline{\Delta}^-)$, where $\underline{\Delta}^{\pm}$ are the half Spin representations of $\mathrm{Spin}(2N)$ of complex dimension 2^{N-1}. The vector spaces $\mathrm{Ker}\, d$ and $\mathrm{Coker}\, d$ are finite dimensional and they are the vector spaces of harmonic half spinors on M. The $\hat{A}$-genus of M, $\hat{A}(M) \in \mathbb{Z}$, is equal to the index of d:

$$(1) \qquad \hat{A}(M) = \mathrm{index}\,(d) = \dim_{\mathbb{C}}(\mathrm{Ker}\, d) - \dim_{\mathbb{C}}(\mathrm{Coker}\, d).$$

We want to keep the geometric information of these vector spaces of harmonic half spinors.

DEFINITION 1 (GEOMETRIC SPIN INDEX). For a closed Riemannian Spin manifold M^{2N}, we define its Spin index $\hat{\mathcal{A}}(M)$ as a super-pair of vector spaces of half spinors:

$$(2) \qquad \hat{\mathcal{A}}(M) = [\mathrm{Ker}\, d \; ; \; \mathrm{Coker}\, d].$$

In (2) the first vector space $\mathrm{Ker}\, d$ is regarded as the "even" part and $\mathrm{Coker}\, d$ is regarded as the "odd" part, and its super-dimension s-$\dim_{\mathbb{C}}\hat{\mathcal{A}}(M)$ is given by the difference of dimensions and is equal to the $\hat{A}$-genus of M above.

We consider a generalization of the above Spin index. For a complex vector bundle F, we let $S^k(F)$ be its k-th symmetric power bundle and let $S_t(F) = \sum_{0 \leq k \in \mathbb{Z}} S^k(F)t^k$ be a formal power series in t with vector bundle coefficients. Let bundles S_{ℓ}, $0 \leq \ell \in \mathbb{Z}$, be defined by

$$(3) \qquad \underline{S}_q = \sum_{0 \leq \ell \in \mathbb{Z}} S_{\ell}q^{\ell} = \bigotimes_{0 < m \in \mathbb{Z}} S_{q^m}\,(T_{\mathbb{C}}M(-m))\,.$$

Here, $T_{\mathbb{C}}M(-m)$ is the complexified cotangent bundle carrying weight n. The parameter q will be regarded as $q = e^{2\pi i \tau}$, $\mathrm{Im}\, \tau > 0$ later. We consider Dirac operators twisted by S_{ℓ}'s:

$$(4) \qquad d_{S_{\ell}}: \Gamma(\underline{\Delta}^+ \otimes S_{\ell}) \to \Gamma(\underline{\Delta}^- \otimes S_{\ell}), \qquad 0 \leq \ell \in \mathbb{Z}.$$

For details of twisted Dirac operators, see §5.3. Note that $S_0 = \mathbb{C}$ is the trivial 1-dimensional vector bundle on M, and d_{S_0} is the usual Dirac operator on M.

DEFINITION 2 (WITTEN GENUS). Let M^{2N} be a closed Riemannian Spin manifold. Its Witten genus, denoted by $\Phi_\theta(M)$, is a super-pair of $\mathbb{Z}$-graded vector spaces of twisted spinors given by

$$(5) \qquad \Phi_\theta(M^{2N}) = \left[\bigoplus_{0 \le \ell \in \mathbb{Z}} (\text{Ker } d_{S_\ell}) q^{\ell - \frac{N}{12}} \; ; \; \bigoplus_{0 \le \ell \in \mathbb{Z}} (\text{Coker } d_{S_\ell}) q^{\ell - \frac{N}{12}} \right].$$

Its graded super-dimension s-dim$_*\Phi_\theta(M^{2N})$ is denoted by $\varphi_\theta(M)$. It is a series of indices of operators given by

$$(6) \qquad \varphi_\theta(M^{2N}) = q^{-\frac{N}{12}} \sum_{0 \le \ell \in \mathbb{Z}} (\text{Index } d_{S_\ell}) q^\ell \in q^{-\frac{N}{12}} \mathbb{Z}[[q]].$$

Note that the lowest weight super-pair (the coefficient of $q^{-N/12}$) in the Witten genus $\Phi_\theta(M)$ is the geometric Spin index $\hat{A}(M)$ in Definition 1. Witten [W3] showed that a formal expression of the S^1-equivariant index of Dirac-like operators on the loop space LM is given by the above formula. By index theory, the numerical Witten genus $\varphi_\theta(M)$ is given by

$$(7) \qquad \varphi_\theta(M^{2N}) = q^{-\frac{N}{12}} \hat{A}(M) \, [\text{ch } \underline{S}_q] \, [M^{2N}].$$

Here, by convention, $\hat{A}(M)$ in the above formula denotes the $\hat{A}$ series which is a polynomial in Pontrjagin classes of M, rather than the integer $\hat{A}$-genus of M.

Both $\hat{A}(M)$ and $\varphi_\theta(M)$ are multiplicative genera on Spin manifolds. From the general theory of multiplicative genera φ, if $P(M) = \prod_{i=1}^N (1 + x_i^2)$ is the formal decomposition of the total Pontryagin class, then the value of the genus on M is given by $\varphi(M^{2N}) = \prod_{i=1}^N x_i / g^{-1}(x_i)$, where $g(X)$ is the logarithm series given by $g(X) = \sum_{n \ge 0} \varphi([\mathbb{C}P^n]) X^{n+1} / (n+1)$. Elliptic genera are those genera whose characteristic power series are of the form $x/g^{-1}(x)$ where $g^{-1}(x)$ is an elliptic function in x vanishing once at $x = 0$. Contrary to our expectation, the Witten genus $\varphi_\theta(M)$ is NOT an elliptic genus. It is a "theta" genus, as our notation $\varphi_\theta(M)$ indicates. To describe this, we recall the Dedekind η-function $\eta(q)$, $q = e^{2\pi i \tau}$, and a theta function $\vartheta_{11}(z; \tau)$ with half integral characteristics:

$$\eta(q) = q^{\frac{1}{24}} \prod_{m=1}^\infty (1 - q^m), \qquad \vartheta_{11}(z; \tau) = \sum_{n \in \mathbb{Z}} e^{\pi i \tau (n + \frac{1}{2})^2} e^{2\pi i (n + \frac{1}{2})(z + \frac{1}{2})}.$$

PROPOSITION 3 (WITTEN GENUS AS A THETA GENUS). *The characteristic power series* $x/g^{-1}(x)$ *of the Witten genus* $\varphi_\theta : \Omega_*^{\text{Spin}} \to \mathbb{Z}[q^{-\frac{1}{6}}][[q]]$ *is given by*

$$g^{-1}(x) = \frac{\vartheta_{11}(\frac{x}{2\pi i}; \tau)}{i \eta(q)} = 2q^{\frac{1}{12}} \sinh \frac{x}{2} \cdot \prod_{m=1}^\infty (1 - q^m e^x)(1 - q^m e^{-x}).$$

PROOF. Let $P(M^{2N}) = \prod_{j=1}^{N}(1 + x_j^2)$ be the formal decomposition of the total Pontryagin class. Then,

$$q^{-\frac{N}{12}} \hat{A}(M) \operatorname{ch} \underline{S}_q = q^{-\frac{N}{12}} \prod_{j=1}^{N} \left(\frac{x_j/2}{\sinh(x_j/2)} \prod_{m=1}^{\infty} \frac{1}{(1 - q^m e^{x_j})(1 - q^m e^{-x_j})} \right).$$

Thus, the characteristic power series $x/g^{-1}(x)$ is given by

$$\frac{x}{g^{-1}(x)} = \frac{x}{2q^{\frac{1}{2}} \sinh(x/2) \cdot \prod_{m=1}^{\infty} \{(1 - q^m e^x)(1 - q^m e^{-x})\}}.$$

To see that $g^{-1}(x)$ is the quotient of the theta function by the eta function, we use the product formula for the theta function $\vartheta_{11}(z; \tau)$ given in §1.2, (6):

$$\vartheta_{11}(z; \tau) = -2q^{\frac{1}{8}} \sin \pi z \cdot \prod_{m=1}^{\infty}(1 - q^m) \cdot \prod_{m=1}^{\infty} \{(1 - q^m e^{2\pi i z})(1 - q^m e^{-2\pi i z})\}.$$

From this formula, we immediately see that

$$\frac{\vartheta_{11}(\frac{x}{2\pi i}, \tau)}{i\eta(q)} = \frac{-2q^{\frac{1}{8}} \cdot \frac{1}{i} \sinh(x/2) \cdot \prod_{m=1}^{\infty}(1 - q^m) \cdot \prod_{m=1}^{\infty} \{(1 - q^m e^x)(1 - q^m e^{-x})\}}{iq^{\frac{1}{24}} \prod_{m=1}^{\infty}(1 - q^m)}$$

$$= 2q^{\frac{1}{2}} \sinh \frac{x}{2} \cdot \prod_{m=1}^{\infty} \{(1 - q^m e^x)(1 - q^m e^{-x})\}.$$

This completes the proof. □

CHAPTER II

VERTEX OPERATOR SUPER ALGEBRAS

§2.1   Vertex operator super algebras: a brief introduction

So far, we have only discussed properties of numerical elliptic genera $\varphi^*_{\mathrm{ell}}(M)$ for $* = R, Q, Q'$ defined in Definition 2 of §1.1 as modular function valued cobordism invariants. We turn our attention to the geometric elliptic genera $\Phi^*_{\mathrm{ell}}(M)$ as defined in Definition 1 of §1.1. The value of this genus is a super-pair of infinite dimensional graded vector spaces. We are interested in internal structures of these graded vector spaces which reflect global properties of the closed Riemannian Spin manifold M. It turns out that the elliptic genus $\Phi^*_{\mathrm{ell}}(M)$ is a module over various infinite dimensional Lie algebras such as Virasoro algebras, affine Lie algebras, (infinite dimensional) Clifford algebras, etc. Although these algebras are rather different in nature, there is an encompassing notion which incorporates all of the above algebras. This is the notion of *vertex operator (super) algebras*. Our result discussed in Chapter 5 says that the geometric elliptic genus $\Phi^*_{\mathrm{ell}}(M)$ of a closed Riemannian Spin manifold M has the structure of a module over a certain vertex operator super algebra $\mathcal{P}_M(\underline{V})$ which depends on the Riemannian structure on M. Recall that each vector of finite weight in $\Phi^*_{\mathrm{ell}}(M)$ is a twisted harmonic spinor which lives in some twisted Spin bundle on M. Our machinery of vertex operator super algebra enables us to creat infinitely many twisted harmonis spinors living in various twisted Spin bundles from any vector in the geometric elliptic genus.

The vertex operator super algebra and its $\mathbb{Z}_2$-twisted module relevant to elliptic genera are constructed as the level 1 spin representations of the orthogonal affine Lie algebras [§2.2].

Before discussing the details on the action of the vertex operator super algebra $\mathcal{P}_M(\underline{V})$ on the elliptic genus $\Phi^*_{\mathrm{ell}}(M)$ in Chapter 5, we briefly describe generalities of vertex operator (super) algebras [FFR, FLM, FHL, Ge]. Although references on vertex operator algebras are readily available in literature, here we give a *very gentle* introduction to vertex operator algebras: we do not cover much materials, but we have given ample details on the materials covered.

Our main object is a half-intgrally and non-negatively graded vector space

$$(1) \qquad\qquad V = \bigoplus_{\substack{n \in \frac{1}{2}\mathbb{Z} \\ n \geq 0}} (V)_n,$$

where $(V)_n = \{0\}$ for $n < 0$. The graded vector space V can be thought of as a

$\mathbb{Z}_2$ graded vector space by letting $V = V_0 \bigoplus V_1$, where

(2)
$$V_0 = \bigoplus_{\substack{n \in \mathbb{Z} \\ n \geq 0}} (V)_n, \qquad V_1 = \bigoplus_{\substack{n \in \mathbb{Z} \\ n \geq 0}} (V)_{n+\frac{1}{2}}.$$

A vector $v \in (V)_n$ has weight n and the weight of v is denoted by $\mathrm{wt}(v)$. We let $|v| = 0, 1$ according as $v \in V_0$ or V_1. So, $|v| \equiv 2\,\mathrm{wt}(v) \mod 2$.

Let ζ be a formal variable and let $V[[\zeta, \zeta^{-1}]]$ be a vector space of formal Laurent series with coefficients in V. A vertex operator super algebra is a quadruple $(V, \mathrm{ad}_\zeta, 1, \omega)$ with the following properties. The weight 0 vector $1 \in (V)_0$ is the vacuum vector, also denoted by Ω, generating a 1-dimensional subspace of $(V)_0$. If $(V)_0 = \mathbb{C}\Omega$, we call the vertex operator super algebra V *connected*. This terminology comes from our geometrical considerations of the algebra. See Chapter 5 for details, where it is explained that a connected Riemannian manifold give rise to connected vertex operator super algebra denoted by $\mathcal{P}_M(\underline{V})$. The main structure of a vertex operator super algebra is an injective linear map

(3)
$$\mathrm{ad}_\zeta : V \to \mathrm{End}[[\zeta, \zeta^{-1}]]$$

(4)
$$\mathrm{ad}_\zeta(v) = \sum_{n \in \mathbb{Z}} \{v\}_n \zeta^{-n-1} \equiv Y(v, \zeta)$$

$Y(v, \zeta)$ is called the *vertex operator* associated to the vector v. Thus to each vector $v \in V$ the map ad_ζ associates an infinitely many component operators $\{v\}_n : V \to V$ for $n \in \mathbb{Z}$ acting on V. For a homogeneous vector $v \in V$, the corresponding operator $\{v\}_n$ lowers weight by $n + 1 - \mathrm{wt}(v)$. These operators are required to satisfy the following Jacobi identity:

Jacobi identity. For any v_1 and v_2 in V and any $m, n, r \in \mathbb{Z}$,

(5)
$$\sum_{0 \leq i \in \mathbb{Z}} (-1)^i \binom{r}{i} \Big(\{v_1\}_{m+r-i} \{v_2\}_{n+i} - (-1)^{|v_1||v_2|+r} \{v_2\}_{n+r-i} \{v_1\}_{m+i} \Big)$$
$$= \sum_{0 \leq k \in \mathbb{Z}} \binom{m}{k} \{\{v_1\}_{r+k} v_2\}_{m+n-k}.$$

On the right hand side, the summation is finite, since V is non-negatively graded. For $1 = \Omega \in (V)_0$ and $\omega \in (V)_2$, the corresponding vertex operators are

(6)
$$Y(\Omega, \zeta) = \mathrm{Id}_V,$$

(7)
$$Y(\omega, \zeta) = \sum_{n \in \mathbb{Z}} D(n) \zeta^{-n-2},$$

where $D(n) = \{\omega\}_{n+1}$ for $n \in \mathbb{Z}$ generate the Virasoro algebra satisfying the following commutation relation for any $m, n \in \mathbb{Z}$:

(8)
$$[D(m), D(n)] = (n - m) D(m + n) + \frac{m^3 - m}{12} \delta_{n+m,0} c \cdot \mathrm{Id}_V.$$

Here, c is a constant called the rank of V. Operators $D(-1)$, $D(0)$ and $D(1)$ generate a copy of the Lie algebra $\mathfrak{sl}_2(\mathbb{C})$. We require that

$$(9) \qquad\qquad [D(-1), Y(v, \zeta)] = -\frac{d}{d\zeta} Y(v, \zeta),$$

$$(10) \qquad\qquad D(0)v = -nv, \qquad \text{if } n \in (V)_n.$$

If all vectors in V have integral weight, then we have vertex operator algebras. This completes the description of the vertex operator super algebra.

Instead of (4), if we write $Y(v, \zeta) = \sum_{m \in \mathbb{Z}} Y_m(v) \zeta^{-m - \text{wt}(v)}$, then

$$(11) \qquad\qquad \text{wt}(Y_m(v)w) = \text{wt}(w) - m.$$

So, $Y_m(v)$ lowers the weight by $m \in \mathbb{Z} + \text{wt}(v)$.

If we specialize the Jacobi identity by letting $r = 0$, we get

Super commutator formula. For any $v_1, v_2 \in V$ and $m, n \in \mathbb{Z}$,

$$(12) \quad \{v_1\}_m \{v_2\}_n - (-1)^{|v_1||v_2|} \{v_2\}_n \{v_1\}_m = \sum_{0 \leq k \in \mathbb{Z}} \binom{m}{k} \{\{v_1\}_k v_2\}_{m+n-k}.$$

Again, the summation on the right hand side is finite. This shows that the operators $\{v\}_n$ are closed under Lie super bracket and $V \otimes \mathbb{C}[t, t^{-1}]$ has a structure of a Lie super algebra.

Another interesting property of vertex operator (super) algebras is its skew symmetry.

Skew Symmetry. For any $u, v \in V$, we have

$$(12') \qquad\qquad Y(u, \zeta)v = e^{-\zeta D(-1)} Y(v, -\zeta)u.$$

This skew symmetry formula has interesting applications. For example, see Lemma 2 below.

For an application to elliptic genera, we need the notion of a $\mathbb{Z}_2$-*twisted module* $W = W_0 \oplus W_1$ over a vertex operator super algebra $V = V_0 \oplus V_1$. For a homogeneous $v \in V$, the vertex operator $Y(v, \zeta)$ defining the module structure is in $\text{End}(W)[[\zeta^{\frac{1}{2}}, \zeta^{-\frac{1}{2}}]]$ and

$$(13) \qquad\qquad Y(v, \zeta) = \sum_{n \in \mathbb{Z} + \frac{1}{2}|v|} \{v\}_n \zeta^{-n-1}.$$

The Jacobi identity (5) for these operators should be modified by letting $m \in \mathbb{Z} + \frac{1}{2}|v_1|$, $n \in \mathbb{Z} + \frac{1}{2}|v_2|$. See §2.2 for an example. For the definition of general G-twisted modules of a vertex operator algebra V admitting an action of a finite group G, see [DM].

Consequences of the commutator formula. As simple consequences of the commutator formula (12), we prove in detail a well known fact that if V is a connected vertex operator super algebra, the vertex operators corresponding to weight 1/2 vectors generate a Clifford algebra, and the vertex operators corresponding to weight 1 vectors generate an affine Lie algebra, under an additional mild assumption.

For a vector $a \in (V)_{\frac{1}{2}}$, the n-th component of the associated vertex operator $\{v\}_n$ lowers the weight by $n + \frac{1}{2}$. If V is connected, then $\{a\}_0 b \in (V)_0 = \mathbb{C}\Omega$ for $a, b \in (V)_{\frac{1}{2}}$. We let

(14) $$\{a\}_0 b \equiv \langle a, b \rangle \Omega, \qquad \text{for } a, b \in (V)_{\frac{1}{2}}.$$

This defines a bilinear pairing $\langle \ , \ \rangle : (V)_{\frac{1}{2}} \times (V)_{\frac{1}{2}} \to \mathbb{C}$.

PROPOSITION 1. *Suppose V is a connected vertex operator super algebra. The above pairing $\langle \ , \ \rangle$ on $(V)_{\frac{1}{2}}$ is a $\mathbb{C}$-linear symmetric bilinear pairing on $(V)_{\frac{1}{2}}$. Furthermore, vertex operators associated to vectors in $(V)_{\frac{1}{2}}$ realize a representation of an infinite dimensional Clifford algebra on V. More precisely, for any $a \in (V)_{\frac{1}{2}}$, let*

(15) $$Y(a, \zeta) = \sum_{n \in \mathbb{Z}} a(n + \tfrac{1}{2}) \zeta^{-n-1}.$$

Then operator $a(n + \frac{1}{2})$ for $n \in \mathbb{Z}$ lowers weight by $n + \frac{1}{2}$ and satisfy the following commutation relation:

(16) $$a(m + \tfrac{1}{2})b(-n - \tfrac{1}{2}) + b(-n - \tfrac{1}{2})a(m + \tfrac{1}{2}) = \langle a, b \rangle \delta_{m,n} \cdot \text{Id}.$$

PROOF. Let $a, b \in (V)_{\frac{1}{2}}$. Since these are vectors in the "odd" part of the vertex operator super algebra V, we have $|a| = |b| = 1$ and the super commutator formula (12) gives

$$\{a\}_m \{b\}_n + \{b\}_n \{a\}_m = \sum_{0 \le k \in \mathbb{Z}} \binom{m}{k} \{\{a\}_k b\}_{m+n-k}.$$

In the summation on the right hand side, only the first term survives nontrivially since $\{a\}_n$ lowers the weight by $n + \frac{1}{2}$ and $\{a\}_k b \in (V)_{-k} = \{0\}$ if $k \ge 1$. Since $\{\{a\}_0 b\}_{m+n} = \langle a, b \rangle \{\Omega\}_{m+n} = \langle a, b \rangle \delta_{m+n,-1} \cdot \text{Id}$ by the definition of the pairing $\langle \ , \ \rangle$ given in (14), the above relation becomes

$$a(m + \tfrac{1}{2})b(n + \tfrac{1}{2}) + b(n + \tfrac{1}{2})a(m + \tfrac{1}{2}) = \langle a, b \rangle \delta_{m+n,-1} \cdot \text{Id},$$

which is equivalent to (16). Here we have put $\{a\}_m = a(m + \frac{1}{2})$. Finally, the pairing is symmetric since the left hand side of the above pairing is symmetric in $a(m + \frac{1}{2})$ and $b(n + \frac{1}{2})$. $\square$

Now, we examine $(V)_1$ consisting of weight 1 vectors. First we show that $(V)_1$ has the structure of a Lie algebra. For convenience we let $\mathfrak{g} = (V)_1$ to indicate that it has the structure of a Lie algebra. Let $x, y \in \mathfrak{g}$. The vertex operator $\{x\}_0 : V \to V$ preserves the weight of V since $\mathrm{wt}(x) = 1$, and $\{x\}_0 y \in \mathfrak{g}$. We use this fact to define the bracket product $[\ ,\ ] : \mathfrak{g} \times \mathfrak{g} \to \mathfrak{g}$ for $\mathfrak{g}$. Namely,

$$(17) \qquad [x, y] \equiv \{x\}_0 y \in \mathfrak{g}, \qquad \text{for any } x, y \in \mathfrak{g}.$$

It turns out that this makes $\mathfrak{g}$ into a Lie algebra.

LEMMA 2. *Let V be a connected vertex operator (super) algebra. Then $\mathfrak{g} = (V)_1$ has the structure of a Lie algebra with respect to the bracket product given in (17).*

PROOF. In the super commutator formula (12), we let $v_1 = x$, $v_2 = y$, $|v_1| = |v_2| = 0$, and $m = n = 0$. We then obtain

$$(18) \qquad [\{x\}_0, \{y\}_0] = \sum_{0 \leq k \in \mathbb{Z}} \binom{0}{k} \{\{x\}_k y\}_{-k} = \{\{x\}_0 y\}_0 .$$

Applying the operators in both sides to $z \in \mathfrak{g}$,

$$\{x\}_0 \{y\}_0 z - \{y\}_0 \{x\}_0 z = \{\{x\}_0 y\}_0 z$$

or equivalently, in terms of the bracket defined in (17),

$$[x, [y, z]] - [y, [x, z]] = [x, [y, z]],$$

which is the Jacobi identity for the bracket product (17) for $\mathfrak{g}$.

In the skew symmetry formula (12'), letting $u = x$ and $v = y$ be weight 1 elements and taking the coefficients of ζ^{-1} of both sides, we get

$$\{u\}_0 v = -\{v\}_0 u - D(-1)\{v\}_1 u - \sum_{m \geq 2} D(-1)^m \left(\frac{\{v\}_m u}{m!} \right).$$

Since the weight of $\{v\}_m u$ is $1 - m < 0$ for $m \geq 2$, the summation above vanishes. Since V is connected, $\{v\}_1 u$ is a constant multiple of the vacuum vector Ω. Since $D(-1)\Omega = 0$, the above formula reduces to $[u, v] = -[v, u]$, which is he skew symmetry of bracket product. $\square$

Now, we consider the commutator formula for the vertex operators corresponding to $x, y \in \mathfrak{g}$. For any $m, n \in \mathbb{Z}$, we have

$$[\{x\}_m, \{y\}_n] = \sum_{0 \leq k \in \mathbb{Z}} \binom{m}{k} \{\{x\}_k y\}_{m+n-k}$$

Since $\{x\}_k y \in (V)_{1-k} = 0$ for $k \geq 2$, only the first two terms can be nontrivial. Using $\{x\}_0 y = [x, y]$, this commutator is equal to

$$= \{[x, y]\}_{m+n} + m\{\{x\}_1 y\}_{m+n-1}.$$

Since $wt(x) = 1$, the operator $\{x\}_1$ lowers the weight by 1 and we have $\{x\}_1 y \in (V)_0 = \mathbb{C}\Omega$ if V is connected. In this case, we let

$$(19) \qquad \{x\}_1 y = \langle x, y \rangle \Omega.$$

This defines a bilinear pairing $\langle \ , \ \rangle : \mathfrak{g} \times \mathfrak{g} \to \mathbb{C}$. From our requirement (6), we have $\{\Omega\}_n = \delta_{n,-1} \mathrm{Id}_V$. So, $\{\{x\}_1 y\}_{m+n-1} = \langle x, y \rangle \{\Omega\}_{m+n-1} = \langle x, y \rangle \delta_{m,-n} \cdot \mathrm{Id}$ and the commutator formula now becomes

$$(20) \qquad [\{x\}_m, \{y\}_n] = \{[x, y]\}_{m+n} + m\langle x, y \rangle \delta_{m,-n} \cdot \mathrm{Id}.$$

This commutation relation defines an affine Lie algebra, if the bilinear pairing $\langle \ , \ \rangle$ is symmetric and $\mathfrak{g}$-invariant. This turns out to be the case.

LEMMA 3. *Let V be a connected vertex operator super algebra. Then the $\mathbb{C}$-bilinear pairing $\langle \ , \ \rangle : \mathfrak{g} \times \mathfrak{g} \to \mathbb{C}$ in (19) is a symmetric $\mathfrak{g}$-invariant pairing: for any $x, y, z \in \mathfrak{g}$, we have*

$$(21) \qquad \langle x, y \rangle = \langle y, x \rangle, \qquad \langle [z, x], y \rangle + \langle x, [z, y] \rangle = 0.$$

PROOF. In the commutator formula (20), we let $m = -n$. Then,

$$(22) \qquad [\{x\}_{-n}, \{y\}_n] = \{[x, y]\}_0 - n\langle x, y \rangle \cdot \mathrm{Id}.$$

Switching x, y, and n, $-n$, we also get

$$[\{y\}_n, \{x\}_{-n}] = \{[y, x]\}_0 + n\langle y, x \rangle \cdot \mathrm{Id}.$$

Since both of the above brackets are anti-symmetric, choosing $n \neq 0$ in the above two formulae, we obtain $\langle x, y \rangle = \langle y, x \rangle$.

Next, we show $\mathfrak{g}$-invariance of the bilinear pairing $\langle \ , \ \rangle$ on $\mathfrak{g}$. Since vertex operators form an associative algebra acting on V, for any $x, y, z \in \mathfrak{g}$ and $n \in \mathbb{Z}$ we have

$$(*) \qquad [\{z\}_0, [\{x\}_n, \{y\}_{-n}]] = [[\{z\}_0, \{x\}_n], \{y\}_{-n}] + [\{x\}_n, [\{z\}_0, \{y\}_{-n}]].$$

For the moment, we call these three terms (I), (II), (III) from left to right. From the commutator formula (20), we observe that for $n \neq 0$,

$$[\{z\}_0, \{x\}_n] = \{\{z\}_0 x\}_n = \{[z, x]\}_n, \qquad [\{z\}_0, \{y\}_n] = \{\{z\}_0 y\}_n = \{[z, y]\}_n.$$

Then, from (22) with n replaced by $-n$,

$$\text{(II)} = [\{[z, x]\}_n, \{y\}_{-n}] = \{[[z, x], y]\}_0 + n\langle[z, x], y\rangle\text{Id},$$
$$\text{(III)} = [\{x\}_n, \{[z, y]\}_{-n}] = \{[x, [z, y]]\}_0 + n\langle x, [z, y]\rangle\text{Id}.$$

Similarly, from (22) with n replaced by $-n$, and noting that the identity operator commutes with any operator,

$$\text{(I)} = [\{z\}_0, [\{x\}_n, \{y\}_{-n}]] = [\{z\}_0, \{[x, y]\}_0 + n\langle x, y\rangle\text{Id}]$$
$$= [\{z\}_0, \{[x, y]\}_0] = \{[z, [x, y]]\}_0,$$

where the last equality comes from (18). Since the Jacobi identity is satisfied in the Lie algebra $\mathfrak{g}$ due to Lemma 2, cancelling those terms arising from this Jacobi identity, we get

$$n\langle[z, x], y\rangle\text{Id} + n\langle x, [z, y]\rangle\text{Id} = 0.$$

Since $n \neq 0$, we get $\mathfrak{g}$-invariance of the pairing $\langle\ ,\ \rangle$ as in (21). $\square$

Combining Lemma 2 and Lemma 3, we get a statement for $(V)_1$ similar to Proposition 1 for $(V)_{\frac{1}{2}}$.

PROPOSITION 4. *For a connected vertex operator (super) algebra V, let $\mathfrak{g} = (V)_1$. Then*
(i) *$\mathfrak{g}$ has a structure of a Lie algebra with trivial center with respect to the bracket product defined by $[x, y] = \{x\}_0 y$ for $x, y \in \mathfrak{g}$.*
(ii) *Vertex operators $Y(x, \zeta) = \sum_{n\in\mathbb{Z}} x(n)\zeta^{-n-1}$ for $x \in \mathfrak{g}$ generate a representation of an affine Lie algebra $\hat{\mathfrak{g}}$ on V with respect to a bilinear symmetric $\mathfrak{g}$-invariant pairing $\langle\ ,\ \rangle$ defined by $\{x\}_1 y = \langle x, y\rangle\Omega$. Namely, for any $x, y \in \mathfrak{g}$ the following identity holds:*

$$(23) \qquad [x(m), y(n)] = [x, y](m + n) + m\langle x, y\rangle\delta_{m,-n} \cdot \text{Id}.$$

Next, we study the collection of operators of the form $\{v\}_0$ for $v \in V$.

LEMMA 5. *For any $v \in V$ and for any $0 \leq m \in \mathbb{Z}$, we have $\{v\}_m\Omega = 0$.*

PROOF. For any $n \in \mathbb{Z}$ and $0 \leq m \in \mathbb{Z}$, the super commutator formula (12) gives

$$\{v\}_m\{\Omega\}_n - \{\Omega\}_n\{v\}_m = \sum_{0 \leq k \in \mathbb{Z}} \{\{v\}_k\Omega\}_{m+n-k} \cdot$$

Since $\{\Omega\}_n = \delta_{n,-1} \cdot \text{Id}_V$, the left hand side vanishes whether $n = -1$ or not. Thus,

$$\{\{v\}_0\Omega\}_{m+n} + \cdots + \binom{m}{k}\{\{v\}_k\Omega\}_{m+n-k} + \cdots + \{\{v\}_m\Omega\}_n = 0$$

When $m = 0$, the above reduces to $\{\{v\}_0\Omega\}_n = 0$ for any $n \in \mathbb{Z}$. Then by induction on $0 \leq m \in \mathbb{Z}$, we have that $\{\{v\}_m\Omega\}_n = 0$ for any $n \in \mathbb{Z}$. This means that $Y(\{v\}_m\Omega, \zeta) = 0$ for any $0 \leq m \in \mathbb{Z}$. By the injectivity of the vertex operator map, $\{v\}_m\Omega = 0$ for any $0 \leq m \in \mathbb{Z}$. This completes the proof of Lemma 5. $\square$

Let W be any module over a vertex operator super algebra V. Let

$$I_W = \text{Ker} \{ \ \}_0 = \{v \in V \mid \{v\}_0 w = 0, \forall w \in W\}.$$

Since $\{v\}_0$ lowers the weight by $1 - \text{wt}(v)$ when $v \in V$ is homogeneous, it is easy to see that I_W is a graded subspace of V. That is,

$$I_W = \bigoplus_{0 \leq n \in \frac{1}{2}\mathbb{Z}} (I_W \cap (V)_n) = \bigoplus_{0 \leq n \in \frac{1}{2}\mathbb{Z}} (I_W)_n.$$

PROPOSITION 6. *In the above situation, the following statements hold:*
(i) *For any $v \in V$ the subspace $I_W \subset V$ is invariant under the operator $\{v\}_0$, that is, $\{v\}_0 I_W \subset I_W$. We also have $\{I_W\}_0 v \subset I_W$ for any $v \in V$.*
(ii) *The quotient graded vector space V/I_W has the structure of a Lie super algebra with respect to the bracket product defined by $[u_1, u_2] = \{v_1\}_0 v_2 \mod I_W$ for any representatives $v_1 \in u_1$, $v_2 \in u_2$, and W is a module over this Lie super algebra which we denote by $\{V\}_0^W$.*

PROOF. For (i), from the super commutator formula (12), for any $v_1, v_2 \in V$,

(*) $$\{v_1\}_0\{v_2\}_0 - (-1)^{|v_1||v_2|}\{v_2\}_0\{v_1\}_0 = \{\{v_1\}_0 v_2\}_0.$$

If $v_2 \in I_W$, then $\{v_2\}_0 = 0$ on W and the left hand side is 0 and we have $\{\{v_1\}_0 v_2\}_0 = 0$, or $\{v_1\}_0 v_2 \in I_W$. This means that $\{v\}_0 I_W \subset I_W$ for any $v \in V$.

Similarly, if $v_1 \in I_W$, then $\{v_1\}_0 = 0$ as an operator on W and the left hand side of the above identity vanishes. So, $\{\{v_1\}_0 v_2\}_0 = 0$ on W and hence $\{v_1\}_0 v_2 \in I_W$. This means $\{I_W\}_0 v \subset I_W$ for any $v \in V$.

(ii) First we show that the bracket product $[u_1, u_2]$ is independent of the choice of the representatives. Let $u_1 = v_1 + (I_W)_s$, $u_2 = v_2 + (I_W)_t$ for some $v_1, v_2 \in V$, where $s = \text{wt}(v_1)$ and $t = \text{wt}(v_2)$. Using (i), we have

$$\{v_1 + (I_W)_s\}_0(v_2 + (I_W)_t) = \{v_1\}_0 v_2 + \{v_1\}_0(I_W)_t + (I_W)_{t+s-1}$$
$$\subset \{v_1\}_0 v_2 + (I_W)_{s+t-1}.$$

Hence the bracket product is well defined in V/I_W. Now, the above commutator together with the one in which v_1 and v_2 are switched, we get $\{\{v_1\}_0 v_2 + (-1)^{|v_1||v_2|}\{v_2\}_0 v_1\}_0 = 0$, which in turn implies $\{v_1\}_0 v_2 + (-1)^{|v_1||v_2|}\{v_2\}_0 v_1 \in I_W$. Hence, modulo I_W, we get $[u_1, u_2] + (-1)^{|u_1||u_2|}[u_2, u_1] = 0$. Here $|u_1|$ and $|u_2|$ are defined by $|u_1| = |v_1|$ and $|u_2| = |v_2|$. We let $\text{ad}(u) = \{u\}_0 = [u, \ ]$ as operators on W. The super commutator formula (*) above modulo I_W gives

$$\text{ad}(u_1)\text{ad}(u_2) - (-1)^{|u_1||u_2|}\text{ad}(u_2)\text{ad}(u_1) = \text{ad}(\text{ad}(u_1)u_2).$$

This is a super Jacobi identity. Thus, V/I_W has a structure of a Lie super-algebra and W is a representation over it through the operator $\{ \ \}_0$. $\square$

Deduction of the Jacobi identity for vertex operator super algebras from the crossing symmetry. The Jacobi identity (5) for vertex operator super algebras can be deduced from a geometric property called crossing symmetry of vertex operators, making the Jacobi identity look more natural.

Let $w \in V$ and $w' \in V^*$, where V^* is the graded dual vector space, i.e., $V^* = \bigoplus_{n \in \frac{1}{2}\mathbb{Z}} (V)_n^*$. We consider three ways the vertex operators associated to $v_1, v_2 \in V$ act on w, and we let the resulting vectors pair with w' as follows:

$$
\begin{array}{ll}
\text{(I')} & \left(Y(v_1, \zeta_1) Y(v_2, \zeta_2) w, w' \right), \\
\text{(II')} & (-1)^{|v_1||v_2|} \left(Y(v_2, \zeta_2) Y(v_1, \zeta_1) w, w' \right), \\
\text{(III')} & \left(Y(Y(v_1, \zeta_1 - \zeta_2) v_2, \zeta_2) w, w' \right).
\end{array}
$$

(24)

These are formal power series. [The notations (I), (II), (III) without primes are preserved for other objects described below.] The sign $(-1)^{|v_1||v_2|}$ in (II) is equal to -1 only when $v_1, v_2 \in V_1$, in other words only when $\mathrm{wt}(v_1), \mathrm{wt}(v_2) \in \mathbb{Z} + \frac{1}{2}$. The reason for this sign is that the vertex operator $Y(v, \zeta)$ is regarded "fermionic" when $\mathrm{wt}(v) \in \mathbb{Z} + \frac{1}{2}$ and any two fermionic vertex operators anticommute. For the above power series, we make the following assumptions:

(i) *The above power series* (I'), (II'), (III') *converge absolutely on the domains* $|\zeta_1| > |\zeta_2| > 0$, $|\zeta_2| > |\zeta_1| > 0$, $|\zeta_2| > |\zeta_1 - \zeta_2| > 0$, *respectively.*

(ii) (Crossing symmetry) *These three functions in* ζ_1, ζ_2 *have analytic continuation to the same meromorphic function on* $\mathbb{P}^1 \times \mathbb{P}^1$ *with possible poles along* $\zeta_1, \zeta_2 = 0, \infty$, *and* $\zeta_1 = \zeta_2$.

The above assumption on the domain of the convergence of the above power series is, in a sense, reasonable from geometric pictures each of the above three actions of vertex operators describe. In (I'), the vertex operator $Y(v_2, \zeta_2)$ acts on w first, then subsequently the vertex operator $Y(v_1, \zeta_1)$ acts. Since w is supposedly sitting at the origin of $\mathbb{P}^1 = \mathbb{C} \cup \{\infty\}$, this successive interaction makes sense when v_2 is located closer to w then v_1, which is exactly the condition $|\zeta_1| > |\zeta_2| > 0$. A similar justification applies to (II'). For (III'), the vector v_1 acts on v_2 first which is located at $\zeta_1 - \zeta_2$ relative to v_2. Then, the resulting vector acts on the vector w located at $0 \in \mathbb{P}^1$. This picture makes sense when ζ_1 is closer to ζ_2 than $0 \in \mathbb{P}^1$, which is the case when $|\zeta_2| > |\zeta_1 - \zeta_2| > 0$.

Let us denote the meromorphic function in (ii) above by $\mathcal{Y}(\zeta_1, \zeta_2)$. Let $f(\zeta_1, \zeta_2)$ be a meromorphic function in ζ_1, ζ_2 with possible poles along $\zeta_1, \zeta_2 = 0, \infty$ and $\zeta_1 = \zeta_2$ in $\mathbb{P}^1 \times \mathbb{P}^1$. So, $f(\zeta_1, \zeta_2)$ is a rational function which belongs to $\mathbb{C}[\zeta_1, \zeta_1^{-1}, \zeta_2, \zeta_2^{-1}, (\zeta_1 - \zeta_2)^{-1}]$. Now, we fix the variable ζ_2 for a moment, and regard the product $\mathcal{Y}(\zeta_1, \zeta_2) \cdot f(\zeta_1, \zeta_2)$ as a meromorphic function in $\zeta = \zeta_1$ with possible poles at $\zeta = 0, \zeta_2, \infty$. We then take residues at these poles. Since the sum of all the residues at poles in $\mathbb{P}^1$ is 0, we get

$$
(25) \quad -\operatorname*{Res}_{\zeta=\infty} \mathcal{Y}(\zeta, \zeta_2) \cdot f(\zeta, \zeta_2) - \operatorname*{Res}_{\zeta=0} \mathcal{Y}(\zeta, \zeta_2) \cdot f(\zeta, \zeta_2) = \operatorname*{Res}_{\zeta=\zeta_2} \mathcal{Y}(\zeta, \zeta_2) \cdot f(\zeta, \zeta_2).
$$

Since we are considering poles at $\zeta = 0, \zeta_2, \infty$, we may let

$$
(26) \quad f(\zeta, \zeta_2) = \zeta^m (\zeta - \zeta_2)^r, \qquad m, r \in \mathbb{Z},
$$

without loss of generality. For convenience, we let $(I) = -\operatorname{Res}_{\zeta=\infty}$, $(II)=\operatorname{Res}_{\zeta=0}$, $(III) = \operatorname{Res}_{\zeta=\zeta_2}$, where residues are taken for the function $\mathcal{Y}(\zeta, \zeta_2) \cdot f(\zeta, \zeta_2)$. To calculate these residues, we need power series expansions around $\zeta = \infty, 0, \zeta_2$, respectively. The power series expansions for $\mathcal{Y}(\zeta, \zeta_2)$ are provided by the power series (24) in view of our assumption (i) above on the domain of the convergence. Although we are assuming the above domain of convergence, in all the known examples, these assumptions are true. In particular, our assumption can be verified for the vertex operator super algebra described in §2.2 which are of interest to us in the context of elliptic genera [FFR].

The calculation of (I). Recall that the residue at $\zeta = \infty$ of a meromorphic function $f(\zeta)$ around $\zeta = \infty$ is given by $\operatorname{Res}_{\zeta=\infty} f(\zeta) = -a_{-1}$, where $f(\zeta) = \sum_{n \in \mathbb{Z}} a_n \zeta^n$ is a power series expansion on a region $|\zeta| > R$ for some $R > 0$. For the meromorphic function $\mathcal{Y}(\zeta, \zeta_2)$, such expansion is given by the first power series (I') in (24) which converges on $|\zeta| > |\zeta_2| > 0$ due to the assumption (i). The power series expansion of $f(\zeta, \zeta_2)$ given in (26) on the same region is given by

$$f(\zeta, \zeta_2) = \zeta^m \cdot \zeta^r \left(1 - \frac{\zeta_2}{\zeta}\right)^r = \zeta^{m+r} \sum_{i \geq 0} \binom{r}{i} (-1)^i \frac{\zeta_2^i}{\zeta^i} = \sum_{i \geq 0} \binom{r}{i} (-1)^i \zeta^{m+r-i} \zeta_2^i.$$

Thus, on $|\zeta| > |\zeta_2| > 0$, the power series expansion of $\mathcal{Y}(\zeta, \zeta_2) \cdot f(\zeta, \zeta_2)$ is given by

$$\mathcal{Y}(\zeta, \zeta_2) \cdot f(\zeta, \zeta_2) = \sum_{p,q \in \mathbb{Z}} (\{v_1\}_p \{v_2\}_q w, w') \zeta^{-p-1} \zeta_1^{-q-1} \cdot \sum_{i \geq 0} \binom{r}{i} (-1)^i \zeta^{m+r-i} \zeta_2^i$$

$$= \sum_{i \geq 0} \sum_{p,q \in \mathbb{Z}} \left(\binom{r}{i}(-1)^i \{v_1\}_p \{v_2\}_q w, w'\right) \zeta^{m+r-i-p-1} \zeta_2^{i-q-1}.$$

The coefficient of ζ^{-1} gives the residue $-\operatorname{Res}_{\zeta=\infty}$. Since $m + r - i - p - 1 = -1$ when $p = m + r - i$, we have

$$(27) \qquad (I) = -\operatorname*{Res}_{\zeta=\infty} = \sum_{q \in \mathbb{Z}} \left(\sum_{i \geq 0} \binom{r}{i} (-1)^i \{v_1\}_{m+r-i} \{v_2\}_q w, w'\right) \zeta_2^{i-q-1}.$$

The calculation of (II). To calculate the residue at $\zeta = 0$, we need the power series expansion of $\mathcal{Y}(\zeta, \zeta_2) \cdot f(\zeta, \zeta_2)$ on a punctured disc centered at $\zeta = 0$. For $\mathcal{Y}(\zeta, \zeta_2)$, this is given by the second power series (II') in (24) which converges on $0 < |\zeta| < |\zeta_2|$ due to the assumption (i). For the function $f(\zeta, \zeta_2)$ in (26), the expansion on this region is given by

$$f(\zeta, \zeta_2) = \zeta^m \cdot \zeta_2^r (-1)^r \left(1 - \frac{\zeta}{\zeta_2}\right)^r = (-1)^r \zeta^m \zeta_2^r \sum_{i \geq 0} \binom{r}{i} (-1)^i \frac{\zeta^i}{\zeta_2^i}$$

$$= (-1)^r \sum_{i \geq 0} \binom{r}{i} (-1)^i \zeta^{m+i} \zeta_2^{r-i}.$$

Thus the power series expansion of $\mathcal{Y}(\zeta,\zeta_2) \cdot f(\zeta,\zeta_2)$ around $\zeta = 0$ is given by

$$\mathcal{Y}(\zeta,\zeta_2)\,f(\zeta,\zeta_2) = \sum_{p,q\in\mathbb{Z}}(\{v_2\}_p\{v_1\}_q w, w')\,\zeta_2^{-p-1}\zeta^{-q-1}(-1)^r\sum_{i\geq 0}\binom{r}{i}(-1)^i\zeta^{m+i}\zeta_2^{r-i}$$

$$= \sum_{p,q\in\mathbb{Z}}\left((-1)^r\sum_{i\geq 0}\binom{r}{i}(-1)^i\{v_2\}_p\{v_1\}_q w, w'\right)\zeta^{m+i-q-1}\zeta_2^{r-i-p-1}.$$

The residue at $\zeta = 0$ is the coefficient of ζ^{-1}. When $m + i - q - 1 = -1$, we have $q = m + i$. Hence,

$$(28) \qquad \text{(II)} = \operatorname*{Res}_{\zeta=0} = \sum_{p\in\mathbb{Z}}\left((-1)^r\sum_{i\geq 0}\binom{r}{i}(-1)^i\{v_2\}_p\{v_1\}_{m+i} w, w'\right)\zeta_2^{r-i-p-1}.$$

The calculation of (III). To calculate the residue at $\zeta = \zeta_2$, we need the convergent power series expansion of $\mathcal{Y}(\zeta,\zeta_2) \cdot f(\zeta,\zeta_2)$ on a punctured disc centered at $\zeta = \zeta_2$. For $\mathcal{Y}(\zeta,\zeta_2)$, it is given by the third power series (III′) in (24), which converges on $0 < |\zeta - \zeta_2| < |\zeta_2|$. For the function $f(\zeta,\zeta_2)$ in (26), using $\zeta = \zeta_2 + (\zeta - \zeta_2) = \zeta_2(1 + \frac{\zeta-\zeta_2}{\zeta_2})$, its power series expansion on $0 < |\zeta - \zeta_2| < |\zeta_2|$ is given by

$$f(\zeta,\zeta_2) = (\zeta - \zeta_2)^r \cdot \zeta_2^m\left(1 + \frac{\zeta-\zeta_2}{\zeta_2}\right)^m = (\zeta - \zeta_2)^r \zeta_2^m \cdot \sum_{k\geq 0}\binom{m}{k}\frac{(\zeta-\zeta_2)^k}{\zeta_2^k}$$

$$= \sum_{k\geq 0}\binom{m}{k}(\zeta - \zeta_2)^{r+k}\zeta_2^{m-k}.$$

Thus, the power series expansion of $\mathcal{Y}(\zeta,\zeta_2) \cdot f(\zeta,\zeta_2)$ on $0 < |\zeta - \zeta_2| < |\zeta_2|$ is

$$\mathcal{Y}(\zeta,\zeta_2) \cdot f(\zeta,\zeta_2)$$

$$= \sum_{p,q\in\mathbb{Z}}\left(\{\{v_1\}_p v_2\}_q w, w'\right)(\zeta - \zeta_2)^{-p-1}\zeta_2^{-q-1} \cdot \sum_{k\geq 0}\binom{m}{k}(\zeta - \zeta_2)^{r+k}\zeta_2^{m-k}$$

$$= \sum_{p,q\in\mathbb{Z}}\left(\sum_{k\geq 0}\{\{v_1\}_p v_2\}_q w, w'\right)(\zeta - \zeta_2)^{r+k-p-1}\zeta_2^{m-k-q-1}.$$

The residue $\operatorname{Res}_{\zeta=\zeta_2}$ is given by the coefficient of $(\zeta-\zeta_2)^{-1}$. When $r+k-p-1 = -1$, we have $p = r + k$. Hence,

$$(29) \qquad \text{(III)} = \operatorname*{Res}_{\zeta=\zeta_2} = \sum_{q\in\mathbb{Z}}\left(\sum_{k\geq 0}\binom{m}{k}\{\{v_1\}_{r+k}v_2\}_q w, w'\right)\zeta_2^{m-k-q-1}.$$

Deduction of the Jacobi identity. Now, we assemble our calculations together to deduce the Jacobi identity (5). Substituting (27), (28), (29) into the residue identity (25), we obtain the identity of the function in ζ_2.

$$\sum_{q\in\mathbb{Z}}\left(\sum_{i\geq 0}\binom{r}{i}(-1)^i\{v_1\}_{m+r-i}\{v_2\}_q w, w'\right)\zeta_2^{i-q-1}$$

$$-(-1)^{r+|v_1||v_2|}\sum_{p\in\mathbb{Z}}\left(\sum_{i\geq 0}\binom{r}{i}(-1)^i\{v_2\}_p\{v_1\}_{m+i}w, w'\right)\zeta_2^{r-i-p-1}$$

$$=\sum_{q\in\mathbb{Z}}\left(\sum_{k\geq 0}\binom{m}{k}\{\{v_1\}_{r+k}v_2\}_q w, w'\right)\zeta_2^{m-k-q-1}.$$

We equate the coefficients of ζ_2^{-n-1} of both sides for each $n \in \mathbb{Z}$. In the first summation we let $q = n + i$, in the second we let $p = n + r - i$, and in the third summation we let $q = m + n - k$. We then obtain

$$\left(\sum_{i\geq 0}\binom{r}{i}(-1)^i\{v_1\}_{m+r-i}\{v_2\}_{n+i}w, w'\right)$$

$$-(-1)^{r+|v_1||v_2|}\left(\sum_{i\geq 0}\binom{r}{i}(-1)^i\{v_2\}_{n+r-i}\{v_1\}_{m+i}w, w'\right)$$

$$=\left(\sum_{k\geq 0}\binom{m}{k}\{\{v_1\}_{r+k}v_2\}_{m+n-k}w, w'\right).$$

Since this identity holds for any $w \in V$, $w' \in V^*$, we get the Jacobi identity (5).

§2.2   The infinite dimensional Clifford modules as vertex operator super algebras and their modules

We construct the level 1 Spin representations for orthogonal affine Lie algebra $\hat{o}(2N)$ and describe the vertex operator super algebra structure and the module structure in these Spin representations. For more details, see [FFR].

We start from the construction of Spin representations for finite dimensional orthogonal Lie algebras of type D. This method generalizes to the construction of Spin representations for infinite dimensional orthogonal affine Lie algebras.

Let $A \cong \mathbb{C}^{2N}$ be a $2N$-dimensional complex vector space which is a complexification of a real $2N$-dimensional vector space $E \cong \mathbb{R}^{2N}$ on which the real orthogonal Lie algebra $o(2N)$ acts. So the complexified Lie algebra $o(2N) \otimes_{\mathbb{R}} \mathbb{C}$ acts on A. We also denote this complexification by $o(2N)$ for convenience. Now suppose A is equipped with a nondegenerate symmetric $\mathbb{C}$-bilinear form $\langle\ ,\ \rangle$ which is the $\mathbb{C}$-linear extension of an $\mathbb{R}$-bilinear nondegenerate pairing $\langle\ ,\ \rangle$ in E. Let $\text{Cliff}(A)$ be the complex Clifford algebra on A, which is a complexification of the real Clifford algebra $\text{Cliff}(E)$, generated by 1 and vectors in A subject to a relation:

$$(1)\qquad\qquad a\cdot b + b\cdot a = \langle a,b\rangle 1,\qquad\text{for any}\quad a,b\in A.$$

In Cliff(A) elements of the form $:ab:=\frac{1}{2}(ab-ba)$ for $a,b\in A$ are closed under the bracket operations: for any $a,b,c,d\in A$ we have

$$(2) \qquad [:ab:,:cd:]=\langle a,d\rangle :bc:+\langle b,c\rangle :ad:-\langle a,c\rangle :bd:-\langle b,d\rangle :ac:.$$

Elements of the form $:ab:$ for $a,b\in A$ form the orthogonal Lie algebra $\mathfrak{o}(2N)$ in the Clifford algebra Cliff(A) with respect to the bracket product (2). This Lie algebra acts on the vector space A by the bracket:

$$(3) \qquad [:ab:,c]=\langle b,c\rangle a-\langle a,c\rangle b.$$

This action is skew in the sense that $\langle[:ab:,c],d\rangle+\langle c,[:ab:,d]\rangle=0$. This also means that $\langle\ ,\ \rangle$ is an $\mathfrak{o}(2N)$-invariant pairing. As usual, this action of $\mathfrak{o}(2N)$ extends to any exterior powers of A. In particular, $\bigwedge^2 A$ is an $\mathfrak{o}(2N)$ module and the map

$$(4) \qquad \Upsilon:\mathfrak{o}(2n)\to\bigwedge^2 A$$

defined by $\Upsilon(:ab:)=a\wedge b$ is an isomorphism of $\mathfrak{o}(2N)$ modules. The pairing on A can be extended to an nondegenerate symmetric invariant pairing on $\mathfrak{o}(2N)$ and also on $\bigwedge^2 A$ by

$$(5) \qquad \begin{aligned} \langle :a_1a_2:,:b_1b_2:\rangle &= \langle a_1,b_2\rangle\langle a_2,b_1\rangle - \langle a_1,b_1\rangle\langle a_2,b_2\rangle \\ &= \langle[:a_1a_2:,b_1],b_2\rangle = \langle a_1\wedge a_2,b_1\wedge b_2\rangle. \end{aligned}$$

Let $A=A^+\oplus A^-$ be a decomposition into maximal isotropic subspaces. Both of these complex vector spaces $A^\pm$ have complex dimension N. The Spinor representation Δ of $\mathfrak{o}(2N)$ is realized on the exterior algebra $\bigwedge^* A^-$, which is the Clifford module for the Clifford algebra Cliff(A) whcih contains $\mathfrak{o}(2N)$. On this space, A^- acts by exterior multiplication and vectors in A^+ acts by derivation. Its direct summands $\Delta^+=\bigwedge^{\text{even}} A^-$, $\Delta^-=\bigwedge^{\text{odd}} A^-$ are the half Spin representations of $\mathfrak{o}(2N)$.

Infinite dimensional Clifford algebras and their modules. Spin representations for the orthogonal affine Lie algebras can be constructed in a way analogous to the finite dimensional case. But this time we use infinite dimensional Clifford algebras Cliff(Z) for $Z=\mathbb{Z}$ or $\mathbb{Z}+\frac{1}{2}$ instead of the above finite dimensional Clifford algebras Cliff(A). To define Cliff(Z), we let

$$(6) \qquad A(\mathbb{Z})=A\otimes\mathbb{C}[t,t^{-1}], \qquad A(\mathbb{Z}+\tfrac{1}{2})=A\otimes\mathbb{C}t^{\frac{1}{2}}[t,t^{-1}].$$

We denote $a\otimes t^m$ for $m\in Z$ by $a(m)$. We can introduce a nondegenerate symmetric bilinear pairing on $A(Z)$ by $\langle a(m),b(n)\rangle=\langle a,b\rangle\delta_{m,-n}$. By definition, the infinite dimensional Clifford algebra Cliff(Z) is generated by the unit 1 and elements $a(m)$ for $a\in A$ and $m\in Z$ subject to a relation:

$$(7) \qquad a(n)\cdot b(m)+b(m)\cdot a(n)=\langle a(n),b(n)\rangle\cdot 1=\langle a,b\rangle\delta_{n+m,0}\cdot 1.$$

The Clifford module $\mathrm{CM}(Z)$ for the Clifford algebra $\mathrm{Cliff}\,(Z)$ can be constructed using the polarization of $A(Z)$. We let

$$
(8) \qquad A(Z)^{\pm} = \begin{cases} A^{\pm} \oplus A \otimes t^{\pm 1}\mathbb{C}[t^{\pm 1}] & \text{if } Z = \mathbb{Z}, \\ A \otimes t^{\pm\frac{1}{2}}\mathbb{C}[t^{\pm 1}] & \text{if } Z = \mathbb{Z} + \frac{1}{2}. \end{cases}
$$

The Clifford module $\mathrm{CM}(Z)$ is realized on the total exterior algebra $\bigwedge^* A(Z)^-$. Let $A(m)$ denote $A \otimes t^m$, $m \in \frac{1}{2}\mathbb{Z}$. For simplicity we let $V = \mathrm{CM}(\mathbb{Z} + \frac{1}{2})$ and $W = \mathrm{CM}(\mathbb{Z})$. Then we have

$$
(9) \qquad \begin{aligned} V &= \bigotimes_{m \in \mathbb{Z}^+} \bigwedge{}^* A(-m - \tfrac{1}{2}), \\ W &= \Delta \otimes \bigotimes_{m \in \mathbb{N}} \bigwedge{}^* A(-m), \end{aligned}
$$

where $\Delta = \bigwedge^* A^-(0)$ is the total Spin representation of the finite dimensional Lie algebra $\mathfrak{o}(2N)$. As in the finite dimensional case, the vectors in $A(Z)^-$ act on V (when $Z = \mathbb{Z} + \frac{1}{2}$) and on W (when $Z = \mathbb{Z}$) by exterior multiplication. The vectors in $A(Z)^+$ act on them by derivation and annihilate the "vacuum" vector 1. This derivation property comes from (7).

With an application to elliptic genera in mind, for a finite dimensional vector space E, we make the following definition:

$$
(10) \qquad \bigwedge{}^*_t E = \mathbb{C} + Et + (\textstyle\bigwedge^2 E)t^2 + \cdots + (\textstyle\bigwedge^e E)t^e, \qquad e = \dim E.
$$

Here t is a formal variable. We define V_q and W_q by

$$
(11) \qquad \begin{aligned} V_q &= \bigotimes_{\ell \in \mathbb{Z}^+} \bigwedge{}^*_{q^{\ell+\frac{1}{2}}} A(-\ell - \tfrac{1}{2}) = \bigoplus_{m \in \frac{1}{2}\mathbb{Z}^+} (V)_m q^m, \\ W_q &= q^{\frac{N}{8}} \Delta \otimes \bigotimes_{\ell \in \mathbb{N}} \bigwedge{}^*_{q^{\ell}} A(-\ell) = \bigoplus_{m \in \mathbb{Z}^+} (W)_{m + \frac{N}{8}} q^{m + \frac{N}{8}}. \end{aligned}
$$

In the context of vertex operator super algebras and their modules, q as in (11) means $q = e^{2\pi i \tau}$, where $\mathrm{Im}\,\tau > 0$. In this expression vectors in the coefficient of q^r have weight r. The reason of the introduction of $q^{\frac{N}{8}}$ for W_q comes from the Virasoro operator $D(0)$ acting on W. Eigenvalues of the operator $-D(0)$ on homogeneous vectors are equal to their weight, which requires the correction above [FFR, p42]. The weight 0 vector space in V_q is 1-dimensional and it is generated by Ω or **vac**.

The vector spaces V and W are $\mathbb{Z}_2$-graded, as $V = V_0 \oplus V_1$, $W = W_0 \oplus W_1$, where V_0 consists of vectors of integral weights and V_1 consists of vectors of half integral weights. Similarly, W_0 consists of vectors whose weights belong to $2\mathbb{Z} + \frac{N}{8}$ and W_1 consists of those vectors whose weights belong to $2\mathbb{Z} + 1 + \frac{N}{8}$.

Representation theoretically V and W are important because V_0, V_1, W_0 and W_1 exhausts all the level 1 irreducible Spin representations of the orthogonal affine Lie algebra $\hat{o}(2N)$. We will describe the action of $\hat{o}(2N)$ on V and on W in terms of the Clifford multiplication later.

The graded vector spaces V_q and W_q play a major role in the construction of elliptic genera which is the index of the "Dirac operator" on free loop spaces twisted by total Spin bundles associated to V, W.

What makes V and W even more interesting is that V has the structure of a vertex operator super algebra and W has the structure of its $\mathbb{Z}_2$-twisted module:

(12)
$$Y(\ ,\zeta) : V \to \text{End}(V)[[\zeta, \zeta^{-1}]],$$
$$Y(\ ,\zeta) : V \to \text{End}(W)[[\zeta^{\frac{1}{2}}, \zeta^{-\frac{1}{2}}]],$$

satisfying the Jacobi identities. We now describe these structure maps.

Construction of the vertex operator super algebra V. The graded vector space V is spanned by elements of the form

(13)
$$a_1(-n_1 - \tfrac{1}{2})a_2(-n_2 - \tfrac{1}{2}) \cdots a_r(-n_r - \tfrac{1}{2})\Omega,$$

where $a_i \in A$, $n_i \in \mathbb{Z}^+$, $1 \le i \le r$. The basic definition is the following. For $a \in A$, $n \in \mathbb{Z}^+$, let

(14) $$Y(a(-n - \tfrac{1}{2})\Omega, \zeta) = \frac{1}{n!}\frac{d^n}{d\zeta^n}\left\{\sum_{m\in\mathbb{Z}} a(m + \tfrac{1}{2})\zeta^{-m-1}\right\} \equiv a^{(n)}(\zeta).$$

For vectors of the form (13), we define the corresponding vertex operator by
(15)
$$Y(a_1(-n_1 - \tfrac{1}{2})a_2(-n_2 - \tfrac{1}{2})\cdots a_r(-n_r - \tfrac{1}{2})\Omega, \zeta) =: a_1^{(n_1)}(\zeta)a_2^{(n_2)}(\zeta)\cdots a_r^{(n_r)}(\zeta):,$$

where : : denotes the *normal ordering* defined by

(16) $$: a_1(m_1)a_2(m_2)\cdots a_r(m_r) := \text{sgn}(\sigma)a_{\sigma(1)}(m_{\sigma(1)})\cdots a_{\sigma(r)}(m_{\sigma(r)}).$$

Here σ is a permutation on r letters such that $m_{\sigma(1)} \le m_{\sigma(2)} \le \cdots \le m_{\sigma(r)}$. Note that any two vectors in $A(\mathbb{Z} + \tfrac{1}{2})^-$ anticommute because this space is isotropic with respect to the pairing (7) extended from A. Similarly for $A(\mathbb{Z} + \tfrac{1}{2})^+$. Thus, different choices of σ with the above property define the same normal ordering.

For each power of ζ, its coefficient in (15) is an infinite sum of elements of the Clifford algebra $\text{Cliff}(\mathbb{Z} + \tfrac{1}{2})$. Its action on the Clifford module V is well defined because only finitely many elements act nontrivialy on any given homogeneous vector in V.

To define vertex operators associated to general vector $v \in V$, we extend the above definition linearly. This completes the construction of the vertex operator map $\text{ad}_\zeta : V \to \text{End}(V)[[\zeta, \zeta^{-1}]]$ for V.

The construction of $\mathbb{Z}_2$-twisted module structure on W. The construction is done in two steps. First we define $a^{(m)}(\zeta) = a_W^{(m)}(\zeta)$ for $a \in A$, $n \in \mathbb{Z}^+$ by

$$(17) \quad a^{(m)}(\zeta) = \frac{1}{m!}\frac{d^m}{d\zeta^m}\left\{\sum_{\ell \in \mathbb{Z}+\frac{1}{2}} a(\ell + \tfrac{1}{2})\zeta^{-\ell-1}\right\} = \frac{1}{m!}\frac{d^m}{d\zeta^m}\left\{\sum_{\ell \in \mathbb{Z}} a(\ell)\zeta^{-\ell-\frac{1}{2}}\right\}.$$

This is the same formula as (14) except that the index set is shifted by $\frac{1}{2}$. We define an operator $\overline{Y}(v, \zeta)$ by a formula similar to (15):

$$(18) \quad \overline{Y}(a_1(-n_1 - \tfrac{1}{2})\cdots a_r(-n_r - \tfrac{1}{2})\Omega, \zeta) =: a_1^{(n_1)}(\zeta)a_2^{(n_2)}(\zeta)\cdots a_r^{(n_r)}(\zeta):.$$

Here we can try to define the normal ordering as before, moving creation operators to the left and annihilation operators to the right. However for the normal ordering for $a(0)$'s, we get different normal ordering than that for finite dimensional Clifford algebras where we defined $:ab: = \frac{1}{2}(ab - ba)$. To adjust this difference, we define the normal ordering for 0-modes as follows:

$$(19) \quad :a_1(0)a_2(0)\cdots a_r(0): = \frac{1}{r!}\sum_{\sigma \in S_r} \operatorname{sgn}(\sigma)a_{\sigma(1)}(0)a_{\sigma(2)}(0)\cdots a_{\sigma(r)}(0).$$

By linearly extending the above definition, we obtain an operator $\overline{Y}(v, \zeta)$ for any vector $v \in V$. Note that the coefficient of any power of ζ in $\overline{Y}(v, \zeta)$ is an infinite sum of elements in $\mathrm{Cliff}(\mathbb{Z})$ which acts on W by Clifford multiplication. So, $\overline{Y}(v, \zeta) \in \mathrm{End}(W)[[\zeta^{\frac{1}{2}}, \zeta^{-\frac{1}{2}}]]$.

To define the vertex operator $Y(v, \zeta)$ acting on W, we need to modify this operator $\overline{Y}(v, \zeta)$ in order that the Jacobi identity is satisfied. Let $A = A^+ \oplus A^-$ be the polarization of A with respect to the nondegenerate symmetric bilinear form $\langle \, , \, \rangle$ on A and let $\{a_1, a_2, \ldots, a_N\}$ and $\{a_1^*, a_2^*, \ldots, a_N^*\}$ be a basis for A^+, A^- such that $\langle a_i, a_j^* \rangle = \delta_{ij}$, $1 \le i, j \le N$. We define an operator $\Delta(\zeta)$ on V by

$$(20) \quad \Delta(\zeta) = \sum_{i=1}^{N} \sum_{0 \le r,s \in \mathbb{Z}} \frac{1}{2}\left(\frac{r-s}{r+s+1}\right)\binom{-\frac{1}{2}}{r}\binom{-\frac{1}{2}}{s} a_i(r + \tfrac{1}{2})a_i^*(s + \tfrac{1}{2})\zeta^{-r-s-1}.$$

Note that $\Delta(\zeta)$ is quadratic and preserves the $\mathbb{Z}_2$ grading of V. Also note that $\Delta(\zeta)$ is of the form

$$(21) \quad \Delta(\zeta) = \frac{1}{8}\sum_{i=1}^{N}\{a_i(\tfrac{1}{2})a_i^*(\tfrac{3}{2}) + a_i^*(\tfrac{1}{2})a_i(\tfrac{3}{2})\}\zeta^{-2} + \sum_{\substack{0 \le r \ne s \in \mathbb{Z} \\ r+s \ge 2}} a_{rs}\zeta^{-r-s-1},$$

where a_{rs} is an operator which lowers the weight by $r+s+1$. So, $\Delta(\zeta)$ annihilates $(V)_0$, $(V)_{\frac{1}{2}}$, $(V)_1$ and $(V)_{\frac{3}{2}}$.

Finally the vertex operator $Y(v, \zeta) \in \mathrm{End}(W)[[\zeta^{\frac{1}{2}}, \zeta^{-\frac{1}{2}}]]$ in (12) is defined by

$$(22) \quad Y(v, \zeta) = \overline{Y}(e^{\Delta(\zeta)}v, \zeta).$$

This finishes the description of the $\mathbb{Z}_2$-twisted module W over V.

One can check that vertex operators $Y(v, \zeta)$ in (15) and (22) satisfy all requirements of a vertex operator super algebra and its $\mathbb{Z}_2$-twisted modules including the Jacobi identity (5) of §2.1.

§2.3 Clifford algebras, Virasoro algebras and affine Lie algebras as subalgebras of vertex operator super algebra V

As we have seen in §2.2, the underlying vector spaces of our vertex operator super algebra V and its $\mathbb{Z}_2$-twisted module W can be constructed as irreducible Clifford modules of the infinite dimensional Clifford algebras $\mathrm{Cliff}(Z)$, where $Z = \mathbb{Z} + \frac{1}{2}$ for V or $\mathbb{Z}$ for W. The vertex operator $Y(v, \zeta)$ associated to $v \in V$ acting on V or on W is an infinite sum of elements of the Clifford algebras. As such, vertex operators may seem to be nothing more than certain complicated operators obtained by Clifford multiplication by elements of Clifford algebras, even if we use new terminologies such as vertex operator super algebras and their $\mathbb{Z}_2$-twisted modules. This is not quite so. The reason for this is that as *infinite sums* of Clifford elements, vertex operators don't belong to the Clifford algebra, and their nature as operators can be very different from Clifford multiplication.

This is similar to the fact that the behaviour of an analytic function, which is an infinite sum of monomials, can be very different from the behavior of polynomials. For example, an analytic function can be doubly periodic like elliptic functions, it can have regular singularities of the form $(z - z_0)^\alpha$ for $\alpha \in \mathbb{C}$, like Gauß hypergeometric functions.

It is known that our vertex operator super algebra V contains not only the Clifford algebra but also the Virasoro algebra described in (8) of §2.1 and an affine Lie algebra. The representations of these algebras can be realized by vertex operators acting on V and on W.

With applications to elliptic genera in mind, we explicitly describe how these algebras arise as subalgebras of vertex operator algebras. It turns out that in gneral,

(i) the vertex operators associated to vectors of weight $1/2$ generate infinite dimensional Clifford algebras,

(ii) the vertex operators associated to vectors of weight 1 generate an affine Lie algebra $\hat{\mathfrak{g}}$,

(iii) the vertex operator associated to a $\mathfrak{g}$-invariant vector ω of weight 2 (the canonical Virasoro element) generates the Virasoro algebra with rank ($=$ central charge) N.

It is an interesting question to see what kind of operators and algebras arise from vertex operators associated to vectors of higher weights. See Chapter 3 for some calculations associated to products of Kähler forms or of Casimir forms.

Before starting calculations of vertex operators in V, we recall the definition of an (untwisted) affine Lie algebra $\hat{\mathfrak{g}}$ associated to a Lie algebra $\mathfrak{g}$ equipped with a $\mathfrak{g}$-invariant symmetric bilinear form $\langle\ ,\ \rangle$. The Lie algebra $\hat{\mathfrak{g}}$ is given by

$$(1) \qquad\qquad \hat{\mathfrak{g}} = \mathfrak{g} \otimes \mathbb{C}[t, t^{-1}] \oplus \mathbb{C}c,$$

where c is a central element, so it acts by multiplication by a constant on any irreducible representation of $\hat{\mathfrak{g}}$. This constant is called the level of the representation. We denote $x \otimes t^m$ by $x(m)$ for $x \in \mathfrak{g}$ and $m \in \mathbb{Z}$. The commutation relation is given by

$$(2) \qquad [x(m), y(n)] = [x, y](m + n) + m\langle x, y\rangle\delta_{m+n,0} \cdot c.$$

Note that this bracket defines a Lie algebra structure on $\hat{\mathfrak{g}}$ because the pairing $\langle\,,\,\rangle$ on $\mathfrak{g}$ is $\mathfrak{g}$-invariant and symmetric.

Now we study vertex operators associated to vectors of low weight. From the construction of V we see that

$$(3)$$
$$(V)_{\frac{1}{2}} = A(-\tfrac{1}{2}), \quad (V)_1 = \textstyle\bigwedge^2 A(-\tfrac{1}{2}), \quad (V)_2 = \textstyle\bigwedge^4 A(-\tfrac{1}{2}) \oplus \left(A(-\tfrac{3}{2}) \otimes A(-\tfrac{1}{2})\right).$$

We know that $o(2N)$ and $\bigwedge^2 A$ are isomorphic $o(2N)$ modules via Υ in (4) of §2.2, and elements of $o(2N) \subset \mathrm{Cliff}(A)$ can be written as $x =: ab := \tfrac{1}{2}(ab - ba)$ where $a, b \in A$. We define $o(2N)$-maps

$$(4) \qquad \Upsilon = \Upsilon_{\frac{1}{2}} : A \to (V)_{\frac{1}{2}} \quad \text{and} \quad \Upsilon = \Upsilon_1 : o(2N) \to (V)_1$$

by $\Upsilon(a) = a(-\tfrac{1}{2})\Omega$ and $\Upsilon(:a_1 a_2:) = a_1(-\tfrac{1}{2})a_2(-\tfrac{1}{2})\Omega$ for $a, a_1, a_2 \in A$. From our construction of vertex operators acting on V and on W given in (15), (18), (22) of §2.2, we have

$$(5) \qquad Y(a(-\tfrac{1}{2})\Omega, \zeta) = a(\zeta) = \sum_{m \in \mathbb{Z}'} a(m + \tfrac{1}{2})\zeta^{-m-1},$$

$$(6) \qquad Y(a_1(-\tfrac{1}{2})a_2(-\tfrac{1}{2})\Omega, \zeta) =: a_1(\zeta)a_2(\zeta) := \sum_{m \in \mathbb{Z}} x(m)\zeta^{-m-1}.$$

Here, $\mathbb{Z}' = \tfrac{1}{2}\mathbb{Z} - \mathbb{Z}$ and $x =: a_1 a_2 :\in o(2N)$. More explicitly, the m-th mode vertex operator $x(m)$ associated to $x =: a_1 a_2 :$ is given by

$$(7) \qquad x(m) = \sum_{k \in \mathbb{Z}} :a_1(k)a_2(m - k):, \qquad m \in \mathbb{Z}.$$

In (6), (7), m runs over $\mathbb{Z}$ whether $x(m)$ is acting on V or on W. For later use, we record the vertex operators for $a \in A$ and for $x =: a_1 a_2 :\in o(2N)$, where $a, a_1, a_2 \in A$:

$$(8) \qquad \begin{aligned} \{\Upsilon(a)\}_m &= \{a(-\tfrac{1}{2})\Omega\}_m = a(m + \tfrac{1}{2}), \\ \{\Upsilon(x)\}_m &= \{a_1(-\tfrac{1}{2})a_2(-\tfrac{1}{2})\Omega\}_m = x(m). \end{aligned}$$

In the first formula, we have $m \in \mathbb{Z}$ when it acts on V and $m \in \mathbb{Z} + \tfrac{1}{2}$ when the operator is acting on W. In the second formula, we always have $m \in \mathbb{Z}$.

In $(V)_2$, there is a 1-dimensional distinguished subspace $\mathbb{C}\omega$. Namely, $o(2N)$-invariant subspace $\left(A(-\tfrac{3}{2}) \otimes A(-\tfrac{1}{2})\right)^{o(2N)}$. We choose its generator as follows:

$$(9) \qquad \omega = -\tfrac{1}{2} \sum_{i=1}^{N} \left\{a_i^*(-\tfrac{3}{2})a_i(-\tfrac{1}{2}) + a_i(-\tfrac{3}{2})a_i^*(-\tfrac{1}{2})\right\} \Omega.$$

We write the corresponding vertex operator as

$$Y(\omega, \zeta) = -\frac{1}{2} \sum_{i=1}^{N} \left\{ : a_i^{*(1)}(\zeta) a_i(\zeta) : + : a_i^{(1)}(\zeta) a_i^*(\zeta) : \right\} - \frac{N}{8} \zeta^{-2} \delta_{Z,\mathbb{Z}}$$

(10)

$$= \sum_{m \in \mathbb{Z}} D^Z(m) \zeta^{-m-2} = \sum_{m \in \mathbb{Z}} \{\omega\}_{m+1} \zeta^{-m-2}.$$

In the above, $\delta_{Z,\mathbb{Z}}$ means that $\delta_{Z,\mathbb{Z}} = 1$ only when $Z = \mathbb{Z}$, that is, it is 1 when the vertex operator acts on W, and 0 when the vertex operator acts on V. The last term in the first line of (10) comes from the fact that $\Delta(\zeta)\omega = -\frac{N}{8}\zeta^{-2}$. For the operator $\Delta(\zeta)$, see (22) of §2.2. For operators $a^{(1)}(\zeta)$'s, see (14) and (17) of §2.2. After some calculation we see that for $Z = \mathbb{Z} + \frac{1}{2}$ or $\mathbb{Z}$ and for $k \in \mathbb{Z}$,

(11)
$$D^Z(k) = \sum_{i=1}^{N} \sum_{m \in \mathbb{Z}} (m - \tfrac{1}{2}k) : a_i(m) a_i^*(k-m) :,$$

except that for $D^{\mathbb{Z}}(0)$ we have to add an extra term $-\frac{N}{8}$ to the above expression. This finishes the description of vertex operators associated to vectors in $(V)_{\frac{1}{2}}$, $(V)_1$, and $\omega \in (V)_2$.

We now turn to the description of commutation relations for these operators. We prepare two lemmas to facilitate our calculation.

LEMMA 1. *Let* $x(0)$ *and* $x(1)$ *be operators of* (7) *for* $x \in \mathfrak{o}(2N)$. *Then*

(i) *The maps* $\Upsilon_{\frac{1}{2}}$ *and* Υ_1 *of* (4) *are* $\mathfrak{o}(2N)$-*module isomorphisms:*

(12) $\Upsilon_{\frac{1}{2}}(x \cdot a) = x(0) \cdot \Upsilon_{\frac{1}{2}}(a)$ *for any* $a \in A,\ x \in \mathfrak{o}(2N)$,

(13) $\Upsilon_1([x, y]) = x(0) \cdot \Upsilon_1(y)$ *for any* $x, y \in \mathfrak{o}(2N)$.

(ii) *The map* $x(1) : (V)_1 \to (V)_0$ *is described by*

(14) $x(1) \cdot \Upsilon_1(y) = \langle x, y \rangle \Omega,$ *for any* $x, y \in \mathfrak{o}(2N)$.

Here, $\langle\ ,\ \rangle$ *is the bilinear pairing which is used to construct the Clifford algebras. In* (i), *the action of* $x \in \mathfrak{o}(2N)$ *on* $a \in A$ *is given by the commutator* $x \cdot a = [x, a]$ *as in* (3) *of* §2.2.

PROOF. Let $x =: a_1 a_2 : \in \mathfrak{o}(2N)$. Then $x(0)$ is given by (7) where $Z = \mathbb{Z} + \frac{1}{2}$ since $x(0)$ is acting on V in the current context. We have

$$x(0) = \sum_{m \in \mathbb{Z} + \frac{1}{2}} : a_1(-m) a_2(m) :$$

$$= \cdots + a_1(-\tfrac{3}{2}) a_2(\tfrac{3}{2}) + a_1(-\tfrac{1}{2}) a_2(\tfrac{1}{2}) - a_2(-\tfrac{1}{2}) a_1(\tfrac{1}{2}) - a_2(-\tfrac{3}{2}) a_1(\tfrac{3}{2}) - \cdots.$$

Since only vectors in $A(\frac{1}{2}) \subset \mathrm{Cliff}(\mathbb{Z}+\frac{1}{2})$ can act nontrivially on $\Upsilon(a) = a(-\frac{1}{2})\Omega$, only the middle two terms in the above expression can act nontrivially on $\Upsilon(a)$. Since $a_i(\frac{1}{2})$'s act as derivations,

$$x(0) \cdot \Upsilon(a) = \{a_1(-\tfrac{1}{2})a_2(\tfrac{1}{2}) - a_2(-\tfrac{1}{2})a_1(\tfrac{1}{2})\} \, a(-\tfrac{1}{2})\Omega$$
$$= \langle a_2, a\rangle a_1(-\tfrac{1}{2})\Omega - \langle a_1, a\rangle a_2(-\tfrac{1}{2})\Omega$$
$$= \Upsilon(\langle a_2, a\rangle a_1 - \langle a_1, a\rangle a_2) = \Upsilon(x \cdot a).$$

Here for the last identity we used (3) of §2.2. For the second identity (13), we let $x =: a_1 a_2 :$ and $y =: b_1 b_2 :$. Then $x(0)\Upsilon(y) = x(0) \cdot b_1(-\frac{1}{2})b_2(-\frac{1}{2})\Omega$. Since only terms with vectors from $A(\frac{1}{2})$ can act nontrivially on $\Upsilon(y)$,

$$x(0)\Upsilon(y) = \{a_1(-\tfrac{1}{2})a_2(\tfrac{1}{2}) - a_2(-\tfrac{1}{2})a_1(\tfrac{1}{2})\} \cdot b_1(-\tfrac{1}{2})b_2(-\tfrac{1}{2})\Omega$$
$$= \langle a_2, b_1\rangle a_1(-\tfrac{1}{2})b_2(-\tfrac{1}{2})\Omega - \langle a_2, b_2\rangle a_1(-\tfrac{1}{2})b_1(-\tfrac{1}{2})\Omega$$
$$\quad - \langle a_1, b_1\rangle a_2(-\tfrac{1}{2})b_2(-\tfrac{1}{2})\Omega + \langle a_1, b_2\rangle a_2(-\tfrac{1}{2})b_1(-\tfrac{1}{2})\Omega$$
$$= \Upsilon(\langle a_2, b_1\rangle : a_1 b_2 : + \langle a_1, b_2\rangle : a_2 b_1 : - \langle a_2, b_2\rangle : a_1 b_1 :$$
$$\quad - \langle a_1, b_1\rangle : a_2 b_2 := \Upsilon([x, y]), \qquad \text{by (2) in §3.2.}$$

For the seond part of Lemma 1, note that $x(1)$ is given by

$$x(1) = \sum_{m \in \mathbb{Z}+\frac{1}{2}} :a_1(m)a_2(1-m) := \cdots + a_1(-\tfrac{1}{2})a_2(\tfrac{3}{2}) + a_1(\tfrac{1}{2})a_2(\tfrac{1}{2}) - a_2(-\tfrac{1}{2})a_1(\tfrac{3}{2}) - \cdots .$$

On $\Upsilon(y)$, only those terms containing vectors from $A(\frac{1}{2})$ can act nontrivially and they act as derivations. Using (5) of §2.2, we have

$$x(1)\Upsilon(y) = a_1(\tfrac{1}{2})a_2(\tfrac{1}{2})b_1(-\tfrac{1}{2})b_2(-\tfrac{1}{2})\Omega$$
$$= a_1(\tfrac{1}{2}) \left(\langle a_2, b_1\rangle b_2(-\tfrac{1}{2})\Omega - b_1(-\tfrac{1}{2})\langle a_2, b_1\rangle\Omega\right)$$
$$= (\langle a_2, b_1\rangle\langle a_1, b_2\rangle - \langle a_1, b_1\rangle\langle a_2, b_2\rangle)\Omega = \langle x, y\rangle\Omega.$$

This completes the proof. $\square$

LEMMA 2. *Let $D(-1) = \{\omega\}_0$ be the 0-th mode Virasoro operator. Then for any vector $v \in V$, we have*

$$(15) \quad Y(D(-1)v, \zeta) = -\frac{d}{d\zeta}Y(v, \zeta), \quad \text{and} \quad \{D(-1)v\}_n = n\{v\}_{n-1}, \quad n \in \tfrac{1}{2}\mathbb{Z}.$$

In the second identity, when the vertex operator act on V, we let $n \in \mathbb{Z}$. When they act on W, we let $n \in \mathbb{Z} + \frac{1}{2}|v|$.

PROOF. In the super commutator formula (12) in §2.1, we let $v_1 = \omega$, $m = 0$, and $v_2 = v$. Multiplying both sides by ζ^{-n-1} and summing over n, we get

$$[D(-1), Y(v, \zeta)] = Y(D(-1)v, \zeta).$$

Combining this with the derivative formula (9) of §2.1, we get Lemma 2. $\square$

Now we are ready to calculate commutation relations among vertex operators. All of these formulae follow from the commutator formula (12) of §2.2 with the help of Lemma 1 and Lemma 2.

PROPOSITION 3. *Let* $a, b \in A$ *and* $x, y \in \mathfrak{o}(2N)$. *For any* $k, \ell \in Z$, $m, n \in Z$, *the following commutation relations hold as operators on* V *and on* W:

(16) $\quad a(\ell)b(k) + b(k)a(\ell) = \langle a, b \rangle \delta_{k+\ell,0} \cdot \mathrm{Id}$,

(17) $\quad [x(m), y(n)] = [x, y](m+n) + m\langle x, y \rangle \delta_{m+n,0} \cdot \mathrm{Id}$,

(18) $\quad [x(m), a(k)] = (x \cdot a)(m+k)$,

(19) $\quad [D(m), D(n)] = (n-m)D(m+n) + \dfrac{m(m^2-1)}{12} N \delta_{m+n,0} \cdot \mathrm{Id}$,

(20) $\quad [D(m), a(k)] = (k + \tfrac{1}{2}m)a(m+k)$,

(21) $\quad [D(m), x(n)] = nx(m+n)$.

In (18) *the action of* $\mathfrak{o}(2N)$ *on* A *is given by the commutator* (3) *of* §2.2.

PROOF. For (16), let $v_1, v_2 \in (V)_{\frac{1}{2}}$, say, $v_1 = a_1(-\frac{1}{2})\Omega$ and $v_2 = a_2(-\frac{1}{2})\Omega$. Since these vectors are "fermionic", that is, $|\Upsilon(a_1)| = |\Upsilon(a_2)| = 1$, the super commutator formula actually gives an anticommutation relation. Thus for $m, \ell \in Z'$, (8) implies

$$[a_1(m + \tfrac{1}{2}), a_2(\ell + \tfrac{1}{2})]_+ = [\{a_1(-\tfrac{1}{2})\Omega\}_m, \{a_2(-\tfrac{1}{2})\Omega\}_\ell]_+$$
$$= \sum_{k \geq 0} \binom{m}{k} \{\{a_1(-\tfrac{1}{2})\Omega\}_k a_2(-\tfrac{1}{2})\Omega\}_{m+\ell-k}$$
$$= \sum_{k \geq 0} \binom{m}{k} \{a_1(k + \tfrac{1}{2})a_2(-\tfrac{1}{2})\Omega\}_{m+\ell-k}.$$

Here $[\ ,\ ]_+$ is an anti-commutator. Since $a_1(k+\tfrac{1}{2})$ lowers weight by $k+\tfrac{1}{2}$, only the term corresponding to $k = 0$ contributes nontrivially to the above summation. The above is then equal to

$$[a_1(m + \tfrac{1}{2}), a_2(\ell + \tfrac{1}{2})]_+ = \langle a_1, a_2 \rangle \{\Omega\}_{m+\ell} = \langle a_1, a_2 \rangle \delta_{m+\ell+1,0} \cdot \mathrm{Id}.$$

In the last identity we used $\{\Omega\}_m = \delta_{m+1,0} \cdot \mathrm{Id}$, which follows from $Y(\Omega, \zeta) = 1 = \sum_{m \in Z} \{\Omega\}_m \zeta^{-m-1}$. Thus we have recovered the anticommutation relations of the Clifford algebra we started with.

For (17) let $v_1 = \Upsilon(x)$ and $v_2 = \Upsilon(y)$ for $x, y \in \mathfrak{o}(2N)$ in the super commutator formula. For $m, n \in Z$ the commutator formula gives

$$[x(m), y(n)] = \sum_{k \geq 0} \binom{m}{k} \{x(k)\Upsilon(y)\}_{m+n-k}.$$

Here $y(n) = \{\Upsilon(y)\}_n$ from (8). Since $x(k)$ lowers weight by k and $\Upsilon(y) \in (V)_1$, only those terms with $k = 0, 1$ contribute nontrivially to the above summation, and we get

$$[x(m), y(n)] = \{x(0)\Upsilon(y)\}_{m+n} + m\{x(1)\Upsilon(y)\}_{m+n-1}$$
$$= [x, y](m+n) + m\langle x, y \rangle \delta_{m+n,0} \cdot \mathrm{Id}.$$

For the last equality above, we used Lemma 1. Thus operators $x(m)$ in (7) for $x \in \mathfrak{o}(2N)$ and $m \in \mathbb{Z}$, generate the orthogonal affine Lie algebra $\hat{\mathfrak{o}}(2N)$, and V and W are its representations of level 1.

Next, for $m \in \mathbb{Z}$ and $n \in \mathbb{Z}'$, the commutator formula gives

$$[x(m), a(n + \tfrac{1}{2})] = \sum_{k \geq 0} \binom{m}{k} \{x(k)a(-\tfrac{1}{2})\Omega\}_{m+n-k} .$$

Only the first term contributes nontrivially since wt $(x(k)a(-\tfrac{1}{2})\Omega) = \tfrac{1}{2} - k$, and we have

$$[x(m), a(n + \tfrac{1}{2})] = \{x(0)a(-\tfrac{1}{2})\Omega\}_{m+n} = \{(x \cdot a)(-\tfrac{1}{2})\Omega\}_{m+n} = (x \cdot a)(m + n + \tfrac{1}{2}).$$

The last two equalities are due to (12) and (8). This proves (18).

The proof for (19) goes as follows. In the commutator formula (12) of §2.2, we let $v_1 = v_2 = \omega$. Since $\{\omega\}_{m+1} = D(m)$ by (10), we have

$$[D(m), D(n)] = [\{\omega\}_{m+1}, \{\omega\}_{n+1}] = \sum_{k \geq 0} \binom{m+1}{k} \{\{\omega\}_k \omega\}_{m+n+2-k} .$$

Since $\{\omega\}_k$ lowers weight by $k - 1$, those terms with $k = 0, 1, 2, 3$ contributes to the above summation nontrivially, and we have

$$\begin{aligned}
[D(m), D(n)] = {} & \{D(-1)\omega\}_{m+n+2} + (m+1)\{D(0)\omega\}_{m+n+1} \\
& + \binom{m+1}{2}\{D(1)\omega\}_{m+n} + \binom{m+1}{3}\{D(2)\omega\}_{m+n-1}.
\end{aligned}$$

From (15), $\{D(-1)\omega\}_{m+n+2} = (m+n+2)\{\omega\}_{m+n+1} = (m+n+2)D(m+n)$. Since $D(0)\omega = -2\omega$, the second term is equal to $-2(m+1)D(m+n)$. We have to calculate the third and the fourth terms. $D(1)$ and $D(2)$ are given by (11) with $Z = \mathbb{Z} + \tfrac{1}{2}$ since they are acting on $\omega \in V$. Due to the form of ω given in (9), in the terms of $D(m)$'s only those terms with vectors from $A(\tfrac{1}{2})$ and $A(\tfrac{3}{2})$ act nontrivially on ω. With this in mind, we write down the relevant terms in the expressions of $D(1)$ and $D(2)$ below:

$$D(1) = \sum_{i=1}^{N} \sum_{n \in \mathbb{Z}+\frac{1}{2}} (n - \tfrac{1}{2}) : a_i(n)a_i^*(1 - n):$$

$$= \sum_{i=1}^{N} \{\cdots - a_i(-\tfrac{1}{2})a_i^*(\tfrac{3}{2}) - a_i^*(-\tfrac{1}{2})a_i(\tfrac{3}{2}) - \ldots\},$$

$$D(2) = \sum_{i=1}^{N} \sum_{n \in \mathbb{Z}+\frac{1}{2}} (n - 1) : a_i(n)a_i^*(2 - n):$$

$$= \sum_{i=1}^{N} \{\cdots - \tfrac{1}{2}a_i(\tfrac{1}{2})a_i^*(\tfrac{3}{2}) - \tfrac{1}{2}a_i^*(\tfrac{1}{2})a_i(\tfrac{3}{2}) - \ldots\}.$$

A simple calculation shows that $D(1)\omega = 0$ and $D(2)\omega = \frac{N}{2}\Omega$. Thus, the third term is 0 and the fourth term is $\binom{m+1}{3}\frac{N}{2}\delta_{m+n,0}$. Combining these calculations, we obtain (19).

Formula (20) is now a routine. For $m \in \mathbb{Z}$ and $n \in \mathbb{Z}'$,

$$[D(m), a(n + \tfrac{1}{2})] = [\{\omega\}_{m+1}, \{a(-\tfrac{1}{2})\Omega\}_n] = \sum_{k \geq 0} \binom{m+1}{k} \{\{\omega\}_k a(-\tfrac{1}{2})\Omega\}_{m+n+1-k}$$

$$= \{D(-1)a(-\tfrac{1}{2})\Omega\}_{m+n+1} + (m+1)\{D(0)a(-\tfrac{1}{2})\Omega\}_{m+n}.$$

We use (15) for the first term. For the second term, since $a(-\tfrac{1}{2})\Omega$ has weight $\tfrac{1}{2}$, we have $D(0)a(-\tfrac{1}{2})\Omega = -\tfrac{1}{2}a(-\tfrac{1}{2})\Omega$. The above formula becomes

$$[D(m), a(n + \tfrac{1}{2})] = (n + m + 1)a(n + m + \tfrac{1}{2}) - \tfrac{1}{2}(m+1)a(m + n + \tfrac{1}{2})$$

$$= (n + \tfrac{1}{2} + \tfrac{m}{2})a(m + n + \tfrac{1}{2}), \qquad n \in \mathbb{Z}'.$$

For formula (21), for $m, n \in \mathbb{Z}$ we have

$$[D(m), x(n)] = [\{\omega\}_{m+1}, x(n)] = \sum_{k \geq 0} \binom{m+1}{k} \{\{\omega\}_k \Upsilon(x)\}_{m+n+1-k}$$

$$= \{D(-1)\Upsilon(x)\}_{m+n+1} + (m+1)\{D(0)\Upsilon(x)\}_{m+n} + \binom{m+1}{2}\{D(1)\Upsilon(x)\}_{m+n-1}.$$

By (15), The first term is equal to $(m + n + 1)x(m + n)$. The second term is equal to $(m+1)\{-\Upsilon(x)\}_{m+n} = -(m+1)x(m+n)$ since $\Upsilon(x)$ has weight 1. For the third term, from (11) we have

$$D^{\mathbb{Z}+\frac{1}{2}}(1) = \sum_{i=1}^{N} \sum_{n \in \mathbb{Z}+\frac{1}{2}} (n - \tfrac{1}{2}) : a_i(n)a_i^*(1 - n) : .$$

In this summation, there are no terms involving vectors from $A(\tfrac{1}{2})$. Thus, $D(1)$ acts trivially in $\Upsilon(x)$. Collecting our calculations together, we get (21). $\square$

As we noted in the proof of (17), V and W are representations of the orthogonal affine Lie algebra $\hat{o}(2N)$. Since $x(m)$'s are quadratic in $a(k)$'s, actions of $x(m)$'s on V and on W preserve their $\mathbb{Z}_2$-gradings. Thus the action of $\hat{o}(2N)$ preserves the vector spaces V_0, V_1, W_0, W_1. It is known that V_0, V_1, W_0 and W_1 exhaust all the irreducible level 1 representations of $\hat{o}(2N)$ [K1].

The subalgebra $o(2N)(0) \subset \hat{o}(2N)$ consisting of 0-mode operators acts on V and W by Clifford multiplication, preserving weights. On the other hand, the original Lie algebra $o(2N)$ itself acts on all the exterior powers of A and their tensor products by extending the its action on A given in (3) of §2.2 and it also acts on finite dimensional Clifford representations Δ by Clifford multiplications. Thus $o(2N)$ also acts on V and W. These two actions of $o(2N)(0)$ and of $o(2N)$ on V and on W may look different, but in fact they are the same.

COROLLARY 4. *The action of the 0-mode of the affine orthogonal Lie algebra* $o(2N)(0) \subset \hat{o}(2N)$ *on* V *and on* W *(by infinite sums of elements in* $\text{Cliff}(Z)$*) is the same as the standard action of* $o(2N)$ *on* V *and on* W.

PROOF. Let $x \in o(2N)$. We calculate the action of $x(0)$ on V and on W. For $a_1, \ldots, a_r \in A$, we have

$$x(0) \cdot (a_1(-n_1)a_2(-n_2) \cdots a_r(-n_r)\Omega)$$

$$= \sum_{i=1}^{r} a_1(-n_1) \cdots [x(0), a_i(-n_i)] \cdots a_r(-n_r)\Omega + a_1(-n_1) \cdots a_r(-n_r)x(0) \cdot \Omega.$$

By (18), $[x(0), a_i(-n_i)] = (x \cdot a_i)(-n_i)$, where $x \cdot a_i$ is the standard action of $o(2N)$ on A. As for $x(0) \cdot \Omega$, we claim that $x(0) \cdot \Omega = x \cdot \Omega$ in V and W. If $x =: a_1 a_2 :$, then $x(0) = \sum_{n \in Z} : a_1(-n)a_2(n) :$. Since the vacuum vectors Ω in V and W are annihilated by Clifford multiplication by vectors in $A(n)$ with $n > 0$, it follows that in V we have $x(0) \cdot \Omega = 0$, and in W we have $x(0) \cdot \Omega =: a_1(0)a_2(0) : \Omega =: a_1 a_2 : \Omega = x \cdot \Omega \in \Delta$. Since $(V)_0 = \mathbb{C}\Omega$ is a trivial representation of $o(2N)$, we have $x \cdot \Omega = 0$ in V. Thus we have proved our claim and the actions of $o(2N)(0)$ and of $o(2N)$ on the vacuum vector Ω of V and of W are the same. Thus the above summation is equal to

$$\sum_{i=1}^{r} a_1(-n_1) \cdots (x \cdot a_i)(-n_i) \cdots a_r(-n_r)\Omega + a_1(-n_1) \cdots a_r(-n_r)x \cdot \Omega$$

$$= x \cdot (a_1(-n_1) \cdots a_r(-n_r)\Omega),$$

where $\cdot$ denotes the action of $x \in o(2N)$ on V and on W. Hence, the actions of $x(0) \in \hat{o}(2N)$ and $x \in o(2N)$ on V and W are the same. $\square$

§2.4   The adjoint operators of vertex operators and vertex operators for adjoint vectors

We study adjoint operators of vertex operators in the vertex operator super algebra V which was constructed in §2.2, §2.3 on Spin representations of orthogonal affine Lie algebras. We study these operators from the point of view of positive definite Hermitian pairings on V. In most existing literature on vertex operator algebras, Hermitian pairings are not usually treated and so one doesn't often see adjoint operators of vertex operators.

There are several reasons why we want to deal with adjoint operators. First of all, for our vertex operator super algebra V the Hermitian pairing is a natural integral part of the structure: V is a total Spin representation of the affine orthogonal Lie algebra $o(\hat{2}N)$, and the very construction of finite dimensional orthogonal Lie algebra $o(2N)$ requires a positive definite pairing on real $2N$-dimensional vector space. Incorporating the complex structure gives us a Hermitian pairing on V. Hermitian pairings give rise to richer structure and convenient tools.

The second reason comes from the essential use of the Hermitian pairings in the unitary representation theory of the Virasoro algebras where the Kac determinant formula play an important role. Since the Virasoro algebra is the most important algebra in the theory of vertex operator algebras, Hermitian pairings should be useful in the general study of the representation theory of vertex operator algebras.

Later in this section we introduce another nondegenerate positive definite Hermitian pairing on V using vertex operators. This pairing is defined by $(u, v) \mapsto Y_k(u^*)v \in \mathbb{C}\Omega$, where $k = \text{wt}(v)$ and u^* is the adjoint vector of u defined later [Definition 22].

Our interest in this Hermitian pairing arises from the fact that its $\mathbb{C}$-linear version arises naturally and it is important. For example, $(u, v) \mapsto Y_{\frac{1}{2}}(u)v \in \mathbb{C}\Omega$ for $u, v \in (V)_{\frac{1}{2}}$ recovers the original pairing $\langle\ ,\ \rangle$ on A (See (14) of §2.1.), and the pairing $(u, v) \mapsto Y_1(u)v \in \mathbb{C}\Omega$ for $u, v \in (V)_1 = \mathfrak{g}$ recovers the $\mathfrak{g}$ invariant pairing of the Lie algebra $\mathfrak{g} = (V)_1$ (See (19) of §2.1.). Properties of this Hermitian pairing are described and proved in terms of Gramians.

A Hermitian pairing in V induced from a pairing in A. To describe in detail the Hermitian pairing on V induced from that on A, we start with a real Euclidean vector space E^{2N} with an isometric complex structure I. Let $\{e_1, \ldots, e_N, e'_1, \ldots, e'_N\}$ be an orthonormal basis of E such that $I(e_j) = e'_j$ and $I(e'_j) = -e_j$ for $1 \leq j \leq N$. Our vector space V is a complexification of a real vector space. Namely,

$$(1) \qquad V = \bigotimes_{0 \leq n \in \mathbb{Z}} \wedge_{\mathbb{C}}^* A(-n - \tfrac{1}{2}) = \left\{ \bigotimes_{0 \leq n \in \mathbb{Z}} \wedge_{\mathbb{R}}^* E(-n - \tfrac{1}{2}) \right\} \otimes \mathbb{C}.$$

Here $E(-m)$ is a copy of the real vector space E carrying weight m. As a complexified vector space, V is equipped with a complex conjugation operator. Let $A = A^+ \oplus A^-$ be the decomposition of the complexification of E into $\pm i$-eigenspaces of the complex structure I. These eigenspaces are maximal isotropic subspaces with respect to the complex linear extension of the pairing $\langle\ ,\ \rangle$. These subspaces A^+ and A^- have bases $\{a_1, \ldots, a_N\}$ and $\{a_1^*, \ldots, a_N^*\}$ where

$$(2) \qquad a_j = \frac{e_j - ie'_j}{\sqrt{2}}, \qquad a_j^* = \frac{e_j + ie'_j}{\sqrt{2}}, \qquad \text{for } 1 \leq j \leq N.$$

We have $\langle a_i, a_j \rangle = \langle a_i^*, a_j^* \rangle = 0$ and $\langle a_i, a_j^* \rangle = \delta_{i,j}$ for any $1 \leq i, j \leq N$. Let $\bar{\ } : A \to A$ be the complex conjugation map with respect to the real structure on V given in (1). Its effect on the above basis vectors is given by

$$(3) \qquad \bar{a}_j = a_j^*, \qquad \overline{a_j^*} = a_j, \qquad 1 \leq j \leq N.$$

We define a Hermitian form $(\ ,\ )$ on A by

$$(4) \qquad (a, b) = \langle \bar{a}, b \rangle$$

for any $a, b \in A$. This Hermitian form is conjugate linear in the first variable. We could consider another Hermitian pairing which is conjugate linear in the second variable. But in our context the one above is more convenient. Note that

$$(4') \qquad \overline{(a, b)} = (\bar{a}, \bar{b}).$$

We extend this pairing to a direct sum $\bigoplus_{n\in\mathbb{Z}+\frac{1}{2}} A(n)$ by letting

(5) $(a(n), b(m)) = (a, b)\delta_{n,m}$,   for any $a, b \in A$.

Note that vectors of different weight pair trivially.

For general vectors in the exterior algebra V, we let $(\Omega, \Omega) = 1$ and for $r, s \geq 0$, $b_1, \ldots, b_r, c_1 \ldots, c_s \in A$, and $n_1, \ldots, n_r, m_1, \ldots, m_s \in \mathbb{Z}_- - \frac{1}{2}$, we let

(6) $\left(b_1(n_1) \cdots b_r(n_r)\Omega, c_1(m_1) \cdots c_s(m_s)\Omega\right) = \delta_{r,s} \det\left((b_j(n_j), c_k(m_k))\right)$.

Elements of the form $a_{i_1}(n_1) \cdots a_{i_r}(n_r)a_{j_1}^*(m_1) \cdots a_{j_s}^*(m_s)\Omega$ form an orthonormal basis of V, where $n_1, \ldots, n_r, m_1, \ldots, m_s$ are non-positive half integers. This completes definition of the Hermitian pairing on V induced from that on A given in (4).

Adjoint operators of vertex operators. Now we calculate adjoint operators of vertex operators of V with respect to the positive definite Hermitian pairing given by (6). For any element $b \in A$, we let

(7) $b(\zeta) = \sum_{m\in\mathbb{Z}} b(m + \tfrac{1}{2})\zeta^{-m-1}$.

The adjoint operator $b(\zeta)^*$ of $b(\zeta)$ is given by the next lemma.

LEMMA 1. *Let $b \in A$. Then the adjoint operator $b(\zeta)^*$ is given by*
(8)
$$b(\zeta)^* = \sum_{m\in\mathbb{Z}} \bar{b}(-m - \tfrac{1}{2})\zeta^{-m-1} = \zeta^{-1}\bar{b}(\zeta^{-1}), \quad \text{that is,} \quad b(m + \tfrac{1}{2})^* = \bar{b}(-m - \tfrac{1}{2}).$$

More explicitly, for any $u, v \in V$, we have $(b(m + \tfrac{1}{2})u, v) = (u, \bar{b}(-m - \tfrac{1}{2})v)$.

PROOF. We show $b(-n)^* = \bar{b}(n)$ for $b \in A$ and $n \in \mathbb{Z} + \frac{1}{2}$. First we deal with the case $n > 0$. We let $-n = n_1 < 0$, $b_1 = b$ and let $b_2, \ldots, b_r, c_1, \ldots, c_s \in A$ and $n_2, \ldots, n_r, m_1, \ldots, m_s \in \mathbb{Z} + \frac{1}{2}$. We may assume that $n_2, \ldots, n_r, m_1, \ldots, m_s$ are all negative half integers. Now we consider a pairing

$(b_1(n_1) \cdots b_r(n_r)\Omega, c_1(m_1) \cdots c_s(m_s)\Omega) = \delta_{r,s} \det\left((b_j(n_j), c_k(m_k))\right)$.

Expanding the determinant on the right along the first column, we get

$\delta_{rs}\sum_{j=1}^{r}(-1)^{j+1}(b_1(n_1), c_j(m_j))(b_2(n_2) \cdots b_r(n_r)\Omega, c_1(m_1) \cdots \widehat{c_j(m_j)} \cdots c_s(m_s)\Omega)$,

where a hat means that that term is omitted. Since

$(b_1(n_1), c_j(m_j)) = (b_1, c_j)\delta_{n_1,m_j} = \langle \bar{b}_1, c_j\rangle\delta_{n_1,m_j}$, moving the summation inside of the parenthesis in the second entry, we have

$$\delta_{rs}\left(b_2(n_2)\cdots b_r(n_r)\Omega,\ \sum_{j=1}^{r}(-1)^{j+1}\langle\bar{b}_1,c_j\rangle\delta_{n_1,m_j}c_1(m_1)\cdots\widehat{c_j(m_j)}\cdots c_s(m_s)\Omega\right).$$

On the other hand, since $\bar{b}(-n_1) = \bar{b}(n)$ with $n > 0$ acts as a derivation on V,

$$\bar{b}_1(-n_1)\cdot(c_1(m_1)\cdots c_s(m_s)\Omega)$$
$$= \sum_{j=1}^{s}(-1)^{j+1}\langle\bar{b}_1,c_j\rangle\delta_{n_1,m_j}c_1(m_1)\cdots\widehat{c_j(m_j)}\cdots c_s(m_s)\Omega.$$

Comparing with the above formula, we see that

$$(b_1(n_1)\cdots b_r(n_r)\Omega,\ c_1(m_1)\cdots c_s(m_s)\Omega)$$
$$= (b_2(n_2)\cdots b_r(n_r)\Omega,\ \bar{b}_1(-n_1)(c_1(m_1)\cdots c_s(n_s)\Omega))\,.$$

Since b_i's and c_j's are arbitrary, elements of the above form run over all the basis vectors of V. Hence $b(n)^* = \bar{b}(-n)$ if $n < 0$ and for any $b \in A$. The case $n > 0$ can be shown in a similar way. $\square$

Applying this lemma repeatedly, we get the next lemma.

LEMMA 2. *For any $b_1,\ldots,b_r \in A$ and $m_1,\ldots,m_r \in \mathbb{Z}+\frac{1}{2}$, we have*

$$(9)\qquad (:b_1(m_1)\cdots b_r(m_r):v,\ u) = (v,\ :\bar{b}_r(-m_r)\cdots\bar{b}(-m_1):u)$$

for any $u,v \in V$.

PROOF. Unraveling the meaning of normal ordering,

$$:b_1(m_1)\cdots b_r(m_r): := \mathrm{sgn}(\sigma)b_{\sigma(1)}(m_{\sigma(1)})\cdots b_{\sigma(r)}(m_{\sigma(r)}),$$

where the permutation σ is such that $m_{\sigma(1)} \leq m_{\sigma(2)} \leq \cdots \leq m_{\sigma(r)}$. Then, by Lemma 1,

$$(:b_1(m_1)\cdots b_r(m_r):v,\ u) = (v,\ \mathrm{sgn}(\sigma)\bar{b}_{\sigma(r)}(-m_{\sigma(r)})\cdots\bar{b}_{\sigma(1)}(-m_{\sigma(1)})u)$$
$$= (v,\ :\bar{b}_r(-m_r)\cdots\bar{b}(-m_1):u).$$

In the right hand side of the first line above, operators are already normally ordered since $-m_{\sigma(r)} \leq \cdots \leq -m_{\sigma(1)}$. So the formula in the second line follows by the definition of normal ordering. $\square$

Collecting operators of various modes, we get the following:

COROLLARY 3. *Let $b_1,\ldots,b_r \in A$. Then*

$$(10)\qquad\begin{aligned}:b_1(\zeta)b_2(\zeta)\cdots b_r(\zeta):^* &= :b_r(\zeta)^*\cdots b_2(\zeta)^*b_1(\zeta)^*:\\ &= \zeta^{-r}:\bar{b}_r(\zeta^{-1})\cdots\bar{b}_1(\zeta^{-1}):\end{aligned}$$

Here, $b(\zeta)^*$'s are as in (8).

In terms of a vertex operator $Y(v,\zeta) = \sum_{m\in\mathbb{Z}+\mathrm{wt}(v)} Y_m(v)\zeta^{-m-\mathrm{wt}(v)}$, the above corollary means the following:

PROPOSITION 4 (ADJOINT OPERATORS OF VERTEX OPERATORS: (I)).
Let $v \in \bigwedge^* A(-\frac{1}{2})\Omega \subset V$. Then

$$(11) \qquad\qquad Y_m(v)^* = Y_{-m}(v^*).$$

Here, if $v = b_1(-\frac{1}{2})\cdots b_r(-\frac{1}{2})\Omega$, then $v^* = \bar{b}_r(-\frac{1}{2})\cdots \bar{b}_1(-\frac{1}{2})\Omega$.

PROOF. We only have to recall that for v as above, its associated vertex operator $Y(v, \zeta)$ is given by

$$Y(v, \zeta) =: b_1(\zeta)\cdots b_r(\zeta): .$$

Thus Corollary 3 gives $Y(v, \zeta)^*$. □

The subspace $\bigwedge^* A(-\frac{1}{2})\Omega$ of V has other nice properties. For example, see Proposition 8 of §3.5.

Specializing this Corollary 4 to a vector $x \in \bigwedge^2 A(-\frac{1}{2})\Omega$, we obtain

COROLLARY 5. Let $x \in o(2N) \otimes \mathbb{C}$. Then,

$$(12) \qquad\qquad x(n)^* = x^*(-n) = -\bar{x}(-n), \qquad for\ any\ n \in \mathbb{Z}.$$

In the above corollary, the negative sign in front of $\bar{x}(-n)$ comes from switching the order of elements in $A(n)$'s under the normal ordering and $\bar{x}$ is the complex conjugate of x taken in $A = E \otimes \mathbb{C}$

Now we consider adjoint operators of general vertex operators. Recall that $b^{(n)}(\zeta)$ is defined in (14) of §2.2. Corresponding to this, for $b \in A$, we let

$$(13) \qquad b^{(n)}(\zeta)^* = \frac{1}{n!}\frac{d^n}{d\zeta^n}b(\zeta)^* = \frac{1}{n!}\frac{d^n}{d\zeta^n}\left(\sum_{m\in\mathbb{Z}}\bar{b}(-m-\frac{1}{2})\zeta^{-m-1}\right).$$

Using the above notation, we have

LEMMA 6. The adjoint of $Y(b(-n-\frac{1}{2})\Omega, \zeta)$ is given by

$$(14) \qquad\qquad Y(b(-n-\frac{1}{2})\Omega, \zeta)^* = \left(b^{(n)}(\zeta)\right)^* = b^{(n)}(\zeta)^*.$$

PROOF. This is a formal calculation :

$$\left(b^{(n)}(\zeta)u, v\right) = \frac{1}{n!}\frac{d^n}{d\zeta^n}\left(b(\zeta)u, v\right) = \frac{1}{n!}\frac{d^n}{d\zeta^n}\left(u, b(\zeta)^*v\right) = \left(u, b^{(n)}(\zeta)^*v\right).$$

Thus we get our conclusion. □

Now we can state the adjoint formula for general vertex operators.

PROPOSITION 7 (ADJOINT OPERATORS OF VERTEX OPERATORS: (II)). *Let*

$$v = b_1(-n_1 - \tfrac{1}{2})b_2(-n_2 - \tfrac{1}{2}) \cdots b_r(-n_r - \tfrac{1}{2})\Omega$$

be a vector in V, *where* $b_1, \ldots, b_r \in A$ *and* $n_1, \ldots, n_r \geq 0$. *Then the adjoint of the vertex operator* $Y(v, \zeta)$ *is given by*

(15) $$Y(v, \zeta)^* =: b_r^{(n_r)}(\zeta)^* \cdots b_1^{(n_1)}(\zeta)^* :$$

PROOF. We recall that $Y(v, \zeta) =: b_1^{(n_1)}(\zeta) \cdots b_r^{(n_r)}(\zeta):$. The above lemmas and corollaries imply that for variables $\zeta_1, \ldots, \zeta_r$, we have

$$: b_1^{(n_1)}(\zeta_1) \ldots b_r^{(n_r)}(\zeta_r) :^* =: b_r^{(n_r)}(\zeta_r)^* \ldots b_1^{(n_1)}(\zeta_1)^* :$$

Letting $\zeta_1 = \cdots = \zeta_r = \zeta$, we get the above formula. $\square$

We write down these operators explicitly for future reference. Let the vector $v \in V$ be as in Proposition 7. Then, letting $|n| = \sum_{j=1}^r n_j$, a simple calculation using $Y(v, \zeta) = \sum_{m \in \mathbb{Z} + \text{wt}(v)} Y_m(v) \zeta^{-m - \text{wt}(v)}$ shows that

(16)

$$Y_m(v) = (-1)^{|\bar{n}|} \sum_{\sum_j(m_j + \frac{1}{2})=m} \prod_{j=1}^r \frac{(m_j + 1)_{n_j}}{n_j!} : b_1(m_1 + \tfrac{1}{2}) \cdots b_r(m_r + \tfrac{1}{2}):,$$

(17)

$$Y_m(v)^* = (-1)^{|\bar{n}|} \sum_{\sum_j(m_j + \frac{1}{2})=m} \prod_{j=1}^r \frac{(m_j + 1)_{n_j}}{n_j!} : \bar{b}_r(-m_r - \tfrac{1}{2}) \cdots \bar{b}_1(-m_1 - \tfrac{1}{2}): .$$

Here $(\alpha)_n = \alpha \cdot (\alpha + 1) \cdots (\alpha + n - 1)$ for any $\alpha \in \mathbb{C}$ and $n \in \mathbb{Z}_+$. It is the Pochhammer symbol. Note that $(m_j + 1)_{n_j}/n_j! = (m_j + n_j)!/(m_j! \, n_j!)$ is a binomial coefficient.

For a homogeneous element $v \in V$, the operator $Y_m(v)$ with $m \in \mathbb{Z} + \text{wt}(v)$, lowers the weight by m. So its adjoint operator $Y_m(v)^*$ raises the weight by m. An operator $Y_{-m}(v)$ also raises the weight by m. We compare these two operators. Especially we compare $Y_m(v^*)v$ and $Y_{-m}(v)^*v$, where $m = \text{wt}(v)$. For the definition of v^*, see Proposition 4 above or Definition 22 below.

LEMMA 8. *Let* $v \in V$ *be a homogeneous vector. Then*

(18) $$Y_{-\text{wt}(v)}(v)^* v = (v, v)\Omega.$$

PROOF. By the reproducing property of vertex operators, $v = Y_{-\text{wt}(v)}(v)\Omega$ for any $v \in V$. Then, by the adjoint property,

$$(v, v) = (Y_{-\text{wt}(v)}(v)\Omega, v) = (\Omega, Y_{-\text{wt}(v)}(v)^* v).$$

Since $Y_{-\text{wt}(v)}(v)^* v$ has weight 0, it must be a constant multiple of Ω. Since $(\Omega, \Omega) = 1$, this constant must be (v, v). $\square$

To calculate $Y_m(v^*)v$, we need to introduce Gramians. The result of calculation is given later in Proposition 23 and Proposition 24.

Gramian Determinants. Earlier in this section, we defined a Hermitian pairing on V induced from a positive definite nondegenerate Hermitian pairing on A. Later we will define another Hermitian pairing on V in terms of vertex operators. To describe this pairing, we need to introduce Gramians. In this subsection we describe Gramians and their properties.

Let $\vec{u}(x) = (u_1(x), \ldots, u_r(x))$ and $\vec{v}(x) = (v_1(x), \ldots, v_r(x))$ be (not necessarily continuous) $\mathbb{R}^r$-valued functions on an interval $[a, b]$. We consider the following determinant of pairings:

$$(19) \qquad G(\vec{u}, \vec{v}) \overset{\text{def}}{=} \det(\langle u_i, v_j \rangle) = \begin{vmatrix} \langle u_1, v_1 \rangle & \langle u_1, v_2 \rangle & \cdots & \langle u_1, v_r \rangle \\ \langle u_2, v_1 \rangle & \langle u_2, v_2 \rangle & \cdots & \langle u_2, v_r \rangle \\ \vdots & \vdots & \ddots & \vdots \\ \langle u_r, v_1 \rangle & \langle u_r, v_2 \rangle & \cdots & \langle u_r, v_r \rangle \end{vmatrix},$$

where the pairing $\langle u_i, v_j \rangle$ is defined in the usual way by the integral over $[a, b]$:

$$\langle u_i, v_j \rangle = \int_a^b u_i(x) \cdot v_j(x) \, dx.$$

The determinant $G(\vec{u}, \vec{u})$ is called a Gramian determinant and it is the discriminant of the quadratic form

$$(20) \qquad (\xi_1, \ldots, \xi_r) \to \int_a^b \left(\sum_{j=1}^r \xi_j u_j(x) \right)^2 dx.$$

An integral representation of the Gramian determinant will be useful. Let $\mathbf{x} = (x_1, x_2, \ldots, x_r)$ be a sequence of r real variables. For an $\mathbb{R}^r$-valued function $\vec{u}(x) = (u_1(x), \ldots, u_r(x))$, let $U(\mathbf{x})$ be a matrix-valued function given by

$$(21) \qquad U(\mathbf{x}) = (u_i(x_j))_{i,j} = \begin{pmatrix} u_1(x_1) & u_1(x_2) & \cdots & u_1(x_r) \\ u_2(x_1) & u_2(x_2) & \cdots & u_2(x_r) \\ \vdots & \vdots & \ddots & \vdots \\ u_r(x_1) & u_r(x_2) & \cdots & u_r(x_r) \end{pmatrix}.$$

For another $\mathbb{R}^r$-valued function $\vec{v}(x)$, we define a matrix-valued function $V(\mathbf{x})$ in r variables in a similar way. The following integral representation of Gramians is known, although a reference is hard to find. So we give its proof here.

PROPOSITION 9 (THE INTEGRAL REPRESENTATION OF GRAMIANS). *Let $\vec{u}(x)$, $\vec{v}(x)$ be $\mathbb{R}^r$-valued functions on $[a, b]$. The Gramian $G(\vec{u}, \vec{v})$ is given by*

$$(22) \qquad G(\vec{u}, \vec{v}) = \det(\langle u_i, v_j \rangle) = \frac{1}{r!} \underbrace{\int_a^b \cdots \int_a^b}_{r} \det U(\mathbf{x}) \cdot \det V(\mathbf{x}) \, d\mathbf{x},$$

where $dx = dx_1 \cdots dx_r$. *In particular,*

(23) $$G(\vec{u}, \vec{u}) = \det(\langle u_i, u_j \rangle) = \frac{1}{r!} \underbrace{\int_a^b \cdots \int_a^b}_{r} (\det U(\mathbf{x}))^2 \, dx \geq 0.$$

PROOF. First note that $\det U(\mathbf{x}) \cdot \det V(\mathbf{x}) = \det (U(\mathbf{x}) \cdot {}^t V(\mathbf{x}))$, and the (i, j) component of the matrix $U(\mathbf{x}) \cdot {}^t V(\mathbf{x})$ is given by

$$\sum_{\ell=1}^r u_i(x_\ell) v_j(x_\ell) = \sum_{\ell=1}^r \rho_{ij}(x_\ell),$$

where $\rho_{ij}(x) = u_i(x) \cdot v_j(x)$. The right hand side of (22) is then given by

$$(*) \quad \frac{1}{r!} \underbrace{\int_a^b \cdots \int_a^b}_{r} \sum_{\sigma \in \mathfrak{S}_r} \mathrm{sgn}(\sigma) \left(\sum_{\ell=1}^r \rho_{1,\sigma(1)}(x_\ell) \right) \cdots \left(\sum_{\ell=1}^r \rho_{r,\sigma(r)}(x_\ell) \right) dx_1 \cdots dx_r.$$

For each σ, the integrand gives r^r terms of the form

$$\rho_{1,\sigma(1)}(x_{\ell_1}) \cdot \rho_{2,\sigma(2)}(x_{\ell_2}) \cdots \rho_{r,\sigma(r)}(x_{\ell_r}).$$

for some permutation $\sigma \in \mathfrak{S}_r$. Suppose that the variables $\{x_{\ell_1}, x_{\ell_2}, \ldots, x_{\ell_r}\}$ appearing in this term are not mutually distinct, say, $x_{\ell_1} = x_{\ell_2} = x_1$. Let $\tau \in \mathfrak{S}_r$ be such that $\tau(1) = 2$, $\tau(2) = 1$, $\tau(j) = j$ for $3 \leq j \leq r$. Then the summation of the above term over $\sigma \in \mathfrak{S}_r$ is given by

$$\sum_{\sigma \in \mathfrak{S}_r} \mathrm{sgn}(\sigma) \rho_{1,\sigma(1)}(x_{\ell_1}) \cdot \rho_{2,\sigma(2)}(x_{\ell_2}) \cdot \rho_{3,\sigma(3)}(x_{\ell_3}) \cdots \rho_{r,\sigma(r)}(x_{\ell_r})$$

$$= \sum_{\sigma \in \mathfrak{S}_r} \mathrm{sgn}(\sigma\tau) \rho_{1,\sigma\tau(1)}(x_{\ell_1}) \cdot \rho_{2,\sigma\tau(2)}(x_{\ell_2}) \cdot \rho_{3,\sigma\tau(3)}(x_{\ell_3}) \cdots \rho_{r,\sigma\tau(r)}(x_{\ell_r})$$

$$= - \sum_{\sigma \in \mathfrak{S}_r} \mathrm{sgn}(\sigma) \rho_{1,\sigma(2)}(x_{\ell_1}) \cdot \rho_{2,\sigma(1)}(x_{\ell_2}) \cdot \rho_{3,\sigma(3)}(x_{\ell_3}) \cdots \rho_{r,\sigma(r)}(x_{\ell_r}).$$

Since by definition $\rho_{1,\sigma(1)}(x_1) = u_1(x_1)v_{\sigma(1)}(x_1)$, $\rho_{2,\sigma(2)}(x_1) = u_2(x_1)v_{\sigma(2)}(x_1)$, and $\rho_{1,\sigma(2)}(x_1) = u_1(x_1)v_{\sigma(2)}(x_1)$, $\rho_{2,\sigma(1)}(x_1) = u_2(x_1)v_{\sigma(1)}(x_1)$, by rearranging terms it immediately follows that $\rho_{1,\sigma(1)}(x_1)\rho_{2,\sigma(2)}(x_1) = \rho_{1,\sigma(2)}(x_1)\rho_{2,\sigma(1)}(x_1)$. So, the first and the last formula above are identical except for the sign. and the above sum must vanish. Hence, afer summing over $\sigma \in \mathfrak{S}_r$, only those monomials in ρ's for which the variables $\{x_{\ell_1}, \ldots, x_{\ell_r}\}$ is a permutation of $\{x_1, x_2, \ldots, x_r\}$ can contribute nontrivially. Hence $(*)$ is equal to

$$\frac{1}{r!} \int_a^b \cdots \int_a^b \sum_{\sigma \in \mathfrak{S}_r} \mathrm{sgn}(\sigma) \sum_{\tau \in \mathfrak{S}_r} \rho_{1,\sigma(1)}(x_{\tau(1)}) \cdots \rho_{r,\sigma(r)}(x_{\tau(r)}) dx_1 \cdots dx_r.$$

For each $\sigma \in \mathfrak{S}_r$, different τ gives the same result after integrating over r variables $x_1, \ldots, x_r$. Thus the summation over $\tau \in \mathfrak{S}_r$ cancels the denominator $r!$, and the above integral is equal to

$$\sum_{\sigma \in \mathfrak{S}_r} \operatorname{sgn}(\sigma) \int_a^b \cdots \int_a^b \rho_{1,\sigma(1)}(x_1) \cdots \rho_{r,\sigma(r)}(x_r)\, dx_1 \cdots dx_r$$

$$= \sum_{\sigma \in \mathfrak{S}_r} \operatorname{sgn}(\sigma) \prod_{j=1}^r \left(\int_a^b \rho_{j,\sigma(j)}(x_j) dx_j \right) = \sum_{\sigma \in \mathfrak{S}_r} \operatorname{sgn}(\sigma) \prod_{j=1}^r \langle u_j, v_{\sigma(j)} \rangle$$

$$= \det(\langle u_i, v_j \rangle)_{i,j} = G(\vec{u}, \vec{v}).$$

This completes the proof of the integration formula. $\square$

We deduce several corollaries to the integral representation.

COROLLARY 10. *For any $\mathbb{R}^r$-valued function $\vec{u}(x) = (u_1(x), \ldots, u_r(x))$ on an interval $[a, b]$, we have*

(24) $$G(\vec{u}, \vec{u}) = \det(\langle u_i, u_j \rangle)_{i,j} \geq 0.$$

Furthermore, we have the following equivalences :
(25)

$$\begin{array}{ccc} \det U(\mathbf{x}) \equiv 0 \text{ as a function} & & \{u_1(x), \ldots, u_r(x)\} \text{ is} \\ \quad & \Longleftrightarrow \quad G(\vec{u}, \vec{u}) = 0 \quad \Longleftrightarrow & \quad \\ \text{in } \mathbf{x} = (x_1, \ldots, x_r) & & \text{linearly dependent.} \end{array}$$

PROOF. The first part follows from the integral representation (23) of Proposition 9. So, we examine the case in which the equality holds. From (23), it is immediate that $G(\vec{u}, \vec{u}) = 0$ if and only if $\det U(\mathbf{x}) \equiv 0$. When the set $\{u_1(x), \ldots, u_r(x)\}$ is a set of linearly dependent functions, then the matrix $(\langle u_i, u_j \rangle)_{i,j}$ has rank less than r. Hence, its determinant vanishes, i.e., $G(\vec{u}, \vec{u}) = 0$. Conversely, suppose $G(\vec{u}, \vec{u}) = 0$. Since the matrix $U = (\langle u_i, u_j \rangle)$ is a real symmetric matrix, it is diagonalizable by an orthogonal matrix. Since the determinant of U vanishes, it has 0 as its eigenvalue. Let $\xi = (\xi_1, \ldots, \xi_r) \neq \vec{0}$ be an eigenvector associated to the eigenvalue 0. Then, $\vec{\xi} U^t \vec{\xi} = \vec{0}$. In terms of the quadratic form (20), this means that $\int_a^b \left(\sum_{j=1}^r \xi_j u_j(x) \right)^2 dx = 0$. Hence, we must have $\sum_{j=1}^r \xi_j u_j(x) \equiv 0$ as a function in x. This means that functions $u_1(x), \ldots, u_r(x)$ are linearly dependent. This completes the proof. $\square$

COROLLARY 11. *Let $\vec{u}(x)$ and $\vec{v}(x)$ be $\mathbb{R}^r$-valued function on an interval $[a, b]$. Then we have*

(26) $$G(\vec{u}, \vec{u}) \cdot G(\vec{v}, \vec{v}) \geq G(\vec{u}, \vec{v})^2.$$

PROOF. We consider the following quadratic form in variables ξ_1 and in ξ_2:

$$\frac{1}{r!} \int_a^b \cdots \int_a^b \left(\xi_1 \det U(\mathbf{x}) + \xi_2 \det V(\mathbf{x}) \right)^2 dx_1 \cdots dx_r \geq 0,$$

where $U(\mathbf{x}) = (u_i(x_j))$ and $V(\mathbf{x}) = (v_i(x_j))$. Evaluating this integral, we get

$$\xi_1^2 G(\vec{u}, \vec{u}) + 2\xi_1 \xi_2 G(\vec{u}, \vec{v}) + \xi_2^2 G(\vec{v}, \vec{v}) \geq 0,$$

which is valid for any reals ξ_1 and ξ_2. If the set of functions $\{u_1(x), \ldots, u_r(x)\}$ is linearly dependent, then $G(\vec{u}, \vec{u}) = 0$, $G(\vec{u}, \vec{v}) = 0$. So (26) is trivially true. If $\{u_1(x), \ldots, u_r(x)\}$ is linearly independent, then $G(\vec{u}, \vec{u}) > 0$ and the nonpositivity of the discriminant of the above quadratic form gives (26) above. $\square$

Next, we examine when equality holds in the Schwarz type inequality (26).

LEMMA 12. *Let* $\vec{u}(x) = (u_1(x), \ldots, u_r(x))$ *and* $\vec{v}(x) = (v_1(x), \ldots, v_r(x))$ *be* $\mathbb{R}^r$*-valued functions on an interval* $[a, b]$. *Suppose that* $\{u_1(x), \ldots, u_r(x)\}$ *and* $\{v_1(x), \ldots, v_r(x)\}$ *are subsets of linearly independent vectors in the function space* $F([a, b]; \mathbb{R}^r)$. *Then the following two statements are equivalent:*

(i) *Two sets of vectors* $\{u_1(x), \ldots, u_r(x)\}$, $\{v_1(x), \ldots, v_r(x)\}$ *span the same* r*-dimensional subspace in the function space* $F([a, b]; \mathbb{R}^r)$. *Namely,*

$$\bigoplus_{j=1}^r \mathbb{R} u_j = \bigoplus_{j=1}^r \mathbb{R} v_j.$$

(ii) *The following relation between determinants holds as functions in* r *independent variables* $x_1, \ldots, x_r$:

$$\begin{vmatrix} u_1(x_1) & \cdots & u_1(x_r) \\ \vdots & \ddots & \vdots \\ u_r(x_1) & \cdots & u_r(x_r) \end{vmatrix} = \xi \begin{vmatrix} v_1(x_1) & \cdots & v_1(x_r) \\ \vdots & \ddots & \vdots \\ v_r(x_1) & \cdots & v_r(x_r) \end{vmatrix},$$

for some nonzero constant ξ.

PROOF. Assuming (i), there exists a nonsingular (constant) matrix A such that $(u_1(x), \ldots, u_r(x)) = (v_1(x), \ldots, v_r(x))A$. Evaluating this identity at $x = x_1, \ldots, x_r$ and assembling them together, we get

$$\begin{pmatrix} u_1(x_1) & \cdots & u_1(x_r) \\ \vdots & \ddots & \vdots \\ u_r(x_1) & \cdots & u_r(x_r) \end{pmatrix} = {}^tA \begin{pmatrix} v_1(x_1) & \cdots & v_1(x_r) \\ \vdots & \ddots & \vdots \\ v_r(x_1) & \cdots & v_r(x_r) \end{pmatrix}.$$

Taking the determinants of both sides, we get (ii) with $\xi = \det A \neq 0$.

Since $\{u_1(x), \ldots, u_r(x)\}$ is a set of linearly independent functions, $\det U(\mathbf{x})$ does not vanish constantly as a function in $\mathbf{x} = (x_1, \ldots, x_r)$ by (25). So, there exists $\mathbf{c} = (c_1, \ldots, c_r)$ such that $\det U(\mathbf{c}) \neq 0$. Since $\xi \neq 0$ in (ii), we also have that $\det V(\mathbf{c}) = (1/\xi) \det U(\mathbf{c}) \neq 0$. Let $U_k(x) = U(c_1, \ldots, c_{k-1}, x, c_{k+1}, \ldots, c_r)$ be the matrix obtained from $U(\mathbf{c})$ by replacing the k-th column by the column vector ${}^t\vec{u}(x)$. We define $V_k(x)$ similarly. Expanding the determinant $\det U_k(x)$ along the k-th column, we get

$$\det U_k(x) = (u_1(x), \ldots, u_r(x)) \begin{pmatrix} \tilde{a}_{1k} \\ \vdots \\ \tilde{a}_{rk} \end{pmatrix},$$

where $(-1)^{i+k}\tilde{a}_{ik}$ is a constant which is the determinant of the matrix obtained by deleting the i-th row and k-th column from $U(\mathbf{c})$. That is, $\tilde{a}_{ik}$ is the (i, k) cofactor of $\det U(\mathbf{c})$. Collecting the above formulae for $1 \le k \le r$, we obtain

$$(\det U_1(x), \ldots, \det U_r(x)) = (u_1(x), \ldots, u_r(x)) \cdot {}^t\mathrm{adj}(U(\mathbf{c})),$$

where $\mathrm{adj}(U(\mathbf{c})) = {}^t(\tilde{a}_{ij})$ is the adjoint matrix of $U(\mathbf{c})$. Since $\det(U) \cdot U^{-1} = \mathrm{adj}(U)$, the above can be rewritten as

$$(\det U_1(x), \ldots, \det U_r(x)) = (u_1(x), \ldots, u_r(x)) \cdot |U(\mathbf{c})| \cdot {}^tU(\mathbf{c})^{-1}.$$

Similarly, applying the same arguments to $V_k(x)$'s, we get

$$(\det V_1(x), \ldots, \det V_r(x)) = (v_1(x), \ldots, v_r(x)) \cdot |V(\mathbf{c})| \cdot {}^tV(\mathbf{c})^{-1}.$$

But from our hypothesis (ii), we have $\det U_k(x) = \xi \det V_k(x)$ for $1 \le k \le r$. So,

$$\begin{aligned}
|U(\mathbf{c})| \cdot (u_1(x), \ldots, u_r(x)) \cdot {}^tU^{-1}(\mathbf{c}) &= (\det U_1(x), \ldots, \det U_k(x)) \\
&= \xi \cdot (\det V_1(x), \ldots, V_r(x)) \\
&= \xi|V(\mathbf{c})| \cdot (v_1(x), \ldots, v_r(x)) \cdot {}^tV^{-1}(\mathbf{c}).
\end{aligned}$$

Sine (ii) implies $|U(\mathbf{c})| = \xi|V(\mathbf{c})| \ne 0$, we finally have

$$(u_1(x), \ldots, u_r(x)) = (v_1(x), \ldots, v_r(x)) \cdot {}^t(U(\mathbf{c})V^{-1}(\mathbf{c})),$$

where $\det U(\mathbf{c})V^{-1}(\mathbf{c}) \ne 0$. Hence, the two sets of linearly independent functions defined on the interval $[a, b]$, $\{u_1(x), \ldots, u_r(x)\}$ and $\{v_1(x), \ldots, v_r(x)\}$, span the same subspace of the function space. $\square$

PROPOSITION 13. *Let* $\vec{u}(x) = (u_1(x), \ldots, u_r(x))$, $\vec{v}(x) = (v_1(x), \ldots, v_r(x))$ *be* $\mathbb{R}^r$-*valued functions on an interval* $[a, b]$. *Then the equality*

$$(*) \qquad\qquad G(\vec{u}, \vec{u}) \cdot G(\vec{v}, \vec{v}) = G(\vec{u}, \vec{v})^2$$

holds if and only if either of the following statements are true:

(i) *Either* $\{u_1(x), \ldots, u_r(x)\}$ *or* $\{v_1(x), \ldots, v_r(x)\}$ *is a set of linearly dependent functions.*

(ii) *Both* $\{u_1(x), \ldots, u_r(x)\}$ *and* $\{v_1(x), \ldots, v_r(x)\}$ *are sets of linearly independent functions and they span the same r-dimensional subspace in the function space* $F([a,b]; \mathbb{R}^r)$, *that is,*

$$\bigoplus_{j=1}^{r} \mathbb{R} u_j = \bigoplus_{j=1}^{r} \mathbb{R} v_j.$$

PROOF. Suppose (i) is the case, say, $\{u_1(x), \ldots, u_r(x)\}$ is a linearly dependent set of functions. Then $G(\vec{u}, \vec{u}) = 0$ and $G(\vec{u}, \vec{v}) = 0$, and the above equality $(*)$ is trivially true. Next, suppose (ii) is the case, then there exists a nonsingular matrix A such that $(u_1, \ldots, u_r) = (v_1, \ldots, v_r)A$. Then

$$G(\vec{u}, \vec{u}) = \det\left({}^t A \begin{pmatrix} v_1 \\ \vdots \\ v_r \end{pmatrix} \cdot (\, v_1 \quad \cdots \quad v_r \,)\, A\right) = \det {}^t A \cdot G(\vec{v}, \vec{v}) \cdot \det A$$

$$= (\det A)^2 G(\vec{v}, \vec{v}),$$

$$G(\vec{u}, \vec{v}) = \det\left({}^t A \begin{pmatrix} v_1 \\ \vdots \\ v_r \end{pmatrix} \cdot (\, v_1 \quad \cdots \quad v_r \,)\right) = \det A \cdot G(\vec{v}, \vec{v}).$$

In the above, pairings are denoted by dots ".". From these formulae, obviously the above equality $(*)$ holds.

Conversely, suppose $(*)$ holds and the sets $\{u_1, \ldots, u_r\}$, $\{v_1, \ldots, v_r\}$ are linearly independent sets. Then from Corollary 10 we have $G(\vec{u}, \vec{u}) > 0$ and $G(\vec{v}, \vec{v}) > 0$. Since the discriminant of the quadratic form

$$\xi^2 G(\vec{u}, \vec{u}) + 2\xi_1 \xi_2 G(\vec{u}, \vec{v}) + \xi_2^2 G(\vec{v}, \vec{v})$$

vanishes by $(*)$, there exists a pair of real number $(\xi_1, \xi_2) \neq (0,0)$ such that the value of the above quadratic form is 0. In terms of the integral representation, this implies that

$$\frac{1}{r!} \int_a^b \cdots \int_a^b \left(\xi_1 \det U(\mathbf{x}) + \xi_2 \det V(\mathbf{x})\right)^2 dx_1 \cdots dx_r = 0.$$

This in turn implies that $\xi_1 \det U(\mathbf{x}) + \xi_2 \det V(\mathbf{x}) = 0$ as functions in r real variables $x_1, \ldots, x_r$. Since $\{u_1, \ldots, u_r\}$ and $\{v_1, \ldots, v_r\}$ are linearly independent sets, by (25) we see that $\det U(\mathbf{x})$ and $\det V(\mathbf{x})$ are not constantly 0. So, if $\xi_1 = 0$, then we must have $\xi_2 \det V(\mathbf{x}) = 0$ which implies that $\xi_2 = 0$. This contradicts to our choice of (ξ_1, ξ_2). Thus $\xi_1 \neq 0$. Similarly, $\xi_2 \neq 0$. Hence we get

$$\det U(\mathbf{x}) = \xi \det V(\mathbf{x}), \qquad \xi \neq 0,$$

where $\xi = -\xi_2/\xi_1$. By Lemma 12 this implies that $\{u_1, \ldots, u_r\}$ and $\{v_1, \ldots, v_r\}$ span the same subspace in the function space. $\square$

Determinants of matrices of binomial coefficients. We apply our previous results of Gramians to specific functions $u_j(x)$ to obtain results which will be applied later to calculate Hermitian pairings of vectors in V defined by vertex operators.

For any non-negative integer n, let $u_n(x)$ be a function on $[0, 1]$ defined by

$$(27) \qquad u_n(x) = \sum_{\ell=0}^{n} \sqrt{2} \binom{n}{\ell} \cos \pi(\ell + 1)x.$$

The exclusion of the constant function from the expression above is merely due to a notational reason. For our purpose, we could use any orthonormal set of functions. The use of cosine functions is merely for convenience.

First we observe an elementary fact.

LEMMA 14. *The set* $\{\sqrt{2}\cos \pi(\ell+1)x\}_{\ell \geq 0}$ *is an orthonormal set of functions on the interval* $[0, 1]$, *that is,*

$$\int_0^1 \sqrt{2}\cos \pi(\ell + 1)x \cdot \sqrt{2}\cos \pi(\ell' + 1)x \, dx = \delta_{\ell,\ell'}, \qquad 0 \leq \ell, \ell' \in \mathbb{Z}.$$

PROOF. This is a straightforward calculation using the addition formula

$$\sqrt{2}\cos \pi(\ell + 1)x \cdot \sqrt{2}\cos \pi(\ell' + 1)x = \cos \pi(\ell + \ell' + 2)x + \cos \pi(\ell - \ell')x,$$

for $\ell, \ell' \geq 0$. If $\ell \neq \ell'$, then the integral over $[0, 1]$ yields 0. When $\ell = \ell'$, then the integral over $[0, 1]$ is 1 since the second term on the right hand side in the above formula is then a constant function 1. $\square$

LEMMA 15. *The set of functions* $\{u_0(x), u_1(x), \ldots, u_n(x), \ldots, \}$ *is a set of linearly independent functions on* $[0, 1]$ *such that for any* $0 \leq m, n \in \mathbb{Z}$,

$$(28) \qquad \langle u_m, u_n \rangle = \int_0^1 u_m(x) \cdot u_n(x) \, dx = \binom{m+n}{n} = \frac{(m+n)!}{m!\,n!} = (m, n).$$

PROOF. Suppose there exists a linear relation of the form

$$c_1 u_{k_1}(x) + c_2 u_{k_2}(x) + \cdots + c_r u_{k_r}(x) \equiv 0,$$

for some $0 \leq k_1 < k_2 < \cdots < k_r$ with $c_r \neq 0$. Since $\sqrt{2}\cos(k_r+1)x$ appears only in $u_{k_r}(x)$ among functions appearing in the above relation, the integration of the above relation against $\sqrt{2}\cos(k_r + 1)x$ over $[0, 1]$ gives $c_r = 0$, contradicting to our hypothesis $c_r \neq 0$. Hence no linear relation of the above form can exist and $\{u_k(x)\}_{k \geq 0}$ is a set of linearly independent functions.

For (28), from Lemma 14, when $m \leq n$, we have

$$\langle u_m, u_n \rangle = \int_0^1 u_m(x)u_n(x) \, dx = \sum_{\ell=0}^{m} \binom{m}{\ell}\binom{n}{\ell} = \sum_{\ell=0}^{m} \binom{m}{m-\ell}\binom{n}{\ell}.$$

This is the coefficient of z^m in the expansion of $(1+z)^m(1+z)^n = (1+z)^{m+n}$, which is equal to $\binom{m+n}{m}$. The case $m > n$ gives the same answer. $\square$

Let r be a positive integer. Let $\mathbb{Z}_r^+$ be the set of r-tuples of non-negative integers. For any $\vec{n} = (n_1, n_2, \ldots, n_r) \in \mathbb{Z}_+^r$, we let

$$(29) \qquad u_{\vec{n}}(\mathbf{x}) = u_{n_1}(x_1) \cdot u_{n_2}(x_2) \cdots u_{n_r}(x_r).$$

This is a function in r variables $\mathbf{x} = (x_1, x_2, \ldots, x_r)$.

COROLLARY 16. *The set of functions $\{u_{\vec{n}}(\mathbf{x})\}_{\vec{n} \in \mathbb{Z}_+^r}$ is a set of linearly independent functions in r variables $x_1, x_2, \ldots, x_r$.*

PROOF. By induction on r. $\square$

For a positive integer r, let $\mathcal{S}_r^+$ denote the set of sequences of strictly increasing r non-negative integers:

$$(30) \qquad \mathcal{S}_r^+ = \{\mathbf{s} = (0 \le n_1 < n_2 < \cdots < n_r) \mid n_j \in \mathbb{Z}\}.$$

We let $\mathcal{S}_0^+ = \emptyset$. For each sequence $\mathbf{s} = (0 \le n_1 < \cdots < n_r) \in \mathcal{S}_r^+$ of length r, let $u_{\mathbf{s}}(x) = (u_{n_1}(x), \ldots, u_{n_r}(x))$. To each such sequence $\mathbf{s}$, we associate a matrix $U_{\mathbf{s}}(\mathbf{x})$ in r real variables $\mathbf{x} = (x_1, \ldots, x_r)$ given by

$$(31) \qquad U_{\mathbf{s}}(\mathbf{x}) = \big(u_{n_i}(x_j)\big)_{i,j} = \begin{pmatrix} u_{n_1}(x_1) & u_{n_1}(x_2) & \cdots & u_{n_1}(x_r) \\ u_{n_2}(x_1) & u_{n_2}(x_2) & \cdots & u_{n_2}(x_r) \\ \vdots & \vdots & \ddots & \vdots \\ u_{n_r}(x_1) & u_{n_r}(x_2) & \cdots & u_{n_r}(x_r) \end{pmatrix}.$$

For two sequences $\mathbf{s}_1 = (0 \le m_1 < \cdots < m_r)$, $\mathbf{s}_2 = (0 \le n_1 < \cdots < n_r)$ of length r in $\mathcal{S}_r^+$, let $G(\mathbf{s}_1, \mathbf{s}_2)$ be the Gramian determinant of pairings $\langle u_{m_i}, u_{n_j} \rangle = (m_i, n_j)$:

$$(32)$$

$$G(\mathbf{s}_1, \mathbf{s}_2) = \det\big((\langle u_{m_i}, u_{n_j} \rangle)\big)_{i,j} = \begin{vmatrix} (m_1, n_1) & (m_1, n_2) & \cdots & (m_1, n_r) \\ (m_2, n_1) & (m_2, n_2) & \cdots & (m_2, n_r) \\ \vdots & \vdots & \ddots & \vdots \\ (m_r, n_1) & (m_r, n_2) & \cdots & (m_r, n_r) \end{vmatrix} \in \mathbb{Z}.$$

By Proposition 9, this determinant can be expressed as an integral over $[0,1]^r$:

$$(33) \qquad G(\mathbf{s}_1, \mathbf{s}_2) = \frac{1}{r!} \underbrace{\int_0^1 \cdots \int_0^1}_{r} \det U_{\mathbf{s}_1}(\mathbf{x}) \cdot \det U_{\mathbf{s}_2}(\mathbf{x}) \, d\mathbf{x}.$$

Applying Corollary 10 and Corollary 11 to our present context, we obtain the next Proposition.

PROPOSITION 17. (i) *For any length r sequence* $s \in S_r^+$, *we have* $G(s, s) > 0$. *In other words, if* $s = (0 \leq n_1 < \cdots < n_r)$, *then*

(34)
$$\begin{vmatrix} (n_1, n_1) & \cdots & (n_1, n_r) \\ \vdots & \ddots & \vdots \\ (n_r, n_1) & \cdots & (n_r, n_r) \end{vmatrix} > 0.$$

(ii) *For any two sequences* $s_1, s_2 \in S_r^+$, *we have*

$$G(s_1, s_1) \cdot G(s_2, s_2) \geq G(s_1, s_2)^2.$$

In other words, if $s_1 = (0 \leq m_1 < \cdots < m_r)$ *and* $s_2 = (0 \leq n_1 < \cdots < n_r)$, *then*
(35)
$$\begin{vmatrix} (m_1, m_1) & \cdots & (m_1, m_r) \\ \vdots & \ddots & \vdots \\ (m_r, m_1) & \cdots & (m_r, m_r) \end{vmatrix} \cdot \begin{vmatrix} (n_1, n_1) & \cdots & (n_1, n_r) \\ \vdots & \ddots & \vdots \\ (n_r, n_1) & \cdots & (n_r, n_r) \end{vmatrix} \geq \begin{vmatrix} (m_1, n_1) & \cdots & (m_1, n_r) \\ \vdots & \ddots & \vdots \\ (m_r, n_1) & \cdots & (m_r, n_r) \end{vmatrix}^2.$$

Here $(m, n) = (m + n)!/(m! \, n!)$ *is the binomial coefficient.*

The inequality in (ii) is somewhat reminiscent of the Schwarz inequality. We will show later that indeed the above inequality can be interpreted in that way.

LEMMA 18. *Let* $s_1, \ldots, s_\ell \in S_r^+$ *be distinct non-negative sequences of length r. Then the functions*

$$\det U_{s_1}(\mathbf{x}), \quad \det U_{s_2}(\mathbf{x}), \quad \ldots, \quad \det U_{s_\ell}(\mathbf{x})$$

are linearly independent as functions in r variables $\mathbf{x} = (x_1, \ldots, x_\ell)$.

PROOF. Suppose there exists a nontrivial linear relation of the form

(∗)
$$\xi_1 \det U_{s_1}(\mathbf{x}) + \xi_2 \det U_{s_2}(\mathbf{x}) + \cdots + \xi_\ell \det U_{s_\ell}(\mathbf{x}) = 0,$$

as a function in r variables $\mathbf{x} = (x_1, \ldots, x_r)$ for some constants $(\xi_1, \ldots, \xi_\ell) \neq (0, \ldots, 0)$. Without loss of generality, we may assume that $\xi_1 \neq 0$. Expanding the above determinants, the linear relation (∗) can be written as a linear relation among the functions $\{u_{\vec{n}}(\mathbf{x})\}_{\vec{n} \in \mathbb{Z}_+^r}$. Let $s_1 = (0 \leq n_1 < \cdots < n_r)$. Since the sequences $s_1, \ldots, s_\ell$ are distinct, the function $u_{n_1}(x_1) u_{n_2}(x_2) \cdots u_{n_r}(x_r)$ appears only in the expansion of $\det U_{s_1}(\mathbf{x})$, so its coefficient in the expansion of the above linear relation is ξ_1. Since the set of functions $\{u_{\vec{n}}(\mathbf{x})\}_{\vec{n} \in \mathbb{Z}_+^r}$ is a set of linearly independent functions by Corollary 16, we must have $\xi_1 = 0$. This is a contradiction to our hypothesis. Hence no linear relations of the form (∗) can exist and functions $\det U_{s_1}(\mathbf{x}), \ldots, \det U_{s_\ell}(\mathbf{x})$ are linearly independent. □

To each sequence $s \in S_r^+$, we associate a symbol $a(s)$ and consider a lattice $L(r)$ and a real vector space $F(r)$ generated by these symbols:

(36)
$$L(r) = \bigoplus_{s \in S_r^+} \mathbb{Z} a(s), \qquad F(r) = \bigoplus_{s \in S_r^+} \mathbb{R} \, a(s) = L(r) \otimes_{\mathbb{Z}} \mathbb{R}.$$

Since there are infinitely many distinct elements in $\mathcal{S}_r^+$, the lattice $L(r)$ has infinite rank and the vector space $F(r)$ is infinite dimensional. We introduce a bilinear pairing in $L(r)$ and in $F(r)$ defined by

$$(37) \qquad \langle a(\mathbf{s}_1), a(\mathbf{s}_2) \rangle = G(\mathbf{s}_1, \mathbf{s}_2) \in \mathbb{Z}.$$

The pairing of general vectors $u = \sum_{j=1}^{\ell} \xi_j a(\mathbf{s}_j)$ and $v = \sum_{k=1}^{\ell'} \eta_k a(\mathbf{s}_k)$ in $L(r)$ or in $F(r)$ for some sequences $\mathbf{s}_j$, $\mathbf{s}_k \in \mathcal{S}_r^+$ can be defined by $\mathbb{Z}$- or $\mathbb{R}$-linearity. In view of the integral representation of Gramian determinants (33), the pairing $\langle u, v \rangle$ for the above u, v can be expressed as follows.

$$(38) \qquad \begin{aligned} \langle u, v \rangle &= \sum_{j=1}^{\ell} \sum_{k=1}^{\ell'} \xi_j \eta_k G(\mathbf{s}_j, \mathbf{s}_k) \\ &= \frac{1}{r!} \underbrace{\int_0^1 \cdots \int_0^1}_{r} \Big(\sum_{j=1}^{\ell} \xi_j \det U_{\mathbf{s}_j}(\mathbf{x}) \Big) \cdot \Big(\sum_{k=1}^{\ell'} \eta_k \det U_{\mathbf{s}_k}(\mathbf{x}) \Big) \, d\mathbf{x}. \end{aligned}$$

PROPOSITION 19. *The bilinear pairing $\langle\ ,\ \rangle : F(r) \times F(r) \to \mathbb{R}$ on the vector space $F(r)$ defined by $\langle a(\mathbf{s}_1), a(\mathbf{s}_2) \rangle = G(\mathbf{s}_1, \mathbf{s}_2) \in \mathbb{Z}$ is a positive definite pairing on $F(r)$. In fact, it defines an integral positive definite pairing on the lattice $L(r)$, $\langle\ ,\ \rangle : L(r) \times L(r) \to \mathbb{Z}$.*

PROOF. Let $u = \sum_{j=1}^{\ell} \xi_j a(\mathbf{s}_j) \in F(r)$ be a general vector in $F(r)$, where $x_1, \ldots, x_\ell \in \mathbb{R}$ and $\mathbf{s}_1, \ldots, \mathbf{s}_\ell \in \mathcal{S}_r^+$ are distinct non-negative sequences of length r. Then by (38), we have

$$\langle u, u \rangle = \frac{1}{r!} \underbrace{\int_0^1 \cdots \int_0^1}_{r} \Big(\sum_{j=1}^{\ell} \xi_j \det U_{\mathbf{s}_j}(\mathbf{x}) \Big)^2 d\mathbf{x} \geq 0.$$

This shows that the pairing is positive semi-definite. To prove that it is actually positive definite, suppose $\langle u, u \rangle = 0$. Then the above integral formula implies that as a function in $\mathbf{x} = (x_1, \ldots, x_r)$ we have $\sum_{j=1}^{\ell} \xi_j \det U_{\mathbf{s}_j}(\mathbf{x}) \equiv 0$. By Lemma 18, the functions $\det U_{\mathbf{s}_1}(\mathbf{x}), \ldots, \det U_{\mathbf{s}_\ell}(\mathbf{x})$ are linearly independent functions in r variables. Hence, the above linear relation among these functions imply that $\xi_1 = \xi_2 = \cdots = \xi_\ell = 0$ and thus $u = 0$ in $F(r)$. This shows that the above pairing is actually positive definite.

Since the pairing on the basis vectors are integer valued, we can restrict our pairing to the lattice $L(r)$ to get $\mathbb{Z}$-valued pairing. The same argument shows that this pairing is also positive definite. $\square$

Positive definite pairings defined by Gramians. Let N be a positive integer and let $\vec{r} = (r_1, \ldots, r_N)$ be a sequence of N positive integers. Let

$$(39) \qquad \mathcal{S}_{\vec{r}}^+ = \mathcal{S}_{r_1}^+ \times \mathcal{S}_{r_2}^+ \times \cdots \times \mathcal{S}_{r_N}^+$$

be a set of N sequences of non-negative integers of length $r_1, \ldots, r_N$, respectively. Any element in this set is of the form $\vec{s} = (s_1, s_2, \ldots, s_N)$ where $s_j \in S^+_{r_j}$. For any two such elements $\vec{s} = (s_1, s_2, \ldots, s_N)$ and $\vec{s'} = (s'_1, s'_2, \ldots, s'_N)$, we define

$$(40) \qquad G(\vec{s}, \vec{s'}) = \prod_{j=1}^{N} G(s_j, s'_j) \in \mathbb{Z}.$$

For each element $\vec{s} \in S^+_{\vec{r}}$, we associate a symbol $a(\vec{s})$. Let $L(\vec{r})$ and $F(\vec{r})$ be a lattice and an $\mathbb{R}$-vector space spanned by $a(\vec{s})$'s. Namely,

$$(41) \qquad L(\vec{r}) = \bigoplus_{\vec{s} \in S^+_{\vec{r}}} \mathbb{Z} a(\vec{s}), \qquad F(\vec{r}) = \bigoplus_{\vec{s} \in S^+_{\vec{r}}} \mathbb{R}\, a(\vec{s}) = L(\vec{r}) \otimes_{\mathbb{Z}} \mathbb{R}.$$

It is easy to see that there is a canonical isomorphism given by

$$(42) \qquad \begin{array}{c} F(\vec{r}) \cong F(r_1) \otimes_{\mathbb{R}} F(r_2) \otimes_{\mathbb{R}} \cdots \otimes_{\mathbb{R}} F(r_N), \\ a(\vec{s}) \leftrightarrow a(s_1) \otimes a(s_2) \otimes \cdots \otimes a(s_N). \end{array}$$

A similar isomorphism exists for the lattice $L(\vec{r})$. Using (40), we define a bilinear pairing in $L(\vec{r})$ and in $F(\vec{r})$ by

$$(43) \qquad \langle a(\vec{s}), a(\vec{s'}) \rangle = G(\vec{s}, \vec{s'}) \qquad \text{for } \vec{s}, \vec{s'} \in S^+_{\vec{r}}.$$

This pairing has an integral representation given by
(44)
$$\langle a(\vec{s}), a(\vec{s'}) \rangle = \frac{1}{\prod_{j=1}^{N} r_j!} \cdot \underbrace{\int_0^1 \cdots \int_0^1}_{|\vec{r}|} \left(\prod_{j=1}^{N} \det U_{s_j}(\mathbf{x}^{(j)}) \right) \cdot \left(\prod_{j=1}^{N} \det U_{s'_j}(\mathbf{x}^{(j)}) \right) \cdot \prod_{j=1}^{N} d\mathbf{x}^{(j)}$$

Here $\mathbf{x}^{(j)} = (x_1^{(j)}, x_2^{(j)}, \ldots, x_N^{(j)})$ is the j-th set of real variables for $1 \leq j \leq N$.

LEMMA 20. *For any distinct elements* $\vec{s}^{(1)}, \vec{s}^{(2)}, \ldots, \vec{s}^{(\ell)}$ *in* $S^+_{\vec{r}}$, *the set of functions* $\{\prod_{j=1}^{N} \det U_{s^{(k)}_j}(\mathbf{x}^{(j)})\}^{\ell}_{k=1}$ *in* $|\vec{r}| = \sum_{j=1}^{N} r_j$ *variables* $\mathbf{x}^{(1)}, \mathbf{x}^{(2)}, \ldots, \mathbf{x}^{(N)}$ *is a set of linearly independent functions over* $\mathbb{R}$ *and also over* $\mathbb{C}$.

PROOF. When $N = 1$, this is Lemma 18. The general case can be proved by induction on N. $\square$

PROPOSITION 21. *Let* $\vec{r} \in \mathbb{N}^N$ *be a sequence of* N *positive integers for a given positive integer* N. *The* $\mathbb{R}$*-bilinear pairing* $\langle\ ,\ \rangle$ *on* $F(\vec{r})$ *defined by (43) is positive definite. Similarly, the pairing* $\langle\ ,\ \rangle$ *on* $L(\vec{r})$ *defined by (43) is an integral positive definite pairing.*

PROOF. Let v be any vector in $F(\vec{r})$. So, it is of the form $v = \sum_{k=1}^{\ell} \xi_k a(\vec{s}^{(k)})$ for some constants $\xi_1, \ldots, \xi_\ell$ and for some elements $\vec{s}^{(1)}, \ldots, \vec{s}^{(\ell)}$ in $S^+_{\vec{r}}$. From the integration formula (44) we have

$$\langle v, v \rangle = \frac{1}{\prod_{j=1}^{N} r_j!} \cdot \underbrace{\int_0^1 \cdots \int_0^1}_{|\vec{r}|} \left(\sum_{k=1}^{\ell} \xi_k \prod_{j=1}^{N} \det U_{s^{(k)}_j}(\mathbf{x}^{(j)}) \right)^2 \cdot \prod_{j=1}^{N} d\mathbf{x}^{(j)} \geq 0.$$

So, $\langle v, v \rangle \geq 0$ for any $v \in F(\vec{r})$. This shows that the pairing is positive-semi definite. To show that it is positive definite, suppose that $\langle v, v \rangle = 0$ for some $v \in F(\vec{r})$. From the above integral representation, we must have

$$\sum_{k=1}^{\ell} \xi_k \prod_{j=1}^{N} \det U_{\mathbf{s}_j^{(k)}}(\mathbf{x}^{(j)}) \equiv 0$$

as a function in $|\vec{r}|$ variables $(\mathbf{x}^{(1)}, \mathbf{x}^{(2)}, \ldots, \mathbf{x}^{(N)})$. From Lemma 20, the functions $\{\prod_{j=1}^{N} \det U_{\mathbf{s}_j^{(k)}}(\mathbf{x}^{(j)})\}_{k=1}^{\ell}$ are linearly independent. Thus, the above linear relation implies that $\xi_1 = \cdots = \xi_\ell = 0$. Hence $v = 0$ in $F(\vec{r})$. This proves that the pairing $\langle\ ,\ \rangle$ on $F(\vec{r})$ is positive definite. $\square$

Vertex operators for adjoint vectors and a Hermitian pairing defined by vertex operators. We will define a new Hermitian pairing on the vertex operator super algebra V in Theorem 25 below. As a preparation, we define the notion of adjoint vectors first.

DEFINITION 22 (ADJOINT VECTORS). We define a conjugate linear involutive isomorphism $* : V \to V$ as follows. If $u = b_1(-n_1 - \frac{1}{2}) \ldots b_r(-n_r - \frac{1}{2})\Omega$ for $b_1, \ldots, b_r \in A$ and for non-negative integers $0 \leq n_1, \ldots, n_r \in \mathbb{Z}$, u^* is defined by

$$(45) \qquad u^* = (-1)^{|\vec{n}|} \bar{b}_r(-n_r - \tfrac{1}{2}) \cdots \bar{b}_1(-n_1 - \tfrac{1}{2})\Omega,$$

where $|\vec{n}| = n_1 + n_2 + \cdots + n_r$. We call u^* the *adjoint vector* of u.

Note that this operation has already appeared in Proposition 4 for vectors in $\bigwedge^* A(-\frac{1}{2})\Omega$. We calculate $Y_{\mathrm{wt}(v)}(u^*)v$ for $u, v \in V$. The formula (16) applied to u^* as in the above Definition 22 gives the operator $Y_m(u^*)$:

$$(46) \qquad Y_m(u^*) = \sum_{\sum(m_j + \frac{1}{2}) = m} \prod_{j=1}^{r} \frac{(m_j + n_j)!}{m_j! \, n_j!} : \bar{b}_r(m_r + \tfrac{1}{2}) \cdots \bar{b}_1(m_1 + \tfrac{1}{2}) : .$$

For any $b \in A$ and for any sequence of distinct non-negative integers $\vec{n} = (n_1, n_2, \ldots, n_k)$, we let

$$(47) \qquad b(\vec{n}) = b(-n_1 - \tfrac{1}{2})b(-n_2 - \tfrac{1}{2}) \cdots b(-n_k - \tfrac{1}{2})$$

be an element in the Clifford algebra $\mathrm{Cliff}(\mathbb{Z} + \frac{1}{2})$. Its weight is $\mathrm{wt}(b(\vec{n})) = \sum_{j=1}^{k}(n_j + \frac{1}{2}) = |\vec{n}| + \frac{k}{2}$. For convenience, we let the element corresponding to an empty sequence be 1 of weight 0, that is, $b(\emptyset) = 1 \in \mathrm{Cliff}(\mathbb{Z} + \frac{1}{2})$.

PROPOSITION 23. *Let* $b \in \{a_1, \ldots, a_N, a_1^*, \ldots, a_N^*\}$ *be a canonical basis vector. Let* $\vec{n} = (n_1, n_2, \ldots, n_r)$, $\vec{m} = (m_1, m_2, \ldots, m_r)$ *be a sequence of strictly increasing non-negative integers of length* r, *that is,* $\vec{n}, \vec{m} \in S_r^+$. *Let* $u = b(\vec{m})\Omega$, $v = b(\vec{n})\Omega$. *Then we have*

$$(48) \qquad Y_{\mathrm{wt}(v)}(u^*)v = G(\vec{m}, \vec{n}) \cdot \Omega.$$

PROOF. First note that wt $(v) = (\sum_{j=1}^{r} n_j) + \frac{r}{2}$. By (46), the vertex operator $Y_{\text{wt}(v)}(u^*)$ is given as follows:

$$Y_{\text{wt}(v)}(u^*) = \sum_{\sum \ell_j = \sum n_j} \prod_{j=1}^{r} \frac{(\ell_j + m_j)!}{\ell_j! \, m_j!} : \bar{b}(\ell_r + \tfrac{1}{2}) \cdots \bar{b}(\ell_1 + \tfrac{1}{2}): .$$

Here ℓ_j's have to be distinct for the term to be nontrivial. Each term of $Y_{\text{wt}(v)}(u^*)v$ to be calculated is of the form

$$: \bar{b}(\ell_r + \tfrac{1}{2}) \cdots \bar{b}(\ell_1 + \tfrac{1}{2}): b(-n_1 - \tfrac{1}{2}) \cdots b(-n_r - \tfrac{1}{2})\Omega,$$

times some coefficient. Since $\sum n_j = \sum \ell_j$, this element is nonzero if and only if $(\ell_1, \ldots, \ell_r)$ is equal to $(n_{\sigma(1)}, \ldots, n_{\sigma(r)})$ for some permutation σ. Since all $\ell_j + \frac{1}{2}$'s are positive, we can drop the normal ordering sign and then the above element is equal to $\text{sgn}(\sigma)\Omega$. Thus, our summation becomes

$$Y_{\text{wt}(v)}(u^*)v = \sum_{\sigma \in \mathfrak{S}_r} \text{sgn}(\sigma) \prod_{j=1}^{r} \frac{(m_j + n_{\sigma(j)})!}{m_j! \, n_{\sigma(j)}!} \cdot \Omega = G(\vec{m}, \vec{n}) \cdot \Omega.$$

This completes the proof. □

By the definition of the Gramian (32) we have $G(\vec{m}, \vec{n}) = G(\vec{n}, \vec{m})$, since $\det A = \det {}^t A$ for any matrix A.

Let $\vec{r} = (r_1, r_2, \ldots, r_{2N}) \in \mathbb{Z}_+^{2N}$ be a sequence of $2N$ non-negative integer. To each such $\vec{r}$, we associate a set of multi-sequences of type $\vec{r}$:

(49) $$\mathcal{S}_{\vec{r}}^+ = \mathcal{S}_{r_1}^+ \times \mathcal{S}_{r_2}^+ \times \cdots \times \mathcal{S}_{r_{2N}}^+.$$

Here, by convention, $\mathcal{S}_0^+ = \emptyset$. To each multi-sequence $\vec{s} = (s_1, s_2, \ldots, s_{2N}) \in \mathcal{S}_{\vec{r}}^+$, we associate a vector $a(\vec{s})$ in the vertex operator super algebra V given by

(50) $$a(\vec{s}) = a_1(s_1) \cdots a_N(s_N) a_{N+1}(s_{n+1}) \cdots a_{2N}(s_{2N}) \cdot \Omega \in V.$$

Here we are using the notation (47), and $a_{N+j} = a_j^* = \bar{a}_j$. If $r_i = 0$, then we set $a_i(\emptyset) = 1$. Note that wt $(a(\vec{s})) = \sum_{j=1}^{2N} |s_j| + \frac{1}{2}|\vec{r}|$.

PROPOSITION 24. Let $\vec{r}, \vec{r'} \in \mathbb{Z}_+^{2N}$ be sequences of $2N$ non-negative integers. Let $\vec{s} \in \mathcal{S}_{\vec{r}}^+$, $\vec{s'} \in \mathcal{S}_{\vec{r'}}^+$ be two multi-sequences of type $\vec{r}$, $\vec{r'}$, respectively. Then

(51) $$Y_{\text{wt}(a(\vec{s'}))}(a(\vec{s})^*)a(\vec{s'}) = \begin{cases} G(\vec{s}, \vec{s'}) \cdot \Omega & \text{if } \vec{r} = \vec{r'}, \\ 0 & \text{if } \vec{r} \neq \vec{r'}. \end{cases}$$

PROOF. Let the types $\vec{r}$ and $\vec{r'}$ be $\vec{r} = (r_1, \ldots, r_{2N})$ and $\vec{r'} = (r'_1, \ldots, r'_{2N})$, respectively. Also, let the mulri-sequences be of the form $\vec{s} = (s_1, s_2, \ldots, s_{2N})$, $\vec{s'} = (s'_1, s'_2, \ldots, s'_{2N})$, where

$$s_j = (0 \leq m_1^{(j)} < m_2^{(j)} < \cdots < m_{r_j}^{(j)}) \in \mathcal{S}_{r_j}^+,$$
$$s'_j = (0 \leq n_1^{(j)} < n_2^{(j)} < \cdots < n_{r_j}^{(j)}) \in \mathcal{S}_{r'_j}^+,$$

for $1 \leq j \leq 2N$. For a sequence of integers $\ell = (\ell_1, \ell_2, \ldots, \ell_r) \in \mathbb{Z}^r$ and $b \in A$, we let

$$b^+(\ell) = b(\ell_1 + \tfrac{1}{2})b(\ell_2 + \tfrac{1}{2}) \cdots b(\ell_r + \tfrac{1}{2}) \in \mathrm{Cliff}(\mathbb{Z} + \tfrac{1}{2}).$$

Please compare with the notation (47). Then $Y_{\mathrm{wt}(a(\vec{s}'))}(a(\vec{s})^*)a(\vec{s}')$ is equal to

$$(*) \quad \sum_{\substack{\Sigma_j |\ell^{(j)}| \\ =\Sigma_j |s_j'|}} \prod_{j=1}^{2N} \left(\prod_{k=1}^{r_j} (\ell_k^{(j)}, m_k^{(j)}) \right) : \bar{a}_{2N}^+(\ell^{(2N)}) \cdots \bar{a}_2^+(\ell^{(2)}) \bar{a}_1^+(\ell^{(1)}) :$$

$$\cdot \, a_1(s_1')a_2(s_2') \cdots a_{2N}(s_{2N}')\Omega,$$

where $\ell^{(j)} = (\ell_1^{(j)}, \ell_2^{(j)}, \ldots, \ell_{r_j}^{(j)}) \in \mathbb{Z}^{r_j}$ is a sequence of r_j integers and the summation above is over all the sequences $\ell^{(1)}, \ldots, \ell^{(2N)} \in \mathbb{Z}^{r_1} \times \cdots \times \mathbb{Z}^{r_{2N}}$ satisfying the above weight condition. As before, $(m, n) = (m + n)!/(m!\,n!)$ denotes the binomial coefficient. If m is negative, the above means $(m + 1)_n/n!$. See (16).

First in the above expression, we observe that for the nontriviality of the action of a component operator $: \bar{a}_{2N}^+(\ell^{(2N)}) \cdots \bar{a}_2^+(\ell^{(2)})\bar{a}_1^+(\ell^{(1)}) :$ on $a(\vec{s}')$, the operator must consist only of annihilation operators. To see this, suupose there was a creation operator in this expression. The normal ordering places the creation operators to the left of all the annihilation operators. But then, due to the weight constraint $\sum |\ell^{(j)}| = \sum |s_j'|$, the product of all the annihilation operators act on $a(\vec{s}')\Omega$ to produce vectors of negative weight, which is nonexistent in V. Hence the action is trivial if the operator contains creation operators. Thus all the operators be annihilation operators, that is, $\ell^{(j)}$'s must be non-negative sequences. Since $\bar{a}_j(\ell + \tfrac{1}{2})$ can pair nontrivially only with $a_j(-\ell - \tfrac{1}{2})$, for the nontriviality of the action of the operator $: \bar{a}_{2N}^+(\ell^{(2N)}) \cdots \bar{a}_2^+(\ell^{(2)})\bar{a}_1^+(\ell^{(1)}) :$ on $a(\vec{s}')\Omega$, we must have that $\ell^{(j)}$ is a permutation of s_j' for all j. Thus, we see that the nontriviality of the action of the vertex operator $(*)$ implies that $\vec{r} = \vec{r}'$ and $\ell^{(j)} = \sigma_j(s_j')$ for some $\sigma_j \in \mathfrak{S}_{r_j}$ for $1 \leq j \leq 2N$. For $\sigma_j \in \mathfrak{S}_{r_j}$, we use the notation

$$a_j^+(\sigma_j(s_j')) = a_j(n_{\sigma_j(1)}^{(j)} + \tfrac{1}{2}) \cdots a_j(n_{\sigma_j(r_j)}^{(j)} + \tfrac{1}{2}).$$

Then the above summation $(*)$ reduces to a summation over all permutations $(\sigma_1, \sigma_2, \ldots, \sigma_{2N}) \in \mathfrak{S}_{r_1} \times \mathfrak{S}_{r_2} \times \cdots \times \mathfrak{S}_{r_{2N}}$:

$$\sum_{(\sigma_1, \ldots, \sigma_{2N})} \prod_{j=1}^{2N} \prod_{k=1}^{r_j} (n_{\sigma_j(k)}^{(j)}, m_k^{(j)}) : \bar{a}_{2N}^+(\sigma_{2N}(s_{2N}')) \cdots \bar{a}_1^+(\sigma_1(s_1')) :$$

$$\cdot \, a_1(s_1') \cdots a_{2N}(s_{2N}')\Omega.$$

Since no two elements inside the normal ordering symbol can pair nontrivially, we can remove the normal ordering symbol. We move $a_2(s_2'), \ldots, a_{2N}(s_{2N}')$ forward to be placed to the left of $\bar{a}_2(\sigma_2(s_2')), \ldots, \bar{a}_{2N}(\sigma_{2N}(s_{2N}'))$, respectively. There is no sign change in this process because products $a_j(s_j')$ pass even number of elements. Hence, the above can be rewritten as

$$\prod_{j=1}^{2N} \left\{ \sum_{\sigma_j \in \mathfrak{S}_{r_j}} \prod_{k=1}^{r_j} (n_{\sigma_j(k)}^{(j)}, m_k^{(j)}) \bar{a}_j(\sigma_j(s_j')) a_j(s_j') \right\} \Omega.$$

From our calculation in Proposition 23, inside of { } is $G(\mathbf{s}_j, \mathbf{s}'_j)$. So, the above is equal to $\{\prod_{j=1}^{2N} G(\mathbf{s}_j, \mathbf{s}'_j)\}\Omega = G(\vec{\mathbf{s}}, \vec{\mathbf{s}}')\Omega$. This completes the proof. $\square$

The vertex operator super algebra can be decomposed according to types of the vectors. For a non-negative sequence $\vec{r} \in \mathbb{Z}_+^r$ of length r, let

$$(52) \qquad\qquad F_{\mathbb{C}}(\vec{r}) = \bigoplus_{\vec{\mathbf{s}} \in \mathcal{S}_{\vec{r}}^+} \mathbb{C}\,a(\vec{\mathbf{s}}) \subset V.$$

This is a subspace of V of infinite dimension. $F_{\mathbb{C}}(\vec{r})$ itself is a graded vector space consisting of vectors of various weights. Since V has a basis consisting of vectors of the form (50), we have the following decomposition:

$$(53) \qquad\qquad V = \bigoplus_{\vec{r} \in \mathbb{Z}_+^{2N}} F_{\mathbb{C}}(\vec{r}).$$

We shall see that this decomposition is an orthogonal decomposition with respect to a Hermitain pairing defined in terms of vertex operators.

THEOREM 25. *The pairing* $(\ ,\ ) : V \otimes_{\mathbb{C}} V \to \mathbb{C}$ *defined by*

$$(54) \qquad\qquad Y_{\mathrm{wt}\,(v)}(u^*)v = (u, v)\Omega$$

is a positive definite Hermitian pairing.

If $\vec{r} \neq \vec{r}'$, *then the vector subspaces* $F_{\mathbb{C}}(\vec{r})$ *and* $F_{\mathbb{C}}(\vec{r}')$ *is orthogonal and* (53) *is an orthogonal decomposition.*

In $F_{\mathbb{C}}(\vec{r})$, *if* $\vec{\mathbf{s}}, \vec{\mathbf{s}}' \in \mathcal{S}_{\vec{r}}^+$ *are multi-sequences of type* $\vec{r}$, *then the pairing of two vectors* $a(\vec{\mathbf{s}})$, $a(\vec{\mathbf{s}}')$ *in* V *is given by the Gramian determinant:*

$$(55) \qquad\qquad (a(\vec{\mathbf{s}}), a(\vec{\mathbf{s}}')) = G(\vec{\mathbf{s}}, \vec{\mathbf{s}}').$$

PROOF. From the definition of the adjoint vector u^* in Definition 22, it is clear that the above pairing is sesquilinear in the first variable. The formula (55) is proved in Proposition 24. Since $G(\vec{\mathbf{s}}, \vec{\mathbf{s}}') \in \mathbb{Z}$, it is in particular a real number. From this it easily follows that $\overline{(u, v)} = (v, u)$.

From Proposition 24, it also follows that the decomposition (53) is an orthogonal decomposition. The restriction of the Hermitian pairing $(\ ,\ )$ on $F_{\mathbb{C}}(\vec{r})$ is positive definite by sesquilinear extension of Proposition 21. Hence the Hermitain pairing is positive definite on the entire vector space V. $\square$

CHAPTER III

G-INVARIANT VERTEX OPERATOR SUPER SUBALGEBRAS

§3.1 V^G as vertex operator super algebra

Vertex operator super algebras don't seem to be abundant in nature. However, given a vertex operator super algebra V, there are several ways to construct new vertex operator super algebras. One method is to take tensor products of V and to consider their vertex operator subalgebras. Another interesting method is to take G-invariant subspace of V when V admits G as its automorphism group.

For the vertex operator super algebra V described in §2.2, G can be any Lie subgroup of $\mathrm{SO}(2N)$. It turns out that the resulting G-invariant subspace V^G always has the structure of a vertex operator super algebra for any Lie subgroup G of $\mathrm{SO}(2N)$. Choosing different G's, we obtain various vertex operator super algebras. All of these algebras V^G are nontrivial, because V^G always contains the canonical Virasoro element $\omega \in V_2$ of weight 2 which is fixed by the entire group $\mathrm{SO}(2N)$.

Of particular interest to us are those vertex operator super algebras V^G corresponding to $G = \mathrm{U}(N), \mathrm{SU}(N), \mathrm{Sp}(N')$, and $\mathrm{Sp}(N') \cdot \mathrm{Sp}(1)$. These Lie groups come from the classification theorem of holonomy groups of Riemannian manifolds [Bes]. We use the vertex operator super algebra V^G for various Lie groups G to investigate the infinite dimensional symmetries in elliptic genera for Kählerian manifolds in Chapter V.

We prepare one lemma. Recall that on the vertex operator super algebra V, $\mathrm{Spin}(2N)$ acts through $\mathrm{SO}(2N)$ preserving the grading of V.

LEMMA 1. *For any element $g \in \mathrm{Spin}(2N)$ and any $v \in V$, as operators acting on V or on W, we have*

$$(1) \qquad g \cdot Y(v, \zeta) \cdot g^{-1} = Y(g \cdot v, \zeta).$$

Consequently, for any Lie subgroup $G \subset \mathrm{Spin}(2N)$, not necessarily connected, and for any G-invariant vector $v \in V^G$ and for any $g \in G$, we have

$$(2) \qquad g \cdot Y(v, \zeta) \cdot g^{-1} = Y(v, \zeta), \quad or \quad g \cdot \{v\}_n = \{v\}_n \cdot g \quad for \ n \in \tfrac{1}{2}\mathbb{Z}.$$

That is, $\{v\}_n$ is a G-equivariant map for any $n \in \tfrac{1}{2}\mathbb{Z}$ if v is G-invariant.

PROOF. In the commutator formula (12) of §2.1, we let $v_1 = \Upsilon(x)$ with $x \in \mathfrak{o}(2N)$, $v_2 = v \in V$, and $m = 0$. Multiplying the identity by ζ^{-n-1} and summing over n, we get

$$(3) \qquad [x(0), Y(v, \zeta)] = Y(x(0)v, \zeta).$$

By exponentiating this Lie algebra identity, we get the first part of the lemma for $g \in \text{Spin}(2N)$ since $\text{Spin}(2N)$ is connected.

For the second part, we simply note that for $v \in V^G$, we have $g \cdot v = v$ for $g \in G$. We then use (1). Note that the group G does not have to be connected since (1) is valid for any element in $\text{Spin}(2N)$. $\square$

THEOREM 2 (V^G AS VERTEX OPERATOR SUPER ALGEBRA). *Let V and W be a vertex operator super algebra and its $\mathbb{Z}_2$-twisted module of §2.2. Let $G \subset \text{Spin}(2N)$ be any Lie subgroup, not necessarily connected. Let V^G be the subspace of G-invariant vectors. Then $V^G \subset V$ has the structure of a vertex operator super algebra, and it has W^G as its $\mathbb{Z}_2$-twisted module. That is, there exists a structure map*

(4)
$$Y(\ , z) : V^G \to \text{End}(V^G)[[z, z^{-1}]],$$
$$Y(\ , z) : V^G \to \text{End}(W^G)[[z^{1/2}, z^{-1/2}]].$$

satisfying Jacobi identities and other requirements for a vertex operator super algebra and its module.

PROOF. We first show that operators $\{v\}_n$ for $v \in V^G$, $n \in \frac{1}{2}\mathbb{Z}$ preserve V^G. Let $w \in V^G$. For any $g \in G$, using (2) above,

$$g \cdot (\{v\}_n(w)) = \{v\}_n(g \cdot w) = \{v\}_n(w),$$

which means $\{v\}_n w$ is G-invariant. Thus $\{v\}_n$ preserves V^G and we have a well defined map

$$Y(\ , z) : V^G \to \text{End}(V^G)[[z, z^{-1}]].$$

Since $1, \omega \in V^{\text{Spin}(2N)}$, we have $1, \omega \in V^G$ for any $G \subset \text{Spin}(2N)$. Furthermore, the Jacobi identity and other requirements for V^G to be a vertex operator super algebra simply follows from the restriction of the structures from V to V^G.

The proof that W^G is a $\mathbb{Z}_2$-twisted module of V^G is similar. $\square$

The next corollary is immediate.

COROLLARY 3. *For any Lie subgroup $G \subset \text{Spin}(2N)$ and $v \in V^G$, the vertex operators $\{v\}_n$ with $n \in \frac{1}{2}\mathbb{Z}$ are G-equivariant maps on V and on its module W:*

(5)
$$\{v\}_n : V \to V \quad \text{with } n \in \mathbb{Z},$$
$$\{v\}_n : W \to W \quad \text{with } n \in \frac{1}{2}\mathbb{Z}.$$

§3.2 Lie subalgebras of Clifford algebras

We apply Theorem 2 of §3.1 with various Lie groups G. Most of the Lie groups of interest to us are connected, and we have $V^G = V^{\mathfrak{g}}$ where $\mathfrak{g} = \text{Lie}(G)$ is the Lie algebra of G. Since the vertex operator super algebra $(V, \text{ad}_\zeta, 1, \omega)$ of §2.2 is constructed as the Clifford module of the Clifford algebra $\text{Cliff}(\mathbb{Z} + \frac{1}{2})$, to understand V^G it is important and convenient to have a description of (the complexification of) the Lie algebra $\mathfrak{g}$ as a Lie subalgebra of $\mathrm{Cliff}(\mathbb{R}^{2N}) \subset \mathrm{Cliff}(\mathbb{C}^{2N})$.

Let $A = \mathbb{C}^{2N}$. In this section, we give a description of the following Lie algebras as subalgebras of real or complex Clifford algebras:

$$(1) \qquad \mathfrak{sp}([\tfrac{N}{2}]) \subset \mathfrak{su}(N) \subset \mathfrak{u}(N) \subset \mathfrak{o}(2N) \supset \mathfrak{o}(k).$$

To describe $\mathfrak{o}(k) \subset \mathfrak{o}(2N)$, the real basis of A is convenient. For other algebras the complex basis of A is useful.

Lie algebras $\mathfrak{o}(k) \subset \mathfrak{o}(2N)$. Let $\mathbb{R}^{2N} = \bigoplus_{i=1}^{2N} \mathbb{R}e_i$ be the real $2N$ dimensional inner product space equipped with an orthonormal basis $\{e_1, e_2, \ldots, e_{2N}\}$. Let $\mathfrak{o}(k)$ acts on $\bigoplus_{i=1}^{k} \mathbb{R}e_i$ and leaves the rest of the basis vectors invariant. In the Clifford algebra $\mathrm{Cliff}(\mathbb{R}^{2N})$, the Lie algebra $\mathfrak{o}(k)$ is described as follows:

LEMMA 1 (LIE ALGEBRA $\mathfrak{o}(k)$). *Let $1 \leq k \leq 2N$. The Lie algebra $\mathfrak{o}(k)$ can be realized in the Clifford algebra $\mathrm{Cliff}(\mathbb{R}^{2N})$ as*

$$(2) \qquad \mathfrak{o}(k) = \bigoplus_{0 \leq i < j \leq k} \mathbb{R} :e_i e_j: .$$

PROOF. From (2) of §2.2, we know that elements of the form $:e_i e_j:$ for $1 \leq i, j \leq 2N$ form a basis of the Lie algebra $\mathfrak{o}(2N)$. Restricting this to the subspace $\mathbb{R}^k \subset \mathbb{R}^{2N}$, we get (2) above. $\square$

Recall that $:ab:$ is defined to be $\frac{1}{2}(ab - ba)$ in the Clifford algebra $\mathrm{Cliff}(A)$, where $a, b \in A$. If $\langle a, b \rangle = 0$, then $:ab: := ab = -ba$ in $\mathrm{Cliff}(A)$ and there is no need to use normal ordering. However, if $\langle a, b \rangle \neq 0$, then $:ab: := ab - \frac{1}{2}\langle a, b \rangle$. The reason of using normal ordering is the property $:ab: := - :ba:$ so that we can define a map $\Upsilon : o(2N) \to \bigwedge^2 A$ by $\Upsilon(:ab:) = a \wedge b$.

To describe other algebras in the list (1) in terms of Clifford elements, it is more convenient to use complex basis of A. Let $E = (\mathbb{R}^{2N}; I, (\ ,\ )) \cong \mathbb{C}^N$ be a complex vector space with nondegenerate Hermitian bilinear pairing $(\ ,\ )$. The unitary group $U(N)$ acts on E preserving the complex structure I and the Hermitian pairing $(\ ,\ )$. Let $\{e_1, e_2, \ldots, e_N\}$ be a Hermitian orthonormal basis of E and let $e'_i = I(e_i)$ for $1 \leq i \leq N$. The underlying real vector space is then given by $E_{\mathbb{R}} = \bigoplus_{i=1}^{N}(\mathbb{R}e_i + \mathbb{R}e'_i)$, and has the nondegenerate symmetric pairing $\langle\ ,\ \rangle_{\mathbb{R}} = \Re(\ ,\ )$, the real part of the Hermitian pairing.

We extend the pairing $\langle\ ,\ \rangle_{\mathbb{R}}$ complex linearly to the complexification $A = E_{\mathbb{R}} \otimes_{\mathbb{R}} \mathbb{C}$ on which I acts as $I \otimes 1$. The resulting $\mathbb{C}$-linear pairing is denoted by $\langle\ ,\ \rangle$. Let $A^{\pm}$ be the $\pm i$ eigenspaces of this map I. These subspaces are isotropic subspaces of A with respect to the $\mathbb{C}$-linear pairing $\langle\ ,\ \rangle = \langle\ ,\ \rangle_{\mathbb{R}} \otimes \mathbb{C}$. We let

$$(3) \qquad a_j = \frac{e_j - ie'_j}{\sqrt{2}}, \qquad a^*_j = \frac{e_j + ie'_j}{\sqrt{2}}, \qquad \text{for } 1 \leq j \leq N.$$

Then $\{a_i, a_2, \ldots, a_N\}$ and $\{a^*_1, a^*_2, \ldots, a^*_N\}$ are bases of A^+ and of A^-, respectively. The $\mathbb{C}$-linear pairings among these basis vectors are given by

$$(4) \qquad \langle a_j, a^*_k \rangle = \delta_{jk}, \qquad \langle a_j, a_k \rangle = 0, \qquad \langle a^*_j, a^*_k \rangle = 0, \qquad 1 \leq j, k \leq N.$$

The relations between complex basis and real basis is given in the next lemma, which can be checked directly.

LEMMA 2. *The following relations hold.*

(a) $:a_j a_k := \frac{1}{2}(:e_j e_k: - :e'_j e'_k: -i:e_j e'_k: -i:e'_j e_k:)$ *for* $1 \leq j < k \leq N$.

(b) $:a^*_j a^*_k := \frac{1}{2}(:e_j e_k: - :e'_j e'_k: +i:e_j e'_k: +i:e'_j e_k:)$ *for* $1 \leq j < k \leq N$.

(c) $:a_j a^*_k := \frac{1}{2}(:e_j e_k: + :e'_j e'_k: +i:e_j e'_k: -i:e'_j e_k:)$ *for* $1 \leq j < k \leq N$.

(d) $:a^*_j a_k := \frac{1}{2}(:e_j e_k: + :e'_j e'_k: -i:e_j e'_k: +i:e'_j e_k:)$ *for* $1 \leq j < k \leq N$.

(e) $:a_j a^*_j := i:e_j e'_j:$ *for* $1 \leq j \leq N$.

From this we have two expressions of $o(2N)$, one in terms of real basis and the other in terms of complex basis.

LEMMA 3. *The orthogonal Lie algebra* $o(2N)$ *has the following descriptions in real and complex Clifford algebras:*

$$
(5) \quad
\left\{
\begin{aligned}
o(2N) &= \bigoplus_{1 \leq j < k \leq N} \mathbb{R}:e_j e_k: \oplus \bigoplus_{1 \leq j < k \leq N} \mathbb{R}:e'_j e'_k: \oplus \bigoplus_{1 \leq j,k \leq N} \mathbb{R}:e_j e'_k:, \\
o(2N) \otimes \mathbb{C} &= \bigoplus_{1 \leq j < k \leq N} \mathbb{C}:a_j a_k: \oplus \bigoplus_{1 \leq j < k \leq N} \mathbb{C}:a^*_j a^*_k: \oplus \bigoplus_{1 \leq j,k \leq N} \mathbb{C}:a_j a^*_k: .
\end{aligned}
\right.
$$

The Cartan subalgebra $t^N \subset o(2N)$ *is described as*

$$
(6) \quad t^N = \bigoplus_{j=1}^{N} \mathbb{R}:e_j e'_j: \quad and \quad t^N \otimes \mathbb{C} = \bigotimes_{j=1}^{N} \mathbb{C}:a_j a^*_j: .
$$

The matrix description of the orthogonal Lie algebras is well known: $o(2N)$ consists of skew matrices in $M_{2N}(\mathbb{R})$. We will translate the matrix description of various Lie algebras in (1) into Clifford description. To do this we first make a correspondence between matrices and Clifford elements for the Lie algebra $o(2N)$. Recall that the Lie algebra $o(2N) \subset \mathrm{Cliff}(\mathbb{R}^{2N})$ acts on $\mathbb{R}^{2N}$ by the formula $[:ab:,c] = \langle b,c \rangle a - \langle a,c \rangle b$. Here a, b, c are elements in $\mathbb{R}^{2N} \subset \mathrm{Cliff}(\mathbb{R}^{2N})$ and $:ab := \frac{1}{2}(ab - ba) \in o(2N)$. Simple calculations show that this action is given as follows:

LEMMA 4 (CLIFFORD ACTIONS OF $o(2N)$).

(1) *The action of* $:e_i e_j:$ *with* $1 \leq i < j \leq N$ *is given by*

$$[:e_i e_j:, e_i] = -e_j, \qquad\qquad [:e_i e_j:, e_j] = e_i,$$
$$[:e_i e_j:, e_k] = 0 \text{ for } k \neq i,j, \quad [:e_i e_j:, e'_k] = 0 \text{ for } 1 \leq k \leq N.$$

(2) *The action of* $:e'_i e'_j:$ *with* $1 \leq i < j \leq N$ *is given by*

$$[:e'_i e'_j:, e'_i] = -e'_j, \qquad\qquad [:e'_i e'_j:, e'_j] = e'_i,$$
$$[:e'_i e'_j:, e'_k] = 0 \text{ for } k \neq i,j, \quad [:e'_i e'_j:, e_k] = 0 \text{ for } 1 \leq k \leq N.$$

(3) *The action of* $:e_i e'_j:$ *with* $1 \leq i < j \leq N$ *is given by*

$$[:e_i e'_j:, e_i] = -e'_j, \qquad\qquad [:e_i e'_j:, e'_j] = e_i,$$
$$[:e_i e'_j:, e_k] = 0 \text{ for } k \neq i, \quad [:e_i e'_j:, e'_k] = 0 \text{ for } k \neq j.$$

Now we make the correspondence between Clifford elements and matrices. To write down the matrix of a linear operator we need to fix a basis. As a basis for $E_{\mathbb{R}} = \mathbb{R}^{2N}$, we use $\{e_1, \ldots, e_N, e'_1, \ldots, e'_N\}$. Let E_{ij} be a $2N \times 2N$ matrix with 1 at (i, j) entry and 0 at other entries. From the above calculation, we immediately see the following correspondence:

LEMMA 5. *In the expression of* $\mathfrak{o}(2N)$ *in terms of Clifford elements given in Lemma 3, the corresponding matrices are given by*

(7)
$$: e_i e_j :\longleftrightarrow E_{i,j} - E_{j,i}, \qquad : e'_i e'_j :\longleftrightarrow E_{i+N,j+N} - E_{j+N,i+N},$$
$$: e_i e'_j :\longleftrightarrow E_{i,j+N} - E_{j+N,i}.$$

Lie algebras $\mathfrak{su}(N) \subset \mathfrak{u}(N)$ **and their complexifications.** Next, we want to find descriptions of $\mathfrak{su}(N)$ and $\mathfrak{u}(N)$ in terms of Clifford elements as a subspace of $\mathfrak{o}(2N)$ given in Lemma 3.

Recall that an $\mathbb{R}$-basis of the Lie algebra $\mathfrak{u}(N)$ consists of skew-adjoint matrices in $M_N(\mathbb{C})$:

(8)
$$\mathfrak{u}(N) = \bigoplus_{1 \leq j < k \leq N} \mathbb{R}(E_{jk} - E_{kj}) \oplus \bigoplus_{1 \leq j < k \leq N} \mathbb{R}(-iE_{jk} - iE_{kj}) \oplus \bigoplus_{j=1}^{N} \mathbb{R}(-iE_{jj}).$$

Also recall that $\mathbb{R}$-basis of $\mathfrak{su}(N) \subset M_n(\mathbb{C})$ consists of traceless skew-adjoint matrices in $M_n(\mathbb{C})$:

(9)
$$\mathfrak{su}(N) = \bigoplus_{1 \leq j < k \leq N} \mathbb{R}(E_{jk} - E_{kj}) \oplus \bigoplus_{1 \leq j < k \leq N} \mathbb{R}(-iE_{jk} - iE_{kj}) \oplus \bigoplus_{j=1}^{N-1} \mathbb{R}(-iE_{jj} + iE_{i+1,i+1}).$$

These matrices act on $E = \bigoplus_{j=1}^{N} \mathbb{C} e_j = \bigoplus_{j=1}^{N} (\mathbb{R} e_j + \mathbb{R} e'_j)$ where $e'_j = I(e_j) = i \cdot e_j$ and I is the complex structure. By checking their effects on the basis $\{e_1, \ldots, e_N, e'_1, \ldots, e'_N\}$, and comparing the result with Lemma 5, we get

PROPOSITION 6. *An* $\mathbb{R}$-*basis of the Lie algebras* $\mathfrak{su}(N) \subset \mathfrak{u}(N) \subset \mathfrak{o}(2N) \subset$ $\mathrm{Cliff}(\mathbb{R}^{2N})$ *is given as follows in terms of Clliford elements:*

(10)
$$\mathfrak{u}(N) = \bigoplus_{1 \leq j < k \leq N} \mathbb{R}(: e_j e_k : + : e'_j e'_k :) \oplus \bigoplus_{1 \leq j < k \leq N} \mathbb{R}(: e_j e'_k : + : e_k e'_j :) \oplus \bigoplus_{j=1}^{N} \mathbb{R}(: e_j e'_j :),$$

(11)
$$\mathfrak{su}(N) = \bigoplus_{1 \leq j < k \leq N} \mathbb{R}(: e_j e_k : + : e'_j e'_k :) \oplus \bigoplus_{1 \leq j < k \leq N} \mathbb{R}(: e_j e'_k : + : e_k e'_j :)$$
$$\oplus \bigoplus_{j=1}^{N-1} \mathbb{R}(: e_j e'_j : - : e_{j+1} e'_{j+1} :).$$

PROOF. We only have to check that $E_{jk} - E_{kj}$ in $M_N(\mathbb{C})$ corresponds to $: e_j e_k : + : e'_j e'_k :$, $-iE_{jk} - iE_{kj}$ corresponds to $: e_j e'_k : + : e_k e'_j :$, and $-iE_{jj}$

corresponds to $: e_j e'_j :$. For example, the second correspondence can be seen as follows. Let f be a linear transformation on $E = \bigoplus_{j=1}^{N} \mathbb{C}e_j = \bigoplus_{j=1}^{N}(\mathbb{R}e_j + \mathbb{R}e'_j)$ corresponding to $-iE_{jk} - iE_{kj} \in M_N(\mathbb{C})$. Then its effect on the real basis is given by

$$f(e_\ell) = \begin{cases} -e'_k, & \text{if } \ell = j, \\ -e'_j, & \text{if } \ell = k, \\ 0, & \text{otherwise.} \end{cases}, \qquad f(e'_\ell) = \begin{cases} -e_k, & \text{if } \ell = j, \\ -e_j, & \text{if } \ell = k, \\ 0, & \text{otherwise.} \end{cases}.$$

Writing f as an element in $M_{2N}(\mathbb{R})$ and comparing with Lemma 5, we get the second correspondence above. Other correspondences can be shown similarly. This proves Proposition 6. $\square$

Next, we express (10) and (11) in terms of complex basis of A. From Lemma 2, we immediately obtain the next lemma.

LEMMA 7. *The following relations hold in* $\mathrm{Cliff}(A)$:
  (1) $: e_j e_k : + : e'_j e'_k := : a_j a^*_k : + : a^*_j a_k :$, *for* $1 \le j < k \le N$.
  (2) $: e_j e'_k : + : e_k e'_j := (-i)(: a_j a^*_k : - : a^*_j a_k :)$, *for* $1 \le j < k \le N$.
  (3) $: e_j e'_j := (-i) : a_j a^*_j :$, *for* $1 \le j \le N$.

Combining Proposition 6 and Lemma 7, we finally get realizations of Lie algebras $\mathfrak{u}(N)$, $\mathfrak{su}(N)$ in complex Clifford algebra.

PROPOSITION 8. *A* $\mathbb{C}$ *basis of* $\mathfrak{su}(N) \otimes \mathbb{C} \subset \mathfrak{u}(N) \otimes \mathbb{C} \subset \mathrm{Cliff}(A)$ *is given by*

$$(12) \qquad \mathfrak{u}(N) \otimes \mathbb{C} = \bigoplus_{1 \le j,k \le N} \mathbb{C} : a_j a^*_k :,$$

$$(13) \qquad \mathfrak{su}(N) \otimes \mathbb{C} = \bigoplus_{1 \le j \ne k \le N} \mathbb{C} : a_j a^*_k : \oplus \bigoplus_{j=1}^{N-1} \mathbb{C}(: a_j a^*_j : - : a_{j+1} a^*_{j+1} :).$$

The Cartan subalgebra of $\mathfrak{u}(N) \otimes \mathbb{C}$ *is given by*

$$(14) \qquad \mathfrak{t}^N_{\mathbb{C}} = \bigoplus_{j=1}^{N} \mathbb{C} : e_j e_j := \bigoplus_{j=1}^{N} \mathbb{C} : a_j a^*_j : .$$

Note the concise description (12) in contrast to (8) of the unitary Lie algebra $\mathfrak{u}(N)$ in terms of the complex basis a_j's and a^*_k's. For later use, we record the correspondence between the matrices in $\mathfrak{u}(N) \subset M_N(\mathbb{C})$ and Clifford elements in $\mathfrak{u}(N) \subset \mathrm{Cliff}(A)$:

$$(15) \qquad \begin{array}{lcll} E_{jk} - E_{kj} & \longleftrightarrow & : a_j a^*_k : + : a^*_j a_k :, & 1 \le j < k \le N, \\ -iE_{jk} - iE_{kj} & \longleftrightarrow & (-1)(: a_j a^*_k : - : a^*_j a_k :), & 1 \le j < k \le N, \\ -iE_{jj} & \longleftrightarrow & (-i) : a_j a^*_j :, & 1 \le j \le N. \end{array}$$

Note that we cannot cancel $i = \sqrt{-1}$ from both side of the correspondence. The multiplication by i in the left hand side comes from the complex structure I on the vector space $E = (\mathbb{R}^{2N}; I)$, whereas on the right hand side the multiplication by i comes from the $\mathbb{C}$ factor in the complexification $E_{\mathbb{R}} \otimes \mathbb{C}$.

Lie algebra $\mathfrak{sp}(N') \subset \mathfrak{o}(4N') \subset \mathrm{Cliff}(\mathbb{R}^{4N'})$ and its complexification. We describe the symplectic Lie algebra $\mathfrak{sp}(N') = \mathrm{Lie}(\mathrm{Sp}(N'))$ in terms of Clifford elements. Again, the complex basis consisting of a_j's and a_j^*'s are convenient to use here also. We begin with generalities concerning quaternionic vector spaces, and describe $\mathfrak{sp}(N')$ in terms of matrices. We then translate this matrix description into Clifford description.

Let $\mathbb{H} = \mathbb{R} + \mathbb{R}i + \mathbb{R}j + \mathbb{R}k$ be the skew field of quaternions, where $ij = k$, $jk = i$, $ki = j$, $i^2 = -1$, $j^2 = -1$, $k^2 = -1$. Let $E = \mathbb{H}^{N'}$ be the N' dimensional quaternionic vector space with basis $\{e_1, e_2, \ldots, e_{N'}\}$. Since quaternions are noncommutative, from now on, we regard $\mathbb{H}^{N'}$ as a right $\mathbb{H}$ vector space. We adopt this convention to accommodate the usual matrix multiplication formalism, as in (21) below. Let $E_{\mathbb{C}}$, $E_{\mathbb{R}}$ be the underlying complex and real vector spaces, where scalars act from the right:

$$E = \bigoplus_{\ell=1}^{N'} e_{\ell}\mathbb{H},$$

(16) $$E_{\mathbb{C}} = \bigoplus_{\ell=1}^{N'} (e_{\ell}\mathbb{C} + e_{\ell+N'}\mathbb{C}), \qquad \text{where } e_{\ell+N'} = e_{\ell} \cdot j,$$

$$E_{\mathbb{R}} = \bigoplus_{\ell=1}^{2N'} (e_{\ell}\mathbb{R} + e_{\ell}'\mathbb{R}), \qquad \text{where } e_{\ell}' = e_{\ell} \cdot i, \quad 1 \leq \ell \leq 2N'.$$

Thus $E_{\mathbb{R}}$ has $\{e_1, e_2, \ldots, e_{N'}\} \cup \{e_1', e_2', \ldots, e_{N'}'\} \cup \{e_{1+N'}, e_{2+N'}, \ldots, e_{2N'}\} \cup \{e_{1+N'}', e_{2+N'}', \ldots, e_{2N'}'\}$ as its basis over $\mathbb{R}$. The following relations exist among these vectors:

(17)
$$
\begin{array}{ccc}
\{e_1, \ldots, e_{N'}\} & \xrightarrow{\;\cdot j\;} & \{e_{1+N'}, \ldots, e_{2N'}\} \\
\downarrow{\scriptstyle \cdot i} & & \downarrow{\scriptstyle \cdot i} \\
\{e_1', \ldots, e_{N'}'\} & \xrightarrow{\;\cdot(-j)\;} & \{e_{1+N'}', \ldots, e_{2N'}'\}
\end{array}
$$

When we regard E as a real vector space $E_{\mathbb{R}}$, the right multiplications by i, j, k causes a $\mathbb{R}$-linear transformations of $E_{\mathbb{R}}$. We denote these linear isomorphisms by I, J, K, which act on $E_{\mathbb{R}}$ from the left. Namely, for any $v \in E = E_{\mathbb{R}}$,

(18) $$I(v) = vi, \qquad J(v) = vj, \qquad K(v) = vk.$$

These linear maps satisfy the following relations :

(19) $\quad I^2 = -1, \quad J^2 = -1, \quad K^2 = -1, \quad IJ = -K, \quad JK = -I, \quad KI = -J.$

The introduction of the negative sign in the last three formulae is due to the fact that I, J, K act from the left on $E_{\mathbb{R}}$, where i, j, k act from the right. This switch of roles necessitates the introduction of inverses and these inverses cause the negative sign in the last three identities.

REMARK. We could define I, J, K by $I(v) = -vi = vi^{-1}$, $J(v) = -vj = vj^{-1}$, $K(v) = -vk = vk^{-1}$. In this case, we have the usual quaternion relations for these maps, $I^2 = -1$, $J^2 = -1$, $K^2 = -1$, $IJ = K$, $JK = I$, $KI = J$.

In $E_{\mathbb{R}}$, one can immediately check that for $1 \leq r \leq N'$,

(20) $J(e_r) = e_{r+N'}, \quad J(e_{r+N'}) = -e_r, \quad J(e'_r) = -e'_{r+N'}, \quad J(e'_{r+N'}) = e'_r.$

The maps I, J and K are $\mathbb{R}$-linear because $\mathbb{R} \subset \mathbb{H}$ is the center of the skew field of quaternions. However, on $E_{\mathbb{C}}$ in (16) the map $J : E_{\mathbb{C}} \to E_{\mathbb{C}}$ is not $\mathbb{C}$ linear, because J doesn't commute with I. This map has the following property :

(21)
$$J\left(\sum_{r=1}^{N'} e_r \alpha_r + \sum_{r=1}^{N'} e_{r+N'} \beta_r\right) = \sum_{r=1}^{N'} e_r(-\bar{\beta}_r) + \sum_{r=1}^{N'} e_{r+N'} \bar{\alpha}_r,$$

or $$J\left[(e_1, \ldots, e_{2N'}) \binom{\alpha}{\beta}\right] = (e_1, \ldots, e_{2N'}) \binom{-\bar{\beta}}{\bar{\alpha}}.$$

Here, $\alpha = {}^t(\alpha_1, \ldots, \alpha_{N'})$ and $\beta = {}^t(\beta_1, \ldots, \beta_{N'})$ denote column vectors with $\alpha_r, \beta_r \in \mathbb{C}$.

As usual, the conjugate of the quaternion $q = \alpha + \beta j$ with $\alpha, \beta \in \mathbb{C}$ is given by $\bar{q} = \bar{\alpha} - \beta j$. The operation of conjugation is an anti-automorphism of $\mathbb{H}$, namely $\overline{q_1 \cdot q_2} = \bar{q}_2 \cdot \bar{q}_1$.

The canonical pairing on $\mathbb{H}^{N'}$ has the following property with respect to the scalar multiplication by $\mathbb{H}$:

(22) $$\langle e_r q_1, e_s q_2 \rangle_{\mathbb{H}} = q_1 \bar{q}_2 \delta_{rs}, \qquad 1 \leq r, s \leq N',$$

where $q_1, q_2 \in \mathbb{H}$. One can easily check that the maps I, J and K on $\mathbb{H}^{N'}$ are isometries. Let $C : \mathbb{H} \to \mathbb{C}$ be the map given by $C(\alpha + \beta j) = \alpha$. We define

(23) $$\langle \, , \, \rangle_{\mathbb{C}} = C(\langle \, , \, \rangle_{\mathbb{H}}), \qquad \langle \, , \, \rangle_{\mathbb{R}} = \Re(\langle \, , \, \rangle_{\mathbb{C}}) = \Re(\langle \, , \, \rangle_{\mathbb{H}}).$$

The first pairing $\langle \, , \, \rangle_{\mathbb{C}}$ defines a nondegenerate Hermitian pairing on $E_{\mathbb{C}}$. It is sesquilinear in the first variable. That is, for any $\alpha, \beta \in \mathbb{C}$, we have

(24) $$\langle e_r \alpha, e_s \beta \rangle_{\mathbb{C}} = \alpha \bar{\beta} \delta_{rs} \qquad \text{for } 1 \leq r, s \leq N = 2N'.$$

Note that over $\mathbb{C}$, E has basis $\{e_1, e_2, \ldots, e_N\}$. The second pairing $\langle \, , \, \rangle_{\mathbb{R}}$ defines a nondegenerate symmetric $\mathbb{R}$-bilinear pairing on $E_{\mathbb{R}}$.

We describe the Lie algebra $\mathfrak{sp}(N')$ as a subalgebra of $\mathfrak{u}(2N')$ first in terms of matrices then in terms of Clifford elements. For this, we have to specify how the group $\mathrm{Sp}(N')$ sits inside $\mathrm{U}(2N')$. To describe an embedding $M_{N'}(\mathbb{H}) \to M_{2N'}(\mathbb{C})$, let $T : \mathbb{H}^{N'} \to \mathbb{H}^{N'}$ be an $\mathbb{H}$-linear map. This map is described by a matrix $Q \in M_{N'}(\mathbb{H})$ by $(T(e_1), T(e_2), \ldots, T(e_{N'})) = (e_1, e_2, \ldots, e_{N'})Q$. Here, we are using right multiplication of $\mathbb{H}$ on E on the right hand side. We let $Q = A + Bj$, where $A, B \in M_{2N'}(\mathbb{C})$, and the (r, s) component is given by

$q_{rs} = a_{rs} + b_{rs}j$ with a_{rs}, $b_{rs} \in \mathbb{C}$. Now we regard E as a complex vector space $E_{\mathbb{C}}$ with basis $\{e_1, e_2, \ldots, e_{N'}, e_{1+N'}, \ldots, e_{2N'}\}$ as in (16) above. We write down the matrix of T with respect to this basis. For $1 \leq s \leq N'$, we have

$$T(e_s) = \sum_{r=1}^{N'} e_r(a_{rs} + b_{rs}j) = \sum_{r=1}^{N'} \{e_r a_{rs} + (e_r j)\bar{b}_{rs}\}.$$

Also, using the right $\mathbb{H}$ linearity of T, we have

$$T(e_{s+N'}) = T(e_s j) = T(e_s)j = \sum_{r=1}^{N'} e_r(a_{rs} + b_{rs}j)j$$

$$= \sum_{r=1}^{N'} e_r(-b_{rs} + j\bar{a}_{rs}) = \sum_{r=1}^{N'} \{e_r(-b_{rs}) + e_{r+N'}\bar{a}_{rs}\}.$$

Combining these two calculations in a matrix form, we have

(25) $(T(e_1), \ldots, T(e_{N'}), T(e_{1+N'}), \ldots, T(e_{2N'}))$

$$= (e_1, \ldots, e_{N'}, e_{1+N'}, \ldots, e_{2N'}) \begin{pmatrix} A & -B \\ B & \bar{A} \end{pmatrix}.$$

This defines a homomorphic embedding of $M_{N'}(\mathbb{H})$ into $M_{2N'}(\mathbb{C})$, sending $Q = A + Bj \in M_{N'}(\mathbb{H})$ to $\begin{pmatrix} A & -B \\ B & \bar{A} \end{pmatrix} \in M_{2N'}(\mathbb{C})$. Since $M_{N'}(\mathbb{H})$ is closed under exponentiation and bracketing, the subset of $M_{2N'}(\mathbb{C})$ consisting of matrices of this block form is also closed under exponentiation and bracketing. For any matrix $Q = A + Bj \in M_{N'}(\mathbb{H})$, its adjoint matrix is given by $Q^* = {}^t\bar{Q} = {}^t\bar{A} - {}^tBj$.

The symplectic group $\mathrm{Sp}(N')$ is, by definition, the set of all invertible $\mathbb{H}$ linear transformation on $E = \mathbb{H}^{N'}$ preserving the quaternionic pairing $\langle \ , \ \rangle_{\mathbb{H}}$ on $\mathbb{H}^{N'}$. In terms of matrices with respect to the basis $\{e_1, e_2, \ldots, e_{N'}\}$, this group is described by

(26) $$\mathrm{Sp}(N') = \{Q \in M_{N'}(\mathbb{H}) \mid Q^*Q = 1_{N'}\}.$$

This is a real Lie group of real dimension $N'(2N' + 1)$. In terms of complex matrices A and B such that $Q = A + Bj$, this condition $Q^*Q = 1_{N'}$ is equivalent to ${}^t\bar{A}A + {}^tB\bar{B} = 1_{N'}$ and ${}^t\bar{A}B = {}^tB\bar{A}$. By a simple calculation, one sees that this condition is precisely equivalent to the condition that the matrix $\begin{pmatrix} A & -B \\ B & \bar{A} \end{pmatrix}$ be unitary. Thus, we have a homomorphic embedding $\mathrm{Sp}(N') \to \mathrm{U}(2N')$. Some more calculations in terms of Lie algebra prove the next lemma.

LEMMA 9. *The Lie algebra* $\mathfrak{sp}(N')$ *is described by*

(27)
$$\mathfrak{sp}(N') = \left\{ \begin{pmatrix} X & -Y \\ \bar{Y} & \bar{X} \end{pmatrix} \middle| {}^t\bar{X} + X = 0, \ {}^tY = Y, \ X, Y \in M_{N'}(\mathbb{C}) \right\} \subset \mathfrak{u}(2N').$$

The conditions simply mean that the matrix of the above form is skew-adjoint. We can give another more intrinsic description of this algebra in terms of the operator J. First, we identify those matrices which commute with J.

LEMMA 10. *Let $E_\mathbb{C}$ be as in (16) and let J be as in (18). Then*

$$\{T \in M_{2N'}(\mathbb{C}) \mid JT = TJ : E_\mathbb{C} \to E_\mathbb{C}\} = \left\{ \begin{pmatrix} A & -B \\ \bar{B} & \bar{A} \end{pmatrix} \middle| A, C \in M_{N'}(\mathbb{C}) \right\}.$$

PROOF. Let $T = \begin{pmatrix} A & B \\ C & D \end{pmatrix} \in M_{2N'}(\mathbb{C})$. Let T also denote the corresponding linear transformation of $E_\mathbb{C}$ acting from the left. Using the second identity in (21), we have

$$JT(e_1, \ldots, e_{2N'}) = J\left((e_1, \ldots, e_{2N'})\begin{pmatrix} A & B \\ C & D \end{pmatrix}\right) = (e_1, \ldots, e_{2N'})\begin{pmatrix} -\bar{C} & -\bar{D} \\ \bar{A} & \bar{B} \end{pmatrix}$$

$$TJ(e_1, \ldots, e_{2N'}) = T\left((e_1, \ldots, e_{2N'})\begin{pmatrix} 0 & -I_{N'} \\ I_{N'} & 0 \end{pmatrix}\right)$$

$$= (e_1, \ldots, e_{2N'})\begin{pmatrix} A & B \\ C & D \end{pmatrix}\begin{pmatrix} 0 & -I_{N'} \\ I_{N'} & 0 \end{pmatrix}$$

$$= (e_1, \ldots, e_{2N'})\begin{pmatrix} B & -A \\ D & -C \end{pmatrix}$$

The equality $JT = TJ$ implies that $B = -\bar{C}$ and $A = \bar{D}$. □

LEMMA 11. *The set of symplectic matrices are precisely those skew-adjoint matrices which commute with J:*

$$(28) \qquad \mathfrak{sp}(N') = \{T \in \mathfrak{u}(2N') \mid TJ = JT : E_\mathbb{C} \to E_\mathbb{C}\}.$$

PROOF. Since $T \in \mathfrak{u}(2N')$, ${}^t\bar{T} + T = 0$. Then Lemma 9 and Lemma 10 imply this lemma. □

The $\mathbb{R}$-basis of this Lie algebra consists of the following elements:

$$(29) \qquad \begin{array}{ll} E_{rs} - E_{sr} + E_{r+N',s+N'} - E_{s+N',r+N'}, & 1 \le r < s \le N', \\ -iE_{rs} - iE_{sr} + iE_{r+N',s+N'} + iE_{s+N',r+N'}, & 1 \le r < s \le N', \\ -iE_{rr} + iE_{r+N',r+N'} & 1 \le r \le N', \\ E_{r,s+N'} - E_{s+N',r} + E_{s,r+N'} - E_{r+N',s}, & 1 \le r < s \le N', \\ -i(E_{r,s+N'} + E_{s+N',r} + E_{s,r+N'} + E_{r+N',s}), & 1 \le r < s \le N', \\ E_{r,r+N'} - E_{r+N',r} & 1 \le r \le N', \\ -i(E_{r,r+N'} + E_{r+N',r}) & 1 \le r \le N'. \end{array}$$

Now we are ready to describe how the Lie algebra $\mathfrak{sp}(N')$ sits inside of $\mathfrak{u}(2N')\otimes\mathbb{C}$ in terms of Clifford elements in $\mathrm{Cliff}(E \otimes_\mathbb{R} \mathbb{C})$. By comparing (29) and (15), and recombining elements, we get the following description:

(30)
$$\mathfrak{sp}(N') \otimes \mathbb{C} = \bigoplus_{1 \le r < s \le N'} \mathbb{C}(: a_r a_s^* : + : a_{r+N'}^* a_{s+N'} :) \oplus \bigoplus_{1 \le r < s \le N'} \mathbb{C}(: a_r^* a_s : + : a_{r+N'} a_{s+N'}^* :)$$

$$\oplus \bigoplus_{1 \le r < s \le N'} \mathbb{C}(: a_r a_{s+N'}^* : + : a_s a_{r+N'}^* :) \oplus \bigoplus_{1 \le r < s \le N'} \mathbb{C}(: a_r^* a_{s+N'} : + : a_s^* a_{r+N'} :)$$

$$\oplus \bigoplus_{1 \le r \le N'} \mathbb{C} : a_r a_{r+N'}^* : \oplus \bigoplus_{1 \le r \le N'} \mathbb{C} : a_r^* a_{r+N'} : \oplus \bigoplus_{1 \le r \le N'} \mathbb{C}(: a_r a_r^* : - : a_{r+N'} a_{r+N'}^* :).$$

The Cartan subalgebra is given by the last summand. Rearranging summands, we finally get the next proposition.

PROPOSITION 12. *The Lie algebra* $\mathfrak{sp}(N') \otimes \mathbb{C} \subset \mathfrak{u}(2N') \otimes \mathbb{C} = \bigoplus_{1 \le , r, s \le N'} \mathbb{C} : a_r a_s^* :$ *is given by*

$$(31) \quad \mathfrak{sp}(N') \otimes \mathbb{C} = \bigoplus_{1 \le r \le s \le N'} \mathbb{C}(: a_r a_{s+N'}^* : + : a_s a_{r+N'}^* :)$$

$$\oplus \bigoplus_{1 \le r \le s \le N'} \mathbb{C}(: a_r^* a_{s+N'} : + : a_s^* a_{r+N'} :) \oplus \bigoplus_{1 \le r, s \le N'} \mathbb{C}(: a_r a_s^* : + : a_{r+N'}^* a_{s+N'} :).$$

We can also give a characterization of $\mathfrak{sp}(N') \otimes \mathbb{C}$ sitting inside of Cliff(A) in terms of the map J. The complex linear extension of this map $J \otimes \mathbb{C}$ on $A = E \otimes \mathbb{C}$ is given as follows. Recall that $A = A^+ \oplus A^-$ has canonical basis $\{a_1, \ldots, a_{N'}, a_{1+N'}, \ldots, a_{2N'}\}$ and $\{a_1^*, \ldots, a_{N'}^*, a_{1+N'}^*, \ldots, a_{2N'}^*\}$ given by (3).

LEMMA 13. *The $\mathbb{C}$ linear map $J \otimes \mathbb{C} : A \to A$ commutes with the action of* $\mathfrak{sp}(N')$. *Furthermore, $J \otimes \mathbb{C}$ interchanges A^+ and A^- and induces $\mathfrak{sp}(N') \otimes \mathbb{C}$ isomorphism $J \otimes \mathbb{C} : A^\pm \xrightarrow{\cong} A^\mp$ between these representations $A^\pm$. This $\mathfrak{sp}(N')$-equivariant isomorphism is described by*

$$(32) \quad J(a_r) = a_{r+N'}^*, \quad J(a_{r+N'}) = -a_r^*, \quad J(a_r^*) = a_{r+N'}, \quad J(a_{r+N'}^*) = -a_r,$$

for $1 \le r \le N'$,

PROOF. Since any element in Sp(N') is a right $\mathbb{H}$-linear map, it commutes with I, J, K given in (18). Identities in (32) follow from (20) applied to (3). $\square$

Since J is an isometry on $\mathbb{H}^{N'}$ with respect to the pairing (20), it is also an isometry on $E_{\mathbb{R}}$ in (21) with respect to the pairing $\langle \ , \ \rangle_{\mathbb{R}}$ given in (23). Thus, $J \otimes \mathbb{C}$ preserves the $\mathbb{C}$-linear pairing on A because it is the $\mathbb{C}$ linear extension of $\langle \ , \ \rangle_{\mathbb{R}}$. Hence J extends to the whole Clifford algebra Cliff(A) by $J(v_1 \ldots v_r) = J(v_1) \ldots J(v_r)$.

LEMMA 14. *The Lie algebra $\mathfrak{u}(2N') \otimes \mathbb{C} \subset$ Cliff(A) is preserved under the map $J \otimes \mathbb{C} :$ Cliff$(A) \to$ Cliff(A).*

PROOF. This is clear from Lemma 13 and the description of the algebra $\mathfrak{u}(2N') \otimes \mathbb{C}$ given in (12). $\square$

Now we state the characterization of $\mathfrak{sp}(N') \otimes \mathbb{C}$ in terms of the map J acting on the Clifford algebra.

PROPOSITION 15. *As a Lie subalgebra of the Clifford algebra* Cliff(A), *the complexified symplectic Lie algebra can be described as follows*:

$$(33) \qquad \mathfrak{sp}(N') \otimes \mathbb{C} = (\mathfrak{u}(2N') \otimes \mathbb{C})^J = \{ x \in \mathfrak{u}(2N') \otimes \mathbb{C} \mid J(x) = x \}.$$

PROOF. Let $x \in \mathfrak{u}(2N') \otimes \mathbb{C}$. Recall that x acts on $a \in A$ by the bracket $x \cdot a = [x, a] = xa - ax$ with respect to the Clifford multiplication. As an endomorphism on A, $x \in \mathfrak{sp}(N') \otimes \mathbb{C}$ if and only if x commutes with J, that is, $[x, J(a)] = J([x, a])$, for any $a \in A$, in view of Lemma 11. But J preserves the Clifford multiplication, so it also preserves the bracket product and $J([x, a]) = [J(x), J(a)]$. Thus, as endomorphisms of A, $[x, \ ]$ and $[J(x), \ ]$ are the same. But the representation of $\mathfrak{u}(2N') \otimes \mathbb{C} \subset \mathfrak{o}(4N') \otimes \mathbb{C}$ on A is faithful. Hence $x = J(x)$. This proves the above proposition. □

§3.3   Commutants of Lie subalgebras in $\mathfrak{o}(2N)$ and Casimir elements

We calculate commutants of those subalgebras of $\mathfrak{o}(2N)$ which are discussed in §3.2. The commutant $\mathfrak{g}'$ of a Lie subalgebra $\mathfrak{g}$ of $\mathfrak{o}(2N)$ is defined by

$$\mathfrak{g}' = \{ X \in \mathfrak{o}(2N) \mid [X, Y] = 0 \text{ for all } Y \in \mathfrak{g} \}.$$

This coincides with the subset of $\mathfrak{g}$-invariant elements of $\mathfrak{o}(2N)$ under the adjoint action. These commutants are subalgebras of $\mathfrak{o}(2N)$ and give rise to infinite dimensional symmetries in elliptic genera, eventually.

We also calculate Casimir elements for these Lie subalgebras $\mathfrak{g}$. Casimir elements live in universal enveloping algebras U$(\mathfrak{g})$. But what we actually calculate is the image of Casimir elements under the map

$$(1) \qquad \mathrm{U}(\mathfrak{g}) \xrightarrow{\cong} S^*(\mathfrak{g}) \to \bigwedge\nolimits^{\mathrm{even}} A(-\tfrac{1}{2}) \subset \mathcal{A} = \bigwedge\nolimits^*[\bigoplus_{n \geq 0} A(-n - \tfrac{1}{2})].$$

Here $S^*(\mathfrak{g})$ is the symmetric algebra of $\mathfrak{g}$ and the above isomorphism is due to Poincaré-Birkhoff-Witt Theorem [Hu]. The images of Casimir elements in $\mathcal{A}$ can be zero. But if they are nonzero, then they are vectors in $\mathcal{A}$ invariant under $\mathfrak{g}$. Note that the action of $\mathfrak{g}$ on A induces an action on $\mathcal{A}$.

We examine $\mathfrak{g}$-invariants $\mathfrak{o}(2N)^{\mathfrak{g}} = \mathfrak{g}'$ for various $\mathfrak{g}$ via the $\mathfrak{o}(2N)$-equivariant map $\Upsilon : \mathfrak{o}(2N) \xrightarrow{\cong} \bigwedge^2 A$ where $\Upsilon(:a_1 a_2:) = a_1 \wedge a_2$ for any $a_1, a_2 \in A$. The Lie algebra $\mathfrak{g}$ we consider is one of the followings:

$$(2) \qquad \mathfrak{sp}(N') \subset \mathfrak{su}(N) \subset \mathfrak{u}(N) \subset \mathfrak{o}(2N) \supset \mathfrak{t}^N.$$

When $\mathfrak{g} = \mathfrak{o}(2N)$, obviously $\mathfrak{o}(2N)^{\mathfrak{o}(2N)} = \{0\}$ because the Lie algebra $\mathfrak{o}(2N)$ has no center.

When $\mathfrak{g} = \mathfrak{u}(N)$, the vector space A is a direct sum of two non-isomorphic irreducible $\mathfrak{u}(N)$-representations A^+ and A^- of the same dimension N. The second exterior power of A decomposes as $\bigwedge^2 A = \bigwedge^2 A \oplus (A^+ \otimes A^-) \oplus \bigwedge^2 A^-$

under the action of $\mathfrak{u}(N)$. The first and the third summand can never have trivial subrepresentations (when $N = 1$, $\bigwedge^2 A^{\pm} = \{0\}$). The representation $A^+ \otimes A^-$ has a 1-dimensional trivial subrepresentation generated by

$$(3) \qquad h = \sum_{r=1}^{N} h_r, \qquad \text{where} \qquad h_r = a_r a_r^* \in A^+ \otimes A^-.$$

By abuse of notation, we also denote $\Upsilon^{-1}(h) \in \mathfrak{o}(2N)$ by h. Thus $\mathfrak{o}(2N)^{\mathfrak{u}(N)} = \mathbb{C}h$, where $h = \sum_{r=1}^{N} : a_r a_r^* : \in \mathrm{Cliff}(A)$.

When $\mathfrak{g} = \mathfrak{su}(N)$, we have the same decomposition of A and of $\bigwedge^2 A$ as for $\mathfrak{u}(N)$ case. In the decomposition of $\bigwedge^2 A$, there is always one dimensional subspace of $\mathfrak{su}(N)$-invariants from the summand $A^+ \otimes A^-$. When $N \geq 3$, there are no more invariants. When $N = 2$, the first and the third summands are trivial 1-dimensional representations. But since $\mathfrak{su}(2) = \mathfrak{sp}(1)$, we include this case in the next symplectic case.

Now under the action on $\mathfrak{sp}(N')$ with $2N' = N$, the representation A splits into a sum of two irreducible $\mathfrak{sp}(N')$-representations A^+ and A^- of complex dimension N, both of which are isomorphic as $\mathfrak{sp}(N')$ representations under the $\mathfrak{sp}(N')$-equivariant isomorphism J (§3.2 Lemma 13). For the second exterior power $\bigwedge^2 A = \bigwedge^2 A^+ \oplus (A^+ \otimes A^-) \oplus \bigwedge^2 A^-$, each summand contains a 1-dimensional trivial representation. To see this, we recall some representation theory of $\mathfrak{sp}(N')$.

If P is the basic N dimensional representation ($P \cong A^+ \cong A^-$), then $\bigwedge^2 P \cong P_2 \oplus \mathbb{C}$, where P_2 is the irreducible representation and the trivial summand is generated by the symplectic form in P, which is $\mathfrak{sp}(N')$-invariant. In our context, the symplectic forms in vector spaces $\bigwedge^2 A^+$ and in $\bigwedge^2 A^-$ are given by $x = \frac{1}{2}(1 \otimes J)h$ and $x^* = \frac{1}{2}(J \otimes 1)h$, respectively, where $1 \otimes J : A^+ \otimes A^- \to \bigwedge^2 A^+$ and $J \otimes 1 : A^+ \otimes A^- \to \bigwedge^2 A^-$ and h is as in (3). These elements are invariant because h is $\mathfrak{sp}(N')$-invariant and J is an $\mathfrak{sp}(N')$-equivariant map. The nontriviality of these elements can be checked by a simple direct calculation. (For more details for the representation theory of $\mathrm{Sp}(N')$, see a subsection on $V^{\mathrm{Sp}(N')}$ in §3.12.) We have

$$\left(\bigwedge^2 A\right)^{\mathfrak{sp}(N')} = \mathbb{C}x \oplus \mathbb{C}h \oplus \mathbb{C}x^*, \qquad \text{where}$$

$$(4) \qquad x = \sum_{r=1}^{N'} a_r a_{r+N'}, \qquad x^* = \sum_{r=1}^{N'} a_r^* a_{r+N'}^*, \qquad h = \sum_{j=1}^{N} a_j a_j^*.$$

Transferring these elements back to $\mathfrak{o}(2N) \subset \mathrm{Cliff}(A)$ via the map Υ, we have

$$\left(\mathfrak{o}(2N) \otimes \mathbb{C}\right)^{\mathfrak{sp}(N')} = \mathbb{C}x \oplus \mathbb{C}h \oplus \mathbb{C}x^* \subset \mathrm{Cliff}(A), \qquad \text{where}$$

$$(5) \qquad x = \sum_{r=1}^{N'} : a_r a_{r+N'} :, \qquad x^* = \sum_{r=1}^{N'} : a_r^* a_{r+N'}^* :, \qquad h = \sum_{j=1}^{N} : a_j a_j^* : .$$

We could check the $\mathfrak{sp}(N')$-invariance of elements in (5) directly using basis vectors of $\mathfrak{sp}(N')$ described in (31) of §3.2 using the bracket formula:

(6) $[:ab:,:cd:] = \langle a,d \rangle :bc: + \langle b,c \rangle :ad: - \langle a,c \rangle :bd: - \langle b,d \rangle :ac:.$

For example, the invariance of x above can be checked as follows. Note that basis vectors of $\mathfrak{sp}(N') \otimes \mathbb{C}$ come in three types as in (31) of §3.2. For one type of elements with indices $1 \le r, s \le N'$, we have

$$[:a_r a_s^*:+ :a_{r+N'}^* a_{s+N'}:, \sum_{p=1}^{N'} :a_p a_{p+N'}:]$$

$$= \sum_{p=1}^{N'} (\langle a_s^*, a_p \rangle :a_r a_{p+N'}: + \langle a_{r+N'}^*, a_{p+N'} \rangle :a_{s+N'} a_p:)$$

$$=:a_r a_{s+N'}: + :a_{s+N'} a_r := 0.$$

For commutators with other types of elements, for $1 \le r \le s \le N'$ we have

$$[:a_r a_{s+N'}^*:+ :a_s a_{r+N'}^*:, \sum_{p=1}^{N'} :a_p a_{p+N'}:]$$

$$= \sum_{p=1}^{N'} (-\langle a_{s+N'}^*, a_{p+N'} \rangle :a_r a_p: - \langle a_{r+N'}^*, a_{p+N'} \rangle :a_s a_p:)$$

$$= - :a_r a_s: - :a_s a_r := 0,$$

$$[:a_r^* a_{s+N'}:+ :a_s^* a_{r+N'}:, \sum_{p=1}^{N'} :a_p a_{p+N'}:]$$

$$= \sum_{p=1}^{N'} (-\langle a_r^*, a_r \rangle :a_{s+N'} a_{p+N'}: - \langle a_s^*, a_p \rangle :a_{r+N'} a_{p+N'}:)$$

$$= - :a_{s+N'} a_{r+N'}: - :a_{r+N'} a_{s+N'} := 0.$$

The $\mathfrak{sp}(N')$-invariance of x^* and h can be checked similarly. Since commutants are always Lie subalgebras, they are closed under Lie bracket. We calculate brackets for x, h, x^*.

LEMMA 1. *The elements* x, h, $x^* \in \mathrm{Cliff}(A)$ *in* (5) *have the following brackets:*

(7) $[h, x] = 2x, \qquad [h, x^*] = -2x^*, \qquad [x, x^*] = -h.$

Thus $\mathfrak{o}(2N)^{\mathfrak{sp}(N')} = \mathbb{C}x \oplus \mathbb{C}h \oplus \mathbb{C}x^* \cong \mathfrak{sp}(1) \otimes \mathbb{C} \cong \mathfrak{su}(2) \otimes \mathbb{C} \cong \mathfrak{sl}_2(\mathbb{C}) \subset \mathfrak{o}(2N).$

PROOF. We check these brackets by direct calculation using (6):

$$[h, x] = \sum_{j=1}^{N} \sum_{p=1}^{N'} [:a_j a_j^*:, :a_p a_{p+N'}:]$$

$$= \sum_{r,p} (\langle a_r^*, a_p \rangle :a_r a_{p+N'}: - \langle a_r^*, a_{p+N'} \rangle :a_r a_p:)$$

$$= \sum_{r=1}^{N'} (:a_r a_{r+N'}: - :a_{r+N'} a_r:) = 2x.$$

Taking complex conjugation $*$ and noting $h^* = -h$, we get $[h, x^*] = -2x^*$. The last bracket can be calculated similarly. $\square$

Under the action of the Cartan subalgebra t^N, the vector space A decomposes as $A = \bigoplus_{r=1}^{N}(u_r \oplus u_r^{-1})$, where u_r is the standard 1 dimensional representation of the r-th summand of t^N generated by the basis vector a_r, and u_r^{-1} is its complex conjugate representation $\mathbb{C}a_r^*$. Thus t^N invariant subspace in $\bigwedge^2 A$ is N-dimensional given by

$$\left(\textstyle\bigwedge^2 A\right)^{t^N} = \bigoplus_{r=1}^{N} \mathbb{C}u_r \cdot u_r^{-1} = \bigoplus_{r=1}^{N} \mathbb{C}a_r\, a_r^* = \bigoplus_{r=1}^{N} \mathbb{C}h_r.$$

Transferring this calculation to $\mathfrak{o}(2N)$ via the $\mathfrak{o}(2N)$-isomorphism Υ, we see that $\left(\mathfrak{o}(2N)\right)^{t^N} = t^N$. This should be the case since t^N is the maximal abelian Lie subalgebra of $\mathfrak{o}(2N)$.

We summarize our calculations of commutants in the next proposition.

PROPOSITION 2. *The commutants of the Lie subalgebras* $\mathfrak{sp}(N') \subset \mathfrak{su}(N) \subset \mathfrak{u}(N) \subset \mathfrak{o}(2N) \supset t^N$ *of* $\mathfrak{o}(2N)$ *are given as follows:*

(8)
$$\mathfrak{o}(2N)^{\mathfrak{u}(N)} = \mathfrak{o}(2N)^{\mathfrak{su}(N)} = \mathbb{C}h, \qquad \mathfrak{o}(2N)^{t^N} = t^N,$$
$$\mathfrak{o}(2N)^{\mathfrak{sp}(N')} = \mathbb{C}x \oplus \mathbb{C}h \oplus \mathbb{C}x^*.$$

Here x, h, x^* *are as in* (4), (5) *satisfying commutation relations of* $\mathfrak{sl}(\mathbb{C})$:

$$[h, x] = 2x, \qquad [h, x^*] = -2x^*, \qquad [x, x^*] = -h.$$

In the above, $\mathfrak{su}(N)$ *is considered for* $N \geq 3$, *and* $\mathfrak{su}(2) = \mathfrak{sp}(1)$.

Next we calculate Casimir elements. For any Lie algebra $\mathfrak{g}$ with nondegenerate symmetric bilinear pairing $\langle\ ,\ \rangle : \mathfrak{g} \otimes \mathfrak{g} \to \mathbb{C}$, the Casimir element $\phi_\mathfrak{g}$ is defined as follows. Let $\{u_1, u_2, \ldots, u_n\}$ be a basis of $\mathfrak{g}$ and let $\{\hat{u}_1, \hat{u}_2, \ldots, \hat{u}_n\}$ be the dual basis of $\mathfrak{g}$ so that $\langle u_i, \hat{u}_j \rangle = \delta_{ij}$ for $1 \leq i, j \leq n$. Then the Casimir element $\phi_\mathfrak{g}$ is defined as an element in the universal enveloping algebra $U(\mathfrak{g})$ given by

(9)
$$\phi_\mathfrak{g} = \sum_{i=1}^{n} u_i \hat{u}_i \in U(\mathfrak{g}).$$

This element is independent of the choice of basis $\{u_1, u_2, \ldots, u_n\}$. By Poincaré-Birkhoff-Witt Theorem, the map from $U(\mathfrak{g})$ to its associated graded algebra which turns out to be the symmetric algebra $S^*(\mathfrak{g})$, is actually an isomorphism $U(\mathfrak{g}) \xrightarrow{\cong} S^*(\mathfrak{g})$. Since $\mathfrak{g} \subset \mathfrak{o}(2N) \xrightarrow{\cong} \bigwedge^2 A$ via the map Υ, we can consider an induced map $S^*(\mathfrak{g}) \to \bigwedge^{\mathrm{even}} A \subset \bigwedge^* A \xrightarrow{\cong} \bigwedge^* A(-\tfrac{1}{2}) \subset A$. Combining these maps, we have a map $U(\mathfrak{g}) \to A$. We let $\phi_\mathfrak{g}$ also denote the image in A of the Casimir element $\phi_\mathfrak{g}$ in $U(\mathfrak{g})$ under this map. We calculate $\phi_\mathfrak{g} \in A$ for

$\mathfrak{g} = \mathfrak{o}(2N)$, $\mathfrak{su}(N)$, $\mathfrak{u}(N)$, $\mathfrak{sp}(N')$, and $\mathfrak{o}(2N)^{\mathfrak{sp}(N')} = \mathfrak{sp}(1)$. Recall that for the Lie algebra $\mathfrak{o}(2N)$ regarded as a Lie subalgebra of the Clifford algebra $\mathrm{Cliff}(A)$, a nondegenerate symmetric pairing is given by

$$(10) \qquad \langle :ab:, :cd:\rangle = \langle a, d\rangle\langle b, c\rangle - \langle a, c\rangle\langle b, d\rangle.$$

From (5) of §3.2, we have

$$\mathfrak{o}(2N) = \bigoplus_{i<j} \mathbb{C}: a_i a_j: \oplus \bigoplus_{i<j} \mathbb{C}: a_i^* a_j^*: \oplus \bigoplus_{1\leq i,j\leq N} \mathbb{C}: a_i a_j^*: .$$

The elements dual to these basis elements are : $a_j^* a_i^*$:, : $a_j a_i$:, and : $a_j a_i^*$:, respectively. In the following calculations, the map $\Upsilon : U(\mathfrak{g}) \to \mathcal{A}$ may not be written explicitly. We let h_i denote both $h_i =: a_i a_i^* : \in \mathfrak{o}(2N)$ and $h_i = a_i(-\frac{1}{2}) \wedge a_i^*(-\frac{1}{2}) \in \mathcal{A}$ for $1 \leq i \leq N$. The Casimir element $\phi_{\mathfrak{o}(2N)}$ is given by

$$(11) \qquad \begin{aligned} \phi_{\mathfrak{o}(2N)} &= \sum_{i<j} :a_i a_j ::a_j^* a_i^*: + \sum_{i<j} :a_i^* a_j^* ::a_j a_i: + \sum_{i,j} :a_i a_j^* ::a_j a_i^*: \\ &= \sum_{i<j} h_i h_j + \sum_{i<j} h_i h_j - \sum_{i,j} h_i h_j = 0 \in \textstyle\bigwedge^4 A. \end{aligned}$$

The last equality is due to $h_i h_j = h_j h_i$ and $h_i^2 = 0$ for $1 \leq i, j \leq N$ in the exterior algebra $\bigwedge^{\mathrm{even}} A$. Note that we are not working in the Clifford algebra $\mathrm{Cliff}(A)$, in which we have different relations, for example $h_i^2 = \frac{1}{4}$.

For the unitary Lie algebra $\mathfrak{u}(N) = \bigoplus_{1\leq i,j\leq N} \mathbb{C}:a_i a_j^*:$, since the element dual to :$a_i a_j^*$: is :$a_j a_i^*$:, the Casimir element $\phi_{\mathfrak{u}(N)}$ is given by

$$(12) \qquad \phi_{\mathfrak{u}(N)} = \sum_{1\leq i,j\leq N} :a_i a_j^* ::a_j a_i^*: := -\sum_{1\leq i,j\leq N} h_i h_j = -h^2 \in \textstyle\bigwedge^4 A,$$

where $h = \sum_{i=1}^{N} h_i$.

Similarly from (13) of §3.2, the Lie algebra $\mathfrak{su}(N) \subset \mathrm{Cliff}(A)$ is given by

$$\mathfrak{su}(N) = \bigoplus_{1\leq i\neq j\leq N} \mathbb{C}: a_i a_j^*: \oplus \bigoplus_{i=1}^{N-1} \mathbb{C}(:a_i a_i^*: - :a_{i+1} a_{i+1}^*:).$$

The vectors in the first summand pairs trivially with the vectors in the second summand. The vector dual to :$a_i a_j^*$: is :$a_j a_i^*$: := $a_j a_i^*$. In the vector space $\bigoplus_{i=1}^{N-1} \mathbb{C}(h_i - h_{i+1})$, the element dual to $h_i - h_{i+1}$ is $h_1 + h_2 + \cdots + h_i - (i/N)h$ since $\langle h_i, h_j\rangle = \delta_{ij}$. So the Casimir element is given by

$$\phi_{\mathfrak{su}(N)} = \sum_{1\leq i\neq j\leq N} (:a_i a_j^*:)(:a_j a_i^*:) + \sum_{i=1}^{N-1} (h_i - h_{i+1})(h_1 + h_2 + \cdots + h_i - (i/N)h).$$

The first summation is equal to $-h^2$ as above. The second summation gives $(\sum_{i=1}^{N} h_i^2) - h^2/N = -h^2/N$, since $h_i^2 = 0$. Hence the Casimir element $\phi_{\mathfrak{su}(N)} \in \mathcal{A}$ for the Lie algebra $\mathfrak{su}(N)$ is given by

$$(13) \qquad \phi_{\mathfrak{su}(N)} = -\left(\frac{N+1}{N}\right) h^2 \in \bigwedge^4 \mathcal{A}.$$

Now we discuss the symplectic Casimir element. First, recall (30) from §3.2:

$$(14)$$
$$\mathfrak{sp}(N') \otimes \mathbb{C} = \bigoplus_{1 \leq r < s \leq N'} \mathbb{C}(:a_r a_s^*: + :a_{r+N'}^* a_{s+N'}:) \oplus \bigoplus_{1 \leq r < s \leq N'} \mathbb{C}(:a_r^* a_s: + :a_{r+N'} a_{s+N'}^*:)$$
$$\oplus \bigoplus_{1 \leq r < s \leq N'} \mathbb{C}(:a_r a_{s+N'}^*: + :a_s a_{r+N'}^*:) \oplus \bigoplus_{1 \leq r < s \leq N'} \mathbb{C}(:a_r^* a_{s+N'}: + :a_s^* a_{r+N'}:)$$
$$\oplus \bigoplus_{1 \leq r \leq N'} \mathbb{C}:a_r a_{r+N'}^*: \oplus \bigoplus_{1 \leq r \leq N'} \mathbb{C}:a_r^* a_{r+N'}: \oplus \bigoplus_{1 \leq r \leq N'} \mathbb{C}(:a_r a_r^*: - :a_{r+N'} a_{r+N'}^*:).$$

Using (10), we see that the elements dual to the above basis vectors are

$$(15) \qquad
\begin{aligned}
&-\tfrac{1}{2}(:a_r^* a_s: + :a_{r+N'} a_{s+N'}^*:), && -\tfrac{1}{2}(:a_r a_s^*: + :a_{r+N'}^* a_{s+N'}:) \\
&-\tfrac{1}{2}(:a_r^* a_{s+N'}: + :a_s^* a_{r+N'}:), && -\tfrac{1}{2}(:a_r a_{s+N'}^*: + :a_s a_{r+N'}^*:) \\
&-:a_r^* a_{r+N'}:, && -:a_r a_{r+N'}^*:, && \tfrac{1}{2}(:a_r a_r^*: - :a_{r+N'} a_{r+N'}^*:).
\end{aligned}$$

Thus the symplectic Casimir element is given by

$$\phi_{\mathfrak{sp}(N')} = -\sum_{1 \leq r < s \leq N'}(:a_r a_s^*: + :a_{r+N'}^* a_{s+N'}:)(:a_r^* a_s: + :a_{r+N'} a_{s+N'}^*:),$$
$$-\sum_{1 \leq r < s \leq N'}(:a_r a_{s+N'}^*: + :a_s a_{r+N'}^*:)(:a_r^* a_{s+N'}: + :a_s^* a_{r+N'}:),$$
$$-2\sum_{r=1}^{N'} :a_r a_{r+N'}^*: :a_r^* a_{r+N'}: + \tfrac{1}{2}\sum_{r=1}^{N'}(:a_r a_r^*: - :a_{r+N'} a_{r+N'}^*:)^2.$$

Let $h_r =: a_r a_r^*:$, $x_r =: a_r a_{r+N'}:$, and $x_r^* =: a_r^* a_{r+N'}^*:$. The above is equal to

$$(16)$$
$$\phi_{\mathfrak{sp}(N')} = -\sum_{1 \leq r < s \leq N'}(h_r + h_{r+N'})(h_s + h_{s+N'}) + 2\sum_{1 \leq r < s \leq N'}(x_r x_s^* + x_r^* x_s) - 3\sum_{r=1}^{N} h_r h_{r+N'}$$
$$= -\tfrac{1}{2}h^2 + 2xx^* \in \bigwedge^4 \mathcal{A}.$$

Finally, for $\mathfrak{sp}(1) \otimes \mathbb{C} = (\mathfrak{o}(2N) \otimes \mathbb{C})^{\mathfrak{sp}(N')} = \mathbb{C}x \oplus \mathbb{C}h \oplus \mathbb{C}x^*$, the only nontrivial pairings among these elements are $\langle x, x^* \rangle = -N'$, $\langle h, h \rangle = N$. Thus the elements dual to x, h, x^* are $-x^*/N'$, h/N, $-x/N'$. So, the Casimir element is

$$(17) \qquad \phi_{\mathfrak{sp}(1)} = -\frac{xx^*}{N'} + \frac{h^2}{N} - \frac{x^*x}{N'} = \frac{1}{N'}(\tfrac{1}{2}h^2 - 2xx^*) \in \bigwedge^4 \mathcal{A}.$$

We collect our calculations of Casimir elements in the next proposition.

PROPOSITION 3. *The Casimir elements* $\phi_\mathfrak{g} \in \bigwedge^4 A$ *for* $\mathfrak{g} = \mathfrak{o}(2N), \mathfrak{u}(N),$ $\mathfrak{su}(N), \mathfrak{sp}(N'),$ *and* $\mathfrak{sp}(1)$ *are given as follows:*

$$
(18) \quad
\begin{aligned}
&\phi_{\mathfrak{o}(2N)} = 0, \qquad \phi_{\mathfrak{u}(N)} = -h^2, \qquad \phi_{\mathfrak{su}(N)} = -(N+1)h^2/N, \quad N \geq 2, \\
&\phi_{\mathfrak{sp}(N')} = -\tfrac{1}{2}h^2 + 2xx^*, \qquad \phi_{\mathfrak{sp}(1)} = (\tfrac{1}{2}h^2 - 2xx^*)/N', \quad N' \geq 1.
\end{aligned}
$$

Here, the last Lie algebra $\mathfrak{sp}(1)$ *is contained in* $\mathfrak{o}(2N)$ *as* $\mathfrak{sp}(1) = \mathfrak{o}(2N)^{\mathfrak{sp}(N')},$ *which is the commutant of* $\mathfrak{sp}(N').$

Note that when $N = 2$ (equivalently $N' = 1$), we have a relation $xx^* = -h^2/2,$ and $\phi_{\mathfrak{sp}(1)} = -(3/2)h^2$ which is the same as $\phi_{\mathfrak{su}(2)}.$ This should be the case since $\mathfrak{su}(2) = \mathfrak{sp}(1).$ The unitary and symplectic Casimir elements h^2 and $\frac{1}{2}h^2 - 2xx^*$ play an important role in our calculation of vertex operators in later sections.

For later use, we describe module structure of some symmetric powers of the Lie algebra $\mathfrak{g} = \mathfrak{sl}_2(\mathbb{C}).$ Let V_n be the unique irreducible representation of $\mathfrak{g}$ of dimension $n.$ Let $\mathfrak{g} = \mathbb{C}x \oplus \mathbb{C}h \oplus \mathbb{C}x^*$ as above.

PROPOSITION 4. *The second and third symmetric powers of* $\mathfrak{g}$ *decomposes under* $\mathfrak{g}$ *as follows:*

(i) $S^2(\mathfrak{g})$ *is a direct sum of* V_5 *and a 1-dimensional trivial representation:*

$$
(19) \quad
\begin{aligned}
S^2(\mathfrak{g}) &= V_5 \oplus \mathbb{C}(\tfrac{1}{2}h^2 - 2xx^*), \qquad where \\
V_5 &= \mathbb{C}x^2 \oplus \mathbb{C}hx \oplus \mathbb{C}(h^2 + 2xx^*) \oplus \mathbb{C}hx^* \oplus \mathbb{C}(x^*)^2
\end{aligned}
$$

is the weight space decomposition of $V_5 \subset S^2(\mathfrak{g}).$

(ii) $S^3(\mathfrak{g})$ *is the sum of two nontrivial irreducible representations given by*

$$
(20) \quad
\begin{aligned}
S^3(\mathfrak{g}) &= V_3 \oplus V_7, \qquad where \\
V_3 &= \mathbb{C}(xh^2 - 4x^2x^*) \oplus \mathbb{C}(h^3 - 4hx x^*) \oplus \mathbb{C}(h^2x^* - 4x(x^*)^2), \\
V_7 &= \mathbb{C}x^3 \oplus \mathbb{C}hx^2 \oplus \mathbb{C}(x^2x^* + xh^2) \oplus \mathbb{C}(h^3 + 6xhx^*) \\
&\quad \oplus \mathbb{C}(h^2x^* + x(x^*)^2) \oplus \mathbb{C}h(x^*)^2 \oplus \mathbb{C}(x^*)^3.
\end{aligned}
$$

The above V_3 *and* V_7 *give weight space decompositions.*

PROOF. Using (7), One can easily check that by successively applying $\operatorname{ad} x$ and $\operatorname{ad} x^*,$ we can generate the whole vector space V_5 from x^2 or from $(x^*)^2$:

$$
\begin{aligned}
(\operatorname{ad} x^*)(x^2) &= 2hx, \qquad \frac{1}{2!}(\operatorname{ad} x^*)^2(x^2) = h^2 + 2xx^*, \\
\frac{1}{3!}(\operatorname{ad} x^*)^3(x^2) &= 2hx^*, \qquad \frac{1}{4!}(\operatorname{ad} x^*)^4(x^2) = (x^*)^2.
\end{aligned}
$$

By applying $*$ and noting that $h^* = -h,$ we get the similar formulae for $\operatorname{ad} x^*.$ So, V_5 given in (i) is irreducible. Since it has codimension 1, there must be one 1-dimensional subspace which must be a trivial representation of $\mathfrak{sl}_2(\mathbb{C}).$ To identify this vector space, note that $\mathbb{C}h^2 \oplus \mathbb{C}xx^*$ is the subspace of $S^2(\mathfrak{g})$ of

h-weight 0. One can check that $\mathrm{Ker}\,(\mathrm{ad}\,x) = \mathrm{Ker}\,(\mathrm{ad}\,x^*) = \mathbb{C}(h^2 - 4xx^*)$. Thus the generator of the trivial representation is $h^2 - 4xx^*$.

(ii) can be checked similarly by direct calculations of effects of $\mathrm{ad}\,x$ and $\mathrm{ad}\,x^*$ given in (7). In fact, we have

$$(\mathrm{ad}\,x^*)(x^3) = 3hx^2, \qquad \frac{1}{2!}(\mathrm{ad}\,x^*)^2(x^3) = 3(x^2x^* + xh^2),$$

$$\frac{1}{3!}(\mathrm{ad}\,x^*)^3(x^3) = h^3 + 6xhx^*, \qquad \frac{1}{4!}(\mathrm{ad}\,x^*)^4(x^3) = 3(h^2x^* + x(x^*)^2),$$

$$\frac{1}{5!}(\mathrm{ad}\,x^*)^5(x^3) = 3h(x^*)^2, \qquad \frac{1}{6!}(\mathrm{ad}\,x^*)^6(x^3) = (x^*)^3.$$

For V_3, we first check that $xh^2 - 4x^2x^*$ is the $\mathrm{ad}\,x$ highest weight vector and its $*$ image $h^2x^* - 4x(x^*)^2$ is the $\mathrm{ad}\,x^*$ lowest weight vector. Then we calculate:

$$(\mathrm{ad}\,x^*)(xh^2 - 4x^2x^*) = h^3 - 4xhx^*, \qquad \frac{1}{2!}(\mathrm{ad}\,x^*)^2(xh^2 - 4x^2x^*) = h^2x^* - 4x(x^*)^2.$$

This calculation shows that V_3 in (ii) is an irreducible representation of $\mathfrak{g}$. $\square$

§3.4 Generating G-invariant vectors

Let V be the vertex operator super algebra of §2.2 on which $\mathrm{SO}(2N)$ acts as the automorphism group. Let G be any Lie subgroup in $\mathrm{Spin}(2N)$ or $\mathrm{SO}(2N)$. To understand the structure of the vertex operator super subalgebra V^G of V consisting of G invariant vectors, we develop a method to produce a large number of vectors in V^G for various G. The calculation of invariants is facilitated by a certain multiplicative structure in V^G and the existence of the vertex operator algebra atructure on V^G.

Let $A \cong \mathbb{C}^{2N}$ be the standard complex $\mathfrak{o}(2N)$-representation. Let $A(-n)$ with $n \in \frac{1}{2}\mathbb{Z}$ be a copy of A carrying weight n. We let

$$(1) \qquad \mathcal{A} = \bigwedge{}^* \Big[\bigoplus_{0 \le n \in \mathbb{Z}} A(-n - \tfrac{1}{2}) \Big]$$

be the exterior algebra on positive weight vectors. Note that this algebra is a subalgebra of the Clifford algebra $\mathrm{Cliff}\,(\mathbb{Z} + \frac{1}{2})$. We consider a map $\Omega = [\ ]$ from $\mathcal{A}$ to V defined by applying elements in $\mathcal{A}$ to the vacuum vector $\Omega \in (V)_0$:

$$(2) \qquad \Omega = [\ ] : \mathcal{A} \xrightarrow{\cong} V, \qquad \Omega(\alpha) = \alpha \cdot \Omega = [\alpha] \in V,$$

for any $\alpha \in \mathcal{A}$. The inverse of this map is denoted by $\Omega^{-1} : V \xrightarrow{\cong} \mathcal{A}$.

Although $\mathcal{A}$ and V are isomorphic as graded vector spaces, they have different algebraic structures. The vector space $\mathcal{A}$ is an exterior algebra contained in the Clifford algebra $\mathrm{Cliff}\,(\mathbb{Z} + \frac{1}{2})$ and has an associative algebra structure. Since $\mathcal{A}$ is not an ideal in the Clifford algebra, it is not a module over the Clifford algebra. On the other hand, the vector space V is a module over the Clifford algebra and in fact it is a module over the vertex operator super algebra V. These two different aspects of the same vector space play a complementary role in our study of V^G.

Now let $G \subset \mathrm{Spin}(2N)$ be a Lie subgroup and let $\mathfrak{g} = \mathrm{Lie}(G)$ be its Lie algebra. We also denote its complexification $\mathfrak{g} \otimes \mathbb{C}$ by $\mathfrak{g}$. When G is connected, dealing with G and dealing with $\mathfrak{g}$ don't make any difference. We compare the action of G and $\mathfrak{g}(0)$ on $\mathcal{A}$ and on V, where $\mathfrak{g}(0)$ is the Lie algebra consisting of 0-th mode operators of the vertex operators associated to $\mathfrak{g} \subset (V)_1$. By Corollary 4 of §2.3, the action of $\mathfrak{g}(0)$ on V coincides with the action of $\mathfrak{g}$ on V induced from $\mathfrak{g}$-action on $\mathcal{A}$.

On the Clifford module V, G and $\mathfrak{g}(0)$ act by Clifford multiplication,

$$(3) \qquad g \cdot [\alpha] = (g \cdot \alpha)\Omega, \qquad x(0) \cdot [\alpha] = (x(0) \cdot \alpha)\Omega,$$

for any $g \in G$, $x(0) \in \mathfrak{g}(0)$ and $\alpha \in \mathcal{A}$. Note that $g \cdot \alpha$ and $x(0) \cdot \alpha$ are no longer in $\mathcal{A}$, although their Clifford multiplication on the vacuum vector Ω is still well defined.

On the other hand, G and $\mathfrak{g}$ act on $\mathcal{A}$ by conjugation and by bracket product, respectively, with respect to the Clifford multiplication. Namely,

$$(4) \qquad g(\alpha) = g \cdot \alpha \cdot g^{-1}, \qquad x(0)(\alpha) = [x(0), \alpha],$$

for any $g \in G$ and $\alpha \in \mathcal{A}$. Here the dot "$\cdot$" denotes the Clifford multiplication. These actions are well defined since the finite dimensional vector space A itself is preserved by conjugation by $g \in G$ and bracket product by $x \in \mathfrak{g}$, and $x(0)$ acts as a derivation on $\mathcal{A}$. A little more examination shows that these actions of G and $\mathfrak{g}(0)$ on $\mathcal{A}$ and on V are in fact the same.

LEMMA 1. *For any Lie subgroup $G \subset \mathrm{Spin}(2N)$, the map $[\ ] : \mathcal{A} \to V$ is a G-isomorphism.*

PROOF. We only have to recall that the vacuum vector $\Omega \in (V)_0$ generates a trivial representation of $\mathrm{Spin}(2N)$. Then, by (3) and (4), for any $g \in G$ and $\alpha \in \mathcal{A}$, we have $[g(\alpha)] = g(\alpha)\Omega = g \cdot \alpha \cdot g^{-1}\Omega = (g \cdot \alpha)\Omega = g \cdot [\alpha]$, since $g^{-1}\Omega = \Omega$ for any $g \in G$. Also, $[x(0)(\alpha)] = (x(0)(\alpha))\Omega = [x(0), \alpha]\Omega = (x(0) \cdot \alpha)\Omega - (\alpha \cdot x(0))\Omega = (x(0) \cdot \alpha)\Omega = x(0) \cdot [\alpha]$, since $x(0)\Omega = 0$. $\square$

Thus, the map $[\ ] : \mathcal{A} \xrightarrow{\cong} V$ induces an isomorphism of G-invariants, $[\ ] : \mathcal{A}^G \xrightarrow{\cong} V^G$. We investigate the structure of G-invariant subspace through the associative algebra structure on $\mathcal{A}$ and also through the module structure on V over the vertex operator super algebra V^G. First we note an elementary fact.

LEMMA 2. *$\mathcal{A}^G$ is a subalgebra of $\mathcal{A}$.*

PROOF. Let $\alpha_1, \alpha_2 \in \mathcal{A}^G$ be two elements in this invariant set. Then for any $g \in G$, we have $g \cdot \alpha_1 \cdot g^{-1} = \alpha_1$ and $g \cdot \alpha_2 \cdot g^{-1} = \alpha_2$ with respect to the Clifford multiplication by (4). Then we have $g \cdot (\alpha_1 \cdot \alpha_2) \cdot g^{-1} = (g\alpha_1 g^{-1}) \cdot (g\alpha_2 g^{-1}) = \alpha_1 \cdot \alpha_2$. So, $\alpha_1 \cdot \alpha_2 \in \mathcal{A}^G$, and $\mathcal{A}^G$ is closed under Clifford multiplication. $\square$

Next we turn our attention to V^G as a vertex operator super algebra [§3.1 Theorem 2]. Since V is naturally equipped with a Hermitian pairing arising from the Hermitian pairing in A [§2.4], we can consider (i) adjoint operators associated to vertex operators in V^G, and (ii) vertex operators associated to adjoint vectors [§2.4 Definition 22]. These operators also preserve V^G, as shown in the next lemma.

LEMMA 3. Let $v \in V^G$ and let $Y(v, \zeta) = \sum_{n \in 1/2\mathbb{Z}} \{v\}_n \zeta^{-n-1}$ be the associated vertex operator. Let $Y(v, \zeta)^* = \sum \{v\}_n^* \zeta^{-n-1}$ and $Y(v^*, \zeta) = \sum \{v^*\}_n \zeta^{-n-1}$ be the adjoint vertex operator and the vertex operator for the adjoint vector, respectively. Then all component operators are $\mathfrak{g}$-equivariant: for any $n \in \mathbb{Z}$,

$$[\mathfrak{g}(0), \{v\}_n] = 0, \qquad [\mathfrak{g}(0), \{v\}_n^*] = 0, \qquad [\mathfrak{g}(0), \{v^*\}_n] = 0.$$

Hence operators $\{v\}_n$, $\{v\}_n^*$, $\{v^*\}_n$ for $n \in \mathbb{Z}$ preserve V^G:

(5) $\qquad \{V^G\}_n V^G \subset V^G, \qquad \{V^G\}_n^* V^G \subset V^G, \qquad \{V^{G^*}\} \subset V^G.$

PROOF. The statement $[\mathfrak{g}(0), \{v\}_n] = 0$ is the Lie algebra version of §3.1 Lemma 1 which shows that $\{v\}_n$ is G-equivariant.

For the second commutator relation, for any $u_1, u_2 \in V$ and $x(0) \in \mathfrak{g}(0)$, we have $(u_1, x(0)\{v\}_n^* u_2) = (\{v\}_n x(0)^* u_1, u_2)$. Since $x(0)^* = -\bar{x}(0)$ by Corollary 5 of §2.4 and since $\mathfrak{g}$ is stable under complex conjugation, the operator $x(0)^*$ is in $\mathfrak{g}(0)$ and it commutes with $\{v\}_n$'s by the first identity. So, this pairing is further equal to $(x(0)^*\{v\}_n u_1, u_2) = (u_1, \{v\}_n^* x(0)u_2)$. Since the Hermitian pairing in V is nondegenerate, as operators on V we have $[x(0), \{v\}_n^*] = 0$ by comparing the first and the last pairings. Since $\{v\}_n^*$ commutes with $\mathfrak{g}(0)$, it preserves the invariant subspace V^G.

Let v^* be the adjoint vector of $v \in V^G$. Applying complex conjugation to the identity $g \cdot v = v$, we have $g \cdot \bar{v} = \bar{v}$, since $g \in G$ is real and $\bar{g} = g$. Since $v^* = \pm\bar{v}$, we have $g \cdot v^* = v^*$. Hence $v^* \in V^G$ when $v \in V^G$. Thus the associated vertex operator $\{v^*\}_*$ commutes with the action of $\mathfrak{g}(0)$ and preserves V^G. $\square$

Since $[\ ] : A \to V$ is an isomorphism of graded vector spaces, for any $\alpha \in A$ we denote the vertex operator associated to $[\alpha] \in V$ by

(6) $$Y(\alpha, \zeta) = Y([\alpha], \zeta) = \sum_{n \in \mathbb{Z}} \{\alpha\}_n \zeta^{-n-1}.$$

Using this notation, previous lemmas can be summarized as follows.

PROPOSITION 5. Let $\alpha_1, \alpha_2 \in A^G$. Then

(1) $[\alpha_1 \cdot \alpha_2] \in V^G$,
(2) $\{\alpha_1\}_n[\alpha_2]$, $\{\alpha_1\}_n^*[\alpha_2]$, $\{\alpha_1^*\}_n[\alpha_2] \in V^G$ for any $n \in \mathbb{Z}$.

Thus the space V^G is a module over A^G via the Clifford multiplication, and over the vertex operator super algebra V_*^G and its dual $(V_*^G)^*$.

We make the following definition to illuminate the interplay between the associative algebra structure on A^G and the vertex operator algebra structure on V^G described in Proposition 5.

DEFINITION 6. Let $N \subset V$ be a subspace of V. Let $\mathrm{Alg}(\Omega^{-1}(N))$ be the subalgebra of $\mathcal{A}$ generated by elements in $\Omega^{-1}(N)$. We then let

$$(7) \qquad\qquad A(N) \equiv \Omega\left(\mathrm{Alg}(\Omega^{-1}(N))\right)$$

and we call it the *algebraic hull* of N.

As a consequence of Proposition 5, we immediately see that if $N \subset V^G$, then its algebraic hull $A(N)$ is also contained in V^G. By applying (2) of Proposition 5 to this algebraic hull, namely by applying vertex operators associated to vectors in $A(N)$ on $A(N)$, we obtain possibly larger G-invariant subspace. This observation gives the next useful corollary.

COROLLARY 7. *Let $N \subset V^G$. Then*

$$(8) \qquad\qquad A\left(\{A(N)\}_* (A(N))\right) \subset V^G.$$

REMARK. (i) From the point of view of the representation theory of a Lie group G or of a Lie algebra $\mathfrak{g}$, the above procedure provides us with a convenient tool to identify invariants in the $\mathfrak{g}$ representation $\mathcal{A}$: instead of decomposing tensor products of representations of $\mathfrak{g}$ and identifying trivial summands, we can use the vertex operator super algebra V^G acting on the associative algebra of $\mathfrak{g}$-invariants to produce new $\mathfrak{g}$-invariants. In a way, this is similar to a situation in algebraic topology in which a finite group G acts on the mod p cohomology ring of a space and the subset of G-invariants is a subring preserved under the action of the Steenrod algebra.

(ii) The above statement is for a subspace of V^G. We could equally start from a subspace M of $\mathcal{A}^G$.

For any vector subspace $M \subset \mathcal{A}$, let $[M] \subset V$ be the corresponding subspace in V. For each $\alpha \in M$, we have corresponding vertex operators $\{\alpha\}_n = \{[\alpha]\}_n$ for each $n \in \mathbb{Z}$ acting on V. We let

$$(9) \qquad\qquad \{M\}[M] = \{\{\alpha\}_n[\beta] \mid \alpha, \beta \in M,\ n \in \mathbb{Z}\} \subset V.$$

Its algebraic hull is then denoted by $A(\{M\}[M])$. For a family of elements $\{\alpha_i\}_{i\in I}$ in $\mathcal{A}$, let $\mathbb{C}\{\alpha_i\}_{i\in I}$ denote the subalgebra of $\mathcal{A}$ generated by these elements.

COROLLARY 8. *Let I be an index set and suppose we have a set of G-invariant elements $\alpha_i \in \mathcal{A}^G$ for any $i \in I$. Then*

$$(10) \qquad A\left(\{M\}[M]\right) \subset V^G, \qquad \text{where} \quad M = \mathbb{C}\{\alpha_i\}_{i\in I}.$$

We use (8) or (10) to identify many elements in V^G for various $G \subset \mathrm{Spin}(2N)$. We are interested in the following groups, where $2N' = N$:

$$(11) \quad \mathrm{SU}(N) \subset \mathrm{Spin}(2N) \supset T^N, \qquad \mathrm{Sp}(N') \subset \mathrm{Sp}(N') \cdot \mathrm{Sp}(1) \subset \mathrm{Spin}(2N).$$

Orthogonal Case: $\mathfrak{o}(2N)$. We define several families of elements in $\mathcal{A}^{\mathfrak{o}(2N)}$ below, and we then apply Corollary 8. Recall that $\dim_{\mathbb{C}} A = 2N$.

We consider following elements in $\mathcal{A}$ defined for $n_1 > n_2 \geq 0$ which are generalizations of the canonical Virasoro element $\omega \in \mathcal{A}_2$:

(12)
$$\omega(n_1, n_2) = -\frac{1}{2} \sum_{j=1}^{N} \{a_j(-n_1 - \tfrac{1}{2}) a_j^*(-n_2 - \tfrac{1}{2}) + a_j^*(-n_1 - \tfrac{1}{2}) a_j(-n_2 - \tfrac{1}{2})\}$$
$$\in A(-n_1 - \tfrac{1}{2}) \wedge A(-n_2 - \tfrac{1}{2}) \subset \mathcal{A}.$$

Since we work in the exterior algebra $\mathcal{A}$, we have $\omega(n_1, n_2) = -\omega(n_2, n_1)$. Furthermore, when $n_1 = n_2 = n$, the above element $\omega(n, n)$ in the exterior algebra $\mathcal{A}$ represents a zero vector in $\bigwedge^2 A(-n - \tfrac{1}{2})$. This is the reason of the restriction $n_1 > n_2$. Note that the element $\omega(1, 0)$ is nothing but the canonical Virasoro element ω of the vertex operator super algebra V. The above element $\omega(n_1, n_2)$ corresponds to the nondegenerate symmetric bilinear pairing $\langle\ ,\ \rangle : A(n_1 + \tfrac{1}{2}) \otimes A(n_2 + \tfrac{1}{2}) \to \mathbb{C}$ for which only nontrivial pairings are given by $\langle a_j(n_1 + \tfrac{1}{2}), a_j^*(n_2 + \tfrac{1}{2})\rangle = -\tfrac{1}{2}$. Due to dimensional reason of vector spaces in which it lives, we have $\omega(n_1, n_2)^{2N+1} = 0$.

We also consider the following top exterior element for $n \geq 0$:

(13)
$$\rho(-n - \tfrac{1}{2}) = a_1(-n - \tfrac{1}{2}) a_1^*(-n - \tfrac{1}{2}) \cdots a_N(-n - \tfrac{1}{2}) a_N^*(-n - \tfrac{1}{2})$$
$$\in \bigwedge^{2N} A(-n - \tfrac{1}{2}).$$

Note that $\rho(-n - \tfrac{1}{2})^2 = 0$ by dimensional reason. Among elements in (12) and in (13), the following relation holds:

(14)
$$\frac{1}{(2N)!} \cdot \omega(n_1, n_2)^{2N} = \frac{1}{2^{2N}} \cdot \rho(-n_1 - \tfrac{1}{2}) \cdot \rho(-n_2 - \tfrac{1}{2}).$$

Let $M_{\mathfrak{o}(2N)}$ be a subspace in $\mathcal{A}$ given by

(15) $$M_{\mathfrak{o}(2N)} = \mathbb{C}\{\omega(n_1, n_2), \rho(-n - \tfrac{1}{2}) \mid n_1 > n_2 \geq 0, n \geq 0\} \subset \mathcal{A}.$$

On an oriented Riemannian manifold M, $\omega(n_1, n_2)$ and $\rho(-n - \tfrac{1}{2})$ correspond to generalized Riemannian tensors and generalized volume forms on M [§5.2].

PROPOSITION 9 ($\mathfrak{o}(2N)$-INVARIANT VECTORS). *For any integers $n_1 > n_2 \geq 0$, $n \geq 0$, vectors $[\omega(n_1, n_2)]$ and $[\rho(-n - \tfrac{1}{2})]$ are $\mathfrak{o}(2N)$-invariant and $M_{\mathfrak{o}(2N)}$ consists of $\mathfrak{o}(2N)$-invariant vectors. The subalgebra $M_{\mathfrak{o}(2N)} \subset \mathcal{A}$ generates the following $\mathfrak{o}(2N)$-invariant subspace of V:*

(16) $$A\left(\{M_{\mathfrak{o}(2N)}\}[M_{\mathfrak{o}(2N)}]\right) \subset V^{\mathfrak{o}(2N)}.$$

PROOF. The $\mathfrak{o}(2N)$-invariance of $\rho(-n - \tfrac{1}{2})$ is clear because it is a vector in the trivial 1-dimensional $\mathfrak{o}(2N)$-representation $\bigwedge^{2N} A(-n - \tfrac{1}{2})$. The existence of the nondegenerate bilinear invariant pairing $\langle \, , \, \rangle : A \otimes A \to \mathbb{C}$ gives us an isomorphism $A^* \cong A$ and we have $(A \otimes A)^{o(2N)} \cong \mathrm{Hom}_{o(2N)}(A, A)$. The latter group is 1-dimensional generated by the identity map by Schur's lemma since A is an irreducible $o(2N)$-representation. Since $\langle a_j, a_k^* \rangle = \delta_{jk}$ and A^+, A^- are isotropic subspaces of A, the element in $A \otimes A$ corresponding to the identity map of A is given by $\sum_{j=1}^{N} a_j \otimes a_j^* + \sum_{j=1}^{N} a_j^* \otimes a_j$. This shows that $\omega(n_1, n_2)$ spans a 1-dimensional invariant subspace in $A(-n_1 - \frac{1}{2}) \otimes A(-n_2 - \frac{1}{2})$. (16) is now a consequence of Corollary 8. $\square$

(Special) Unitary Case: $\mathfrak{u}(N)$, $\mathfrak{su}(N)$. When $\mathfrak{g} = \mathfrak{u}(N) \subset o(2N)$, the vector space A splits into the direct sum of two irreducible $\mathfrak{u}(N)$-representations A^+ and A^-, which are conjugate to each other. We consider the following elements in $\mathcal{A}$ corresponding to the pairing pairing $\langle \, , \, \rangle : A^+ \otimes A^- \to \mathbb{C}$ for $n_1 > n_2 \geq 0$:

(17)
$$\omega_+(n_1, n_2) = -\sum_{j=1}^{N} a_j(-n_1 - \tfrac{1}{2}) a_j^*(-n_2 - \tfrac{1}{2}) \in A^+(-n_1 - \tfrac{1}{2}) \wedge A^-(-n_2 - \tfrac{1}{2}),$$

$$\omega_-(n_1, n_2) = -\sum_{j=1}^{N} a_j^*(-n_1 - \tfrac{1}{2}) a_j(-n_2 - \tfrac{1}{2}) \in A^-(-n_1 - \tfrac{1}{2}) \wedge A^+(-n_2 - \tfrac{1}{2}).$$

When $n_1 = n_2 = n$, we have $\omega_+(n, n) = -\omega_-(n, n)$ and it is equal to $h(-n - \frac{1}{2})$ given in (21) below. Observe that for the $\mathfrak{u}(N)$ case, the element $\omega(n_1, n_2)$ is the average of these two elements. The difference of the above two elements is denoted by $\kappa(n_1, n_2)$ and we call it a *generalized Kähler form*:

(18)
$$\omega(n_1, n_2) = \tfrac{1}{2}\{\omega_+(n_1, n_2) + \omega_-(n_1, n_2)\},$$
$$\kappa(n_1, n_2) = i\{\omega_+(n_1, n_2) - \omega_-(n_1, n_2)\}.$$

The form $\kappa(n_1, n_2)$ does correspond to a generalization of Kähler forms on complex Kähler manifolds. Since $\dim_{\mathbb{C}} A^+ = \dim_{\mathbb{C}} A^- = N$, by dimensional reason we have $\omega_\pm(n_1, n_2)^{N+1} = 0$.

Similarly, $\rho(-n - \frac{1}{2}) \in \bigwedge^{2N} A$ in (13) is a product of two top exterior powers $\rho_\pm(-n - \frac{1}{2})$ in $A^\pm$ given by

(19)
$$\rho_+(-n - \tfrac{1}{2}) = a_1(-n - \tfrac{1}{2}) a_2(-n - \tfrac{1}{2}) \cdots a_N(-n - \tfrac{1}{2}) \in \bigwedge{}^{N} A^+(-n - \tfrac{1}{2}),$$
$$\rho_-(-n - \tfrac{1}{2}) = a_1^*(-n - \tfrac{1}{2}) a_2^*(-n - \tfrac{1}{2}) \cdots a_N^*(-n - \tfrac{1}{2}) \in \bigwedge{}^{N} A^-(-n - \tfrac{1}{2}).$$

As with the case for $\rho(-n - \frac{1}{2})$, these elements satisfy $\rho_\pm(-n - \frac{1}{2})^2 = 0$ by dimensional reason. The exact relation to $\rho(-n - \frac{1}{2})$ is given as follows:

(20)
$$\rho_+(-n - \tfrac{1}{2}) \rho_-(-n - \tfrac{1}{2}) = (-1)^{\frac{N(N-1)}{2}} \rho(-n - \tfrac{1}{2}).$$

The unitary Lie algebra $\mathfrak{u}(N)$ contains a 1-dimensional center $\mathbb{C}h \subset o(2N)$, where $h = \sum_{r=1}^{N} : a_r a_r^* :$ commutes with both $\mathfrak{su}(N)$ and $\mathfrak{u}(N)$. We consider related elements in $\mathcal{A}$. For each $n \geq 0$, let

(21)
$$h(-n - \tfrac{1}{2}) = \sum_{j=1}^{N} a_j(-n - \tfrac{1}{2})a_j^*(-n - \tfrac{1}{2}) \in A^+(-n - \tfrac{1}{2}) \wedge A^-(-n - \tfrac{1}{2})$$
$$\subset \textstyle\bigwedge^2 A(-n - \tfrac{1}{2}) \subset \mathcal{A}.$$

Again, by dimensional reason we have $h(-n - \tfrac{1}{2})^{N+1} = 0$.

Using elements in (17), (19), (20), we consider the following subspaces of $\mathcal{A}$:

(22)
$$M_{\mathfrak{u}(N)} = \mathbb{C}\{\omega_\pm(n_1, n_2), h(-n - \tfrac{1}{2}) \mid n_1 > n_2 \geq 0, n \geq 0\} \subset \mathcal{A},$$
$$M_{\mathfrak{su}(N)} = \mathbb{C}\{\omega_\pm(n_1, n_2), \rho_\pm(-n - \tfrac{1}{2}), h(-n - \tfrac{1}{2}) \mid n_1 > n_2 \geq 0, n \geq 0\} \subset \mathcal{A}.$$

PROPOSITION 10 ($\mathfrak{u}(N)$- AND $\mathfrak{su}(N)$-INVARIANT VECTORS). *For any integers $n_1 > n_2 \geq 0$ and $n \geq 0$, the vectors $[\omega_\pm(n_1, n_2)]$, $[\rho_\pm(-n - \tfrac{1}{2})]$, $[h(-n - \tfrac{1}{2})]$ are $\mathfrak{su}(N)$-invariant. Among these elements, only $[\omega_\pm(n_1, n_2)]$ and $[h(-n - \tfrac{1}{2})]$ are $\mathfrak{u}(N)$-invariant. Thus $[M_{\mathfrak{u}(N)}] \subset (V)^{\mathfrak{u}(N)}$ and $[M_{\mathfrak{su}(N)}] \subset (V)^{\mathfrak{su}(N)}$.*

The following relations hold among these elements:

(23)
$$\frac{1}{N!}\omega_+(n_1, n_2)^N = (-1)^{\frac{N(N+1)}{2}} \rho_+(-n_1 - \tfrac{1}{2})\rho_-(-n_2 - \tfrac{1}{2}),$$
$$\frac{1}{N!}\omega_-(n_1, n_2)^N = (-1)^{\frac{N(N+1)}{2}} \rho_-(-n_1 - \tfrac{1}{2})\rho_+(-n_2 - \tfrac{1}{2}),$$
$$\frac{1}{N!}h(-n - \tfrac{1}{2})^N = \rho(-n - \tfrac{1}{2}).$$

Invariant subspaces $V^{\mathfrak{u}(N)}$ and $V^{\mathfrak{su}(N)}$ contain the following subspaces generated by $M_{\mathfrak{u}(N)}$ and $M_{\mathfrak{su}(N)}$, respectively:

(24)
$$A\left(\{M_{\mathfrak{u}(N)}\}[M_{\mathfrak{u}(N)}]\right) \subset V^{\mathfrak{u}(N)},$$
$$A\left(\{M_{\mathfrak{su}(N)}\}[M_{\mathfrak{su}(N)}]\right) \subset V^{\mathfrak{su}(N)}.$$

PROOF. We only have to show $\mathfrak{u}(N)$- and $\mathfrak{su}(N)$-invariance of elements in (22). Under the action of Lie algebras $\mathfrak{u}(N)$ and $\mathfrak{su}(N)$, the vector space A is a sum of two irreducible representations $A = A^+ \oplus A^-$. The existence of the nondegenerate $\mathbb{C}$-bilinear pairing on A shows that $(A^+)^* = A^-$ and $(A^-)^* = A^+$. Therefore, there are exactly one 1-dimensional invariant subspaces in $A^+ \otimes A^- = \text{Hom}(A^-, A^-)$, and in $A^- \otimes A^+ = \text{Hom}(A^+, A^-)$. Consequently, $A^+(-n_1 - \tfrac{1}{2}) \wedge A^-(-n_2 - \tfrac{1}{2})$, $A^-(-n_1 - \tfrac{1}{2}) \wedge A^+(-n_2 - \tfrac{1}{2})$, and $A^+(-n - \tfrac{1}{2}) \wedge A^-(-n - \tfrac{1}{2})$ have 1-dimensional $\mathfrak{u}(N)$- and $\mathfrak{su}(N)$-invariant subspaces spanned by $\omega_+(n_1, n_2)$, $\omega_-(n_1, n_2)$, and $h(-n - \tfrac{1}{2})$, respectively. The vectors $\rho_\pm(-n - \tfrac{1}{2})$ are $\mathfrak{su}(N)$-invariant because they belong to the trivial $\mathfrak{su}(N)$-representation $\bigwedge^N A^\pm(-n-\tfrac{1}{2})$. However these top exterior powers are not trivial representations of $\mathfrak{u}(N)$, and $\rho_\pm(-n - \tfrac{1}{2})$ do not belong to $V^{\mathfrak{u}(N)}$.

(23) is by direct calculation. (24) is a consequence of Corollary 8. □

Symplectic Case: $\mathfrak{sp}(N')$, $N = 2N'$. When $\mathfrak{g} = \mathfrak{su}(N)$ with $N = 2N'$, the representation A splits into a direct sum of two irreducible representations A^+ and A^-. Under the action of the Lie algebra $\mathfrak{g} = \mathfrak{su}(N)$, these two representations $A^\pm$ are dual to each other. However, under the symplectic Lie algebra $\mathfrak{sp}(N') \subset \mathfrak{su}(N)$, these representations are isomorphic to each other, and consequently they are self-dual representations. The $\mathfrak{sp}(N')$-isomorphism between $A^\pm$ is given by the symplectic structure map $J : A^\pm \xrightarrow{\cong} A^\mp$. We recall that on the basis of A the map J is given by

(25) $\quad J(a_r) = a^*_{r+N'}, \quad J(a_{r+N'}) = -a^*_r, \quad J(a^*_r) = a_{r+N'}, \quad J(a^*_{r+N'}) = -a_r.$

We define the following elements:

(26)
$$x\left(-n - \tfrac{1}{2}\right) = \sum_{j=1}^{N'} a_j\left(-n - \tfrac{1}{2}\right)a_{j+N'}\left(-n - \tfrac{1}{2}\right) \in \textstyle\bigwedge^2 A^+\left(-n - \tfrac{1}{2}\right),$$

$$x^*\left(-n - \tfrac{1}{2}\right) = \sum_{j=1}^{N'} a^*_j\left(-n - \tfrac{1}{2}\right)a^*_{j+N'}\left(-n - \tfrac{1}{2}\right) \in \textstyle\bigwedge^2 A^-\left(-n - \tfrac{1}{2}\right).$$

These elements are related to $h(-n - \tfrac{1}{2})$ of (21) through the symplectic structure J, as is shown in Lemma 11 below. Like $h(-n - \tfrac{1}{2})$, these elements are such that $x(-n - \tfrac{1}{2})^{N'+1} = 0$ and $x^*(-n - \tfrac{1}{2})^{N'+1} = 0$. We also consider the following related elements $\chi(n_1, n_2)$ and $\chi^*(n_1, n_2)$ defined by

(27)
$$\chi(n_1, n_2) = -\sum_{j=1}^{N'} a_j\left(-n_1 - \tfrac{1}{2}\right)a_{j+N'}\left(-n_2 - \tfrac{1}{2}\right)$$
$$+ \sum_{j=1}^{N'} a_{j+N'}\left(-n_1 - \tfrac{1}{2}\right)a_j\left(-n_2 - \tfrac{1}{2}\right) \in A^+\left(-n_1 - \tfrac{1}{2}\right) \otimes A^+\left(-n_2 - \tfrac{1}{2}\right),$$

$$\chi^*(n_1, n_2) = -\sum_{j=1}^{N'} a^*_j\left(-n_1 - \tfrac{1}{2}\right)a^*_{j+N'}\left(-n_2 - \tfrac{1}{2}\right)$$
$$+ \sum_{j=1}^{N'} a^*_{j+N'}\left(-n_1 - \tfrac{1}{2}\right)a^*_j\left(-n_2 - \tfrac{1}{2}\right) \in A^-\left(-n_1 - \tfrac{1}{2}\right) \otimes A^-\left(-n_2 - \tfrac{1}{2}\right).$$

These elements are nilpotent: we have $\chi(n_1, n_2)^{N+1} = 0$ and $\chi^*(n_1, n_2)^{N+1} = 0$.

The relationships among these elements in terms of the symplectic structure map J are given in the next lemma.

LEMMA 11. *Let* $J : A \to A$ *be the symplectic map given by* (25). *Then the following relations hold among elements* (17), (21), (26), *and* (27) *in* $\mathcal{A}$:

(28)
$$(1 \otimes J)h(-n - \tfrac{1}{2}) = 2x(-n - \tfrac{1}{2}), \qquad (J \otimes 1)h(-n - \tfrac{1}{2}) = -2x^*(-n - \tfrac{1}{2}),$$
$$(J \otimes J)h(-n - \tfrac{1}{2}) = -h(-n - \tfrac{1}{2}), \qquad (J \otimes J)\omega_+(n_1, n_2) = \omega_-(n_1, n_2),$$
$$(1 \otimes J)\omega_+(n_1, n_2) = \chi(n_1, n_2), \qquad (J \otimes 1)\omega_+(n_1, n_2) = -\chi^*(n_1, n_2).$$

PROOF. By direct calculation using (25). $\square$

We let $M_{\mathfrak{sp}(N')}$ be the subalgebra of $\mathcal{A}$ generated by the following elements:

$$
\text{(29)} \quad
\begin{aligned}
M_{\mathfrak{sp}(N')} = \mathbb{C}\{ &\omega_\pm(n_1, n_2), \chi(n_1, n_2), \chi^*(n_1, n_2), x(-n - \tfrac{1}{2}), x^*(-n - \tfrac{1}{2}), \\
&h(-n - \tfrac{1}{2}) \mid n_1 > n_2 \geq 0, n \geq 0\}.
\end{aligned}
$$

PROPOSITION 12 ($\mathfrak{sp}(N')$-INVARIANT VECTORS). *Let* $2N' = N$. *The following vectors in* V *are* $\mathfrak{sp}(N')$-*invariant*:

$$
[\omega_\pm(n_1, n_2)], \ [\chi(n_1, n_2)], \ [\chi^*(n_1, n_2)], \ [x(-n - \tfrac{1}{2})], \ [x^*(-n - \tfrac{1}{2})], \ [h(-n - \tfrac{1}{2})].
$$

Thus $[M_{\mathfrak{sp}(N')}] \subset (V)^{\mathfrak{sp}(N')}$.
   There exist the following relations among these elements:

$$
\text{(30)} \quad
\begin{aligned}
\frac{1}{N!}\chi(n_1, n_2)^N &= (-1)^{N'} \rho_+(-n_1 - \tfrac{1}{2})\rho_+(-n_2 - \tfrac{1}{2}), \\
\frac{1}{N!}\chi^*(n_1, n_2)^N &= (-1)^{N'} \rho_-(-n_1 - \tfrac{1}{2})\rho_-(-n_2 - \tfrac{1}{2}), \\
\frac{1}{(N!)}x(-n - \tfrac{1}{2})^{N'} &= (-1)^{\frac{N'(N'-1)}{2}} \rho_+(-n - \tfrac{1}{2}), \\
\frac{1}{(N!)}x^*(-n - \tfrac{1}{2})^{N'} &= (-1)^{\frac{N'(N'-1)}{2}} \rho_-(-n - \tfrac{1}{2}), \\
x(-n - \tfrac{1}{2})^{N'+1} = x(-n - \tfrac{1}{2})^{N'+1} &= 0, \quad \chi(n_1, n_2)^{N+1} = \chi^*(n_1, n_2)^{N+1} = 0.
\end{aligned}
$$

$M_{\mathfrak{sp}(N')}$ *generates the following subspace of* $V^{\mathfrak{sp}(N')}$:

$$
\text{(31)} \quad A\left(\{M_{\mathfrak{sp}(N')}\}[M_{\mathfrak{sp}(N')}]\right) \subset V^{\mathfrak{sp}(N')}.
$$

PROOF. Under the action of the Lie algebra $\mathfrak{sp}(N') \subset \mathfrak{u}(N)$, A splits into a direct sum of two irreducible self-dual isomorphic representations A^+ and A^-. Thus, there are exactly one 1-dimensional invariant subspace in each of $A^+ \otimes A^+$, $A^+ \otimes A^-$, $A^- \otimes A^-$ corresponding to the invariant nondegenerate pairing $\langle \, , \, \rangle$ of A. So, invariant subspaces in $A^+(-n_1 - \tfrac{1}{2}) \wedge A^-(-n_2 - \tfrac{1}{2})$, $A^-(-n_1 - \tfrac{1}{2}) \wedge A^+(-n_2 - \tfrac{1}{2})$, and $A^+(-n - \tfrac{1}{2}) \wedge A^-(-n - \tfrac{1}{2})$ are generated by $\omega_+(n_1, n_2)$, $\omega_-(n_1, n_2)$, and $h(-n - \tfrac{1}{2})$, respectively. For invariants in other spaces, we apply $\mathfrak{sp}(N')$-equivariant map $J \otimes 1$ and $1 \otimes J$ described in (25) to the above invariants to get $\chi(n_1, n_2)$, $\chi^*(n_1, n_2)$, $x(-n - \tfrac{1}{2})$, and $x^*(-n - \tfrac{1}{2})$ as in (28). So, these vectors are also $\mathfrak{sp}(N')$-invariant.
   (30) can be checked directly in the exterior algebra $\mathcal{A}$. (31) is a consequence of Corollary 8. $\square$

Quaternionic Case: $\mathfrak{sp}(N') \oplus \mathfrak{sp}(1)$, $N = 2N'$. Now we consider the case $\mathfrak{g} = \mathfrak{sp}(N') \oplus \mathfrak{sp}(1) \subset \mathfrak{o}(2N)$. The second Lie algebra $\mathfrak{sp}(1) \subset \mathfrak{o}(2N)$ is the commutant of $\mathfrak{sp}(N')$ in $\mathfrak{o}(2N)$: $\mathfrak{sp}(1) = \mathfrak{o}(2N)^{\mathfrak{sp}(N')}$. Since we have considered $\mathfrak{sp}(N')$-invariants above, we consider those $\mathfrak{sp}(N')$-invariant vectors which are also invariant under the action of the Lie algebra $\mathfrak{sp}(1) \otimes \mathbb{C} = \mathbb{C}\{h\}_0 \oplus \mathbb{C}\{x\}_0 \oplus \mathbb{C}\{x^*\}_0$ spanned by 0-modes of vertex operators associated to $\mathfrak{sp}(N')$-invariant vectors $h, x, x^* \in (V)_1$. These vertex operators satisfy the following commutation relations [§3.3 Lemma 1]:

$$[\{h\}_0, \{x\}_0] = 2\{x\}_0, \quad [\{h\}_0, \{x^*\}_0] = -2\{x^*\}_0, \quad [\{x\}_0, \{x^*\}_0] = -\{h\}_0.$$

We recall that the quadratic Casimir element for the $\mathfrak{sp}(1)$ above lives in the universal enveloping algebra and is given by $\phi = \frac{1}{2}\{h\}_0^2 - 2\{x\}_0\{x^*\}_0$ up to a constant multiple [§3.3 Proposition 3]. This element commutes with all the elements in the enveloping algebra of $\mathfrak{sp}(N') \otimes \mathbb{C}$ and of $\mathfrak{sp}(1) \otimes \mathbb{C}$.

In our present context, we consider the following elements for each $n \geq 0$:

$$(32) \qquad \phi(-n - \tfrac{1}{2}) = \tfrac{1}{2}h^2(-n - \tfrac{1}{2}) - 2x(-n - \tfrac{1}{2})x^*(-n - \tfrac{1}{2})$$
$$\in \bigwedge\nolimits^4 A(-n - \tfrac{1}{2}).$$

These elements correspond to the above Casimir element ϕ. Let

$$(33) \qquad M_{\mathfrak{sp}(N')\oplus\mathfrak{sp}(1)} = \mathbb{C}\{\omega(n_1, n_2), \rho(-n - \tfrac{1}{2}), \phi(-n - \tfrac{1}{2})\} \subset \mathcal{A}$$

be a subalgebra of $\mathcal{A}$ generated by elements in (12), (13), and (32).

PROPOSITION 13 ($\mathfrak{sp}(N') \oplus \mathfrak{sp}(1)$-INVARIANT VECTORS). *For any integers $n_1 > n_2 \geq 0$ and $n \geq 0$, vectors $[\omega(n_1, n_2)]$, $[\rho(-n - \tfrac{1}{2})]$, and $[\phi(-n - \tfrac{1}{2})]$ are $\mathfrak{sp}(N') \oplus \mathfrak{sp}(1)$-invariant. Thus, $[M_{\mathfrak{sp}(N')\oplus\mathfrak{sp}(1)}] \subset (V)^{\mathfrak{sp}(N')\oplus\mathfrak{sp}(1)}$. The subalgebra $M_{\mathfrak{sp}(N')\oplus\mathfrak{sp}(1)} \subset \mathcal{A}$ generates the following subspace of $\mathfrak{sp}(N') \oplus \mathfrak{sp}(1)$-invariants:*

$$(34) \qquad A\left(\{M_{\mathfrak{sp}(N')\oplus\mathfrak{sp}(1)}\}[M_{\mathfrak{sp}(N')\oplus\mathfrak{sp}(1)}]\right) \subset V^{\mathfrak{sp}(N')\oplus\mathfrak{sp}(1)}.$$

PROOF. Since vectors $\omega(n_1, n_2)$ and $\rho(-n - \tfrac{1}{2})$ are $\mathfrak{o}(2N)$-invariant by Proposition 9, they are also $\mathfrak{sp}(N') \oplus \mathfrak{sp}(1)$-invariant. Hence all we need to demonstrate is the invariance of $\phi(-n - \tfrac{1}{2})$ under the Lie algebra $\mathfrak{sp}(1)$ generated by the above vertex operators. If K is a 3-dimensional irreducible representation of $\mathfrak{sp}(1)$, then, there is a unique $\mathfrak{sp}(1)$-isomorphism $K \rightarrow \mathfrak{sp}(1) \otimes \mathbb{C}$ up to a constant multiple: the one which maps an h-highest weight vector to x. By Poincaré-Birkhoff-Witt Theorem, the symmetric algebra $S^*(K)$ on K is isomorphic to the associated graded algebra of the universal enveloping algebra of $\mathfrak{sp}(1) \otimes \mathbb{C}$ as representations of $\mathfrak{sp}(1)$. The element in $S^2(K)$ corresponding to the Casimir element ϕ is, therefore, invariant under $\mathfrak{sp}(1)$. Now let $K = \mathbb{C}h(-n - \tfrac{1}{2}) \oplus \mathbb{C}x(-n - \tfrac{1}{2}) \oplus \mathbb{C}x^*(-n - \tfrac{1}{2})$. The element corresponding to the Casimir element ϕ is exactly $\phi(-n - \tfrac{1}{2})$. Thus, $\phi(-n - \tfrac{1}{2})$ is $\mathfrak{sp}(1)$-invariant. Since $h(-n - \tfrac{1}{2})$, $x(-n - \tfrac{1}{2})$, $x^*(-n - \tfrac{1}{2})$ are all $\mathfrak{sp}(N')$-invariant by Proposition 12, $\phi(-n - \tfrac{1}{2})$ is $\mathfrak{sp}(N') \oplus \mathfrak{sp}(1)$-invariant. $\square$

Maximal Torus Case: t^N. The Cartan subalgebra $t^N \subset o(2N)$ is given by $t^N = \bigoplus_{j=1}^{N} \mathbb{C} : a_j a_j^* :$. For integers $n_1 > n_2 \geq 0$, $n \geq 0$, and $1 \leq j \leq N$, we let

$$
\begin{aligned}
\omega_j(n_1, n_2) &= -\tfrac{1}{2}\{a_j(-n_1 - \tfrac{1}{2})a_j^*(-n_2 - \tfrac{1}{2}) + a_j^*(-n_1 - \tfrac{1}{2})a_j(-n_2 - \tfrac{1}{2})\} \\
&\in A(-n_1 - \tfrac{1}{2}) \wedge A(-n_2 - \tfrac{1}{2}),
\end{aligned}
$$

(35)
$$
\begin{aligned}
\delta_j(n_1, n_2) &= -a_j(-n_1 - \tfrac{1}{2})a_j^*(-n_2 - \tfrac{1}{2}) + a_j^*(-n_1 - \tfrac{1}{2})a_j(-n_2 - \tfrac{1}{2}) \\
&\in A(-n_1 - \tfrac{1}{2}) \wedge A(-n_2 - \tfrac{1}{2}),
\end{aligned}
$$
$$
\begin{aligned}
h_j(-n - \tfrac{1}{2}) &= a_j(-n - \tfrac{1}{2})a_j^*(-n - \tfrac{1}{2}) \\
&\in A^+(-n - \tfrac{1}{2}) \wedge A^-(-n - \tfrac{1}{2}).
\end{aligned}
$$

Since these elements are in the exterior algebra A, simple calculations show that there are following relations among these elements:

(36) $\quad h_j(-n - \tfrac{1}{2})^2 = 0, \quad \omega_j(n_1, n_2)\delta_j(n_1, n_2) = 0, \quad \delta_j(n_1, n_2)^2 = 4\omega_j(n_1, n_2)^2.$

In terms of these elements, let $(M_{t^N} \subset A$ be a subalgebra given by
(37)
$$
M_{t^N} = \mathbb{C}\{\omega_j(n_1, n_2), \delta(n_1, n_2), h_j(-n - \tfrac{1}{2}) \mid n_1 > n_2 \geq 0, n \geq 0, 1 \leq j \leq N\}.
$$

PROPOSITION 14 (t^N-INVARIANT VECTORS IN V). *For $1 \leq j \leq N$ and for integers $n_1 > n_2 \geq 0$ and $n \geq 0$, vectors $[\omega_j(n_1, n_2)]$, $[\delta_j(n_1, n_2)]$, $[h_j(-n - \tfrac{1}{2})]$ are t^N-invariant. Thus $[M_{t^N}] \subset (V)^{t^N}$, and the subalgebra $M_{t^N} \subset A$ generates the following t^N-invariant subspace of V:*

(38)
$$
A(\{M_{t^N}\}[M_{t^N}]) \subset V^{t^N}.
$$

PROOF. Under the action of t^N, we have $(\bigwedge^2 A)^{t^N} = \bigoplus_{j=1}^{N} \mathbb{C} a_j a_j^*$. From this, invariance of $h_j(-n - \tfrac{1}{2})$'s follow. In $A(-n_1 - \tfrac{1}{2}) \wedge A(-n_2 - \tfrac{1}{2})$, the invariant subspace is given by

$$
\bigoplus_{j=1}^{N} \mathbb{C} a_j(-n_1 - \tfrac{1}{2})a_j^*(-n_2 - \tfrac{1}{2}) \oplus \bigoplus_{j=1}^{N} \mathbb{C} a_j^*(-n_1 - \tfrac{1}{2})a_j(-n_2 - \tfrac{1}{2}).
$$

By considering suitable linear combinations, we see that $\omega_j(n_1, n_2)$, $\delta_j(n_1, n_2)$ are t^N-invariant.

(35) is a consequence of Corollary 8. $\square$

Note that elements in (35) are related to other elements by

(39)
$$
\omega(n_1, n_2) = \sum_{j=1}^{N} \omega_j(n_1, n_2), \qquad \kappa(n_1, n_2) = i\sum_{j=1}^{N} \delta_j(n_1, n_2),
$$
$$
h(-n - \tfrac{1}{2}) = \sum_{j=1}^{N} h_j(-n - \tfrac{1}{2}).
$$

§3.5   Vertex operators associated to G-invariant vectors of low weight

Some vectors in V of low weight are of particular interest: vertex operators associated to vectors of weight 1/2 generate Clifford algebras, vertex operators for vectors of weight 1 generate affine Lie algebras, and the vertex operator associated to the canonical Virasoro element $[\omega]$ of weight 2 generates the Virasoro algebra [§2.3]. Below is a list of vectors of low weight in $\mathcal{A}$ which we encountered in §3.4, with simplified names:

(1)
$$\omega(1,0) = \omega, \qquad \omega_\pm(1,0) = \omega_\pm, \qquad \chi(1,0) = \chi, \qquad \chi^*(1,0) = \chi^*,$$
$$h(-\tfrac{1}{2}) = h, \qquad x(-\tfrac{1}{2}) = x, \qquad x^*(-\tfrac{1}{2}) = x^*,$$
$$\omega_j(1,0) = \omega_j \qquad \delta_j(1,0) = \delta_j, \qquad h_j(-\tfrac{1}{2}) = h_j.$$

Note that $[\omega], [\omega_\pm], [\chi], [\chi^*], [\omega_j], [\delta_j] \in (V)_2$ and $[h], [h_j], [x], [x^*] \in (V)_1$. We recall their explicit expressions in the exterior algebra $\mathcal{A}$:

(2)
$$\omega = -\tfrac{1}{2}\sum_{j=1}^{N}\left\{a_j(-\tfrac{3}{2})a_j^*(-\tfrac{1}{2}) + a_j^*(-\tfrac{3}{2})a_j(-\tfrac{1}{2})\right\},$$

$$\omega_+ = -\sum_{j=1}^{N}a_j(-\tfrac{3}{2})a_j^*(-\tfrac{1}{2}), \qquad \omega_- = -\sum_{j=1}^{N}a_j^*(-\tfrac{3}{2})a_j(-\tfrac{1}{2}),$$

$$\chi = -\sum_{j=1}^{N'}a_j(-\tfrac{3}{2})a_{j+N'}(-\tfrac{1}{2}) + \sum_{j=1}^{N'}a_{j+N'}(-\tfrac{3}{2})a_j(-\tfrac{1}{2}),$$

$$\chi^* = -\sum_{j=1}^{N'}a_j^*(-\tfrac{3}{2})a_{j+N'}^*(-\tfrac{1}{2}) + \sum_{j=1}^{N'}a_{j+N'}^*(-\tfrac{3}{2})a_j^*(-\tfrac{1}{2}),$$

$$x = \sum_{j=1}^{N'}a_j(-\tfrac{1}{2})a_{j+N'}(-\tfrac{1}{2}), \qquad x^* = \sum_{j=1}^{N'}a_j^*(-\tfrac{1}{2})a_{j+N'}^*(-\tfrac{1}{2}),$$

$$h = \sum_{j=1}^{N}a_j(-\tfrac{1}{2})a_j^*(-\tfrac{1}{2}).$$

Virasoro operators. Next we describe relations among the above elements through Virasoro operators. Recall that Virasoro operators $\{D(n)\}_{n\in\mathbb{Z}}$ acting on V are components of the vertex operator $Y([\omega], \zeta) = \sum_{n\in\mathbb{Z}} D(n)\zeta^{-n-2}$, and their explicit expressions are given by

(3)
$$D^{\mathbb{Z}+\frac{1}{2}}(k) = \sum_{1\leq i\leq N}\sum_{n\in\mathbb{Z}+\frac{1}{2}}(n - \tfrac{1}{2}k) :a_i(n)a_i^*(k-n): \quad \text{for } k \in \mathbb{Z}.$$

Note that 0-th Virasoro operator $D(0)$ is given correctly by the above formula without any corrections [See §3.3 (10)]. To remove normal ordering, we consider three cases: $k < 0$, $k = 0$, and $k > 0$.

When $k > 0$, Virasoro operators have the following expresson without normal ordering symbols:

(4)
$$D^{\mathbb{Z}+\frac{1}{2}}(k) = -\sum_{1 \leq i \leq N} \sum_{\substack{n > 0 \\ n \in \mathbb{Z}+\frac{1}{2}}} (n - \tfrac{k}{2})\{a_i(k-n)a_i^*(n) + a_i^*(k-n)a_i(n)\}$$

$$-\sum_{1 \leq i \leq N} \sum_{\substack{\frac{k}{2} < n < 0 \\ n \in \mathbb{Z}+\frac{1}{2}}} (n - \tfrac{k}{2})\{a_i(k-n)a_i^*(n) + a_i^*(k-n)a_i(n)\}, \qquad k < 0$$

In the first summation, $k - n < 0$ and $n > 0$. In the second summation, both $k - n$ and n are negative.

When $k = 0$, we have

(5)
$$D^{\mathbb{Z}+\frac{1}{2}}(0) = -\sum_{1 \leq i \leq N} \sum_{\substack{n > 0 \\ n \in \mathbb{Z}+\frac{1}{2}}} n\{a_i(-n)a_i^*(n) + a_i^*(-n)a_i(n)\}.$$

When $k > 0$, the expression of $D(k)$ is given as follows:

(6)
$$D^{\mathbb{Z}+\frac{1}{2}}(k) = -\sum_{1 \leq i \leq N} \sum_{n > k} (n - \tfrac{k}{2})\{a_i(k-n)a_i^*(n) + a_i^*(k-n)a_i(n)\}$$

$$-\sum_{1 \leq i \leq N} \sum_{\frac{k}{2} < n < k} (n - \tfrac{k}{2})\{a_i(k-n)a_i^*(n) + a_i^*(k-n)a_i(n)\}, \qquad k > 0$$

PROPOSITION 1. *Virasoro operators $D(-1)$ and $D(-2)$ act on vectors in (2) in the following way*:

$$D(-2)\Omega = [\omega] = \tfrac{1}{2}([\omega_+] + [\omega_-]), \qquad D(-1)[\omega] = -2[\omega(2,0)],$$

$$D(-1)[\omega_+] = -2[\omega_+(2,0)] + [h(-\tfrac{3}{2})],$$

$$D(-1)[\omega_-] = -2[\omega_-(2,0)] - [h(-\tfrac{3}{2})],$$

(7)
$$D(-1)[\omega_+ - \omega_-] = 2[h(-\tfrac{3}{2})] - 2[\omega_+(2,0) - \omega_-(2,0)],$$

$$D(-1)[\chi] = -2[\chi(2,0)], \qquad D(-1)[\chi^*] = -2[\chi^*(2,0)],$$

$$D(-1)[h] = [\omega_+] - [\omega_-], \qquad D(-1)[h_j] = [\delta_j], \quad 1 \leq j \leq N,$$

$$D(-1)[x] = [\chi], \qquad D(-1)[x^*] = [\chi^*].$$

PROOF. Since elements in (2) are constructed from vectors in $A(-\tfrac{3}{2})$ and in $A(-\tfrac{1}{2})$, in the expression of $D(k)$ those terms with vectors from $A(n + \tfrac{1}{2})$ with $n \geq 2$ act trivially on elements in (2). The relevant terms for $D(-1)$ are

$$D(-1) = -\sum_{j=1}^{N}(\cdots + 2a_j(-\tfrac{5}{2})a_j^*(\tfrac{3}{2}) + a_j(-\tfrac{3}{2})a_j^*(\tfrac{1}{2})$$
$$+ a_j^*(-\tfrac{3}{2})a_j(\tfrac{1}{2}) + 2a_j^*(-\tfrac{5}{2})a_j(\tfrac{3}{2}) + \cdots).$$

Thus, applying this operator to $[\omega_+]$, we get

$$D(-1)[\omega_+] = D(-1)(-\sum_{1 \leq i \leq N} a_i(-\tfrac{3}{2})a_i^*(-\tfrac{1}{2})\Omega)$$

$$= \sum_i \{2a_i(-\tfrac{5}{2})a_i^*(-\tfrac{1}{2}) - a_i^*(-\tfrac{3}{2})a_i(-\tfrac{3}{2})\}\Omega = -2[\omega_+(2,0)] + [h(-\tfrac{3}{2})].$$

Taking complex conjugation $*$ and using $D(-1)^* = D(-1)$, $h^* = -h$ and $\omega_+^* = \omega_-$, we get the corresponding formula $D(-1)[\omega_-] = -[h(-\tfrac{3}{2})] - 2[\omega_-(2,0)]$. From these, we have formulae for $D(-1)[\omega]$ and for $D(-1)[\omega_+ - \omega_-]$ in (7). Similarly

$$D(-1)[\chi] = 2\sum_{j=1}^{N'}\{a_j(-\tfrac{5}{2})a_{j+N'}(-\tfrac{1}{2}) - a_{j+N'}(-\tfrac{5}{2})a_j(-\tfrac{1}{2})\}\Omega$$

$$- \sum_{j=1}^{N'}\{a_{j+N'}(-\tfrac{3}{2})a_j(-\tfrac{3}{2}) + a_j(-\tfrac{3}{2})a_{j+N'}(-\tfrac{3}{2})\}\Omega = -2[\chi(2,0)].$$

Again applying $*$, we get $D(-1)[\chi^*] = -2[\chi^*(2,0)]$.

For the remaining four identities we only have to use the middle two terms in the above explicit expression of $D(-1)$ because other terms annihilate vectors of weight 1 in V, and we get

$$D(-1)[h] = D(-1)(\sum_{j=1}^{N} a_j(-\tfrac{1}{2})a_j^*(-\tfrac{1}{2})\Omega)$$

$$= -\sum_{j=1}^{N} a_j(-\tfrac{3}{2})a_j^*(-\tfrac{1}{2})\Omega + \sum_{j=1}^{N} a_j^*(-\tfrac{3}{2})a_j(-\tfrac{1}{2})\Omega = [\omega_+] - [\omega_-].$$

Similarly, we have $D(-1)[h_j] = [\delta_j]$ for $1 \leq j \leq N$. Furthermore,

$$D(-1)[x] = D(-1)(\sum_{j=1}^{N'} a_j(-\tfrac{1}{2})a_{j+N'}(-\tfrac{1}{2})\Omega)$$

$$= -\sum_{j=1}^{N'} a_j(-\tfrac{3}{2})a_{j+N'}(-\tfrac{1}{2})\Omega + \sum_{j=1}^{N'} a_{j+N'}(-\tfrac{3}{2})a_j(-\tfrac{1}{2})\Omega = \chi.$$

Taking complex conjugation $*$ of this identity, we also have $D(-1)[x^*] = [\chi^*]$. $\square$

Among elements in Proposition 1, $[\omega]$ generates the Virasoro algebra. To calculate vertex operators generated by other elements, we recall the identity $Y(D(-1)v, \zeta) = -\dfrac{d}{d\zeta}Y(v, \zeta)$ [§2.1 (9)]. From this we have

$$\{D(-1)v\}_{n+1} = (n+1)\{v\}_n$$

for $n \in \mathbb{Z}$. This implies the following corollary of Proposition 1:

COROLLARY 2. *Vertex operators generated by* $[\omega_+ - \omega_-]$, $[\chi]$, $[\chi^*]$ *and* $[\delta_j]$ *are given as follows:*

(8)
$$\{\omega_+ - \omega_-\}_{n+1} = (n+1)h(n), \qquad \{\delta_j\}_{n+1} = (n+1)h_j(n),$$
$$\{\chi\}_{n+1} = (n+1)x(n), \qquad \{\chi^*\}_{n+1} = (n+1)x^*(n).$$

This describes vertex operators associated to some vectors of weight 2 in V. To calculate vertex operators for other vectors, we prepare a few lemmas.

LEMMA 3. *Let* $a(n) \in \mathrm{Cliff}\,(\mathbb{Z} + \frac{1}{2})$ *and* $x = b_1(-n_1)b_2(-n_2)\cdots b_r(-n_r) \in \mathcal{A}$ *for some vectors* a, b_1, b_2, $\ldots$, $b_r \in A$ *and half integers* n, $n_1, \ldots$, $n_r \in \mathbb{N} - \frac{1}{2}$. *Then the graded commutator* $[a(n), x]_{\pm} = a(n)\, x - (-1)^r x\, a(n)$ *in the Clifford algebra* $\mathrm{Cliff}\,(\mathbb{Z} + \frac{1}{2})$ *is given by*

$$[a(n), x]_{\pm} = \sum_{j=1}^{r} \delta_{n,n_j}\langle a, b_j\rangle b_1(-n_1)\cdots \widehat{b_j(-n_j)} \cdots b_r(-n_r) \in \mathcal{A}.$$

PROOF. This is straightforward using the Clifford relations in $\mathrm{Cliff}\,(\mathbb{Z} + \frac{1}{2})$: $[a(n), b_j(-n_j)]_{+} = \langle a, b_j\rangle \delta_{n,n_j}$ for $1 \leq j \leq r$. $\square$

Let **Vir**$^-$ be a subspace of the Virasoro algebra **Vir** spanned by "negative half" of Virasoro operators:

(9)
$$\mathbf{Vir}^- = \bigoplus_{k \leq 1} \mathbb{C}D(k) \subset \mathbf{Vir}.$$

This is a Lie subalgebra of the Virasoro algebra, which can be easily seen from the commutation relation of the Virasoro algebra given in §2.1 (8).

PROPOSITION 4 (ACTION OF **Vir**$^-$ ON $\mathcal{A}$). *For any* $k \in \mathbb{Z}$, $k \leq 1$, *the map*

(10)
$$\mathrm{ad}\, D(k) : \mathcal{A} \to \mathcal{A} \subset \mathrm{Cliff}\,(\mathbb{Z} + \tfrac{1}{2})$$

defined by $(\mathrm{ad}\, D(k))(x) = [D(k), x]$ *preserves the subalgebra* $\mathcal{A}$, *and* $\mathcal{A}$ *is a representation of the Lie algebra* **Vir**$^-$.

PROOF. The restriction $k \leq 1$ implies that $D(k)$ is an infinite sum of elements of the form $a(m)b(n)$, where either $m < 0$, $n < 0$ or $m < 0$, $n > 0$. This is clear from (4), (5), (6) above. If both m, n are negative, then $[a(m)b(n), x] = 0$ for any $x \in \mathcal{A}$ because $\mathcal{A}$ is an exterior subalgebra of the Clifford algebra $\mathrm{Cliff}\,(\mathbb{Z} + \frac{1}{2})$ generated by elements of non-negative weight. If $m < 0$, $n > 0$, then $[a(m)b(n), x] = a(m)[b(n), x]_{\pm} \in \mathcal{A}$ by Lemma 3. Thus $[D(k), x] \in \mathcal{A}$ whenever $x \in \mathcal{A}$ and $k \leq 1$. $\square$

Virasoro operators $D(-1)$, $D(0)$, and $D(1)$ are rather special because they annihilate the vacuum vector $\Omega \in V$, and they are closed under bracket products forming a copy of $\mathfrak{sl}_2(\mathbb{C})$. This fact implies the following lemma:

LEMMA 5. *Let* $k = 0, \pm 1$. *Suppose in the vertex operator super algebra* V *we have* $D(k)[x] = [y]$ *for some* $x, y \in \mathcal{A}$. *Then in the algebra* $\mathcal{A}$, $[D(k), x] = y \in \mathcal{A}$.

PROOF. Since $D(k)\Omega = 0 \in V$ for $k = 0, \pm 1$, we have $[D(k), x]\Omega = D(k)x\Omega - xD(k)\Omega = D(k)[x] = [y] = y\Omega$. Since $[D(k), x] \in \mathcal{A}$ due to Proposition 4 and since the map $[\ ] = \Omega : \mathcal{A} \to V$ is an isomorphism, we must have $[D(k), x] = y$. $\square$

Now we state our formula for the action the subalgebra $\mathfrak{sl}_2(\mathbb{C})$ spanned by Virasoro operators $D(-1)$, $D(0)$, and $D(1)$ on product elements in V.

PROPOSITION 6 (ACTION OF $\mathfrak{sl}_2(\mathbb{C})$ ON V). *Let* $k = 0, \pm 1$. *Suppose we have* $D(k)[x_\ell] = [y_\ell]$ *for* $x_\ell, y_\ell \in \mathcal{A}$, $1 \leq \ell \leq r$. *Then* $D(k)$ *acts as a derivation on* $\mathcal{A}$:

$$(11) \qquad D(k)[x_1 x_2 \cdots x_r] = \sum_{\ell=1}^{r} [x_1 \cdots x_{\ell-1} y_\ell x_{\ell+1} \cdots x_r].$$

PROOF. By Lemma 5, $D(k)[x_\ell] = [y_\ell]$ implies $[D(k), x_\ell] = y_\ell \in \mathcal{A}$. Then

$$D(k)[x_1 x_2 \cdots x_r] = D(k)x_1 x_2 \cdots x_r \Omega = \sum_{\ell=1}^{r} x_1 \cdots x_{\ell-1}[D(k), x_\ell]x_{\ell+1} \cdots x_r \Omega$$
$$= \sum_{\ell=1}^{r} x_1 \cdots x_{\ell-1} y_\ell x_{\ell+1} \cdots x_r \Omega = \sum_{\ell=1}^{r} [x_1 \cdots x_{\ell-1} y_\ell x_{\ell+1} \cdots x_r].$$

This proves our formula. $\square$

Proposition 6 allows us to calculate the effect of $D(-1)$ on product elements in V if we know the effect of $D(-1)$ on individual elements. Applying Proposition 6 to results of Proposition 1, we immediately obtain the next corollary.

COROLLARY 7. *We have the following identities:*

$$(12)$$
$$D(-1)[\tfrac{1}{2}h^2] = [h(\omega_+ - \omega_-)], \qquad\qquad D(-1)[xx^*] = [\chi x^*] + [x\chi^*],$$
$$D(-1)[\tfrac{1}{2}x^2] = [x\chi], \qquad\qquad D(-1)[\tfrac{1}{2}(x^*)^2] = [x^*\chi^*],$$
$$D(-1)[hx] = [x(\omega_+ - \omega_-)] + [h\chi], \qquad D(-1)[hx^*] = [x^*(\omega_+ - \omega_-)] + [h\chi^*].$$

Further application of Proposition 6 allows us to describe the action of $D(-1)$ on the subspace of V corresponding to the subalgebra $\mathbb{C}\{x, h, x^*\} \subset \bigwedge^{\text{even}} A(-\tfrac{1}{2})$ generated by x, h, and x^*.

Next we discuss the action of $D(k)$ for positive k's. It turns out that the operator $D(k)$ for any positive k acts trivially on the subspace of V corresponding to the subalgebra $\bigwedge^* A(-\tfrac{1}{2})$ of $\mathcal{A}$.

PROPOSITION 8 (CONFORMAL HIGHEST WEIGHT VECTORS). *The subspace* $[\bigwedge^* A(-\tfrac{1}{2})] \subset V$ *consists of conformal highest weight vectors. Namely, for any* $x \in \bigwedge^* A(-\tfrac{1}{2})$, *we have* $D(k)[x] = 0 \in V$ *for any* $k \geq 1$.

PROOF. First observe that any vector $x \in \bigwedge^* A(-\tfrac{1}{2})$ is a product of vectors in $A(-\tfrac{1}{2})$. To prove this proposition, we simply note that in the expression of $D(k)$ for a positive k given in (6), each term has a vector from $A(n + \tfrac{1}{2})$ with $n \geq 1$ which annihilates $[x]$. So, $D(k)$ annihilates $[x]$ for any $k \geq 1$. $\square$

ALTERNATE PROOF OF PROPOSITION 8. We could argue differently as follows. The vector space $V_{\frac{1}{2}} = [A(-\frac{1}{2})]$ of weight $\frac{1}{2}$ is annihilated by Virasoro operators $D(k)$ with $k \geq 1$ by weight reason. Then Proposition 6 applied to products of weight $\frac{1}{2}$ vectors show that such vectors are always annihilated by $D(k)$ for all $k \geq 1$. So, all the vectors in $\wedge^* A(-\frac{1}{2})$ are conformal highest weight vectors. $\square$

In general, vertex operators associated to conformal highest weight vectors have simple commutation relations with Virasoro generators.

PROPOSITION 9. Let $[x] \in (V)_r$ be a conformal highest weight vector, that is, $D(k)[x] = 0$ for all $k \geq 1$. Let $Y([x], \zeta) = \sum_{n \in \mathbb{Z}+r} x(n) \zeta^{-n-r}$ be its vertex operator, where $x(n) = \{x\}_{n+r-1}$ lowers weight by n. Then

(i) $x(n)$'s have the following commutation relation with Virasoro operators:

(13) $[D(m), x(n)] = (n - (r-1)m)x(m+n)$ for $m \in \mathbb{Z}, n \in \mathbb{Z}+r$.

(ii) $x(n)$'s have the following effect on the vacuum vector Ω:

(14) $x(n)\Omega = 0$ for all $n > -r$ with $n \in \mathbb{Z}+r$,   and $x(-r)\Omega = [x]$.

PROOF. We use the commutator formula for vertex operators. Since the Virasoro operator $D(m)$ is given by $\{\omega\}_{n+1}$, we have

$$[D(m), x(n)] = [\{\omega\}_{m+1}, \{x\}_{n+r-1}]$$
$$= \sum_{0 \leq k \in \mathbb{Z}} \binom{m+1}{k} \{\{\omega\}_k[x]\}_{m+n+r-k}$$
$$= \{D(-1)[x]\}_{m+n+r} + (m+1)\{D(0)[x]\}_{m+n+r-1}$$
$$= (m+n+r)\{x\}_{m+n+r-1} - r(m+1)\{x\}_{m+n+r-1}$$
$$= (n+m-rm)x(m+n).$$

This proves (13). For (ii), from the reproducing property of vertex operators, we have $\lim_{\zeta \to 0} Y([x], \zeta)\Omega = [x]$, which means that $x(-r)\Omega = [x]$, where $[x] \in (V)_r$. Now we let $n = -r$ in (13) and apply it to Ω. We get $D(m)x(-r)\Omega - x(-r)D(m)\Omega = -(r+m(r-1))x(m-r)\Omega$. The left hand side is 0 for $m > 0$ because $[x]$ and Ω are conformal highest weight vectors. Except for the case in which $r = \frac{1}{2}$ and $m = 1$, the coefficient of $x(m-r)$ in the right hand side is nonzero for any $r \in \frac{1}{2}\mathbb{Z}_+$ and any $m \in \mathbb{N}$, and it follows that $x(m-r)\Omega = 0$ in this case. When $r = \frac{1}{2}$ and $m = 1$, we have $x(m-r)\Omega = x(\frac{1}{2})\Omega = 0$ by weight reason. This proves (14). $\square$

We can combine Proposition 8 and Proposition 9 to obtain the next corollary.

COROLLARY 10. Let $Y([x], \zeta) = \sum_{n \in \mathbb{Z}+r} x(n)\zeta^{-n-r}$ be the vertex operator associated to $x \in \wedge^{2r} A(-\frac{1}{2})$ with $r \in \frac{1}{2}\mathbb{Z}_+$. Then $[x] \in V_r$ is a conformal highest weight vector and we have

(15) $[D(m), x(n)] = (n - (r-1)m)\, x(m+n)$ for $m, n \in \mathbb{Z}$.

Some other commutation relations. In this subsection, we calculate the commutator $[\{h\}_m, \{h^2\}_n]$ of vertex operators associated to $h \in \mathcal{A}_1$ and $h^2 \in \mathcal{A}_2$.

PROPOSITION 11. *The action of vertex operators $\{h\}_k$ for $k \geq 0$ on $[h^2]$ is given by*

$$\{h\}_0[h^2] = 0, \qquad \{h\}_1[h^2] = 2(N-1)[h], \qquad \{h\}_k[h^2] = 0, \quad k \geq 2.$$

The commutator between vertex operators $\{h\}_m$ and $\{h^2\}_n$ is given by

$$(16) \qquad [\{h\}_m, \{h^2\}_{n+1}] = 2m(N-1)\{h\}_{m+n} \quad \text{for } m, n \in \mathbb{Z}.$$

PROOF. Since $h = \sum_{j=1}^{N} a_j(-\frac{1}{2})a_j^*(-\frac{1}{2}) \in \mathcal{A}_1$ from (2), we have

$$\{h\}_k = \sum_{j=1}^{N} \Big\{ \sum_{m_1+m_2=k-1} : a_j(m_1 + \tfrac{1}{2})a_j^*(m_2 + \tfrac{1}{2}): \Big\},$$

$$[h^2] = \sum_{1 \leq j \neq k \leq N} a_j(-\tfrac{1}{2})a_j^*(-\tfrac{1}{2})a_k(-\tfrac{1}{2})a_k^*(-\tfrac{1}{2})\Omega.$$

We note that operators $a_j(m_1 + \frac{1}{2})$ and $a_j^*(m_2 + \frac{1}{2})$ annihilate $[h^2]$ when $m_1 \geq 1$ and $m_2 \geq 1$. Thus for nontriviality of the action, we need $m_1 \leq 0$ and $m_2 \leq 0$, which implies $m_1 + m_2 \leq 0$. However, when $k \geq 2$, there are no terms in the expression for $\{h\}_k$ satisfying this condition $m_1 + m_2 \leq 0$. Hence $\{h\}_k[h^2] = 0$ when $k \geq 2$. When $k = 0$, relevant terms in the expression of $\{h\}_0$ are given by

$$\{h\}_0 = \sum_{p=1}^{N} (\cdots - a_p(-\tfrac{1}{2})a_p(\tfrac{1}{2}) + a_p(-\tfrac{1}{2})a_p^*(\tfrac{1}{2}) + \cdots).$$

The action of each of these terms on $[h^2]$ is then calculated as follows:

$$\{h\}_0[h^2] = \sum_{1 \leq j \neq k \leq N} \{a_j^*(-\tfrac{1}{2})a_j(-\tfrac{1}{2})a_k(-\tfrac{1}{2})a_k^*(-\tfrac{1}{2}) + a_k^*(-\tfrac{1}{2})a_j(-\tfrac{1}{2})a_j^*(-\tfrac{1}{2})a_k(-\tfrac{1}{2})\}\Omega$$

$$+ \sum_{1 \leq j \neq k \leq N} \{a_j(-\tfrac{1}{2})a_j^*(-\tfrac{1}{2})a_k(-\tfrac{1}{2})a_k^*(-\tfrac{1}{2}) + a_k(-\tfrac{1}{2})a_j(-\tfrac{1}{2})a_j^*(-\tfrac{1}{2})a_k^*(-\tfrac{1}{2})\}\Omega = 0.$$

For $\{h\}_1$, relevant terms are $\{h\}_1 = \sum_{p=1}^{N}(\cdots + a_p(\tfrac{1}{2})a_p(\tfrac{1}{2}) + \cdots)$. Its action on $[h^2]$ is given by

$$\{h\}_1[h^2] = \sum_{1 \leq j \neq k \leq N} \{a_k(-\tfrac{1}{2})a_k^*(-\tfrac{1}{2}) + a_j(-\tfrac{1}{2})a_j^*(-\tfrac{1}{2})\}\Omega$$

$$= 2 \sum_{1 \leq j \neq k \leq N} a_k(-\tfrac{1}{2})a_k^*(-\tfrac{1}{2})\Omega = 2(N-1)[h].$$

This proves the formulae for the action of $\{h\}_k$ on $[h^2]$ for $k \geq 0$. For the commutation relation (16) between $\{h\}_m$ and $\{h^2\}_n$, we apply the commutator formula and we use the above result on $\{h\}_k[h^2]$ for $k \geq 0$. We have

$$[\{h\}_m, \{h^2\}_{n+1}] = \sum_{0 \leq k \in \mathbb{Z}} \binom{m}{k} \{\{h\}_k[h^2]\}_{m+n-k} = 2m(N-1)\{h\}_{m+n}.$$

This completes the proof. □

The above commutator formula will be used to study algebras arising from Kähler forms which act on elliptic genera for Kähler manifolds.

§3.6 Unitary Virasoro algebras

For the unitary Lie algebra $\mathfrak{u}(N) \subset \mathfrak{o}(2N)$, the $\mathfrak{u}(N)$-invariant subspace of weight 1 in the vertex operator super algebra V is 1-dimensional spanned by h, and a $\mathfrak{u}(N)$-invariant unitary Casimir element $[\phi_{\mathfrak{u}(N)}] \in (V)_2^{\mathfrak{u}(N)}$ is given by $[\phi_{\mathfrak{u}(N)}] = -[h^2]$ [§3.4 Proposition 10]. We show that vertex operators associated to the canonical Virasoro element $[\omega]$ and the unitary Casimir element $[h^2]$, together with a 1-dimensional center, close under Lie bracket [Theorem 1 below]. Let these vertex operators be
(1)
$$Y([\omega], \zeta) = \sum_{n \in \mathbb{Z}} D(n)\zeta^{-n-2} \quad \text{and} \quad Y([h^2], \zeta) = -Y([\phi_{\mathfrak{u}(N)}], \zeta) = \sum_{n \in \mathbb{Z}} H(n)\zeta^{-n-2}.$$

We call this Lie algebra the *unitary Virasoro algebra* denoted by $\mathbf{Vir}_{\mathfrak{u}(N)}$:

$$(2) \qquad \mathbf{Vir}_{\mathfrak{u}(N)} = \bigoplus_{m \in \mathbb{Z}} \mathbb{C}D(m) \oplus \bigoplus_{n \in \mathbb{Z}} \mathbb{C}H(n) \oplus \mathbb{C}c,$$

where c is the central element which acts as constants on its irreducible representations. The structure of $\mathbf{Vir}_{\mathfrak{u}(N)}$ is described as follows:

THEOREM 1 (UNITARY VIRASORO ALGEBRA (I)). *Let* $H(m) = \{h^2\}_{m+1}$ *and* $D(m) = \{\omega\}_{m+1}$ *for* $m \in \mathbb{Z}$. *Then the following commutation relations hold for any* $m, n \in \mathbb{Z}$:

$$(3) \qquad [D(m), D(n)] = (n-m)D(m+n) + \frac{m(m^2-1)}{12}\delta_{m+n,0} N \cdot \mathrm{Id},$$

$$(4) \qquad [D(m), H(n)] = (n-m)H(m+n),$$

$$[H(m), H(n)] = 4(N-1)(n-m)D(m+n) - 2(N-2)(n-m)H(m+n)$$
$$(5) \qquad\qquad\qquad + \frac{m(m^2-1)}{3}N(N-1)\delta_{m+n,0} \cdot \mathrm{Id}.$$

Formula (3) is the usual commutation relation for the Virasoro algebra and (4) is a consequence of Corollary 10 of §3.5. Formula (5) is a result of a long calculation which is described below. To understand the structure of the unitary Virasoro algebra, we reformulate this algebra by choosing a different set of basis vectors of the algebra to simplify commutation relations. Note that formulae (4) and (5) are "mixed": they are expressed in terms of both operators $D(m)$ and $H(m)$. However, we can choose certain linear combinations of ω and h^2 in such a way that associated vertex operators have "closed" commutation relations. It turns out that such linear combinations are unique, and they are given by

$$(6) \qquad \theta = \frac{1}{2N}(2\omega - h^2), \qquad \lambda = \frac{1}{2N}\left(2(N-1)\omega + h^2\right).$$

Note that $\theta + \lambda = \omega$, and we have decomposed the canonical Virasoro element ω into two parts using h^2. Let the corresponding vertex operators be

$$(7) \qquad Y([\theta], \zeta) = \sum_{n \in \mathbb{Z}} \Theta(n)\zeta^{-n-2}, \qquad Y([\lambda], \zeta) = \sum_{n \in \mathbb{Z}} \Lambda(n)\zeta^{-n-2},$$

$$\Theta(n) = \frac{1}{2N}\{2D(n) - H(n)\}, \quad \Lambda(n) = \frac{1}{2N}\{2(N-1)D(n) + H(n)\}.$$

Commutation relations of the unitary Virasoro algebra with respect to the above operators are now described as follows:

THEOREM 2 (UNITARY VIRASORO ALGEBRA (II)). *The unitary Virasoro algebra given in terms of operators in* (7)

$$(2') \qquad \mathbf{Vir}_{u(N)} = \bigoplus_{m \in \mathbb{Z}} \mathbb{C}\Theta(m) \oplus \bigoplus_{n \in \mathbb{Z}} \mathbb{C}\Lambda(n) \oplus \mathbb{C}c$$

is a direct sum of two commuting Virasoro algebras with central charges 1 *and* $N - 1$, *respectively, and their commutation relations are given as follows:*

$$(8) \qquad [\Theta(m), \Theta(n)] = (n - m)\Theta(m + n) + \frac{m(m^2 - 1)}{12}\delta_{m+n,0}\cdot\mathrm{Id},$$

$$(9) \qquad [\Lambda(m), \Lambda(n)] = (n - m)\Lambda(m + n) + \frac{m(m^2 - 1)}{12}\delta_{m+n,0}(N - 1)\cdot\mathrm{Id},$$

$$(10) \qquad [\Theta(m), \Lambda(n)] = 0, \qquad \Theta(m) + \Lambda(m) = D(m).$$

The rest of this section is devoted to the proof of Theorem 1 and Theorem 2. We first prove (5) in Theorem 1. The rest follows from this. The proof of the formula (5) is by lengthy calculation. Later we will give another proof in the framework of Kähler forms in §3.11.

From the commutator formula of vertex operators, we have

$$[H(m), H(n)] = [\{h^2\}_{m+1}, \{h^2\}_{n+1}] = \sum_{0 \leq k \in \mathbb{Z}} \binom{m+1}{k}\left\{\{h^2\}_k[h^2]\right\}_{m+n+2-k}$$

$$(11) \qquad = \left\{\{h^2\}_0[h^2]\right\}_{m+n+2} + (m+1)\left\{\{h^2\}_1[h^2]\right\}_{m+n+1}$$

$$+ \binom{m+1}{2}\left\{\{h^2\}_2[h^2]\right\}_{m+n} + \binom{m+1}{3}\left\{\{h^2\}_3[h^2]\right\}_{m+n-1}.$$

We have to calculate these terms. Since $h = \sum_{j=1}^{N} a_j(-\frac{1}{2})a_j^*(-\frac{1}{2})$, we have

(12) $$[h^2] = \sum_{1 \leq j \neq k \leq N} a_j(-\frac{1}{2})a_j^*(-\frac{1}{2})a_k(-\frac{1}{2})a_k^*(-\frac{1}{2})\Omega.$$

The associated vertex operator $\{h^2\}_k$ for $k \in \mathbb{Z}$ is given by

(13) $$\{h^2\}_k = \sum_{\substack{1 \leq p \neq q \leq N \\ m_1 + m_2 + m_3 + m_4 = k+1}} :a_p(m_1 - \frac{1}{2})a_p^*(m_2 - \frac{1}{2})a_q(m_3 - \frac{1}{2})a_q^*(m_4 - \frac{1}{2}): .$$

Our calculation of $\{h^2\}_k[h^2]$ for relevant k is summarized as follows:

PROPOSITION 3. *The action of the vertex operator $\{h^2\}_k$ on $[h^2] \in V$ for $0 \leq k \leq 3$ is given as follows:*

(14) $$\begin{aligned}
\{h^2\}_0[h^2] &= 4(N-1)D(-1)[\omega] - 2(N-2)D(-1)[h^2], \\
\{h^2\}_1[h^2] &= -8(N-1)[\omega] + 4(N-2)[h^2], \\
\{h^2\}_2[h^2] &= 0, \\
\{h^2\}_3[h^2] &= 2N(N-1)\Omega.
\end{aligned}$$

Here $D(-1)[\omega] = -2[\omega(2,0)]$ and $D(-1)[h^2] = 2[h \cdot (\omega_+ - \omega_-)]$.

For the definitions of $\omega(n_1, n_2)$ and $\omega_\pm$, see (12) and (17) of §3.4 and (1) of §3.5.

Assuming Proposition 3 for a moment, we can prove Theorem 1 and, with a help of Lemma 4, also Theorem 2.

(PROOF OF THEOREM 1 ASSUMING PROPOSITION 3). We continue the calculation in (11) using formuae in (14). We have

$$\begin{aligned}
[H(m), H(n)] = {}& 4(N-1)\{D(-1)[\omega]\}_{m+n+2} - 2(N-2)\{D(-1)[h^2]\}_{m+n+2} \\
& - 8(N-1)(m+1)\{\omega\}_{m+n+1} + 4(N-2)(m+1)\{h^2\}_{m+n+1} \\
& + \binom{m+1}{3} 2N(N-1)\{\Omega\}_{m+n-1}.
\end{aligned}$$

Since $\{D(-1)[x]\}_{r+1} = (r+1)\{x\}_r$, $\{\omega\}_{m+n+1} = D(m+n)$, $\{h^2\}_{m+n+1} = H(m+n)$, and $\{\Omega\}_{m+n-1} = \delta_{m+n,0} \cdot \mathrm{Id}$, we have

$$\begin{aligned}
[H(m), H(n)] = {}& 4(N-1)(n-m)D(m+n) - 2(N-2)(n-m)H(m+n) \\
& + \frac{m(m^2-1)}{3}N(N-1)\delta_{m+n,0} \cdot \mathrm{Id}.
\end{aligned}$$

This proves formula (5). Formulae (3) and (4) are already known. This completes the proof of Theorem 1. □

To explain the reason of the choice of vectors in (6), we let $v = a[\omega] + b[h^2]$ for some $a, b \in \mathbb{C}$ and we calculate commutation relations for components of the associated vertex operator.

LEMMA 4. *The commutation relation for components of the vertex operator* $Y(v, \zeta) = \sum_{n \in \mathbb{Z}} \{v\}_n \zeta^{-n-1}$ *associated to* $v = a[\omega] + b[h^2]$ *is given by*

$$(15) \quad [\{v\}_{m+1}, \{v\}_{n+1}] = (n-m)2\{a - b(N-2)\}\{v\}_{m+n+1}$$

$$+ \frac{m(m^2 - 1)}{12} \{a^2 + 4b^2(N-1)\} N \cdot \delta_{m+n,0} \cdot \mathrm{Id}$$

$$+ (n-m)\{-a^2 + 2ab(N-2) + 4b^2(N-1)\} D(m+n).$$

PROOF. This is a straightforward calculation using Theorem 1. Note that $\{v\}_{n+1} = aD(n) + bH(n)$. Thus,

$$[\{v\}_{m+1}, \{v\}_{n+1}] = a^2[D(m), D(n)] + ab[D(m), H(n)] + ab[H(m), D(n)]$$

$$+ b^2[H(m), H(n)] = (n-m)\{a^2 + 4b^2(N-1)\} D(m+n)$$

$$+ (n-m)\{2ab - 2b^2(N-2)\} H(m+n)$$

$$+ \frac{m(m^2 - 1)}{12} \{a^2 + 4b^2(N-1)\} N \cdot \delta_{m+n,0} \cdot \mathrm{Id}.$$

Since $bH(m+n) = \{v\}_{m+n+1} - aD(m+n)$, eliminating operators $H(m)$ from the above expression, we get (15). $\square$

(PROOF OF THEOREM 2). If we want the commutation relation (15) to be "closed", we want the coefficient of $D(m+n)$ to be 0 for any $m, n \in \mathbb{Z}$. For this to be the case, we must have $a = -2b$ or $a = (2N - 2)b$. In these two cases, the coefficient of $\{v\}_{m+n+1}$ is either $(n-m) \cdot (-2bN)$ or $(n-m) \cdot 2bN$. To normalize this coeffient to be $(n-m)$ as in the Virasoro algebra case, we let $b = -1/(2N)$ for the case $a = -2b$, and $b = 1/(2N)$ for the case $a = b(2N - 2)$. This way, we arrive at expressions in (6) for vectors θ and λ.

Corresponding commutation formulae are then immediate from Lemma 4 for both choices of a and b. The central charges of the Virasoro algebras generated by operators $\Theta(m)$ and $\Lambda(n)$ turn out to be 1 and $N - 1$, respectively. Note that their sum is N which is the central charge for the Virasoro algebra generated by $D(n)$'s. This should be the case since $\omega = \theta + \lambda$.

We can easily show that operators $\Theta(m)$ and $\Lambda(n)$ commute using (7) and Theorem 1. This finishes the proof of Theorem 2. $\square$

The rest of this section is devoted to the proof of Proposition 3.

Calculation of $\{h^2\}_0[h^2]$. To calculate $\{h^2\}_0[h^2]$, we have to calculate the following:

$$(16) \quad \sum_{\substack{j \neq k, p \neq q \\ m_1 + m_2 + m_3 + m_4 = 1}} : a_p(m_1 - \tfrac{1}{2}) a_p^*(m_2 - \tfrac{1}{2}) a_q(m_3 - \tfrac{1}{2}) a_q^*(m_4 - \tfrac{1}{2}) :$$

$$\cdot a_j(-\tfrac{1}{2}) a_j^*(-\tfrac{1}{2}) a_k(-\tfrac{1}{2}) a_k^*(-\tfrac{1}{2}) \Omega.$$

Since h^2 is formed out of vectors in $A(-\tfrac{1}{2})$, all integers m_j must be less than or equal to 1 for nontriviality of the action of the above normal ordered product. Since the sum of m_i's must be 1, we see that (m_1, m_2, m_3, m_4) must be a permutation of $(1, 1, 1, -2)$, $(1, 1, -1, 0)$, $(1, 0, 0, 0)$.

Case (I): $(m_1, m_2, m_3, m_4) \in$ permutations of $(1, 1, 1, -2)$. Since $a_p(\frac{1}{2})$ can pair nontrivially only with $a_p^*(-\frac{1}{2})$ for any p, for nontriviality of the action we must have $(p, q) = (j, k)$ or (k, j). Since $j \neq k$, the condition $p \neq q$ is automatically satisfied. For these (p, q) there is no need to use normal ordering because there are no possible "internal" pairings among operators within the normal ordering symbol, and we can remove them without affecting the formula. Writing out all possible permutations in this case, we have

$$\sum_{j \neq k} \{ a_j(\tfrac{1}{2}) a_j^*(\tfrac{1}{2}) a_k(\tfrac{1}{2}) a_k^*(-\tfrac{5}{2}) + a_k(\tfrac{1}{2}) a_k^*(\tfrac{1}{2}) a_j(\tfrac{1}{2}) a_j^*(-\tfrac{5}{2})$$
$$+ a_j(\tfrac{1}{2}) a_j^*(\tfrac{1}{2}) a_k(-\tfrac{5}{2}) a_k^*(\tfrac{1}{2}) + a_k(\tfrac{1}{2}) a_k^*(\tfrac{1}{2}) a_j(-\tfrac{5}{2}) a_j^*(\tfrac{1}{2})$$
$$+ a_j(\tfrac{1}{2}) a_j^*(-\tfrac{5}{2}) a_k(\tfrac{1}{2}) a_k^*(\tfrac{1}{2}) + a_k(\tfrac{1}{2}) a_k^*(-\tfrac{5}{2}) a_j(\tfrac{1}{2}) a_j^*(\tfrac{1}{2})$$
$$+ a_j(-\tfrac{5}{2}) a_j^*(\tfrac{1}{2}) a_k(\tfrac{1}{2}) a_k^*(\tfrac{1}{2}) + a_k(-\tfrac{5}{2}) a_k^*(\tfrac{1}{2}) a_j(\tfrac{1}{2}) a_j^*(\tfrac{1}{2}) \}$$
$$\cdot a_j(-\tfrac{1}{2}) a_j^*(-\tfrac{1}{2}) a_k(-\tfrac{1}{2}) a_k^*(-\tfrac{1}{2}) \Omega.$$

After simplifying this, we arrive at the following expression:

$$2 \sum_{1 \leq j \neq k \leq N} \{ a_k^*(-\tfrac{5}{2}) a_k(-\tfrac{1}{2}) \Omega + a_k(-\tfrac{5}{2}) a_k^*(-\tfrac{1}{2}) \Omega + a_j^*(-\tfrac{5}{2}) a_j(-\tfrac{1}{2}) \Omega + a_j(-\tfrac{5}{2}) a_j^*(-\tfrac{1}{2}) \Omega \}.$$

The summation of the first two terms over $j \neq k$ gives $-4(N - 1)\omega(2, 0)$. The summation of the last two terms gives the same result, and the above sum is equal to $-8(N - 1)[\omega(2, 0)]$. See (12) of §3.4 for the definition of $\omega(2, 0)$.

Case (II): $(m_1, m_2, m_3, m_4) \in$ permutations of $(1, 0, 0, 0)$. Removing the normal ordering symbol, we see that in this case (16) is equal to

$$\sum_{j \neq k, p \neq q} \{ -a_p^*(-\tfrac{1}{2}) a_q(-\tfrac{1}{2}) a_q^*(-\tfrac{1}{2}) a_p(\tfrac{1}{2}) + a_p(-\tfrac{1}{2}) a_q(-\tfrac{1}{2}) a_q^*(-\tfrac{1}{2}) a_p^*(\tfrac{1}{2})$$
$$- a_p(-\tfrac{1}{2}) a_p^*(-\tfrac{1}{2}) a_q^*(-\tfrac{1}{2}) a_q(\tfrac{1}{2}) + a_p(-\tfrac{1}{2}) a_p^*(-\tfrac{1}{2}) a_q(-\tfrac{1}{2}) a_q^*(\tfrac{1}{2}) \}$$
$$\cdot a_j(-\tfrac{1}{2}) a_j^*(-\tfrac{1}{2}) a_k(-\tfrac{1}{2}) a_k^*(-\tfrac{1}{2}) \Omega.$$

For the first two terms, p must be equal to j or k for nontriviality of the action. For the last two terms, q must be equal to j or k by the same reason. After doing some calculations and using the notation $h_j = a_j(-\frac{1}{2}) a_j^*(-\frac{1}{2})$, we see that the above is equal to

$$\sum_{j \neq k \neq q} \{ -h_j h_q h_k - h_k h_q h_j + h_j h_q h_k + h_k h_q h_j \} \Omega$$
$$+ \sum_{j \neq k \neq p} \{ -h_p h_j h_k - h_p h_k h_j + h_p h_j h_k + h_p h_k h_j \} \Omega = 0.$$

Case (III): $(m_1, m_2, m_3, m_4) \in$ permutations of $(1, 1, -1, 0)$. Calculation is the longest for this case. There are 12 different permutations for the above

4-tuple. Writing these out, we see that in this case (16) is equal to

$$\sum_{j\neq k,p\neq q} \Big\{ :a_p(\tfrac{1}{2})a_p^*(\tfrac{1}{2})a_q(-\tfrac{3}{2})a_q^*(-\tfrac{1}{2}): + :a_p(\tfrac{1}{2})a_p^*(\tfrac{1}{2})a_q(-\tfrac{1}{2})a_q^*(-\tfrac{3}{2}): $$
$$+ :a_p(-\tfrac{3}{2})a_p^*(-\tfrac{1}{2})a_q(\tfrac{1}{2})a_q^*(\tfrac{1}{2}): + :a_p(-\tfrac{1}{2})a_p^*(-\tfrac{3}{2})a_q(\tfrac{1}{2})a_q^*(\tfrac{1}{2}): $$
$$+ :a_p(\tfrac{1}{2})a_p^*(-\tfrac{3}{2})a_q(\tfrac{1}{2})a_q^*(-\tfrac{1}{2}): + :a_p(-\tfrac{3}{2})a_p^*(\tfrac{1}{2})a_q(\tfrac{1}{2})a_q^*(-\tfrac{1}{2}): $$
$$+ :a_p(\tfrac{1}{2})a_p^*(-\tfrac{3}{2})a_q(-\tfrac{1}{2})a_q^*(\tfrac{1}{2}): + :a_p(-\tfrac{3}{2})a_p^*(\tfrac{1}{2})a_q(-\tfrac{1}{2})a_q^*(\tfrac{1}{2}): $$
$$+ :a_p(\tfrac{1}{2})a_p^*(-\tfrac{1}{2})a_q(\tfrac{1}{2})a_q^*(-\tfrac{3}{2}): + :a_p(-\tfrac{1}{2})a_p^*(\tfrac{1}{2})a_q(\tfrac{1}{2})a_q^*(-\tfrac{3}{2}): $$
$$+ :a_p(\tfrac{1}{2})a_p^*(-\tfrac{1}{2})a_q(-\tfrac{3}{2})a_q^*(\tfrac{1}{2}): + :a_p(-\tfrac{1}{2})a_p^*(\tfrac{1}{2})a_q(-\tfrac{3}{2})a_q^*(\tfrac{1}{2}): \Big\}$$
$$\cdot a_j(-\tfrac{1}{2})a_j^*(-\tfrac{1}{2})a_k(-\tfrac{1}{2})a_k^*(-\tfrac{1}{2})\Omega.$$

Inside of { }, the middle four terms and the last four terms give rise to the same summation over $p \neq q$. First we calculate summation of the first four terms. For the first two terms, p must be equal to j or k for nontriviality of the action. For the next two terms, q must be equal to j or k for nontriviality. Then the action of the sum of the first four terms gives rise to

$$\sum_{j\neq k}\Big[\sum_{q\neq j}\{a_q(-\tfrac{3}{2})a_q^*(-\tfrac{1}{2})a_k(-\tfrac{1}{2})a_k^*(-\tfrac{1}{2})\Omega + a_q(-\tfrac{1}{2})a_q^*(-\tfrac{3}{2})a_k(-\tfrac{1}{2})a_k^*(-\tfrac{1}{2})\Omega\}$$
$$+\sum_{q\neq k}\{a_q(-\tfrac{3}{2})a_q^*(-\tfrac{1}{2})a_j(-\tfrac{1}{2})a_j^*(-\tfrac{1}{2})\Omega + a_q(-\tfrac{1}{2})a_q^*(-\tfrac{3}{2})a_k(-\tfrac{1}{2})a_k^*(-\tfrac{1}{2})\Omega\}$$
$$+\sum_{p\neq j}\{a_p(-\tfrac{3}{2})a_p^*(-\tfrac{1}{2})a_k(-\tfrac{1}{2})a_k^*(-\tfrac{1}{2})\Omega + a_p(-\tfrac{1}{2})a_p^*(-\tfrac{3}{2})a_k(-\tfrac{1}{2})a_k^*(-\tfrac{1}{2})\Omega\}$$
$$+\sum_{p\neq k}\{a_p(-\tfrac{3}{2})a_p^*(-\tfrac{1}{2})a_j(-\tfrac{1}{2})a_j^*(-\tfrac{1}{2})\Omega + a_p(-\tfrac{1}{2})a_p^*(-\tfrac{3}{2})a_j(-\tfrac{1}{2})a_j^*(-\tfrac{1}{2})\Omega\}\Big].$$

In the first line above, if $q = k$, then inside is 0 because $\mathcal{A}$ is an exterior algebra. So summation over q can be taken to be over $q \neq k, j$. Similar remarks apply to the other three lines which follow. But then, each of the above four lines denotes the same summation, and the above is equal to

$$4\sum_{p\neq k}\sum_{j\neq p,k}\{a_p(-\tfrac{3}{2})a_p^*(-\tfrac{1}{2}) + a_p(-\tfrac{1}{2})a_p^*(-\tfrac{3}{2})\}\,a_k(-\tfrac{1}{2})a_k^*(-\tfrac{1}{2})\Omega.$$

Since terms to be summed are independent of j, the summation over j gives a factor of $(N-2)$ because $j \neq p, k$. If $p = k$, then each term is zero. Thus, we may assume that the summation is over all possible p and k between 1 and N, and we can sum over p, k independently. Then the sum is equal to $4(N-2)(-\omega_+ + \omega_-)h$.

Next, the contribution from the last eight terms is twice the contribution from the middle four terms. Removing normal ordering symbols, the sum of the last eight terms in the above expression is given by

$$2\sum_{j\neq k}\sum_{p\neq q}\{-a_p^*(-\tfrac{3}{2})a_q^*(-\tfrac{1}{2})a_p(\tfrac{1}{2})a_q(\tfrac{1}{2}) + a_p(-\tfrac{3}{2})a_q^*(-\tfrac{1}{2})a_p^*(\tfrac{1}{2})a_q(\tfrac{1}{2})$$
$$+a_p^*(-\tfrac{3}{2})a_q(-\tfrac{1}{2})a_p(\tfrac{1}{2})a_q^*(\tfrac{1}{2}) - a_p(-\tfrac{3}{2})a_q(-\tfrac{1}{2})a_p^*(\tfrac{1}{2})a_q^*(\tfrac{1}{2})\}$$
$$\cdot a_j(-\tfrac{1}{2})a_j^*(-\tfrac{1}{2})a_k(-\tfrac{1}{2})a_k^*(-\tfrac{1}{2})\Omega.$$

For nontriviality of the action, we must have $(p, q) = (j, k)$ or (k, j). After some calculation of pairings, this is equal to

$$
2 \sum_{\substack{j \neq k}} \{ a_j^*(-\tfrac{3}{2}) a_j(-\tfrac{1}{2}) a_k^*(-\tfrac{1}{2}) a_k(-\tfrac{1}{2}) \Omega + a_k^*(-\tfrac{3}{2}) a_k(-\tfrac{1}{2}) a_j^*(-\tfrac{1}{2}) a_j(-\tfrac{1}{2}) \Omega \\
+ a_j(-\tfrac{3}{2}) a_j^*(-\tfrac{1}{2}) a_k^*(-\tfrac{1}{2}) a_k(-\tfrac{1}{2}) \Omega + a_k(-\tfrac{3}{2}) a_k^*(-\tfrac{1}{2}) a_j^*(-\tfrac{1}{2}) a_j(-\tfrac{1}{2}) \Omega \\
+ a_j^*(-\tfrac{3}{2}) a_j(-\tfrac{1}{2}) a_k(-\tfrac{1}{2}) a_k^*(-\tfrac{1}{2}) \Omega + a_k^*(-\tfrac{3}{2}) a_k(-\tfrac{1}{2}) a_j(-\tfrac{1}{2}) a_j^*(-\tfrac{1}{2}) \Omega \\
+ a_j(-\tfrac{3}{2}) a_j^*(-\tfrac{1}{2}) a_k(-\tfrac{1}{2}) a_k^*(-\tfrac{1}{2}) \Omega + a_k(-\tfrac{3}{2}) a_k^*(-\tfrac{1}{2}) a_j(-\tfrac{1}{2}) a_j^*(-\tfrac{1}{2}) \Omega \}.
$$

In this summation, the first four terms cancel with the last four terms by a trivial reason of signs arising from rearranging orders of products. Thus, the sum above is 0, and the contribution from the case (III) is $4(N-2)[(-\omega_+ + \omega_-)h]$.

Collecting the contributions from (I), (II) and (III), we have the first formula in (14).

Calculation of $\{h^2\}_1[h^2]$. Now, we move on to the proof of the second formula in (14). We have to calculate $\{h^2\}_1[h^2]$ given explicitly as follows:

$$
(17) \qquad \sum_{\substack{j \neq k, p \neq q \\ m_1 + m_2 + m_3 + m_4 = 2}} : a_p(m_1 - \tfrac{1}{2}) a_p^*(m_2 - \tfrac{1}{2}) a_q(-\tfrac{1}{2}) a_q^*(m_4 - \tfrac{1}{2}) : \\
\cdot a_j(-\tfrac{1}{2}) a_j^*(-\tfrac{1}{2}) a_k(-\tfrac{1}{2}) a_k^*(-\tfrac{1}{2}) \Omega.
$$

In the above summation, indices j, k, p, q run from 1 to N. As in the case of (16), all integers m_j must be less than or equal to 1, otherwise at least one of $a(m - \tfrac{1}{2})$'s annihilate the vacuum vector Ω. Since the sum of m_j's must be 2, (m_1, m_2, m_3, m_4) must be a permutation of either $(1, 1, 1, -1)$ or $(1, 1, 0, 0)$. As before we divide our calculation into two cases.

Case (I): $(m_1, m_2, m_3, m_4) \in$ permutations of $(1, 1, 1, -1)$. To simplify our argument, we define a few terminologies. Recall that complex conjugation $*$ acts on V. (Note that complex conjugation operator $*$ is not the same as the operation to get the adjoint vector as defined in §2.4 Definition 22.) We say that a vector $v \in V$ is $*$-invariant if it is invariant under $*$. Two elements $\rho_1, \rho_2 \in \mathcal{A}$ are called $*$-related if $\rho_1^* = \rho_2$. There are four permutations of $(1, 1, 1, -1)$ and writing these out, we get

$$
(18) \quad \sum_{p \neq q} \{ : a_p(\tfrac{1}{2}) a_p^*(\tfrac{1}{2}) a_q(\tfrac{1}{2}) a_q^*(-\tfrac{3}{2}) : + : a_p(\tfrac{1}{2}) a_p^*(\tfrac{1}{2}) a_q(-\tfrac{3}{2}) a_q^*(\tfrac{1}{2}) : \\
+ : a_p(\tfrac{1}{2}) a_p^*(-\tfrac{3}{2}) a_q(\tfrac{1}{2}) a_q^*(\tfrac{1}{2}) : + : a_p(-\tfrac{3}{2}) a_p^*(\tfrac{1}{2}) a_q(\tfrac{1}{2}) a_q^*(\tfrac{1}{2}) : \} \\
\cdot \sum_{j \neq k} a_j(-\tfrac{1}{2}) a_j^*(-\tfrac{1}{2}) a_k(-\tfrac{1}{2}) a_k^*(-\tfrac{1}{2}) \Omega.
$$

Note that $[h^2] = \sum_{j \neq k} a_j(-\tfrac{1}{2}) a_j^*(-\tfrac{1}{2}) a_k(-\tfrac{1}{2}) a_k^*(-\tfrac{1}{2})$ is $*$-invariant since $h^* = -h$. Among four terms inside of braces, the first term and the second term are $*$-related and the third term and the fourth term are $*$-related. We also note that after taking summation over p, q, the first and the third sums are equal. The second and the fourth sums are also equal. Thus essentially we only have to calculate the first sum. For its nontriviality for each given (j,k), the pair of indices (p,q) must be equal to either (j,k) or (k,j). Then, after removing the normal ordering symbols, the first sum is equal to

$$\sum_{j\neq k} \{:a_j(\tfrac{1}{2})a_j^*(\tfrac{1}{2})a_k(\tfrac{1}{2})a_k^*(-\tfrac{3}{2}): + :a_k(\tfrac{1}{2})a_k^*(\tfrac{1}{2})a_j(\tfrac{1}{2})a_j^*(-\tfrac{3}{2}):\}$$
$$\cdot a_j(-\tfrac{1}{2})a_j^*(-\tfrac{1}{2})a_k(-\tfrac{1}{2})a_k^*(-\tfrac{1}{2})\Omega$$
$$= \sum_{j\neq k} \{a_k^*(-\tfrac{3}{2})a_k(-\tfrac{1}{2}) + a_j^*(-\tfrac{3}{2})a_j(-\tfrac{1}{2})\}\,\Omega.$$

The first term above is independent of j and summation over j gives a factor $(N-1)$. The second term gives the same sum. Thus the above summation is equal to $2(N-1)\sum_{k=1}^N a_k^*(-\tfrac{3}{2})a_k(-\tfrac{1}{2})\Omega$.

The second sum in (18) is obtained by applying $*$(=complex conjugation) to the first sum and this is equal to $2(N-1)\sum_{k=1}^N a_k(-\tfrac{3}{2})a_k^*(-\tfrac{1}{2})\Omega$. The third and the fourth sums in (18) are the same as the above sums. So (18) is equal to

$$4(N-1)\sum_{k=1}^N \{a_k(-\tfrac{3}{2})a_k^*(-\tfrac{1}{2}) + a_k^*(-\tfrac{3}{2})a_k(-\tfrac{1}{2})\}\,\Omega = -8(N-1)\omega.$$

Case (II): $(m_1, m_2, m_3, m_4) \in$ permutations of $(1,1,0,0)$. There are six permutations for $(1,1,0,0)$. Writing these out, we have the following:

$$(19) \quad \sum_{p\neq q} \{:a_p(\tfrac{1}{2})a_p^*(\tfrac{1}{2})a_q(-\tfrac{1}{2})a_q(-\tfrac{1}{2}): + :a_p(-\tfrac{1}{2})a_p^*(-\tfrac{1}{2})a_q(\tfrac{1}{2})a_q(\tfrac{1}{2}):$$
$$+ :a_p(\tfrac{1}{2})a_p^*(-\tfrac{1}{2})a_q(\tfrac{1}{2})a_q(-\tfrac{1}{2}): + :a_p(\tfrac{1}{2})a_p^*(-\tfrac{1}{2})a_q(-\tfrac{1}{2})a_q(\tfrac{1}{2}):$$
$$+ :a_p(-\tfrac{1}{2})a_p^*(\tfrac{1}{2})a_q(\tfrac{1}{2})a_q(-\tfrac{1}{2}): + :a_p(-\tfrac{1}{2})a_p^*(\tfrac{1}{2})a_q(-\tfrac{1}{2})a_q(\tfrac{1}{2}):\}$$
$$\cdot \sum_{j\neq k} a_j(-\tfrac{1}{2})a_j^*(-\tfrac{1}{2})a_k(-\tfrac{1}{2})a_k^*(-\tfrac{1}{2})\Omega.$$

Note that the first and the second terms give rise to the same result after summing over p and q with $p \neq q$. Also note that the third and the sixth terms are $*$-related, and the fourth and the fifth terms are $*$-related. Now for the first sum, for nontriviality of the action, p must be equal to j or k for each given (j,k). Thus the first term gives rise to

$$\sum_{j\neq k} \{\sum_{q\neq j} :a_j(\tfrac{1}{2})a_j^*(\tfrac{1}{2})a_q(-\tfrac{1}{2})a_q^*(-\tfrac{1}{2}): + \sum_{q\neq k} :a_k(\tfrac{1}{2})a_k^*(\tfrac{1}{2})a_q(-\tfrac{1}{2})a_q^*(-\tfrac{1}{2}):\}$$
$$\cdot a_j(-\tfrac{1}{2})a_j^*(-\tfrac{1}{2})a_k(-\tfrac{1}{2})a_k^*(-\tfrac{1}{2})\Omega$$
$$= \sum_{q\neq k}\sum_{j\neq k,q} a_q(-\tfrac{1}{2})a_q^*(-\tfrac{1}{2})a_k(-\tfrac{1}{2})a_k^*(-\tfrac{1}{2})\Omega$$
$$+ \sum_{q\neq j}\sum_{k\neq j,q} a_q(-\tfrac{1}{2})a_q^*(-\tfrac{1}{2})a_j(-\tfrac{1}{2})a_j^*(-\tfrac{1}{2})\Omega.$$

For the first sum, summing over j gives a factor $(N-2)$. If $q = k$, the term inside is 0 in the Clifford algebra $\mathrm{Cliff}(\mathbb{Z}+\tfrac{1}{2})$. So we may assume that summation is over independent indices $1 \leq q, k \leq N$. The second sum is the same as the first sum. Thus, the above is equal to

$$2(N - 2) \sum_{q,k=1}^{N} a_q(-\tfrac{1}{2})a_q^*(-\tfrac{1}{2})a_k(-\tfrac{1}{2})a_k^*(-\tfrac{1}{2})\Omega = 2(N - 2)[h^2].$$

As pointed out earlier, the second sum in (19) is the same as the first one.

For the third sum, given (j, k), for nontriviality of the action the pair of indices (p, q) must be equal to either (j, k) or (k, j). Thus the third term gives rise to

$$\sum_{j \neq k} \{ : a_j(\tfrac{1}{2})a_j^*(-\tfrac{1}{2})a_k(\tfrac{1}{2})a_k^*(-\tfrac{1}{2}) : + : a_k(\tfrac{1}{2})a_k^*(-\tfrac{1}{2})a_j(\tfrac{1}{2})a_j^*(-\tfrac{1}{2}): \}$$
$$\cdot a_j(-\tfrac{1}{2})a_j^*(-\tfrac{1}{2})a_k(-\tfrac{1}{2})a_k^*(-\tfrac{1}{2})\Omega.$$

Removing normal ordering symbols and calculating some pairings, we see that the above is equal to

$$\sum_{j \neq k} a_j(-\tfrac{1}{2})a_j^*(-\tfrac{1}{2})a_k(-\tfrac{1}{2})a_k^*(-\tfrac{1}{2})\Omega + \sum_{j \neq k} a_j(-\tfrac{1}{2})a_j^*(-\tfrac{1}{2})a_k(-\tfrac{1}{2})a_k^*(-\tfrac{1}{2})\Omega = 2[h^2].$$

Similarly, for nontriviality of the fourth term, given (j, k), we must have that $(p, q) = (j, k)$ or (k, j), and we have

$$\sum_{j \neq k} \{ a_j(\tfrac{1}{2})a_j^*(-\tfrac{1}{2})a_k(-\tfrac{1}{2})a_k^*(\tfrac{1}{2}) : + a_k(\tfrac{1}{2})a_k^*(-\tfrac{1}{2})a_j(-\tfrac{1}{2})a_j^*(\tfrac{1}{2}) : \}$$
$$a_j(-\tfrac{1}{2})a_j^*(-\tfrac{1}{2})a_k(-\tfrac{1}{2})a_k^*(-\tfrac{1}{2})\Omega$$
$$= -\sum_{j \neq k} a_j(-\tfrac{1}{2})a_j^*(-\tfrac{1}{2})a_k(-\tfrac{1}{2})a_k^*(-\tfrac{1}{2})\Omega - \sum_{j \neq k} a_j(-\tfrac{1}{2})a_j^*(-\tfrac{1}{2})a_k(-\tfrac{1}{2})a_k^*(-\tfrac{1}{2})\Omega$$
$$= -2[h^2].$$

Since $[(h^2)^*] = [(-h)^2] = [h^2]$, using the $*$-relation, we see that the fifth term and the sixth term give rise to sums $-2[h^2]$ and $2[h^2]$, respectively. Combining all the calculations above, we see that (19) is equal to $4(N - 2)[h^2]$.

Thus, cases (I) and (II) give that $\{h^2\}_1[h^2] = -8(N - 1)[\omega] + 4(N - 2)[h^2]$, which is the second formula in (14).

Calculation of $\{h^2\}_2[h^2]$. Next we calculate $\{h^2\}_2[h^2]$ given by

$$(20) \qquad \sum_{\substack{p \neq q, j \neq k \\ m_1+m_2+m_3+m_4=3}} : a_p(m_1 - \tfrac{1}{2})a_p^*(m_2 - \tfrac{1}{2})a_q(m_3 - \tfrac{1}{2})a_q^*(m_4 - \tfrac{1}{2}):$$
$$a_j(-\tfrac{1}{2})a_j^*(-\tfrac{1}{2})a_k(-\tfrac{1}{2})a_k^*(-\tfrac{1}{2})\Omega.$$

As before, for nontriviality of the action, integers m_j must be less than or equal to 1. Since the sum of m_j's must be 3, integers (m_1, m_2, m_3, m_4) must be a permutation of $(1, 1, 1, 0)$. There are four permutations of this. Writing these out, we see that (20) is equal to

(21)
$$\{h^2\}_2[h^2] = \sum_{j \neq k} \sum_{p \neq q} \{ :a_p(\tfrac{1}{2})a_p^*(\tfrac{1}{2})a_q(\tfrac{1}{2})a_q^*(-\tfrac{1}{2}): + :a_p(\tfrac{1}{2})a_p^*(\tfrac{1}{2})a_q(-\tfrac{1}{2})a_q^*(\tfrac{1}{2}):$$
$$+ :a_p(\tfrac{1}{2})a_p^*(-\tfrac{1}{2})a_q(\tfrac{1}{2})a_q^*(\tfrac{1}{2}): + :a_p(-\tfrac{1}{2})a_p^*(\tfrac{1}{2})a_q(\tfrac{1}{2})a_q^*(\tfrac{1}{2}): \}$$
$$\cdot a_j(-\tfrac{1}{2})a_j^*(-\tfrac{1}{2})a_k(-\tfrac{1}{2})a_k^*(-\tfrac{1}{2})\Omega.$$

We note that after summing over p and q, the first and the third terms give rise to the same sums, and similarly the second term and the fourth term give rise to the same sums. We also note that the first and the second terms are $*$-related. So we only have to calculate the sum of the first term. For nontriviality of this sum, given (j, k), the pair of indices (p, q) must be equal to either (j, k) or (k, j). The first sum is then equal to

$$\sum_{j \neq k} \{ :a_j(\tfrac{1}{2})a_j^*(\tfrac{1}{2})a_k(\tfrac{1}{2})a_k^*(-\tfrac{1}{2}): + :a_k(\tfrac{1}{2})a_k^*(\tfrac{1}{2})a_j(\tfrac{1}{2})a_j^*(-\tfrac{1}{2}): \}$$
$$\cdot a_j(-\tfrac{1}{2})a_j^*(-\tfrac{1}{2})a_k(-\tfrac{1}{2})a_k^*(-\tfrac{1}{2})\Omega.$$
$$= \sum_{j \neq k} a_k^*(-\tfrac{1}{2})a_k(-\tfrac{1}{2})\Omega + \sum_{j \neq k} a_j^*(-\tfrac{1}{2})a_j(-\tfrac{1}{2})\Omega.$$

In the first sum above, summation over j gives a factor $(N - 1)$ and changing the order of a^*'s and a's, the sum is equal to $-(N - 1)[h]$. The second summation have the same sum, and the above is equal to $-2(N - 1)[h]$.

The second sum in (21) is calculated by applying $*$ to the first sum and thus it is equal to $-2(N - 1)[h^*] = 2(N - 1)[h]$. Since the third and the fourth are the same as the first and the second, we see that (21) is equal to 0. This proves the third formula in Proposition 3.

Calculation of $\{h^2\}_3[h^2]$. Finally we calculate $\{h^2\}_3[h^2]$ given by

(22)
$$\sum_{\substack{p \neq q, j \neq k \\ m_1 + m_2 + m_3 + m_4 = 4}} :a_p(m_1 - \tfrac{1}{2})a_p^*(m_2 - \tfrac{1}{2})a_q(m_3 - \tfrac{1}{2})a_q^*(m_4 - \tfrac{1}{2}):$$
$$\cdot a_j(-\tfrac{1}{2})a_j^*(-\tfrac{1}{2})a_k(-\tfrac{1}{2})a_k^*(-\tfrac{1}{2})\Omega.$$

As before, for nontriviality of the action, integers m_j must be less than or equal to 1 for all j. Since the sum of m_j's must be 4, there is only one possibility for m_j's: namely $(m_1, m_2, m_3, m_4) = (1, 1, 1, 1)$. In this case, given (j, k), for nontriviality of the result, indices (p, q) must be equal to (j, k) or (k, j). Then, (22) is equal to

$$\sum_{j \neq k} \{ :a_j(\tfrac{1}{2})a_j^*(\tfrac{1}{2})a_k(\tfrac{1}{2})a_k^*(\tfrac{1}{2}): + :a_k(\tfrac{1}{2})a_k^*(\tfrac{1}{2})a_j(\tfrac{1}{2})a_j^*(\tfrac{1}{2}): \}$$
$$\cdot a_j(-\tfrac{1}{2})a_j^*(-\tfrac{1}{2})a_k(-\tfrac{1}{2})a_k^*(-\tfrac{1}{2})\Omega.$$

The first term and the second term give rise to the same sum after summing over j and k, and (22) is equal to $2 \sum_{j \neq k} \Omega = 2(N^2 - N)\Omega$. This proves the last formula in Proposition 3 and its proof is complete.

§3.7   Vertex operators generated by complex volume forms

Let E^{2N} be a real $2N$-dimensional vector space with an inner product $\langle\ ,\ \rangle$ and an isometric complex structure I. Let $E(-n - \frac{1}{2})$ be a copy of E carrying weight $n + \frac{1}{2}$. Let $E^{2N}(-n - \frac{1}{2}) \otimes \mathbb{C} = A(-n - \frac{1}{2}) = A^+(-n - \frac{1}{2}) \oplus A^-(-n - \frac{1}{2})$ be the I-eigenspace decomposition of the complexification A with a basis

$$A^+(-n - \tfrac{1}{2}) = \bigoplus_{j=1}^{N} \mathbb{C}a_j(-n - \tfrac{1}{2}), \qquad A^-(-m - \tfrac{1}{2}) = \bigoplus_{j=1}^{N} \mathbb{C}a_j^*(-n - \tfrac{1}{2}).$$

We consider top exterior powers $\bigwedge^N A^\pm(-n - \frac{1}{2}) = \mathbb{C}\rho_\pm(-n - \frac{1}{2})$ spanned by

(1)
$$\begin{aligned}
\rho_+(-n - \tfrac{1}{2}) &= a_1(-n - \tfrac{1}{2})a_2(-n - \tfrac{1}{2})\cdots a_N(-n - \tfrac{1}{2}), \\
\rho_-(-n - \tfrac{1}{2}) &= a_1^*(-n - \tfrac{1}{2})a_2^*(-n - \tfrac{1}{2})\cdots a_N^*(-n - \tfrac{1}{2}).
\end{aligned}$$

Vectors $\rho_+(-n - \frac{1}{2})$ and $\rho_-(-n - \frac{1}{2})$ for $n \geq 0$ should be regarded as complex volume forms in the "holomorphic" vector space $A^+(-n - \frac{1}{2})$ and in the "anti-holomorphic" vector space $A^-(-n - \frac{1}{2})$, respectively. Our interest in complex volume forms comes from the fact that these vectors are invariant under the standard action of the group $\mathrm{SU}(N)$.

We calculate the algebra generated by vertex operators associated to complex volume forms. These algebras have geometric consequences on elliptic genera of special Kähler manifolds. Calculation of commutation relations among the associated vertex operators

(2)
$$Y([\rho_\pm(-n - \tfrac{1}{2})], z) = \sum_{k\in\mathbb{Z}}\{\rho_\pm(-n - \tfrac{1}{2})\}_k z^{-k-1}$$

becomes exceedingly difficult as n gets larger and larger since the weight of $\rho_\pm(-n-\frac{1}{2})$, which is $(n+\frac{1}{2})N$, also increases. However, when $n = 0$, it turns out that commutation relations are simple among operators $\{\rho_\pm(-\frac{1}{2})\}_*$, although crossing commutation relations turn out to be unexpectedly complicated. In this section, we present our partial calculation [Proposition 3, Proposition 5].

The vertex operator $Y([\rho_+(-n - \frac{1}{2})], z)$ is given by

$$Y([\rho_+(-n - \tfrac{1}{2})], z) = : a_1^{(n)}(z)a_2^{(n)}(z)\cdots a_N^{(n)}(z): , \qquad \text{where}$$

$$a_j^{(n)}(z) = \frac{1}{n!}\frac{d^n}{dz^n}\left[\sum_{m\in\mathbb{Z}}a_j(m + \tfrac{1}{2})z^{-m-1}\right] = (-1)^n\sum_{m\in\mathbb{Z}}\binom{m+n}{n}a_j(m + \tfrac{1}{2})z^{-m-n-1}.$$

The component operators $\{\rho_+(-n-\frac{1}{2})\}_k$ for $k \in \mathbb{Z}$ are given explicitly as follows:
(3)

$$\{\rho_+(-n - \tfrac{1}{2})\}_k = (-1)^{nN}\sum_{\substack{\sum_{j=1}^{N}m_j \\ =k+1-(n+1)N}}\prod_{j=1}^{N}\binom{m_j + n}{n}\, :a_1(m_1 + \tfrac{1}{2})\cdots a_N(m_N + \tfrac{1}{2}): .$$

Complex conjugation of (3) yields the component operator $\{\rho_-(-n-\frac{1}{2})\}_k$:
(4)

$$\{\rho_-(-n-\tfrac{1}{2})\}_k = (-1)^{nN} \sum_{\substack{\sum_{j=1}^N m_j \\ =k+1-(n+1)N}} \prod_{j=1}^N \binom{m_j+n}{n} : a_1^*(m_1+\tfrac{1}{2})\cdots a_N^*(m_N+\tfrac{1}{2}): \, .$$

LEMMA 1. *Let $k \geq 0$. In the summation on the right hand side of (3), no term with $m_j \leq -n-1$ for all $1 \leq j \leq N$ can appear. That is, for at least one j, we must have $m_j \geq -n$.*

PROOF. This is clear because if $\overrightarrow{m} = (m_1, \ldots, m_N)$ is such that $m_j \leq -n-1$ for all $1 \leq j \leq N$, then, $\sum_{j=1}^N m_j \leq -(n+1)N < (k+1) - (n+1)N$ for $k \geq 0$. Thus the condition on summation indices is not satisfied and $\overrightarrow{m}$ with the stated property cannot appear in the summation. $\square$

The same statement holds for $\{\rho_-(-n-\frac{1}{2})\}_k$ for $k \geq 0$.

A simple application of this observation to the case $n = 0$ yields the next result.

LEMMA 2. *The action of components of the vertex operator $Y([\rho_\pm(-\frac{1}{2}), z)$ has the following properties:*
(5)

$$\{\rho_+(-\tfrac{1}{2})\}_k[\rho_+(-\tfrac{1}{2})] = 0 \quad \text{and} \quad \{\rho_-(-\tfrac{1}{2})\}_k[\rho_-(-\tfrac{1}{2})] = 0 \quad \text{for any } k \geq 0.$$

PROOF. We apply Lemma 1 with $n = 0$ to the vertex operator $\{\rho_+(-\frac{1}{2})\}_k = \sum :a_1(m_1+\frac{1}{2})\cdots a_N(m_N+\frac{1}{2}):$. We then see that we must have $m_j \geq 0$ for at least one j. But then, $a_j(m_j+\frac{1}{2})$ annihilates $[\rho_+(-\frac{1}{2})] = a_1(-\frac{1}{2})\cdots a_N(-\frac{1}{2})\Omega$. Hence $\{\rho_+(-\frac{1}{2})\}_k[\rho_+(-\frac{1}{2})] = 0$. We can prove the second identity in a similar way. $\square$

We now calculate commutation relations of vertex operators associated to complex volume forms. Let $[\ ,\ ]_-$ and $[\ ,\ ]_+$ denote commutator and anti-commutator, respectively. Together, $[\ ,\ ]_\pm$ denotes a graded commutator: for any elements a, b in the graded vector space, we have $[a, b]_\pm = ab - (-1)^{|a||b|}ba$, where $|a|$, $|b|$ are the weights of a and $|b|$.

PROPOSITION 3 (VERTEX OPERATORS FOR VOLUME FORMS: (I)). *Let $\rho_\pm = \rho_\pm(-\frac{1}{2})$ be complex volume forms in $\bigwedge^N A^\pm(-\frac{1}{2})$. For any $m, n \in \mathbb{Z}$, we have*
(6)

$$[\{\rho_+\}_m, \{\rho_+\}_n]_\pm = 0, \quad [\{\rho_-\}_m, \{\rho_-\}_n]_\pm = 0, \quad \text{where } \pm = \begin{cases} - & \text{if } N \text{ is even,} \\ + & \text{if } N \text{ is odd.} \end{cases}$$

Thus, the vertex operator $Y([\rho_+], z) = \sum_{k\in\mathbb{Z}}\{\rho_+\}_k z^{-k-1}$ generates an infinite dimensional abelian Lie algebra when N is even, and an infinite dimensional exterior algebra when N is odd. Similar statement holds for the vertex operator $Y([\rho_-], z)$.

PROOF. The commutator formula for vertex operators gives

$$[\{\rho_+\}_m, \{\rho_+\}_n]_\pm = \sum_{k \geq 0} \binom{m}{k} \{\{\rho_+\}_k[\rho_+]\}_{m+n-k} \quad \text{for any} \quad m, n \in \mathbb{Z}.$$

Due to Lemma 2, $\{\rho_+\}_k[\rho_+] = 0$ for $k \geq 0$. So, each term in the above summation is zero. This proves our commutator formula for $\{\rho_+\}_*$. Similarly for $\{\rho_-\}_*$'s. $\square$

Next, we consider crossing commutation relations between $\{\rho_+\}_*$ and $\{\rho_-\}_*$. These turn out to be highly nontrivial and we only have partial calculations. Let $\rho_\pm = \rho_\pm(-\frac{1}{2}) \in \bigwedge^N A^\pm(-\frac{1}{2})$. Then

(7)
$$\{\rho_-\}_k[\rho_+] = \sum_{\substack{\sum_{j=1}^N m_j \\ =k+1-N}} :a_1^*(m_1 + \tfrac{1}{2}) \cdots a_N^*(m_N + \tfrac{1}{2}): a_1(-\tfrac{1}{2}) \cdots a_N(-\tfrac{1}{2})\Omega,$$

$$\{\rho_+\}_k[\rho_-] = \sum_{\substack{\sum_{j=1}^N m_j \\ =k+1-N}} :a_1(m_1 + \tfrac{1}{2}) \cdots a_N(m_N + \tfrac{1}{2}): a_1^*(-\tfrac{1}{2}) \cdots a_N^*(-\tfrac{1}{2})\Omega.$$

For notations of elements appearing in the next proposition, see (2) of §3.5.

PROPOSITION 4. *Elements* $\{\rho_\pm\}_k[\rho_\mp]$ *are given as follows for* $k \geq N - 4$:

(i) $\qquad\qquad \{\rho_+\}_k[\rho_-] = \{\rho_-\}_k[\rho_+] = 0 \quad$ *for* $k \geq N$,

(ii) $\qquad (-1)^{\frac{N(N-1)}{2}} \{\rho_+\}_{N-1}[\rho_-] = (-1)^{\frac{N(N-1)}{2}} \{\rho_-\}_{N-1}[\rho_+] = \Omega,$

(iii) $(-1)^{\frac{N(N-1)}{2}} \{\rho_+\}_{N-2}[\rho_-] = [h], \qquad (-1)^{\frac{N(N-1)}{2}} \{\rho_-\}_{N-2}[\rho_+] = -[h],$

(iv) $\quad \begin{cases} (-1)^{\frac{N(N-1)}{2}} \{\rho_+\}_{N-3}[\rho_-] = \frac{1}{2}[h^2] - [\omega_+], \\ (-1)^{\frac{N(N-1)}{2}} \{\rho_-\}_{N-3}[\rho_+] = \frac{1}{2}[h^2] - [\omega_-], \end{cases}$

(v) $\quad \begin{cases} (-1)^{\frac{N(N-1)}{2}} \{\rho_+\}_{N-4}[\rho_-] = -[\omega_+(2,0)] + \frac{1}{6}[h^3] - [\omega_+ \cdot h], \\ (-1)^{\frac{N(N-1)}{2}} \{\rho_-\}_{N-4}[\rho_+] = -[\omega_-(2,0)] - \frac{1}{6}[h^3] + [\omega_- \cdot h]. \end{cases}$

PROOF. We prove these identities for $\{\rho_-\}_*[\rho_+]$. Identities for $\{\rho_+\}_*[\rho_-]$ follow immediately by applying complex conjugation operator $*$, and by using $h^* = -h$ and $\omega_\pm^* = \omega_\mp$.

(i) When $k \geq N$, the corresponding condition on m_j's is $m_1 + \cdots + m_N = 1 - N + k \geq 1$. So, there exists at least one j for which $m_j \geq 1$. But then $a_j^*(m_j + \frac{1}{2})$ annihilates $[\rho_+]$. Hence $\{\rho_-\}_k[\rho_+] = 0$ in this case.

(ii) When $k = N - 1$, the condition on m_j's is $m_1 + \cdots + m_N = 0$. For nontriviality of the action, all m_j must be non-positive. The only possibility is $m_1 = \cdots = m_N = 0$, and the summation reduces to a single term and we have

$$\{\rho_-\}_{N-1}[\rho_+] =: a_1^*(\tfrac{1}{2}) \cdots a_N^*(\tfrac{1}{2}): a_1(-\tfrac{1}{2}) \cdots a_N(-\tfrac{1}{2}) \cdot \Omega = (-1)^{\frac{N(N-1)}{2}} \Omega.$$

(iii) When $k = N - 2$, the condition on m_j's is $m_1 + \cdots + m_N = -1$. Since m_j's must be non-positive for nontriviality of the action of vertex operators, the only possibility is that $(m_1, \ldots, m_N)$ is a permutation of $(-1, 0, \ldots, 0)$. Then

$$\{\rho_-\}_{N-2}[\rho_+] = \sum_{i=1}^{N} :a_1^*(\tfrac{1}{2}) \cdots a_{i-1}^*(\tfrac{1}{2})a_i^*(-\tfrac{1}{2})a_{i+1}^*(\tfrac{1}{2}) \cdots a_N^*(\tfrac{1}{2}):$$
$$\cdot a_1(-\tfrac{1}{2}) \cdots a_N(-\tfrac{1}{2})\Omega.$$

Removing the normal ordering symbol and then reversing the order of operators a_j^*, we see that the sign we get is (-1) to the $N(N-1)/2$-th power and all terms cancel out except $a_i^*(-\tfrac{1}{2})a_i(-\tfrac{1}{2})\Omega$. So the above is equal to

$$(-1)^{N(N-1)/2} \sum_{j=1}^{N} a_i^*(-\tfrac{1}{2})a_i(-\tfrac{1}{2})\Omega = (-1)^{N(N-1)/2}[-h].$$

This proves (iii).

(iv) When $k = N - 3$, a similar consideration shows that $(m_1, \ldots, m_N)$ must be a permutation of $(-2, 0, \ldots, 0)$ or $(-1, -1, 0, \ldots, 0)$. In the first case, the corresponding summation is given by

$$\sum_{i=1}^{N} :a_1^*(\tfrac{1}{2}) \cdots a_{i-1}^*(\tfrac{1}{2})a_i^*(-\tfrac{3}{2})a_{i+1}^*(\tfrac{1}{2}) \cdots a_N^*(\tfrac{1}{2}): a_1(-\tfrac{1}{2}) \cdots a_N(-\tfrac{1}{2}) \cdot \Omega$$
$$= (-1)^{\frac{N(N-1)}{2}} \sum_{i=1}^{N} a_i^*(-\tfrac{3}{2})a_i(-\tfrac{1}{2})\Omega = -(-1)^{\frac{N(N-1)}{2}}[\omega_-(1,0)].$$

In the second case, the relevant summation is given by

$$\sum_{1<i<j<N} :a_1^*(\tfrac{1}{2}) \cdots a_i^*(-\tfrac{1}{2}) \cdots a_j^*(-\tfrac{1}{2}) \cdots a_N^*(\tfrac{1}{2}): a_1(-\tfrac{1}{2}) \cdots a_N(-\tfrac{1}{2})\Omega$$
$$= (-1)^{\frac{N(N-1)}{2}} \sum_{1<i<j<N} a_i^*(-\tfrac{1}{2})a_i(-\tfrac{1}{2})a_j^*(-\tfrac{1}{2})a_j(-\tfrac{1}{2})\Omega = \tfrac{1}{2}(-1)^{\frac{N(N-1)}{2}}[h^2].$$

From these calculations, formula (iv) follows.

(v) When $k = N-4$, a term in the summation (7) can be nontrivial only when $(m_1, \ldots, m_N)$ is a permutation of (a) $(-3, 0, \ldots, 0)$, (b) $(-1, -1, -1, 0 \ldots, 0)$, or (c) $(-2, -1, 0, \ldots, 0)$. Summation of relevant terms in the case (a) is given by

$$\sum_{j=1}^{N} :a_1^*(\tfrac{1}{2}) \cdots a_j^*(-\tfrac{5}{2}) \cdots a_N^*(\tfrac{1}{2}): a_1(-\tfrac{1}{2}) \cdots a_N(-\tfrac{1}{2})\Omega$$
$$= (-1)^{\frac{N(N-1)}{2}} \sum_{j=1}^{N} a_j^*(-\tfrac{5}{2})a_j(-\tfrac{1}{2})\Omega = (-1)^{\frac{N(N-1)}{2}}[-\omega_-(2,0)].$$

For the case (b), we have a summation over $1 \le i < j < k \le N$ of $\binom{N}{3}$ terms and it is given by

$$\sum_{1 \le i < j < k \le N} : a_1^*(\tfrac{1}{2}) \cdots a_i^*(-\tfrac{1}{2}) \cdots a_j^*(-\tfrac{1}{2}) \cdots a_k^*(-\tfrac{1}{2}) \cdots a_N^*(\tfrac{1}{2}): a_1(-\tfrac{1}{2}) \cdots a_N(-\tfrac{1}{2})\Omega$$

$$= (-1)^{\frac{N(N-1)}{2}} \sum_{1 \le i < j < k \le N} a_i^*(-\tfrac{1}{2})a_i^{\zeta} - \tfrac{1}{2})a_j^*(-\tfrac{1}{2})a_j^{\zeta} - \tfrac{1}{2})a_k^*(-\tfrac{1}{2})a_k^{\zeta} - \tfrac{1}{2})\Omega$$

$$= \tfrac{1}{6}(-1)^{\frac{N(N-1)}{2}} \sum_{1 \le i,j,k \le N} a_i^*(-\tfrac{1}{2})a_i(-\tfrac{1}{2})a_j^*(-\tfrac{1}{2})a_j(-\tfrac{1}{2})a_k^*(-\tfrac{1}{2})a_k(-\tfrac{1}{2})\Omega$$

$$= (-1)^{\frac{N(N-1)}{2}}(-\tfrac{1}{6})[h^3].$$

In the last case (c), the corresponding summation is calculated as follows:

$$\sum_{1 \le i < j \le N} : a_1^*(\tfrac{1}{2}) \cdots a_i^*(-\tfrac{3}{2}) \cdots a_j^*(-\tfrac{1}{2}) \cdots a_N^*(\tfrac{1}{2}): a_1(-\tfrac{1}{2}) \cdots a_N(-\tfrac{1}{2})\Omega$$

$$+ \sum_{1 \le i < j \le N} : a_1^*(\tfrac{1}{2}) \cdots a_i^*(-\tfrac{1}{2}) \cdots a_j^*(-\tfrac{3}{2}) \cdots a_N^*(\tfrac{1}{2}): a_1(-\tfrac{1}{2}) \cdots a_N(-\tfrac{1}{2})\Omega$$

$$= (-1)^{\frac{N(N-1)}{2}} \sum_{1 \le i < j \le N} a_i^*(-\tfrac{3}{2})a_i(-\tfrac{1}{2})a_j^*(-\tfrac{1}{2})a_j(-\tfrac{1}{2})\Omega$$

$$+ (-1)^{\frac{N(N-1)}{2}} \sum_{1 \le i < j \le N} a_i^*(-\tfrac{1}{2})a_i(-\tfrac{1}{2})a_j^*(-\tfrac{3}{2})a_j(-\tfrac{1}{2})\Omega$$

$$= (-1)^{\frac{N(N-1)}{2}} \sum_{1 \le i,j \le N} a_i^*(-\tfrac{3}{2})a_i(-\tfrac{1}{2})a_j^*(-\tfrac{1}{2})a_j(-\tfrac{1}{2})\Omega = (-1)^{\frac{N(N-1)}{2}} [\omega_- \cdot h].$$

Combining calculations in cases (a), (b), (c), we obtain formula (v). □

REMARK. Note that the above calculations for $(-1)^{\frac{N(N-1)}{2}}\{\rho_-\}_k[\rho_+]$ and $(-1)^{\frac{N(N-1)}{2}}\{\rho_-\}_k[\rho_+]$ with $k \ge N - 4$ don't have N in them. A moment's reflection on calculations shows that this is the case in general because nontrivial results come from those terms which arise due to the difference between N and k, although it is rather difficult to see a general pattern for general k.

The above calculations determine graded commutators $[\{\rho_-\}_m, \{\rho_+\}_n]_\pm$ for $m, n \in \mathbb{Z}$ for $N \le 4$ using the commutator formula:

$$[\{\rho_-\}_m, \{\rho_+\}_n]_\pm = \{\rho_-\}_m\{\rho_+\}_n - (-1)^{N-1}\{\rho_+\}_n\{\rho_-\}_m$$

(8)
$$= \sum_{k \ge 0} \binom{m}{k}\{\{\rho_-\}_k[\rho_+]\}_{m+n-k}.$$

PROPOSITION 5 (VERTEX OPERATORS FOR VOLUME FORMS: (II)). *Graded commutators $[\{\rho_-\}_m, \{\rho_+\}_n]_\pm$ for vertex operators associated to volume forms $[\rho_\pm]$ are given as follows for $0 \le N \le 4$:*

(9) $[\{\rho_-\}_m, \{\rho_+\}_n]_- = h(m+n) - m \cdot \mathrm{Id} \cdot \delta_{m+n,0}$ *when $N = 2$,*

(10)

$$[\{\rho_-\}_m, \{\rho_+\}_n]_+ = -\frac{1}{2}\{h^2\}_{m+n} + D(m+n-1)$$

$$-\frac{(n-m)}{2}h(m+n-1) - \frac{m(m-1)}{2}\cdot \mathrm{Id}\cdot\delta_{m+n-1,0} \quad \text{when} \quad N=3,$$

(11)

$$[\{\rho_-\}_m, \{\rho_+\}_n]_- = \frac{1}{2}\{h(-\tfrac{3}{2})\}_{m+n} - \frac{1}{6}\{h^3\}_{m+n} + \{\omega\cdot h\}_{m+n}$$

$$-\frac{(n-m)}{4}\{h^2\}_{m+n-1} + \frac{(n-m)}{2}D(m+n-2)$$

$$-\frac{1}{2}\left\{\binom{m}{2}+\binom{n}{2}\right\}h(m+n-2)+\binom{m}{3}\delta_{m+n-2,0}\cdot\mathrm{Id} \quad \text{when} \quad N=4.$$

PROOF. This follows from Proposition 4 and the commutator formula, except that we symmetrize our calculation. When $N=2$, formula (10) is a straightforward consequence of the commutator formula. For the other two cases, we calculate commutation relations in two different ways, switching ρ_- and ρ_+. We then take their average. The reason for doing this is that the commutator formula for vertex operators isn't symmetric in two entries of the graded commutator.

When $N=3$, from the commutator formula we have

$$[\{\rho_-\}_m, \{\rho_+\}_n]_+ = -\tfrac{1}{2}\{h^2\}_{m+n}+\{\omega_-\}_{m+n}+m\cdot h(m+n-1)-\tfrac{m(m-1)}{2}\cdot\mathrm{Id}\cdot\delta_{m+n,1}.$$

Switching $\rho_\pm$ and applying the commutator formula again, we get

$$[\{\rho_+\}_n, \{\rho_-\}_m]_+ = -\tfrac{1}{2}\{h^2\}_{m+n}+\{\omega_+\}_{m+n}-n\cdot h(m+n-1)-\tfrac{m(m-1)}{2}\cdot\mathrm{Id}\cdot\delta_{m+n,1}.$$

Since we are calculating anticommutators here, the above two are equal, although the expressions look different. The average of the above two gives the result for $N=3$ since $\omega=\frac{1}{2}(\omega_+ + \omega_-)$.

For the case $N=4$, the commutator formula applied in two ways give

$$[\{\rho_-\}_m, \{\rho_+\}_n] = -\{\omega_-(2,0)\}_{m+n} - \tfrac{1}{6}\{h^3\}_{m+n} + \{\omega_-\cdot h\}_{m+n}$$

$$+m\left(\tfrac{1}{2}\{h^2\}_{m+n-1}-\{\omega_-\}_{m+n-1}\right)-\binom{m}{2}\{h\}_{m+n-2}+\binom{m}{3}\cdot\mathrm{Id}\cdot\delta_{m+n-2,0},$$

$$[\{\rho_+\}_n, \{\rho_-\}_m] = -\{\omega_+(2,0)\}_{m+n} + \tfrac{1}{6}\{h^3\}_{m+n} - \{\omega_+\cdot h\}_{m+n}$$

$$+n\left(\tfrac{1}{2}\{h^2\}_{m+n-1}-\{\omega_+\}_{m+n-1}\right)+\binom{n}{2}\{h\}_{m+n-2}+\binom{n}{3}\cdot\mathrm{Id}\cdot\delta_{m+n-2,0}.$$

The only difference of the above two formulae is sign. To simplify the difference of the above two, we use the following formulae:

$$\{\omega_+(2,0)-\omega_-(2,0)\}_{m+n} = \{h(-\tfrac{3}{2})\}_{m+n}-\tfrac{1}{2}\{D(-1)D(-1)[h]\}_{m+n}$$

$$= \{h(-\tfrac{3}{2})\}_{m+n}-\frac{(m+n)(m+n-1)}{2}\{h\}_{m+n-2},$$

$-m\{\omega_-\}_{m+n-1} + n\{\omega_+\}_{m+n-1}$

$$= \frac{m+n}{2}\{\omega_+ - \omega_-\}_{m+n-1} + \frac{n-m}{2}\{\omega_+ + \omega_-\}_{m+n-1}$$

$$= \frac{m+n}{2}\{D(-1)[h]\}_{m+n-1} + (n-m)\{\omega\}_{m+n-1}$$

$$= \frac{(m+n)(m+n-1)}{2}h(m+n-2) + (n-m)D(m+n-2).$$

Here, we used relations $D(-1)(h) = [\omega_+] - [\omega_-]$ and $D(-1)[\omega_\pm] = -2[\omega_\pm(2,0)] \pm [h(-\frac{3}{2})]$ from Proposition 1 of §3.5 and $2[\omega] = [\omega_+] + [\omega_-]$ by definition. Also note that $\binom{n}{3}\delta_{m+n-2,0} = -\binom{m}{3}\delta_{m+n-2,0}$. Our formula follows from these.   □

Next, we calculate commutation relations between vertex operators associated to complex volume forms and those associated to $[h]$, $[h^2]$, $[\omega]$ in V.

LEMMA 6. *The action of vertex operators $\{h\}_k$ and $\{h^2\}_k$ for $k \geq 0$ on complex volume forms $[\rho_\pm]$ is described as follows:*

$$(13) \qquad \begin{aligned} \{h\}_0[\rho_\pm] &= \pm N[\rho_\pm], \qquad \{h\}_k[\rho_\pm] = 0 \ \text{for} \ k \geq 1, \\ \{h^2\}_k[\rho_\pm] &= 0 \ \text{for} \ k \geq 0. \end{aligned}$$

PROOF. Since $\rho_\pm$ is a product of vectors in $A(-\frac{1}{2})$, vectors in $A(n+\frac{1}{2})$ with $n > 0$ annihilate $[\rho_\pm]$ by Clifford multiplication. Now operators $\{h\}_k$ with $k \geq 0$ are given by

$$\{h\}_k = \sum_{j=1}^{N}\Big\{ \sum_{m_1+m_2=k-1} :a_j(m_1 + \tfrac{1}{2})a_j^*(m_2 + \tfrac{1}{2}): \Big\}.$$

Emphasizing those terms in $\{h\}_0$ which act nontrivially on $[\rho_\pm]$, we have

$$\{h\}_0 = \sum_{j=1}^{N}\big(\cdots - a_j^*(-\tfrac{1}{2})a_j(\tfrac{1}{2}) + a_j(-\tfrac{1}{2})a_j^*(\tfrac{1}{2}) + \cdots\big).$$

The action of $\{h\}_0$ on $[\rho_\pm]$ is then given by

$$\{h\}_0[\rho_+] = \sum_{j=1}^{N}(-1)^{j-1}a_j(-\tfrac{1}{2})a_1(-\tfrac{1}{2})\cdots\widehat{a_j(-\tfrac{1}{2})}\cdots a_N(-\tfrac{1}{2})\cdot\Omega = N[\rho_+],$$

$$\{h\}_0[\rho_-] = -\sum_{j=1}^{N}(-1)^{j-1}a_j^*(-\tfrac{1}{2})a_1^*(-\tfrac{1}{2})\cdots\widehat{a_j^*(-\tfrac{1}{2})}\cdots a_N^*(-\tfrac{1}{2})\cdot\Omega = -N[\rho_-].$$

This proves the first identity. When $k > 0$, in the expression of $\{h\}_k$ above we have $m_1 + m_2 \geq 0$. The operator $:a_j(m_1 + \frac{1}{2})a_j^*(m_2 + \frac{1}{2}):$ annihilates $[\rho_\pm]$ if $m_1 \geq 1$ or $m_2 \geq 1$. When $m_1 = m_2 = 0$, it is easy to see that the corresponding operator also annihilates $[\rho_\pm]$. Hence all operators appearing in the expression of $\{h\}_k$ for any $k \geq 1$ act trivially on $[\rho_\pm]$. This proves the second statement.

For the third identity, we first note that

$$\{h^2\}_k = \sum_{1 \leq i \neq j \leq N} \sum_{\substack{m_1+m_2+m_3 \\ +m_4=k+1}} : a_i(m_1 - \tfrac{1}{2})a_i^*(m_2 - \tfrac{1}{2})a_j(m_3 - \tfrac{1}{2})a_j^*(m_4 - \tfrac{1}{2}): \, .$$

We consider $\{h^2\}_k[\rho_+]$. Since $\rho_+ \in \bigwedge^N A^+(-\tfrac{1}{2})$, vectors $a_i^*(m_2 - \tfrac{1}{2})$ and $a_j^*(m_4 - \tfrac{1}{2})$ can act nontrivially on $[\rho_+]$ only when $m_2, m_4 \leq 1$. Similarly, $a_i(m_1 - \tfrac{1}{2})$ and $a_j(m_3 - \tfrac{1}{2})$ can act nontrivially on $[\rho_+]$ only when $m_1, m_3 \leq -1$. Here note that when $m_1 = 0$ or $m_3 = 0$, both $a_i(-\tfrac{1}{2})$ and $a_j(-\tfrac{1}{2})$ annihilate $[\rho_+]$. Hence for nontriviality of the action of $\{h^2\}_k$ on $[\rho_+]$, a condition $m_1 + m_2 + m_3 + m_4 \leq 0$ is necessary. However, when $k \geq 0$, $\{h^2\}_k$ doesn't have a term satisfying this condition on m_j's. Hence, $\{h^2\}_k[\rho_+] = 0$ for $k \geq 0$. By taking complex conjugate, we also have $\{h^2\}_k[\rho_-] = 0$ for $k \geq 0$ since $h^* = -h$, $(\rho_+)^* = \rho_-$. $\square$

Now we apply the commutator formula for vertex operators to calculate commutation relations of $\{\rho_\pm\}_*$ with other operators. Let

$$Y([\rho_\pm], z) = \sum_{n \in \mathbb{Z} - \frac{N}{2}} \rho_\pm(n) z^{-n - \frac{N}{2}}.$$

Then the operator $\rho_\pm(n) = \{\rho_\pm\}_{n+\frac{N}{2}-1}$ lowers weight by n.

PROPOSITION 7. *For any* $m, n \in \mathbb{Z}$, *we have*

$$
\begin{aligned}
& [h(m), \rho_\pm(n)] = \pm N \cdot \rho_\pm(n + m), \\
(14) \quad & [D(m), \rho_\pm(n)] = \left(n - (\tfrac{N}{2} - 1)m\right)\rho_\pm(n + m), \\
& [\{h^2\}_{m+1}, \rho_\pm(n)] = 0.
\end{aligned}
$$

PROOF. For the first identity, the commutator formula and (13) says that

$$[h(m), \rho_\pm(n)] = [\{h\}_m, \{\rho_\pm\}_{n+\frac{N}{2}-1}] = \sum_{k \geq 0} \binom{m}{k} \{\{h\}_k[\rho_\pm]\}_{m+n+\frac{N}{2}-1-k}$$

$$= \{\{h\}_0[\rho_\pm]\}_{m+n+\frac{N}{2}-1} = \pm N \{\rho_\pm\}_{m+n+\frac{N}{2}-1} = \pm N \rho_\pm(m + n).$$

For the second commutation relation, from Proposition 8 of §3.5 we know that $[\rho_\pm] \in (V)_{\frac{N}{2}}$ are conformal highest weight vectors. So, by Proposition 9 of §3.5 we have the second formula.

The last commutation relation follows easily since $\{h^2\}_k[\rho_\pm] = 0$ for $k \geq 0$ by (13). This completes the proof of Proposition 7. $\square$

§3.8   Symplectic Virasoro algebras

Throughout this section, N denotes an even positive integer, $N = 2N'$. Among Lie algebras we have discussed so far, the symplectic Lie algebra $\mathfrak{sp}(N') \subset \mathfrak{u}(N)$ is the most interesting one. Although there are many $\mathfrak{sp}(N')$-invariant vectors in the vertex operator super algebra V [§3.4 Proposition 12], there are two distinguished $\mathfrak{sp}(N')$-invariant vectors of weight 2: the canonical Virasoro element $[\omega]$ and the symplectic Casimir element $[\phi_{\mathfrak{sp}(N')}]$ given by

$$(1) \qquad [\phi_{\mathfrak{sp}(N')}] = -\frac{1}{2}[h^2] + 2[xx^*] \in (V)_2.$$

For notations, see (2) of §3.5. Let associated vertex operators be

$$(2) \qquad Y([\omega], \zeta) = \sum_{n \in \mathbb{Z}} D(n)\zeta^{-n-2}, \qquad Y([\phi_{\mathfrak{sp}(N')}], \zeta) = \sum_{n \in \mathbb{Z}} Q(n)\zeta^{-n-2}.$$

It turns out that within the vector space of operators spanned by components of these vertex operators and a 1-dimensional center, Lie bracket closes. We call this Lie algebra *the symplectic Virasoro algebra* and we denote it by $\mathbf{Vir}_{\mathfrak{sp}(N')}$:

$$(3) \qquad \mathbf{Vir}_{\mathfrak{sp}(N')} = \bigoplus_{n \in \mathbb{Z}} \mathbb{C}D(n) \oplus \bigoplus_{n \in \mathbb{Z}} \mathbb{C}Q(n) \oplus \mathbb{C}\mathrm{Id},$$

where Id is the identity operator. We calculate commutation relations of this Lie algebra, and study its structure.

THEOREM 1 (SYMPLECTIC VIRASORO ALGEBRA). *Commutation relations of the symplectic Virasoro algebra* $\mathbf{Vir}_{\mathfrak{sp}(N')}$ *is described as follows:*

$$(4) \qquad [D(m), D(n)] = (n - m)D(m + n) + \frac{m(m^2 - 1)}{12}N \cdot \delta_{m+n,0} \cdot \mathrm{Id},$$

$$(5) \qquad [D(m), Q(n)] = (n - m)Q(m + n),$$

$$(6) \quad [Q(m), Q(n)] = 3(n - m)(N + 1)D(m + n) + (n - m)(N - 2)Q(m + n)$$
$$+ \frac{m(m^2 - 1)}{4}N(N + 1) \cdot \delta_{m+n,0} \cdot \mathrm{Id}.$$

In view of (6), operators $Q(m)$ don't close among themselves under Lie bracket even with the central extension. As with the unitary Virasoro algebra, we choose a "canonical" set of basis vectors in the symplectic Virasoro algebra so that commutation relations simplify and "close" within smaller subalgebras. We use notation $\phi_{\mathfrak{sp}}$ to denote $\phi_{\mathfrak{sp}(N')}$ for simplicity when $N = 2N'$ is understood. Let

$$(7) \qquad [\sigma] = \frac{1}{N + 4}(3[\omega] + [\phi_{\mathfrak{sp}}]), \qquad [\tau] = \frac{1}{N + 4}((N + 1)[\omega] - [\phi_{\mathfrak{sp}}]),$$
$$[\sigma] + [\tau] = [\omega].$$

The third relation in (7) shows that elements σ and τ give a decomposition of the canonical Virasoro element ω using the symplectic Casimir element ϕ_{sp}. We consider vertex operators associated to $[\sigma]$ and $[\tau] \in V_2$:

$$(8) \qquad Y([\sigma], \zeta) = \sum_{n \in \mathbb{Z}} S(n) \zeta^{-n-2}, \qquad Y([\tau], \zeta) = \sum_{n \in \mathbb{Z}} T(n) \zeta^{-n-2}.$$

Components of these vertex operators are preferred canonical basis vectors for the symplectic Virasoro algebra. In terms of $D(n)$'s and $Q(n)$'s, they are

$$(9) \qquad S(n) = \frac{1}{N+4}(3D(n) + Q(n)), \qquad T(n) = \frac{1}{N+4}((N+1)D(n) - Q(n)),$$
$$S(n) + T(n) = D(n).$$

The symplectic Virasoro algebra can also be given as follows:

$$(3') \qquad \mathbf{Vir}_{\mathfrak{sp}(N')} = \bigoplus_{n \in \mathbb{Z}} \mathbb{C} S(n) \oplus \bigoplus_{n \in \mathbb{Z}} \mathbb{C} T(n) \oplus \mathbb{C} \mathrm{Id}.$$

The reason why we call these basis vectors canonical is explained by corresponding commutation relations among new basis vectors.

THEOREM 2 (CANONICAL COMMUTATION RELATIONS FOR $\mathbf{Vir}_{\mathfrak{sp}(N')}$). *With respect to the canonical basis $(3')$ of $\mathbf{Vir}_{\mathfrak{sp}(N')}$, commutation relations of the symplectic Virasoro algebra are described by the following formulae:*

$$(10) \, [S(m), S(n)] = (n - m)S(m + n) + \frac{m(m^2 - 1)}{12} \left\{ \frac{3N}{(N+4)} \right\} \delta_{m+n,0} \cdot \mathrm{Id},$$

(11)

$$[T(m), T(n)] = (n - m)T(m + n) + \frac{m(m^2 - 1)}{12} \left\{ \frac{N(N+1)}{(N+4)} \right\} \delta_{m+n,0} \cdot \mathrm{Id},$$

$$(12) \qquad [T(m), S(n)] = 0.$$

Two subspaces $\hat{S} = \bigoplus_{n \in \mathbb{Z}} \mathbb{C} S(n) \oplus \mathbb{C} \mathrm{Id}$, $\hat{T} = \bigoplus_{n \in \mathbb{Z}} \mathbb{C} T(n) \oplus \mathbb{C} \mathrm{Id}$ of the symplectic Virasoro algebra are closed under Lie bracket by Theorem 2 and so they are actually Lie subalgebras. Furthermore, these Lie subalgebras $\hat{S}$ and $\hat{T}$ commute, and hence they are in fact ideals of $\mathbf{Vir}_{\mathfrak{sp}(N')}$. Recall that a similar situation ocurred for the unitary Virasoro algebra [§3.6 Theorem 2].

Note that the sum of central charges for subalgebras $\hat{S}$, $\hat{T}$ is $\frac{3N}{N+4} + \frac{N(N+1)}{N+4} = N$. This has to be the case because operators $S(n) + T(n) = D(n)$ generate the canonical Virasoro algebra with central charge N.

REMARKS. (i) When $N = 2$, vectors $[h^2]$ and $[xx^*]$ are not linearly independent in V_2 and we have a linear relation $[h^2] + 2[xx^*] = 0$. So, $[\phi_{\mathrm{sp}}] = -\frac{1}{2}[h^2] + 2[xx^*] = -\frac{3}{2}[h^2]$ when $N = 2$. This implies $[\sigma] = [\theta]$, $[\tau] = [\lambda]$, where $[\theta]$, $[\lambda] \in (V)_2$ generate the unitary Virasoro algebra [§3.6 (6)]. On operator level, we then have $Q(n) = -\frac{3}{2}H(n)$, $S(n) = \Theta(n)$, and $T(n) = \Lambda(n)$. Thus, when $N = 2$, the symplectic Virasoro algebra coincides with the unitary Virasoro algebra. This is not surprising because $\mathfrak{sp}(1) = \mathfrak{su}(2)$.

(ii) For higher $N = 2N'$ with $N' > 1$, symplectic Virasoro algebras and unitary Virasoro algebras are different. We can show that $\mathbb{C}[\omega] \oplus \mathbb{C}[h^2] = (V)_2^{\mathfrak{u}(N)}$ and $\mathbb{C}[\omega] \oplus \mathbb{C}[\phi_{\mathfrak{sp}}] = (V)_2^{\mathrm{Sp}(N')\mathrm{Sp}(1)}$, which are proved in §3.12. So, vectors generating unitary or symplectic Virasoro algebras can be realized as invariant subspaces of some Lie subalgebras of $\mathfrak{o}(2N)$. Since there doesn't seem to exist any Lie subalgebra $\mathfrak{g}$ of $\mathfrak{o}(2N)$ such that $(V)_2^{\mathfrak{g}} = \mathbb{C}[\omega] \oplus \mathbb{C}[h^2] \oplus \mathbb{C}[\phi_{\mathfrak{sp}}]$, it seems that it may not be fruitful to consider the structure of a Lie algebra generated by vertex operators associated to these three vectors of weight 2.

To prove Theorem 1, we have to calculate the action of vertex operators $\{\phi_{\mathfrak{sp}(N')}\}_*$ on the vector $[\phi_{\mathfrak{sp}(N')}]$. This is given in the next theorem.

THEOREM 3. Let $\phi_{\mathfrak{sp}} = \phi_{\mathfrak{sp}(N')}$. The following identities hold:

(13) $\{\phi_{\mathfrak{sp}}\}_0[\phi_{\mathfrak{sp}}] = 3(N + 1)D(-1)[\omega] + (N - 2)D(-1)[\phi_{\mathfrak{sp}}]$,

(14) $\{\phi_{\mathfrak{sp}}\}_1[\phi_{\mathfrak{sp}}] = -6(N + 1)[\omega] - 2(N - 2)[\phi_{\mathfrak{sp}}]$,

(15) $\{\phi_{\mathfrak{sp}}\}_2[\phi_{\mathfrak{sp}}] = 0$,

(16) $\{\phi_{\mathfrak{sp}}\}_3[\phi_{\mathfrak{sp}}] = \frac{3}{2}N(N + 1)\Omega$.

Among the above identities, it is easy to see the vanishing of $\{\phi_{\mathfrak{sp}}\}_2[\phi_{\mathfrak{sp}}]$ by checking invariant vectors. To see this, since $[\phi_{\mathfrak{sp}}] \in (V)_2^{\mathrm{Sp}(N')\mathrm{Sp}(1)}$ [§3.4 Proposition 13], $\{\phi_{\mathfrak{sp}}\}_2[\phi_{\mathfrak{sp}}]$ is also $\mathrm{Sp}(N')\mathrm{Sp}(1)$-invariant because in general $(V)^{\mathfrak{g}}$ is a vertex operator super algebra for any Lie subalgebra $\mathfrak{g} \subset \mathfrak{o}(2N)$ [§2.1]. Note that $\{\phi_{\mathfrak{sp}}\}_2[\phi_{\mathfrak{sp}}]$ has weight 1. But there is no nontrivial invariant vector of weight 1 from our calculation of $(V)^{\mathrm{Sp}(N')\mathrm{Sp}(1)}$ for low weight vectors in §3.12. Thus, we must have $\{\phi_{\mathfrak{sp}}\}_2[\phi_{\mathfrak{sp}}] = 0$. For other formulae, we must actually calculate. But comparing (13) and (14), we observe that $\{\phi_{\mathfrak{sp}}\}_0[\phi_{\mathfrak{sp}}] = -\frac{1}{2}D(-1)(\{\phi_{\mathfrak{sp}}\}_1[\phi_{\mathfrak{sp}}])$. This type of formula is also true for the unitary Casimir element $[\phi_{\mathfrak{u}(N)}]$ [§3.6 Proposition 3] and for the canonical Virasoro element $[\omega]$. We don't have an intrinsic explanation of this identity, except by direct calculation. Even so, we will prove similar results in a more general context later in Theorem 24 of §3.11.

We can now prove Theorem 1 from Theorem 3.

(PROOF OF THEOREM 1 ASSUMING THEOREM 3). Formula (4) is the usual commutation relation for the Virasoro algebra. For (6), from the general commutator formula for vertex operators,

$$[Q(m), Q(n)] = [\{\phi_{\mathfrak{sp}}\}_{m+1}, \{\phi_{\mathfrak{sp}}\}_{n+1}] = \sum_{0 \leq k \in \mathbb{Z}} \binom{m+1}{k} \{\{\phi_{\mathfrak{sp}}\}_k \phi_{\mathfrak{sp}}\}_{m+n+2-k}$$

$$= \{\{\phi_{\mathfrak{sp}}\}_0 \phi_{\mathfrak{sp}}\}_{m+n+2} + (m+1) \{\{\phi_{\mathfrak{sp}}\}_1 \phi_{\mathfrak{sp}}\}_{m+n+1}$$

$$+ \binom{m+1}{2} \{\{\phi_{\mathfrak{sp}}\}_2 \phi_{\mathfrak{sp}}\}_{m+n} + \binom{m+1}{3} \{\{\phi_{\mathfrak{sp}}\}_3 \phi_{\mathfrak{sp}}\}_{m+n-1} .$$

Substituting results in Theorem 3 into the above formula, we have

$$[Q(m), Q(n)] = 3(N+1)\{D(-1)[\omega]\}_{m+n+2} + (N-2)\{D(-1)[\phi_{\mathrm{sp}}]\}_{m+n+2}$$
$$- 6(N+1)(m+1)\{[\omega]\}_{m+n+1} - 2(m+1)(N-2)\{\phi_{\mathrm{sp}}\}_{m+n+1}$$
$$+ \binom{m+1}{3}\frac{3}{2}N(N+1)\{\Omega\}_{m+n-1}.$$

Recall that $\{D(-1)v\}_{n+1} = (n+1)\{v\}_n$ for any $v \in V$ [see §2.3 Lemma 2], and $\{\Omega\}_{m+n-1} = \delta_{m+n,0} \cdot \mathrm{Id}$. By definition $\{\omega\}_{n+1} = D(n)$ and $\{\phi_{\mathrm{sp}}\}_{n+1} = Q(n)$, and the above can be simplified and we obtain

$$[Q(m), Q(n)] = 3(n-m)(N+1)D(m+n) + (n-m)(N-2)Q(m+n)$$
$$+ \frac{m(m^2-1)}{4}N(N+1)\delta_{m+n,0} \cdot \mathrm{Id}.$$

This proves (6). For (5), since ϕ_{sp} is in the vector space $\bigwedge^* A(-\frac{1}{2})$, the symplectic Casimir element $[\phi_{\mathrm{sp}}]$ is a conformal highest weight vector by Proposition 8 of §3.5. Hence Corollary 10 of §3.5 implies (5). This completes the proof of Theorem 1. $\square$

Although commutation relations in Theorem 1 basically describes the structure of the symplectic Virasoro algebra, these formulae are not very enlightening. To understand intrinsic structure more explicitly, we need to reformulate the structure theorem. Since the symplectic Virasoro algebra is generated by vertex operators associated to $[\omega]$ and $[\phi_{\mathrm{sp}}]$, we take a linear combination of these vectors which best clarifies the structure of $\mathbf{Vir}_{\mathrm{sp}}$.

LEMMA 4. *Let $v = a[\omega] + b[\phi_{\mathrm{sp}}] \in V_2$ for $a, b \in \mathbb{C}$. Let the associated vertex operator be $Y(v, \zeta) = \sum_{n \in \mathbb{Z}} \{v\}_{n+1} \zeta^{-n-2}$. Then the commutation relation for $\{v\}_*$ is given by*

$$[\{v\}_{m+1}, \{v\}_{n+1}] = (n-m)\{2a + (N-2)b\}\{v\}_{m+n+1}$$
$$- (n-m)\{a^2 + ab(N-2) - 3b^2(N+1)\}D(m+n)$$
$$+ \frac{m(m^2-1)}{12}N\left(a^2 + 3b^2(N+1)\right)\delta_{m+n,0} \cdot \mathrm{Id}.$$

PROOF. Since $\{v\}_{n+1} = aD(n) + bQ(n)$, we have

$$[\{v\}_{m+1}, \{v\}_{n+1}] = a^2[D(m), D(n)] + ab[D(m), Q(n)]$$
$$+ ab[Q(m), D(n)] + b^2[Q(m), Q(n)].$$

We then apply Theorem 1 and eliminate $Q(n)$'s from the result using $bQ(n) = \{v\}_{n+1} - aD(n)$ to get the above identity. $\square$

From the formula in Lemma 4, vertex operators associated to v close under Lie bracket if the coefficient of $D(m+n)$ vanishes. This happens when a and b

satisfy $a^2 + ab(N-2) - 3b^2(N+1) = (a - 3b)(a + b(N+1)) = 0$, or $a = 3b$, $a = -b(N+1)$. The same condition on a, b can be also obtained when we express the commutation relation of $\{v\}_n$'s in terms of $\{v\}_n$'s and $Q(n)$'s and then set the coefficient of $Q(n)$ zero.

When $a = 3b$, we have $v = b(3[\omega] + [\phi_{sp}])$, and from Lemma 4 components $\{v\}_*$ of the associated vertex operator satisfy
(18)

$$[\{v\}_{m+1}, \{v\}_{n+1}] = (n-m)(N+4)b\{v\}_{m+n+1} + \frac{m(m^2-1)}{4}N(N+4)b^2\delta_{m+n,0}\cdot\text{Id}.$$

Similarly, when $a = -b(N+1)$, we have $v = -b((N+1)[\omega] - [\phi_{sp}])$, and the commutator for component operators $\{v\}_*$ is given by

$$(19) \quad [\{v\}_{m+1}, \{v\}_{n+1}] = -(n-m)(N+4)b\{v\}_{m+n+1}$$
$$+ \frac{m(m^2-1)}{12}N(N+1)(N+4)b^2\delta_{m+n,0}\cdot\text{Id}.$$

To put commutation relations into "canonical" forms, we choose b so that coefficients of $\{v\}_{m+n+1}$ in the right hand sides of (18) and (19) become $(n-m)$. Thus, in (18) we choose $b = 1/(N+4)$, and in (19) we let $b = -1/(N+4)$. With these choices of a and b, we arrive at elements in (7) and associated vertex operators satisfy commutation relations (10), (11), as can be seen from (18) and (19). Finally to see (12), we prepare one lemma.

LEMMA 5. *Lie subalgebras*

$$\hat{S} = \bigoplus_{n\in\mathbb{Z}}\mathbb{C}S(n) \oplus \mathbb{C}\text{Id}, \qquad \hat{T} = \bigoplus_{n\in\mathbb{Z}}\mathbb{C}T(n) \oplus \mathbb{C}\text{Id}$$

of the symplectic Virasoro algebra are invariant under Lie bracket with Virasoro operators: we have
(20)

$$[D(m), S(n)] = (n-m)S(m+n) + \frac{m(m^2-1)}{12}\left\{\frac{3N}{N+4}\right\}\delta_{m+n,0}\cdot\text{Id},$$

$$[D(m), T(n)] = (n-m)T(m+n) + \frac{m(m^2-1)}{12}\left\{\frac{N(N+1)}{N+4}\right\}\delta_{m+n,0}\cdot\text{Id}.$$

PROOF. Using (9) and Theorem 1, we have

$$[D(m), S(n)] = \frac{3}{N+4}[D(m), D(n)] + \frac{1}{N+4}[D(m), Q(n)]$$

$$= \frac{3}{N+4}\left((n-m)D(m+n) + \frac{m(m^2-1)}{12}N\delta_{m+n,0}\cdot\text{Id}\right)$$

$$+\frac{1}{N+4}(n-m)Q(m+n) = (n-m)S(m+n) + \frac{m(m^2-1)}{4}\left\{\frac{N}{N+4}\right\}\delta_{m+n,0}\cdot\text{Id}.$$

This proves the first formula. The second formula can be shown similarly. $\square$

To see (12), since $T(m) = D(m) - S(m)$, we have

$$[T(m), S(n)] = [D(m), S(n)] - [S(m), S(n)]$$

$$= (n-m)S(m+n) + \frac{m(m^2-1)}{12}\left\{\frac{3N}{N+4}\right\}\delta_{m+n,0}\cdot\text{Id}$$

$$- (n-m)S(m+n) - \frac{m(m^2-1)}{12}\left\{\frac{3N}{(N+4)}\right\}\delta_{m+n,0}\cdot\text{Id} = 0.$$

This completes the proof of Theorem 2. Theorem 3 remains to be shown. Its proof occupies the rest of this section.

Proof of the formula for $\{\phi_{sp}\}_0[\phi_{sp}]$

Recall that $[\phi_{sp}] = -\frac{1}{2}[h^2] + 2[xx^*]$, where

(21)
$$[h^2] = \sum_{1\le j\ne k\le N} a_j(-\tfrac{1}{2})a_j^*(-\tfrac{1}{2})a_k(-\tfrac{1}{2})a_k^*(-\tfrac{1}{2})\Omega,$$

(22)
$$[xx^*] = \sum_{r,s=1}^{N'} a_r(-\tfrac{1}{2})a_{r+N'}(-\tfrac{1}{2})a_s^*(-\tfrac{1}{2})a_{s+N'}^*(-\tfrac{1}{2})\Omega.$$

Corresponding vertex operators are given by

(23)
$$\{h^2\}_k = \sum_{\substack{1\le p\ne q\le N \\ m_1+m_2+m_3+m_4=k+1}} :a_p(m_1-\tfrac{1}{2})a_p^*(m_2-\tfrac{1}{2})a_q(m_3-\tfrac{1}{2})a_q^*(m_4-\tfrac{1}{2}):,$$

(24)
$$\{xx^*\}_k = \sum_{\substack{1\le p,q\le N' \\ n_1+n_2+n_3+n_4=k+1}} :a_p(n_1-\tfrac{1}{2})a_{p+N'}(n_2-\tfrac{1}{2})a_q^*(n_3-\tfrac{1}{2})a_{q+N'}^*(-\tfrac{1}{2}): .$$

Calculation of $\{\phi_{sp}\}_0[\phi_{sp}]$ depends on the next proposition.

PROPOSITION 6. *The action of vertex operators $\{h^2\}_0$, $\{xx^*\}_0$ on vectors $[h^2]$, $[xx^*] \in V_2$ is given as follows:*

(25)
$$\{h^2\}_0[h^2] = 4(N-1)D(-1)[\omega] - 2(N-2)D(-1)[h^2],$$

(26)
$$\{h^2\}_0[xx^*] = -2D(-1)[\omega] + D(-1)[h^2] + 2D(-1)[xx^*],$$

(27)
$$\{xx^*\}_0[h^2] = -2D(-1)[\omega] + D(-1)[h^2] + 2D(-1)[xx^*],$$

(28)
$$\{xx^*\}_0[xx^*] = \tfrac{N}{2}D(-1)[\omega] + \tfrac{1}{2}D(-1)[h^2] + \tfrac{N}{2}D(-1)[xx^*].$$

Here, $D(-1)[\omega] = -2[\omega(2,0)]$, $D(-1)[h^2] = 2[h(\omega_+ - \omega_-)]$ and $D(-1)[xx^*] = x\chi^* + \chi x^*$.

REMARK. In this proposition, note that $\{h^2\}_0[xx^*] = \{xx^*\}_0[h^2]$. It is not clear to me what is behind this symmetry. It is interesting to find for which $u, v \in V$, we have the symmetry property $\{u\}_k v = \{v\}_k u$. For more information, see Proposition 7 below and Lemma 23 of §3.11.

(PROOF OF (13) IN THEOREM 3 ASSUMING PROPOSITION 6). This is a simple calculation. First expanding ϕ_{sp}, we have

$$\{\phi_{sp}\}_0[\phi_{sp}] = \tfrac{1}{4}\{h^2\}_0[h^2] - \{h^2\}_0[xx^*] - \{xx^*\}_0[h^2] + 4\{xx^*\}_0[xx^*].$$

Using Proposition 6 to rewrite each of these terms, we have

$$\{\phi_{sp}\}_0[\phi_{sp}] = 3(N+1)D(-1)[\omega] + (N-2)D(-1)\left(-\tfrac{1}{2}[h^2] + 2[xx^*]\right)$$
$$= 3(N+1)D(-1)[\omega] + (N-2)D(-1)[\phi_{sp}].$$

This proves (13). $\square$

The formula (25) has been calculated in Proposition 3 of §3.6. We show three other identities.

Calculation of $\{h^2\}_0[xx^*]$. We have to calculate the following object:

$$(*_0) \quad \{h^2\}_0[xx^*] = \sum_{\substack{1 \leq p \neq q \leq N \\ m_1+m_2+m_3+m_4=1}} :a_p(m_1 - \tfrac{1}{2})a_p^*(m_2 - \tfrac{1}{2})a_q(m_3 - \tfrac{1}{2})a_q^*(m_4 - \tfrac{1}{2}): $$
$$\cdot \sum_{1 \leq r,s \leq N'} a_r(-\tfrac{1}{2})a_{r+N'}(-\tfrac{1}{2})a_s^*(-\tfrac{1}{2})a_{s+N'}^*(-\tfrac{1}{2})\Omega.$$

Observe that all m_j's must be less than or equal to 1, otherwise those terms with at least one $m_j > 1$ annihilate Ω. Since the sum of m_j's must be 1, (m_1, m_2, m_3, m_4) must be a permutation of $(1, 1, 1, -2)$, $(1, 0, 0, 0)$ or $(1, 1, -1, 0)$. We examine each case separately.

Case (I): $(m_1, m_2, m_3, m_4) \in \{\text{permutations of } (1, 1, 1, -2)\}$. There are four permutations. Writing out relevant terms in $(*_0)$, we have

$$\sum_{p \neq q} \{ :a_p(\tfrac{1}{2})a_p^*(\tfrac{1}{2})a_q(\tfrac{1}{2})a_q^*(-\tfrac{5}{2}): + :a_p(\tfrac{1}{2})a_p^*(\tfrac{1}{2})a_q(-\tfrac{5}{2})a_q^*(\tfrac{1}{2}):$$
$$+ :a_p(\tfrac{1}{2})a_p^*(-\tfrac{5}{2})a_q(\tfrac{1}{2})a_q^*(\tfrac{1}{2}): + :a_p(-\tfrac{5}{2})a_p^*(\tfrac{1}{2})a_q(\tfrac{1}{2})a_q^*(\tfrac{1}{2}): \}$$
$$\sum_{1 \leq r,s \leq N'} a_r(-\tfrac{1}{2})a_{r+N'}(-\tfrac{1}{2})a_s^*(-\tfrac{1}{2})a_{s+N'}^*(-\tfrac{1}{2})\Omega.$$

We can remove normal ordering symbols because there are no possible nontrivial pairings among vectors in each monomial inside of { }. Note that the first and the third terms give rise to the same contributions after summing over $p \neq q$. Similarly, the second term and the fourth term give rise to the same contributions. Thus the above summation is twice of that of the first two terms. Given (r, s), the action of the vertex operator corresponding to (p, q) on the term in $[xx^*]$ corresponding to (r, s) is nontrivial only when $\{p, q\} = \{s, s+N'\}$ and $p \in \{r, r+N'\}$. These imply that $r = s$. Letting $(p, q) = (r, r+N')$, $(r+N', r)$ and summing over $1 \leq r \leq N'$, we see that the above is equal to

$$2\sum_{r=1}^{N'} \{ a_r(\tfrac{1}{2})a_r^*(\tfrac{1}{2})a_{r+N'}(\tfrac{1}{2})a_{r+N'}^*(-\tfrac{5}{2}) + a_{r+N'}(\tfrac{1}{2})a_{r+N'}^*(\tfrac{1}{2})a_r(\tfrac{1}{2})a_r^*(-\tfrac{5}{2})$$
$$+ a_r(\tfrac{1}{2})a_r^*(\tfrac{1}{2})a_{r+N'}(-\tfrac{5}{2})a_{r+N'}^*(\tfrac{1}{2}) + a_{r+N'}(\tfrac{1}{2})a_{r+N'}^*(\tfrac{1}{2})a_r(-\tfrac{5}{2})a_r^*(\tfrac{1}{2}) \}$$
$$\cdot a_r(-\tfrac{1}{2})a_{r+N'}(-\tfrac{1}{2})a_s^*(-\tfrac{1}{2})a_{s+N'}^*(-\tfrac{1}{2})\Omega.$$

After taking care of the pairing, we see that this is further equal to

$$-2\sum_{r=1}^{N'} a^*_{r+N'}(-\tfrac{5}{2})a_{r+N'}(-\tfrac{1}{2})\Omega + a^*_r(-\tfrac{5}{2})a_r(-\tfrac{1}{2})\Omega + a_{r+N'}(-\tfrac{5}{2})a^*_{r+N'}(-\tfrac{1}{2})\Omega$$

$$+a_r(-\tfrac{5}{2})a^*_r(-\tfrac{1}{2})\Omega = -2\sum_{r=1}^{N}\{a^*_r(-\tfrac{5}{2})a_r(-\tfrac{1}{2})\Omega + a_r(-\tfrac{5}{2})a^*_r(-\tfrac{1}{2})\Omega\} = 4\omega(2,0).$$

This finishes the calculation for the Case (I).

Case (II): $(m_1, m_2, m_3, m_4) \in \{$permutations of $(1,0,0,0)\}$. There are four permutations. Relevant terms in $(*_0)$ are

$$\sum_{p\neq q}\{:a_p(\tfrac{1}{2})a^*_p(-\tfrac{1}{2})a_q(-\tfrac{1}{2})a^*_q(-\tfrac{1}{2}): + :a_p(-\tfrac{1}{2})a^*_p(\tfrac{1}{2})a_q(-\tfrac{1}{2})a^*_q(-\tfrac{1}{2}):$$

$$+ :a_p(-\tfrac{1}{2})a^*_p(-\tfrac{1}{2})a_q(\tfrac{1}{2})a^*_q(-\tfrac{1}{2}): + :a_p(-\tfrac{1}{2})a^*_p(-\tfrac{1}{2})a_q(-\tfrac{1}{2})a^*_q(\tfrac{1}{2}):\}$$

$$\cdot \sum_{1\leq r,s\leq N'} a_r(-\tfrac{1}{2})a_{r+N'}(-\tfrac{1}{2})a^*_s(-\tfrac{1}{2})a^*_{s+N'}(-\tfrac{1}{2})\Omega.$$

We observe that the first and the third summation over p, q are the same and the second and the fourth summations over p, q are the same. So, we consider contribution from the first two terms. For nontriviality of the action of a monomial in $\{h^2\}_0$ corresponding to (p, q) on the vector in $[xx^*]$ corresponding to (r, s), in the first term above we must have $p \in \{s, s + N'\}$, in the second term $p \in \{r, r + N'\}$. Writing out two possibilities for p for the first and the second terms and taking care of signs when removing normal ordering symbols, we see that the above is twice the contribution from the first and the second term:

$$2\sum_{r,s=1}^{N'}\Big\{-\sum_{q\neq s}a^*_s(-\tfrac{1}{2})a_s(\tfrac{1}{2})h_q - \sum_{q\neq s+N'}a^*_{s+N'}(-\tfrac{1}{2})a_{s+N'}(\tfrac{1}{2})h_q$$

$$+ \sum_{q\neq r}a_r(-\tfrac{1}{2})a^*_r(\tfrac{1}{2})h_q + \sum_{q\neq r+N'}a_{r+N'}(-\tfrac{1}{2})a^*_{r+N'}(\tfrac{1}{2})h_q\Big\}$$

$$\cdot a_r(-\tfrac{1}{2})a_{r+N'}(-\tfrac{1}{2})a^*_s(-\tfrac{1}{2})a^*_{s+N'}(-\tfrac{1}{2})\Omega.$$

Here $h_q = a_q(-\tfrac{1}{2})a^*_q(-\tfrac{1}{2}) \in \mathcal{A}_1$ for $1 \leq q \leq N = 2N'$. We examine each of the above four sums closely. After calculating actions, we see that the first summation over q above is equal to

$$-2\sum_{r,s=1}^{N'}\sum_{q\neq s}h_q a_r(-\tfrac{1}{2})a_{r+N'}(-\tfrac{1}{2})a^*_s(-\tfrac{1}{2})a^*_{s+N'}(-\tfrac{1}{2})\Omega.$$

When $q = s$, the term inside is 0. So we may include this term and then the summation is over independent indices r, s, q. Summation over q gives $h \in \mathcal{A}_1$, and summations over r, s give $x, x^* \in \mathcal{A}_1$, respectively. So the above is equal to $-2[xhx^*]$. Contribution from the second term is the same as that from the first term. Contributions from the third and the fourth terms turns out to be the negative of contribution from the first two terms, as signs in front of these terms indicate. Thus, the total contribution is 0, and hence (II) $= 0$.

Case (III): $(m_1, m_2, m_3, m_4) \in \{$permutations of $(1, 1, -1, 0)\}$. Relevant terms in $(*_0)$ corresponding to twelve permutations are the followings:

$$\sum_{p \neq q} \{ : a_p(\tfrac{1}{2}) a_p^*(\tfrac{1}{2}) a_q(-\tfrac{3}{2}) a_q^*(-\tfrac{1}{2}) : + : a_p(\tfrac{1}{2}) a_p^*(\tfrac{1}{2}) a_q(-\tfrac{1}{2}) a_q^*(-\tfrac{3}{2}) :$$

$$+ : a_p(-\tfrac{3}{2}) a_p^*(-\tfrac{1}{2}) a_q(\tfrac{1}{2}) a_q^*(\tfrac{1}{2}) : + : a_p(-\tfrac{1}{2}) a_p^*(-\tfrac{3}{2}) a_q(\tfrac{1}{2}) a_q^*(\tfrac{1}{2}) :$$

$$+ : a_p(\tfrac{1}{2}) a_p^*(-\tfrac{3}{2}) a_q(\tfrac{1}{2}) a_q^*(-\tfrac{1}{2}) : + : a_p(-\tfrac{3}{2}) a_p^*(\tfrac{1}{2}) a_q(-\tfrac{1}{2}) a_q^*(\tfrac{1}{2}) :$$

$$+ : a_p(\tfrac{1}{2}) a_p^*(-\tfrac{3}{2}) a_q(-\tfrac{1}{2}) a_q^*(\tfrac{1}{2}) : + : a_p(-\tfrac{3}{2}) a_p^*(\tfrac{1}{2}) a_q(\tfrac{1}{2}) a_q^*(-\tfrac{1}{2}) :$$

$$+ : a_p(\tfrac{1}{2}) a_p^*(-\tfrac{1}{2}) a_q(\tfrac{1}{2}) a_q^*(-\tfrac{3}{2}) : + : a_p(-\tfrac{1}{2}) a_p^*(\tfrac{1}{2}) a_q(\tfrac{1}{2}) a_q^*(-\tfrac{3}{2}) :$$

$$+ : a_p(\tfrac{1}{2}) a_p^*(-\tfrac{1}{2}) a_q(-\tfrac{3}{2}) a_q^*(\tfrac{1}{2}) : + : a_p(-\tfrac{1}{2}) a_p^*(\tfrac{1}{2}) a_q(-\tfrac{3}{2}) a_q^*(\tfrac{1}{2}) : \}$$

$$\cdot \sum_{1 \leq r, s \leq N'} a_r(-\tfrac{1}{2}) a_{r+N'}(-\tfrac{1}{2}) a_s^*(-\tfrac{1}{2}) a_{s+N'}^*(-\tfrac{1}{2}) \Omega.$$

In the above, p and q run from 1 to N. Contribution from the first four terms will be denoted by (i), contribution from the next four terms is denoted by (ii), and contribution from the last four terms will be called (iii). Note that after summing over $p \neq q$, (ii)=(iii), which can be seen easily by switching p and q.

For (i), the first and the third terms give rise to the same contributions which can be seen by switching p and q. Similarly, the second term and the fourth terms contribute equally. So, we concentrate on the first two terms of (i). For the first term, its action on the vector in $[xx^*]$ corresponding to (r, s) is nontrivial only when p satisfies $p = s = r$ or $p = s + N' = r + N'$. Thus, only terms with $r = s$ in $[xx^*]$ contribute the above summation and p can be either r or $r + N'$. Similar situation occurs for the second term; only those terms in $[xx^*]$ with $r = s$ give rise to nontrivial contribution and then p must be either r or $r + N'$. Writing out those terms which contribute nontrivially, we see that (i) is equal to twice of the contribution from the first two terms:

$$(i) = 2 \sum_{r=1}^{N'} \Big\{ \sum_{q \neq r} : a_r(\tfrac{1}{2}) a_r^*(\tfrac{1}{2}) a_q(-\tfrac{3}{2}) a_q^*(-\tfrac{1}{2}) :$$

$$+ \sum_{q \neq r+N'} : a_{r+N'}(\tfrac{1}{2}) a_{r+N'}^*(\tfrac{1}{2}) a_q(-\tfrac{3}{2}) a_q^*(-\tfrac{1}{2}) : + \sum_{q \neq r} : a_r(\tfrac{1}{2}) a_r^*(\tfrac{1}{2}) a_q(-\tfrac{1}{2}) a_q^*(-\tfrac{3}{2}) :$$

$$+ \sum_{q \neq r+N'} : a_{r+N'}(\tfrac{1}{2}) a_{r+N'}^*(\tfrac{1}{2}) a_q(-\tfrac{1}{2}) a_q^*(-\tfrac{3}{2}) : \Big\}$$

$$\cdot a_r(-\tfrac{1}{2}) a_{r+N'}(-\tfrac{1}{2}) a_r^*(-\tfrac{1}{2}) a_{r+N'}^*(-\tfrac{1}{2}) \Omega.$$

We can simply remove normal ordering symbols without any changes. After computing actions and combining the first and the third summations over $q \neq r$, and the second and the fourth summations over $q \neq r + N'$, we have

$$(i) = -2 \sum_{r=1}^{N'} \sum_{q \neq r} \{ a_q(-\tfrac{3}{2}) a_q^*(-\tfrac{1}{2}) - a_q^*(-\tfrac{3}{2}) a_q(-\tfrac{1}{2}) \} \, a_{r+N'}(-\tfrac{1}{2}) a_{r+N'}^*(-\tfrac{1}{2}) \Omega$$

$$- 2 \sum_{r=1}^{N'} \sum_{q \neq r+N'} \{ a_q(-\tfrac{3}{2}) a_q^*(-\tfrac{1}{2}) - a_q^*(-\tfrac{3}{2}) a_q(-\tfrac{1}{2}) \} \, a_r(-\tfrac{1}{2}) a_r^*(-\tfrac{1}{2}) \Omega.$$

If there were no restrictions on summations over q, the above summation would be equal to $[2(\omega_+ - \omega_-)h]$. However, the restriction on q requires correction terms corresponding to $q = r$ and $q = r + N'$, which are given as follows:

$$2\sum_{r=1}^{N'} \left\{ a_r(-\tfrac{3}{2})a_r^*(-\tfrac{1}{2}) - a_r^*(-\tfrac{3}{2})a_r(-\tfrac{1}{2}) \right\} a_{r+N'}(-\tfrac{1}{2})a_{r+N'}^*(-\tfrac{1}{2})\Omega$$

$$+ 2\sum_{r=1}^{N'} \left\{ a_{r+N'}(-\tfrac{3}{2})a_{r+N'}^*(-\tfrac{1}{2}) - a_{r+N'}^*(-\tfrac{3}{2})a_{r+N'}(-\tfrac{1}{2}) \right\} a_r(-\tfrac{1}{2})a_r^*(-\tfrac{1}{2})\Omega.$$

As we shall see later, these correction terms cancel with correction terms arising from (ii) and (iii).

For (ii), we first examine summation over p and q. For nontriviality of the action of operators in (ii) corresponding to (p, q) on the term in $[xx^*]$ corresponding to (r, s), the index (p, q) must satisfy the following restrictions:

(a) For the first term : $(p, q) = (s, s + N'), (s + N', s)$.
(b) For the second term : $(p, q) = (r, r + N'), (r + N', r)$.
(c) For the third term : $(p, q) = (s, r), (s + N', r), (s, r + N'), (s + N', r + N')$.
(d) For the fourth term : $(p, q) = (r, s), (r + N', s), (r, s + N'), (r + N', s + N')$.

Here, indices s, r run from 1 to N', whereas p, q run from 1 to N. The restriction $p \neq q$ imposes a restriction $r \neq s$ for cases (c) and (d). Writing out operators satisfying the above restrictions but without the restriction $p \neq q$, we see that the corresponding summation is given by

$$\sum_{r,s=1}^{N'} \{ :a_s(\tfrac{1}{2})a_s^*(-\tfrac{3}{2})a_{s+N'}(\tfrac{1}{2})a_{s+N'}^*(-\tfrac{1}{2}): \; + \; :a_{s+N'}(\tfrac{1}{2})a_{s+N'}^*(-\tfrac{3}{2})a_s(\tfrac{1}{2})a_s^*(-\tfrac{1}{2}):$$
$$+ :a_r(-\tfrac{3}{2})a_r^*(\tfrac{1}{2})a_{r+N'}(-\tfrac{1}{2})a_{r+N'}^*(\tfrac{1}{2}): \; + \; :a_{r+N'}(-\tfrac{3}{2})a_{r+N'}^*(\tfrac{1}{2})a_r(-\tfrac{1}{2})a_r^*(\tfrac{1}{2}):$$
$$+ :a_s(\tfrac{1}{2})a_s^*(-\tfrac{3}{2})a_r(-\tfrac{1}{2})a_r^*(\tfrac{1}{2}): \; + \; :a_{s+N'}(\tfrac{1}{2})a_{s+N'}^*(-\tfrac{3}{2})a_r(-\tfrac{1}{2})a_r^*(\tfrac{1}{2}):$$
$$+ :a_s(\tfrac{1}{2})a_s^*(-\tfrac{3}{2})a_{r+N'}(-\tfrac{1}{2})a_{r+N'}^*(\tfrac{1}{2}): \; + \; :a_{s+N'}(\tfrac{1}{2})a_{s+N'}^*(-\tfrac{3}{2})a_{r+N'}(-\tfrac{1}{2})a_{r+N'}^*(\tfrac{1}{2}):$$
$$+ :a_r(-\tfrac{3}{2})a_r^*(\tfrac{1}{2})a_s(\tfrac{1}{2})a_s^*(-\tfrac{1}{2}): \; + \; :a_{r+N'}(-\tfrac{3}{2})a_{r+N'}^*(\tfrac{1}{2})a_s(\tfrac{1}{2})a_s^*(-\tfrac{1}{2}):$$
$$+ :a_r(-\tfrac{3}{2})a_r^*(\tfrac{1}{2})a_{s+N'}(\tfrac{1}{2})a_{s+N'}^*(-\tfrac{1}{2}): \; + \; :a_{r+N'}(-\tfrac{3}{2})a_{r+N'}^*(\tfrac{1}{2})a_{s+N'}(\tfrac{1}{2})a_{s+N'}^*(-\tfrac{1}{2}): \}$$
$$\cdot a_r(-\tfrac{1}{2})a_{r+N'}(-\tfrac{1}{2})a_s^*(-\tfrac{1}{2})a_{s+N'}^*(-\tfrac{1}{2})\Omega.$$

Since our summation in (ii) actually imposes $p \neq q$, when taking summation over indices r and s, we must impose $r \neq s$ for the fifth, eighth, ninth and twelfth terms in the above summation over r, s. Calculating the action of operators with these restrictions, we see that (ii) is given by

$$\sum_{r,s} a_s^*(-\tfrac{3}{2})a_{s+N'}^*(-\tfrac{1}{2})a_r(-\tfrac{1}{2})a_{r+N'}(-\tfrac{1}{2})\Omega$$

$$- \sum_{r,s} a_{s+N'}^*(-\tfrac{3}{2})a_s^*(-\tfrac{1}{2})a_r(-\tfrac{1}{2})a_{r+N'}(-\tfrac{1}{2})\Omega$$

$$+ \sum_{r,s} \{ a_r(-\tfrac{3}{2})a_{r+N'}(-\tfrac{1}{2})a_s^*(-\tfrac{1}{2})a_{s+N'}^*(-\tfrac{1}{2})\Omega$$

$$-\sum_{r,s} a_{r+N'}(-\tfrac{3}{2})a_r(-\tfrac{1}{2})a_s^*(-\tfrac{1}{2})a_{s+N'}^*(-\tfrac{1}{2})\Omega$$

$$-\sum_{r\neq s} a_s^*(-\tfrac{3}{2})a_{s+N'}^*(-\tfrac{1}{2})a_r(-\tfrac{1}{2})a_{r+N'}(-\tfrac{1}{2})\Omega$$

$$-\sum_{r,s} a_{s+N'}^*(-\tfrac{3}{2})a_s^*(-\tfrac{1}{2})a_r(-\tfrac{1}{2})a_{r+N'}(-\tfrac{1}{2})\Omega$$

$$-\sum_{r,s} a_s^*(-\tfrac{3}{2})a_{s+N'}^*(-\tfrac{1}{2})a_r(-\tfrac{1}{2})a_{r+N'}(-\tfrac{1}{2})\Omega$$

$$+\sum_{r\neq s} a_{s+N'}^*(-\tfrac{3}{2})a_s^*(-\tfrac{1}{2})a_r(-\tfrac{1}{2})a_{r+N'}(-\tfrac{1}{2})\Omega$$

$$-\sum_{r\neq s} a_r(-\tfrac{3}{2})a_{r+N'}(-\tfrac{1}{2})a_s^*(-\tfrac{1}{2})a_{s+N'}^*(-\tfrac{1}{2})\Omega$$

$$+\sum_{r,s} a_{r+N'}(-\tfrac{3}{2})a_r(-\tfrac{1}{2})a_s^*(-\tfrac{1}{2})a_{s+N'}^*(-\tfrac{1}{2})\Omega$$

$$-\sum_{r,s} a_r(-\tfrac{3}{2})a_{r+N'}(-\tfrac{1}{2})a_s^*(-\tfrac{1}{2})a_{s+N'}^*(-\tfrac{1}{2})\Omega$$

$$+\sum_{r\neq s} a_{r+N'}(-\tfrac{3}{2})a_r(-\tfrac{1}{2})a_s^*(-\tfrac{1}{2})a_{s+N'}^*(-\tfrac{1}{2})\Omega.$$

If we ignore the condition $r \neq s$ in some of the above summations, we see that two consecutive terms pair up to give $-[\chi^*x]$, $-[\chi x^*]$, $[\chi^*x]$, $[\chi^*x]$, $[\chi x^*]$, $[\chi x^*]$, respectively, and the sum would be $[\chi^*x]+[\chi x^*]$. However, due to the restriction $r \neq s$ in some of the above summations, we need correction terms. With these correction terms, (ii) is given by

$$(ii) = [\chi^*x]+[\chi x^*]-\sum_{r=1}^{N'}\{a_r(-\tfrac{3}{2})a_r^*(-\tfrac{1}{2}) - a_r^*(-\tfrac{3}{2})a_r(-\tfrac{1}{2})\}a_{r+N'}(-\tfrac{1}{2})a_{r+N'}^*(-\tfrac{1}{2})\Omega$$
$$-\sum_{r=1}^{N'}\{a_{r+N'}(-\tfrac{3}{2})a_{r+N'}^*(-\tfrac{1}{2}) - a_{r+N'}^*(-\tfrac{3}{2})a_{r+N'}(-\tfrac{1}{2})\}\,a_r(-\tfrac{1}{2})a_r^*(-\tfrac{1}{2})\Omega.$$

(iii) gives the same sum as (ii). Thus, (i) + (ii) + (iii) = (i) + 2(ii), and when we calculate this sum, we find that correction terms from (i), (ii), (iii) precisely cancel with each other. Thus, (III) $= 2[(\omega_+ -\omega_-)h]+2[\chi^*x]+2[\chi x^*]$. Collecting (I), (II), (III), we finally obtain

$$\{h^2\}_0[xx^*] = 4[\omega(2,0)] + 2[(\omega_+ - \omega_-)h] + 2[\chi^*x] + 2[\chi x^*]$$
$$= -2D(-1)[\omega] + D(-1)[h^2] + 2D(-1)[xx^*].$$

Here $D(-1)[\omega] = -2[\omega(2,0)]$, $D(-1)[h^2] = 2[h(\omega_+ - \omega_-)]$, and $D(-1)[xx^*] = [\chi x^*] + [x\chi^*]$ from Proposition 1 and Proposition 7 both of §3.5. This proves formula (26).

Calculation of $\{xx^*\}_0[h^2]$. Our object of calculation is the following:

$$(*_1)\quad \{xx^*\}_0[h^2] = \sum_{\substack{1\le p,q\le N' \\ n_1+n_2+n_3+n_4=1}} :a_p(n_1 - \tfrac{1}{2})a_{p+N'}(n_2 - \tfrac{1}{2})a_q^*(n_3 - \tfrac{1}{2})a_{q+N'}^*(-\tfrac{1}{2}):$$
$$\cdot \sum_{1\le j\ne k\le N} a_j(-\tfrac{1}{2})a_j^*(-\tfrac{1}{2})a_k(-\tfrac{1}{2})a_k^*(-\tfrac{1}{2})\Omega.$$

As in the previous case, all n_j's must be less than or equal to 1, otherwise we get a term with trivial contribution. For the sum of n_j's to be 1, (n_1, n_2, n_3, n_4) must be a permutation of $(1,1,1,-2)$, $(1,0,0,0)$, $(1,1,-1,0)$. We examine each case separately.

Case (I): $(n_1, n_2, n_3, n_4) \in \{\text{permutations of } (1,1,1,-2)\}$. There are four permutations and relevant terms in $(*_1)$ are given by

$$\sum_{p,q=1}^{N'} \{:a_p(\tfrac{1}{2})a_{p+N'}(\tfrac{1}{2})a_q^*(\tfrac{1}{2})a_{q+N'}^*(-\tfrac{5}{2}): + :a_p(\tfrac{1}{2})a_{p+N'}(\tfrac{1}{2})a_q^*(-\tfrac{5}{2})a_{q+N'}^*(\tfrac{1}{2}):$$
$$+ :a_p(\tfrac{1}{2})a_{p+N'}(-\tfrac{5}{2})a_q^*(\tfrac{1}{2})a_{q+N'}^*(\tfrac{1}{2}): + :a_p(-\tfrac{5}{2})a_{p+N'}(\tfrac{1}{2})a_q^*(\tfrac{1}{2})a_{q+N'}^*(\tfrac{1}{2}):\}$$
$$\cdot \sum_{j\ne k} a_j(-\tfrac{1}{2})a_j^*(-\tfrac{1}{2})a_k(-\tfrac{1}{2})a_k^*(-\tfrac{1}{2})\Omega.$$

Note that the first and the third terms are $*$-related (complex conjugate of each other). Also, the second and the fourth terms are $*$-related. The last part $[h^2]$ is $*$-invariant. So we calculate the summation of the first two terms. For nontriviality of the action of the first term on $[h^2]$, given (j,k), we must have $\{p, p+N'\} = \{j, k\}$ and $q \in \{j, k\}$. Since $1 \le p, q \le N'$, this happens only when either $p = q = j$, $k = j + N'$ or $p = q = k$, $j = k + N'$. Note also that we can remove normal ordering symbols without any changes. By interchanging j, k, we see that contributions from these two cases are the same. Hence contribution from the first term is

$$2\sum_{j=1}^{N'} a_j(\tfrac{1}{2})a_{j+N'}(\tfrac{1}{2})a_j^*(\tfrac{1}{2})a_{j+N'}^*(-\tfrac{5}{2}) \cdot a_j(-\tfrac{1}{2})a_j^*(-\tfrac{1}{2})a_{j+N'}(-\tfrac{1}{2})a_{j+N'}^*(-\tfrac{1}{2})\Omega$$
$$= -2\sum_{j=1}^{N'} a_{j+N'}^*(-\tfrac{5}{2})a_{j+N'}(-\tfrac{1}{2})\Omega.$$

As for the second term, nontriviality of its action on $[h^2]$ imposes that either $p = q = j$, $k = j + N'$ or $p = q = k$, $j = k + N'$. A similar calculation shows that contribution from the second term is equal to $-2\sum_{j=1}^{N'} a_j^*(-\tfrac{5}{2})a_j(-\tfrac{1}{2})\Omega$. Thus, the summation of the first two terms is $-2\sum_{j=1}^{N} a_j^*(-\tfrac{5}{2})a_j(-\tfrac{1}{2})\Omega$. The contribution from the last two terms is obtained by applying $*$ to this sum. Altogether, we get

$$(\text{I}) = -2\sum_{j=1}^{N} \{a_j(-\tfrac{5}{2})a_j^*(-\tfrac{1}{2}) + a_j^*(-\tfrac{5}{2})a_j(-\tfrac{1}{2})\}\Omega = 4\omega(2,0) = -2D(-1)[\omega].$$

Case (II): $(n_1, n_2, n_3, n_4) \in \{$permutations of $(1, 0, 0, 0)\}$. There are four permutations and relevant terms in $(*_1)$ are the followings:

$$(\mathrm{II}) = \sum_{p,q=1}^{N'} \{ :a_p(\tfrac{1}{2})a_{p+N'}(-\tfrac{1}{2})a_q^*(-\tfrac{1}{2})a_{q+N'}^*(-\tfrac{1}{2}):$$

$$+ :a_p(-\tfrac{1}{2})a_{p+N'}(\tfrac{1}{2})a_q^*(-\tfrac{1}{2})a_{q+N'}^*(-\tfrac{1}{2}): + :a_p(-\tfrac{1}{2})a_{p+N'}(-\tfrac{1}{2})a_q^*(\tfrac{1}{2})a_{q+N'}^*(-\tfrac{1}{2}):$$

$$+ :a_p(-\tfrac{1}{2})a_{p+N'}(-\tfrac{1}{2})a_q^*(-\tfrac{1}{2})a_{q+N'}^*(\tfrac{1}{2}): \} \cdot \sum_{j \neq k} a_j(-\tfrac{1}{2})a_j^*(-\tfrac{1}{2})a_k(-\tfrac{1}{2})a_k^*(-\tfrac{1}{2})\Omega.$$

We note that contributions from the first and the third terms are *-related, those from the second and the fourth terms are also *-related. So, we concentrate on contributions from the first two terms. For nontriviality of the action of the operator corresponding to (p, q) on the vector in $[h^2]$ corresponding to (j, k), we must have either $j = p$ or $k = p$. Explicitly writing out relevant terms, we see that these two cases give rise to the same summation and contribution from the first term is then equal to

$$-2 \sum_{p,q=1}^{N'} \sum_{k \neq p} a_p(-\tfrac{1}{2})a_{p+N'}(-\tfrac{1}{2})a_q^*(-\tfrac{1}{2})a_{q+N'}^*(-\tfrac{1}{2})a_k(-\tfrac{1}{2})a_k^*(-\tfrac{1}{2})\Omega.$$

In this summation, the inside term is 0 when $k = p$. So, we can include this term into the summation without changing the sum, and then independent summations over p, q, k give x, x^*, h, respectively. Thus, the above is equal to $-2[xhx^*]$.

For the contribution from the second term in (II), given (p, q), we must have that either $j = p + N'$ or $k = p + N'$ for nontriviality of the action. A similar calculation as above shows that contribution from the second term is the same as the first one, namely $-2[xhx^*]$. Together, we have $-4[xhx^*]$. Since the contribution from the last two terms is obtained by applying $*$ to the contribution from the first two terms, it is equal to $-4[xhx^*]^* = -4[x^*h^*x] = 4[x^*hx]$, since $h^* = -h$. This cancels the contribution from the first two terms. Thus (II) = 0.

Case (III): $(n_1, n_2, n_3, n_4) \in \{$permutations of $(1, 1, -1, 0)\}$. There are twelve permutations. Relevant partial summation in $(*_1)$ is the following:

$$\sum_{p,q=1}^{N'} \{ :a_p(\tfrac{1}{2})a_{p+N'}(\tfrac{1}{2})a_q^*(-\tfrac{3}{2})a_{q+N'}^*(-\tfrac{1}{2}): + :a_p(\tfrac{1}{2})a_{p+N'}(\tfrac{1}{2})a_q^*(-\tfrac{1}{2})a_{q+N'}^*(-\tfrac{3}{2}):$$

$$+ :a_p(-\tfrac{3}{2})a_{p+N'}(-\tfrac{1}{2})a_q^*(\tfrac{1}{2})a_{q+N'}^*(\tfrac{1}{2}): + :a_p(-\tfrac{1}{2})a_{p+N'}(-\tfrac{3}{2})a_q^*(\tfrac{1}{2})a_{q+N'}^*(\tfrac{1}{2}):$$

$$+ :a_p(\tfrac{1}{2})a_{p+N'}(-\tfrac{3}{2})a_q^*(\tfrac{1}{2})a_{q+N'}^*(-\tfrac{1}{2}): + :a_p(\tfrac{1}{2})a_{p+N'}(-\tfrac{3}{2})a_q^*(-\tfrac{1}{2})a_{q+N'}^*(\tfrac{1}{2}):$$

$$+ :a_p(-\tfrac{3}{2})a_{p+N'}(\tfrac{1}{2})a_q^*(\tfrac{1}{2})a_{q+N'}^*(-\tfrac{1}{2}): + :a_p(-\tfrac{3}{2})a_{p+N'}(\tfrac{1}{2})a_q^*(-\tfrac{1}{2})a_{q+N'}^*(\tfrac{1}{2}):$$

$$+ :a_p(\tfrac{1}{2})a_{p+N'}(-\tfrac{1}{2})a_q^*(\tfrac{1}{2})a_{q+N'}^*(-\tfrac{3}{2}): + :a_p(\tfrac{1}{2})a_{p+N'}(-\tfrac{1}{2})a_q^*(-\tfrac{3}{2})a_{q+N'}^*(\tfrac{1}{2}):$$

$$+ :a_p(-\tfrac{1}{2})a_{p+N'}(\tfrac{1}{2})a_q^*(\tfrac{1}{2})a_{q+N'}^*(-\tfrac{3}{2}): + :a_p(-\tfrac{1}{2})a_{p+N'}(\tfrac{1}{2})a_q^*(-\tfrac{3}{2})a_{q+N'}^*(\tfrac{1}{2}): \}$$

$$\cdot \sum_{j \neq k} a_j(-\tfrac{1}{2})a_j^*(-\tfrac{1}{2})a_k(-\tfrac{1}{2})a_k^*(-\tfrac{1}{2})\Omega.$$

We divide the above sum into three groups. Contribution from the first four terms is denoted by (i), that from the next four terms by (ii), that from the last four terms by (iii). We calculate (i), (ii), and (iii) separately.

For the first term in (i), given (p, q), for nontriviality of the action on $[h^2]$, we must have either $(j, k) = (p, p + N')$ or $(p + N', p)$. These two cases give the same result by symmetry between j, k. After some calculation, contribution from the first term is

$$2\sum_{p,q=1}^{N'} a_q^*(-\tfrac{3}{2})a_{q+N'}^*(-\tfrac{1}{2})a_p(-\tfrac{1}{2})a_{p+N'}(-\tfrac{1}{2})\Omega = 2x\Big\{\sum_{q=1}^{N'} a_q^*(-\tfrac{3}{2})a_{q+N'}^*(-\tfrac{1}{2})\Omega\Big\}.$$

Ccontribution from the second term is the same as that from the first term except that we replace $a_q^*(-\tfrac{3}{2})a_{q+N'}^*(-\tfrac{1}{2})$ by $a_q^*(-\tfrac{1}{2})a_{q+N'}^*(-\tfrac{3}{2})$. Thus, the contribution from the first two terms in (III) is given by

$$2x\Big\{\sum_{q=1}^{N'} a_q^*(-\tfrac{3}{2})a_{q+N'}^*(-\tfrac{1}{2})\Omega + \sum_{q=1}^{N'} a_q^*(-\tfrac{1}{2})a_{q+N'}^*(-\tfrac{3}{2})\Omega\Big\} = -2[x\chi^*].$$

Contributions from the third and the fourth terms are obtained by applying the conjugation operation $*$ to contributions from the first two terms, and is equal to $-2[x\chi^*]^* = -2[x^*\chi]$. Altogether (i) is given by

$$(\mathrm{i}) = -2[x\chi^*] - 2[x^*\chi] = -2D(-1)[xx^*].$$

Next we look at the first term in (ii). Given (p, q), for nontriviality of the action on $[h^2]$, we must have $(j, k) = (p, q), (q, p)$ if $p \neq q$. If $p = q$, then either $j = p$, $k \neq p$ or $k = p$, $j \neq p$. The first two cases with $p \neq q$ give rise to the same summations and together their contribution is

$$2x^*\sum_{p=1}^{N'} a_{p+N'}(-\tfrac{3}{2})a_p(-\tfrac{1}{2})\Omega - 2\sum_{p=1}^{N'} a_{p+N'}(-\tfrac{3}{2})a_{p+N'}^*(-\tfrac{1}{2})a_p(-\tfrac{1}{2})a_p^*(-\tfrac{1}{2})\Omega.$$

The second summation in the above formula is there because we have to subtract terms corresponding to $p = q$. Contributions from the third and the fourth cases with $p = q$ are the same, which can be seen by switching j and k. Altogether contribution from the first term for the case $p = q$ is

$$-2h \cdot \sum_{p=1}^{N'} a_{p+N'}(-\tfrac{3}{2})a_{p+N'}^*(-\tfrac{1}{2})\Omega + 2\sum_{p=1}^{N'} a_{p+N'}(-\tfrac{3}{2})a_{p+N'}^*(-\tfrac{1}{2})a_p(-\tfrac{1}{2})a_p^*(-\tfrac{1}{2})\Omega.$$

Here, the second summation is there due to the condition $k \neq p$, and it corresponds to terms with $k = p$. Thus, the total contribution from the first term in

(ii) is the sum of the above two formulae, and we see that "correction" terms precisely cancel with each other and contribution from the first term in (ii) is

$$2x^* \cdot \sum_{p=1}^{N'} a_{p+N'}(-\tfrac{3}{2})a_p(-\tfrac{1}{2})\Omega - 2h \cdot \sum_{p=1}^{N'} a_{p+N'}(-\tfrac{3}{2})a_{p+N'}^*(-\tfrac{1}{2})\Omega.$$

For the second term in (ii), for nontriviality of the action on $[h^2]$, we must have either $(j,k) = (p, q+N')$ or $(q+N', p)$. Contributions from these two cases are the same by symmetry and together, we obtain $2x^* \cdot \sum_{p=1}^{N'} a_{p+N'}(-\tfrac{3}{2})a_p(-\tfrac{1}{2})\Omega$.

Contribution from the third term is similar. Possible contribution comes from $(j,k) = (p+N', q), (q, p+N')$ for given (p,q), and the summation over p, q gives $-2x^* \cdot \sum_{p=1}^{N'} a_p(-\tfrac{3}{2})a_{p+N'}(-\tfrac{1}{2})\Omega$.

Contribution from the fourth term is similar to that for the first term. Given (p,q), possible contribution comes from $(j,k) = (p+N', q+N'), (q+N', p+N')$ when $p \neq q$. When $p = q$, we have either $j = p+N'$, $k \neq p+N'$ or $k = p+N'$, $j \neq p+N'$. For the first two cases, writing the summation over $p \neq q$ as $\sum_{p,q} - \sum_{p=q}$, we get

$$-2x^* \cdot \sum_{p=1}^{N'} a_p(-\tfrac{3}{2})a_{p+N'}(-\tfrac{1}{2})\Omega + 2\sum_{p=1}^{N'} a_p(-\tfrac{3}{2})a_{p+N'}(-\tfrac{1}{2})a_p^*(-\tfrac{1}{2})a_{p+N'}^*(-\tfrac{1}{2})\Omega.$$

The two cases for $p = q$ give rise to the same contributions which can be seen by switching k and j. Writing $\sum_{k\neq p+N'}$ as $\sum_{k,p} - \sum_{k=p+N'}$, contribution from these two cases is

$$-2h \cdot \sum_{p=1}^{N'} a_p(-\tfrac{3}{2})a_p^*(-\tfrac{1}{2})\Omega + 2 \sum_{p=1}^{N'} a_p(-\tfrac{3}{2})a_p^*(-\tfrac{1}{2})a_{p+N'}(-\tfrac{1}{2})a_{p+N'}^*(-\tfrac{1}{2})\Omega.$$

Again, correction terms in the last two formulae exactly cancel with each other. Altogether contribution from the fourth term in (ii) is given by

$$-2x^* \cdot \sum_{p=1}^{N'} a_p(-\tfrac{3}{2})a_{p+N'}(-\tfrac{1}{2})\Omega - 2h \cdot \sum_{p=1}^{N'} a_p(-\tfrac{3}{2})a_p^*(-\tfrac{1}{2})\Omega.$$

Collecting our calculations for (ii), we finally get

$$\text{(ii)} = 4x^* \cdot \sum_{p=1}^{N'} a_{p+N'}(-\tfrac{3}{2})a_p(-\tfrac{1}{2})\Omega - 2h \cdot \sum_{p=1}^{N'} a_{p+N'}(-\tfrac{3}{2})a_{p+N'}^*(-\tfrac{1}{2})\Omega$$
$$- 4x^* \cdot \sum_{p=1}^{N'} a_p(-\tfrac{3}{2})a_{p+N'}(-\tfrac{1}{2})\Omega - 2h \cdot \sum_{p=1}^{N'} a_p(-\tfrac{3}{2})a_p^*(-\tfrac{1}{2})\Omega$$
$$= 4[x^*\chi] - 2h \cdot \sum_{p=1}^{N} a_p(-\tfrac{3}{2})a_p^*(-\tfrac{1}{2})\Omega = 4[x^*\chi] + 2[h\omega_+].$$

As for (iii), examining terms in (ii) and (iii), we see that (iii) is $*$-related to (ii), and hence (iii) $=$ (ii)$^* = 4[x\chi^*] - 2[h\omega_-]$, since $h^* = -h$ and $\omega_+^* = \omega_-$. Thus, contribution from Case (III) is

$$(\text{III}) = (-2[x\chi^*] - 2[x^*\chi]) + (4[x^*\chi] + 2[h\omega_+]) + (4[x\chi^*] - 2[h\omega_-])$$
$$= 2[x\chi^*] + 2[x^*\chi] + 2[h(\omega_+ - \omega_-)] = 2D(-1)[xx^*] + D(-1)[h^2].$$

Now, (I), (II), (III) together gives

$$\{xx^*\}_0[h^2] = (\text{I}) + (\text{II}) + (\text{III}) = -2D(-1)[\omega] + D(-1)[h^2] + 2D(-1)[xx^*].$$

This proves the third formula (27) in Proposition 6.

Calculation of $\{xx^*\}_0[xx^*]$. We now calculate the last formula in Proposition 6. The following object is what we have to deal with:

$$(*_2) \qquad \sum_{\substack{1 \le p,q \le N' \\ n_1+n_2+n_3+n_4=1}} :a_p(n_1 - \tfrac{1}{2})a_{p+N'}(n_2 - \tfrac{1}{2})a_q^*(n_3 - \tfrac{1}{2})a_{q+N'}^*(n_4 - \tfrac{1}{2}):$$
$$\cdot \sum_{1 \le r,s \le N'} a_r(-\tfrac{1}{2})a_{r+N'}(-\tfrac{1}{2})a_s^*(-\tfrac{1}{2})a_{s+N'}^*(-\tfrac{1}{2})\Omega.$$

Arguing as before, we see that (n_1, n_2, n_3, n_4) must be a permutation of either $(1,1,1,-2)$, $(1,0,0,0)$ or $(1,1,-1,0)$.

Case (I): $(n_1, n_2, n_3, n_4) \in \{\text{permutations of } (1,1,1,-2)\}$. There are four permutations and relevant terms in $(*_2)$ are given by

$$\sum_{p,q=1}^{N'} \{:a_p(\tfrac{1}{2})a_{p+N'}(\tfrac{1}{2})a_q^*(\tfrac{1}{2})a_{q+N'}^*(-\tfrac{5}{2}): + :a_p(\tfrac{1}{2})a_{p+N'}(\tfrac{1}{2})a_q^*(-\tfrac{5}{2})a_{q+N'}^*(\tfrac{1}{2}):$$
$$+ :a_p(\tfrac{1}{2})a_{p+N'}(-\tfrac{5}{2})a_q^*(\tfrac{1}{2})a_{q+N'}^*(\tfrac{1}{2}): + :a_p(-\tfrac{5}{2})a_{p+N'}(\tfrac{1}{2})a_q^*(\tfrac{1}{2})a_{q+N'}^*(\tfrac{1}{2}):\}$$
$$\cdot \sum_{1 \le r,s \le N'} a_r(-\tfrac{1}{2})a_{r+N'}(-\tfrac{1}{2})a_s^*(-\tfrac{1}{2})a_{s+N'}^*(-\tfrac{1}{2})\Omega.$$

We call this summation (I). Given (r, s), for nontriviality of actions of the operator in $\{xx^*\}_0$ corresponding to (p, q) on the vector in $[xx^*]$ corresponding to (r, s), we must have $p = s$, $q = r$ for each of the above terms. After calculating the action, (I) is equal to

$$\sum_{r,s=1}^{N'} \{a_{r+N'}^*(-\tfrac{5}{2})a_{r+N'}(-\tfrac{1}{2}) + a_r^*(-\tfrac{5}{2})a_r(-\tfrac{1}{2}) + a_{r+N'}(-\tfrac{5}{2})a_{r+N'}^*(-\tfrac{1}{2})$$
$$+ a_r(-\tfrac{5}{2})a_r^*(-\tfrac{1}{2})\} \Omega = N' \sum_{r=1}^{N} \{a_r(-\tfrac{5}{2})a_r^*(-\tfrac{1}{2}) + a_r^*(-\tfrac{5}{2})a_r(-\tfrac{1}{2})\}\Omega$$
$$= -N\omega(2,0) = \tfrac{N}{2}D(-1)[\omega].$$

The first equality is due to absence of index s in the first summation. The last identity is due to $D(-1)[\omega] = -2[\omega(2,0)]$. This finishes the case (I).

Case (II): $(n_1, n_2, n_3, n_4) \in \{$permutations of $(1, 0, 0, 0)\}$. There are four permutations and relevant terms in $(*_2)$ are

$$(\text{II}) = \sum_{p,q=1}^{N'} \{ :a_p(\tfrac{1}{2})a_{p+N'}(-\tfrac{1}{2})a_q^*(-\tfrac{1}{2})a_{q+N'}^*(-\tfrac{1}{2}):$$

$$+ :a_p(-\tfrac{1}{2})a_{p+N'}(\tfrac{1}{2})a_q^*(-\tfrac{1}{2})a_{q+N'}^*(-\tfrac{1}{2}): + :a_p(-\tfrac{1}{2})a_{p+N'}(-\tfrac{1}{2})a_q^*(\tfrac{1}{2})a_{q+N'}^*(-\tfrac{1}{2}):$$

$$+ :a_p(-\tfrac{1}{2})a_{p+N'}(-\tfrac{1}{2})a_q^*(-\tfrac{1}{2})a_{q+N'}^*(\tfrac{1}{2}): \}$$

$$\cdot \sum_{1 \le r,s \le N'} a_r(-\tfrac{1}{2})a_{r+N'}(-\tfrac{1}{2})a_s^*(-\tfrac{1}{2})a_{s+N'}^*(-\tfrac{1}{2})\Omega.$$

Note that contributions from the third and the fourth terms are obtained by applying $*$ to those from the first and the second terms. So calculation of summations of the first two terms is sufficient for us. For the first two operators in $\{xx^*\}_0$ corresponding to (p, q) to act nontrivialy on the term in $[xx^*]$ corresponding to (r, s), we must have $p = s$. After calculating some pairings, we see that contribution from the first two terms of (II) is

$$\{ \sum_{q=1}^{N'} a_q^*(-\tfrac{1}{2})a_{q+N'}^*(-\tfrac{1}{2}) \} \{ \sum_{r=1}^{N'} a_r(-\tfrac{1}{2})a_{r+N'}(-\tfrac{1}{2}) \} \{ \sum_{s=1}^{N'} a_{s+N'}(-\tfrac{1}{2})a_{s+N'}^*(-\tfrac{1}{2}) \}\Omega$$

$$- \{ \sum_{q=1}^{N'} a_q^*(-\tfrac{1}{2})a_{q+N'}^*(-\tfrac{1}{2}) \} \{ \sum_{r=1}^{N'} a_r(-\tfrac{1}{2})a_{r+N'}(-\tfrac{1}{2}) \} \{ \sum_{s=1}^{N'} a_s(-\tfrac{1}{2})a_s^*(-\tfrac{1}{2}) \}\Omega$$

$$= -x^*x \sum_{s=1}^{N} a_s(-\tfrac{1}{2})a_s^*(-\tfrac{1}{2})\Omega = -[xhx^*].$$

Contribution from the third and the fourth terms is then obtained by applying $*$ to the above result and it is equal to $-[xhx^*]^* = [xhx^*]$ since $h^* = -h$. Collecting our calculations, we have $(\text{II}) = 0$.

Case (III): $(n_1, n_2, n_3, n_4) \in \{$permutations of $(1, 1, -1, 0)\}$. There are twelve permutations and relevant terms in $(*_2)$ are given as follows:

$$\sum_{p,q=1}^{N'} \{ :a_p(\tfrac{1}{2})a_{p+N'}(\tfrac{1}{2})a_q^*(-\tfrac{3}{2})a_{q+N'}^*(-\tfrac{1}{2}): + :a_p(\tfrac{1}{2})a_{p+N'}(\tfrac{1}{2})a_q^*(-\tfrac{1}{2})a_{q+N'}^*(-\tfrac{3}{2}):$$

$$+ :a_p(-\tfrac{3}{2})a_{p+N'}(-\tfrac{1}{2})a_q^*(\tfrac{1}{2})a_{q+N'}^*(\tfrac{1}{2}): + :a_p(-\tfrac{1}{2})a_{p+N'}(-\tfrac{3}{2})a_q^*(\tfrac{1}{2})a_{q+N'}^*(\tfrac{1}{2}):$$

$$+ :a_p(\tfrac{1}{2})a_{p+N'}(-\tfrac{3}{2})a_q^*(\tfrac{1}{2})a_{q+N'}^*(-\tfrac{1}{2}): + :a_p(\tfrac{1}{2})a_{p+N'}(-\tfrac{3}{2})a_q^*(-\tfrac{1}{2})a_{q+N'}^*(\tfrac{1}{2}):$$

$$+ :a_p(-\tfrac{3}{2})a_{p+N'}(\tfrac{1}{2})a_q^*(\tfrac{1}{2})a_{q+N'}^*(-\tfrac{1}{2}): + :a_p(-\tfrac{3}{2})a_{p+N'}(\tfrac{1}{2})a_q^*(-\tfrac{1}{2})a_{q+N'}^*(\tfrac{1}{2}):$$

$$+ :a_p(\tfrac{1}{2})a_{p+N'}(-\tfrac{1}{2})a_q^*(\tfrac{1}{2})a_{q+N'}^*(-\tfrac{3}{2}): + :a_p(\tfrac{1}{2})a_{p+N'}(-\tfrac{1}{2})a_q^*(-\tfrac{3}{2})a_{q+N'}^*(\tfrac{1}{2}):$$

$$+ :a_p(-\tfrac{1}{2})a_{p+N'}(\tfrac{1}{2})a_q^*(\tfrac{1}{2})a_{q+N'}^*(-\tfrac{3}{2}): + :a_p(-\tfrac{1}{2})a_{p+N'}(\tfrac{1}{2})a_q^*(-\tfrac{3}{2})a_{q+N'}^*(\tfrac{1}{2}): \}$$

$$\cdot \sum_{1 \le r,s \le N'} a_r(-\tfrac{1}{2})a_{r+N'}(-\tfrac{1}{2})a_s^*(-\tfrac{1}{2})a_{s+N'}^*(-\tfrac{1}{2})\Omega.$$

We call this sum (III). As before, contribution from the first four terms is called (i), that from the next four terms is called (ii), that from the last four terms is denoted by (iii).

We look at (i). Given a pair of indices (r, s), for nontriviality of the action of four operators in (i) corresponding to (p, q) on the vector in $[xx^*]$ corresponding to (r, s), we must have $p = s$ for the first two terms, and $q = r$ for the next two terms. After calculating annihilation operators, we see that the first two summations in (i) become independent of s. Summation over s gives us a factor N', summation over r gives x, and finally summation over q gives χ^*. Thus contribution from the first two terms is $N'[x\chi^*]$. Similarly next two summations in (i) become independent of r after taking care of annihilation operators, and summation over r gives a factor N', summation over s gives x^*, and summation over q gives χ. So, contribution from the third and the fourth terms is $N'[x^*\chi]$. Together, we have (i) $= N'[x\chi^*] + N'[x^*\chi]$.

For (ii), for a given pair of indices (r, s), nontriviality of the action of four terms in (ii) corresponding to (p, q) on the term in $[xx^*]$ corresponding to (r, s) requires that $p = s$, $q = r$ for all the four terms in (ii). After taking care of annihilation operators, the four summations over r, s combine together into a single summation over r, s, where now r, s range from 1 to $N = 2N'$. Then contribution from (ii) can be easily seen to be $[h\omega_+]$.

Since by inspection, (iii) $=$ (ii)*, we have (iii) $= [h^*\omega_+^*] = -[h\omega_-]$. Thus,

$$(III) = N'([\chi^*x] + [\chi x^*]) + [h(\omega_+ - \omega_-)] = \tfrac{N}{2}D(-1)[xx^*] + \tfrac{1}{2}D(-1)[h^2].$$

Now collecting summations (I), (II), (III), we finally obtain

$$\{xx^*\}_0[xx^*] = \tfrac{N}{2}D(-1)[\omega] + \tfrac{1}{2}D(-1)[h^2] + \tfrac{N}{2}D(-1)[xx^*].$$

This completes the proof (28) and hence of Proposition 6.

Proof of the formula for $\{\phi_{\mathfrak{sp}}\}_1[\phi_{\mathfrak{sp}}]$

We show (14) of Theorem 3. The proof is by rather long but somewhat subtle calculations as in the calculation for (13). Recall that $[\phi_{\mathfrak{sp}}] = -\tfrac{1}{2}[h^2] + 2[xx^*] \in \bigwedge^*A(-\tfrac{1}{2})\Omega \subset V$, and $\{\phi_{\mathfrak{sp}}\}_1 = -\tfrac{1}{2}\{h^2\}_1 + 2\{xx^*\}_1 : V_* \to V_*$ is a weight preserving vertex operator. From (21), (22), (23), and (24), these vectors and operators are given by

$$[h^2] = \sum_{1 \le j \ne k \le N} a_j(-\tfrac{1}{2})a_j^*(-\tfrac{1}{2})a_k(-\tfrac{1}{2})a_k^*(-\tfrac{1}{2})\Omega,$$

$$[xx^*] = \sum_{r,s=1}^{N'} a_r(-\tfrac{1}{2})a_{r+N'}(-\tfrac{1}{2})a_s^*(-\tfrac{1}{2})a_{s+N'}^*(-\tfrac{1}{2})\Omega,$$

(29)

$$\{h^2\}_1 = \sum_{\substack{1 \le p \ne q \le N \\ m_1+m_2+m_3+m_4=2}} : a_p(m_1 - \tfrac{1}{2})a_p^*(m_2 - \tfrac{1}{2})a_q(m_3 - \tfrac{1}{2})a_q^*(m_4 - \tfrac{1}{2}): ,$$

$$\{xx^*\}_1 = \sum_{\substack{1 \le p,q \le N' \\ n_1+n_2+n_3+n_4=2}} : a_p(n_1 - \tfrac{1}{2})a_{p+N'}(n_2 - \tfrac{1}{2})a_q^*(n_3 - \tfrac{1}{2})a_{q+N'}^*(-\tfrac{1}{2}): .$$

Calculation of $\{\phi_{sp}\}_1[\phi_{sp}]$ is immediate once the next proposition is proved.

PROPOSITION 7. *The action of vertex operators* $\{h^2\}_1$, $\{xx^*\}_1$ *on vectors* $[h^2]$, $[xx^*]$ *is described as follows:*

$$(30) \qquad \{h^2\}_1[h^2] = -8(N-1)[\omega] + 4(N-2)[h^2],$$

$$(31) \qquad \{h^2\}_1[xx^*] = 4[\omega] - 2[h^2] - 4[xx^*],$$

$$(32) \qquad \{xx^*\}_1[h^2] = 4[\omega] - 2[h^2] - 4[xx^*],$$

$$(33) \qquad \{xx^*\}_1[xx^*] = -N[\omega] - [h^2] - N[xx^*].$$

Note that (30) has been shown in Proposition 3 of §3.6. We prove other formulae below. But before that, we use them to prove (14).

(PROOF OF (14) OF THEOREM 3 ASSUMING PROPOSITION 7). We have

$$\{\phi_{sp}\}_1[\phi_{sp}] = \tfrac{1}{4}\{h^2\}_1[h^2] - \{h^2\}_1[xx^*] - \{xx^*\}_1[h^2] + 4\{xx^*\}_1[xx^*]$$
$$= -6(N+1)[\omega] + (N-2)[h^2] - 4(N-2)[xx^*] = -6(N+1)[\omega] - 2(N-2)[\phi_{sp}].$$

This gives (14). Note that although these elements are linear combinations of $[\omega]$, $[h^2]$, $[xx^*]$, elements of interest are actually can be written as a linear combination of $[\omega]$, $[\phi_{sp}]$ only. $\square$

Now we turn to the proof of (31), (32) and (33).

Calculation of $\{h^2\}_1[xx^*]$. We must calculate the following summation:

$$(*_3) \quad \{h^2\}_1[xx^*] = \sum_{\substack{1 \le p \ne q \le N \\ m_1+m_2+m_3+m_4=2}} : a_p(m_1 - \tfrac{1}{2})a_p^*(m_2 - \tfrac{1}{2})a_q(m_3 - \tfrac{1}{2})a_q^*(m_4 - \tfrac{1}{2}) :$$
$$\sum_{1 \le r,s \le N'} a_r(-\tfrac{1}{2})a_{r+N'}(-\tfrac{1}{2})a_s^*(-\tfrac{1}{2})a_{s+N'}^*(-\tfrac{1}{2})\Omega.$$

For the action of operators in $\{h^2\}_1$ on $[xx^*]$ to be nontrivial, each m_j's must be less than or equal to 1, otherwise those terms with at least one m_j bigger than 1 annihilate $[xx^*]$. Since the sum of m_j's must be 2, we see that (m_1, m_2, m_3, m_4) must be a permutation of $(1,1,1,-1)$ or $(1,1,0,0)$. These two cases are treated separately.

Case (I): $(m_1, m_2, m_3, m_4) \in \{\text{permutations of } (1,1,1,-1)\}$. There are four permutations, and relevant terms in $(*_3)$ are

$$(I) = \sum_{p \ne q} \{ :a_p(\tfrac{1}{2})a_p^*(\tfrac{1}{2})a_q(\tfrac{1}{2})a_q^*(-\tfrac{3}{2}): + :a_p(\tfrac{1}{2})a_p^*(\tfrac{1}{2})a_q(-\tfrac{3}{2})a_q^*(\tfrac{1}{2}):$$
$$+ :a_p(\tfrac{1}{2})a_p^*(-\tfrac{3}{2})a_q(\tfrac{1}{2})a_q^*(\tfrac{1}{2}): + :a_p(-\tfrac{3}{2})a_p^*(\tfrac{1}{2})a_q(\tfrac{1}{2})a_q^*(\tfrac{1}{2}): \}$$
$$\sum_{1 \le r,s \le N'} a_r(-\tfrac{1}{2})a_{r+N'}(-\tfrac{1}{2})a_s^*(-\tfrac{1}{2})a_{s+N'}^*(-\tfrac{1}{2})\Omega.$$

Note that contributions from the first and the second terms are *-related, which can be seen by switching p, q. We also see that the first and the third terms give rise to the same sums and the second and the fourth terms also give rise to the same sums. So, we only have to calculate the contribution from the first term. For nontriviality of the action of the first term corresponding to (p, q) on the term in $[xx^*]$ corresponding to (r, s), we must have $s = r$ and given s, (p, q) must be either $(s, s + N')$ or $(s + N', s)$. After some calculation, we find that contribution from the first term is exactly $[\omega_-]$. Thus, the contribution from the second term is $[\omega_-]^* = [\omega_+]$. As noted above, contributions from the third and the fourth terms are the same as those from the first and the second terms. Hence we have $(\mathrm{I}) = 2([\omega_+] + [\omega_-]) = 4[\omega]$.

Case (II): $(m_1, m_2, m_3, m_4) \in \{\text{permutations of } (1, 1, 0, 0)\}$. There are six permutations and the relevant terms in $(*_3)$ are

$$(\mathrm{II}) = \sum_{p \neq q} \{ :a_p(\tfrac{1}{2})a_p^*(\tfrac{1}{2})a_q(-\tfrac{1}{2})a_q^*(-\tfrac{1}{2}): \ + \ :a_p(-\tfrac{1}{2})a_p^*(-\tfrac{1}{2})a_q(\tfrac{1}{2})a_q^*(\tfrac{1}{2}):$$
$$+ :a_p(\tfrac{1}{2})a_p^*(-\tfrac{1}{2})a_q(\tfrac{1}{2})a_q^*(-\tfrac{1}{2}): \ + \ :a_p(\tfrac{1}{2})a_p^*(-\tfrac{1}{2})a_q(-\tfrac{1}{2})a_q^*(\tfrac{1}{2}):$$
$$+ :a_p(-\tfrac{1}{2})a_p^*(\tfrac{1}{2})a_q(\tfrac{1}{2})a_q^*(-\tfrac{1}{2}): \ + \ :a_p(-\tfrac{1}{2})a_p^*(\tfrac{1}{2})a_q(-\tfrac{1}{2})a_q^*(\tfrac{1}{2}): \}$$
$$\cdot \sum_{1 \leq r, s \leq N'} a_r(-\tfrac{1}{2})a_{r+N'}(-\tfrac{1}{2})a_s^*(-\tfrac{1}{2})a_{s+N'}^*(-\tfrac{1}{2})\Omega.$$

Under summation over p and q, the first and the second terms give rise to the same sums, contributions from the third and the sixth terms are $*$-related, and sums for the fourth and the fifth terms are also $*$-related and in fact they are equal, which can be seen by symmetry between p and q. So, we calculate contributions from the first, the third and the fourth terms.

For the first term, given (r, s), for nontriviality of the action on the term in $[xx^*]$ corresponding to (r, s), we must have $p = s$, $s + N'$ and $p = r$, $r + N'$ at the same time. This implies that $r = s$ and $p = r$, $r + N'$. Thus contribution from the first term is

$$\sum_{r=1}^{N'} \Big\{ \sum_{q \neq r} :a_r(\tfrac{1}{2})a_r^*(\tfrac{1}{2})a_q(-\tfrac{1}{2})a_q^*(-\tfrac{1}{2}): + \sum_{q \neq r+N'} :a_{r+N'}(\tfrac{1}{2})a_{r+N'}^*(\tfrac{1}{2})a_q(-\tfrac{1}{2})a_q^*(-\tfrac{1}{2}): \Big\}$$
$$\cdot a_r(-\tfrac{1}{2})a_{r+N'}(-\tfrac{1}{2})a_r^*(-\tfrac{1}{2})a_{r+N'}^*(-\tfrac{1}{2})\Omega$$
$$= -\sum_{q \neq r} a_q(-\tfrac{1}{2})a_q^*(-\tfrac{1}{2})a_{r+N'}(-\tfrac{1}{2})a_{r+N'}^*(-\tfrac{1}{2})\Omega$$
$$- \sum_{q \neq r+N'} a_q(-\tfrac{1}{2})a_q^*(-\tfrac{1}{2})a_r(-\tfrac{1}{2})a_r^*(-\tfrac{1}{2})\Omega.$$

Separating out those cases for which $q = r$ or $q = r + N'$, and noting that the summation over q gives the element $h \in \mathcal{A}$, we see that the above is equal to

$$- h \cdot \Big(\sum_{r=1}^{N'} \{ a_{r+N'}(-\tfrac{1}{2})a_{r+N'}^*(-\tfrac{1}{2}) + a_r(-\tfrac{1}{2})a_r^*(-\tfrac{1}{2}) \} \Big) \Omega$$
$$+ 2 \sum_{r=1}^{N'} a_r(-\tfrac{1}{2})a_r^*(-\tfrac{1}{2})a_{r+N'}(-\tfrac{1}{2})a_{r+N'}^*(-\tfrac{1}{2})\Omega$$
$$= -[h^2] + 2 \sum_{r=1}^{N'} a_r(-\tfrac{1}{2})a_r^*(-\tfrac{1}{2})a_{r+N'}(-\tfrac{1}{2})a_{r+N'}^*(-\tfrac{1}{2})\Omega.$$

For the third term, given (r, s), nontriviality of the action of the operator corresponding to (p, q) on the (r, s) term in $[xx^*]$ requires that $(p, q) = (s, s+N')$ or $(s + N', s)$. For both of these cases, contributions can be easily seen to be $[xx^*]$. So, together we get $2[xx^*]$.

We have to be careful in the calculation of contribution from the fourth term. Given (r, s), nontriviality of the action requires that $p = s$ or $s + N'$, and $q = r$ or $r + N'$. Since $p \neq q$, all the possible (p, q) is (s, r) with $r \neq s$, $(s, r + N')$, $(s + N', r)$, $(s + N', r + N')$ with $r \neq s$. Thus contribution from the fourth term in (II) is given by

$$\left\{ \sum_{r \neq s} (: a_s(\tfrac{1}{2})a_s^*(-\tfrac{1}{2})a_r(-\tfrac{1}{2})a_r^*(\tfrac{1}{2}): + : a_{s+N'}(\tfrac{1}{2})a_{s+N'}^*(-\tfrac{1}{2})a_{r+N'}(-\tfrac{1}{2})a_{r+N'}^*(\tfrac{1}{2}):) \right.$$

$$\left. + \sum_{r,s} (: a_s(\tfrac{1}{2})a_s^*(-\tfrac{1}{2})a_{r+N'}(-\tfrac{1}{2})a_{r+N'}^*(\tfrac{1}{2}): + : a_{s+N'}(\tfrac{1}{2})a_{s+N'}^*(-\tfrac{1}{2})a_r(-\tfrac{1}{2})a_r^*(\tfrac{1}{2}):) \right\}$$

$$\cdot a_r(-\tfrac{1}{2})a_{r+N'}(-\tfrac{1}{2})a_s^*(-\tfrac{1}{2})a_{s+N'}(-\tfrac{1}{2})\Omega.$$

If there were no restrictions on r, s, each term would give $-[xx^*]$. However, for the first two sums, we need correction terms. Taking these into account, the whole contribution from the fourth term in (II) is

$$-4[xx^*] + 2\sum_{r=1}^{N'} a_r(-\tfrac{1}{2})a_{r+N'}(-\tfrac{1}{2})a_r^*(-\tfrac{1}{2})a_{r+N'}^*(-\tfrac{1}{2})\Omega.$$

Recalling $*$-relations among the six terms in (II) explained right after the formula for (II), we see that contribution from Case (II) is

$$-2[h^2] - 4\sum_{r=1}^{N'} a_r(-\tfrac{1}{2})a_{r+N'}(-\tfrac{1}{2})a_r^*(-\tfrac{1}{2})a_{r+N'}^*(-\tfrac{1}{2})\Omega + 4[xx^*] - 8[xx^*]$$
$$+ 4\sum_{r=1}^{N'} a_r(-\tfrac{1}{2})a_{r+N'}(-\tfrac{1}{2})a_r^*(-\tfrac{1}{2})a_{r+N'}^*(-\tfrac{1}{2})\Omega = -2[h^2] - 4[xx^*].$$

Collecting (I) and (II), we finally have

$$\{h^2\}_1[xx^*] = (I) + (II) = 4[\omega] - 2[h^2] - 4[xx^*].$$

This proves (31) of Proposition 7.

Calculation of $\{xx^*\}_1[h^2]$. This time, we must calculate the following:

$$(*4) \quad \{xx^*\}_1[h^2] = \sum_{\substack{1 \leq r,s \leq N' \\ n_1+n_2+n_3+n_4=2}} : a_r(n_1 - \tfrac{1}{2})a_{r+N'}(n_2 - \tfrac{1}{2})a_s^*(n_3 - \tfrac{1}{2})a_{s+N'}^*(-\tfrac{1}{2}):$$
$$\cdot \sum_{1 \leq j \neq k \leq N} a_j(-\tfrac{1}{2})a_j^*(-\tfrac{1}{2})a_k(-\tfrac{1}{2})a_k^*(-\tfrac{1}{2})\Omega.$$

As before, we consider two cases depending on whether (n_1, n_2, n_3, n_4) is a permutation of $(1, 1, 1, -1)$ or of $(1, 1, 0, 0)$, which are only possibilities for n_j's for possible nontrivial contribution to $(*4)$.

Case (I): $(n_1, n_2, n_3, n_4) \in \{$permutations of $(1, 1, 1, -1)\}$. There are four different permutations and summation of relevant four terms in $(*_4)$ are given by

$$\sum_{r,s=1}^{N'} \{ :a_r(\tfrac{1}{2})a_{r+N'}(\tfrac{1}{2})a_s^*(\tfrac{1}{2})a_{s+N'}^*(-\tfrac{3}{2}): + :a_r(\tfrac{1}{2})a_{r+N'}(\tfrac{1}{2})a_s^*(-\tfrac{3}{2})a_{s+N'}^*(\tfrac{1}{2}): $$
$$+ :a_r(\tfrac{1}{2})a_{r+N'}(-\tfrac{3}{2})a_s^*(\tfrac{1}{2})a_{s+N'}^*(\tfrac{1}{2}): + :a_r(-\tfrac{3}{2})a_{r+N'}(\tfrac{1}{2})a_s^*(\tfrac{1}{2})a_{s+N'}^*(\tfrac{1}{2}): \}$$
$$\cdot \sum_{1 \le j \ne k \le N} a_j(-\tfrac{1}{2})a_j^*(-\tfrac{1}{2})a_k(-\tfrac{1}{2})a_k^*(-\tfrac{1}{2})\Omega.$$

We call this summation (I). We observe that contributions from the first and the third terms are $*$-related and contributions from the second and the fourth terms are also $*$-related.

For the first term, for nontriviality of the action of the operator in $\{xx^*\}_1$ corresponding to (r, s) on the term in $[h^2]$ corresponding to (j, k), we must have $s \in \{j, k\}$ and $\{r, r + N'\} = \{j, k\}$. This is possible only when $r = s$ and then, given $1 \le r \le N'$, (j, k) must be $(r, r + N')$, $(r + N', r)$. After summing over r, these two cases give the same contribution and together we have $-2 \sum_{r=1}^{N'} a_{r+N'}^*(-\tfrac{3}{2})a_{r+N'}(-\tfrac{1}{2})\Omega$.

For contribution from the second term, nontriviality of the action of the operator in $\{xx^*\}_1$ corresponding to (p, q) on the term in $[h^2]$ corresponding to (j, k) again requires that $r = s$ and, given r, $(j, k) = (r, r + N')$ or $(r + N', r)$. Similar calculation shows that contribution from the second term is given by $-2 \sum_{r=1}^{N'} a_r^*(-\tfrac{3}{2})a_r(-\tfrac{1}{2})\Omega$. Thus contribution from the first two terms is the sum of the above two, and this is equal to $2[\omega_-]$. For contribution from the last two terms we apply $*$, and it is given by $[\omega_-]^* = [\omega_+]$. Since $[\omega_+] + [\omega_-] = 2[\omega]$, we see that (I) $= 4[\omega]$.

Case (II): $(n_1, n_2, n_3, n_4) \in \{$permutations of $(1, 1, 0, 0)\}$. There are six permutations. The summation of relevenat terms in $(*_4)$ is

$$\sum_{r,s=1}^{N'} \{ :a_r(\tfrac{1}{2})a_{r+N'}(\tfrac{1}{2})a_s^*(-\tfrac{1}{2})a_{s+N'}^*(-\tfrac{1}{2}): + :a_r(-\tfrac{1}{2})a_{r+N'}(-\tfrac{1}{2})a_s^*(\tfrac{1}{2})a_{s+N'}^*(\tfrac{1}{2}):$$
$$+ :a_r(\tfrac{1}{2})a_{r+N'}(-\tfrac{1}{2})a_s^*(\tfrac{1}{2})a_{s+N'}^*(-\tfrac{1}{2}): + :a_r(\tfrac{1}{2})a_{r+N'}(-\tfrac{1}{2})a_s^*(-\tfrac{1}{2})a_{s+N'}^*(\tfrac{1}{2}):$$
$$+ :a_r(-\tfrac{1}{2})a_{r+N'}(\tfrac{1}{2})a_s^*(\tfrac{1}{2})a_{s+N'}^*(-\tfrac{1}{2}): + :a_r(-\tfrac{1}{2})a_{r+N'}(\tfrac{1}{2})a_s^*(-\tfrac{1}{2})a_{s+N'}^*(\tfrac{1}{2}): \}$$
$$\cdot \sum_{j \ne k} a_j(-\tfrac{1}{2})a_j^*(-\tfrac{1}{2})a_k(-\tfrac{1}{2})a_k^*(-\tfrac{1}{2})\Omega.$$

We call this summation (II). We examine contribution from the first term. Given the operator in $\{xx^*\}_1$ corresponding to (r, s), only those terms in $[h^2]$ with $(j, k) = (r, r + N')$ or $(r + N', r)$ contribute nontrivially. Contributions fom these two terms are the same by symmetry of indices, and together contribution from the first term of (II) is equal to

$$2 \sum_{r,s=1}^{N'} :a_s(\tfrac{1}{2})a_{r+N'}(\tfrac{1}{2})a_s^*(-\tfrac{1}{2})a_{s+N'}^*(-\tfrac{1}{2}): a_r(-\tfrac{1}{2})a_r^*(-\tfrac{1}{2})a_{r+N'}(-\tfrac{1}{2})a_{r+N'}^*(-\tfrac{1}{2})\Omega$$
$$= 2 \Big(\sum_{s=1}^{N'} a_s^*(-\tfrac{1}{2})a_{s+N'}^*(-\tfrac{1}{2}) \Big) \Big(\sum_{r=1}^{N'} a_r(-\tfrac{1}{2})a_{r+N'}(-\tfrac{1}{2}) \Big) \Omega = 2[xx^*].$$

Contribution from the second term is the same as that from the first term, as can be seen by switching r and s.

For contribution from the third term, nontriviality of the action of the operator in $\{xx^*\}_1$ corresponding to (r,s) on the term in $[h^2]$ corresponding to (j,k) requires that $r=j,k$ and $s=j,k$. So, $(r,s)=(j,j),\ (j,k),\ (k,j),\ (k,k)$. Since j,k range from 1 to N and r,s range from 1 to N', for some (j,k), there are no corresponding pairs (r,s) satisfying above identities. So, we rephrase the above in terms of (r,s). Thus, given (r,s), if $r=s$, then either $j=r,k\neq r$ or $k=r, j\neq r$. If $r\neq s$, then $(j,k)=(r,s)$ or (s,r). By symmetry of indices, we see that contributions from the two cases for $r=s$ are the same, and contributions from the two cases for $r\neq s$ are also the same. Thus, contribution from the third term is equal to

$$2\sum_{r=1}^{N'} :a_r(\tfrac{1}{2})a_{r+N'}(-\tfrac{1}{2})a_r^*(\tfrac{1}{2})a_{r+N'}^*(-\tfrac{1}{2}): \sum_{k\neq r} a_r(-\tfrac{1}{2})a_r^*(-\tfrac{1}{2})a_k(-\tfrac{1}{2})a_k^*(-\tfrac{1}{2})\Omega$$

$$+\ 2\sum_{r\neq s} :a_r(\tfrac{1}{2})a_{r+N'}(-\tfrac{1}{2})a_s^*(\tfrac{1}{2})a_{s+N'}^*(-\tfrac{1}{2}): a_r(-\tfrac{1}{2})a_r^*(-\tfrac{1}{2})a_s(-\tfrac{1}{2})a_s^*(-\tfrac{1}{2})\Omega$$

$$=-2\sum_{k\neq r} a_k(-\tfrac{1}{2})a_k^*(-\tfrac{1}{2})a_{r+N'}(-\tfrac{1}{2})a_{r+N'}^*(-\tfrac{1}{2})\Omega$$
$$-\ 2\sum_{r\neq s} a_r(-\tfrac{1}{2})a_{r+N'}(-\tfrac{1}{2})a_s^*(-\tfrac{1}{2})a_{s+N'}^*(-\tfrac{1}{2})\Omega.$$

If there were no restrictions such as $k\neq r$ and $r\neq s$, summation over k in the first sum directly above would have given $h\in\mathcal{A}$, and the second summation over r,s would have given $-2[xx^*]$. Separating correction terms for the cases with $r=k$ or $r=s$, we have

$$=-2h\cdot\left(\sum_{r=1}^{N'} a_{r+N'}(-\tfrac{1}{2})a_{r+N'}^*(-\tfrac{1}{2})\right)\Omega$$
$$-\ 2\sum_{r=1}^{N'} a_r(-\tfrac{1}{2})a_{r+N'}(-\tfrac{1}{2})a_r^*(-\tfrac{1}{2})a_{r+N'}^*(-\tfrac{1}{2})\Omega - 2[xx^*]$$
$$+\ 2\sum_{r=1}^{N'} a_r(-\tfrac{1}{2})a_{r+N'}(-\tfrac{1}{2})a_r^*(-\tfrac{1}{2})a_{r+N'}^*(-\tfrac{1}{2})\Omega$$
$$=-2[xx^*]-2h\cdot\sum_{r=1}^{N'} a_{r+N'}(-\tfrac{1}{2})a_{r+N'}^*(-\tfrac{1}{2})\Omega.$$

Note that the summation above is the "second half" of $h\in\mathcal{A}$.

Now we look at contribution from the fourth term. For nontriviality of the action, we need $r=j,k$ and $s+N'=j,k$. This happens when $(j,k)=(r,s+N')$ or $(s+N',r)$. By symmetry of indices, we see that contributions from these two cases are the same, and contribution from the fourth term is

$$2\sum_{r,s=1}^{N'} :a_r(\tfrac{1}{2})a_{r+N'}(-\tfrac{1}{2})a_s^*(-\tfrac{1}{2})a_{s+N'}^*(\tfrac{1}{2}): a_r(-\tfrac{1}{2})a_r^*(-\tfrac{1}{2})a_{s+N'}(-\tfrac{1}{2})a_{s+N'}^*(-\tfrac{1}{2})\Omega$$
$$=-2\sum_{r,s=1}^{N'} a_r(-\tfrac{1}{2})a_{r+N'}(-\tfrac{1}{2})a_s^*(-\tfrac{1}{2})a_{s+N'}^*(-\tfrac{1}{2})\Omega = -2[xx^*].$$

Contribution from the fifth term is ∗-related to that from the fourth term and is equal to $-2[xx^*]^* = -2[xx^*]$.

Calculation of contribution from the sixth term is similar to that for the third term. Nontriviality of the action requires that $r + N' = j, k$ and $s + N' = j, k$. Rephrasing this as in the third term case in terms of (r, s), we have that if $r = s$, then either $j = r + N', k \neq r + N'$ or $k = r + N', j \neq r + N'$. If $r \neq s$, then $(j, k) = (r + N', s + N')$ or $(s + N', r + N')$. By symmetry of indices, contributions from the two cases for $r = s$ are the same, and those from the two cases for $r \neq s$ are also the same. Thus, contribution from the sixth term is

$$2\sum_{k \neq r+N'} :a_r(-\tfrac{1}{2})a_{r+N'}(\tfrac{1}{2})a_r^*(-\tfrac{1}{2})a_{r+N'}^*(\tfrac{1}{2}): a_{r+N'}(-\tfrac{1}{2})a_{r+N'}^*(-\tfrac{1}{2})a_k(-\tfrac{1}{2})a_k^*(-\tfrac{1}{2})\Omega$$

$$+2\sum_{r \neq s}:a_r(-\tfrac{1}{2})a_{r+N'}(\tfrac{1}{2})a_s^*(-\tfrac{1}{2})a_{s+N'}^*(\tfrac{1}{2}):a_{r+N'}(-\tfrac{1}{2})a_{r+N'}^*(-\tfrac{1}{2})a_{s+N'}(-\tfrac{1}{2})a_{s+N'}^*(-\tfrac{1}{2})\Omega$$

$$= -2\sum_{k \neq r+N'} a_k(-\tfrac{1}{2})a_k^*(-\tfrac{1}{2})a_r(-\tfrac{1}{2})a_r^*(-\tfrac{1}{2})\Omega$$

$$- 2\sum_{r \neq s} a_r(-\tfrac{1}{2})a_{r+N'}(-\tfrac{1}{2})a_s^*(-\tfrac{1}{2})a_{s+N'}^*(-\tfrac{1}{2})\Omega.$$

As before, to deal with restrictions on summation indices, we separate out the cases $k = r + N'$ and $r = s$. The above is then equal to

$$- 2h \cdot \sum_{r=1}^{N'} a_r(-\tfrac{1}{2})a_r^*(-\tfrac{1}{2})\Omega + 2\sum_{r=1}^{N'} a_r(-\tfrac{1}{2})a_r^*(-\tfrac{1}{2})a_{r+N'}(-\tfrac{1}{2})a_{r+N'}^*(-\tfrac{1}{2})\Omega$$

$$- 2[xx^*] + 2\sum_{r=1}^{N'} a_r(-\tfrac{1}{2})a_{r+N'}(-\tfrac{1}{2})a_r^*(-\tfrac{1}{2})a_{r+N'}^*(-\tfrac{1}{2})\Omega$$

$$= -2h \cdot \sum_{r=1}^{N'} a_r(-\tfrac{1}{2})a_r^*(-\tfrac{1}{2})\Omega - 2[xx^*].$$

This is the contribution from the sixth term. We note that contributions from the third term and the sixth term are complementary and their sum is $-4[xx^*] - 2[h^2]$. Contributions from other terms cancel and $(II) = -2[h^2] - 4[xx^*]$.

Together with (I), we have

$$\{xx^*\}_1[h^2] = (I) + (II) = 4[\omega] - 2[h^2] - 4[xx^*].$$

This proves (32) of Proposition 7.

Calculation of $\{xx^*\}_1[xx^*]$. To complete the proof of Proposition 7, we have to calculate

(∗5) $$\{xx^*\}_1[xx^*] = \sum_{\substack{1 \leq p,q \leq N' \\ n_1+n_2+n_3+n_4=2}} :a_p(n_1 - \tfrac{1}{2})a_{p+N'}(n_2 - \tfrac{1}{2})a_q^*(n_3 - \tfrac{1}{2})a_{q+N'}^*(-\tfrac{1}{2}):$$
$$\cdot \sum_{1 \leq r,s \leq N'} a_r(-\tfrac{1}{2})a_{r+N'}(-\tfrac{1}{2})a_s^*(-\tfrac{1}{2})a_{s+N'}^*(-\tfrac{1}{2})\Omega.$$

As before, nontrivial contributions arise only when (n_1, n_2, n_3, n_4) is a permutation of $(1, 1, 1, -1)$ or of $(1, 1, 0, 0)$. For other quadruples, corresponding operators annihilate $[xx^*]$.

Case (I): $(n_1, n_2, n_3, n_4) \in \{$permutations of $(1, 1, 1, -1)\}$. There are four permutations, and they give rise to the following partial summation of $(*_5)$:

$$\sum_{p,q=1}^{N'} \{ :a_p(\tfrac{1}{2})a_{p+N'}(\tfrac{1}{2})a_q^*(\tfrac{1}{2})a_{q+N'}^*(-\tfrac{3}{2}): + :a_p(\tfrac{1}{2})a_{p+N'}(\tfrac{1}{2})a_q^*(-\tfrac{3}{2})a_{q+N'}^*(\tfrac{1}{2}):$$
$$+ :a_p(\tfrac{1}{2})a_{p+N'}(-\tfrac{3}{2})a_q^*(\tfrac{1}{2})a_{q+N'}^*(\tfrac{1}{2}): + :a_p(-\tfrac{3}{2})a_{p+N'}(\tfrac{1}{2})a_q^*(\tfrac{1}{2})a_{q+N'}^*(\tfrac{1}{2}):\}$$
$$\cdot \sum_{1 \leq r,s \leq N'} a_r(-\tfrac{1}{2})a_{r+N'}(-\tfrac{1}{2})a_s^*(-\tfrac{1}{2})a_{s+N'}^*(-\tfrac{1}{2})\Omega.$$

We call this sum (I). For nontriviality of the action on $[xx^*]$, we must have $p = s$ and $q = r$. After simplifying each term, we obtain

$$\text{(I)} = \sum_{1 \leq r,s \leq N'} \{ a_{r+N'}^*(-\tfrac{3}{2})a_{r+N'}(-\tfrac{1}{2}) + a_r^*(-\tfrac{3}{2})a_r(-\tfrac{1}{2}) + a_{r+N'}(-\tfrac{3}{2})a_{r+N'}^*(-\tfrac{1}{2})$$
$$+ a_r(-\tfrac{3}{2})a_r^*(-\tfrac{1}{2})\}\Omega = N' \sum_{p=1}^{N} \{ a_p^*(-\tfrac{3}{2})a_p(-\tfrac{1}{2}) + a_p(-\tfrac{3}{2})a_p^*(-\tfrac{1}{2})\}\Omega = -N[\omega].$$

Case (II): $(n_1, n_2, n_3, n_4) \in \{$permutations of $(1, 1, 0, 0)\}$. The relevant six permutations give rise to the following partial summation of $(*_5)$:

$$\sum_{p,q=1}^{N'} \{ :a_p(\tfrac{1}{2})a_{p+N'}(\tfrac{1}{2})a_q^*(-\tfrac{1}{2})a_{q+N'}^*(-\tfrac{1}{2}): + :a_p(-\tfrac{1}{2})a_{p+N'}(-\tfrac{1}{2})a_q^*(\tfrac{1}{2})a_{q+N'}^*(\tfrac{1}{2}):$$
$$+ :a_p(\tfrac{1}{2})a_{p+N'}(-\tfrac{1}{2})a_q^*(\tfrac{1}{2})a_{q+N'}^*(-\tfrac{1}{2}): + :a_p(\tfrac{1}{2})a_{p+N'}(-\tfrac{1}{2})a_q^*(-\tfrac{1}{2})a_{q+N'}^*(\tfrac{1}{2}):$$
$$+ :a_p(-\tfrac{1}{2})a_{p+N'}(\tfrac{1}{2})a_q^*(\tfrac{1}{2})a_{q+N'}^*(-\tfrac{1}{2}): + :a_p(-\tfrac{1}{2})a_{p+N'}(\tfrac{1}{2})a_q^*(-\tfrac{1}{2})a_{q+N'}^*(\tfrac{1}{2}):\}$$
$$\cdot \sum_{1 \leq r,s \leq N'} a_r(-\tfrac{1}{2})a_{r+N'}(-\tfrac{1}{2})a_s^*(-\tfrac{1}{2})a_{s+N'}^*(-\tfrac{1}{2})\Omega.$$

We call this summation (II). For a given pair (r, s), a little examination shows that nontrivial contribution results from the first operator corresponding to (p, q) only when $p = s$. In this case, contribution from the first term is

$$\sum_{r,s,q=1}^{N'} :a_s(\tfrac{1}{2})a_{s+N'}(\tfrac{1}{2})a_q^*(-\tfrac{1}{2})a_{q+N'}^*(-\tfrac{1}{2}): a_r(-\tfrac{1}{2})a_{r+N'}(-\tfrac{1}{2})a_s^*(-\tfrac{1}{2})a_{s+N'}^*(-\tfrac{1}{2})\Omega$$
$$= -\sum_{r,s,q=1}^{N'} a_q^*(-\tfrac{1}{2})a_{q+N'}^*(-\tfrac{1}{2})a_r(-\tfrac{1}{2})a_{r+N'}(-\tfrac{1}{2})\Omega = -N'[xx^*].$$

Contribution from the second term is $*$-related to that of the first term. So it is the same as above and together we have $-2N'[xx^*] = -N[xx^*]$.

From summation of the remaining four terms, nontrivial contributions result only when $p = s$ and $q = r$. And simple calculations show that contributions from these four terms are given by

$$-\Big(\sum_{r=1}^{N'} a_{r+N'}(-\tfrac{1}{2})a_{r+N'}^*(-\tfrac{1}{2})\Big)^2 \Omega,$$

$$-\Big(\sum_{r=1}^{N'} a_r(-\tfrac{1}{2})a_r^*(-\tfrac{1}{2})\Big)\Big(\sum_{s=1}^{N'} a_{s+N'}(-\tfrac{1}{2})a_{s+N'}^*(-\tfrac{1}{2})\Big)\Omega,$$

$$-\Big(\sum_{r=1}^{N'} a_r(-\tfrac{1}{2})a_r^*(-\tfrac{1}{2})\Big)\Big(\sum_{s=1}^{N'} a_{s+N'}(-\tfrac{1}{2})a_{s+N'}^*(-\tfrac{1}{2})\Big)\Omega,$$

$$-\Big(\sum_{r=1}^{N'} a_r(-\tfrac{1}{2})a_r^*(-\tfrac{1}{2})\Big)^2\Omega.$$

Thus, contribution from the last four terms is the sum of the above four vectors and it is equal to $-[h^2]$. Hence altogether, (II) $= -N[xx^*] - [h^2]$.

Collecting our calculations, we now have

$$\{xx^*\}_1[xx^*] = (\mathrm{I}) + (\mathrm{II}) = -N[\omega] - [h^2] - N[xx^*].$$

This proves (33) and completes the proof of Proposition 7.

Proof of the formula for $\{\phi_{sp}\}_3[\phi_{sp}]$

We finally prove the last identity (16) in Theorem 3. We list the objects relevant for the calculation of $\{\phi_{sp}\}_3[\phi_{sp}]$, where $\phi_{sp} = -\tfrac{1}{2}h^2 + 2xx^*$.

(34)

$$[h^2] = \sum_{1 \le j \ne k \le N} a_j(-\tfrac{1}{2})a_j^*(-\tfrac{1}{2})a_k(-\tfrac{1}{2})a_k^*(-\tfrac{1}{2})\Omega,$$

$$[xx^*] = \sum_{r,s=1}^{N'} a_r(-\tfrac{1}{2})a_{r+N'}(-\tfrac{1}{2})a_s^*(-\tfrac{1}{2})a_{s+N'}^*(-\tfrac{1}{2})\Omega,$$

$$\{h^2\}_3 = \sum_{\substack{1 \le p \ne q \le N \\ m_1+m_2+m_3+m_4=4}} :a_p(m_1 - \tfrac{1}{2})a_p^*(m_2 - \tfrac{1}{2})a_q(m_3 - \tfrac{1}{2})a_q^*(m_4 - \tfrac{1}{2}):,$$

$$\{xx^*\}_3 = \sum_{\substack{1 \le p,q \le N' \\ n_1+n_2+n_3+n_4=4}} :a_p(n_1 - \tfrac{1}{2})a_{p+N'}(n_2 - \tfrac{1}{2})a_q^*(n_3 - \tfrac{1}{2})a_{q+N'}^*(-\tfrac{1}{2}): .$$

In the last two identities of (34), we observe that summation indices m_j's and n_j's must be less than or equal to 1, otherwise the corresponding operators annihilate $[h^2]$ or $[xx^*]$. Since sums of m_j's and n_j's must be 4 in both cases, the only possibility for these indices is $(m_1, m_2, m_3, m_4) = (1,1,1,1)$ and $(n_1, n_2, n_3, n_4) = (1,1,1,1)$. Thus $\{h^2\}_3$ and $\{xx^*\}_3$ consist only of summations over p, q.

PROPOSITION 8. *The action of vertex operators $\{h^2\}_3$, $\{xx^*\}_3$ on $[h^2]$, $[xx^*]$ is described as follows:*

(35) $$\{h^2\}_3[h^2] = 2N(N-1)\Omega,$$

(36) $$\{h^2\}_3[xx^*] = -N\Omega,$$

(37) $$\{xx^*\}_3[h^2] = -N\Omega,$$

(38) $$\{xx^*\}_3[xx^*] = N'^2\Omega = \tfrac{1}{4}N^2\Omega.$$

PROOF. Although (35) was proved in Proposition 3 of §3.6, we include it here for completeness' sake. So,

$$\{h^2\}_3[h^2] = \sum_{1 \le p \ne q \le N} : a_p(\tfrac{1}{2})a_p^*(\tfrac{1}{2})a_q(\tfrac{1}{2})a_q^*(\tfrac{1}{2}) : \sum_{j \ne k} a_j(-\tfrac{1}{2})a_j^*(-\tfrac{1}{2})a_k(-\tfrac{1}{2})a_k^*(-\tfrac{1}{2})\Omega.$$

Given the term in $[h^2]$ corresponding to (j, k), for nontrivial contribution the index (p, q) for the operator must be either (j, k) or (k, j), and the contributions from these two cases are the same, as can be seen by switching j and k. So, the above is equal to

$$2\sum_{j \ne k} : a_j(\tfrac{1}{2})a_j^*(\tfrac{1}{2})a_k(\tfrac{1}{2})a_k^*(\tfrac{1}{2}) : a_j(-\tfrac{1}{2})a_j^*(-\tfrac{1}{2})a_k(-\tfrac{1}{2})a_k^*(-\tfrac{1}{2})\Omega$$

$$= 2\sum_{1 \le j \ne k \le N} \Omega = 2N(N-1)\Omega.$$

For (36), letting $(m_1, m_2, m_3, m_4) = (1, 1, 1, 1)$, we have

$$\{h^2\}_3[xx^*] = \sum_{\substack{1 \le p \ne q \le N \\ 1 \le r,s \le N'}} : a_p(\tfrac{1}{2})a_p^*(\tfrac{1}{2})a_q(\tfrac{1}{2})a_q^*(\tfrac{1}{2}) : a_r(-\tfrac{1}{2})a_{r+N'}(-\tfrac{1}{2})a_s^*(-\tfrac{1}{2})a_{s+N'}^*(-\tfrac{1}{2})\Omega.$$

Given (r, s), for nontriviality of the action of the operator corresponding to (p, q), we must have $(p, q) = (s, s + N'), (s + N', s)$ and at the same time $(p, q) = (r, r + N'), (r + N', r)$. This happens only when $r = s$ and in this case, $(p, q) = (r, r+N'), (r+N', r)$. These two cases contribute equally by symmetry of indices, and the above is equal to

$$2\sum_{r=1}^{N'} : a_r(\tfrac{1}{2})a_r^*(\tfrac{1}{2})a_{r+N'}(\tfrac{1}{2})a_{r+N'}^*(\tfrac{1}{2}) : a_r(-\tfrac{1}{2})a_{r+N'}(-\tfrac{1}{2})a_r^*(-\tfrac{1}{2})a_{r+N'}^*(-\tfrac{1}{2})\Omega$$

$$= 2\sum_{r=1}^{N'}(-\Omega) = -2N'\Omega = -N\Omega.$$

The calculation of $\{xx^*\}_3[h^2]$ in (37) is essentially the same as above and gives the same result.

For (38), we proceed as follows. The object we must calculate is the following:

$$\{xx^*\}_3[xx^*] = \sum_{1 \le p \ne q \le N'} : a_p(\tfrac{1}{2})a_{p+N'}(\tfrac{1}{2})a_q^*(\tfrac{1}{2})a_{q+N'}^*(\tfrac{1}{2}) :$$
$$\sum_{1 \le r,s \le N'} a_r(-\tfrac{1}{2})a_{r+N'}(-\tfrac{1}{2})a_s^*(-\tfrac{1}{2})a_{s+N'}^*(-\tfrac{1}{2})\Omega.$$

Given (r, s), for nontriviality of the action of the operator corresponding to (p, q), we must have $p = s$ and $q = r$. Thus, the above is equal to

$$\sum_{r,s=1}^{N'} : a_s(\tfrac{1}{2})a_{s+N'}(\tfrac{1}{2})a_r^*(\tfrac{1}{2})a_{r+N'}^*(\tfrac{1}{2}) : a_r(-\tfrac{1}{2})a_{r+N'}(-\tfrac{1}{2})a_s^*(-\tfrac{1}{2})a_{s+N'}^*(-\tfrac{1}{2})\Omega$$

$$= \sum_{1 \le r,s \le N'} \Omega = N'^2\Omega = \tfrac{1}{4}N^2\Omega.$$

Here $N = 2N'$. This completes the proof of Proposition 8. $\Box$

Now, (16) of Theorem 3 can be finally proved.

(PROOF OF (16) OF THEOREM 3). Using Proposition 8, we get

$$\{\phi_{\mathfrak{sp}}\}_3[\phi_{\mathfrak{sp}}] = \tfrac{1}{4}\{h^2\}_3[h^2] - \{h^2\}_3[xx^*] - \{xx^*\}_3[h^2] + 4\{xx^*\}_3[xx^*]$$
$$= \tfrac{1}{2}N(N-1)\Omega + N\Omega + N\Omega + N^2\Omega = \tfrac{3}{2}N(N+1)\Omega.$$

This proves (16) and finishes the proof of Theorem 3. $\Box$

Now everything in this section has at last been proved.

§3.9   Kähler forms on quaternionic vector spaces

On a Kähler manifold with an integrable almost complex structure I, a basic geometric objet is a Kähler form κ_I on M^{2N}. This is a real closed differential 2-form on M of type $(1,1)$. In this section, we consider an analogue of a Kähler form for a complex vector space. We also call this object a Kähler form and denote it by ω_I; it is an alternating 2-form on the vector space. Its construction is modeled on restriction of a Kähler form κ_I on a Kähler manifold to a tangent space.

We are particularly interested in Kähler forms for vector spaces equipped with more than one complex structures. A quaternionic vector space is such a vector space with complex structures I, J, K satisfying quaternionic relations. For a quaternionic vector space, there is a real 2-dimensional family of complex structures, and consequently there is a real 2-dimensional family of Kähler forms. Among these, ω_I, ω_J, ω_K play important roles. This situation is a local version of a hyperkähler manifold whose holonomy group is contained in the symplectic group $\mathrm{Sp}(N')$. An object of interest for a quaternionic vector space is the quaternion-Kähler form $\omega_{\mathrm{Q\text{-}K}}$ given by $\omega_I^2 + \omega_J^2 + \omega_K^2$. This 4-form is important geometrically for a quaternion-Kähler manifold whose holonomy group is contained in $\mathrm{Sp}(N') \cdot \mathrm{Sp}(1) \subset \mathrm{SO}(4N')$. It turns out that the quaternion-Kähler form coincides, up to a constant multiple, with Casimir elements for Lie algebras $\mathfrak{g} = \mathfrak{sp}(N')$, $\mathfrak{sp}(N') \oplus \mathfrak{sp}(1)$, and $\mathfrak{sp}(1)$, where $\mathfrak{sp}(1) = \mathfrak{o}(4N')^{\mathfrak{sp}(N')}$.

Finally, we consider a generalization of Kähler forms and quaternion-Kähler forms to a context of vertex operator super algebras.

Kähler forms associated to complex structures on vector spaces. To define a Kähler form on a complex vector space, we start with a real $2N$-dimensional Euclidean vector space $E = (E^{2N}, \langle\, , \,\rangle)$ where $\langle\, , \,\rangle : E \otimes E \to \mathbb{R}$ is an inner product. Let $I : E \to E$ be a complex structure on E compatible with the inner product, that is, I is an isometry of E satisfying $I^2 = -1$. Let $\{e_1, e_2, \ldots, e_N, e'_1, e'_2, \ldots, e'_N\}$ be an orthonormal basis for E where $e'_j = I(e_j)$, $1 \le j \le N$. Its complexification $A = E \otimes \mathbb{C}$ splits into a direct sum $A^+ \oplus A^-$ of $\pm i$-eigenspaces of I. Complex bases for $A^{\pm}$ and pairings among the basis vectors are given as follows:

$$A^+ : \{a_1, a_2, \ldots, a_N\}, \qquad A^- : \{a_1^*, a_2^*, \ldots, a_N^*\}$$

(1)
$$a_j = \frac{e_j - ie_j'}{\sqrt{2}}, \qquad a_j^* = \frac{e_j + ie_j'}{\sqrt{2}},$$

$$\langle a_j, a_k \rangle = 0, \quad \langle a_j^*, a_k^* \rangle = 0, \quad \langle a_j, a_k^* \rangle = \delta_{j,k}, \qquad 1 \le j, k \le N.$$

Later we will compare I with other complex structures on E in terms of the above basis. Using the $\mathbb{C}$-linear pairing $\langle\ ,\ \rangle$ on A, we can identify A with its dual $\mathrm{Hom}(A, \mathbb{C})$. So, we have $a_j(a_k) = 0$, $a_j^*(a_k^*) = 0$, $a_j(a_k^*) = \delta_{jk} = a_j^*(a_k)$. In other words, the dual basis of $\{a_1, a_2, \ldots, a_N, a_1^*, a_2^*, \ldots, a_N^*\}$ is given by $\{a_1^*, a_2^*, \ldots, a_N^*, a_1, a_2, \ldots, a_N\}$. Thus, we can obtain dual basis vectors by applying complex conjugation operator $*$. To define a Kähler form on the vector space A associated to the complex structure I, we recall the following well known lemma.

LEMMA 1. *Let* $(E; \langle\ ,\ \rangle)$ *be a Euclidean vector space with an isometric complex structure* I. *Then there exists a unique Hermitian pairing* $(\ ,\ )$ *on* E *whose real part is the given inner product* $\langle\ ,\ \rangle$ *and which is conjugate linear in the second variable. It is given by*

(2)
$$(u, v) = \langle u, v \rangle - i\omega_I(u, v),$$

where the imaginary part ω_I *is an alternating 2-form on* E *given by*

(3)
$$\omega_I(u, v) = \langle Iu, v \rangle = -\langle u, Iv \rangle, \qquad u, v \in E.$$

There exists another unique Hermitian pairing $(\ ,\ )'$ *with the same properties except that it is conjugate linear in the first variable. It is given by*

(4)
$$(u, v)' = \langle u, v \rangle + i\omega_I(u, v),$$

The complex linear extension of ω_I *to* $A = E \otimes \mathbb{C}$ *is called the Kähler form associated to the complex structure* I *on* E.

PROOF. The above pairing $(\ ,\ )$ can be easily checked to be a Hermitian pairing using the fact that I is an isometry. For example, $(u, I(v)) = \langle u, I(v) \rangle - i\omega_I(u, I(v)) = \langle u, I(v) \rangle + i\langle u, I^2(v) \rangle = \langle u, I(v) \rangle - i\langle u, v \rangle$, from the definition of ω_I and using $I^2 = -1$. The first term is equal to $-\omega_I(u, v)$. So, factoring out $-i$, we see that $(u, I(v))$ is equal to $-i(\langle u, v \rangle - i\omega_I(u, v)) = -i(u, v)$. Another identity $(v, u) = \overline{(u, v)}$ can be proved similarly.

To see the uniqueness of ω_I with the stated property, let a Hermitian pairing satisfying requirements be given by $(u, v) = \langle u, v \rangle - i\omega'(u, v)$ for some pairing $\omega : E \times E \to \mathbb{R}$. By comparing (Iu, v) and $i(u, v)$, we can see that ω' must satisfy $\omega'(u, v) = \langle Iu, v \rangle = -\langle u, Iv \rangle$. Thus, imaginary part of a Hermitian pairing is uniquely determined by the real part of the pairing. This proves uniqueness.

The second part can be proved similarly. $\square$

The Kähler form ω_I depends on the complex structure I. This dependence will turn out to be linear as is proved in Theorem 15.

For the next lemma, recall that an alternating form defined on A is called real if it is invariant under complex conjugation with respect to $\mathbb{C}$ in $A = E \otimes \mathbb{C}$.

LEMMA 2. *Let* $(E^{2N}; \langle \ , \ \rangle, I)$ *be a Hermitian vector space. The Kähler form* ω_I *associated to the isometric complex structure* I *is a real 2-form given by*

$$(5) \qquad \omega_I = -i \sum_{j=1}^{N} a_j \wedge a_j^* = \sum_{j=1}^{N} e_j \wedge e_j' \in \wedge^2 A \cong \wedge^2 A^*.$$

The Kähler form ω_I *is independent of the choice of orthonormal basis of* A.

PROOF. Since $I(a_j) = ia_j$, $I(a_j^*) = -ia_j^*$, formula (2) for ω_I and pairing relations in (1) gives

$$\omega_I(a_j, a_k) = 0, \quad \omega_I(a_j, a_k^*) = i\delta_{jk}, \quad \omega_I(a_j^*, a_k) = -i\delta_{jk}, \quad \omega_I(a_j^*, a_k^*) = 0.$$

This implies that $\omega_I = -i\sum_{j=1}^{N} a_j \wedge a_j^*$ in terms of dual basis which has been identified with the standard basis through $*$. Rewriting in terms of e_j's using (1), we get the above expression. $\square$

REMARK. In the above definition of the Kähler form ω_I for a Hermitian vector space $(E^{2N}; \langle \ , \ \rangle, I)$, we could use $-\omega_I$ as the Kähler form. The reason of our choice of ω_I is that the highest exterior power gives rise to the correct orientation of the complex vector space E. That is,

$$(6) \qquad \omega_I^N = N!(e_1 \wedge e_1') \wedge (e_2 \wedge e_2') \wedge \cdots \wedge (e_N \wedge e_N').$$

General complex structures on E. Now we consider an arbitrary isometric complex structure $\mathcal{J}$ on $(E; \langle \ , \ \rangle)$ and study its properties. We reprove statements similar to those in the previous subsection in this slightly more genral context. One simple observation for an isometric complex structure $\mathcal{J}$ is that $\mathcal{J}$ maps a vector to a vector perpendicular to it.

LEMMA 3. *Let* $\mathcal{J}$ *be an isometric complex structure on a Euclidean space* $(E, \langle \ , \ \rangle)$. *Then for any vector* $v \in E$, *we have*

$$(7) \qquad \langle v, \mathcal{J}(v) \rangle = 0.$$

In fact, the above is true for any vectors in A *when the pairing and the complex structure* $\mathcal{J}$ *is extended* $\mathbb{C}$-*linearly to* A.

PROOF. Since $\mathcal{J}$ is an isometry with $\mathcal{J}^2 = -1$, for any vector $v \in E$ we have $\langle v, \mathcal{J}(v) \rangle = \langle \mathcal{J}(v), \mathcal{J}^2(v) \rangle = -\langle \mathcal{J}(v), v \rangle = -\langle v, \mathcal{J}(v) \rangle$. The last equality is due to the symmetry of the pairing $\langle \ , \ \rangle$. Thus $\langle v, \mathcal{J}(v) \rangle = 0$. Extending this calculation complex linearly and noting that $\mathcal{J}$ commutes with multiplication by $\mathbb{C}$, we see that (7) holds for any v in the complexification A. $\square$

A complex structure $\mathcal{J}$ induces a splitting of $A = E \otimes \mathbb{C}$ into $\pm i$ eigenspaces of $\mathcal{J}$, which we denote by $A_{\mathcal{J}}^+$, $A_{\mathcal{J}}^-$:

$$(8) \qquad A = A_{\mathcal{J}}^+ \oplus A_{\mathcal{J}}^-.$$

LEMMA 4. *Let $\mathcal{J}$ be an arbitrary isometric complex structure on a Euclidean space $(E; \langle\ ,\ \rangle)$ Then eigensubspaces $A_{\mathcal{J}}^{\pm}$ are conjugate to each other, and they are maximal isotropic subspaces with respect to the $\mathbb{C}$-linear pairing $\langle\ ,\ \rangle$:*

$$(9) \qquad \overline{A_{\mathcal{J}}^{+}} = A_{\mathcal{J}}^{-}, \qquad \text{and} \qquad \langle A_{\mathcal{J}}^{+}, A_{\mathcal{J}}^{+} \rangle = 0, \quad \langle A_{\mathcal{J}}^{-}, A_{\mathcal{J}}^{-} \rangle = 0.$$

PROOF. For $v \in A_{\mathcal{J}}^{+}$, we have $\mathcal{J}(v) = iv$. Since $\mathcal{J}$ is a real map commuting with conjugation, we have $\mathcal{J}(\bar{v}) = \overline{\mathcal{J}(v)} = -i\bar{v}$, and $\bar{v} \in A_{\mathcal{J}}^{-}$. Thus $A_{\mathcal{J}}^{\pm}$ are conjugate to each other in $A = E \otimes \mathbb{C}$.

To see that $A_{\mathcal{J}}^{+}$ is isotropic, since $\mathcal{J}$ is isometric and $\langle\ ,\ \rangle$ is $\mathbb{C}$-linear, for any $a_1, a_2 \in A_{\mathcal{J}}^{+}$ we have $\langle a_1, a_2 \rangle = \langle \mathcal{J}(a_1), \mathcal{J}(a_2) \rangle = \langle ia_1, ia_2 \rangle = -\langle a_1, a_2 \rangle$, since $\langle\ ,\ \rangle$ is $\mathbb{C}$-linear in both variables. Thus, $\langle a_1, a_2 \rangle = 0$, Similarly for $A_{\mathcal{J}}^{-}$. $\square$

For any complex structure $\mathcal{J}$, we define maps $L_{\mathcal{J}}^{\pm} : A \to A$ by the following formulae:

$$(10) \qquad L_{\mathcal{J}}^{+}(a) = \frac{a - i\mathcal{J}(a)}{2}, \qquad L_{\mathcal{J}}^{-}(a) = \frac{a + i\mathcal{J}(a)}{2}, \qquad a \in A.$$

These maps are in fact projections onto $A_{\mathcal{J}}^{\pm}$ as is shown in the next lemma.

LEMMA 5. *Let $\mathcal{J}$ be an arbitrary complex structure on E. Then the associated maps $L_{\mathcal{J}}^{\pm} : A \to A$ defined above have the following properties.*

$$L_{\mathcal{J}}^{+} + L_{\mathcal{J}}^{-} = \mathrm{Id},$$
$$(11) \qquad (L_{\mathcal{J}}^{+})^{2} = L_{\mathcal{J}}^{+}, \qquad (L_{\mathcal{J}}^{-})^{2} = L_{\mathcal{J}}^{-}, \qquad L_{\mathcal{J}}^{+} \cdot L_{\mathcal{J}}^{-} = L_{\mathcal{J}}^{-} \cdot L_{\mathcal{J}}^{+} = 0,$$
$$L_{\mathcal{J}}^{+}(A) \subset A_{\mathcal{J}}^{+}, \qquad L_{\mathcal{J}}^{-}(A) \subset A_{\mathcal{J}}^{-}.$$

Thus, $L_{\mathcal{J}}^{+}, L_{\mathcal{J}}^{-}$ are projection operators onto $A_{\mathcal{J}}^{+}, A_{\mathcal{J}}^{-}$, respectively.

PROOF. The first identity is obvious. To see that $L_{\mathcal{J}}^{+}(A) \subset A_{\mathcal{J}}^{+}$, let $a \in A$ be an arbitrary element in A. Then, $\mathcal{J}(L_{\mathcal{J}}^{+}(a)) = \frac{1}{2}\mathcal{J}(a - i\mathcal{J}(a)) = \frac{1}{2}(\mathcal{J}(a) + ia) = \frac{1}{2}i(a - i\mathcal{J}(a)) = iL_{\mathcal{J}}^{+}(a)$. Thus, $L_{\mathcal{J}}^{+}(a) \in A_{\mathcal{J}}^{+}$. Similarly, one can show that $L_{\mathcal{J}}^{-}(A) \subset A_{\mathcal{J}}^{-}$.

To see that $L_{\mathcal{J}}^{+}$ is an idempotent, for any $a \in A$, $(L_{\mathcal{J}}^{+})^{2}(a) = L_{\mathcal{J}}^{+}(L_{\mathcal{J}}^{+}(a)) = \frac{1}{2}(L_{\mathcal{J}}^{+}(a) - i \cdot iL_{\mathcal{J}}^{+}(a)) = L_{\mathcal{J}}^{+}(a)$. The second equality is because $\mathcal{J}(L_{\mathcal{J}}^{+}(a)) = iL_{\mathcal{J}}^{+}(a)$. This shows that $L_{\mathcal{J}}^{+}$ is an idempotent. The case for $L_{\mathcal{J}}^{-}$ is similar. To see the complementarity property, we apply $L_{\mathcal{J}}^{+}$ to the first identity in (11). We get $(L_{\mathcal{J}}^{+})^{2} + L_{\mathcal{J}}^{+} \cdot L_{\mathcal{J}}^{-} = L_{\mathcal{J}}^{+}$. Thus, $L_{\mathcal{J}}^{+} \cdot L_{\mathcal{J}}^{-} = 0$ since $(L_{\mathcal{J}}^{+})^{2} = L_{\mathcal{J}}^{+}$. Similarly, one can show that $L_{\mathcal{J}}^{-} \cdot L_{\mathcal{J}}^{+} = 0$. This completes the proof of Lemma 5. $\square$

For an arbitrary isometric complex structure $\mathcal{J}$ on a Euclidean vector space $(E; \langle\ ,\ \rangle)$, $\mathcal{J}$ defines a Hermitian pairing $(\ ,\ )$ on E whose real part is the given inner product $\langle\ ,\ \rangle$ and whose imaginary part is the Kähler form $\omega_{\mathcal{J}}$ which is conjugate linear in the second variable:

$$(12) \qquad \begin{aligned} (u, v) &= \langle u, v \rangle - i\omega_{\mathcal{J}}(u, v), \\ \text{where} \quad \omega_{\mathcal{J}}(u, v) &= \langle \mathcal{J}(u), v \rangle = -\langle u, \mathcal{J}(v) \rangle. \end{aligned}$$

This is exactly Lemma 1, with I replaced by $\mathcal{J}$. We extend the alternating form $\omega_{\mathcal{J}}$ complex linearly to $A \otimes A$. The Kähler form associated to an isometric complex structure $\mathcal{J}$ is, by definition, this extension $\omega_{\mathcal{J}}$.

Let $\{a_{\mathcal{J}}(1), \ldots, a_{\mathcal{J}}(N)\}$ and $\{a_{\mathcal{J}}^*(1), \ldots, a_{\mathcal{J}}^*(N)\}$ be basis of isotropic subspaces $A_{\mathcal{J}}^{\pm}$ with the following properties:

$$(13) \qquad \langle a_{\mathcal{J}}(j), a_{\mathcal{J}}^*(k) \rangle = \delta_{jk}, \qquad a_{\mathcal{J}}^*(k) = \overline{a_{\mathcal{J}}(k)}, \qquad \text{for} \quad 1 \leq j, k \leq N.$$

Here, complex conjugation is taken with respect to the conjugation in $\mathbb{C}$ of $A = E \otimes \mathbb{C}$. We call such a basis of $A_{\mathcal{J}}^{\pm}$ a basis *adapted* to the complex structure $\mathcal{J}$ of E. Such a basis can be constructed by first choosing a Hermitian orthonormal basis $\{e_1, \ldots, e_N, e_1', \ldots, e_N'\}$ of E^{2N} such that $\mathcal{J}(e_j) = e_j'$ for $1 \leq j \leq N$. Then we apply the construction of (1) to obtain a complex basis of A. The next lemma can be shown in exactly the same way as the "standard" complex structure I shown in Lemma 2.

PROPOSITION 6. *Let $\mathcal{J}$ be an isometric complex structure on $(E; \langle\ ,\ \rangle)$. For any basis adapted to the complex structure $\mathcal{J}$, we have*

$$(14) \qquad\qquad \omega_{\mathcal{J}} = -i \sum_{j=1}^{N} a_{\mathcal{J}}(j) \wedge a_{\mathcal{J}}^*(j).$$

The Kähler form $\omega_{\mathcal{J}}$ is independent of the choice of basis $\{a_{\mathcal{J}}, a_{\mathcal{J}}^*\}_{1 \leq j \leq N}$ adapted to $\mathcal{J}$.

Note that independence of $\omega_{\mathcal{J}}$ of the choice of basis should be clear since its definition (12) doesn't involve basis vectors.

Kähler forms on quaternionic vector spaces and quaternion-Kähler forms. Let $\mathbb{H}^{N'}$ be a quaternionic vector space on which $\mathbb{H}$ acts on the right. Let I, J, K be a linear map acting on $\mathbb{H}^{N'}$ on the left corresponding to the multiplication by i, j, k on the right. So, for any vector $v \in \mathbb{H}^{N'}$, $I(v) = vi$, $J(v) = vj$ and $K(v) = vk$. These maps have the following properties:

$$(15) \qquad\qquad I^2 = J^2 = K^2 = -1, \qquad IJ = -JI = -K.$$

REMARK. Note that I, J, K act on the left and i, j, k act on the right. This switch from right to left necessitates inversion of the action: we could have defined endomorphisms I, J, K by $I(v) = v \cdot i^{-1} = -v \cdot i$, etc. This results in familiar quaternion relations among I, J, K with $IJ = K$. If we define endomorphisms I, J, K as above without inversions, we get a negative sign in $IJ = -K$. But our formulation is equivalent to this.

Modeling on this case, suppose that a real Euclidean vector space $(E^{4N'}; \langle\ ,\ \rangle)$ admits linear isometries I, J, K having the above quaternionic relations (15).

LEMMA 7. *Let I, J, K be isometric complex structures on $(E^{4N'}; \langle \ , \ \rangle)$ as above. For any nonzero vector $v \in E$, the set of vectors $\{v, I(v), J(v), K(v)\}$ is a set of mutually orthogonal vectors.*

PROOF. Since complex structures I, J and K are isometries, Lemma 3 implies that $\langle v, I(v) \rangle = \langle v, J(v) \rangle = \langle v, K(v) \rangle = 0$. Now, since I is an isometry, we have $\langle I(v), J(v) \rangle = \langle I^2(v), IJ(v) \rangle = \langle v, K(v) \rangle = 0$, using relations (7), (15). Other pairings can be shown to vanish by similar arguments. $\square$

When a Euclidean vector space $(E^{4N'}; \langle \ , \ \rangle)$ has a quaternionic structure I, J, K satisfying (15), we can in fact produce a real 2-dimensional family of complex structures on E.

PROPOSITION 8. *Let (I, J, K) be an isometric quaternionic structure on a Euclidean space $(E; \langle \ , \ \rangle)$. Then an endomorphism of E given by*

$$(16) \qquad \mathcal{J} = \alpha I + \beta J + \gamma K, \quad \text{where} \quad \alpha, \beta, \gamma \in \mathbb{R}, \quad \alpha^2 + \beta^2 + \gamma^2 = 1,$$

is also an isometric complex structure on E:

$$(17) \qquad \mathcal{J}^2 = -1 \quad \text{and} \quad \langle \mathcal{J}(u), \mathcal{J}(v) \rangle = \langle u, v \rangle \quad \text{for any} \quad u, v \in E.$$

PROOF. We have $\mathcal{J}^2 = \alpha^2 I^2 + \beta^2 J^2 + \gamma^2 K^2 + \alpha\beta(IJ + JI) + \beta\gamma(JK + KJ) + \alpha\gamma(IK + KI)$. Using quaternion relations (15) among I, J, K, we see that all the crossing terms vanish and remaining terms give $\mathcal{J}^2 = -\alpha^2 - \beta^2 - \gamma^2 = -1$. Thus $\mathcal{J}$ is a complex structure. To see that it is an isometry of $(E; \langle \ , \ \rangle)$, we do a similar calculation:

$$\langle \mathcal{J}(u), \mathcal{J}(v) \rangle = \alpha^2 \langle I(u), I(v) \rangle + \beta^2 \langle J(u), J(v) \rangle + \gamma^2 \langle K(u), K(v) \rangle$$
$$+ \alpha\beta(\langle I(u), J(v) \rangle + \langle J(u), I(v) \rangle) + \beta\gamma(\langle J(u), K(v) \rangle + \langle K(u), J(v) \rangle)$$
$$+ \gamma\alpha(\langle K(u), I(v) \rangle + \langle I(u) + K(v) \rangle).$$

Now, from the first three term, we obtain $(\alpha^2 + \beta^2 + \gamma^2)\langle u, v \rangle = \langle u, v \rangle$ since I, J, K are isometries. To take care of remaining terms, note that $\langle I(u), J(v) \rangle = \langle u, K(v) \rangle$ and $\langle J(u), I(v) \rangle = -\langle u, K(v) \rangle$. Thus, $\langle I(u), J(v) \rangle + \langle J(u), I(v) \rangle = 0$. Similar relations hold for other pairs (J, K) and (K, I). Thus the last three terms vanish and we have proved (17). $\square$

When a vector space E admits endomorphisms I, J, K satisfying quaternion relations (15), this vector space is a module over the the skew field of quaternions $\mathbb{H}$ acting on the right. Thus dimension of E must be divisible by 4. Since $\dim_{\mathbb{R}} E = 2N$, we let $N = 2N'$ for some $N' \in \mathbb{N}$. Let $\{a_1, a_2, \ldots, a_N\} \cup \{a_1^*, a_2^*, \ldots, a_N^*\}$ be a basis of $A = E \otimes \mathbb{C}$ adapted to the complex structure I given in (1). We calculate the effect of other complex structures J, K on these basis vectors. For this, we choose our Hermitian orthonormal basis $\{e_1, \ldots, e_{4N'}\}$ in such a way that (see (20) of §3.2)

$$(18) \qquad \begin{aligned} I(e_j) &= e_{j+N}, & 1 \le j \le N, \\ J(e_r) &= e_{r+N'}, & J(e_{N+r}) = -e_{N+N'+r}, & 1 \le r \le N'. \end{aligned}$$

LEMMA 9. *With respect to the standard basis* (1), *the complex structures* I, J *and* K *have the following effect:*

(19)
$$I(a_r) = ia_r, \qquad I(a_{r+N'}) = ia_{r+N'}, \quad I(a_r^*) = -ia_r^*, \qquad I(a_{r+N'}^*) = -ia_{r+N'},$$
$$J(a_r) = a_{r+N'}^*, \quad J(a_{r+N'}) = -a_r^*, \quad J(a_r^*) = a_{r+N'}, \quad J(a_{r+N'}^*) = -a_r,$$
$$K(a_r) = ia_{r+N'}^*, \ K(a_{r+N'}) = -ia_r^*, \ K(a_r^*) = -ia_{r+N'}, \ K(a_{r+N'}^*) = ia_r.$$

PROOF. The effect of I is due to the construction of these vectors. The effect of J has been calculated in Lemma 13 of §3.2. Then the effect of K can be calculated by combining these calculations since $K = JI$ from (15). □

Using Lemma 9, we can calculate the effect of a general complex structure $\mathcal{J}$ in Proposition 8 on basis vectors of E given in (1), (18).

PROPOSITION 10. *Let* $(E^{4N'}; \langle \ , \ \rangle, I, J, K)$ *be a quaterninic Euclidean vector space. Let* $\mathcal{J} = \alpha I + \beta J + \gamma K$ *be an isometric complex structure on* E *with* $\alpha^2 + \beta^2 + \gamma^2 = 1$, $\alpha, \beta, \gamma \in \mathbb{R}$. *Then the action of* $\mathcal{J}$ *on basis vectors* $\{a_1, \ldots, a_N, a_1^*, \ldots, a_N^*\}$ *adapted to* I *is given by*

(20)
$$\mathcal{J}(a_r) = i\alpha a_r + (\beta + i\gamma)a_{r+N'}^*, \qquad \mathcal{J}(a_{r+N'}) = i\alpha a_{r+N'} - (\beta + i\gamma)a_r^*,$$
$$\mathcal{J}(a_r^*) = -i\alpha a_r^* + (\beta - i\gamma)a_{r+N'}, \qquad \mathcal{J}(a_{r+N'}^*) = -i\alpha a_{r+N'}^* - (\beta - i\gamma)a_r.$$

Here $1 \le r \le N'$. *Since* $\mathcal{J}$ *is real, we have* $\mathcal{J}(a_r^*) = \overline{\mathcal{J}(a_r)}$ *and* $\mathcal{J}(a_{r+N'}^*) = \overline{\mathcal{J}(a_{r+N'})}$.

PROOF. By straightforward calculation. □

For later use, we calculate some pairings involving a general isometric complex structure $\mathcal{J}$ on E.

COROLLARY 11. *Let* $\mathcal{J}$ *be an isometric complex structure on* $(E; \langle \ , \ \rangle, I)$ *given in* (16), *and let* $\{a_1, \ldots, a_N, a_1^*, \ldots, a_N^*\}$ *be a basis of* A *given in* (1) *adapted to the given complex structure* I. *Then*

(21)
$$\langle \mathcal{J}(a_j), a_k^* \rangle = i\alpha \delta_{jk}, \qquad \langle a_j, \mathcal{J}(a_k^*) \rangle = -i\alpha \delta_{jk}, \qquad 1 \le j, k \le N.$$

PROOF. This follows from (1), (20). For example, for the first pairing, when $1 \le j \le N'$, $\mathcal{J}(a_j) = i\alpha a_j + (\beta + i\gamma)a_{j+N'}^*$. Among vectors in A^-, only a_j^* can pair nontrivially with $\mathcal{J}(a_j)$ and then the value of the pairing is $i\alpha$. The case in which $N' \le j \le N$ is the same. The other identity can be proved similarly. □

We calculate pairings among vectors defined by projection operators of (10).

LEMMA 12. *Let* $(E^{4N'}; \langle \ , \ \rangle, I, J, K)$ *be a quaternionic Euclidean vector space. Let* $\mathcal{J}$ *be an isometric complex structure on* E *given in* (16) *and let* $L_{\mathcal{J}}^{\pm}$ *be projection operators defined in* (10). *With notations as before, we then have*

(22)
$$\langle L_{\mathcal{J}}^+(a_j), L_{\mathcal{J}}^-(a_k^*) \rangle = \left\langle \frac{a_j - i\mathcal{J}(a_j)}{2}, \frac{a_k^* + i\mathcal{J}(a_k^*)}{2} \right\rangle = \left(\frac{1+\alpha}{2} \right) \delta_{jk},$$
$$\langle L_{\mathcal{J}}^+(a_j^*), L_{\mathcal{J}}^-(a_k) \rangle = \left\langle \frac{a_j^* - i\mathcal{J}(a_j^*)}{2}, \frac{a_k + i\mathcal{J}(a_k)}{2} \right\rangle = \left(\frac{1-\alpha}{2} \right) \delta_{jk}.$$

PROOF. For the first identity, we note that

$$4\langle L_{\mathcal{J}}^+(a_j), L_{\mathcal{J}}^-(a_k^*)\rangle = \langle a_j, a_k^*\rangle + \langle \mathcal{J}(a_j), \mathcal{J}(a_k^*)\rangle - i\langle \mathcal{J}(a_j), a_k^*\rangle + i\langle a_j, \mathcal{J}(a_k^*)\rangle.$$

Since $\mathcal{J}$ is an isometry, the second term is equal to $\langle a_j, a_k^*\rangle = \delta_{jk}$. Using Corollary 11 to take care of the last two terms, we see that the above is equal to $\delta_{jk} + \delta_{jk} + \alpha\delta_{jk} + \alpha\delta_{jk} = 2(1+\alpha)\delta_{jk}$. The second identity can be proved in a similar way. □

We give a description of a basis of A adapted to any complex structure $\mathcal{J}$ of (16) in terms of the basis $\{a_1, \ldots, a_N\} \cup \{a_1^*, \ldots, a_N^*\}$ in (1) adapted to a given complex structure I. The basis is constructed from an $\mathbb{H}$-basis of E satisfying (18), (19).

PROPOSITION 13. Let $\mathcal{J} = \alpha I + \beta J + \gamma K$ with $\alpha^2 + \beta^2 + \gamma^2 = 1$ be an isometric complex structure on a quaternionic Euclidean vector space $(E^{4N'}; \langle\,,\,\rangle; I, J, K)$. A basis of isotropic subspaces $A_{\mathcal{J}}^\pm$ adapted to $\mathcal{J}$ is given as follows:
    Case (I) : $1 < \alpha \leq 1$. A basis adapted to $\mathcal{J}$ is given by

$$A_{\mathcal{J}}^+ = \bigoplus_{j=1}^N \mathbb{C}\sqrt{\frac{2}{1+\alpha}}L_{\mathcal{J}}^+(a_j) = \bigoplus_{j=1}^N \mathbb{C}\frac{a_j - i\mathcal{J}(a_j)}{\sqrt{2(1+\alpha)}},$$

(23)
$$A_{\mathcal{J}}^- = \bigoplus_{j=1}^N \mathbb{C}\sqrt{\frac{2}{1+\alpha}}L_{\mathcal{J}}^-(a_j^*) = \bigoplus_{j=1}^N \mathbb{C}\frac{a_j^* + i\mathcal{J}(a_j^*)}{\sqrt{2(1+\alpha)}},$$

$$\left\langle \sqrt{\frac{2}{1+\alpha}}L_{\mathcal{J}}^+(a_j), \sqrt{\frac{2}{1+\alpha}}L_{\mathcal{J}}^-(a_k^*) \right\rangle = \delta_{jk}.$$

Case (II) : $-1 \leq \alpha < 1$. A basis adapted to $\mathcal{J}$ is given by

$$A_{\mathcal{J}}^+ = \bigoplus_{j=1}^N \mathbb{C}\sqrt{\frac{2}{1-\alpha}}L_{\mathcal{J}}^+(a_j^*) = \bigoplus_{j=1}^N \mathbb{C}\frac{a_j^* - i\mathcal{J}(a_j^*)}{\sqrt{2(1-\alpha)}},$$

(24)
$$A_{\mathcal{J}}^- = \bigoplus_{j=1}^N \mathbb{C}\sqrt{\frac{2}{1-\alpha}}L_{\mathcal{J}}^-(a_j) = \bigoplus_{j=1}^N \mathbb{C}\frac{a_j + i\mathcal{J}(a_j)}{\sqrt{2(1-\alpha)}},$$

$$\left\langle \sqrt{\frac{2}{1-\alpha}}L_{\mathcal{J}}^+(a_j^*), \sqrt{\frac{2}{1-\alpha}}L_{\mathcal{J}}^-(a_k) \right\rangle = \delta_{jk}.$$

PROOF. The pairing formula comes from Lemma 12. For the proof of the rest, we only have to prove the linear independence of vectors shown. But this follows from the nondegenerate nature of the pairings among vectors in question. □

Proposition 13 can be restated in a slightly different point of view. Let ι^+ : $A^+ \hookrightarrow A$, $\iota^- : A^- \hookrightarrow A$ be inclusion maps. Cases (I), (II) in the next corollary correspond to Cases (I), (II) in Proposition 13.

COROLLARY 14. (I) *When* $-1 < \alpha \leq 1$, *the following maps are isomorphisms*:

$$L_{\mathcal{J}}^{+} \circ \iota^{+} : A^{+} \xrightarrow{\cong} A_{\mathcal{J}}^{+}, \qquad L_{\mathcal{J}}^{-} \circ \iota^{-} : A^{-} \xrightarrow{\cong} A_{\mathcal{J}}^{-}.$$

(II) *When* $-1 \leq \alpha < 1$, *the following compositions are isomorphisms*:

$$L_{\mathcal{J}}^{+} \circ \iota^{-} : A^{-} \xrightarrow{\cong} A_{\mathcal{J}}^{+}, \qquad L_{\mathcal{J}}^{-} \circ \iota^{+} : A^{+} \xrightarrow{\cong} A_{\mathcal{J}}^{-}.$$

REMARK. In (I) of Proposition 13 and Corollary 14, we had to exclude the case $\alpha = -1$. We check what happens in this case. When $\alpha = -1$, then $\mathcal{J} = -I$. Thus, its eigenspaces are given by $A_{-I}^{+} = A^{-}$, $A_{-I}^{-} = A^{+}$. Thus, the composition $L_{-I}^{+} \circ \iota^{+} : A^{+} \to A \to A_{-I}^{+} = A^{-}$ is a zero map. The other composite map is also zero. This is why we have to exclude this case from (I).

Finally, we calculate the Kähler form $\omega_{\mathcal{J}}$ associated to an isometric complex structure $\mathcal{J}$ of (16). Recall that elements h, x, x^{*} in the exterior algebra $\bigwedge^{*}A$ are defined by

$$(25) \qquad h = \sum_{j=1}^{N} a_j a_j^{*}, \qquad x = \sum_{r=1}^{N'} a_r a_{r+N'}, \qquad x^{*} = \sum_{r=1}^{N'} a_r^{*} a_{r+N'}^{*}.$$

THEOREM 15 (KÄHLER FORMS ON QUATERNIONIC SPACES). *Suppose we have a quaternionic Euclidean vector space* $(E; \langle \ , \ \rangle, I, J, K)$. *For any isometric complex structure* $\mathcal{J} = \alpha I + \beta J + \gamma K$ *with* $\alpha^2 + \beta^2 + \gamma^2 = 1$, $\alpha, \beta, \gamma \in \mathbb{R}$ *on* E, *the associated Kähler form* $\omega_{\mathcal{J}}$ *depends linearly on* $\mathcal{J}$ *and is given by*

$$(26) \qquad \begin{aligned} \omega_{\mathcal{J}} &= -i\alpha h + (\beta - i\gamma)x + (\beta + i\gamma)x^{*} \\ &= -i\alpha h + \beta(x + x^{*}) - i\gamma(x - x^{*}). \end{aligned}$$

In particular, Kähler forms associated to complex structures I, J, K *are given as follows*:

$$(27) \qquad \omega_I = -ih, \qquad \omega_J = x + x^{*}, \qquad \omega_K = -i(x - x^{*}).$$

In terms of these Kähler forms, the Kähler form $\omega_{\mathcal{J}}$ *associated to the isometric complex structure* $\mathcal{J} = \alpha I + \beta J + \gamma K$ *is given by*

$$(28) \qquad \omega_{\mathcal{J}} = \alpha \omega_I + \beta \omega_J + \gamma \omega_K.$$

PROOF. By Lemma 2, we know that $\omega_I = -ih$. So, the case in which $\alpha = 1$, $\beta = \gamma = 0$ has been checked. For other cases, we use the basis of $A_{\mathcal{J}}^{\pm}$ adapted to the complex structure $\mathcal{J}$ corresponding to the case $-1 \leq \alpha < 1$. From Proposition 6 and Case (II) of Proposition 13,

$$i\omega_{\mathcal{J}} = \sum_{j=1}^{N} \sqrt{\frac{2}{1-\alpha}} L_{\mathcal{J}}^{+}(a_j^{*}) \cdot \sqrt{\frac{2}{1-\alpha}} L_{\mathcal{J}}^{-}(a_j)$$

$$= \frac{1}{2(1-\alpha)} \sum_{j=1}^{N} \left(a_j^{*} - i\mathcal{J}(a_j^{*})\right) \left(a_j + i\mathcal{J}(a_j)\right).$$

We multiply $2(1-\alpha)$ on both sides, and separate the sum $\sum_{j=1}^{N}$ into $\sum_{j=1}^{N'}$ plus $\sum_{j=N'+1}^{N}$. After some calculation using Proposition 10, we obtain

$$2(1-\alpha)i\omega_J = (\beta^2+\gamma^2-\alpha^2+2\alpha-1)h+2i(1-\alpha)(\beta-i\gamma)x+2i(1-\alpha)(\beta+i\gamma)x^*,$$

using elements described in (25). Since $\alpha^2+\beta^2+\gamma^2 = 1$, the coefficient of the first term is equal to $2\alpha(1-\alpha)$. Since $\alpha \neq 1$, we divide the whole equation by $2i(1-\alpha)$. We then obtain

$$\begin{aligned}
\omega_J &= -i\alpha h + (\beta - i\gamma)x + (\beta + i\gamma)x^* \\
&= -i\alpha h + \beta(x + x^*) - i\gamma(x - x^*).
\end{aligned}$$

In this formula, when $\alpha = 1$ and $\beta = \gamma = 0$, we recover $\omega_I = -ih$. So $\alpha = 1$ case is included in the above general case. The rest of the statement is clear. □

Recall that the symplectic Casimir element $\phi_{\mathfrak{sp}(N')}$ is given by $\phi_{\mathfrak{sp}(N')} = -\frac{1}{2}h^2 + 4xx^*$ [§3.8]. There is a simple relationship between the symplectic Casimir element and Kähler forms for complex structures I, J, K.

DEFINITION 16 (QUATERNION-KÄHLER FORMS). Let $(E^{4N'}; \langle\ ,\ \rangle, I, J, K)$ be a quaternionic Euclidean vector space, and $A = E \otimes \mathbb{C}$. The *quaternion-Kähler form* $\omega_{Q\text{-}K}$ is defined by

$$(29) \qquad \omega_{Q\text{-}K} = \omega_I^2 + \omega_J^2 + \omega_K^2 \in \textstyle\bigwedge^4 A.$$

Next corollary identifies the quaternion-Kähler form above with the symplectic Casimir element, up to a constant multiple.

COROLLARY 17. *With the notations as in Definition 16, in $\bigwedge^4 A$ we have*

$$(30) \qquad \omega_{Q\text{-}K} = \omega_I^2 + \omega_J^2 + \omega_K^2 = -h^2 + 4xx^* = 2\phi_{\mathfrak{sp}(N')}.$$

PROOF. Straightforward calculation using (23). □

We recall that the Lie algebra $\mathfrak{sl}_2(\mathbb{C}) = \mathbb{C}x \oplus \mathbb{C}h \oplus \mathbb{C}x^* \subset \bigwedge^2 A \cong \mathfrak{o}(2N) \subset \mathrm{Cliff}(A)$ has the following commutation relations:

$$(31) \qquad [h, x] = 2x, \qquad [h, x^*] = -2x^*, \qquad [x, x^*] = -h.$$

Here brackets are calculated as commutators in the Clifford algebra $\mathrm{Cliff}(A)$. These relations are not symmetric among x, h, x^*. However, we can have symmetric commutation relations if we use Kähler forms $\omega_I, \omega_J, \omega_K$ which span the same vector space as x, h, x^*.

LEMMA 18. *The Lie algebra* $\mathfrak{sl}_2(\mathbb{C}) = \mathbb{C}\omega_I \oplus \mathbb{C}\omega_J \oplus \omega_K$ *has following commutation relations:*

$$(32) \qquad [\omega_I, \omega_J] = 2\omega_K, \qquad [\omega_J, \omega_K] = 2\omega_I, \qquad [\omega_K, \omega_I] = 2\omega_J.$$

In general, if $\mathcal{J}_1 = a_1\omega_I + b_1\omega_J + c_1\omega_K$ *and* $\mathcal{J}_2 = a_2\omega_I + b_2\omega_J + c_2\omega_K$, *then the commutator of the corresponding Kähler forms is given by*

$$(33) \qquad [\omega_{\mathcal{J}_1}, \omega_{\mathcal{J}_2}] = 2 \begin{vmatrix} \omega_I & \omega_J & \omega_K \\ a_1 & b_1 & c_1 \\ a_2 & b_2 & c_2 \end{vmatrix}$$

PROOF. (32) is shown by direct calculation. For example,

$$[\omega_I, \omega_J] = [-ih, x + x^*] = -i[h, x] - i[h, x^*] = -2ix + 2ix^* = -2i(x - x^*) = 2\omega_K.$$

Other commutation relations can be shown similarly. Formula (33) can be proved by using (32). $\square$

Generalized Kähler forms and generalized quaternion-Kähler forms on quaternionic vector spaces. Let (E^{2N}, I) be a complex vector space, and for $n \in \mathbb{Z}$ let $A(-n - \frac{1}{2})$ be a copy of its complexification $E^{2N} \otimes \mathbb{C}$ carrying weight $n + \frac{1}{2}$ and a complex structure $I \otimes 1$. We define a generalized Kähler form $\omega_I(-n - \frac{1}{2}) \in \bigwedge^2 A(-n - \frac{1}{2})$ by

$$(34) \qquad \omega_I(-n - \tfrac{1}{2}) = -i \sum_{j=1}^{N} a_j(-n - \tfrac{1}{2})a_j^*(-n - \tfrac{1}{2}) = -ih(-n - \tfrac{1}{2}) \in \mathcal{A}.$$

If $(E^{4N'}; I, J, K)$ is a quaternionic vector space with complex structures I, J, K, then we can consider three generalized Kähler forms $\omega_I(-n - \frac{1}{2})$, $\omega_J(-n - \frac{1}{2})$, $\omega_K(-n - \frac{1}{2})$. With respect to the basis $a_i(-n - \frac{1}{2})$'s and $a_j^*(-n - \frac{1}{2})$'s of $A(-n - \frac{1}{2})$ adapted to the complex structure I, these forms are given by

$$(35) \qquad \begin{cases} \omega_I(-n - \tfrac{1}{2}) = -ih(-n - \tfrac{1}{2}), \\ \omega_J(-n - \tfrac{1}{2}) = x(-n - \tfrac{1}{2}) + x^*(-n - \tfrac{1}{2}), \\ \omega_K(-n - \tfrac{1}{2}) = -i\big(x(-n - \tfrac{1}{2}) - x^*(-n - \tfrac{1}{2})\big). \end{cases}$$

For $n \geq 0$, these forms belong to the vertex operator super algebra V under the isomorphism $[\ ] : \mathcal{A} \xrightarrow{\cong} V$, and has weight $2(n + \frac{1}{2})$. The generalized Kähler form $[\omega_{\text{Q-K},(n)}] \in \bigwedge^4 A(-n - \frac{1}{2}) \subset (V)_{4(N+\frac{1}{2})}$, $n \geq 0$, is then defined by

$$(36) \qquad \begin{aligned} \omega_{\text{Q-K},(n)} &= \omega_I(-n - \tfrac{1}{2})^2 + \omega_J(-n - \tfrac{1}{2})^2 + \omega_K(-n - \tfrac{1}{2})^2 \\ &= -h(-n - \tfrac{1}{2})^2 + 4x(-n - \tfrac{1}{2})x^*(-n - \tfrac{1}{2}). \end{aligned}$$

§3.10   Unitary Virasoro algebras in quaternionic vector spaces

Let $\phi_{\mathrm{u}(N)} = -[h^2] = \omega_I^2 \in (V)_2$ be the Casimir element for the unitary Lie algebra $\mathrm{u}(N)$ acting on a Hermitian vector space $(E^{2N}; \langle\ ,\ \rangle, I)$ [§3.4 Proposition 10]. Associated vertex operators $\{\omega_I^2\}_*$ satisfy the following commutation relation [§3.6 Theorem 1]:

$$[\{\omega_I^2\}_{m+1}, \{\omega_I^2\}_{n+1}] = 4(N-1)(n-m)D(m+n)$$
$$+ 2(N-2)(n-m)\{\omega_I^2\}_{m+n+1} + \frac{m(m^2-1)}{3}N(N-1)\delta_{m+n,0}\cdot \mathrm{Id},$$

for any $m, n \in \mathbb{Z}$. Here we used $\omega_I = -ih$ from Theorem 15 of §3.9. Since the Virasoro operator $D(m+n)$ is independent of the isometric complex structure I on $(E^{2N}, \langle\ ,\ \rangle)$, we also have a similar identity for any isometric complex structure J on E. We apply this observation to a quaternionic Hermitian vector space with three isometric complex structures I, J, K. We then deduce some consequences concerning commutation relations of vertex operators associated to products of Kähler forms.

Let $(E^{4N'}; \langle\ ,\ \rangle; I, J, K)$ be a quaternionic Hermitian vector space with three isometric complex structures I, J, K satisfying quaternion relations (15) of §3.9. By Proposition 8 of §3.9, we actually have a real 2-dimensional family of complex structures given by

$$\mathcal{J} = aI + bJ + cK, \qquad a^2 + b^2 + c^2 = 1, \quad a, b, c \in \mathbb{R},$$

with corresponding Kähler forms given by [§3.9 Theorem 15]:

$$\omega_{\mathcal{J}} = a\omega_I + b\omega_J + c\omega_K.$$

Each complex structure $\mathcal{J}$ determine a unitary group $\mathrm{U}_{\mathcal{J}}(2N') \subset \mathrm{O}(4N')$ of $\mathcal{J}$-linear isometries of $(E; \langle\ ,\ \rangle)$ after fixing a real orthonormal basis of $E^{4N'}$. This subgroup depends on the choice of a complex structure $\mathcal{J}$. The Casimir element of the associated Lie algebra $\mathrm{u}_{\mathcal{J}}(2N') = \mathrm{u}(2N', \mathcal{J})$ is given by

$$\phi_{\mathrm{u}(2N', \mathcal{J})} = \omega_{\mathcal{J}}^2.$$

From the above discussion, the vertex operator associated to this Casimir element must satisfy the same commutation relation above.

THEOREM 1. *Let* $\mathcal{J} = aI + bJ + cK$ *with* $a^2 + b^2 + c^2 = 1$, $a, b, c \in \mathbb{R}$. *Then the vertex operator associated to the Casimir element* $\omega_{\mathcal{J}}^2$ *for the unitary Lie algebra* $\mathrm{u}(2N', \mathcal{J})$ *satisfy the following commutation relation:*

$(*_{\mathcal{J}})$ $[\{\omega_{\mathcal{J}}^2\}_{m+1}, \{\omega_{\mathcal{J}}^2\}_{n+1}] = 4(N-1)(n-m)D(m+n)$

$$+ 2(N-2)(n-m)\{\omega_{\mathcal{J}}^2\}_{m+n+1} + \frac{m(m^2-1)}{3}N(N-1)\delta_{m+n,0}\cdot \mathrm{Id},$$

Since the above identity holds for any real a, b, c such that $a^2 + b^2 + c^2 = 1$, we can deduce some implications on commutators of some related vertex operators.

First, if two of these three parameters a, b, c are zero, we see that we recover the original commutation relation for vertex operators $\{\omega_I^2\}_*$, $\{\omega_J^2\}_*$, $\{\omega_K^2\}_*$. We record these here.

PROPOSITION 2. *The following commutation relations hold:*

(1)
$$[\{\omega_I^2\}_{m+1}, \{\omega_I^2\}_{n+1}] = 4(N-1)(n-m)D(m+n)$$
$$+2(N-2)(n-m)\{\omega_I^2\}_{m+n+1} + \frac{m(m^2-1)}{3}N(N-1)\delta_{m+n,0} \cdot \mathrm{Id},$$

(2)
$$[\{\omega_J^2\}_{m+1}, \{\omega_J^2\}_{n+1}] = 4(N-1)(n-m)D(m+n)$$
$$+2(N-2)(n-m)\{\omega_J^2\}_{m+n+1} + \frac{m(m^2-1)}{3}N(N-1)\delta_{m+n,0} \cdot \mathrm{Id},$$

(3)
$$[\{\omega_K^2\}_{m+1}, \{\omega_K^2\}_{n+1}] = 4(N-1)(n-m)D(m+n)$$
$$+2(N-2)(n-m)\{\omega_K^2\}_{m+n+1} + \frac{m(m^2-1)}{3}N(N-1)\delta_{m+n,0} \cdot \mathrm{Id}.$$

Next, we study implications of Theorem 1 when complex structure $\mathcal{J}$ is of the form $\mathcal{J} = aI + bJ$, $a^2 + b^2 = 1$. In this case, corresponding Kähler form is given by $\omega_{\mathcal{J}} = a\omega_I + b\omega_J$ with $a^2 + b^2 = 1$. From Theorem 1, the difference $(*_{\mathcal{J}}) - (*_J)$ is given by

$$[\{\omega_{\mathcal{J}}^2\}_{m+1}, \{\omega_{\mathcal{J}}^2\}_{n+1}] - [\{\omega_J^2\}_{m+1}, \{\omega_J^2\}_{n+1}] = 2(N-1)(n-m)\{\omega_{\mathcal{J}}^2 - \omega_J^2\}_{m+n+1}.$$

Since $\omega_{\mathcal{J}}^2 = a^2\omega_I^2 + (1-a^2)\omega_J^2 + 2ab\omega_I\omega_J = a^2(\omega_I^2 - \omega_J^2) + 2ab\omega_I\omega_J + \omega_J^2$, expanding the above identity, we obtain

$$a^4[\{\omega_I^2 - \omega_J^2\}_{m+1}, \{\omega_I^2 - \omega_J^2\}_{n+1}] + 4a^2(1-a^2)[\{\omega_I\omega_J\}_{m+1}, \{\omega_I\omega_J\}_{n+1}]$$
$$+ 2a^3b\left([\{\omega_I^2 - \omega_J^2\}_{m+1}, \{\omega_I\omega_J\}_{n+1}] + [\{\omega_I\omega_J\}_{m+1}, \{\omega_I^2 - \omega_J^2\}_{n+1}]\right)$$
$$+ a^2\left([\{\omega_I^2 - \omega_J^2\}_{m+1}, \{\omega_J^2\}_{n+1}] + [\{\omega_J^2\}_{m+1}, \{\omega_I^2 - \omega_J^2\}_{n+1}]\right)$$
$$+ 2ab\left([\{\omega_I\omega_J\}_{m+1}, \{\omega_J^2\}_{n+1}] + [\{\omega_J^2\}_{m+1}, \{\omega_I\omega_J\}_{n+1}]\right)$$
$$= 2(N-2)(n-m)\left(a^2\{\omega_I^2 - \omega_J^2\}_{m+n+1} + 2ab\{\omega_I\omega_J\}_{m+n+1}\right).$$

The coefficients a, b can be thought of as functions on a circle $S^1 = \{(x,y) \in \mathbb{R}^2 \mid x^2 + y^2 = 1\}$. As for the linear (in)dependence of the polynomial functions on S^1, we have the following well known fact.

LEMMA 3. *The ring of complex valued functions on S^1 contains a ring of polynomial functions given by*

$$\mathbb{C}[x,y]/(x^2 + y^2 - 1) \cong \bigoplus_{j=0}^{\infty} \mathbb{C}x^j \oplus \bigoplus_{j=0}^{\infty} \mathbb{C}x^j y.$$

In particular, these functions $\{x^j\}_{j\geq 0} \cup \{x^j y\}_{j\geq 0}$ are linearly independent.

PROOF. If the above polynomial functions are not independent, then we have nontrivial polynomials $f(x)$ and $g(x)$ such that $f(x) + y \cdot g(x) = 0$ on S^1. When

$(x, y) \in S^1$, then $(x, -y) \in S^1$. So, we must have $f(x) - y \cdot g(x) = 0$ for any $(x, y) \in S^1$. From these two formulae, we have $f(x) = g(x) = 0$, contradicting our assumption. □

Going back to our previous calculation, since functions $a^4, a^3 b, a^2, ab$ are linearly independent on S^1, their coefficients must be equal. Hence, we have

(4) $\qquad [\{\omega_I^2 - \omega_J^2\}_{m+1}, \{\omega_I^2 - \omega_J^2\}_{n+1}] = 4[\{\omega_I \omega_J\}_{m+1}, \{\omega_I \omega_J\}_{n+1}],$

(5) $\qquad [\{\omega_I^2 - \omega_J^2\}_{m+1}, \{\omega_I \omega_J\}_{n+1}] + [\{\omega_I \omega_J\}_{m+1}, \{\omega_I^2 - \omega_J^2\}_{n+1}] = 0,$

(6) $\quad [\{\omega_I^2 - \omega_J^2\}_{m+1}, \{\omega_J^2\}_{n+1}] + [\{\omega_J^2\}_{m+1}, \{\omega_I^2 - \omega_J^2\}_{n+1}]$
$$+ 4[\{\omega_I \omega_J\}_{m+1}, \{\omega_I \omega_J\}_{n+1}] = 2(N-2)(n-m)\{\omega_I^2 - \omega_J^2\}_{m+n+1},$$

(7) $\quad [\{\omega_I \omega_J\}_{m+1}, \{\omega_J^2\}_{n+1}] + [\{\omega_J^2\}_{m+1}, \{\omega_I \omega_J\}_{n+1}]$
$$= 2(N-2)(n-m)\{\omega_I \omega_J\}_{m+n+1}.$$

Here, two of the above equations are redundant. In fact, (6) = (1) − (2) − (4), and (5) = (7)′ − (7), where (7)′ is the identity corresponding to (7) when (I, J) are switched. Applying the same calculation to (J, K), (K, I), we obtain

PROPOSITION 4. *We have following identities among vertex operators associated to products of Kähler forms of a quaternionic Hermitian vector space* $(E^{4N'}; \langle \, , \, \rangle; I, J, K)$:

(8) $\begin{cases} [\{\omega_I^2 - \omega_J^2\}_{m+1}, \{\omega_I^2 - \omega_J^2\}_{n+1}] = 4[\{\omega_I \omega_J\}_{m+1}, \{\omega_I \omega_J\}_{n+1}], \\ [\{\omega_I \omega_J\}_{m+1}, \{\omega_J^2\}_{n+1}] + [\{\omega_J^2\}_{m+1}, \{\omega_I \omega_J\}_{n+1}] \\ \qquad\qquad = 2(N-2)(n-m)\{\omega_I \omega_J\}_{m+n+1}, \end{cases}$

(9) $\begin{cases} [\{\omega_J^2 - \omega_K^2\}_{m+1}, \{\omega_J^2 - \omega_K^2\}_{n+1}] = 4[\{\omega_J \omega_K\}_{m+1}, \{\omega_J \omega_K\}_{n+1}], \\ [\{\omega_J \omega_K\}_{m+1}, \{\omega_K^2\}_{n+1}] + [\{\omega_K^2\}_{m+1}, \{\omega_J \omega_K\}_{n+1}] \\ \qquad\qquad = 2(N-2)(n-m)\{\omega_J \omega_K\}_{m+n+1}, \end{cases}$

(10)
$\begin{cases} [\{\omega_K^2 - \omega_I^2\}_{m+1}, \{\omega_K^2 - \omega_I^2\}_{n+1}] = 4[\{\omega_K \omega_I\}_{m+1}, \{\omega_K \omega_I\}_{n+1}], \\ [\{\omega_K \omega_I\}_{m+1}, \{\omega_I^2\}_{n+1}] + [\{\omega_I^2\}_{m+1}, \{\omega_K \omega_I\}_{n+1}] \\ \qquad\qquad = 2(N-2)(n-m)\{\omega_K \omega_I\}_{m+n+1}. \end{cases}$

By switching (I, J, K) *by* $(J, I, -K)$ *in* (8), (I, J, K) *by* $(-I, K, J)$ *in* (9), *and* (I, J, K) *by* $(K, -J, I)$ *in* (10), *respectively, we also obtain other identities.*

If course, we can obtain (9), (10) from (8) by cyclically permuting I, J, K.

Now, we finally let our complex structure $\mathcal{J}$ to be arbitrary. Let $\mathcal{J} = aI + bJ + cK$ with $a^2 + b^2 + c^2 = 1$. The corresponding Kähler form is $\omega_{\mathcal{J}} = a\omega_I + b\omega_J + c\omega_K$. We examine $(*_{\mathcal{J}}) - (*_K)$. The Casimir element $\omega_{\mathcal{J}}^2$ is given by

$$\omega_{\mathcal{J}}^2 = a^2 \omega_I^2 + b^2 \omega_J^2 + c^2 \omega_K + 2ab\omega_I \omega_J + 2bc\omega_J \omega_K + 2ca\omega_K \omega_I$$
$$= a^2(\omega_I^2 - \omega_J^2) + b^2(\omega_J^2 - \omega_K^2) + \omega_K^2 + 2ab\omega_I \omega_J + 2bc\omega_J \omega_K + 2ca\omega_K \omega_I.$$

Here, we can regard a, b, c as functions on the sphere S^2. As for linear independence of polynomial functions on S^2, we have the next lemma.

LEMMA 5. *The following ring is a subring of complex valued functions on* $S^2 = \{(x, y, z) \in \mathbb{R}^3 \mid x^2 + y^2 + z^2 = 1\}.$

$$\mathbb{C}[x, y, z]/(x^2 + y^2 + z^2 - 1) \cong \bigoplus_{i,j \geq 0} \mathbb{C}x^i y^j \oplus \bigoplus_{i,j \geq 0} \mathbb{C}x^i y^j z,$$

In particular, functions $\{x^i y^j\}_{i,j \geq 0} \cup \{x^i y^j z\}_{i,j \geq 0}$ *are linearly independent.*

Substituting the above ω_J^2 into the identity $(*_J) - (*_K)$, and expanding the result into a linear combination of $\{a^i b^j\}_{i,j \geq 0} \cup \{a^i b^j c\}_{i,j \geq 0}$ and equating each coefficient to be zero, we get many identities. Many of them can be derived from our previous identities in Propositions 2 and 4. But there are some new identities. By looking at the coefficient of ab in the expansion, we obtain

PROPOSITION 6. *The following identity holds among vertex operators associated to certain products of Kähler forms* $\omega_I, \omega_J, \omega_K$:

(11) $[\{\omega_K^2\}_{m+1}, \{\omega_I \omega_J\}_{n+1}] + [\{\omega_I \omega_J\}_{m+1}, \{\omega_K^2\}_{n+1}]$

$\qquad + 2[\{\omega_J \omega_K\}_{m+1}, \{\omega_K \omega_I\}_{n+1}] + 2[\{\omega_K \omega_I\}_{m+1}, \{\omega_J \omega_K\}_{n+1}]$

$\qquad\qquad = 2(N - 2)(n - m)\{\omega_I \omega_J\}_{m+n+1}.$

By replacing (I, J, K) *by* (J, K, I), (K, I, J), $(-I, K, J)$, $(K, -J, I)$, $(J, I, -K)$, *we obtain other identities of similar type.*

§3.11   Vertex operators for products of Kähler forms and their commutators

For a compact Kähler manifold M without boundary of complex dimension N, all powers of the Kähler class up to the N-th power, $[\Omega], [\Omega^2], \ldots, [\Omega^N]$, are nontrivial cohomology classes of M [GH]. We considered Kähler forms for complex vector spaces in §3.9. In this section, we calculate vertex operators associated to powers of Kähler forms, and their action on products of Kähler forms.

We are particularly interested in the case when a vector space has the structure of a Hermitian quaternionic vector space. In this case, we have a real 2-dimensional family of isometric complex structures on the vector space, and hence a real 2-dimensional family of corresponding Kähler forms.

Let $E^{4N'}$ with $N = 2N'$ be a quaternionic vector space with complex structures I, J, K satisfying quaternion relations $I^2 = J^2 = -1$, $IJ = -JI = -K$. Let $A = E \otimes \mathbb{C} = A^+ \oplus A^-$ be the decomposition of the complexification A into eigenspaces with respect to the complex structure I. As before, $A(-n - \frac{1}{2})$ with $n \geq 0$ denotes a copy of A carrying weight $n + \frac{1}{2}$, and $\{a_1(-n - \frac{1}{2}), \ldots, a_N(-n - \frac{1}{2})\}$, $\{a_1^*(-n - \frac{1}{2}), \ldots, a_N^*(-n - \frac{1}{2})\}$ are bases for $A^+(-n - \frac{1}{2})$, $A^-(-n - \frac{1}{2})$, respectively [§3.9 (1)]. The Kähler form ω_I for the complex structure I is then given by $\omega_I = -i \sum_{j=1}^{N} a_j(-\frac{1}{2}) a_j^*(-\frac{1}{2}) \in \bigwedge^* A(-\frac{1}{2})$.

In terms of the above basis, Kähler forms corresponding to complex structures I, J, K have been given by Theorem 15 of §3.9:

$$\omega_I = -ih, \qquad \omega_J = x + x^*, \qquad \omega_K = -i(x - x^*), \qquad \text{where}$$

(1)
$$h = \sum_{j=1}^{N} a_j(-\tfrac{1}{2})a_j^*(-\tfrac{1}{2}),$$

$$x = \sum_{r=1}^{N'} a_r(-\tfrac{1}{2})a_{r+N'}(-\tfrac{1}{2}), \qquad x^* = \sum_{r=1}^{N'} a_r^*(-\tfrac{1}{2})a_{r+N'}^*(-\tfrac{1}{2}).$$

Note that these Kähler forms are *real* forms, and hence they are invariant under complex conjugation $*$: $\omega_I^* = \omega_I$, $\omega_J^* = \omega_J$, $\omega_K^* = \omega_K$. Furthermore, there exists a cyclic symmetry among them [§3.9 Lemma 18]:

(2)
$$[\omega_I, \omega_J] = 2\omega_K, \qquad [\omega_J, \omega_K] = 2\omega_I, \qquad [\omega_K, \omega_J] = 2\omega_I.$$

Due to this symmetric nature of commutators for $\omega_I, \omega_J, \omega_K$, calculations with vertex operators can be greatly facilitated for Kähler forms.

The $\mathbb{C}$-vector space spanned by $\omega_I, \omega_J, \omega_K$ is the same as the $\mathbb{C}$-vector space spanned by x, h, x^*. If we use I as the standard complex structure, then x is a "holomorphic" form and x^* is an "anti-holomorphic" form. It is well known that on a compact hyperkähler manifold M, there exists a holomorphic 2-form called the *complex symplectic form* whose j-th power generates a nontrivial cohomology class in $H^{0,2j}(M; \mathbb{C})$ for $1 \leq j \leq N'$. The above element x is the vector space analogue of the complex symplectic form.

Let V be the vertex operator super algebra of §2.2 constructed from a Euclidean vector space E with a quaternionic Hermitian structure (namely, I, J, K are isometries). We calculate vertex operators associated to elements in the algebra generated by Kähler forms ω_I, ω_J, ω_K in $\bigwedge^* A(-\tfrac{1}{2}) \subset \mathcal{A} \cong V$. This algebra is the same as the algebra generated by x, h, x^*. We call this algebra hyperkähler algebra and denote it by $\mathcal{K}$.

LEMMA 1. *In the hyperkähler algebra* $\mathcal{K} \subset \bigwedge^* A(-\tfrac{1}{2})$, *we have* $x^i \cdot (x^*)^j \cdot h^k = 0$ *if* $2i + k > N$ *or* $2j + k > N$.

PROOF. This is obvious, since $x^i \cdot (x^*)^j \cdot h^k \in \bigwedge^{2i+k} A^+(-\tfrac{1}{2}) \otimes \bigwedge^{2j+k} A^-(-\tfrac{1}{2})$ and $\bigwedge^j A^{\pm}(-\tfrac{1}{2}) = \{0\}$ for $j > N$. $\square$

Vertex operators for $S^2(\mathbb{C}[x] \oplus \mathbb{C}[h] \oplus \mathbb{C}[x^*])$. We calculate vertex operators associated to low weight vectors in the algebra $\mathcal{K}$. For a vector $v \in \mathbb{C}x \oplus \mathbb{C}x^* \oplus \mathbb{C}h$, its vertex operator is denoted by $Y([v], z) = \sum_{n \in \mathbb{Z}} v(n)z^{-n-1}$. The operator $v(n)$ lowers weight by n. We will need the following result.

LEMMA 2. *Let* $u, v \in \mathcal{K}$ *be vectors of weight 1 in the hyperkähler algebra. The* *action of $v(-1)$ on u is given as follows:*

$$x(-1)[x] = [x^2], \qquad\qquad x(-1)[x^*] = [xx^*] + [\omega_+],$$
$$x(-1)[h] = [xh] + D(-1)[x], \qquad x^*(-1)[x] = [x^*x] + [\omega_-],$$
$$x^*(-1)[x^*] = [(x^*)^2], \qquad\qquad x^*(-1)[h] = [x^*h] - D(-1)[x^*],$$
$$h(-1)[x] = [hx] - D(-1)[x], \qquad h(-1)[x^*] = [hx^*] + D(-1)[x^*],$$
$$h(-1)[h] = [h^2] - 2[\omega].$$

PROOF. This follows from simple calculations using

$$x(-1) = \sum_{r=1}^{N'}\left(\cdots + a_r\left(-\tfrac{3}{2}\right)a_{r+N'}\left(\tfrac{1}{2}\right) + a_r\left(-\tfrac{1}{2}\right)a_{r+N'}\left(-\tfrac{1}{2}\right) - a_{r+N'}\left(-\tfrac{3}{2}\right)a_r\left(\tfrac{1}{2}\right) + \cdots\right),$$

$$x^*(-1) = \sum_{r=1}^{N'}\left(\cdots + a_r^*\left(-\tfrac{3}{2}\right)a_{r+N'}^*\left(\tfrac{1}{2}\right) + a_r^*\left(-\tfrac{1}{2}\right)a_{r+N'}^*\left(-\tfrac{1}{2}\right) - a_{r+N'}^*\left(-\tfrac{3}{2}\right)a_r^*\left(\tfrac{1}{2}\right) + \cdots\right),$$

$$h(-1) = \sum_{j=1}^{N}\left(\cdots + a_j\left(-\tfrac{3}{2}\right)a_j^*\left(\tfrac{1}{2}\right) + a_j\left(-\tfrac{1}{2}\right)a_j^*\left(-\tfrac{1}{2}\right) - a_j^*\left(-\tfrac{3}{2}\right)a_j\left(\tfrac{1}{2}\right) + \cdots\right),$$

applied to vectors $[x]$, $[h]$, $[x^*]$. □

To calculate vertex operators for these vectors, we apply a special case of the Jacobi identity for vertex operator algebras.

JACOBI IDENTITY (SPECIAL CASE). Let $x \in (V)_1$ and $v \in V$. Then
(3)
$$\{x(-m)v\}_n = \sum_{i \geq 0} \binom{m+i-1}{i}\left(x(-m-i)\{v\}_{n+i} - (-1)^m\{v\}_{n-m-i}x(i)\right).$$

The above formula with $m = 1$ applied to a vector v in the hyperkähler algebra $\mathcal{K}$ yields the following result:

PROPOSITION 3 (VERTEX OPERATORS FOR WEIGHT 2 VECTORS). *Let $n \in \mathbb{Z}$. Vertex operators associated to weight 2 vectors in $\mathcal{K}$ are given as follows:*

$$\{x^2\}_{n+1} = \sum_{0 \leq i \in \mathbb{Z}} x(-i-1)x(n+i+1) + \sum_{0 \leq i \in \mathbb{Z}} x(n-i)x(i),$$

$$\{h^2\}_{n+1} = \sum_{0 \leq i \in \mathbb{Z}} h(-i-1)h(n+i+1) + \sum_{0 \leq i \in \mathbb{Z}} h(n-i)h(i) + 2D(n),$$

$$\{(x^*)^2\}_{n+1} = \sum_{0 \leq i \in \mathbb{Z}} x^*(-i-1)x^*(n+i+1) + \sum_{0 \leq i \in \mathbb{Z}} x^*(n-i)x^*(i),$$

$$\{xh\}_{n+1} = \sum_{0 \leq i \in \mathbb{Z}} h(-i-1)x(n+i+1) + \sum_{0 \leq i \in \mathbb{Z}} x(n-i)h(i) + (n+1)x(n),$$

$$\{x^*h\}_{n+1} = \sum_{0 \leq i \in \mathbb{Z}} x^*(-i-1)h(n+i+1) + \sum_{0 \leq i \in \mathbb{Z}} h(n-i)x^*(i) + (n+1)x^*(n),$$

$$\{xx^*\}_{n+1} = \sum_{0 \leq i \in \mathbb{Z}} x(-i-1)x^*(n+i+1) + \sum_{0 \leq i \in \mathbb{Z}} x^*(n-i)x(i) - \left(\tfrac{n+1}{2}\right)h(n) - \tfrac{1}{2}D(n).$$

Vertex operators for $S^2\left(\mathbb{C}[\omega_I] \oplus \mathbb{C}[\omega_J] \oplus \mathbb{C}[\omega_K]\right)$. We recall that on a quaternionic Hermitian vector space $E^{4N'}$, isometric complex structures on E of the form $\mathcal{J} = aI + bJ + cK$ with $a^2 + b^2 + c^2 = 1$ form a real 2-dimensional family. Corresponding Kähler forms are of the form $\omega_{\mathcal{J}} = a\omega_I + b\omega_J + c\omega_K$ [§3.9 Theorem 15]. Since every complex structure is equally as good as the other, we should discuss Kähler forms in a coordinate free way. With this in mind, we first calculate vertex operators of ω_I, ω_J, ω_k, rather than of h, x, x^*.

LEMMA 4. *In terms of Kähler forms* ω_I, ω_J, ω_K, *we have*

$$\omega_I(-1)\Omega = [\omega_I], \qquad \omega_J(-1)\Omega = [\omega_J], \qquad \omega_K(-1)\Omega = [\omega_K],$$

$$\begin{cases} \omega_I(-1)[\omega_I] = [\omega_I^2] + 2[\omega], \\ \omega_J(-1)[\omega_J] = [\omega_J^2] + 2[\omega], \\ \omega_K(-1)[\omega_K] = [\omega_K^2] + 2[\omega], \end{cases} \qquad \begin{cases} \omega_I(-1)[\omega_J] + \omega_J(-1)[\omega_I] = 2[\omega_I\omega_J], \\ \omega_J(-1)[\omega_K] + \omega_K(-1)[\omega_J] = 2[\omega_J\omega_K], \\ \omega_K(-1)[\omega_I] + \omega_I(-1)[\omega_K] = 2[\omega_K\omega_I], \end{cases}$$

PROOF. The first three identities follow from the reproducing property (creation property) of vertex operators. Other identities follow from straightforward calculations using (1) and Lemma 2. For example,

$$\omega_J(-1)[\omega_J] = \big(x(-1) + x^*(-1)\big)\big([x] + [x^*]\big)$$
$$= [x^2] + [xx^*] + [\omega_+] + [xx^*] + [\omega_-] + [(x^*)^2] = [(x + x^*)^2] + 2[\omega].$$

Another example is that

$$\omega_J(-1)[\omega_K] + \omega_K(-1)[\omega_J] = \big(x(-1) + x^*(-1)\big)[-i(x - x^*)]$$
$$- i\big(x(-1) - x^*(-1)\big)[x + x^*] = -2i\big(x(-1)[x] - x^*(-1)[x^*]\big)$$
$$= -2i\big([x^2] - [(x^*)^2]\big) = 2[\omega_J\omega_K].$$

Other formulae follow similarly. □

The above relations can be combined into a single relation in terms of a general isometric complex structure $\mathcal{J}$.

PROPOSITION 5. *Let* $\mathcal{J} = aI + bJ + cK$ *be a general isometric complex structure on a quaternionic Hermitian vector space* E, *and let* $\omega_{\mathcal{J}}$ *be the corresponding Kähler form. Then*

(4) $$\omega_{\mathcal{J}}(-1)[\omega_{\mathcal{J}}] = [\omega_{\mathcal{J}}^2] + 2[\omega].$$

Let $\mathcal{J}_i = a_i I + b_i J + c_i K$, $i = 1, 2$, *be two isometric complex structures on* E *and let* $\mathcal{J}_1 \cdot \mathcal{J}_2 = a_1 a_2 + b_1 b_2 + c_1 c_2 \in \mathbb{R}$. *Then*

(5) $$\omega_{\mathcal{J}_1}(-1)[\omega_{\mathcal{J}_2}] + \omega_{\mathcal{J}_2}(-1)[\omega_{\mathcal{J}_1}] = 2[\omega_{\mathcal{J}_1}\omega_{\mathcal{J}_2}] + 4(\mathcal{J}_1 \cdot \mathcal{J}_2)[\omega].$$

PROOF. For the first part, using Lemma 4, we have

$$\omega_{\mathcal{J}}(-1)[\omega_{\mathcal{J}}] = a^2[\omega_I^2] + b^2[\omega_J^2] + c^2[\omega_K^2]$$
$$+ 2(a^2 + b^2 + c^2)[\omega] + 2ab[\omega_I\omega_J] + 2bc[\omega_J\omega_K] + 2ac[\omega_K\omega_I] = [\omega_{\mathcal{J}}^2] + 2[\omega].$$

For the second part, we may assume that $\mathcal{J}_1 \neq \mathcal{J}_2$. When $\mathcal{J}_1 = -\mathcal{J}_2$, formula (5) is a consequence of formula (4). When $\mathcal{J}_1 \neq -\mathcal{J}_2$, $\mathcal{J}_1 + \mathcal{J}_2 = (a_1 + a_2)I + (b_1 + b_2)J + (c_1 + c_2)K$ is a nontrivial endomorphism of E, but is not a complex structure. Since $(a_1 + a_2)^2 + (b_1 + b_2)^2 + (c_1 + c_2)^2 = 2(1 + \mathcal{J}_1 \cdot \mathcal{J}_2) > 0$, by normalizing $\mathcal{J}_1 + \mathcal{J}_2$, we see that $\mathcal{J} = (\mathcal{J}_1 + \mathcal{J}_2)/\sqrt{2(1 + \mathcal{J}_1 \cdot \mathcal{J}_2)}$ is an isometric complex structure. Applying the first identity to this complex structure, we have

$$\frac{(\omega_{\mathcal{J}_1}(-1) + \omega_{\mathcal{J}_2}(-1))}{2(1 + \mathcal{J}_1 \cdot \mathcal{J}_2)}[\omega_{\mathcal{J}_1} + \omega_{\mathcal{J}_2}] = \frac{[(\omega_{\mathcal{J}_1} + \omega_{\mathcal{J}_2})^2]}{2(1 + \mathcal{J}_1 \cdot \mathcal{J}_2)} + 2[\omega].$$

Expanding this identity and using (4), we obtain (5). □

We apply the Jacobi identity (3) to the formula (5) of Proposition 5.

COROLLARY 6. *With notations as above, the vertex operator associated to a product of Kähler forms $[\omega_{\mathcal{J}_1} \cdot \omega_{\mathcal{J}_2}] \in (V)_2$ is given by*

$$\{\omega_{\mathcal{J}_1} \cdot \omega_{\mathcal{J}_2}\}_{n+1} = \sum_{\substack{i+j=n \\ i,j \geq 0}} \tfrac{1}{2}\left(\omega_{\mathcal{J}_1}(i) \cdot \omega_{\mathcal{J}_2}(j) + \omega_{\mathcal{J}_2}(j) \cdot \omega_{\mathcal{J}_1}(i)\right)$$

$$+ \sum_{0 \leq i \in \mathbb{Z}} \omega_{\mathcal{J}_1}(-i-1) \cdot \omega_{\mathcal{J}_2}(n+i+1) + \sum_{0 \leq i \in \mathbb{Z}} \omega_{\mathcal{J}_2}(-i-1) \cdot \omega_{\mathcal{J}_1}(n+i+1) - 2(\mathcal{J}_1 \cdot \mathcal{J}_2)D(n).$$

PROOF. All we have to do is to rearrange summations appearing in the Jacobi identity. From (5) of Proposition 5, for each $n \in \mathbb{Z}$, we have

$$\{\omega_{\mathcal{J}_1}\omega_{\mathcal{J}_2}\}_{n+1} = \tfrac{1}{2}\left(\{\omega_{\mathcal{J}_1}(-1)[\omega_{\mathcal{J}_2}]\}_{n+1} + \{\omega_{\mathcal{J}_2}(-1)[\omega_{\mathcal{J}_1}]\}_{n+1}\right) - 2(\mathcal{J}_1 \cdot \mathcal{J}_2)\{\omega\}_{n+1}.$$

From the Jacobi identity (3), we have

$$\{\omega_{\mathcal{J}_1}(-1)[\omega_{\mathcal{J}_2}]\}_{n+1} = \sum_{j \geq 0}\left(\omega_{\mathcal{J}_1}(-1-i)\omega_{\mathcal{J}_2}(n+i+1) + \omega_{\mathcal{J}_2}(n-i)\omega_{\mathcal{J}_1}(i)\right)$$

$$= \sum_{\substack{i+j=n \\ i,j \geq 0}}\omega_{\mathcal{J}_2}(j)\omega_{\mathcal{J}_1}(i) + \sum_{i \geq 0}\omega_{\mathcal{J}_1}(-1-i)\omega_{\mathcal{J}_2}(n+i+1) + \sum_{i \geq 0}\omega_{\mathcal{J}_2}(-1-i)\omega_{\mathcal{J}_1}(n+i+1).$$

Similarly, switching $\mathcal{J}_1$ and $\mathcal{J}_2$, we also have

$$\{\omega_{\mathcal{J}_2}(-1)[\omega_{\mathcal{J}_1}]\}_{n+1} = \sum_{\substack{i+j=n \\ i,j \geq 0}}\omega_{\mathcal{J}_1}(i)\omega_{\mathcal{J}_2}(j) + \sum_{i \geq 0}\omega_{\mathcal{J}_2}(-1-i)\omega_{\mathcal{J}_1}(n+i+1)$$
$$+ \sum_{i \geq 0}\omega_{\mathcal{J}_1}(-1-i)\omega_{\mathcal{J}_2}(n+i+1).$$

From these formulae, Corollary 6 follows. □

Specializing $\mathcal{J}_i$'s, we have the following formulae for products of $\omega_I, \omega_J, \omega_K$.

COROLLARY 7 (VERTEX OPERATORS FOR QUADRATIC KÄHLER FORMS). *Let $E^{4N'}$ be a quaternionic Hermitian vector space and let V be the associated vertex operator super algebra. Let $\omega_I, \omega_J, \omega_K$ be Kähler forms associated to isometric complex structures I, J, K in $E^{4N'}$. Then*

$$\{\omega_I^2\}_{n+1} = \sum_{\substack{i+j=n \\ i,j \geq 0}} \omega_I(i) \cdot \omega_I(j) + 2\sum_{0 \leq i \in \mathbb{Z}} \omega_I(-i-1) \cdot \omega_I(n+i+1) - 2D(n),$$

$$\{\omega_J^2\}_{n+1} = \sum_{\substack{i+j=n \\ i,j \geq 0}} \omega_J(i) \cdot \omega_J(j) + 2\sum_{0 \leq i \in \mathbb{Z}} \omega_J(-i-1) \cdot \omega_J(n+i+1) - 2D(n),$$

$$\{\omega_K^2\}_{n+1} = \sum_{\substack{i+j=n \\ i,j \geq 0}} \omega_K(i) \cdot \omega_K(j) + 2\sum_{0 \leq i \in \mathbb{Z}} \omega_K(-i-1) \cdot \omega_K(n+i+1) - 2D(n),$$

$$\{\omega_I \cdot \omega_J\}_{n+1} = \sum_{\substack{i+j=n \\ i,j \geq 0}} \omega_I(i) \cdot \omega_J(j) - (n+1)\omega_K(n)$$
$$+ \sum_{0 \leq i \in \mathbb{Z}} \omega_I(-i-1) \cdot \omega_J(n+i+1) + \sum_{0 \leq i \in \mathbb{Z}} \omega_J(-i-1) \cdot \omega_I(n+i+1),$$

$$\{\omega_J \cdot \omega_K\}_{n+1} = \sum_{\substack{i+j=n \\ i,j \geq 0}} \omega_J(i) \cdot \omega_K(j) - (n+1)\omega_I(n)$$
$$+ \sum_{0 \leq i \in \mathbb{Z}} \omega_J(-i-1) \cdot \omega_K(n+i+1) + \sum_{0 \leq i \in \mathbb{Z}} \omega_K(-i-1) \cdot \omega_J(n+i+1),$$

$$\{\omega_K \cdot \omega_I\}_{n+1} = \sum_{\substack{i+j=n \\ i,j \geq 0}} \omega_K(i) \cdot \omega_I(j) - (n+1)\omega_J(n)$$
$$+ \sum_{0 \leq i \in \mathbb{Z}} \omega_K(-i-1) \cdot \omega_I(n+i+1) + \sum_{0 \leq i \in \mathbb{Z}} \omega_I(-i-1) \cdot \omega_K(n+i+1).$$

PROOF. We apply Corollary 6. All we need is to take care of the first summation in the formula. Since $[\omega_{\mathcal{J}_1}(i), \omega_{\mathcal{J}_2}(j)] = [\omega_{\mathcal{J}_1}, \omega_{\mathcal{J}_2}](i+j)$ when $i, j \geq 0$ from affine Lie algebra commutation relations (7) below, we have

$$\tfrac{1}{2}\left(\omega_{\mathcal{J}_1}(i) \cdot \omega_{\mathcal{J}_2}(j) + \omega_{\mathcal{J}_2}(j) \cdot \omega_{\mathcal{J}_1}(i)\right) = \omega_{\mathcal{J}_1}(i) \cdot \omega_{\mathcal{J}_2}(j) - \tfrac{1}{2}[\omega_{\mathcal{J}_1}, \omega_{\mathcal{J}_2}](i+j).$$

So, we simply apply commutators

$$[\omega_I, \omega_J] = 2\omega_K, \qquad [\omega_J, \omega_K] = 2\omega_I, \qquad [\omega_K, \omega_J] = 2\omega_I,$$

to the right hand side of the above identity to get our results.   □

Commutators for vertex operators associated to products of Kähler forms. The vector space $S^2\left(\mathbb{C}[\omega_I] \oplus \mathbb{C}[\omega_J] \oplus \mathbb{C}[\omega_K]\right)$ of the second symmetric power is 6-dimensional. There are $6 + \binom{6}{2} = 21$ different commutators of associated vertex operators. However, due to cyclically symmetric nature of Kähler forms $\omega_I, \omega_J, \omega_K$, we only have to calculate 7 of them. Rest of commutators follow by cyclic permutation of I, J, K. To calculate these commutators, we need some preparations.

Let $E^{2N} = \bigoplus_{j=1}^{2N} \mathbb{R}e_j$ and $A = E^{2N} \otimes \mathbb{C}$. Recall that the orthogonal Lie algebra $\mathfrak{o}(2N) \otimes \mathbb{C}$ can be realized as a Lie subalgebra of the Clifford algebra $\mathrm{Cliff}(A)$. It consists of elements of the form $:rs := \frac{1}{2}(rs - sr) \in \mathrm{Cliff}(A)$ for $r, s \in A$. In $\mathfrak{o}(2N) \otimes \mathbb{C}$ with $N = 2N'$, we have elements h, x, x^* given by

$$h = \sum_{j=1}^{N} :a_j a_j^*:, \qquad x = \sum_{r=1}^{N'} :a_r a_{r+N'}:, \qquad x^* = \sum_{r=1}^{N'} :a_r^* a_{r+N'}^*: .$$

Kähler forms ω_I, ω_J, ω_K corresponding to complex structures I, J, K are given by $\omega_I = -ih$, $\omega_J = x + x^*$, $\omega_K = -i(x - x^*)$. We also recall that an invariant bilinear pairing in $\mathfrak{o}(2N) \subset \mathrm{Cliff}(A)$ is given by

$$\langle :r_1 r_2:, :s_1 s_2: \rangle = \langle r_1, s_2 \rangle \langle r_2, s_1 \rangle - \langle r_1, s_1 \rangle \langle r_2, s_2 \rangle.$$

Invariant pairings among the above Kähler forms are given as follows.

LEMMA 8. *Let* $\langle\ ,\ \rangle$ *be the invariant pairing on the Lie algebra* $\mathfrak{o}(4N')$ *given above and let* ω_I, ω_J, ω_K *be Kähler forms associated to isometric complex structures* I, J, K *in the quaternionic Hermitian vector space* $E^{4N'}$. *Then*

(6)
$$\langle \omega_I, \omega_I \rangle = \langle \omega_J, \omega_J \rangle = \langle \omega_K, \omega_K \rangle = -N,$$
$$\langle \omega_I, \omega_J \rangle = \langle \omega_J, \omega_K \rangle = \langle \omega_K, \omega_I \rangle = 0.$$

PROOF. This can be checked by direct calculation using the fact that

$$\langle h, h \rangle = N, \qquad \langle x, x \rangle = 0, \qquad \langle x^*, x^* \rangle = 0,$$
$$\langle h, x \rangle = 0, \qquad \langle h, x^* \rangle = 0, \qquad \langle x, x^* \rangle = -N'.$$

If we note that these three Kähler forms are cyclically symmetric, our calculation is more immediate. $\square$

Recall that ω_I, ω_J, and ω_K span a Lie subalgebra $\mathfrak{sl}_2(\mathbb{C}) = \mathfrak{sp}(1) \otimes \mathbb{C}$ of $\mathfrak{o}(4N')$ [§3.9 Lemma 18]. We consider the action of the associated affine Lie algebra $\widehat{\mathfrak{sl}}_2(\mathbb{C})$ on the vertex operator super algebra V.

LEMMA 9. *The vector space spanned by the vacuum vector* Ω *is a trivial 1-dimensional representation of* $\widehat{\mathfrak{sl}}_2(\mathbb{C})(0) = \mathbb{C}\omega_I(0) \oplus \mathbb{C}\omega_J(0) \oplus \mathbb{C}\omega_K(0)$. *Namely,* $\omega_I(0)\Omega = 0$, $\omega_J(0)\Omega = 0$, *and* $\omega_K(0)\Omega = 0$.

PROOF. This is well known but we give a simple proof here showing that commutation relations force any 1-dimensional representation to be trivial. Let

$\omega_I(0)\Omega = a\Omega$, $\omega_J(0)\Omega = b\Omega$, $\omega_K(0)\Omega = c\Omega$ for some $a, b, c \in \mathbb{C}$. Then, from the commutation relation $[\omega_I(0), \omega_J(0)] = 2\omega_K(0)$, we see that $ab - ba = 2c$, or $c = 0$. Similarly, by sytmmetry of I, J and K, we can also deduce $a = b = 0$. □

Recall that the commutator of the affine Lie algebra $\hat{o}(2N)$ is given by

(7) $$[x(m), y(n)] = [x, y](m + n) + m\langle x, y\rangle\delta_{m,-n} \cdot \mathrm{Id},$$

for any $x, y \in \mathfrak{o}(2N)$. Lemma 8 and Lemma 9 imply the following lemma.

LEMMA 10. *In the vertex operator super algebra V, we have*

$$\begin{array}{lll}
\omega_I(0)[\omega_I] = 0, & \omega_I(0)[\omega_J] = 2[\omega_K], & \omega_I(0)[\omega_K] = -2[\omega_J], \\
\omega_J(0)[\omega_I] = -2[\omega_K], & \omega_J(0)[\omega_J] = 0, & \omega_J(0)[\omega_K] = 2[\omega_I], \\
\omega_K(0)[\omega_I] = 2[\omega_J], & \omega_K(0)[\omega_J] = -2[\omega_I], & \omega_K(1)[\omega_K] = 0.
\end{array}$$

$$\begin{array}{lll}
\omega_I(1)[\omega_I] = -N\Omega, & \omega_I(1)[\omega_J] = 0, & \omega_I(1)[\omega_K] = 0, \\
\omega_J(1)[\omega_I] = 0, & \omega_J(1)[\omega_J] = -N\Omega, & \omega_J(1)[\omega_K] = 0, \\
\omega_K(1)[\omega_I] = 0, & \omega_K(1)[\omega_J] = 0, & \omega_K(1)[\omega_K] = -N\Omega.
\end{array}$$

PROOF. We only have to note that, for any isometric complex structure $\mathcal{J}_1$ and $\mathcal{J}_2$ on E, Lemma 9 and formula (7) with $x(m) = \omega_{\mathcal{J}_1}(m)$ for $m = 0, 1$ and $y(-1) = \omega_{\mathcal{J}_2}(-1)$ imply that

$$\omega_{\mathcal{J}_1}(0)[\omega_{\mathcal{J}_2}] = [\omega_{\mathcal{J}_1}, \omega_{\mathcal{J}_2}]\Omega = [\omega_{\mathcal{J}_1}, \omega_{\mathcal{J}_2}](-1)\Omega,$$
$$\omega_{\mathcal{J}_1}(1)[\omega_{\mathcal{J}_2}] = [\omega_{\mathcal{J}_1}(1), \omega_{\mathcal{J}_2}(-1)]\Omega = [\omega_{\mathcal{J}_1}, \omega_{\mathcal{J}_2}](0)\Omega + \langle\omega_{\mathcal{J}_1}, \omega_{\mathcal{J}_2}\rangle\Omega$$
$$= \langle\omega_{\mathcal{J}_1}, \omega_{\mathcal{J}_2}\rangle\Omega.$$

Results above now follow from Lemma 8 and commutation relations $[\omega_I, \omega_J] = 2\omega_K$, $[\omega_J, \omega_K] = 2\omega_I$, $[\omega_K, \omega_I] = 2\omega_J$. □

The action of Kähler forms on the canonical Virasoro element $\omega = D(-2)\Omega$ is given as follows:

LEMMA 11. *Let $k \geq 0$. Then for any isometric complex structure $\mathcal{J}$ in a quaternionic Hermitian vector space E, we have*

$$\omega_{\mathcal{J}}(k)[\omega] = \begin{cases} -[\omega_{\mathcal{J}}], & \text{if } k = 1, \\ 0, & \text{if } k \neq 1. \end{cases}$$

PROOF. Since $k \geq 0$, we have $\omega_{\mathcal{J}}(k)\Omega = 0$. Now,

$$\omega_{\mathcal{J}}(k)[\omega] = \omega_{\mathcal{J}}(k)D(-2)\Omega = -[D(-2), \omega_{\mathcal{J}}(k)]\Omega = -k\omega_{\mathcal{J}}(k - 2)\Omega.$$

The last equality is due to Proposition 3 of §2.3. This can be nonzero only when $k = 1$ and in this case, the above is equal to $-[\omega_{\mathcal{J}}]$ using the creation property of vertex operators. □

Since vertex operators associated to products of Kähler forms are given as a sum of products of vertex operators associated to individual Kähler forms as in Corollary 6 and Corollary 7, the next calculation is fundamental for our purpose.

PROPOSITION 12. *The action of vertex operators* $\omega_I(k)$, $k \geq 0$ *on products of Kähler forms on V is given as follows:*

$$
\begin{cases}
\omega_I(0)[\omega_I^2] = 0, \quad \omega_I(0)[\omega_J^2] = 4[\omega_J\omega_K], \quad \omega_I(0)[\omega_K^2] = -4[\omega_J\omega_K], \\
\quad \omega_I(0)[\omega_K\omega_I] = -2[\omega_I\omega_J], \quad \omega_I(0)[\omega_I\omega_J] = 2[\omega_K\omega_I], \\
\qquad\qquad \omega_I(0)[\omega_J\omega_K] = 2[\omega_K^2] - 2[\omega_J^2], \\
\omega_I(1)[\omega_I^2] = -2(N-1)[\omega_I], \quad \omega_I(1)[\omega_J^2] = -2[\omega_I], \quad \omega_I(1)[\omega_K^2] = -2[\omega_I], \\
\quad \omega_I(1)[\omega_K\omega_I] = -(N-2)[\omega_K], \quad \omega_I(1)[\omega_I\omega_J] = -(N-2)[\omega_J], \\
\qquad\qquad \omega_I(1)[\omega_J\omega_K] = 0, \\
\omega_I(k)[\omega_I^2] = 0, \quad \omega_I(k)[\omega_J^2] = 0, \quad \omega_I(k)[\omega_K^2] = 0, \\
\omega_I(k)[\omega_J\omega_K] = 0, \quad \omega_I(k)[\omega_K\omega_I] = 0, \quad \omega_I(k)[\omega_I\omega_J] = 0,
\end{cases} \quad \text{for } k \geq 2.
$$

For the operators $\omega_J(k)$, *we have*

$$
\begin{cases}
\omega_J(0)[\omega_I^2] = -4[\omega_K\omega_I], \quad \omega_J(0)[\omega_J^2] = 0, \quad \omega_J(0)[\omega_K^2] = 4[\omega_K\omega_I], \\
\quad \omega_J(0)[\omega_K\omega_I] = 2[\omega_I^2] - 2[\omega_K^2], \quad \omega_J(0)[\omega_I\omega_J] = -2[\omega_J\omega_K], \\
\qquad\qquad \omega_J(0)[\omega_J\omega_K] = 2[\omega_I\omega_J], \\
\omega_J(1)[\omega_I^2] = -2[\omega_J], \quad \omega_J(1)[\omega_J^2] = -2(N-1)[\omega_J], \quad \omega_J(1)[\omega_K^2] = -2[\omega_J], \\
\quad \omega_J(1)[\omega_K\omega_I] = 0, \quad \omega_J(1)[\omega_I\omega_J] = -(N-2)[\omega_I], \\
\qquad\qquad \omega_J(1)[\omega_J\omega_K] = -(N-2)[\omega_K], \\
\omega_J(k)[\omega_I^2] = 0, \quad \omega_J(k)[\omega_J^2] = 0, \quad \omega_J(k)[\omega_K^2] = 0, \\
\omega_J(k)[\omega_J\omega_K] = 0, \quad \omega_J(k)[\omega_K\omega_I] = 0, \quad \omega_J(k)[\omega_I\omega_J] = 0,
\end{cases} \quad \text{for } k \geq 2.
$$

For the operators $\omega_K(k)$ *'s, we have*

$$
\begin{cases}
\omega_K(0)[\omega_I^2] = 4[\omega_I\omega_J], \quad \omega_K(0)[\omega_J^2] = -4[\omega_I\omega_J], \quad \omega_K(0)[\omega_K^2] = 0, \\
\quad \omega_K(0)[\omega_K\omega_I] = 2[\omega_J\omega_K], \quad \omega_K(0)[\omega_I\omega_J] = 2[\omega_J^2] - 2[\omega_I^2], \\
\qquad\qquad \omega_K(0)[\omega_J\omega_K] = -2[\omega_K\omega_I], \\
\omega_K(1)[\omega_I^2] = -2[\omega_K], \quad \omega_K(1)[\omega_J^2] = -2[\omega_K], \quad \omega_K(1)[\omega_K^2] = -2(N-1)[\omega_K], \\
\quad \omega_K(1)[\omega_K\omega_I] = -(N-2)[\omega_I], \quad \omega_K(1)[\omega_I\omega_J] = 0, \\
\qquad\qquad \omega_K(1)[\omega_J\omega_K] = -(N-2)[\omega_J], \\
\omega_K(k)[\omega_I^2] = 0, \quad \omega_K(k)[\omega_J^2] = 0, \quad \omega_K(k)[\omega_K^2] = 0, \\
\omega_K(k)[\omega_J\omega_K] = 0, \quad \omega_K(k)[\omega_K\omega_I] = 0, \quad \omega_K(k)[\omega_I\omega_J] = 0,
\end{cases} \quad \text{for } k \geq 2.
$$

PROOF. Formulae for the action of operators $\omega_J(k)$, $\omega_K(k)$ on products of Kähler forms can be obtained by cyclically permuting (I, J, K) in the formulae for $\omega_I(k)$'s. So, we only have to prove formulae for operators $\omega_I(k)$. For the proof, we use Lemma 4, Lemma 9, Lemma 10, and Lemma 11, together with commutators $[\omega_I, \omega_J] = 2\omega_K$, $[\omega_J, \omega_K] = 2\omega_I$, $[\omega_K, \omega_I] = 2\omega_J$.

Now, we begin our calculation. For the first formula, using Lemma 4 and Lemma 11, we have

$$\omega_I(0)[\omega_I^2] = \omega_I(0)\,(\omega_I(-1)\omega_I(-1)\Omega - 2[\omega]) = 0,$$

since $\omega_I(0)$ and $\omega_I(-1)$ commute. For the next one,

$$\omega_I(0)[\omega_J^2] = \omega_I(0)\,(\omega_J(-1)\omega_J(-1)\Omega - 2[\omega]) = 2\omega_K(-1)\omega_J(-1)\Omega$$
$$+\omega_J(-1)\omega_I(0)\omega_J(-1)\Omega = 2\omega_K(-1)\omega_J(-1)\Omega + 2\omega_J(-1)\omega_K(-1)\Omega = 4[\omega_J\omega_K].$$

Commutation relations among ω_I, ω_J, ω_K remain invariant when we replace J and K and change I by $-I$. So, the formula for $\omega_I(0)[\omega_K^2]$ is a consequence of this observation. We go to the next one.

$$\omega_I(0)[\omega_K\omega_I] = \tfrac{1}{2}\omega_I(0)\,(\omega_K(-1)\omega_I(-1)\Omega + \omega_I(-1)\omega_K(-1)\Omega)$$
$$= \tfrac{1}{2}\,(-2\omega_J(-1)\omega_I(-1)\Omega - 2\omega_I(-1)\omega_J(-1)\Omega) = -2[\omega_I\omega_J].$$

The formula for $\omega_I(0)[\omega_I\omega_J]$ follows by switching J and K, I and $-I$ in the above formula. Next,

$$\omega_I(0)[\omega_J\omega_K] = \tfrac{1}{2}\omega_I(0)\,(\omega_J(-1)\omega_K(-1)\Omega + \omega_K(-1)\omega_J(-1)\Omega)$$
$$= \tfrac{1}{2}\,(2\omega_K(-1)\omega_K(-1)\Omega + \omega_J(-1)\omega_I(0)\omega_K(-1)\Omega - \omega_J(-1)\omega_J(-1)\Omega$$
$$+\omega_K(-1)\omega_I(0)\omega_J(-1)\Omega) = \tfrac{1}{2}\,(2[\omega_K^2] - 2[\omega_J^2] - 2\omega_J(-1)\omega_J(-1)\Omega$$
$$+2\omega_K(-1)\omega_K(-1)\Omega) = 2[\omega_K^2] - 2[\omega_J^2].$$

This takes care of the action of the operator $\omega_I(0)$. Next, we calculate the effect of operator $\omega_I(1)$ on products of Kähler forms. For the first two formulae,

$$\omega_I(1)[\omega_I^2] = \omega_I(1)\,(\omega_I(-1)\omega_I(-1)\Omega - 2[\omega])$$
$$- N[\omega_I] + \omega_I(-1)\omega_I(1)\omega_I(-1)\Omega + 2[\omega_I] = -2(N-1)[\omega_I],$$
$$\omega_I(1)[\omega_J^2] = \omega_I(1)\,(\omega_J(-1)\omega_J(-1)\Omega - 2[\omega]) = 2\omega_K(0)\omega_J(-1)\Omega$$
$$+\omega_J(-1)\omega_I(1)\omega_J(-1)\Omega + 2[\omega] = -4[\omega_I] + 2\omega_J(-1)\omega_K(0)\Omega + 2[\omega_I] = -2[\omega_I].$$

The formula for $\omega_I(1)[\omega_K^2]$ can be obtained by switching J and K, I and $-I$ in the above second formula. We go to the next one.

$$\omega_I(1)[\omega_K\omega_I] = \tfrac{1}{2}\omega_I(1)\,(\omega_K(-1)\omega_I(-1)\Omega + \omega_I(-1)\omega_K(-1)\Omega)$$
$$= \tfrac{1}{2}\,(-2\omega_J(0)\omega_I(-1)\Omega + \omega_K(-1)\omega_I(1)\omega_I(-1)\Omega - N[\omega_K]$$
$$+\omega_I(-1)\omega_I(1)\omega_K(-1)\Omega) = \tfrac{1}{2}\,(4[\omega_K] - N[\omega_K] - N[\omega_K]) = -(N-2)[\omega_K].$$

Formula for $\omega_I(1)[\omega_I\omega_J]$ can be obtained by switching (I, J, K) by $(-I, K, J)$ in the above. Finally, on $[\omega_J\omega_K]$ we have

$$\omega_I(1)[\omega_J\omega_K] = \tfrac{1}{2}\omega_I(1)\left(\omega_J(-1)\omega_K(-1)\Omega + \omega_K(-1)\omega_J(-1)\Omega\right)$$
$$= \tfrac{1}{2}\left(2\omega_K(0)\omega_K(-1)\Omega + \omega_J(-1)\omega_I(1)\omega_K(-1)\Omega - 2\omega_J(0)\omega_J(-1)\Omega\right.$$
$$\left. +\omega_K(-1)\omega_I(1)\omega_J(-1)\Omega\right) = 0$$

For the action of operators $\omega_I(k)$ for $k \geq 0$, we recall that $D(k)$ for $k \geq 1$ acts trivially on $[\bigwedge^* A(-\tfrac{1}{2})]$ in V because $[\bigwedge^* A(-\tfrac{1}{2})] \subset V$ consists of conformal highest weight vectors [§3.5 Proposition 8]. If $v \in S^2 (\mathbb{C}[\omega_I] \oplus \mathbb{C}[\omega_J] \oplus \mathbb{C}[\omega_K]) \subset [\bigwedge^* A(-\tfrac{1}{2})]$, then $D(1)v = 0$. From our calculations above, $\omega_I(1)v \in \mathbb{C}[\omega_I] \oplus \mathbb{C}[\omega_J] \oplus \mathbb{C}[\omega_K]$. It follows that $D(1)(\omega_I(1)v) = 0$ also. Hence we have $\omega_I(2)v = [D(1), \omega_I(1)]v = 0$ for any $v \in S^2 (\mathbb{C}[\omega_I] \oplus \mathbb{C}[\omega_J] \oplus \mathbb{C}[\omega_K])$. For $k \geq 3$, $\omega_I(k)v$ has negative weight, so it must be zero. $\square$

Since an observation we made in the proof of Proposition 12 is very useful, we state it explicitly as a lemma.

LEMMA 13. *Let* $(I', J', K') = (-I, K, J)$, $(K, -J, I)$, $(J, I, -K)$. *Then the corresponding Kähler forms* $\omega_{I'}, \omega_{J'}, \omega_{K'}$ *satisfy the same commutation relations as* $\omega_I, \omega_J, \omega_K$. *Namely,*

$$[\omega_{I'}, \omega_{J'}] = 2\omega_{K'}, \qquad [\omega_{J'}, \omega_{K'}] = 2\omega_{I'}, \qquad [\omega_{K'}, \omega_{I'}] = 2\omega_{J'}.$$

For any isometric complex structures $\mathcal{J}_1$, $\mathcal{J}_2$, $\mathcal{J}_3$, $\mathcal{J}_4$ in the quaternionic Hermitian vector space E, the commutator formula for vertex operators gives

$$[\{\omega_{\mathcal{J}_1}\omega_{\mathcal{J}_2}\}_{m+1}, \{\omega_{\mathcal{J}_3}\omega_{\mathcal{J}_4}\}_{n+1}] = \sum_{0 \leq k \in \mathbb{Z}} \binom{m+1}{k} \{\{\omega_{\mathcal{J}_1}\omega_{\mathcal{J}_2}\}_k[\omega_{\mathcal{J}_3}\omega_{\mathcal{J}_4}]\}_{m+n+2-k} \cdot$$

Our next task is to calculate $\{\omega_{\mathcal{J}_1}\omega_{\mathcal{J}_2}\}_k[\omega_{\mathcal{J}_3}\omega_{\mathcal{J}_4}]$ for $k \geq 0$. To do this, we prepare some lemmas.

LEMMA 14. *The vector* $\omega_I(-2)[\omega_I] \in V_3$ *is in the image of* $D(-1)$ *operator. More precisely,*

$$\begin{cases} -2\omega_I(-2)[\omega_I] = D(-1)[\omega_I^2] + 2D(-1)[\omega], \\ -2\omega_J(-2)[\omega_J] = D(-1)[\omega_J^2] + 2D(-1)[\omega], \\ -2\omega_K(-2)[\omega_K] = D(-1)[\omega_K^2] + 2D(-1)[\omega]. \end{cases}$$

PROOF. We apply the identity $[D(-1), \omega_I(-1)] = -\omega_I(-2)$ on $[\omega_I]$ to obtain

$$D(-1)\omega_I(-1)[\omega_I] - \omega_I(-1)D(-1)\omega_I(-1)\Omega = -\omega_I(-2)[\omega_I].$$

Since $D(-1)\omega_I(-1)\Omega = -\omega_I(-2)\Omega$ and $\omega_I(-1)$, $\omega_I(-2)$ commute, we see that $-\omega_I(-1)D(-1)\omega_I(-1)\Omega = \omega_I(-2)\omega_I(-1)\Omega = \omega_I(-2)[\omega_I]$. Using Lemma 4, we get the first identity. Other identities follow by cyclically permuting I, J, K. $\square$

LEMMA 15. *The following identities hold:*

$$[\omega_I\omega_J] = \omega_I(-1)\omega_J(-1)\Omega - \omega_K(-2)\Omega$$
$$= \omega_J(-1)\omega_I(-1)\Omega + \omega_K(-2)\Omega,$$
$$[\omega_J\omega_K] = \omega_J(-1)\omega_K(-1)\Omega - \omega_I(-2)\Omega$$
$$= \omega_K(-1)\omega_J(-1)\Omega + \omega_I(-2)\Omega,$$
$$[\omega_K\omega_I] = \omega_K(-1)\omega_I(-1)\Omega - \omega_J(-2)\Omega$$
$$= \omega_I(-1)\omega_K(-1)\Omega + \omega_J(-2)\Omega.$$

PROOF. This follows from (2), Lemma 4, and (7). □

LEMMA 16. *The effect of the Virasoro operator $D(-1)$ on quadratic Kähler forms is given as follows:*

$$-D(-1)[\omega_I\omega_J] = \omega_I(-2)[\omega_J] + \omega_J(-2)[\omega_I]$$
$$= \omega_I(-1)\omega_J(-2)\Omega + \omega_J(-1)\omega_I(-2)\Omega,$$
$$-D(-1)[\omega_J\omega_K] = \omega_J(-2)[\omega_K] + \omega_K(-2)[\omega_J]$$
$$= \omega_J(-1)\omega_K(-2)\Omega + \omega_K(-1)\omega_J(-2)\Omega,$$
$$-D(-1)[\omega_K\omega_I] = \omega_K(-2)[\omega_I] + \omega_I(-2)[\omega_K]$$
$$= \omega_I(-1)\omega_K(-2)\Omega + \omega_K(-1)\omega_I(-2)\Omega.$$

PROOF. For the first identity, since $[\omega_I\omega_J] = \omega_I(-1)\omega_J(-1)\Omega - \omega_K(-2)\Omega$ from Lemma 15, the commutation relation (7) gives

$$D(-1)[\omega_I\omega_J] = -\omega_I(-2)\omega_J(-1)\Omega - \omega_I(-1)\omega_J(-2)\Omega + 2\omega_K(-3)\Omega,$$
$$= -(\omega_J(-1)\omega_I(-2)\Omega + 2\omega_K(-3)\Omega)$$
$$\quad - (\omega_I(-1)\omega_J(-2)\Omega - 2\omega_K(-3)\Omega)$$
$$= -\omega_J(-1)\omega_I(-2)\Omega - \omega_I(-1)\omega_J(-2)\Omega.$$

Other identities follow by cyclic permutations of (I, J, K). □

LEMMA 17. *The following identities hold:*

$$\omega_I(-1)[\omega_J\omega_K] - \omega_J(-1)[\omega_K\omega_I] = \tfrac{1}{2}D(-1)\left([\omega_I^2] + [\omega_J^2] - 2[\omega_K^2]\right),$$
$$\omega_J(-1)[\omega_K\omega_I] - \omega_K(-1)[\omega_I\omega_J] = \tfrac{1}{2}D(-1)\left([\omega_J^2] + [\omega_K^2] - 2[\omega_I^2]\right),$$
$$\omega_K(-1)[\omega_I\omega_J] - \omega_I(-1)[\omega_J\omega_K] = \tfrac{1}{2}D(-1)\left([\omega_K^2] + [\omega_I^2] - 2[\omega_J^2]\right).$$

PROOF. We prove the first identity. Using Lemmas 4, 14, and 15, we have

$$\omega_I(-1)[\omega_J\omega_K] - \omega_J(-1)[\omega_K\omega_I]$$
$$= \omega_I(-1)\left(\omega_J(-1)\omega_K(-1)\Omega - \omega_I(-2)\Omega\right)$$
$$\quad - \omega_J(-1)\left(\omega_I(-1)\omega_K(-1)\Omega + \omega_J(-2)\Omega\right)$$
$$= [\omega_I(-1), \omega_J(-1)]\omega_K(-1)\Omega - \omega_I(-2)\omega_I(-1)\Omega - \omega_J(-2)\omega_J(-1)\Omega$$
$$= 2\omega_K(-2)\omega_K(-1)\Omega + \tfrac{1}{2}D(-1)([\omega_I^2] + [\omega_J^2]) + 2D(-1)[\omega]$$
$$= \tfrac{1}{2}D(-1)\left([\omega_I^2] + [\omega_J^2] - 2[\omega_K^2]\right).$$

The other two follows by cyclically permuting (I, J, K).   □

LEMMA 18. *The following identities hold:*

$$\omega_I(-1)\left([\omega_J^2] - [\omega_K^2]\right) - \omega_J(-1)[\omega_I\omega_J] + \omega_K(-1)[\omega_K\omega_I] = -3D(-1)[\omega_J\omega_K],$$

$$\omega_J(-1)\left([\omega_K^2] - [\omega_I^2]\right) - \omega_K(-1)[\omega_J\omega_K] + \omega_I(-1)[\omega_I\omega_J] = -3D(-1)[\omega_K\omega_I],$$

$$\omega_K(-1)\left([\omega_I^2] - [\omega_J^2]\right) - \omega_I(-1)[\omega_K\omega_I] + \omega_J(-1)[\omega_J\omega_K] = -3D(-1)[\omega_I\omega_J].$$

PROOF. For the first identity, we calculate each term as follows.

$$\begin{aligned}
\omega_I(-1)\left([\omega_J^2] - [\omega_K^2]\right) &= \omega_I(-1)\left(\omega_J(-1)\omega_J(-1)\Omega - \omega_K(-1)\omega_K(-1)\Omega\right) \\
&= \left(2\omega_K(-2)\omega_J(-1)\Omega + \omega_J(-1)\omega_I(-1)\omega_J(-1)\Omega\right) \\
&\quad + \left(2\omega_J(-2)\omega_K(-1)\Omega - \omega_K(-1)\omega_I(-1)\omega_K(-1)\Omega\right),
\end{aligned}$$

Here, we used Lemma 14. Lemma 15 implies

$$\begin{aligned}
\omega_J(-1)[\omega_I\omega_J] &= \omega_J(-1)\left(\omega_I(-1)\omega_J(-1)\Omega - \omega_K(-2)\Omega\right) \\
&= \omega_J(-1)\omega_I(-1)\omega_J(-1)\Omega - \omega_J(-1)\omega_K(-2)\Omega, \\
\omega_K(-1)[\omega_K\omega_I] &= \omega_K(-1)\left(\omega_I(-1)\omega_K(-1)\Omega + \omega_J(-2)\Omega\right) \\
&= \omega_K(-1)\omega_I(-1)\omega_K(-1)\Omega + \omega_K(-1)\omega_J(-2)\Omega.
\end{aligned}$$

Thus, after some cancellation, Lemma 16 implies that

$$\begin{aligned}
&\omega_I(-1)\left([\omega_I^2] - [\omega_K^2]\right) - \omega_J(-1)[\omega_I\omega_J] + \omega_K[\omega_K\omega_I] \\
&= 2\omega_K(-2)\omega_J(-1)\Omega + 2\omega_J(-2)\omega_K(-1)\Omega + \omega_J(-1)\omega_K(-2)\Omega + \omega_K(-1)\omega_J(-2)\Omega \\
&= -2D(-1)[\omega_J\omega_K] - D(-1)[\omega_J\omega_K] = -3D(-1)[\omega_J\omega_K].
\end{aligned}$$

This completes the proof of the first identity. Other two identities can be proved similarly. Or, we can simply cyclically permute (I, J, K).   □

Calculation of $\{\omega_I^2\}_*[\omega_I^2]$. Corollary 7 gives us expressions of operators $\{\omega_I^2\}_*$. Since $\omega_I(k)$ acts trivially on $[\omega_I^2]$ for $k \geq 2$ by Proposition 12, and $D(k)$ also acts trivially on a conformal highest weight vector $[\omega_I^2]$ for $k \geq 1$ by Proposition 8 of §3.5, we have

$$\begin{aligned}
\{\omega_I^2\}_0[\omega_I^2] &= \left(2\omega_I(-1)\omega_I(0) + 2\omega_I(-2)\omega_I(1) - 2D(-1)\right)[\omega_I^2] \\
&= -4(N-1)\omega_I(-2)[\omega_I] - 2D(-1)[\omega_I^2] \\
&= 4(N-1)D(-1)[\omega] + 2(N-2)D(-1)[\omega_I^2].
\end{aligned}$$

For the last identity, we used Lemma 14.

$$\begin{aligned}
\{\omega_I^2\}_1[\omega_I^2] &= \left(\omega_I(0)\omega_I(0) + 2\omega_I(-1)\omega_I(1) - 2D(0)\right)[\omega_I^2] \\
&= -4(N-1)([\omega_I^2] + 2[\omega]) + 4[\omega_I^2] = -4(N-2)[\omega_I^2] - 8(N-1)[\omega].
\end{aligned}$$

For the second identity, we used Lemma 4.

$$\{\omega_I^2\}_2[\omega_I^2] = 2\omega_I(0)\omega_I(1)[\omega_I^2] = -4(N-1)\omega_I(0)[\omega_I] = 0.$$

$$\{\omega_I^2\}_3[\omega_I^2] = \omega_I(1)\omega_I(1)[\omega_I^2] = -2(N-1)\omega_I(1)\omega_I(-1)\Omega = 2N(N-1)\Omega.$$

Calculation of $\{\omega_I^2\}_*[\omega_J^2]$. This calculation is similar to the previous one. Lemma 4, Corollary 7, Lemma 10, Proposition 12, and Lemma 14 imply that

$$\{\omega_I^2\}_0[\omega_J^2] = (2\omega_I(-1)\omega_I(0) + 2\omega_I(-2)\omega_I(1) - 2D(-1))\,[\omega_J^2]$$
$$= 8\omega_I(-1)[\omega_J\omega_K] - 4\omega_I(-2)[\omega_I] - 2D(-1)[\omega_J^2]$$
$$= 8\omega_I(-1)[\omega_J\omega_K] + 2D(-1)\left([\omega_I^2] - [\omega_J^2] + 2[\omega]\right).$$

The second equality is due to Proposition 12, the third one is due to Lemma 14.

$$\{\omega_I^2\}_1[\omega_J^2] = (\omega_I(0)\omega_I(0) + 2\omega_I(-1)\omega_I(1) - 2D(0))\,[\omega_I^2]$$
$$= (8[\omega_K^2] - 8[\omega_J^2]) - 4\omega_I(-1)[\omega_I] + 4[\omega_J^2]$$
$$= -4\left([\omega_I^2] + [\omega_J^2] - 2[\omega_K^2] + 2[\omega]\right).$$

The second equality is due to Proposition 12, the third one is due to Lemma 10.

$$\{\omega_I^2\}_2[\omega_J^2] = 2\omega_I(0)\omega_I(1)[\omega_J^2] = -4\omega_I(0)[\omega_I] = 0,$$
$$\{\omega_I^2\}_3[\omega_J^2] = \omega_I(1)\omega_I(1)[\omega_J^2] = -2\omega_I(1)[\omega_I] = 2N\Omega.$$

In the last two calculations, the second equalities are due to Proposition 12, the third equalities are due to Lemma 10. By switching (I, J, K) and $(-I, K, J)$, we obtain formulae for $\{\omega_J^2\}_*[\omega_I^2]$. A cyclic permutation of (I, J, K) also gives $\{\omega_I^2\}_*[\omega_K^2]$.

By applying lemmas and propositions above, we obtain following numerous results. For these calculations, we do not explicitly cite results we used.

Calculation of $\{\omega_I^2\}_*[\omega_I\omega_J]$.

$$\{\omega_I^2\}_0[\omega_I\omega_J] = (2\omega_I(-1)\omega_I(0) + 2\omega_I(-2)\omega_I(1) - 2D(-1))\,[\omega_I\omega_J]$$
$$= 4\omega_I(-1)[\omega_K\omega_I] - 2(N-2)\omega_I(-2)[\omega_J] - 2D(-1)[\omega_I\omega_J],$$
$$\{\omega_I^2\}_1[\omega_I\omega_J] = (\omega_I(0)\omega_I(0) + 2\omega_I(-1)\omega_I(1) - 2D(0))\,[\omega_I\omega_J]$$
$$= 2\omega_I(0)[\omega_K\omega_I] - 2(N-2)\omega_I(-1)[\omega_J] + 4[\omega_I\omega_J]$$
$$= -2(N-2)\omega_I(-1)[\omega_J]$$
$$= -2(N-2)[\omega_I\omega_J] + 2(N-2)D(-1)[\omega_K],$$
$$\{\omega_I^2\}_2[\omega_I\omega_J] = 2\omega_I(1)\omega_I(0)[\omega_I\omega_J] = 4\omega_I(1)[\omega_K\omega_I]$$
$$= -4(N-2)[\omega_K],$$
$$\{\omega_I^2\}_3[\omega_I\omega_J] = \omega_I(1)\omega_I(1)[\omega_I\omega_J] = -(N-2)\omega_I(1)[\omega_J] = 0.$$

By switching (I, J, K) and $(-I, K, J)$ in the above calculation, we also obtain the formulae for $\{\omega_J^2\}_*[\omega_I\omega_J]$. Cyclic permutations of (I, J, K) yield other formulae.

Calculation of $\{\omega_I\omega_J\}_*[\omega_I^2]$.

$$\{\omega_I\omega_J\}_0[\omega_I^2] = (\omega_I(-1)\omega_J(0) + \omega_I(-2)\omega_J(1) + \omega_J(-1)\omega_I(0) + \omega_J(-2)\omega_I(1))\,[\omega_I^2]$$
$$= -4\omega_I(-1)[\omega_K\omega_I] - 2\omega_I(-2)[\omega_J] - 2(N-1)\omega_J(-2)[\omega_I],$$

$$\{\omega_I\omega_J\}_1[\omega_I^2] = (\omega_I(0)\omega_J(0) - \omega_K(0) + \omega_I(-1)\omega_J(1) + \omega_J(-1)\omega_I(1))\,[\omega_I^2]$$
$$= -4\omega_I(0)[\omega_K\omega_I] - 4[\omega_I\omega_J] - 2\omega_I(-1)[\omega_J] - 2(N-1)\omega_J(-1)[\omega_I]$$
$$= 8[\omega_I\omega_J] - 4[\omega_I\omega_J] - 2[\omega_I\omega_J]$$
$$\qquad - 2\omega_K(-2)\Omega - 2(N-1)[\omega_I\omega_J] + 2(N-1)\omega_K(-2)\Omega$$
$$= -2(N-2)[\omega_I\omega_J] - 2(N-2)D(-1)[\omega_K],$$

$$\{\omega_I\omega_J\}_2[\omega_I^2] = (\omega_I(0)\omega_J(1) + \omega_I(1)\omega_J(0) - 2\omega_K(1))\,[\omega_I^2]$$
$$= -2\omega_I(0)[\omega_J] - 4\omega_I(1)[\omega_K\omega_I] + 4[\omega_K]$$
$$= -4[\omega_K] + 4(N-2)[\omega_K] + 4[\omega_K] = 4(N-2)[\omega_K],$$

$$\{\omega_I\omega_J\}_3[\omega_I^2] = \omega_I(1)\omega_J(1)[\omega_I^2] = -2\omega_I(1)[\omega_J] = 0.$$

Calculation of $\{\omega_I^2\}_*[\omega_J\omega_K]$.

$$\{\omega_I^2\}_0[\omega_J\omega_K] = (2\omega_I(-1)\omega_I(0) + 2\omega_I(-2)\omega_I(1) - 2D(-1))\,[\omega_J\omega_K]$$
$$= 4\omega_I(-1)\,([\omega_K^2] - [\omega_J^2]) - 2D(-1)[\omega_J\omega_K],$$

$$\{\omega_I^2\}_1[\omega_J\omega_K] = (\omega_I(0)\omega_I(0) + 2\omega_I(-1)\omega_I(1) - 2D(0))\,[\omega_J\omega_K]$$
$$= 2\omega_I(0)\,([\omega_K^2] - [\omega_J^2]) + 4[\omega_J\omega_K]$$
$$= -12[\omega_J\omega_K],$$

$$\{\omega_I^2\}_2[\omega_J\omega_K] = 2\omega_I(0)\omega_I(1)[\omega_J\omega_K] = 0,$$

$$\{\omega_I^2\}_3[\omega_J\omega_K] = \omega_I(1)\omega_I(1)[\omega_J\omega_K] = 0.$$

Calculation of $\{\omega_J\omega_K\}_*[\omega_I^2]$.

$$\{\omega_J\omega_K\}_0[\omega_I^2] = (\omega_J(-1)\omega_K(0) + \omega_J(-2)\omega_K(1) + \omega_K(-1)\omega_J(0)$$
$$\qquad\qquad +\omega_K(-2)\omega_J(1))\,[\omega_I^2]$$
$$= 4\omega_J(-1)[\omega_I\omega_J] - 2\omega_J(-2)[\omega_K] - 4\omega_K(-1)[\omega_K\omega_I] - 2\omega_K(-2)[\omega_J]$$
$$= 4\omega_J(-1)[\omega_I\omega_J] - 4\omega_K(-1)[\omega_K\omega_I] + 2D(-1)[\omega_J\omega_K],$$

$$\{\omega_J\omega_K\}_1[\omega_I^2] = (\omega_J(0)\omega_K(0) - \omega_I(0) + \omega_J(-1)\omega_K(1) + \omega_K(-1)\omega_J(1))\,[\omega_I^2]$$
$$= 4\omega_J(0)[\omega_I\omega_J] - 2\omega_J(-1)[\omega_K] - 2\omega_K(-1)[\omega_J]$$
$$= -8[\omega_J\omega_K] - 4[\omega_J\omega_K]$$
$$= -12[\omega_J\omega_K],$$

$$\{\omega_J\omega_K\}_2[\omega_I^2] = (\omega_J(1)\omega_K(0) + \omega_J(0)\omega_K(1) - 2\omega_I(1))\,[\omega_I^2]$$
$$= 4\omega_J(1)[\omega_I\omega_J] - 2\omega_J(0)[\omega_K] + 4(N-1)[\omega_I]$$
$$= -4(N-2)[\omega_I] - 4[\omega_I] + 4(N-1)[\omega_I] = 0,$$

$$\{\omega_J\omega_K\}_3[\omega_I^2] = \omega_J(1)\omega_K(1)[\omega_I^2] = -2\omega_J(1)[\omega_K] = 0.$$

Calculation of $\{\omega_I\omega_J\}_*[\omega_I\omega_J]$.

$$\{\omega_I\omega_J\}_0[\omega_I\omega_J] = (\omega_I(-1)\omega_J(0) + \omega_I(-2)\omega_J(1) + \omega_J(-1)\omega_I(0)$$
$$+\omega_J(-2)\omega_I(1))\,[\omega_I\omega_J]$$
$$= -2\omega_I(-1)[\omega_J\omega_K] - (N-2)\omega_I(-2)[\omega_I] + 2\omega_J(-1)[\omega_K\omega_I]$$
$$- (N-2)\omega_J(-2)[\omega_J]$$
$$= -2\omega_I(-1)[\omega_J\omega_K] + 2\omega_J(-1)[\omega_K\omega_I]$$
$$+ \tfrac{N-2}{2}D(-1)\left([\omega_I^2] + [\omega_J^2] + 4[\omega]\right)$$
$$= D(-1)\left(\tfrac{N-4}{2}\left([\omega_I^2] + [\omega_J^2]\right) + 2[\omega_K^2] + 2(N-2)[\omega]\right),$$

$$\{\omega_I\omega_J\}_1[\omega_I\omega_J] = (\omega_I(0)\omega_J(0) - \omega_K(0) + \omega_I(-1)\omega_J(1) + \omega_J(-1)\omega_I(1))\,[\omega_I\omega_J]$$
$$= -2\omega_I(0)[\omega_J\omega_K] - 2\left([\omega_J^2] - [\omega_I^2]\right) - (N-2)\omega_I(-1)[\omega_I]$$
$$- (N-2)\omega_J(-1)[\omega_J]$$
$$= (-4[\omega_K^2] + 4[\omega_J^2]) - 2[\omega_J^2] + 2[\omega_I^2] - (N-2)([\omega_I^2] + 2[\omega])$$
$$- (N-2)([\omega_J^2] + 2[\omega])$$
$$= -(N-4)[\omega_I^2] - (N-4)[\omega_J^2] - 4[\omega_K^2] - 4(N-2)[\omega],$$

$$\{\omega_I\omega_J\}_2[\omega_I\omega_J] = (\omega_I(0)\omega_J(1) + \omega_I(1)\omega_J(0) - 2\omega_K(1))\,[\omega_I\omega_J]$$
$$= -(N-2)\omega_I(0)[\omega_I] - 2\omega_I(1)[\omega_J\omega_K] = 0,$$

$$\{\omega_I\omega_J\}_3[\omega_I\omega_J] = (\omega_I(2)\omega_J(0) + \omega_I(1)\omega_J(1))\,[\omega_I\omega_J]$$
$$= -2\omega_I(2)[\omega_J\omega_K] - (N-2)\omega_I(1)[\omega_I]$$
$$= N(N-2)\Omega.$$

Calculation of $\{\omega_I\omega_J\}_*[\omega_K\omega_I]$.

$$\{\omega_I\omega_J\}_0[\omega_K\omega_I] = (\omega_I(-1)\omega_J(0) + \omega_I(-2)\omega_J(1) + \omega_J(-1)\omega_I(0)$$
$$+\omega_J(-2)\omega_I(1))\,[\omega_K\omega_I]$$
$$=2\omega_I(-1)\left([\omega_I^2] - [\omega_K^2]\right) - 2\omega_J(-1)[\omega_I\omega_J] - (N-2)\omega_J(-2)[\omega_K],$$

$$\{\omega_I\omega_J\}_1[\omega_K\omega_I] = (\omega_I(0)\omega_J(0) - \omega_K(0) + \omega_I(-1)\omega_J(1) + \omega_J(-1)\omega_I(1))\,[\omega_K\omega_I]$$
$$= \omega_I(0)\left(2[\omega_I^2] - 2[\omega_K^2]\right) - 2[\omega_J\omega_K] - (N-2)\omega_J(-1)[\omega_K]$$
$$= 6[\omega_J\omega_K] - (N-2)\omega_J(-1)[\omega_K]$$
$$= -(N-8)[\omega_J\omega_K] - (N-2)D(-1)[\omega_I],$$

$$\{\omega_I\omega_J\}_2[\omega_K\omega_I] = (\omega_I(0)\omega_J(1) + \omega_I(1)\omega_J(0) - 2\omega_K(1))\,[\omega_K\omega_I]$$
$$= \omega_I(1)\left(2[\omega_I^2] - 2[\omega_K^2]\right) + 2(N-2)[\omega_I]$$
$$= -4(N-1)[\omega_I] + 4[\omega_I] + 2(N-2)[\omega_I]$$
$$= -2(N-2)[\omega_I],$$

$$\{\omega_I\omega_J\}_3[\omega_K\omega_I] = \omega_I(1)\omega_I(1)[\omega_K\omega_I] = 0.$$

By switching (I, J, K) by $(-I, K, J)$, we obtain the formula for $\{\omega_K\omega_I\}_*[\omega_I\omega_J]$. Collecting our calculations so far, we obtain the next proposition.

PROPOSITION 19. *The action of vertex operators associated to products of Kähler forms on products of Kähler forms are given as follows. Other formulae are obtained by cyclically permuting* (I, J, K) *or by replacing* (I, J, K) *with* $(-I, K, J)$ *or by doing both of these operations:*

$$
\begin{cases}
\{\omega_I^2\}_0[\omega_I^2] = 4(N-1)D(-1)[\omega] + 2(N-2)D(-1)[\omega_I^2], \\
\qquad = -\tfrac{1}{2}D(-1)\left(\{\omega_I^2\}_1[\omega_I^2]\right) \\
\{\omega_I^2\}_1[\omega_I^2] = -8(N-1)[\omega] - 4(N-2)[\omega_I^2], \\
\{\omega_I^2\}_2[\omega_I^2] = 0, \\
\{\omega_I^2\}_3[\omega_I^2] = 2N(N-1)\Omega.
\end{cases}
$$

$$
\begin{cases}
\{\omega_I^2\}_0[\omega_J^2] = 8\omega_I(-1)[\omega_J\omega_K] + 2D(-1)\left([\omega_I^2] - [\omega_J^2] + 2[\omega]\right), \\
\{\omega_I^2\}_1[\omega_J^2] = -4\left([\omega_I^2] + [\omega_J^2] - 2[\omega_K^2] + 2[\omega]\right), \\
\{\omega_I^2\}_2[\omega_J^2] = 0, \\
\{\omega_I^2\}_3[\omega_J^2] = 2N\Omega.
\end{cases}
$$

$$
\begin{cases}
\{\omega_I^2\}_0[\omega_I\omega_J] = 4\omega_I(-1)[\omega_K\omega_I] - 2(N-2)\omega_I(-2)[\omega_J] - 2D(-1)[\omega_I\omega_J], \\
\{\omega_I^2\}_1[\omega_I\omega_J] = -2(N-2)[\omega_I\omega_J] + 2(N-2)D(-1)[\omega_K], \\
\{\omega_I^2\}_2[\omega_I\omega_J] = -4(N-2)[\omega_K], \\
\{\omega_I^2\}_3[\omega_I\omega_J] = 0.
\end{cases}
$$

$$
\begin{cases}
\{\omega_I\omega_J\}_0[\omega_I^2] = -4\omega_I(-1)[\omega_K\omega_I] - 2\omega_I(-2)[\omega_J] - 2(N-1)\omega_J(-2)[\omega_I], \\
\{\omega_I\omega_J\}_1[\omega_I^2] = -2(N-2)[\omega_I\omega_J] - 2(N-2)D(-1)[\omega_K], \\
\{\omega_I\omega_J\}_2[\omega_I^2] = 4(N-2)[\omega_K], \\
\{\omega_I\omega_J\}_3[\omega_I^2] = 0.
\end{cases}
$$

$$
\begin{cases}
\{\omega_I^2\}_0[\omega_J\omega_K] = 4\omega_I(-1)\left([\omega_K^2] - [\omega_J^2]\right) - 2D(-1)[\omega_J\omega_K], \\
\{\omega_I^2\}_1[\omega_J\omega_K] = -12[\omega_J\omega_K], \\
\{\omega_I^2\}_2[\omega_J\omega_K] = 0, \\
\{\omega_I^2\}_3[\omega_J\omega_K] = 0.
\end{cases}
$$

$$
\begin{cases}
\{\omega_J\omega_K\}_0[\omega_I^2] = 4\omega_J(-1)[\omega_I\omega_J] - 4\omega_K(-1)[\omega_K\omega_I] + 2D(-1)[\omega_J\omega_K], \\
\{\omega_J\omega_K\}_1[\omega_I^2] = -12[\omega_J\omega_K], \\
\{\omega_J\omega_K\}_2[\omega_I^2] = 0, \\
\{\omega_J\omega_K\}_3[\omega_I^2] = 0.
\end{cases}
$$

$$
\begin{cases}
\{\omega_I\omega_J\}_0[\omega_I\omega_J] = \tfrac{N-4}{2}D(-1)\left([\omega_I^2] + [\omega_J^2]\right) + 2[\omega_K^2] + 2(N-2)[\omega] \\
\qquad = -\tfrac{1}{2}D(-1)\left(\{\omega_I\omega_J\}_1[\omega_I\omega_J]\right), \\
\{\omega_I\omega_J\}_1[\omega_I\omega_J] = -(N-4)[\omega_I^2] - (N-4)[\omega_J^2] - 4[\omega_K^2] - 4(N-2)[\omega], \\
\{\omega_I\omega_J\}_2[\omega_I\omega_J] = 0, \\
\{\omega_I\omega_J\}_3[\omega_I\omega_J] = N(N-2)\Omega.
\end{cases}
$$

$$\begin{cases} \{\omega_I\omega_J\}_0[\omega_K\omega_I] = 2\omega_I(-1)\left([\omega_I^2]-[\omega_K^2]\right) - 2\omega_J(-1)[\omega_I\omega_J] \\ \qquad\qquad\qquad - (N-2)\omega_J(-2)[\omega_K], \\ \{\omega_I\omega_J\}_1[\omega_K\omega_I] = -(N-8)[\omega_J\omega_K] - (N-2)D(-1)[\omega_I], \\ \{\omega_I\omega_J\}_2[\omega_K\omega_I] = -2(N-2)[\omega_I], \\ \{\omega_I\omega_J\}_3[\omega_K\omega_I] = 0. \end{cases}$$

The above formulae together with their derivatives give complete information on the action of $\{\omega_{J_1}\omega_{J_2}\}_k$ on $[\omega_{J_3}\omega_{J_4}]$ for $k \geq 0$, for any isometric complex structures J_1, J_2, J_3, J_4 on the quaternionic Hermitian vector space E.

As a corollary to our calculation, we can now easily calculate commutators of the symplectic Virasoro algebra which was calculated in §3.8 in an elementary but tediously long way.

We recall that the Casimir element $\phi_{\mathfrak{sp}(N')}$ for the symplectic Lie algebra $\mathfrak{sp}(N')$ is given by Corollary 17 of §3.9:

$$2[\phi_{\mathfrak{sp}(N')}] = [\omega_I^2] + [\omega_J^2] + [\omega_K^2] \in (V)_2.$$

COROLLARY 20 (SYMPLECTIC VIRASORO ALGEBRA). *The action of symplectic vertex operator $\{\phi_{\mathfrak{sp}(N')}\}_k$ on the symplectic Casimir element $[\phi_{\mathfrak{sp}(N')}]$ is given as follows for $k \geq 0$:*

$$\begin{aligned} \{\phi_{\mathfrak{sp}(N')}\}_0[\phi_{\mathfrak{sp}(N')}] &= 3(N+1)D(-1)[\omega] + (N-2)D(-1)[\phi_{\mathfrak{sp}(N')}] \\ &= -\tfrac{1}{2}D(-1)\left(\{\phi_{\mathfrak{sp}(N')}\}_1[\phi_{\mathfrak{sp}(N')}]\right), \\ \{\phi_{\mathfrak{sp}(N')}\}_1[\phi_{\mathfrak{sp}(N')}] &= -6(N+1)[\omega] - 2(N-2)[\phi_{\mathfrak{sp}(N')}], \\ \{\phi_{\mathfrak{sp}(N')}\}_2[\phi_{\mathfrak{sp}(N')}] &= 0, \\ \{\phi_{\mathfrak{sp}(N')}\}_3[\phi_{\mathfrak{sp}(N')}] &= \tfrac{3}{2}N(N+1)\Omega, \\ \{\phi_{\mathfrak{sp}(N')}\}_k[\phi_{\mathfrak{sp}(N')}] &= 0 \quad \text{for } k \geq 4. \end{aligned}$$

PROOF. First note that $4\{\phi_{\mathfrak{sp}(N')}\}_k[\phi_{\mathfrak{sp}(N')}] = \{\omega_I^2+\omega_J^2+\omega_K^2\}_k[\omega_I^2+\omega_J^2+\omega_K^2]$ for any $k \in \mathbb{Z}$. The formula for the case $k \geq 4$ follow by weight reason. From the first two sets of formulae in Proposition 19, we get, for $k = 0$, that

$$\begin{aligned} \{\omega_I^2\}_0[\omega_I^2] &= 4(N-1)D(-1)[\omega] + 2(N-2)D(-1)[\omega_I^2], \\ \{\omega_J^2\}_0[\omega_J^2] &= 4(N-1)D(-1)[\omega] + 2(N-2)D(-1)[\omega_J^2], \\ \{\omega_K^2\}_0[\omega_K^2] &= 4(N-1)D(-1)[\omega] + 2(N-2)D(-1)[\omega_K^2], \\ \{\omega_I^2\}_0[\omega_J^2] &= 8\omega_I(-1)[\omega_J\omega_K] + 2D(-1)\left([\omega_I^2]-[\omega_J^2]+2[\omega]\right), \\ \{\omega_J^2\}_0[\omega_K^2] &= 8\omega_J(-1)[\omega_K\omega_I] + 2D(-1)\left([\omega_J^2]-[\omega_K^2]+2[\omega]\right), \\ \{\omega_K^2\}_0[\omega_I^2] &= 8\omega_K(-1)[\omega_I\omega_J] + 2D(-1)\left([\omega_K^2]-[\omega_I^2]+2[\omega]\right), \\ \{\omega_J^2\}_0[\omega_I^2] &= -8\omega_J(-1)[\omega_K\omega_I] + 2D(-1)\left([\omega_J^2]-[\omega_I^2]+2[\omega]\right), \\ \{\omega_K^2\}_0[\omega_J^2] &= -8\omega_K(-1)[\omega_I\omega_J] + 2D(-1)\left([\omega_K^2]-[\omega_J^2]+2[\omega]\right), \\ \{\omega_I^2\}_0[\omega_K^2] &= -8\omega_I(-1)[\omega_J\omega_K] + 2D(-1)\left([\omega_I^2]-[\omega_K^2]+2[\omega]\right). \end{aligned}$$

From this we immediately obtain that

$$4\{\phi_{\mathfrak{sp}(N')}\}_0[\phi_{\mathfrak{sp}(N')}] = 12(N+1)D(-1)[\omega] + 4(N-2)D(-1)[\phi_{\mathfrak{sp}(N')}].$$

Note that those terms without $D(-1)$ cancel out completely and we are only left with terms which are $D(-1)$ images.

For $\{\phi_{\mathfrak{sp}(N')}\}_1[\phi_{\mathfrak{sp}(N')}]$, from Proposition 19 we have

$$\{\omega_I^2\}_1[\omega_I^2] = -8(N-1)[\omega] - 4(N-2)[\omega_I^2],$$
$$\{\omega_J^2\}_1[\omega_J^2] = -8(N-1)[\omega] - 4(N-2)[\omega_J^2],$$
$$\{\omega_K^2\}_1[\omega_K^2] = -8(N-1)[\omega] - 4(N-2)[\omega_K^2],$$
$$\{\omega_I^2\}_1[\omega_J^2] = -4\left([\omega_I^2] + [\omega_J^2] - 2[\omega_K^2] + 2[\omega]\right),$$
$$\{\omega_J^2\}_1[\omega_K^2] = -4\left([\omega_J^2] + [\omega_K^2] - 2[\omega_I^2] + 2[\omega]\right),$$
$$\{\omega_K^2\}_1[\omega_I^2] = -4\left([\omega_K^2] + [\omega_I^2] - 2[\omega_J^2] + 2[\omega]\right),$$
$$\{\omega_J^2\}_1[\omega_I^2] = -4\left([\omega_I^2] + [\omega_J^2] - 2[\omega_K^2] + 2[\omega]\right),$$
$$\{\omega_K^2\}_1[\omega_J^2] = -4\left([\omega_J^2] + [\omega_K^2] - 2[\omega_I^2] + 2[\omega]\right),$$
$$\{\omega_I^2\}_1[\omega_K^2] = -4\left([\omega_K^2] + [\omega_I^2] - 2[\omega_J^2] + 2[\omega]\right).$$

From these, we immediately have

$$\begin{aligned}
4\{\phi_{\mathfrak{sp}(N')}\}_1[\phi_{\mathfrak{sp}(N')}] &= \{\omega_I^2 + \omega_J^2 + \omega_K^2\}_1[\omega_I^2 + \omega_J^2 + \omega_K^2] \\
&= -24(N+1)[\omega] - 8(N-2)[\phi_{\mathfrak{sp}(N')}].
\end{aligned}$$

Other identities follow in a similar fashion. $\square$

Next, we calculate $\{\omega_I^2 - \omega_J^2\}_*[\omega_I^2 - \omega_J^2]$. This calculation will be useful later.

COROLLARY 21. *For $k \geq 0$, $\{\omega_I^2 - \omega_J^2\}_k[\omega_I^2 - \omega_J^2]$ is given by*

$$\begin{aligned}
\{\omega_I^2 - \omega_J^2\}_0[\omega_I^2 - \omega_J^2] &= 8(N-2)D(-1)[\omega] + 2(N-4)D(-1)\left([\omega_I^2] + [\omega_J^2]\right) \\
&\quad + 8D(-1)[\omega_K^2] \\
&= -\tfrac{1}{2}D(-1)\left(\{\omega_I^2 - \omega_J^2\}_1[\omega_I^2 - \omega_J^2]\right),
\end{aligned}$$
$$\{\omega_I^2 - \omega_J^2\}_1[\omega_I^2 - \omega_J^2] = -16(N-2)[\omega] - 4(N-4)\left([\omega_I^2] + [\omega_J^2]\right) - 16[\omega_K^2],$$
$$\{\omega_I^2 - \omega_J^2\}_2[\omega_I^2 - \omega_J^2] = 0, \qquad \{\omega_I^2 - \omega_J^2\}_3[\omega_I^2 - \omega_J^2] = 4N(N-2)\Omega,$$
$$\{\omega_I^2 - \omega_J^2\}_k[\omega_I^2 - \omega_J^2] = 0 \quad \text{for} \quad k \geq 4.$$

Results of calculations for $\{\omega_J^2 - \omega_K^2\}_k[\omega_J^2 - \omega_K^2]$, $\{\omega_K^2 - \omega_I^2\}_k[\omega_K^2 - \omega_I^2]$ for $k \geq 0$ can be obtained by cyclically permuting (I, J, K) in the above formulae.

PROOF. From Proposition 19, we have

$$\{\omega_I^2\}_0[\omega_I^2] = 4(N-1)D(-1)[\omega] + 2(N-2)D(-1)[\omega_I^2],$$
$$\{\omega_J^2\}_0[\omega_J^2] = 4(N-1)D(-1)[\omega] + 2(N-2)D(-1)[\omega_J^2],$$
$$\{\omega_I^2\}_0[\omega_J^2] = 8\omega_I(-1)[\omega_J\omega_K] + 2D(-1)\left([\omega_I^2] - [\omega_J^2] + 2[\omega]\right),$$
$$\{\omega_J^2\}_0[\omega_I^2] = -8\omega_J(-1)[\omega_K\omega_I] + 2D(-1)\left([\omega_J^2] - [\omega_I^2] + 2[\omega]\right).$$

These formulae immediately give us

$$\{\omega_I^2 - \omega_J^2\}_0[\omega_I^2 - \omega_J^2] = 8(N-2)D(-1)[\omega] + 2(N-2)D(-1)\left([\omega_I^2] + [\omega_J^2]\right)$$
$$- 8\omega_I(-1)[\omega_J\omega_K] + 8\omega_J(-1)[\omega_K\omega_I].$$

Now, Lemma 17 expresses the last two terms as an image of $D(-1)$ operator. Other identities follow directly from Proposition 19. $\square$

Comparing the calculation in Corollary 21 with the calculation in Proposition 19 of $\{\omega_I\omega_J\}_*[\omega_I\omega_J]$, we see that these are almost the same. From this observation, we obtain the next corollary using the commutator formula for vertex operators. Note that this result is already known to us as Proposition 4 of §3.10.

COROLLARY 22. *For any* $m, n \in \mathbb{Z}$, *we have*

$$[\{\omega_I^2 - \omega_J^2\}_{m+1}, \{\omega_I^2 - \omega_J^2\}_{n+1}] = 4[\{\omega_I\omega_J\}_{m+1}, \{\omega_I\omega_J\}_{n+1}],$$
$$[\{\omega_J^2 - \omega_K^2\}_{m+1}, \{\omega_J^2 - \omega_K^2\}_{n+1}] = 4[\{\omega_J\omega_K\}_{m+1}, \{\omega_J\omega_K\}_{n+1}],$$
$$[\{\omega_K^2 - \omega_I^2\}_{m+1}, \{\omega_K^2 - \omega_I^2\}_{n+1}] = 4[\{\omega_K\omega_I\}_{m+1}, \{\omega_K\omega_I\}_{n+1}].$$

PROOF. For $k \geq 0$, we have

$$\{\omega_I^2 - \omega_J^2\}_k[\omega_I^2 - \omega_J^2] = 4\{\omega_I\omega_J\}_k[\omega_I\omega_J],$$
$$\{\omega_J^2 - \omega_K^2\}_k[\omega_J^2 - \omega_K^2] = 4\{\omega_J\omega_K\}_k[\omega_J\omega_K],$$
$$\{\omega_K^2 - \omega_I^2\}_k[\omega_K^2 - \omega_I^2] = 4\{\omega_K\omega_I\}_k[\omega_K\omega_I].$$

The commutator formula for vertex operators implies the above result. $\square$

As we can see from Proposition 19, Corollary 20, and Corollary 21, equations

$$\{v\}_0[v] = -\tfrac{1}{2}D(-1)\left(\{v\}_1[v]\right),$$
$$\{v\}_2[v] = 0,$$

hold for $v = \omega_I^2, \omega_J^2, \omega_K^2, \omega_I\omega_J, \omega_J\omega_K, \omega_K\omega_I, \phi_{\mathfrak{sp}(N')}, \omega_I^2 - \omega_J^2, \omega_J^2 - \omega_K^2, \omega_K^2 - \omega_I^2$, in the vector space $S^2(W)$, where $W = \mathbb{C}[\omega_I] \oplus \mathbb{C}[\omega_J] \oplus \mathbb{C}[\omega_K]$. Actually it turns out the above quadratic identities hold for any vector v in $S^2(W)$. To see this, we need to do some more calculations using Proposition 19.

Calculations of $\{\omega_I^2\}_*[\omega_J^2] + \{\omega_J^2\}_*[\omega_I^2]$. Lemma 17 implies that

$$\{\omega_I^2\}_0[\omega_J^2] + \{\omega_J^2\}_0[\omega_I^2] = 8\omega_I(-1)[\omega_J\omega_K] + 2D(-1)\left([\omega_I^2] - [\omega_J^2] + 2[\omega]\right)$$
$$- 8\omega_J(-1)[\omega_K\omega_I] + 2D(-1)\left([\omega_J^2] - [\omega_I^2] + 2[\omega]\right)$$
$$= 4D(-1)\left([\omega_I^2] + [\omega_J^2] - 2[\omega_K^2] + 2[\omega]\right),$$
$$\{\omega_I^2\}_1[\omega_J^2] + \{\omega_J^2\}_1[\omega_I^2] = -8\left([\omega_I^2] + [\omega_J^2] - 2[\omega_K^2] + 2[\omega]\right),$$
$$\{\omega_I^2\}_2[\omega_J^2] + \{\omega_J^2\}_2[\omega_I^2] = 0,$$
$$\{\omega_I^2\}_3[\omega_J^2] + \{\omega_J^2\}_3[\omega_I^2] = 4N\Omega.$$

Calculation of $\{\omega_I^2\}_*[\omega_I\omega_J] + \{\omega_I\omega_J\}_*[\omega_I^2]$. **Lemma 16 gives**

$$\{\omega_I^2\}_0[\omega_I\omega_J] + \{\omega_I\omega_J\}_0[\omega_I^2] = 4\omega_I(-1)[\omega_K\omega_I] - 2(N-2)\omega_I(-2)[\omega_J]$$
$$- 2D(-1)[\omega_I\omega_J] - 4\omega_I(-1)[\omega_K\omega_I] - 2\omega_I(-2)[\omega_J]$$
$$- 2(N-1)\omega_J(-2)[\omega_I]$$
$$= -2(N-1)\left\{\omega_I(-2)[\omega_J] + \omega_J(-2)[\omega_I]\right\}$$
$$- D(-1)[\omega_I\omega_J]$$
$$= 2(N-1)D(-1)[\omega_I\omega_J] - 2D(-1)[\omega_I\omega_J]$$
$$= 2(N-2)D(-1)[\omega_I\omega_J],$$
$$\{\omega_I^2\}_1[\omega_I\omega_J] + \{\omega_I\omega_J\}_1[\omega_I^2] = -2(N-2)[\omega_I\omega_J] + 2(N-2)D(-1)[\omega_K]$$
$$- 2(N-2)[\omega_I\omega_J] - 2(N-2)D(-1)[\omega_K]$$
$$= -4(N-2)[\omega_I\omega_J],$$
$$\{\omega_I^2\}_2[\omega_I\omega_J] + \{\omega_I\omega_J\}_2[\omega_I^2] = -4(N-2)[\omega_K] + 4(N-2)[\omega_K] = 0,$$
$$\{\omega_I^2\}_3[\omega_I\omega_J] + \{\omega_I\omega_J\}_3[\omega_I^2] = 0.$$

Calculation of $\{\omega_I^2\}_*[\omega_J\omega_K] + \{\omega_J\omega_K\}_*[\omega_I^2]$. **Lemma 18 yields**

$$\{\omega_I^2\}_0[\omega_J\omega_K] + \{\omega_J\omega_K\}_0[\omega_I^2] = 4\omega_I(-1)\left([\omega_K^2] - [\omega_J^2]\right) - 2D(-1)[\omega_J\omega_K]$$
$$4\omega_J(-1)[\omega_I\omega_J] - 4\omega_K(-1)[\omega_K\omega_I] + 2D(-1)[\omega_J\omega_K]$$
$$= 12D(-1)[\omega_J\omega_K],$$
$$\{\omega_I^2\}_1[\omega_J\omega_K] + \{\omega_J\omega_K\}_1[\omega_I^2] = -24[\omega_J\omega_K],$$
$$\{\omega_I^2\}_2[\omega_J\omega_K] + \{\omega_J\omega_K\}_2[\omega_I^2] = 0,$$
$$\{\omega_I^2\}_3[\omega_J\omega_K] + \{\omega_J\omega_K\}_3[\omega_I^2] = 0.$$

Calculation of $\{\omega_I\omega_J\}_*[\omega_K\omega_I] + \{\omega_K\omega_I\}_*[\omega_I\omega_J]$. **Again Lemma 18 gives**

$$\{\omega_I\omega_J\}_0[\omega_K\omega_I] + \{\omega_K\omega_I\}_0[\omega_I\omega_J] = 2\omega_I(-1)\left([\omega_I^2] - [\omega_K^2]\right) - 2\omega_J(-1)[\omega_I\omega_J]$$
$$- (N-2)\omega_J(-2)[\omega_K] - 2\omega_I(-1)\left([\omega_I^2] - [\omega_J^2]\right)$$
$$2\omega_K(-1)[\omega_K\omega_I] - (N-2)\omega_K(-2)[\omega_J]$$
$$= 2\omega_I(-1)\left([\omega_I^2] - [\omega_K^2]\right) - 2\omega_J(-1)[\omega_I\omega_J]$$
$$+ 2\omega_K(-1)[\omega_K\omega_I] + (N-2)D(-1)[\omega_J\omega_K]$$
$$= -6D(-1)[\omega_J\omega_K] + (N-2)D(-1)[\omega_J\omega_K]$$
$$= (N-8)D(-1)[\omega_J\omega_K],$$
$$\{\omega_I\omega_J\}_1[\omega_K\omega_I] + \{\omega_K\omega_I\}_1[\omega_I\omega_J] = -(N-8)[\omega_J\omega_K] - (N-2)D(-1)[\omega_I]$$
$$- (N-8)[\omega_J\omega_K] + (N-2)D(-1)[\omega_I]$$
$$= -2(N-8)[\omega_J\omega_K],$$
$$\{\omega_I\omega_J\}_2[\omega_K\omega_I] + \{\omega_K\omega_I\}_2[\omega_I\omega_J] = 0,$$
$$\{\omega_I\omega_J\}_3[\omega_K\omega_I] + \{\omega_K\omega_I\}_3[\omega_I\omega_J] = 0.$$

By replacing (I, J, K) by $(-I, K, J)$, $(K, -J, I)$, or $(J, I, -K)$, or performing cyclic permutation of (I, J, K), or doing both of these operations, we obtain all the other calculations of $\{\omega_{\mathcal{J}_1}\omega_{\mathcal{J}_2}\}_k[\omega_{\mathcal{J}_3}\omega_{\mathcal{J}_4}] + \{\omega_{\mathcal{J}_3}\omega_{\mathcal{J}_4}\}_k[\omega_{\mathcal{J}_1}\omega_{\mathcal{J}_2}]$ for any complex structures $\mathcal{J}_i \in \{I, J, K\}$ for $1 \leq i \leq 4$, and for $k \geq 0$.

LEMMA 23. *For any complex structures $\mathcal{J}_i \in \{I, J, K\}$ for $1 \leq i \leq 4$ on the quaternionic Hermitian vector space $E^{4N'}$, we have*

$$\{\omega_{\mathcal{J}_1}\omega_{\mathcal{J}_2}\}_0[\omega_{\mathcal{J}_3}\omega_{\mathcal{J}_4}] + \{\omega_{\mathcal{J}_3}\omega_{\mathcal{J}_4}\}_0[\omega_{\mathcal{J}_1}\omega_{\mathcal{J}_2}]$$
$$= -\tfrac{1}{2}D(-1)\left(\{\omega_{\mathcal{J}_1}\omega_{\mathcal{J}_2}\}_1[\omega_{\mathcal{J}_3}\omega_{\mathcal{J}_4}] + \{\omega_{\mathcal{J}_3}\omega_{\mathcal{J}_4}\}_1[\omega_{\mathcal{J}_1}\omega_{\mathcal{J}_2}]\right),$$

Furthermore, we have

$$\{\omega_{\mathcal{J}_1}\omega_{\mathcal{J}_2}\}_2[\omega_{\mathcal{J}_3}\omega_{\mathcal{J}_4}] + \{\omega_{\mathcal{J}_3}\omega_{\mathcal{J}_4}\}_2[\omega_{\mathcal{J}_1}\omega_{\mathcal{J}_2}] = 0.$$

PROOF. From calculations above and Proposition 19, we see that for all possible choices of complex structures $\mathcal{J}_1, \mathcal{J}_2, \mathcal{J}_3, \mathcal{J}_4$ from $\{I, J, K\}$, the above identities hold.   □

Now, we are ready to state our Theorem.

THEOREM 24. *Let $v \in S^2\left(\mathbb{C}[\omega_I] \oplus \mathbb{C}[\omega_J] \oplus \mathbb{C}[\omega_K]\right)$ be any quadratic Kähler form on the quaternionic Hermitian vector space $E^{4N'} \otimes \mathbb{C}$. Then*

$$\{v\}_0[v] = -\tfrac{1}{2}D(-1)\left(\{v\}_1[v]\right),$$
$$\{v\}_1[v] \in S^2\left(\mathbb{C}[\omega_I] \oplus \mathbb{C}[\omega_J] \oplus \mathbb{C}[\omega_K]\right) \oplus \mathbb{C}[\omega],$$
$$\{v\}_2[v] = 0,$$
$$\{v\}_3[v] \in \mathbb{C}\Omega.$$

More specifically, if we let

$$v = a[\omega_I^2] + b[\omega_J^2] + c[\omega_K^2] + d[\omega_I\omega_J] + e[\omega_J\omega_K] + f[\omega_K\omega_I]$$

for $a, b, c, d, e, f \in \mathbb{C}$. Then, we have

$$
\begin{aligned}
\{v\}_1[v] = &-4\left\{2(a+b+c)^2 + 2(N-2)(a^2+b^2+c^2)\right.\\
&\left.+(N-2)(d^2+e^2+f^2)\right\}[\omega]\\
&-\left(4(N-2)a^2 + (N-4)d^2 + 4e^2 + (N-4)f^2 + 8ab - 16bc + 8ca\right)[\omega_I^2]\\
&-\left(4(N-2)b^2 + (N-4)d^2 + (N-4)e^2 + 4f^2 + 8ab + 8bc - 16ca\right)[\omega_J^2]\\
&-\left(4(N-2)c^2 + 4d^2 + (N-4)e^2 + (N-4)f^2 - 16ab + 8bc + 8ca\right)[\omega_K^2]\\
&-\left(2(N-8)ef + 4(N-2)ad + 24cd + 4(N-2)bd\right)[\omega_I\omega_J]\\
&-\left(2(N-8)df + 4(N-2)be + 24ae + 4(N-2)ce\right)[\omega_J\omega_K]\\
&-\left(2(N-8)de + 4(N-2)cf + 24bf + 4(N-2)af\right)[\omega_K\omega_I],
\end{aligned}
$$
$$
\begin{aligned}
\{v\}_3[v] = &\left\{2N(N-1)(a^2+b^2+c^2) + 4N(ab+bc+ca)\right.\\
&\left.+N(N-2)(d^2+e^2+f^2)\right\}\Omega.
\end{aligned}
$$

PROOF. The formulae for $\{v\}_0[v]$ and for $\{v\}_2[v]$ follow from Lemma 23. Other formulae follow by direct calculations using most of our previous calculations.   □

COROLLARY 25. *Let* $v \in S^2 (\mathbb{C}[\omega_I] \oplus \mathbb{C}[\omega_J] \oplus \mathbb{C}[\omega_K])$. *Then associated vertex operators* $\{v\}_*$ *satisfy following commutation relations:*

$$[\{v\}_{m+1}, \{v\}_{n+1}] = \frac{(m-n)}{2} \{\{v\}_1[v]\}_{m+n+1} + \binom{m+1}{3} (\{v\}_3[v], \Omega) \delta_{m+n,0} \cdot \mathrm{Id}.$$

Here, the pairing $(,)$ *on* V *is such that* $(\Omega, \Omega) = 1$, *and explicit form of the vector* $\{v\}_1[v]$ *and* $\{v\}_3[v]$ *is given in Theorem 24.*

PROOF. From the commutator formula,

$$[\{v\}_{m+1}, \{v\}_{n+1}] = \sum_{0 \leq k \in \mathbb{Z}} \binom{m+1}{k} \{\{v\}_k[v]\}_{m+n+2-k}$$

$$= \{\{v\}_0[v]\}_{m+n+2} + (m+1) \{\{v\}_1[v]\}_{m+n+1} + \binom{m+1}{3} \{\{v\}_3[v]\}_{m+n-1}$$

$$= \frac{(m-n)}{2} \{\{v\}_1[v]\}_{m+n+1} + \binom{m+1}{3} (\{v\}_3[v], \Omega) \delta_{m+n,0} \cdot \mathrm{Id}.$$

Here, we used formulae $\{D(-1)u\}_{n+1} = (n+1)\{u\}_n$ for $u \in V$, and $\{\Omega\}_{n-1} = \delta_{n,0} \cdot \mathrm{Id}$ for $n \in \mathbb{Z}$. □

The formula in Corollary 25 deals with commutators of vertex operators associated to a single quadratic Kähler form. Next, we calculate commutators between vertex operators associated to Kähler forms and vertex operators associated to quadratic Kähler forms.

PROPOSITION 26. *For any* $m, n \in \mathbb{Z}$, *we have*

$$[\{\omega_I\}_m, \{\omega_I^2\}_{n+1}] = -2m(N-1)\{\omega_I\}_{m+n},$$

$$[\{\omega_I\}_m, \{\omega_J^2\}_{n+1}] = 4\{\omega_J \omega_K\}_{m+n+1} - 2m\{\omega_I\}_{m+n},$$

$$[\{\omega_I\}_m, \{\omega_K^2\}_{n+1}] = -4\{\omega_J \omega_K\}_{m+n+1} - 2m\{\omega_I\}_{m+n},$$

$$[\{\omega_I\}_m, \{\omega_K \omega_I\}_{n+1}] = -2\{\omega_I \omega_J\}_{m+n+1} - m(N-2)\{\omega_K\}_{m+n},$$

$$[\{\omega_I\}_m, \{\omega_I \omega_J\}_{n+1}] = 2\{\omega_K \omega_I\}_{m+n+1} - m(N-2)\{\omega_J\}_{m+n},$$

$$[\{\omega_I\}_m, \{\omega_J \omega_K\}_{n+1}] = 2\{\omega_K^2\}_{m+n+1} - 2\{\omega_J^2\}_{m+n+1}.$$

Commutation relations of quadratic Kähler operators with other Kähler operators $\{\omega_J\}_m$, $\{\omega_K\}_m$, $m \in \mathbb{Z}$, *can be obtained by cyclically permuting* (I, J, K) *in the above formulae.*

PROOF. From Proposition 12, we see that for any isometric complex structure $\mathcal{J}_1, \mathcal{J}_2, \mathcal{J}_3 \in \{I, J, K\}$ and for $k \geq 2$, we have $\{\omega_{\mathcal{J}_1}\}_k[\omega_{\mathcal{J}_2} \omega_{\mathcal{J}_3}] = 0$. So the commutator formula for vertex operators implies that

$$[\{\omega_{\mathcal{J}_1}\}_m, \{\omega_{\mathcal{J}_2} \omega_{\mathcal{J}_3}\}_{n+1}] = \sum_{0 \leq k \in \mathbb{Z}} \binom{m}{k} \{\{\omega_{\mathcal{J}_1}\}_k[\omega_{\mathcal{J}_2} \omega_{\mathcal{J}_3}]\}_{m+n+1-k}$$

$$= \{\{\omega_{\mathcal{J}_1}\}_0[\omega_{\mathcal{J}_2} \omega_{\mathcal{J}_3}]\}_{m+n+1} + m \{\{\omega_{\mathcal{J}_1}\}_1[\omega_{\mathcal{J}_2} \omega_{\mathcal{J}_3}]\}_{m+n}.$$

Now, Proposition 12 gives above formulae. □

§3.12   Determination of *G*-invariant vectors of low weight in $(V)_*$

In §3.4, we described some major algebra generators of the algebra of *G*-invariants V^G in the vertex operator super algebra V. Here, we precisely determine a part of the algebra V^G of weight up to 2 for groups $G = O(2N)$, $SO(2N)$, $U(N)$, $SU(N)$, $Sp(N')$, and $Sp(N') \cdot Sp(1)$. We will explicitly see that we don't see any generators other than those described in §3.4 up to weight 2.

Let E^{2N} be a $2N$-dimensional Euclidean vector space with an inner product $\langle\ ,\ \rangle$. Let $A = E \otimes \mathbb{C}$ be its complexification. We recall that our vertex operator super algebra V and its low weight part is given by

$$(1) \qquad V = \bigotimes_{n \geq 0} \wedge^*_{q^{n+\frac{1}{2}}} A(-n - \tfrac{1}{2}) \cdot \Omega.$$

$$(2) \qquad \begin{aligned} (V)_0 &= \mathbb{C}\Omega, \qquad (V)_{\frac{1}{2}} = A(-\tfrac{1}{2}), \qquad (V)_1 = \wedge^2 A(-\tfrac{1}{2}), \\ (V)_{\frac{3}{2}} &= \wedge^3 A(-\tfrac{1}{2}) \oplus A(-\tfrac{3}{2}), \\ (V)_2 &= \wedge^4 A(-\tfrac{1}{2}) \oplus \left(A(-\tfrac{3}{2}) \otimes A(-\tfrac{1}{2})\right). \end{aligned}$$

Since we will deal with Kähler manifolds M^{2N} in Chapter V, we also write down decompositions of low weight homogeneous parts of V under the unitary group $U(N)$. For this, we assume that E has an isometric complex structure I. The complexification A splits into I-eigenspaces $A = A^+ \oplus A^-$ where I acts by multiplication by i and $-i$ on A^+ and on A^-, respectively. Since I acts by isometry, we can introduce a Hermitian pairing on E^{2N} [§3.9 Lemma 1] and the group preserving the Hermitian pairing is $U(N)$ after fixing a Hermitian orthonormal basis of E. Note that the standard $U(N)$-representation A decomposes as $A = A^+ \oplus A^-$ where $\overline{A^+} = A^-$. In terms of A^+ and A^-, we have the following decomposition of $(V)_*$ for $* \leq 2$ under the unitary group $U(N)$ (here we mostly omitted the vacuum vector Ω from our notation for simplicity):

$$(3)$$
$$(V)_0 = \mathbb{C}\Omega, \qquad (V)_{\frac{1}{2}} = A^+(-\tfrac{1}{2}) \oplus A^-(-\tfrac{1}{2}),$$
$$(V)_1 = \wedge^2 A^+(-\tfrac{1}{2}) \oplus \left(A^+(-\tfrac{1}{2}) \otimes A^-(-\tfrac{1}{2})\right) \oplus \wedge^2 A^-(-\tfrac{1}{2}),$$
$$(V)_{\frac{3}{2}} = \wedge^3 A^+(-\tfrac{1}{2}) \oplus \left(\wedge^2 A^+(-\tfrac{1}{2}) \otimes A^-(-\tfrac{1}{2})\right) \oplus \left(A^+(-\tfrac{1}{2}) \otimes \wedge^2 A^-(-\tfrac{1}{2})\right)$$
$$\oplus \wedge^3 A^-(-\tfrac{1}{2}) \oplus A^+(-\tfrac{3}{2}) \oplus A^-(-\tfrac{3}{2}),$$
$$(V)_2 = \left(A^+(-\tfrac{3}{2}) \otimes A^+(-\tfrac{1}{2})\right) \oplus \left(A^-(-\tfrac{3}{2}) \otimes A^-(-\tfrac{1}{2})\right) \oplus \left(A^+(-\tfrac{3}{2}) \otimes A^-(-\tfrac{1}{2})\right)$$
$$\oplus \left(A^-(-\tfrac{3}{2}) \otimes A^+(-\tfrac{1}{2})\right) \oplus \wedge^4 A^+(-\tfrac{1}{2}) \oplus \left(\wedge^3 A^+(-\tfrac{1}{2}) \otimes A^-(-\tfrac{1}{2})\right)$$
$$\left(\wedge^2 A^+(-\tfrac{1}{2}) \otimes \wedge^2 A^-(-\tfrac{1}{2})\right) \oplus \left(A^+(-\tfrac{1}{2}) \otimes \wedge^3 A^-(-\tfrac{1}{2})\right) \oplus \wedge^4 A^-(-\tfrac{1}{2}).$$

We use the same basis as before for A^+ and for A^-:

$$(4) \qquad A^+ = \bigoplus_{j=1}^{N} \mathbb{C}a_j, \qquad A^- = \bigoplus_{j=1}^{N} \mathbb{C}a_j^*.$$

See §3.9 (1) for details. We recall that we defined an exterior algebra $\mathcal{A}$ by

$$(5) \qquad \mathcal{A} = \bigotimes_{n \geq 0} \textstyle\bigwedge^* A(-n - \tfrac{1}{2}) \subset \mathrm{Cliff}\,(\mathbb{Z} + \tfrac{1}{2}).$$

As a vector space, $\mathcal{A}$ is isomorphic to V via $[\ ] : \mathcal{A} \to V$ given by $[\alpha] = \alpha \cdot \Omega$ for $\alpha \in \mathcal{A}$, where Ω is the vacuum vector. We also recall definitions of important elements in $\mathcal{A}$. These are as follows:

$$\omega = -\tfrac{1}{2} \sum_{j=1}^{N} \left\{ a_j(-\tfrac{3}{2}) a_j^*(-\tfrac{1}{2}) + a_j^*(-\tfrac{3}{2}) a_j(-\tfrac{1}{2}) \right\} \in \mathcal{A}_2,$$

$$\omega_+ = -\sum_{j=1}^{N} a_j(-\tfrac{3}{2}) a_j^*(-\tfrac{1}{2}) \in \mathcal{A}_2, \qquad \omega_- = -\sum_{j=1}^{N} a_j^*(-\tfrac{3}{2}) a_j(-\tfrac{1}{2}) \in \mathcal{A}_2,$$

$$\chi = -\sum_{j=1}^{N'} a_j(-\tfrac{3}{2}) a_{j+N'}(-\tfrac{1}{2}) + \sum_{j=1}^{N'} a_{j+N'}(-\tfrac{3}{2}) a_j(-\tfrac{1}{2}) \in \mathcal{A}_2,$$

$$(6) \qquad \chi^* = -\sum_{j=1}^{N'} a_j^*(-\tfrac{3}{2}) a_{j+N'}^*(-\tfrac{1}{2}) + \sum_{j=1}^{N'} a_{j+N'}^*(-\tfrac{3}{2}) a_j^*(-\tfrac{1}{2}) \in \mathcal{A}_2,$$

$$h = \sum_{j=1}^{N} a_j(-\tfrac{1}{2}) a_j^*(-\tfrac{1}{2}) \in \mathcal{A}_1,$$

$$x = \sum_{j=1}^{N'} a_j(-\tfrac{1}{2}) a_{j+N'}(-\tfrac{1}{2}) \in \mathcal{A}_1, \qquad x^* = \sum_{j=1}^{N'} a_j^*(-\tfrac{1}{2}) a_{j+N'}^*(-\tfrac{1}{2}) \in \mathcal{A}_1,$$

$$\rho(-\tfrac{1}{2}) = a_1(-\tfrac{1}{2}) a_1^*(-\tfrac{1}{2}) \cdots a_N(-\tfrac{1}{2}) a_N^*(-\tfrac{1}{2}) \in \mathcal{A}_N,$$

$$\rho^+ = a_1(-\tfrac{1}{2}) a_2(-\tfrac{1}{2}) \cdots a_N(-\tfrac{1}{2}) \in \mathcal{A}_{\frac{N}{2}}, \qquad \rho^- = a_1^*(-\tfrac{1}{2}) a_2^*(-\tfrac{1}{2}) \cdots a_N^*(-\tfrac{1}{2}) \in \mathcal{A}_{\frac{N}{2}}.$$

Invariant vectors $(V)_*^{\mathrm{O}(2N)}$ **of low weight.** Since E^{2N} is a Euclidean vector space, the orthogonal group $\mathrm{O}(2N)$ acts on it once we choose an orthonormal basis of E. The invariant subspace $V_*^{\mathrm{O}(2N)}$ of low weight is given as follows:

PROPOSITION 1 (INVARIANT VECTORS $(V)_*^{\mathrm{O}(2N)}$ OF LOW WEIGHT). *Let $G = \mathrm{O}(2N)$ with $N \geq 2$, or $\mathrm{SO}(2N)$ with $N \geq 3$. Then*

$$(V)_0^G = \mathbb{C}\Omega, \qquad (V)_{\frac{k}{2}}^G = \{0\}, \quad 1 \leq k \leq 3, \qquad (V)_2^G = \mathbb{C}\omega.$$

If $G = \mathrm{SO}(4)$, then we have

$$(V)_0^G = \mathbb{C}\Omega, \qquad (V)_{\frac{k}{2}}^G = \{0\}, \quad 1 \leq k \leq 3, \qquad (V)_2^G = \mathbb{C}\rho(-\tfrac{1}{2}) \oplus \mathbb{C}\omega.$$

PROOF. We only have to note that when $G = \mathrm{O}(4)$, the volume element $\rho(-\tfrac{1}{2}) \in \bigwedge^4 A(-\tfrac{1}{2})$ isn't invariant since it changes its sign for elements with $\det = -1$. The rest is straightforward using the fact that for the group $\mathrm{SO}(2N)$, representations $A, \bigwedge^2 A, \ldots, \bigwedge^{N-1} A$ are all irreducible representations and $\bigwedge^N A$ is a direct sum of non-isomorphic irreducible representations [BD, p274]. □

Invariant vectors $(V)_*^{U(N)}$, $(V)_*^{SU(N)}$ **of low weight.** Under the action of the unitary group $U(N)$, the vector space A splits into a direct sum of irreducible representations $A^+ \oplus A^-$ each of which has complex dimension N, and they are conjugate to each other, $\overline{A^+} = A^-$. Representations $\mathbb{C}$, A^+, $\bigwedge^2 A^+$, ..., $\bigwedge^N A^+$ are all distinct irreducible representations of $U(N)$ [BD]. For the calculation of $U(N)$-invariants of their tensor products, a basic fact is

$$(7) \qquad (A^+ \otimes A^-)^{U(N)} = \mathbb{C} \cdot h, \quad \text{where} \quad h = \sum_{j=1}^{N} a_j \otimes a_j^*.$$

In the exterior algebra $\bigwedge^* A$, we can consider powers of h. It turns out that these exhaust all the $U(N)$-invariants of $\bigwedge^* A \otimes \bigwedge^* A$, as is shown in the next lemma.

LEMMA 2. *Let* $0 \le i, j \le N$. *Then*

$$\left(\textstyle\bigwedge^i A^+ \otimes \bigwedge^j A^-\right)^{U(N)} \cong \mathrm{Hom}_{U(N)}\left(\textstyle\bigwedge^i A^-, \bigwedge^j A^-\right) \cong \begin{cases} \mathbb{C} h^i, & \text{if } i = j \\ \{0\}, & \text{otherwise.} \end{cases}$$

Here $h^{N+1} = 0$.

PROOF. Since $\bigwedge^i A^{\pm}$ for $0 \le i \le N$ are non-isomorphic irreducible $U(N)$-representations, by Shur's lemma, $\mathrm{Hom}_{U(N)}(\bigwedge^i A^-, \bigwedge^i A^-)$ is nontrivial only when $i = j$, and in this case it is 1-dimensional spanned by the identity map. Converting constant multiples of this identity map to elements in the tensor product, we see that the corresponding elements are spanned by

$$\sum_{1 \le k_1 < \cdots < k_i \le N} a_{k_1} a_{k_2} \cdots a_{k_i} a_{k_1}^* a_{k_2}^* \cdots a_{k_i}^* \in \textstyle\bigwedge^i A^+ \otimes \bigwedge^i A^-.$$

On the other hand, $h^i \in \bigwedge^i A^+ \otimes \bigwedge^i A^-$ is a constant multiple of the above element. Hence the $U(N)$-invariant subspace in $\bigwedge^i A^+ \otimes \bigwedge^i A^-$ is spanned over $\mathbb{C}$ by h^i. □

Lemma 2 facilitates the calculation of $U(N)$-invariant vectors of low weight in V. Recall that for an element $v \in \bigotimes_{n \ge 0} \bigwedge^* A(-n - \frac{1}{2})$, the corresponding vector in the vertex operator super algebra V is denoted by $[v] = v \cdot \Omega \in V$.

PROPOSITION 3 (INVARIANTS $V_*^{U(N)}$ OF LOW WEIGHT). $U(N)$-*invariant vectors in* V *of weight up to 2 is described as follows:*

$$(V)_0^{U(N)} = \mathbb{C}\Omega, \qquad\qquad (V)_{\frac{1}{2}}^{U(N)} = \{0\},$$

$$(V)_1^{U(N)} = \mathbb{C}[h], \qquad\qquad (V)_{\frac{3}{2}}^{U(N)} = \{0\},$$

$$(V)_2^{U(N)} = \begin{cases} \mathbb{C}[\omega_+] \oplus \mathbb{C}[\omega_-], & N = 1 \\ \mathbb{C}[\omega_+] \oplus \mathbb{C}[\omega_-] \oplus \mathbb{C}[h^2], & N \ge 2. \end{cases}$$

Here, $h^2 = 0$ if $N = 1$.

PROOF. Since $\bigwedge^i A^\pm$ are irreducible U(N)-representations for $1 \leq i \leq N$, we have $(\bigwedge^i A^\pm)^{\mathrm{U}(N)} = \{0\}$ for $i \geq 1$. For invariants of tensor products, we only have to deal with tensor products of at most two representations since we are dealing with vectors of weight at most 2 (see (3)). For these representations, Lemma 2 and definitions of $\omega_\pm$ proves Proposition 3. $\square$

For the case of SU(N)-invariants, one notable difference is that the top exterior powers $\bigwedge^N A^\pm$ are trivial 1-dimensional representations for SU(N) generated by complex volume forms $\rho_\pm$:

$$(8) \qquad \rho_+ = a_1 a_2 \cdots a_N \in \bigwedge^N A^+, \qquad \rho_- = a_1^* a_2^* \cdots a_N^* \in \bigwedge^N A^-,$$

while representations $A^+, \bigwedge^2 A^+, \ldots, \bigwedge^{N-1} A^+$ are still non-isomorphic irreducible representations of SU(N). A lemma similar to Lemma 2 for the group SU(N) can be stated as follows.

LEMMA 4. *Let $1 \leq i, j \leq N$. Then SU(N)-invariants of tensor products of two irreducible representations are given as follows:*

$$\left(\bigwedge^i A^+ \otimes \bigwedge^j A^-\right)^{\mathrm{SU}(N)} \cong \mathrm{Hom}_{\mathrm{SU}(N)}\left(\bigwedge^i A^-, \bigwedge^j A^-\right) \cong \begin{cases} \mathbb{C}h^i, & \text{if } i = j \\ \mathbb{C}\rho_+, & \text{if } (i,j) = (N,0) \\ \mathbb{C}\rho_-, & \text{if } (i,j) = (0,N) \\ \{0\}, & \text{otherwise.} \end{cases}$$

PROOF. The proof is the same as before. Just note that the top exterior powers $\bigwedge^N A^\pm$ are trivial representations. Thus for $(i,j) = (0,N), (N,0)$, tensor products have invariants generated by $\rho_\pm$. $\square$

Using this lemma, we can write down the vector space of SU(N)-invariants of low weight. We note that the only difference from the unitary group case up to weight 2 is additions of complex volume forms $\rho_\pm$.

PROPOSITION 5 (INVARIANTS $V^{\mathrm{SU}(N)}$ OF LOW WEIGHT). *Let $N \geq 2$. Then SU(N)-invariants in V of weight up to 2 are given as follows:*

$$(V)_0^{\mathrm{SU}(N)} = \mathbb{C}\Omega, \qquad (V)_{1/2}^{\mathrm{SU}(N)} = \{0\},$$

$$(V)_1^{\mathrm{SU}(N)} = \begin{cases} \mathbb{C}[h], & \text{if } N \neq 2 \\ \mathbb{C}[h] \oplus \mathbb{C}[\rho_+] \oplus \mathbb{C}[\rho_-], & \text{if } N = 2, \end{cases}$$

$$(V)_{3/2}^{\mathrm{SU}(N)} = \begin{cases} \{0\}, & \text{if } N \neq 3 \\ \mathbb{C}[\rho_+] \oplus \mathbb{C}[\rho_-], & \text{if } N = 3, \end{cases}$$

$$(V)_2^{\mathrm{SU}(N)} = \begin{cases} \mathbb{C}[\omega_+] \oplus \mathbb{C}[\omega_-] \oplus \mathbb{C}[h^2], & \text{if } N \neq 4 \\ \mathbb{C}[\omega_+] \oplus \mathbb{C}[\omega_-] \oplus \mathbb{C}[h^2] \oplus \mathbb{C}[\rho_+] \oplus \mathbb{C}[\rho_-], & \text{if } N = 4. \end{cases}$$

PROOF. The proof is similar to the unitary group case using Lemma 4. $\square$

Invariant vectors $V_*^{\mathrm{Sp}(N')}$, $V_*^{\mathrm{Sp}(N')\mathrm{Sp}(1)}$ **of low weight.** We have seen that the orthogonal Lie algebra $\mathfrak{o}(4N') \otimes \mathbb{C}$ can be realized as a Lie subalgebra of the Clifford algebra Cliff A [§3.2 Lemma 3]:

$$(9) \qquad \mathfrak{o}(4N') \otimes \mathbb{C} = \bigoplus_{1 \leq j < k \leq 2N'} (\mathbb{C} : a_j a_k : \oplus \mathbb{C} : a_j^* a_k^* :) \oplus \bigoplus_{1 \leq j,k \leq 2N'} \mathbb{C} : a_j a_k^* : .$$

It contains the Lie subalgebra $\mathfrak{sp}(N') \otimes \mathbb{C}$ and its commutant in $\mathfrak{o}(4N')$, $\mathfrak{sp}(1) \otimes \mathbb{C} = \big(\mathfrak{o}(4N') \oplus \mathbb{C}\big)^{\mathfrak{sp}(N')}$ [§3.3 Lemma 1] described by

$$(10) \qquad \begin{aligned} \mathfrak{sp}(N') \otimes \mathbb{C} &= \bigoplus_{1 \leq r,s \leq N'} \mathbb{C}(:a_r a_s^*: + :a_{r+N'}^* a_{s+N'}:) \\ &\oplus \bigoplus_{1 \leq r \leq s \leq N'} \mathbb{C}(:a_r a_{s+N'}^*: + :a_s a_{r+N'}^*:) \oplus \bigoplus_{1 \leq r \leq s \leq N'} \mathbb{C}(:a_r^* a_{s+N'}: + :a_s^* a_{r+N'}:), \end{aligned}$$

$$\mathfrak{sp}(1) \otimes \mathbb{C} = \mathfrak{sl}_2(\mathbb{C}) = \mathbb{C}x \oplus \mathbb{C}x^* \oplus \mathbb{C}h.$$

In the description of $\mathfrak{sp}(1)$, elements x, x^*, h satisfy usual commutation relations $[h, x] = 2x$, $[h, x^*] = -2x^*$, $[x, x^*] = -h$, and they are given by

$$(11) \qquad x = \sum_{r=1}^{N'} :a_r a_{r+N'}:, \qquad x^* = \sum_{r=1}^{N'} :a_r^* a_{r+N'}^*:, \qquad h = \sum_{j=1}^{N} :a_j a_j^*: .$$

Note that after removing the normal ordering symbols, we have $x = \sum_r a_r a_{r+N'}$ and $x^* = \sum_r a_r^* a_{r+N'}^*$ in the Clifford algebra. Also note that Clifford multiplication by x on the Clifford module $\bigwedge^* A^+$ is actually the same as exterior multiplication. Similarly, Clifford multiplication by x^* on $\bigwedge^* A^-$ is the same as exterior multiplication. To consider the subalgebra of invariants in the exterior algebra $\bigwedge^* A$, we introduce the following elements in $\bigwedge^* A$:

$$(12)$$

$$\bar{x} = \sum_{r=1}^{N'} a_r a_{r+N'} \in \bigwedge^* A^+, \qquad \bar{x}^* = \sum_{r=1}^{N'} a_r^* a_{r+N'}^* \in \bigwedge^* A^-, \qquad \bar{h} = \sum_{j=1}^{N} a_j a_j^* \in \bigwedge^* A.$$

Note that elements x, x^*, h are in the Clifford algebra Cliff (A) rather than in the Clifford module $\bigwedge^* A$. Before, we have seen that these two Lie subalgebras $\mathfrak{sp}(N') \otimes \mathbb{C}$ and $\mathfrak{sp}(1) \otimes \mathbb{C}$ commute in $\mathfrak{o}(4N') \otimes \mathbb{C}$ [§3.3 (4)]. We also recall that the quaternionic structure map $J : A \to A$ in the complex vector space A is given by

$$(13) \quad J(a_r) = a_{r+N'}^*, \quad J(a_{r+N'}) = -a_r^*, \quad J(a_r^*) = a_{r+N'}, \quad J(a_{r+N'}^*) = -a_r,$$

and it induces an $\mathfrak{sp}(N')$-equivariant isomorphism $J : \bigwedge^* A^+ \xrightarrow{\cong} \bigwedge^* A^-$ of $\mathfrak{sp}(N')$-representations [§3.2 Lemma 13]. Since these are dual to each other, we see that both of these representations are self-dual. On the spin representation $\bigwedge^* A^\pm$, the Lie algebra $\mathfrak{sp}(N')$ acts preserving the exterior grading, although the action of $\mathfrak{sp}(1) \otimes \mathbb{C} = \mathfrak{sl}_2(\mathbb{C})$ does not, it changes exterior degrees. As for the action of $\mathfrak{sl}_2(\mathbb{C})$ on $\bigwedge^* A^+ \cong \bigwedge^* A^-$, we have the following lemma.

LEMMA 6. *For each $0 \leq k \leq N = 2N'$, the k-th exterior power $\bigwedge^k A^+ \cong \bigwedge^k A^-$ is an eigenspace of the operator $h \in \mathfrak{sl}_2(\mathbb{C})$ with an eigenvalue $k - N'$.*

PROOF. Since $\bigwedge^k A^+$ is spanned by elements of the form $v = a_{i_1} a_{i_2} \cdots a_{i_k} \Omega$ where Ω is the vacuum vector on which A^- acts trivially, we only have to show that the above vector is an eigenvector with an eigenvalue $k - N'$. To show this, we first note that for any $1 \leq j \leq N$, we have $:a_j a_j^* : \Omega = (a_j a_j^* - \frac{1}{2})\Omega = -\frac{1}{2}\Omega$. Here, we recall that normal ordering in Cliff A is defined by $:ab := \frac{1}{2}(ab - ba)$ for $a, b \in A$. So, if $j \notin \{i_1, i_2, \ldots, i_k\}$, then $:a_j a_j^* :$ commutes with $a_{i_1}, \ldots, a_{i_k}$ and we have $(:a_j a_j^* :)v = a_{i_1} \cdots a_{i_k}(:a_j a_j^* :)\Omega = -\frac{1}{2}v$. If $j \in \{i_1, \ldots, i_k\}$, say, $j = i_\ell$, then,

$$(:a_j a_j^* :)v = a_{i_1} \cdots a_{i_{\ell-1}}(:a_{i_\ell} a_{i_\ell}^* : a_{i_\ell})a_{i_{\ell+1}} \cdots a_{i_k}\Omega = \frac{1}{2}v,$$

since $:a_{i_\ell} a_{i_\ell}^* : a_{i_\ell} = (\frac{1}{2} - a_{i_\ell}^* a_{i_\ell})a_{i_\ell} = \frac{1}{2}a_{i_\ell}$. Collecting our calculations, we have

$$h \cdot v = \sum_{j \in \{i_1, \ldots, i_k\}} \tfrac{1}{2}v - \sum_{j \notin \{i_1, \ldots, i_k\}} \tfrac{1}{2}v = \tfrac{1}{2}kv - (N' - \tfrac{1}{2}k)v = (k - N')v.$$

This proves the lemma. $\square$

We say a vector in the Clifford module $\bigwedge^* A^+$ is primitive if it is annihilated by Clifford multiplication by x^*. Similarly, a vector in $\bigwedge^* A^-$ is called primitive if it is annihilated by Clifford multiplication by x. In each homogeneous part of $\bigwedge^* A^\pm$, the subspace of primitive vectors with respect to the action of $\mathfrak{sp}(1) \otimes \mathbb{C}$ is a representation of $\mathfrak{sp}(N') \otimes \mathbb{C}$ since actions of these two Lie subalgebras commute. Let P_j be the subspace of primitive vectors in $\bigwedge^j A^+ \cong \bigwedge^j A^-$ for $1 \leq j \leq N'$:

(14) $$P_j = \{v \in \textstyle\bigwedge^j A^+ \mid x^* \cdot v = 0\}.$$

It is well known that these representations $P_1, P_2, \ldots, P_{N'}$ of $\mathfrak{sp}(N')$ have the correct highest weights and dimensions to be irreducible representations. It is also well known that the complex representation ring for the group $\text{Sp}(N')$ is a polynomial ring generated by these representations [BD]. Using $\mathfrak{sl}_2(\mathbb{C})$-representation theory, we have the next lemma.

LEMMA 7. (i) *For $0 \leq k \leq N'$, Clifford multiplication by x on $\bigwedge^{k-2} A^+$ is injective and we have an $\mathfrak{sp}(N')$-isomorphism*

$$\textstyle\bigwedge^k A^+ \cong P_k \oplus (\bigwedge^{k-2} A^+)x, \qquad 0 \leq k \leq N'.$$

(ii) *For $1 \leq j \leq N'$, the following map of Clifford multiplication by x^j is a $\mathfrak{sp}(N')$-isomorphism:*

$$x^j : \textstyle\bigwedge^{N'-j} A^+ \xrightarrow{\cong} \bigwedge^{N'+j} A^+, \qquad 1 \leq j \leq N'.$$

We have similar statements for $\bigwedge^* A^-$ with x^* in place of x.

REMARK. As we noted above, Clifford multiplication by x^j on the Clifford module $\bigwedge^* A^+ \cdot \Omega$ is the same as exterior multiplication by $\bar{x}^j$ on the exterior algebra $\bigwedge^* A$.

PROOF. We only have to show that the above maps are $\mathfrak{sp}(N')$-equivariant. But this is the case because multiplication by powers of x is $\mathfrak{sp}(N')$-equivariant due to the fact that $[\mathfrak{sp}(N'), \mathfrak{sp}(1)] = 0$ as subalgebras in $\mathfrak{o}(4N')$. $\square$

Here are some examples of decompositions of $\bigwedge^* A^+$ under $\mathfrak{sp}(N') \otimes \mathbb{C}$. Note that $\dim A^+ = N = 2N'$ and powers of x span trivial $Sp(N')$-representations.
(15)

Sp(1)	Sp(2)	Sp(3)	Sp(4)
			$\bigwedge^8 A^+ \cong \mathbb{C}x^4$
		$\bigwedge^6 A^+ \cong \mathbb{C}x^3$	$\bigwedge^7 A^+ \cong P_1 \cdot x^3$
$\bigwedge^4 A^+ \cong \mathbb{C}x^2$	$\bigwedge^5 A^+ \cong P_1 \cdot x^2$	$\bigwedge^6 A^+ \cong P_2 \cdot x^2 \oplus \mathbb{C}x^3$
$\bigwedge^2 A^+ \cong \mathbb{C}x$	$\bigwedge^3 A^+ \cong P_1 \cdot x$	$\bigwedge^4 A^+ \cong P_2 \cdot x \oplus \mathbb{C}x^2$	$\bigwedge^5 A^+ \cong P_3 \cdot x \oplus P_1 \cdot x^2$
$A^+ \cong P_1$	$\bigwedge^2 A^+ \cong P_2 \oplus \mathbb{C}x$	$\bigwedge^3 A^+ \cong P_3 \oplus P_1 \cdot x$	$\bigwedge^4 A^+ \cong P_4 \oplus P_2 \cdot x \oplus \mathbb{C}x^2$
$\mathbb{C}$	$A^+ \cong P_1$	$\bigwedge^2 A^+ \cong P_2 \oplus \mathbb{C}x$	$\bigwedge^3 A^+ \cong P_3 \oplus P_1 \cdot x$
	$\mathbb{C}$	$A^+ \cong P_1$	$\bigwedge^2 A^+ \cong P_2 \oplus \mathbb{C}x$
		$\mathbb{C}$	$A^+ \cong P_1$
			$\mathbb{C}$

Decompositions for A^- is similar with x replaced by x^*.

For calculation of $\mathfrak{sp}(N')$-invariants in V up to weight 2, next lemma will be useful. Recall that $\bar{x}$ and $\bar{x}^*$ in (12) denote elements in the exterior algebra $\bigwedge^* A^\pm$.

LEMMA 8. (i) *Representations $\bigwedge^* A^\pm$ have the following $\mathfrak{sp}(N')$-invariants:*

$$(\bigwedge^k A^+)^{\mathfrak{sp}(N')} = \begin{cases} \mathbb{C}\bar{x}^j, & \text{if } k = 2j, 1 \leq j \leq N' \\ \{0\}, & \text{if } k \text{ is odd.} \end{cases}$$

$$(\bigwedge^k A^-)^{\mathfrak{sp}(N')} = \begin{cases} \mathbb{C}\bar{x}^{*j}, & \text{if } k = 2j, 1 \leq j \leq N' \\ \{0\}, & \text{if } k \text{ is odd.} \end{cases}$$

(ii) *There are no nonzero $\mathfrak{sp}(N')$-invariants in the tensor product of exterior powers of total degree 3, that is, if $i + j = 3$, then*

$$(\bigwedge^i A^\pm \otimes \bigwedge^j A^\pm)^{\mathfrak{sp}(N')} = \{0\}.$$

(iii) *For $\mathfrak{sp}(N')$-invariants of tensor products of exterior powers of total degee 4,* *we have the following results:*

$$
(A^- \otimes \textstyle\bigwedge^3 A^+)^{\mathfrak{sp}(N')} = \begin{cases} \{0\}, & \text{if } N' = 1 \\ \mathbb{C}\bar{x}\,\bar{h}, & \text{if } N' \geq 2, \end{cases}
$$

$$
(A^+ \otimes \textstyle\bigwedge^3 A^-)^{\mathfrak{sp}(N')} = \begin{cases} \{0\}, & \text{if } N' = 1 \\ \mathbb{C}\bar{x}^*\,\bar{h}, & \text{if } N' \geq 2, \end{cases}
$$

$$
(\textstyle\bigwedge^2 A^- \otimes \textstyle\bigwedge^2 A^+)^{\mathfrak{sp}(N')} = \begin{cases} \mathbb{C}\bar{x}\,\bar{x}^*, & \text{if } N' = 1 \\ \mathbb{C}\bar{x}\,\bar{x}^* \oplus \mathbb{C}\bar{h}^2, & \text{if } N' \geq 2. \end{cases}
$$

PROOF. For (i), we proceed by induction on degree. Due to part (ii) of Lemma 7, we may assume that $k \leq N'$. First we note that $\bigwedge^0 A^+ = \mathbb{C}$ and $\bigwedge^1 A^+ = A^+$ which is irreducible, so $(\bigwedge^i A^+)^{\mathfrak{sp}(N')}$ is isomorphic to $\mathbb{C} \cdot 1$ when $i = 0$, and to $\{0\}$ when $i = 1$. Now let $2 \leq k \leq N'$ and assume that the conclusion on $(\bigwedge^j A^+)^{\mathfrak{sp}(N')}$ holds for $j \leq k-1$. Then by (i) of Lemma 7, we have $\bigwedge^k A^+ \cong P_k \oplus x \cdot (\bigwedge^{k-2} A^+)$. Since P_k is $\mathfrak{sp}(N')$-irreducible for $2 \leq k \leq N'$, there is no $\mathfrak{sp}(N')$-invariants in P_k. Thus, $\mathfrak{sp}(N')$-invariants in $\bigwedge^k A^+$ is isomorphic to $\mathfrak{sp}(N')$-invariants in $\bigwedge^{k-2} A^+$ via exterior multiplication by $\bar{x}$. Using inductive hypothesis, we have

$$
(\textstyle\bigwedge^k A^+)^{\mathfrak{sp}(N')} \cong x \cdot (\textstyle\bigwedge^{k-2} A^+)^{\mathfrak{sp}(N')} = \begin{cases} \mathbb{C}\bar{x} \cdot \bar{x}^{j-1}, & \text{if } k - 2 = 2(j-1), j \geq 0 \\ \{0\} & \text{if } k - 2 \text{ is odd.} \end{cases}
$$

Similarly for the case $\bigwedge^* A^-$.

For (ii), when $i = 0$ or $j = 0$, this is (i). We consider invariants in $A^- \otimes \bigwedge^2 A^+$. This representation is isomorphic to $P_1 \otimes \mathbb{C}\bar{x}$ if $N' = 1$ and to $P_1 \otimes (P_2 \oplus \mathbb{C}x)$ if $N' \geq 2$. Since $\mathfrak{sp}(N')$-invariants in $(P_1 \otimes P_2) \cong \text{Hom}(P_1, P_2)$ consist only of a zero vector, there are no nontrivial invariants in these representations. Since $A^+ \cong A^-$ as $\mathfrak{sp}(N')$-representations, other vanishing results follow.

For the first equation in (iii), first note that $\bigwedge^3 A^+ = \{0\}$ when $N' = 1$, $\bigwedge^3 A^+ \cong x \cdot P_1$ when $N' = 2$, and $\bigwedge^3 A^+ \cong P_3 \oplus x \cdot P_1$ when $N' \geq 3$. Since $A^- \cong (A^+)^*$,

$$
(A^- \otimes \textstyle\bigwedge^3 A^+)^{\mathfrak{sp}(N')} \cong \text{Hom}_{\mathfrak{sp}(N')}(A^+, \textstyle\bigwedge^3 A^+) \cong \text{Hom}_{\mathfrak{sp}(N')}(P_1, x \cdot P_1)
$$

$$
= \bar{x} \cdot \text{Hom}_{\mathfrak{sp}(N')}(P_1, P_1) \cong \bar{x} \cdot (A^- \otimes A^+)^{\mathfrak{sp}(N')} = \mathbb{C}\bar{x}\,\bar{h}.
$$

The second equation in (iii) is obtained by taking complex conjugation of the first equation and by noting that $h^* = -h$.

For the third equation in (iii), we observe that $\bigwedge^2 A^+ = \mathbb{C}\bar{x}$ when $N' = 1$, and $\bigwedge^2 A^+ \cong P_1 \oplus \mathbb{C}\bar{x}$ when $N' \geq 2$. By taking complex conjugates or duals, we also have $\bigwedge^2 A^- = \mathbb{C}\bar{x}^*$ when $N' = 1$, and $\bigwedge^2 A^- \cong P_2 \oplus \mathbb{C}\bar{x}^*$ when $N' \geq 2$. Since $\bar{x}$ and $\bar{x}^*$ are $\mathfrak{sp}(N')$-invariant, when $N' = 1$ we have $(\bigwedge^2 A^- \otimes \bigwedge^2 A^+)^{\mathfrak{sp}(N')} = \mathbb{C}\bar{x}^* \bar{x}$. When $N' \geq 2$, we have

$$
(\textstyle\bigwedge^2 A^- \otimes \textstyle\bigwedge^2 A^+)^{\mathfrak{sp}(N')} \cong (P_2 \otimes P_2)^{\mathfrak{sp}(N')} \oplus \mathbb{C}\bar{x}\,\bar{x}^* \cong \text{Hom}_{\mathfrak{sp}(N')}(P_2, P_2) \oplus \mathbb{C}\bar{x}\,\bar{x}^*.
$$

The last isomorphism is because P_2 is self-conjugate. Since P_2 is irreducible, the above vector space of invariants has dimension 2. On the other hand, the vectors $\bar{x}\bar{x}^*$, $\bar{h}^2$ are nontrivial $\mathfrak{sp}(N')$-invariants in this representation since $\bar{x}, \bar{x}^*, \bar{h}$ are $\mathfrak{sp}(N')$-invariants in the exterior algebra $\bigwedge^* A$. The vectors $\bar{x}\bar{x}^*$, $\bar{h}^2$ are distinct invariants when $N' \geq 2$ in $\bigwedge^2 A^- \otimes \bigwedge^2 A^+$. This is obvious when one writes out expressions of these elements, which are given by

$$\bar{x}\bar{x}^* = \sum_{r,s=1}^{N'} a_r a_{r+N'} a_s^* a_{s+N'}^*, \qquad \bar{h}^2 = 2\sum_{1\leq j<k\leq N} a_j a_j^* a_k a_k^*.$$

Hence, $(\bigwedge^2 A^- \otimes \bigwedge^2 A^+)^{\mathfrak{sp}(N')}$ is spanned by $\bar{x}\bar{x}^*$ and $\bar{h}^2$ when $N' \geq 2$. This completes the proof of (iii). $\square$

Symplectic Lie algebras in (10) act on Spin representation $\bigwedge^* A^{\pm}$ of $\mathfrak{o}(2N)$ through Clifford multiplication. On the other hand, the action of $\mathfrak{sp}(N')$ on $V = \bigotimes_{n\geq 0} \bigwedge^* A(-n - \frac{1}{2}) \cdot \Omega$ arises as 0-th vertex operators of vectors in the following subspace of $\bigwedge^2 A(-\frac{1}{2}) = (V)_1$:

$$(16) \qquad \bigoplus_{1\leq r\leq s\leq N'} \mathbb{C}([a_r(-\tfrac{1}{2})a_{s+N'}^*(-\tfrac{1}{2})] + [a_s(-\tfrac{1}{2})a_{r+N'}^*(-\tfrac{1}{2})])$$
$$\oplus \bigoplus_{1\leq r\leq s\leq N'} \mathbb{C}([a_r^*(-\tfrac{1}{2})a_{s+N'}(-\tfrac{1}{2})] + [a_s^*(-\tfrac{1}{2})a_{r+N'}(-\tfrac{1}{2})])$$
$$\oplus \bigoplus_{1\leq r,s\leq N'} \mathbb{C}([a_r(-\tfrac{1}{2})a_s^*(-\tfrac{1}{2})] + [a_{r+N'}^*(-\tfrac{1}{2})a_{s+N'}(-\tfrac{1}{2})]).$$

Note that (10) and (16) correspond to each other under the isomorphism $A \cong A(-\frac{1}{2})$. As we have shown in Corollary 4 of §2.3, these two actions of Lie algebras are the same.

We can now describe the subspace of invariant vectors $(V)_*^{\mathrm{Sp}(N')}$ of low weight.

PROPOSITION 9 (INVARIANTS $(V)_*^{\mathrm{Sp}(N')}$ OF LOW WEIGHT). *Let $N' \geq 1$. Then $\mathrm{Sp}(N')$-invariant vectors of weight up to 2 in the vertex operator super algebra V is given as follows:*

$$(V)_0^{\mathrm{Sp}(N')} = \mathbb{C}\Omega, \qquad\qquad (V)_{1/2}^{\mathrm{Sp}(N')} = \{0\},$$

$$(V)_1^{\mathrm{Sp}(N')} = \mathbb{C}[x] \oplus \mathbb{C}[x^*] \oplus \mathbb{C}[h], \qquad (V)_{3/2}^{\mathrm{Sp}(N')} = \{0\},$$

$$(V)_2^{\mathrm{Sp}(N')} = \begin{cases} (\mathbb{C}[\chi] \oplus \mathbb{C}[\chi^*] \oplus \mathbb{C}[\omega_+ - \omega_-]) \oplus \mathbb{C}[\omega] \oplus \mathbb{C}[h^2], & \text{if } N' = 1 \\[2mm] (\mathbb{C}[\chi] \oplus \mathbb{C}[\chi^*] \oplus \mathbb{C}[\omega_+ - \omega_-]) \oplus \mathbb{C}[\omega] \oplus \mathbb{C}[h^2 - 4xx^*] \\ \oplus\, (\mathbb{C}[x^2] \oplus \mathbb{C}[xh] \oplus \mathbb{C}[h^2 + 2xx^*] \oplus \mathbb{C}[x^* h] \oplus \mathbb{C}[(x^*)^2]), & \text{if } N' \geq 2. \end{cases}$$

When $N' = 1$, we have $[h^2] + 2[xx^*] = [x^2] = [xh] = 0$, so are their conjugates.

PROOF. This follows from Lemma 8. Note that we have written $\mathbb{C}[\omega_+] \oplus \mathbb{C}[\omega_-]$ as a direct sum $\mathbb{C}[\omega_+ - \omega_-] \oplus \mathbb{C}[\omega]$. Similarly, for invariants $[h^2]$, $[xx^*]$

in $\Lambda^2 A^+(-\frac{1}{2}) \otimes \Lambda^2 A^-(-\frac{1}{2})$, we have taken certain linear combinations of these elements. The relation $h^2 + 2xx^* = 0$ when $N' = 1$ can be easily read off from explicit forms of these elements in the proof of Lemma 8.   □

Representation of the Lie algebra $\mathfrak{sl}_2(\mathbb{C})$ on V given by

$$(17) \qquad \mathfrak{sl}_2(\mathbb{C}) = \mathbb{C}[x](0) \oplus \mathbb{C}[x^*](0) \oplus \mathbb{C}[h](0),$$

where $[x](0)$, $[x^*](0)$, and $[h](0)$ are 0-modes of vertex operators associated to vectors $[x]$, $[x^*]$, and $[h]$, respectively. The action of $\mathfrak{sl}_2(\mathbb{C})$ on V commutes with the action of $\mathfrak{sp}(N')$, and hence $\mathfrak{sl}_2(\mathbb{C})$ acts on the space of invariants $(V)_*^{\mathfrak{sp}(N')}$. This action is described by the next lemma.

LEMMA 10. (i) *Following* $\mathfrak{sl}_2(\mathbb{C})$-*representations are isomorphic to the irreducible adjoint representation of* $\mathfrak{sl}_2(\mathbb{C})$:

$$\mathbb{C}[x] \oplus \mathbb{C}[x^*] \oplus \mathbb{C}[h], \qquad \mathbb{C}[\chi] \oplus \mathbb{C}[\chi^*] \oplus \mathbb{C}[\omega_+ - \omega_-].$$

(ii) *Let* $N' \geq 2$. *The following* $\mathfrak{sl}_2(\mathbb{C})$-*representation is isomorphic to the unique nontrivial irreducible 5-dimensional summand of the second symmetric power* $S^2(\mathfrak{sl}_2(\mathbb{C}))$ *of the adjoint representation of* $\mathfrak{sl}_2(\mathbb{C})$:

$$\mathbb{C}[x^2] \oplus \mathbb{C}[xh] \oplus \mathbb{C}[h^2 + 2xx^*] \oplus \mathbb{C}[x^* h] \oplus \mathbb{C}[(x^*)^2].$$

(iii) *The vector* ω *is* $\mathfrak{o}(2N)$-*invariant. For* $N' \geq 1$, *the* $\mathfrak{sp}(N')$-*Casimir element* $\phi_{\mathfrak{sp}(N')} = -\frac{1}{2}[h^2] + 2[xx^*]$ *is* $\mathfrak{sp}(N')$-*invariant.*

PROOF. Part (ii) and part (iii) has been taken care of in Propositions 3 and 4 of §3.3. Note that the representation $S^2(\mathfrak{sl}_2(\mathbb{C}))$ is a direct sum of the 5-dimensional representation in (ii) and a 1-dimensional trivial representation $\mathbb{C}[\phi_{\mathfrak{sp}(N')}]$. For the second representation in (i), from Proposition 1 of §3.5,

$$(18) \qquad D(-1)[x] = [\chi], \qquad D(-1)[x^*] = [\chi^*], \qquad D(-1)[h] = [\omega_+] - [\omega_-].$$

Since the action of Virasoro operators and the action of the Lie algebra $\mathfrak{o}(2N)$ commute on V [§2.3 (23)], it follows that in particular $D(-1)$ is an $\mathfrak{sl}_2(\mathbb{C})$-equivariant map since $\mathfrak{sl}_2(\mathbb{C}) \subset \mathfrak{o}(2N)$. Since $D(-1)$ gives an $\mathfrak{sl}_2(\mathbb{C})$-equivariant isomorphism from the first representation to the second one in (i), the second representation is also irreducible.   □

Lemma 10 immediately calculates $\mathrm{Sp}(N') \cdot \mathrm{Sp}(1)$-invariants in V of low weight.

PROPOSITION 11 (INVARIANTS $(V)_*^{\mathrm{Sp}(N')\mathrm{Sp}(1)}$ OF LOW WEIGHT). *Let* $N' \geq 1$. *Then* $\mathrm{Sp}(N') \cdot \mathrm{Sp}(1)$-*invariant vectors of weight up to 2 in the vertex operator super algebra* V *is given by*

$$(V)_0^{\mathrm{Sp}(N')\mathrm{Sp}(1)} = \mathbb{C}\Omega, \qquad (V)_{k/2}^{\mathrm{Sp}(N')\mathrm{Sp}(1)} = \{0\}, \quad 1 \leq k \leq 3,$$

$$(V)_2^{\mathrm{Sp}(N')\mathrm{Sp}(1)} = \mathbb{C}[\omega] \oplus \mathbb{C}[\phi_{\mathfrak{sp}(N')}].$$

Here, $\phi_{\mathfrak{sp}(N')} = -\frac{1}{2}h^2 + 2xx^*$ is the symplectic Casimir element in $\mathcal{A}$ for $N' \geq 1$. When $N' = 1$, $\phi_{\mathfrak{sp}(N')} = 3xx^* = -\frac{3}{2}h^2 \neq 0$.

REMARK. When the vector space A carries a quaternionic structure and a symplectic scalar product, we can consider three different complex structures I, J, K on $A(-\frac{1}{2})$, and the corresponding Kähler forms ω_I, ω_J, ω_K on $A(-\frac{1}{2})$. These Kähler forms are $\mathfrak{sp}(N')$-invariant. In Theorem 15 and Corollary 16 of §3.9, we showed that

$$(19) \qquad \begin{array}{l} \omega_I = -ih, \qquad \omega_J = x + x^*, \qquad \omega_K = -i(x - x^*), \\[1mm] \omega_I^2 + \omega_J^2 + \omega_K^2 = -h^2 + 4xx^* = 2\phi_{\mathfrak{sp}(N')} = -N\phi_{\mathfrak{sp}(1)}, \end{array}$$

where $\mathfrak{sp}(1) = \mathfrak{o}(2N)^{\mathfrak{sp}(N')}$.

CHAPTER IV

GEOMETRIC STRUCTURE IN VECTOR SPACES AND REDUCTION OF STRUCTURE GROUPS ON MANIFOLDS

§4.1 Some geometric structures on vector spaces

Each tangent space of a given manifold can have various geometric structures which can be systematically defined on the whole tangent bundle. Global existence of a smoothly defined particular geometric structure on the tangent spaces depends on global properties of the given manifold. The more structure the tangent bundle has, the more special the manifold is. Accordingly, the structure group which preserves a particular geometric structure on the manifold reduces to a smaller group.

Before we discuss global existence of a particular geometric structure on a manifold, we discuss in detail interesting geometric structures on a vector space. The following is a list of structures of interest for a real vector space E^n of real dimension n. For a more comprehensive treatment of geometric structures, see §4.3.

(I)	Euclidean scalar product structure.
(II)	Complex structure when $n = 2N$.
(III)	Hermitian scalar product structure when $n = 2N$.
(IV)	Special Hermitian scalar product structure when $n = 2N$.
(V)	Quaternionic vector space structure, when $n = 4N'$.
(VI)	Symplectic scalar product structure when $n = 4N'$.
(VII)	Pseudo-symplectic structure when $n = 4N'$.

We discuss these structures in detail one by one. For the definition of pseudo-symplectic structure, see (VII) below. This terminology is only for the use of this paper and it is not commonly used. Central theme of our discussion is a group of automorphisms of E^n preserving a given geometric structure. We are following the Weyl's dictum ; *"Whenever you have to do with a structure endowed entity, try to determine its group of automorphisms"* [Wey]. The automorphism groups for standard model spaces with various structures are of particular importance. These are vector spaces $\mathbb{R}^n$ with the standard basis $\{e_1, e_2, \ldots, e_n\}$ equipped with geometric structures in a standard way.

Given E^n equipped with one of the above geometric structures (I) to (VII) which we denote by Σ, we can consider the set of all linear isomorphisms $\xi :$ $(\mathbb{R}^n, \Sigma_0) \rightarrow (E^n, \Sigma)$ which preserve the standard structure Σ_0 in $\mathbb{R}^n$ and a given structure Σ in E^n. This set of maps is the set of preferred frames in

E^n equipped with a particular geometric structure Σ. This set of frames is canonically determined for each given geometric structure Σ in E and plays an important role in the reduction of structure groups of manifolds. We will discuss this one by one for the above listed geometric structures. It is also important to note that any frame in E^n can be used to transport any geometric structures in $\mathbb{R}^n$ to E^n.

We use the following notation for the set of all frames in E^n.

$$(1) \qquad \mathcal{F}(E^n) = \{\xi : \mathbb{R}^n \to E^n \mid \xi \text{ is a linear isomorphism}\}$$

Since $\mathbb{R}^n$ has the standard basis $\{e_1, \ldots, e_n\}$, the above set is the same as the set of n-tuples of linearly independent set of vectors $\{v_1, \ldots, v_n\}$ in E^n, where $v_j = \xi(e_j)$. When E^n has an orientation, we denote the set of all frames in the orientation class by $\mathcal{F}^+(E^n)$. These spaces $\mathcal{F}(E^n)$ and $\mathcal{F}^+(E^n)$ are homogeneous spaces of the groups $\mathrm{GL}(E^n)$ and $\mathrm{GL}^+(E^n)$, respectively, with trivial isotropy groups. In each of the geometric structures we deal with, a certain group acts on the set of preferred set of frames compatible with a given geometric structure, both transitively and effectively.

Given a type of geometric structure in the above list, we consider the set of all inequivalent such geometric structures in E^n. We use structured frames to classify all the possible such geometric structures. In fact, for each type of the above geometric structures, the space of inequivalent geometric structures of a given type in E^n turns out to be a homogeneous space: an orbit space of $\mathcal{F}(E^n)$ under the action of the automorphism group of the given geometric structure. Conversely, we can define geometric structures in this way. See §4.3 for this approach.

(I) Euclidean scalar product structures. The group of automorphisms of a vector space E^n equipped with a Euclidean scalar product $\langle\ ,\ \rangle_E$ consists of those linear maps which preserve the inner product. This is the orthogonal group of E^n with respect to $\langle\ ,\ \rangle_E$:

$$(2) \quad \mathrm{Aut}(E^n; \langle\ ,\ \rangle_E) = \mathrm{O}(E^n) = \left\{ T \in \mathrm{GL}(E^n) \,\middle|\, \begin{array}{l} \langle T(u), T(v)\rangle_E = \langle u, v\rangle_E \\ \text{for any } u, v \in E^n \end{array} \right\}.$$

The standard model space is $\mathbb{R}^n$ together with the standard scalar product $\langle\ ,\ \rangle_0$ for which the standard basis $\{e_1, \ldots, e_n\}$ forms an orthonormal basis. Obviously, the automorphism group of the Euclidean vector space $\mathbb{R}^n$ is the standard orthogonal group $\mathrm{O}(n)$.

The set of preferred frames for the Euclidean vector space E^n consists of those frames $\xi : \mathbb{R}^n \to E^n$ which preserve inner products in $\mathbb{R}^n$ and in E^n, i.e., those maps which are isometries. These maps ξ have the property $\langle \xi(u), \xi(v)\rangle_E = \langle u, v\rangle_0$. Note that vectors $\xi(e_j)$, $1 \leq j \leq n$, form an orthonormal basis in E^n. We denote the set of preferred frames in E^n by $\mathcal{F}(E^n; \langle\ ,\ \rangle_E)$. Thus,

$$(3) \qquad \mathcal{F}(E^n; \langle\ ,\ \rangle_E) = \{(v_1, v_2, \ldots, v_n) \mid \langle v_i, v_j\rangle_E = \delta_{ij}\}.$$

Since every orthonormal basis is related to the other such basis by an orthogonal transformations, the above space is a principal homogeneous space for the group $O(n)$ and for any orthonormal frame ξ, we have

$$(4) \qquad \mathcal{F}(E^n; \langle \ , \ \rangle_E) = \xi \cdot O(n) = \mathrm{Aut}(E^n; \langle \ , \ \rangle_E) \cdot \xi.$$

In the formula in the middle above, the dot denotes the matrix multiplication of a row vector $\xi = \{v_1, \ldots, v_n\}$ with an element in $O(n)$, where in the last formula $T : E^n \to E^n$ acts on the frame ξ by $T(\xi) = (T(v_1), \ldots, T(v_n))$. The actions of the above automorphism groups are effective and transitive. Note the difference between the left and the right actions in (4).

Now, we consider the moduli space of the Euclidean scalar products in E^n. Given any frame $\xi : \mathbb{R}^n \to E^n$ in E^n, we can transport the standard inner product $\langle \ , \ \rangle_0$ in $\mathbb{R}^n$ to E^n by decreeing that the basis $\{\xi(e_1), \ldots, \xi(e_n)\}$ forms an orthonormal basis of E^n. Thus, we have a map from $\mathcal{F}(E^n)$ to the moduli space of inner products in E^n. This map is surjective because for any inner product in E^n, we can construct an orthonormal basis for E^n and this frame recovers the original inner product in E^n we started from by the above correspondence. We check the fibres of this map. Suppose two frames $\xi, \eta \in \mathcal{F}(E^n)$ define the same inner product in E^n. With respect to this same inner product, these two frames are orthonormal frames. Thus they are related by an orthogonal matrix. So, $\eta \in \xi \cdot O(n)$. Thus, we have the following lemma.

LEMMA 1. *The moduli space of inner products in E^n is described by the bijection*

$$(5) \qquad \mathcal{F}(E^n)/O(n) \xrightarrow[\text{onto}]{1:1} \{\text{Euclidean scalar products in } E^n\},$$

where for any chosen frame ξ, we can identify the above space with a homogeneous space $\mathrm{GL}_n(\mathbb{R})/O(n)$, and the coset $\xi \cdot O(n) \subset \mathcal{F}(E^n)$ is precisely the set of all orthonormal frames for the corresponding inner product $\langle \ , \ \rangle_\xi$.

(II) Complex structures. Let E^{2N} be a real $2N$ dimensional vector space with a complex structure I. So, $I : E^{2N} \to E^{2N}$ is a linear isomorphism satisfying $I^2 = -1$. The automorphism group of the complex structure I is described by

$$(6) \qquad \mathrm{Aut}(E^{2N}; I) = \{T \in \mathrm{GL}(E^{2N}) \mid T \circ I = I \circ T\}.$$

The standard model space is $(\mathbb{R}^{2N}; I_0)$ with standard basis $\{e_1, e_2, \ldots, e_{2N}\}$ on which the standard complex structure I_0 acts by $I_0(e_j) = e_{j+N}$ and $I_0(e_{j+N}) = -e_j$ for $1 \leq j \leq N$. We can identify this vector space with $\oplus_{j=1}^N \mathbb{C}e_j$, where the multiplication by i is given by I. Under this identification, those linear isomorphisms on $\mathbb{R}^{2N}$ commuting with I_0 can be identified with elements in $\mathrm{GL}_N(\mathbb{C})$ by definition. This is the automorphism group of $(\mathbb{R}^{2N}; I_0)$.

The set of preferred frames in E^{2N} are those linear isomorphisms $\xi : \mathbb{R}^{2N} \to E^{2N}$ which commute with complex structures on these vector spaces. We denote this set of preferred frames by $\mathcal{F}(E^{2N}; I)$:

$$(7) \qquad \mathcal{F}(E^{2N}; I) = \{\xi \in \mathcal{F}(E^{2N}) \mid \xi \circ I_0 = I \circ \xi\}.$$

We observe that $\mathcal{F}(E^{2N}; I) = \xi \cdot \mathrm{GL}_N(\mathbb{C}) = \mathrm{Aut}(E^{2N}; I) \cdot \xi$.

We now turn to the description of the moduli space of inequivalent complex structures on a given vector space E^{2N}. Let $\xi : \mathbb{R}^{2N} \to E^{2N}$ be any frame in E^{2N}. We can transport the standard complex structure I_0 on $\mathbb{R}^{2N}$ to E^{2N} via this map. Namely, if the frame ξ is given by $\xi = (v_1, v_2, \ldots, v_{2N})$ where $v_j = \xi(e_j)$, then a linear map $I : E^{2N} \to E^{2N}$ given by $I(v_j) = v_{j+N}$ and $I(v_{j+N}) = -v_j$ for $1 \leq j \leq N$ defines a complex structure on E^{2N}. This defines a map from the space of all frames $\mathcal{F}(E^{2N})$ to the moduli space of complex structures in E^{2N}. This map is surjective because for a given complex structure I on E^{2N}, one can always choose a $\mathbb{C}$-basis of E^{2N}, say, $\{v_1, \ldots, v_N\}$, where multiplication by i is given by the structure map I. Then we can extend it to an $\mathbb{R}$-basis of E^{2N} by letting $v_{j+N} = I(v_j)$ for $1 \leq j \leq N$. This frame recovers the given complex structure we started from by the above correspondence.

Next, we check fibres of this map from $\mathcal{F}(E^{2N})$ to the moduli space of complex structures on E^{2N}. Suppose frames ξ and $\eta \in \mathcal{F}(E^{2N})$ define the same complex structure I in E^{2N}. So, $\xi \circ I_0 = I \circ \xi$, $\eta \circ I_0 = I \circ \eta$ for the same I. Then $(\xi^{-1} \circ \eta) \circ I_0 = \xi^{-1} \circ I \circ \eta = I_0 \circ (\xi^{-1} \circ \eta)$ and the map $\xi^{-1} \circ \eta : \mathbb{R}^{2N} \to \mathbb{R}^{2N}$ preserves the standard complex structure I_0 in $\mathbb{R}^{2N}$. Thus $\xi^{-1} \circ \eta \in \mathrm{GL}_N(\mathbb{C})$, or $\eta \in \xi \cdot \mathrm{GL}_N(\mathbb{C})$. Thus, two frames in $\mathcal{F}(E^{2N})$ defines the same complex structure if and only if they belong to the same $\mathrm{GL}_N(\mathbb{C})$-coset in $\mathcal{F}(E^{2N})$. Thus we obtain the next lemma.

LEMMA 2. *On a real $2N$ dimensional vector space E^{2N}, the moduli space of complex structures on E^{2N} is described by the following bijection:*

$$(8) \qquad \mathcal{F}(E^{2N})/\mathrm{GL}_N(\mathbb{C}) \xrightarrow[\mathrm{onto}]{1:1} \{\text{complex structures on } E^{2N}\}$$

For any chosen frame $\xi \in \mathcal{F}(E^{2N})$, we can identify the above space with the homogeneous space $\mathrm{GL}_{2N}(\mathbb{R})/\mathrm{GL}_N(\mathbb{C})$, and the coset $\xi \cdot \mathrm{GL}_N(\mathbb{C}) \subset \mathcal{F}(E^{2N})$ is precisely the set of all preferred frames $\mathcal{F}(E^{2N}; I_\xi)$ defined in (7) compatible with the complex structure I_ξ on E^{2N}. We have $\xi \cdot \mathrm{GL}_N(\mathbb{C}) = \mathrm{Aut}(E^{2N}; I_\xi) \cdot \xi$.

(III) Hermitian structures on complex vector spaces. Let $\langle \, , \, \rangle_E$ be a Hermitian scalar product on a given complex vector space $(E^{2N}; I)$. We use Hermitian pairings which are sesqui-linear in the first (or second) entry. The automorphism group of this structure is given by

$$(9) \qquad \mathrm{Aut}(E^{2N}; I, \langle \, , \, \rangle) = \left\{ T \in \mathrm{Aut}(E^{2N}; I) \,\middle|\, \begin{matrix} \langle T(u), T(v) \rangle = \langle u, v \rangle \, for \\ \forall u, \forall v \in E^{2N} \end{matrix} \right\}.$$

The standard model of a Hermitian vector space is $\mathbb{R}^{2N}$ equipped with the standard basis $\{e_1, \ldots, e_{2N}\}$, the standard complex structure I_0, and the standard Hermitian pairing $\langle \, , \, \rangle_0$ for which the basis $\{e_1, \ldots, e_N\}$ is a Hermitian orthonormal basis. The automorphism group for the standard Hermitian structure is the unitary group $U(N)$.

The set of preferred frames in a Hermitian vector space consists of those maps $\xi : \mathbb{R}^{2N} \to E^{2N}$ which preserve the complex structures I_0, I and the Hermitian pairings $\langle \, , \, \rangle_0$, $\langle \, , \, \rangle_E$. We call such frames Hermitian (orthonormal) frames. Thus,

$$(10) \qquad \mathcal{F}(E^{2N}; I, \langle \, , \, \rangle_E) = \left\{ \xi \in \mathcal{F}(E^{2N} : I) \;\middle|\; \begin{array}{c} \langle \xi(u), \xi(v) \rangle_E = \langle u, v \rangle_0 \\ \text{for any } u, v \in \mathbb{R}^{2N} \end{array} \right\}.$$

In other words, the preferred frames for the Hermitian structure are those of the form $\xi = (v_1, \ldots, v_N, I(v_1), \ldots, I(v_N))$ where $\{v_1, v_2, \ldots, v_N\}$ is an Hermitian orthonormal frame for the given Hermitian pairing in E^{2N}. Let ξ and η be such frames. Then $\xi^{-1} \circ \eta$, $\eta \circ \xi^{-1}$ preserve structures in $\mathbb{R}^{2N}$ and in E^{2N}, so they are elements in $U(N)$, $\mathrm{Aut}(E^{2N}; I, \langle \, , \, \rangle_E)$, respectively. Hence, we see that the set of preferred frames is a principal homogeneous space given by $\mathcal{F}(E^{2N}; I, \langle \, , \, \rangle_E) = \xi \cdot U(N) = \mathrm{Aut}(E^{2N}; I, \langle \, , \, \rangle_E) \cdot \xi$. Note that the actions of these groups are effective.

Now we describe the moduli space of all inequivalent Hermitian structures compatible with a given complex structure I in E^{2N}. As before, this space can be described in terms of frames. Let $\xi \in \mathcal{F}(E^{2N}; I)$ be a $\mathbb{C}$-linear frame $\xi : \mathbb{R}^{2N} \to E^{2N}$ commuting with complex structure I_0 in $\mathbb{R}^{2N}$ and I in E^{2N}. We can introduce a Hermitian pairing in E^{2N} by regarding ξ as an isometry between these two vector spaces so that we have $\langle u, v \rangle_E = \langle \xi^{-1}(u), \xi^{-1}(v) \rangle_{\mathbb{R}^{2N}}$. In other words, given ξ, if $\xi = (v_1, \ldots, v_N, I(v_1), \ldots, I(v_N))$, then we decree that $\{v_1, \ldots, v_N\}$ is a Hermitian orthonormal basis. This defines a Hermitian pairing in E^{2N}. This defines a map from the set of complex frames $\mathcal{F}(E^{2N}; I)$ to the set of Hermitian structures compatible with the given complex structure I in E^{2N}. This map is surjective since for any Hermitian structure in E^{2N} one can choose a Hermitian orthonormal basis and we can extend it to a frame over $\mathbb{R}$ using I. This frame recovers the given Hermitian structure we started from.

Now several complex frames can define the same Hermitian structure in E^{2N}. Let ξ and $\eta \in \mathcal{F}(E^{2N}; I)$ be two such complex frames. Then the composition $\xi^{-1} \circ \eta : \mathbb{R}^{2N} \to E^{2N} \to \mathbb{R}^{2N}$ is an isometry of $(\mathbb{R}^{2N}; I_0, \langle \, , \, \rangle_0)$, so it must be an element in the unitary group. Hence $\eta \in \xi \cdot U(N)$ and ξ and η belong to the same coset of $U(N)$. Hence we obtain the following lemma.

LEMMA 3. *Let I be a complex structure in E^{2N}. The space of compatible Hermitian structures in E^{2N} is described by the following bijection:*

$$(11) \qquad \mathcal{F}(E^{2N}; I)/U(N) \xrightarrow[\text{onto}]{1:1} \{\text{Hermitian structures on } (E^{2N}; I)\}.$$

A choice of a frame $\xi \in \mathcal{F}(E^{2N}; I)$ allows us to identify the above space with a homogeneous space $\mathrm{GL}_N(\mathbb{C})/U(N)$. The coset $\xi \cdot U(N) \subset \mathcal{F}(E; I)$ is the set *of all the Hermitian orthonormal frames $\mathcal{F}(E^{2N}; I, \langle \ , \ \rangle_\xi)$ for the corresponding Hermitian pairing $\langle \ , \ \rangle_\xi$, and this can be identified with* $\mathrm{Aut}(E^{2N}; I, \langle \ , \ \rangle_\xi) \cdot \xi$.

(IV) Special Hermitian structures on Hermitian vector spaces. First we explain what we mean by a special Hermitian structure. This is a Hermitian structure together with an element α of unit length in the highest complex exterior power $\bigwedge_{\mathbb{C}}^N(E)$ of a complex vector space $(E^{2N}; I)$. An automorphism of of E^{2N} with a special Hermitian structure is an automorphism of the underlying Hermitian structure $(E^{2N}; I, \langle \ , \ \rangle_E)$ which preserves the element $\alpha \in \bigwedge_{\mathbb{C}}^N(E)$. Thus,

$$(12) \quad \mathrm{Aut}(E^{2N} : I, \langle \ , \ \rangle_E, \alpha) = \{T \in \mathrm{Aut}(E^{2N}; I, \langle \ , \ \rangle_E) \mid (\bigwedge_{\mathbb{C}}^N T)(\alpha) = \alpha\}.$$

The standard model space of a special Hermitian structure is $(\mathbb{R}^{2N}, I_0, \langle \ , \ \rangle_0, \alpha_0)$ with $\alpha_0 = e_1 \wedge e_2 \wedge \cdots \wedge e_N$, where $\{e_1, \ldots, e_N\}$ is the standard Hermitian orthonormal frame. An automorphism of the standard special Hermitian vector space is an element in the unitary group which preserves the element α_0. Since the action of a unitary transformation T on α_0 is by multiplication by its determinant $\det T$, the automorphism group of the special Hermitian structure is the special unitary group $\mathrm{SU}(N)$.

The preferred set of frames for a special Hermitian structure in E^{2N} is the set of Hermitian orthonormal frames $\xi : \mathbb{R}^{2N} \to E^{2N}$ which preserve the preferred complex N-forms α_0 and α in $\bigwedge_{\mathbb{C}}^N \mathbb{C}^N$, $\bigwedge_{\mathbb{C}}^N E$, respectively. We call these frames special Hermitian frames. This set is denoted by

$$(13) \quad \mathcal{F}(E^{2N}; I, \langle \ , \ \rangle_E, \alpha) = \{\xi \in \mathcal{F}(E^{2N}; I, \langle \ , \ \rangle_E) \mid (\bigwedge_{\mathbb{C}}^N \xi)(\alpha_0) = \alpha\}.$$

The condition $(\bigwedge_{\mathbb{C}}^N \xi)(\alpha_0) = \alpha$ is the same as $\xi(e_1) \wedge \cdots \wedge \xi(e_N) = \alpha \in \bigwedge_{\mathbb{C}}^N(E)$. Let ξ and η be any two special Hermitian frames. Then, the compositions $\xi^{-1} \circ \eta$ and $\eta \circ \xi^{-1}$ preserve the special Hermitian structures in $\mathbb{R}^{2N}$ and in E^{2N} respectively, and they must be elements in the their automorphism groups, namely $\mathrm{SU}(N)$ and $\mathrm{Aut}(E^{2N}; I, \langle \ , \ \rangle, \alpha)$, respectively. Thus we see that the set of special Hermitian frames is a homogeneous space of the form

$$(14) \quad \mathcal{F}(E^{2N}; I, \langle \ , \ \rangle, \alpha) = \xi \cdot \mathrm{SU}(N) = \mathrm{Aut}(E^{2N}; I, \langle \ , \ \rangle, \alpha) \cdot \xi,$$

where the groups act transitively and effectively.

We consider the moduli space of special Hermitian structures compatible with a given Hermitian vector space $(E^{2N}; I, \langle \ , \ \rangle_E)$. By the very definition, this space is precisely the set $S(\bigwedge_{\mathbb{C}}^N E)$ of unit vectors in the top exterior power of E^{2N} which can be identified with a circle S^1 as a space, although not canonically. We reformulate this statement in terms of frames.

Let $\xi = (v_1, \ldots, v_N, I(v_1), \ldots, I(v_N)) \in \mathcal{F}(E^{2N}; I, \langle \ , \ \rangle)$ be a Hermitian orthonormal frame, so that $\{v_1, \ldots, v_N\}$ forms a Hermitian orthonormal basis of E^{2N}. Then, the element $\alpha_\xi = v_1 \wedge v_2 \wedge \cdots \wedge v_N$ in $\bigwedge_{\mathbb{C}}^N E$ defines a special Hermitian structure in E^{2N}. This defines a map from $\mathcal{F}(E^{2N}; I, \langle \ , \ \rangle)$ to the set of the special Hermitian structures subordinate to the given Hermitian structure. This map is surjective because given a special Hermitian structure α, one can choose a Hermitian orthonormal frame $\{v_1, v_2, \ldots, v_N\}$ so that $v_1 \wedge \cdots \wedge v_N$ gives the preferred element α, by multiplying by a constant $z \in \mathbb{C}$ with $|z| = 1$ on a basis vector, if necessary. Now let ξ and η be two Hermitian frames giving rise to the same special Hermitian structure on E^{2N}. Then, the composition $\xi^{-1} \circ \eta : \mathbb{R}^{2N} \to E^{2N} \to \mathbb{R}^{2N}$ preserves the standard special Hermitian structure in $\mathbb{R}^{2N}$. Thus, this element belongs to $\mathrm{SU}(N)$ and we have $\eta \in \xi \cdot \mathrm{SU}(N)$. This gives the following lemma.

LEMMA 4. *The moduli space of special Hermitian structures in a Hermitian vector space $(E^{2N}; I, \langle \,, \, \rangle_E)$ has the following description:*
(15)
$$\mathcal{F}(E^{2N}; I, \langle \,, \, \rangle_E)/\mathrm{SU}(N) \xrightarrow[\text{onto}]{1:1} S\left(\bigwedge_{\mathbb{C}}^{N} E\right) \cong \left\{ \begin{array}{c} \text{special Hermitian structures} \\ \text{subordinate to } (E^{2N}; I, \langle \,, \, \rangle_E) \end{array} \right\}$$

Any Hermitian frame ξ allows us to identify the above space as $\mathrm{U}(N)/\mathrm{SU}(N) = S^1$ using α_ξ. The coset $\xi \cdot \mathrm{SU}(N)$ is precisely the set of all preferred frames $\mathcal{F}(E^{2N}; I, \langle \,, \, \rangle_E, \alpha_\xi)$ corresponding to the special Hermitian structure α_ξ associated to the Hermitian frame ξ.

(V) Quaternionic structures. A quaternionic structure in a real vector space E consists of endomorphisms I, J of E such that $I^2 = J^2 = -1$, $IJ = -JI$. We let $JI = K$. This defines a right module structure in E over the skew field of quaternions $\mathbb{H} = \mathbb{R} \oplus \mathbb{R}i \oplus \mathbb{R}j \oplus \mathbb{R}k$ by letting maps I, J, K acting on E from the left correspond to the right multiplications by i, j, k on vectors in E. Note that $K(u) = JI(u) = u \cdot ij = u \cdot k$. This explains our choice of $K = JI$ instead of $K = IJ = -JI$. We then see that the real dimension of E must be divisible by 4, say, $n = 2N = 4N'$. The automorphism group of a quaternion structure consists of elements in $\mathrm{GL}(E^{4N'})$ which commute with I, J.

(16) $\mathrm{Aut}(E^{4N'}; I, J) = \{T \in \mathrm{GL}(E^{4N'}) \mid T \circ I = I \circ T, \ T \circ J = J \circ T \}$.

These are $\mathbb{H}$-linear maps with respect to the right $\mathbb{H}$-module structure on $E^{4N'}$.

The standard model space is $\mathbb{R}^{4N'} = \bigoplus_{\ell=1}^{4N'} \mathbb{R}e_j$ on which endomorphisms I_0, J_0 act by

$$
\begin{aligned}
&& I_0(e_j) = e_{j+N}, && I_0(e_{k+N}) = -e_k, && 1 \le k \le N, \\
(17) && J_0(e_r) = e_{r+N'}, && J_0(e_{r+N'}) = -e_r, && 1 \le r \le N', \\
&& J_0(e_{r+N}) = -e_{r+N+N'}, && J_0(e_{r+N+N'}) = e_{r+N}, && 1 \le r \le N'.
\end{aligned}
$$

It is straightforward to check that $I_0^2 = -1$, $J_0^2 = -1$, $I_0 J_0 = -J_0 I_0$. We regard $(\mathbb{R}^{4N'}; I_0, J_0)$ as a right $\mathbb{H}$-vector space with basis $\{e_1, e_2, \ldots, e_{N'}\}$. The automorphism group of a quaternionic vector space $(\mathbb{R}^{4N'}; I_0, J_0)$ is $\mathrm{GL}_{N'}(\mathbb{H})$, by definition.

The set of preferred frames for the quaternionic vector space $(E^{4N'}; I, J)$ consists of those frames $\xi : \mathbb{R}^{4N'} \to E^{4N'}$ which preserve the quaternionic structures in each of these vector spaces. We call such frames quaternionic frames. The totality of such frames is denoted by $\mathcal{F}(E^{4N'}; I, J)$. Thus,

$$(18) \qquad \mathcal{F}(E^{4N'}; I, J) = \{\xi \in \mathcal{F}(E^{4N'}) \mid I \circ \xi = \xi \circ I_0, \ J \circ \xi = \xi \circ J_0\}.$$

In other words, letting $\xi(e_j) = v_j$ for $1 \leq j \leq 4N'$, this set consists of those $\mathbb{R}$-bases $\{v_1, \ldots, v_{4N'}\}$ of $E^{4N'}$ such that

$$(19) \qquad \begin{aligned} I(v_j) &= v_{j+N}, & I(v_{j+N}) &= -v_j, & 1 \leq j \leq N, \\ J(v_r) &= v_{r+N'}, & J(v_{r+N'}) &= -v_r, & 1 \leq r \leq N', \\ J(v_{r+N}) &= -v_{r+N+N'}, & J(v_{r+N+N'}) &= v_{r+N}, & 1 \leq r \leq N'. \end{aligned}$$

On this space of quaternionic frames, the groups $\mathrm{GL}_{N'}(\mathbb{H})$, $\mathrm{Aut}(E^{4N'}; I, J)$ act from the right and from the left, respectively. These actions are transitive and effective and for any quaternionic frame ξ, we have

$$(20) \qquad \mathcal{F}(E^{4N'}; I, J) = \xi \cdot \mathrm{GL}_{N'}(\mathbb{H}) = \mathrm{Aut}(E^{4N'}; I, J) \cdot \xi.$$

This gives a principal homogeneous space structure on the set of quaternionic frames in two different ways.

Next, we study the set of inequivalent quaternionic structures on real $4N'$ dimensional real vector space $E^{4N'}$. First, we observe that any frame $\xi \in \mathcal{F}(E^{4N'})$ can be used to introduce a quaternionic structure on $E^{4N'}$, as follows. Let $\xi = (v_1, \ldots, v_{4N'})$ be any frame in $E^{4N'}$. It is a linear isomorphism $\xi : \mathbb{R}^{4N'} \to E^{4N'}$. We define endomorphisms I_ξ, J_ξ in $E^{4N'}$ by transporting I_0 and J_0 from $\mathbb{R}^{4N'}$. Namely, by the formula in (19). We can readily check that $I_\xi^2 = -1$, $J_\xi^2 = -1$, $I_\xi J_\xi = -J_\xi I_\xi$. This defines a map from $\mathcal{F}(E^{4N'})$ to the set of quaternionic structures in $E^{4N'}$. This map is surjective because given any quaternionic structures I, J in $E^{4N'}$, one can choose a right $\mathbb{H}$ vector space basis $\{v_1, \ldots, v_{N'}\}$ of $E^{4N'}$. One can then extend this $\mathbb{H}$-basis to a $\mathbb{R}$-basis of $E^{4N'}$ using I and J as in (19). This frame recovers the original quaternionic structure by the above correspondence. We check the fibres of this map from $\mathcal{F}(E^{4N'})$. Suppose two frames ξ and η define the same quaternionic structure on $E^{4N'}$. Then, the composition $\xi^{-1} \circ \eta : \mathbb{R}^{4N'} \to \mathbb{R}^{4N'}$ preserves the quaternionic structure in the standard model space, so this must be an element in $\mathrm{GL}_{N'}(\mathbb{H})$. Hence $\eta \in \xi \cdot \mathrm{GL}_{N'}(\mathbb{H})$. Thus, we have obtained the following lemma.

LEMMA 5. *The moduli space of inequivalent quaternionic structures in real $4N'$ dimensional vector space $E^{4N'}$ is described by the bijection*

$$(21) \qquad \mathcal{F}(E^{4N'})/\mathrm{GL}_{N'}(\mathbb{H}) \xrightarrow[\text{onto}]{1:1} \{\text{quaternionic structures in } E^{4N'}\}.$$

For any frame $\xi \in \mathcal{F}(E^{4N'})$, let I_ξ, J_ξ be the corresponding quaternionic structures in $E^{4N'}$. Then the coset $\xi \cdot \mathrm{GL}_{N'}(\mathbb{H})$ is precisely the set of all preferred frames for this quaternionic structure. Through ξ, the above moduli space can be identified with $GL_{4N'}(\mathbb{R})/\mathrm{GL}_{N'}(\mathbb{H})$.

(VI) Symplectic scalar product structures. We give a brief introduction to symplectic scalar products and their relation to Euclidean pairings on a quaternionic vector space $E^{4N'}$ equipped with endomorphism I, J with $I^2 = J^2 = -1$, $IJ = -JI$. We let $JI = K$. This defines a right $\mathbb{H}$ module structure on $E^{4N'}$ by letting $I(u) = u \cdot i$, $J(u) = u \cdot j$, $K(u) = u \cdot k$. A symplectic scalar product $\langle \ , \ \rangle_{\mathbb{H}} : E^{4N'} \times E^{4N'} \to \mathbb{H}$ is a pairing which is $\mathbb{H}$-sesqui-linear in the first variable in the sense that $\langle uq_1, vq_2 \rangle = \bar{q}_1 \langle u, v \rangle q_2$ for any $u, v \in E^{4N'}$ and $q_1, q_2 \in \mathbb{H}$. The reason of the above behavior of q_1 and q_2 comes from the following matrix representation:

(22)

$$\langle (u_1, \ldots, u_{N'}) \begin{pmatrix} q_1 \\ \vdots \\ q_{N'} \end{pmatrix}, (v_1, \ldots, v_{N'}) \begin{pmatrix} q'_1 \\ \vdots \\ q'_{N'} \end{pmatrix} \rangle = (\bar{q}_1, \ldots, \bar{q}_{N'})(\langle u_i, v_j \rangle) \begin{pmatrix} q'_1 \\ \vdots \\ q'_{N'} \end{pmatrix}.$$

The above form of matrix multiplication is the reason of our choice of right $\mathbb{H}$ module structure on $E^{4N'}$ instead of left $\mathbb{H}$-module structure. The right $\mathbb{H}$ module structure leads to our convention $JI = K$. The relationship between symplectic scalar products and Euclidean scalar products is given as follows:

LEMMA 6. (a) *Let $\langle \ , \ \rangle_{\mathbb{H}}$ be a symplectic scalar product on a quaternionic vector space $E^{4N'}$, as above. Then, its real part $\langle \ , \ \rangle_{\mathbb{R}} = \Re\langle \ , \ \rangle_{\mathbb{H}}$ is a Euclidean scalar product on $E^{4N'}$ on which I and J act as isometries.*

(b) *Conversely, let $\langle \ , \ \rangle_{\mathbb{R}}$ be a Euclidean pairing on $E^{4N'}$ on which I, J act as isometries. Then the pairing*

(23) $\langle u, v \rangle_{\mathbb{H}} = \langle u, v \rangle_{\mathbb{R}} + i\langle I(u), v \rangle_{\mathbb{R}} + j\langle J(u), v \rangle_{\mathbb{R}} + k\langle JI(u), v \rangle_{\mathbb{R}} \in \mathbb{H}$

is a symplectic pairing on $E^{4N'}$ satisfying $\langle uq_1, vq_2 \rangle_{\mathbb{H}} = \bar{q}_1 \cdot \langle u, v \rangle_{\mathbb{H}} \cdot q_2$ for any $u, v \in E$ and $q_1, q_2 \in \mathbb{H}$.

PROOF. From our definition of the Euclidean pairing,

$$\langle I(u), I(v) \rangle_{\mathbb{R}} = \Re\langle ui, vi \rangle_{\mathbb{H}} = \Re\{(-i)\langle u, v \rangle_{\mathbb{H}} i\} = \Re\langle u, v \rangle_{\mathbb{H}}$$
$$= \langle u, v \rangle_{\mathbb{R}},$$

where the third equality is due to the fact that for any nonzero $q \in \mathbb{H}^*$, the conjugation of $\mathbb{H}$ by q keeps the real part invariant. This shows that I acts as an isometry on the Euclidean scalar product $\langle \ , \ \rangle_{\mathbb{R}}$. Similarly we can show that J also acts as an isometry.

The second part of the statement can be checked directly. For example, letting $\langle \ , \ \rangle = \langle \ , \ \rangle_{\mathbb{R}}$, (23) says that $\langle u, I(v) \rangle_{\mathbb{H}}$ is equal to

$$\langle u, I(v) \rangle + i\langle I(u), I(v) \rangle + j\langle J(u), I(v) \rangle + k\langle JI(u), I(v) \rangle$$
$$= -\langle I(u), v \rangle + i\langle u, v \rangle + j\langle JI(u), v \rangle - k\langle J(u), v \rangle = \langle u, v \rangle_{\mathbb{H}} i.$$

Other cases can be checked by similar calculations. $\square$

REMARK. Given a Euclidean scalar product $\langle \, , \, \rangle = \langle \, , \, \rangle_{\mathbb{R}}$ on $E^{4N'}$ for which I, J are isometries, we could define $\langle \, , \, \rangle_{\mathbb{H}}$ by

$$(24) \qquad \langle u, v \rangle'_{\mathbb{H}} = \langle u, v \rangle + i\langle u, I(v) \rangle + j\langle u, J(v) \rangle + k\langle u, IJ(v) \rangle \in \mathbb{H}.$$

This pairing is $\mathbb{H}$-sesquilinear in the second variable and satisfies $\langle uq_1, vq_2 \rangle'_{\mathbb{H}} = q_1 \langle u, v \rangle'_{\mathbb{H}} \bar{q}_2$ for any $q_1, q_2 \in \mathbb{H}$ and any $u, v \in E^{4N'}$. A statement similar to the above Lemma also holds. In fact, we have $\langle u, v \rangle'_{\mathbb{H}} = \overline{\langle u, v \rangle_{\mathbb{H}}} \in \mathbb{H}$. Since we have at least two choices of symplectic pairings for a single Euclidean scalar product, we regard this Euclidean scalar product more fundamental and from now on, by a symplectic scalar product, we mean a Euclidean pairing for which I and J are isometries.

Now, suppose that a quaternionic vector space $(E^{4N'}; I, J)$ carries a compatible Euclidean scalar product $\langle \, , \, \rangle_E$ so that both of the complex structures I and J are isometries. Then, the (left) automorphism group of this structure is defined by

(25)

$$\text{Aut}(E^{4N'}; I, J, \langle \, , \, \rangle_E) = \left\{ T \in \text{GL}(E^{4N'}) \,\middle|\, \begin{array}{c} T \circ I = I \circ T, \ T \circ J = J \circ T, \\ \langle T(u), T(v) \rangle_E = \langle u, v \rangle_E, \\ \forall u, \forall v \in E^{4N'} \end{array} \right\}.$$

This group acts on $E^{4N'}$ from the left. So, it acts on the set of compatible quaternionic frames from the left.

The standard model space for the symplectic scalar product structure is $\mathbb{R}^{4N'} = \bigoplus_{j=1}^{4N'} \mathbb{R}e_j$ equipped with the standard quaternionic structure I_0 and J_0 and the standard Euclidean pairing $\langle \, , \, \rangle_0$. It is straightforward to check that I_0 and J_0 are isometries, because they map a basis vector to another basis vector up to a sign. The (right) automorphism group of this structure is the symplectic group $\text{Sp}(N')$ consisting of $\mathbb{H}$-linear isomorphisms preserving the symplectic scalar product. So, each element in $\text{Sp}(N')$ must preserve the Euclidean scalar product and the quaternionic structure. We have

$$(26) \qquad\qquad O(4N') \cap \text{GL}_{N'}(\mathbb{H}) = \text{Sp}(N').$$

Since I_0 is an isometry, we can think of $(\langle \, , \, \rangle_0, I_0)$ as a Hermitian pairing compatible with the quaternionic structure. Since the automorphism group of the Hermitian pairing is $U(N)$, we also have the following identity among Lie groups:

$$(27) \qquad\qquad U(2N') \cap \text{GL}_{N'}(\mathbb{H}) = \text{Sp}(N').$$

Given a symplectic scalar product on a quaternionic vector space $E^{4N'}$, the set of preferred frames $\xi : \mathbb{R}^{4N'} \to E^{4N'}$ consists of those frames which are $\mathbb{H}$-linear and preserve symplectic scalar product structures. We call such frames symplectic frames and their totality is denoted by $\mathcal{F}(E^{4N'}; I, J, \langle \ , \ \rangle_E)$. So, it is given by

$$(28) \quad \mathcal{F}(E^{4N'}; I, J, \langle \ , \ \rangle_E) = \left\{ \xi \in \mathcal{F}(E^{4N'}) \ \middle| \ \begin{array}{l} I \circ \xi = \xi \circ I_0, \ J \circ \xi = \xi \circ J_0, \\ \langle \xi(u), \xi(v) \rangle_E = \langle u, v \rangle_0, \ \text{for} \\ \text{any } u, v \in \mathbb{R}^{4N'} \end{array} \right\}.$$

The groups $\mathrm{Sp}(N')$ and $\mathrm{Aut}(E^{4N'}; I, J, \langle \ , \ \rangle_E)$ act on this set of preferred frames transitively and effectively from the right and from the left, respectively, and for any symplectic frame ξ, we have

$$(29) \qquad \mathcal{F}(E^{4N'}; I, J, \langle \ , \ \rangle_E) = \xi \cdot \mathrm{Sp}(N') = \mathrm{Aut}(E^{4N'}; I, J, \langle \ , \ \rangle_E) \cdot \xi.$$

Now, we describe the moduli space of symplectic scalar product structures in a real $4N'$ dimensional vector space $E^{4N'}$. We consider this in the following two contexts.

(i) The space of symplectic scalar products which is compatible to a given quaternionic vector space structure in $E^{4N'}$.

(ii) The space of quaternionic vector space structures I, J in $E^{4N'}$ which are compatible with a given Euclidean scalar product in $E^{4N'}$.

Of course, other contexts are possible, but we restrict ourselves to the above two cases because $\mathrm{Sp}(N') = \mathrm{O}(4N') \cap \mathrm{GL}_{N'}(\mathbb{H})$.

CASE (i). Suppose that $E^{4N'}$ has a quaternionic vector space structure, namely, endomorphisms I and J on $E^{4N'}$ such that $I^2 = J^2 = -1$, $IJ = -JI$. Let $\mathcal{F}(E^{4N'}; I, J)$ be the set of quaternionic frames in $E^{4N'}$. Any such frame ξ is of the form
(30)
$$\xi = (v_1, \ldots, v_{N'}, J(v_1), \ldots, J(v_{N'}), I(v_1), \ldots, I(v_{N'}), IJ(v_1), \ldots, IJ(v_{N'})).$$

We can introduce a Euclidean scalar product in $E^{4N'}$ using this frame ξ. Namely, we introduce it by decreeing that ξ is an isometry between $\mathbb{R}^{4N'}$ and $E^{4N'}$ so that the above basis vectors form an orthonormal frame in $E^{4N'}$. Then, one can easily see that I and J are isometries with respect to this Euclidean scalar product. This defines a map from the set of symplectic frames $\mathcal{F}(E^{4N'}; I, J)$ to the set of compatible symplectic scalar products. This map is surjective, since we can always choose an orthonormal $\mathbb{H}$ basis $\{v_1, \ldots, v_{N'}\}$ of $E^{4N'}$, and then extend this $\mathbb{H}$-basis to a symplectic frame which is an $\mathbb{R}$-basis of $E^{4N'}$ using the action of I and J as in (30). Now let ξ and η be two quaternionic frames in $E^{4N'}$ such that the corresponding symplectic scalar products in $E^{4N'}$ are the same. Then, the composition $\xi^{-1} \circ \eta : \mathbb{R}^{4N'} \to \mathbb{R}^{4N'}$ is an $\mathbb{H}$ linear map preserving the standard quaternionic structure and the standard symplectic scalar product. Hence, it must be an element in $\mathrm{Sp}(N')$ and we have $\eta \in \xi \cdot \mathrm{Sp}(N')$. Thus, we have the following lemma.

LEMMA 7. *Let $(E^{4N'}; I, J)$ be a quaternionic vector space. The moduli space of all symplectic scalar products in $E^{4N'}$ compatible with the given quaternionic structure is described by the following bijective map :*

$$(31) \quad \mathcal{F}(E^{4N'} : I, J)/\mathrm{Sp}(N') \xrightarrow[\text{onto}]{1:1} \left\{ \begin{array}{l} \text{compatible symplectic scalar product} \\ \text{structures in } (E^{4N'}; I, J) \end{array} \right\}.$$

Any choice of a symplectic frame identifies the above space with a homogeneous space $\mathrm{GL}_{N'}(\mathbb{H})/\mathrm{Sp}(N')$. For any quaternionic frame $\xi \in \mathcal{F}(E^{4N'} : I, J)$, the coset $\xi \cdot \mathrm{Sp}(N')$ is precisely the set of all the preferred frames $\mathcal{F}(E^{4N'}; I, J, \langle \, , \, \rangle_\xi)$ for the corresponding symplectic pairing $\langle \, , \, \rangle_\xi$.

CASE (ii). Suppose that $E^{4N'}$ is a real $4N'$ dimensional vector space equipped with a Euclidean scalar product $\langle \, , \, \rangle_E$. We describe the totality of inequivalent quaternionic structures (I, J) in $E^{4N'}$ which are compatible with the given Euclidean structure in the sense that I and J are isometries of $E^{4N'}$. Let $\mathcal{F}(E^{4N'}; \langle \, , \, \rangle_E)$ be the set of orthonormal frames in $E^{4N'}$. Any frame $\xi \in \mathcal{F}(E^{4N'}; \langle \, , \, \rangle_E)$ defines an isometry $\xi : \mathbb{R}^{4N'} \to E^{4N'}$ and we use this map to transport the standard quaternionic structure in $\mathbb{R}^{4N'}$ to $E^{4N'}$. So, we let $I = \xi \circ I_0 \circ \xi^{-1}$, $J = \xi \circ J_0 \circ \xi^{-1}$. Obviously these endomorphisms of $E^{4N'}$ define a quaternionic structure in $E^{4N'}$, since it is immediate to check that $I^2 = J^2 = -1$, $IJ = -JI$. Furthermore, I, J are isometries. To see this, for any $u, v \in E^{4N'}$, from the definition of I, J, we have $\langle I(u), I(v) \rangle_E = \langle \xi \circ I_0 \circ \xi^{-1}(u), \xi \circ I_0 \circ \xi^{-1}(v) \rangle_E = \langle I_0 \circ \xi^{-1}(u), I_0 \circ \xi^{-1}(v) \rangle_0$, where we have used the fact that ξ is an isometry between $\mathbb{R}^{4N'}$ and $E^{4N'}$. Since I_0 is an isometry in $\mathbb{R}^{4N'}$, this is equal to $\langle \xi^{-1}(u), \xi^{-1}(v) \rangle_0 = \langle u, v \rangle_E$, again since ξ is an isometry. This shows that I is an isometry on $(E^{4N'}; \langle \, , \, \rangle_E)$. Similarly, one can show that J is also an isometry. Thus, I, J gives a quaternionic structure compatible with the given Euclidean structure $\langle \, , \, \rangle_E$. This way, we have defined a map from the set of Euclidean frames $\mathcal{F}(E^{4N'}; \langle \, , \, \rangle_E)$ to the moduli space of quaternionic structures in $E^{4N'}$ compatible with the given Euclidean structure $\langle \, , \, \rangle_E$. This map is surjective, since given a symplectic orthonormal $\mathbb{H}$-basis, we can always extend this to an orthonormal $\mathbb{R}$-basis using isometries I and J as in (30).

We look at the fibres of this map from $\mathcal{F}(E^{4N'}; \langle \, , \, \rangle_E)$. For two orthonormal frames $\xi, \eta \in \mathcal{F}(E^{4N'}; \langle \, , \, \rangle_E)$, suppose that the corresponding quaternionic structures in $E^{4N'}$ are the same. Then the composition $\xi^{-1} \circ \eta : \mathbb{R}^{4N'} \to \mathbb{R}^{4N'}$ preserves the standard symplectic pairing $\langle \, , \, \rangle_0$ and the standard quaternionic structure I_0, J_0. Hence, this composition must lie in the group $\mathrm{Sp}(N')$, or $\eta \in \xi \cdot \mathrm{Sp}(N')$, and η and ξ lie in the same coset of $\mathrm{Sp}(N')$. Thus, we have the following lemma.

LEMMA 8. *Let $(E^{4N'}; \langle \, , \, \rangle_E)$ be a real $4N'$-dimensional Euclidean vector space. The moduli space of quaternionic structures in $E^{4N'}$ compatible with the given Euclidean structure is described by the following bijective correspondence:*

$$(32) \quad \mathcal{F}(E^{4N'}; \langle \, , \, \rangle_E)/\mathrm{Sp}(N') \xrightarrow[\text{onto}]{1:1} \left\{ \begin{array}{l} \text{quaternionic structures compatible} \\ \text{with } (E^{4N'}; \langle \, , \, \rangle_E) \end{array} \right\}.$$

Any choice of an $\mathbb{R}$ orthonormal frame ξ allows us to identify the above moduli space with a homogeneous space $O(4N')/\mathrm{Sp}(N')$. Each coset $\xi \cdot \mathrm{Sp}(N')$ is precisely the set of all the preferred frames $\mathcal{F}(E^{4N'}; I_\xi, J_\xi, \langle\ ,\ \rangle_E)$ for the corresponding quaternionic structure I_ξ and J_ξ.

(VII) Pseudo-symplectic structures. Suppose that a real $4N'$ dimensional Euclidean vector space $(E^{4N'}; \langle\ ,\ \rangle_E)$ has a compatible quaternionic structure (I, J) so that I, J act as isometries. The automorphism group of this structure was considered in (VI). Let $L = \mathbb{R}I \oplus \mathbb{R}J \oplus \mathbb{R}K$ be a vector subspace of the endomorphism ring $\mathrm{End}(E^{4N'})$. The structure we consider in this subsection is $(E^{4N'}; \langle\ ,\ \rangle_E, L)$, where we do not specify isometric quaternionic structures I, J, K forming a basis of L. We call this structure a pseudo-symplectic structure. An element of the (left) automorphism group for this structure is only required to preserve $\langle\ ,\ \rangle_E$ and the vector space L and not required to preserve a specific isometric quaternionic structure I, J, K, which form a basis of L. So,

$$(33) \qquad \mathrm{Aut}(E^{4N'}; \langle\ ,\ \rangle_E, L) = \{T \in O(E^{4N'}; \langle\ ,\ \rangle_E) \mid T \circ L \circ T^{-1} = L\}.$$

Obviously this group strictly contains the (left) automorphism group for the structure $(E^{4N'}; I, J, \langle\ ,\ \rangle_E)$, which is isomorphic to the (right) automorphism group $\mathrm{Sp}(N')$, although not canonically.

The standard model space is $\mathbb{R}^{4N'}$ equipped with the standard Euclidean scalar product and the standard compatible quaternionic structure together with a subspace of endomorphisms $L_0 = \mathbb{R}I_0 \oplus \mathbb{R}J_0 \oplus \mathbb{R}K_0$ spanned by I_0, J_0, K_0 as in (17). We determine the automorphism group for this structure. First of all, the symplectic group $\mathrm{Sp}(N')$ acting on $\mathbb{R}^{4N'}$ from the left preserves the Euclidean scalar product and the quaternionic structure. So, $\mathrm{Sp}(N')$ is a subgroup of this automorphism group. On the other hand, the group of unit quaternions $\mathrm{Sp}(1)$ acts on $\mathbb{R}^{4N'} = \mathbb{H}^{N'}$ from the right by $u \cdot (a + bi + cj + dk) = au + bI_0(u) + cJ_0(u) + dK_0(u)$, $a^2 + b^2 + c^2 + d^2 = 1$, where $u \in E^{4N'}$. One can easily check that this action preserves the standard Euclidean pairing $\langle\ ,\ \rangle_E$. Since the conjugation of $\mathbb{H}$ by a nonzero quaternion preserves $\mathbb{R}i \oplus \mathbb{R}j \oplus \mathbb{R}k \subset \mathbb{H}$, we see that $q \cdot (\mathbb{R}I_0 \oplus \mathbb{R}J_0 \oplus \mathbb{R}K_0) \cdot q^{-1} = \mathbb{R}I_0 \oplus \mathbb{R}J_0 \oplus \mathbb{R}K_0$ as endomorphisms acting on $\mathbb{R}^{4N'}$. Thus, $\mathrm{Sp}(1) \subset \mathrm{SO}(4N')$ also preserve the structure of $(\mathbb{R}^{4N'}; \langle\ ,\ \rangle_0, L_0)$. This $\mathrm{Sp}(1)$ is the commutant of $\mathrm{Sp}(N')$ in $\mathrm{SO}(4N')$ and not a subgroup of $\mathrm{Sp}(N')$. Their intersection is $\{\pm 1\}$. We let $\mathrm{Sp}(N') \cdot \mathrm{Sp}(1) = \mathrm{Sp}(N') \times_{\{\pm 1\}} \mathrm{Sp}(1)$ be the subgroup of $\mathrm{SO}(4N')$ generated by $\mathrm{Sp}(N')$ and $\mathrm{Sp}(1)$. What we have seen so far is that the group of automorphisms of $(\mathbb{R}^{4N'}; \langle\ ,\ \rangle_0, L_0)$ contains $\mathrm{Sp}(N') \cdot \mathrm{Sp}(1)$. It is certainly strictly contained in $\mathrm{SO}(4N')$, since L_0 is not preserved by the entire group $\mathrm{SO}(4N')$. It turns out that the following lemma is true, whose proof we omit here. See (VII) of §4.2 and Lemma 8 of §4.3.

LEMMA 9. *The group $\mathrm{Sp}(N') \cdot \mathrm{Sp}(1)$ is a maximal subgroup of $\mathrm{SO}(4N')$ and it is precisely the automorphism group of $(\mathbb{R}^{4N'}; \langle\ ,\ \rangle_0, L_0)$.*

The preferred set of frames for the structured vector space $(E^{4N'}; \langle\ ,\ \rangle_E, L)$

consists of those frames $\xi : \mathbb{R}^{4N'} \to E^{4N'}$ which are structure preserving. Thus,

$$(34) \qquad \mathcal{F}(E^{4N'}; \langle\ ,\ \rangle, L) = \{\xi \in \mathcal{F}(E^{4N'}; \langle\ ,\ \rangle_E) \mid \xi^{-1} \circ L \circ \xi = L_0 \}.$$

We call a frame in this set a pseudo-symplectic frame. On this set of preferred frames, the group $\mathrm{Sp}(N') \cdot \mathrm{Sp}(1)$ acts effectively from the right. Actually, this action is transitive. To see this, let ξ, η be in $\mathcal{F}(E^{4N'}; \langle\ ,\ \rangle_E, L)$. Then the composition $\xi^{-1} \circ \eta$ is a structure preserving map from $(\mathbb{R}^{4N'}; \langle\ ,\ \rangle_0, L_0)$ to itself. Hence it is an element in $\mathrm{Sp}(N') \cdot \mathrm{Sp}(1)$. Thus, $\eta \in \xi \cdot (\mathrm{Sp}(N') \cdot \mathrm{Sp}(1))$ and the action is transitive. The similar statement is true for the left actions of the group $\mathrm{Aut}(E^{4N'}; \langle\ ,\ \rangle_E, L)$ and for any pseudo-symplectic frame ξ, we have

$$(35) \qquad \mathcal{F}(E^{4N'}; \langle\ ,\ \rangle_E, L) = \xi \cdot (\mathrm{Sp}(N') \cdot \mathrm{Sp}(1)) = \mathrm{Aut}(E^{4N'}; \langle\ ,\ \rangle_E, L) \cdot \xi.$$

Now, we consider the moduli space of the inequivalent structures of the type $(E^{4N'}; \langle\ ,\ \rangle_E, L)$ compatible with a given Euclidean structure $\langle\ ,\ \rangle_E$ of $E^{4N'}$. We recall that $\mathcal{F}(E^{4N'}; \langle\ ,\ \rangle_E)$ denotes the set of all Euclidean orthonormal frames. We can use any orthonormal frames to transport the standard quaternionic structure I_0, J_0, K_0 in $\mathbb{R}^{4N'}$ and also the vector space L_0 for $\mathbb{R}^{4N'}$ into $E^{4N'}$, giving rise to a quaternionic structure I, J, K and a vector space L for $E^{4N'}$. This defines a map from $\mathcal{F}(E^{4N'}; \langle\ ,\ \rangle_E)$ to the set of subspaces of $\mathrm{End}(E^{4N'})$ of the form $L = \mathbb{R}I \oplus \mathbb{R}J \oplus \mathbb{R}K$, where I, J, K are isometries defining a quaternionic structure on $E^{4N'}$. This map is surjective, since for any choice of L and for any choice of quaternionic isometries spanning I, J, K for L (there are many choices of I, J, K for a given L and any two choices are related by an element in the group $\mathrm{SO}(3)$ as in Lemma 17 in §4.2), we can choose an orthonormal $\mathbb{H}$-basis which we can extend to an $\mathbb{R}$-orthonormal basis using the isometries I, J as in (30). This orthonormal frame in $E^{4N'}$ recovers the original vector space L we started from. Now, let ξ and η be orthonormal frames which give rise to the same vector spaces $L_\xi = L_\eta$ where $L_\xi = \mathbb{R}I_\xi \oplus \mathbb{R}J_\xi \oplus \mathbb{R}K_\xi$ and $L_\eta = \mathbb{R}I_\eta \oplus \mathbb{R}J_\eta \oplus \mathbb{R}K_\eta$. Then, the composition $\xi^{-1} \circ \eta : \mathbb{R}^{4N'} \to \mathbb{R}^{4N'}$ preserves the standard structures in $(\mathbb{R}^{4N'}; \langle\ ,\ \rangle, L_0)$. So, it must be an element in $\mathrm{Sp}(N')\mathrm{Sp}(1)$, or $\eta \in \xi \cdot (\mathrm{Sp}(N') \cdot \mathrm{Sp}(1))$. This proves the following lemma.

LEMMA 10. *Let $(E^{4N'}; \langle\ ,\ \rangle_E)$ be a real $4N'$-dimensional Euclidean vector space. Then, the moduli space of the inequivalent pseudo-symplectic structures in $E^{4N'}$ compatible with the given Euclidean structure $\langle\ ,\ \rangle_E$ is described by* (36)

$$\mathcal{F}(E^{4N'}; \langle\ ,\ \rangle_E)/\mathrm{Sp}(N') \cdot \mathrm{Sp}(1) \xrightarrow[\text{onto}]{1:1} \left\{ \begin{array}{l} L = \mathbb{R}I \oplus \mathbb{R}J \oplus \mathbb{R}K \in \mathrm{End}(E^{4N'}) \\ \text{for some isometric quaternionic} \\ \text{structure } (I, J) \text{ on } E^{4N'} \end{array} \right\}$$

Any orthonormal frame ξ in $E^{4N'}$ allows us to identify the above space with a homogeneous space $\mathrm{O}(4N')/\mathrm{Sp}(N') \cdot \mathrm{Sp}(1)$ and the coset $\xi \cdot (\mathrm{Sp}(N') \cdot \mathrm{Sp}(1))$ is precisely the set of all the pseudo-symplectic frames for the corresponding structure $L_\xi = \mathbb{R}I_\xi \oplus \mathbb{R}J_\xi \oplus \mathbb{R}K_\xi$.

§4.2 Moduli of geometric structures as G-orbits in representations

Let E be a real n dimensional vector space. Moduli spaces of certain types of geometric structures in E naturally have the structure of homogeneous spaces. By choosing a basis of E, we can identify E with $\mathbb{R}^n$. The following interesting cases are well known and are discussed in the previous section §4.1.

(I) $\qquad \mathrm{GL}_n(\mathbb{R}^n)/\mathrm{O}(n) \cong \{$ Euclidean scalar products in $\mathbb{R}^n \}$

(II) $\qquad \mathrm{GL}_{2N}(\mathbb{R})/\mathrm{GL}_N(\mathbb{C}) \cong \{$ complex structures in $\mathbb{R}^{2N} \}$

(III) $\qquad \mathrm{GL}_{4N'}(\mathbb{R})/\mathrm{GL}_{N'}(\mathbb{H}) \cong \{$ quaternionic structures in $\mathbb{R}^{4N'} \}$

(IV) $\qquad \mathrm{O}(4N')/\mathrm{Sp}(N') \cong \{$ symplectic scalar products in $(\mathbb{R}^{4N'}, g_0)\}$

(V) $\qquad \mathrm{GL}_N(\mathbb{C})/\mathrm{U}(N) \cong \{$ Hermitian structures on $\mathbb{C}^N \}$

(VI) $\qquad \mathrm{GL}_N(\mathbb{C})/\mathrm{SU}(N) \cong \{$ special Hermitian structures in $\mathbb{C}^N \}$

(VII)$\mathrm{O}(4N')/(\mathrm{Sp}(N') \cdot \mathrm{Sp}(1)) \cong \{$ pseudo-symplectic structures in $(\mathbb{R}^{4N'}; g_0)\}$

In the above, g_0 denotes the standard Euclidean structure. We can make similar statements for the abstract vector space E without chosen basis, but the above isomorphisms with specific homogeneous spaces depend on the bases chosen. In (VII) above, by pseudo-symplectic structures we mean a three dimensional subspace of endomorphisms $L = \mathbb{R}I \oplus \mathbb{R}J \oplus \mathbb{R}K \subset \mathrm{End}(E)$ generated by isometries (I, J, K) of a Euclidean vector space $(E, \langle\,,\,\rangle)$ satisfying the quaternion relations, $I^2 = J^2 = -1$, $IJ = -JI = -K$.

In this section, we look at these homogeneous moduli spaces above from a different point of view: we identify these moduli spaces with G-orbits in some representation spaces for appropriate groups G. This point of view of orbits becomes important when we consider global geometric objects on manifolds. See §4.5.

Each choice of a basis of E gives rise to a standard choices of each geometric structure.In what follows, we deal with moduli spaces of geometric structures in a vector space $E = \bigoplus_{j=1}^n \mathbb{R}e_j$ equipped with a basis $\{e_1, \ldots, e_n\}$ and we consider the realization of these moduli spaces as G-orbits. For the coordinate free description of the moduli spaces of geometric structures, see §4.3.

(0) Induced Matrix Representations over $\mathbb{R}$ and over $\mathbb{C}$. With a chosen basis in E, we can identify $E \otimes_{\mathbb{R}} E$, $E \otimes_{\mathbb{R}} E^*$, $E^* \otimes_{\mathbb{R}} E^*$ with the vector space of matrices $\mathrm{M}_n(\mathbb{R})$. Given a complex structure in E and a complex basis, we can also identify the above types of tensor products over $\mathbb{C}$ including $\overline{E}^* \otimes_{\mathbb{C}} E^*$ with $\mathrm{M}_N(\mathbb{C})$, where $n = 2N$. Translating our orbit results obtained later into the matrix context, we can obtain interesting results on matrices. See Corollaries 7, 9, 11, 13, 15, 21.

Matrix representations over $\mathbb{R}$. Let $E = \bigoplus_{j=1}^n \mathbb{R}e_j$, $E^* = \bigoplus_{j=1}^n \mathbb{R}e_j^*$, where $\{e_1^*, \ldots, e_n^*\}$ is the dual basis of $\{e_1, \ldots, e_n\}$. In the current and the next subsection, we make some preparations in order to translate our results on moduli spaces of geometric structures into the matrix context. Our discussion here is in slightly more general terms than actually necessary.

Let G be a group and let $\rho : G \to GL(E)$ be a representation of G. Since E has a standard basis, we can represent $\rho(g)$ by a matrix M_g defined by

$$(1) \qquad (\rho(g)(e_1), \ldots, \rho(g)(e_n)) = (e_1, \ldots, e_n) M_g, \qquad M_g \in M_n(\mathbb{R}).$$

The dual representation $\rho^* : G \to GL(E^*)$ is defined by

$$(2) \qquad (\rho^*(g)f)(v) = f(\rho(g^{-1})(v)), \qquad \text{for } f \in E^*, \ v \in E, \ g \in G.$$

The next lemma is elementary, but we give a proof.

LEMMA 1. *Let $\rho : G \to GL(E)$ be a representation, where E^n is equipped with a basis $\{e_1, \ldots, e_n\}$. The matrix representation of the dual representation $\rho^* : G \to GL(E^*)$ with respect to the dual basis $\{e_1^*, \ldots, e_n^*\}$ is given by*

$$(\rho^*(g)e_1^*, \ldots, \rho^*(g)e_n^*) = (e_1^*, \ldots, e_n^*)\, {}^t M_g^{-1},$$

where M_g is the matrix representing $\rho(g)$ with respect to $\{e_1, \ldots, e_n\}$.

PROOF. For $g \in G$, let $(\rho^*(g)e_1^*, \ldots, \rho^*(g)e_n^*) = (e_1^*, \ldots, e_n^*)A_g$. Then, regarding the matrix multiplications as dual pairings, we have

$$\begin{pmatrix} \rho^*(g)e_1^* \\ \vdots \\ \rho^*(g)e_n^* \end{pmatrix} (e_1 \ldots e_n) = {}^t A_g \begin{pmatrix} e_1^* \\ \vdots \\ e_n^* \end{pmatrix} (e_1 \ldots e_n) = {}^t A_g$$

On the other hand, from the definition of the action of ρ^*,

$$\begin{pmatrix} \rho^*(g)e_1^* \\ \vdots \\ \rho^*(g)e_n^* \end{pmatrix} (e_1 \ldots e_n) = \begin{pmatrix} e_1^* \\ \vdots \\ e_n^* \end{pmatrix} (\rho(g^{-1})e_1 \ldots \rho(g^{-1})e_n)$$

$$= \begin{pmatrix} e_1^* \\ \vdots \\ e_n^* \end{pmatrix} (e_1 \ldots e_n) \, M_g^{-1} = M_g^{-1}.$$

Thus, the matrix A_g representing $\rho^*(g)$ is ${}^t M_g^{-1}$. This proves the lemma. □

From now on, we often use $\rho_E(g)$ to denote the matrix M_g associated to a fixed basis of E. For example, Lemma 1 says that $\rho^*(g) = {}^t \rho(g)^{-1}$.

Let $\rho_E : G \to GL(E)$ and $\rho_F : G \to GL(F)$ be two representations of G on vector spaces E^n and F^m of real dimension n and m, respectively. We choose bases for these vector spaces and let $E = \bigoplus_{j=1}^n \mathbb{R}e_j$, $F = \bigoplus_{j=1}^m \mathbb{R}f_j$. With respect to these bases, $\rho_E(g)$, $\rho_F(g)$ can be thought of as elements in $M_n(\mathbb{R})$, $M_m(\mathbb{R})$, respectively. Now, any element in $E \otimes F$ can be written as $\sum_{i=1}^n \sum_{j=1}^m a_{ij} e_i \otimes f_j$. To this element, we can associate a matrix $A = (a_{ij})$ of size $n \times m$. The group G acts diagonally on this tensor product $E \otimes F$.

LEMMA 2. *Through the identification of $E^n \otimes F^m$ with $\mathrm{M}_{n \times m}(\mathbb{R})$, the representation $\rho_{E \otimes F} : G \to \mathrm{GL}(E \otimes F)$ is described by*

$$\rho_{E \otimes F}(g)\left(\sum_{ij} a_{ij} e_i \otimes f_j\right) = \sum_{ij} b_{ij} e_i \otimes f_j,$$

where $B = (b_{ij}) = \rho_E(g) \cdot A \cdot {}^t\rho_F(g)$, $A = (a_{ij})$.

PROOF. If $v = \sum_{i=1}^n \sum_{j=1}^m a_{ij} e_i \otimes f_j$, then regarding the matrix multiplications as tensor products we can rewrite the vector $v \in E \otimes F$ as

$$v = (e_1 \ldots e_n) A \begin{pmatrix} f_1 \\ \vdots \\ f_m \end{pmatrix}.$$

Applying $\rho_{E \otimes F}(g) = \rho_E(g) \otimes \rho_F(g)$, we get

$$\rho_{E \otimes F}(g)v = (\rho_E(g)e_1 \ldots \rho_E(g)e_n) A \begin{pmatrix} \rho_F(g)f_1 \\ \vdots \\ \rho_F(g)f_m \end{pmatrix}$$

$$= (e_1 \ldots e_n) \rho_E(g)A\,{}^t\rho_F(g) \begin{pmatrix} f_1 \\ \vdots \\ f_m \end{pmatrix}.$$

Thus the matrix corresponding to the vector $\rho_{E \times F}(g)v$ is $\rho_E(g)A\,{}^t\rho_F(g)$. $\quad\square$

Given a basis $\{e_1, \ldots, e_n\}$ in E, its dual basis in E^* is denoted by $\{e_1^*, \ldots e_n^*\}$. From Lemma 2, we have the following corollary.

COROLLARY 3. *Let $E = \bigotimes_{j=1}^n \mathbb{R}e_j$ be a representation of G. The diagonal action of G on $E^* \otimes E^*$ is given by*

$$\rho_{E^* \otimes E^*}(g)\left(\sum_{ij} a_{ij} e_i^* \otimes e_j^*\right) = \sum_{ij} b_{ij} e_i^* \otimes e_j^*, \qquad g \in G,$$

where $B = (b_{ij}) = \rho^*(g)A\,{}^t\rho^*(g)$, $A = (a_{ij})$ and $\rho^*(g) = {}^t\rho(g)^{-1}$.
Similarly, the diagonal action of G on $E \otimes E^*$ is given by

$$\rho_{E \otimes E^*}(g)\left(\sum_{ij} a_{ij} e_i \otimes e_j^*\right) = \sum_{ij} b_{ij} e_i \otimes e_j^*, \qquad g \in G,$$

where $B = (b_{ij}) = \rho(g)A\,{}^t\rho^*(g)$, $A = (a_{ij})$.

Matrix representations over $\mathbb{C}$. Now, let E be a complex N dimensional vector space with a chosen basis $\{e_1, \ldots . e_N\}$, $E = \bigoplus_{j=1}^{N} \mathbb{C}e_j$. This complex vector space can be thought of as real $n = 2N$ dimensional vector space with basis $\{e_1, \ldots, e_N, e_{N+1}, \ldots, e_{2N}\}$ where $e_{N+j} = ie_j$, $1 \leq j \leq N$. We consider its dual vector spaces, E^* consisting of $\mathbb{C}$-linear maps with dual basis $\{e_1^*, \ldots, e_N^*\}$, and $\overline{E}^*$ consisting of conjugate linear maps with dual basis $\{\bar{e}_1^*, \ldots, \bar{e}_N^*\}$. Let E be a representation of G via $\rho : G \to \mathrm{GL}_{\mathbb{C}}(E)$. Since E is equipped with a basis, we have a matrix representation of ρ. We let

(3) $(\rho(g)e_1, \ldots, \rho(g)e_N) = (e_1, \ldots, e_N)M_g$, $M_g \in \mathrm{M}_N(\mathbb{C})$, for $g \in G$.

Now G acts on E^* and on $\overline{E}^*$ via $\rho^* : G \to \mathrm{GL}(E^*)$ and $\bar{\rho}^* : G \to \mathrm{GL}(\overline{E}^*)$ defined by the following formulae :

(4) $(\rho^*(g)f)(v) = f(\rho(g^{-1})v)$, $(\bar{\rho}^*(g)\bar{f})(v) = \bar{f}(\rho(g^{-1})v)$,

where $g \in G$, $f \in E^*$, $\bar{f} \in \overline{E}^*$, $v \in E$. The matrix representations of ρ^* and $\bar{\rho}^*$ are given in the next lemma.

LEMMA 4. *Let* $M_g \in \mathrm{GL}_N(\mathbb{C})$ *be the matrix corresponding to* $\rho(g)$. *Then the matrices corresponding to operators* $\rho^*(g)$ *and* $\bar{\rho}^*(g)$ *are given by*

$$(\rho^*(g)e_1^*, \ldots, \rho^*(g)e_N^*) = (e_1^*, \ldots, e_N^*)\,{}^t M_g^{-1},$$
$$(\bar{\rho}^*(g)\bar{e}_1^*, \ldots, \bar{\rho}^*(g)\bar{e}_N^*) = (\bar{e}_1^*, \ldots, \bar{e}_n^*)\,{}^t\overline{M}_g^{-1}.$$

PROOF. The proof of the first identity is the same as the real case. The proof of the second one is almost the same except for the conjugation operation involved. Here is the detail. Let $(\bar{\rho}^*(g)\bar{e}_1^*, \ldots, \bar{\rho}^*(g)\bar{e}_N^*) = (\bar{e}_1^*, \ldots, \bar{e}_N^*)A$ for some $A \in \mathrm{GL}_N(\mathbb{C})$. Then, regarding matrix multiplications as tensor products, and using the definition of the action of $\bar{\rho}^*$,

$${}^tA = {}^tA \begin{pmatrix} \bar{e}_1^* \\ \vdots \\ \bar{e}_N^* \end{pmatrix} (e_1 \ldots e_N) = \begin{pmatrix} \bar{\rho}^*(g)\bar{e}_1^* \\ \vdots \\ \bar{\rho}^*(g)\bar{e}_N^* \end{pmatrix} (e_1 \ldots e_N)$$

$$= \begin{pmatrix} \bar{e}_1^* \\ \vdots \\ \bar{e}_N^* \end{pmatrix} (\rho(g^{-1})e_1 \ldots \rho(g^{-1})e_N) = \begin{pmatrix} \bar{e}_1^* \\ \vdots \\ \bar{e}_N^* \end{pmatrix} ((e_1 \ldots e_N)M_{g^{-1}}) = \overline{M}_g^{-1}.$$

Here in the last identity, we have to conjugate the coefficient matrix M_g^{-1} since $\bar{e}_j^*$'s are conjugate linear. This, $A = {}^t\overline{M}_g^{-1}$ and we have $\bar{\rho}^*(g) = {}^t\overline{\rho(g)}^{-1}$. This completes the proof of the second identity. $\square$

The tensor product $\overline{E}^* \otimes E^*$ has a basis $\{\bar{e}_j^* \otimes e_k^*\}_{jk}$ and can be identified with the vector space of pairing in E which are sesquilinear in the first variable. Also, we can identify this tensor product with the vector space of matrices $\mathrm{M}_N(\mathbb{C})$ by taking the coefficients with respect to the above basis. The group G acts diagonally on the tensor product representation. This action is described in the next lemma which is a consequence of Lemma 2.

LEMMA 5. *The diagonal action $\bar{\rho}^* \otimes \rho^*$ of the group G on $\overline{E}^* \otimes E^*$ is given by*

$$\rho_{\overline{E}^* \otimes E^*}(g)\left(\sum_{ij} a_{ij}\bar{e}_i^* \otimes e_j^*\right) = \sum_{ij} b_{ij}\bar{e}_i^* \otimes e_j^*, \qquad g \in G,$$

where $B = (b_{ij}) = \bar{\rho}^*(g)A\,{}^t\rho^*(g)$, $A = (a_{ij})$, and $\bar{\rho}^*(g) = {}^t\overline{\rho(g)}^{-1}$, $\rho^*(g) = {}^t\rho(g)^{-1}$.

(I)   The moduli space of Euclidean scalar products as an orbit. Let E be a real n dimensional vector space with a chosen basis $\{e_1, \ldots, e_n\}$. The vector space $E^* \otimes E^*$ is the vector space of bilinear pairings on E. This space can be identified with $M_n(\mathbb{R})$ by associating the matrix $A = (a_{ij})$ to an element $\sum_{ij} a_{ij}e_i^* \otimes e_j^*$. Under this identification, symmetric bilinear pairings correspond to symmetric matrices and Euclidean scalar products(=inner products) in E correspond to positive definite symmetric matrices. The standard Euclidean scalar product associated to the chosen basis $\{e_1, \ldots, e_N\}$ is given by

$$(5) \qquad\qquad h_0 = \sum_{j=1}^{n} e_j^* \otimes e_j^* \in E^* \otimes E^*.$$

The matrix corresponding to h_0 is the identity matrix I_n.

From now on in this section, we regard $E = \bigoplus_{j=1}^{n} \mathbb{R}e_j$ as a representation ρ_E of the group $G = \mathrm{GL}_n(\mathbb{R})$ by

$$(6) \qquad (\rho_E(g)e_1, \ldots, \rho_E(g)e_n) = (e_1, \ldots, e_n)g, \qquad \text{for any } g \in \mathrm{GL}_n(\mathbb{R}).$$

Then, $E^* \otimes E^*$ is a representation of $G = \mathrm{GL}_n(\mathbb{R})$ via $\rho_{E^* \otimes E^*} = \rho_{E^*} \otimes \rho_{E^*}$, and we consider the orbit of h_0 under this action of G.

Using the first part of Corollary 3 and Lemma 1, the matrix corresponding to an element $\rho_{E^* \otimes E^*}(g)h_0 \in E^* \otimes E^*$ is given by $\rho^*(g)I_n\,{}^t\rho^*(g) = {}^tM_g^{-1} \cdot M_g^{-1}$ where $M_g = g \in \mathrm{GL}_n(\mathbb{R})$ is the matrix representing $\rho(g)$ on E. This matrix is a positive definite symmetric matrix. The isotropy subgroup of $G = \mathrm{GL}_n(\mathbb{R})$ at h_0 consists of those elements $g \in G$ such that $\rho^*(g) \cdot {}^t\rho^*(g) = I_n$. Thus, $\rho^*(g) = {}^t\rho(g)^{-1}$ must be an orthogonal matrix, which in turn imply that $\rho(g)$ must be also an orthogonal matrix. So, $g \in O(n)$, since ρ_E is really an identity map. Hence the orbit $\mathrm{GL}_n(\mathbb{R}) \cdot h_0$ is isomorphic to $\mathrm{GL}_n(\mathbb{R})/O(n)$. Under the identification $E^* \otimes E^* \cong M_n(\mathbb{R})$, this orbit consists of positive definite symmetric matrices. We show that in fact all the positive definite symmetric matrices are in the orbit and hence the group $\mathrm{GL}_n(\mathbb{R})$ acts transitively on this set. To see this, let $A = (a_{ij})$ be a positive definite symmetric matrix. The corresponding element $h = \sum_{ij} a_{ij}e_i^* \otimes e_j^*$ defines a positive definite symmetric bilinear pairing h on E, that is, an inner product in E. By the Gram-Schmidt process, we can find an orthonormal basis $\{v_1, \ldots, v_n\}$ of E with respect to this inner product, so we have $h(v_i, v_j) = \delta_{ij}$. Then a map $\xi : (\mathbb{R}^n, h_0) \to (\mathbb{R}^n, h)$ defined by $\xi(e_j) = v_j$, $1 \leq j \leq n$, is an isometry between two Euclidean spaces. Expressing $\{v_1, \ldots, v_N\}$ in terms of the originally chosen basis $\{e_1, \ldots, e_n\}$, we obtain a matrix $g = M_g \in GL_n(\mathbb{R})$ such that $\xi = \rho_E(g)$. Namely, $(v_1, \ldots, v_n) = (\xi(e_1), \ldots, \xi(e_n)) = (e_1, \ldots, e_n)M_g$. Recall that ρ_E is really an identity map. But then, $(\rho_{E^* \otimes E^*}(g)(h_0))(v_i, v_j) = h_0\left(\rho_E(g^{-1})v_i, \rho_E(g^{-1})v_j\right) = h_0(\xi^{-1}(v_1), \xi^{-1}(v_j)) = h_0(e_i, e_j) = \delta_{ij}$. Thus, the basis $\{v_1, \ldots, v_n\}$ is an orthonormal basis for a pairing given by $\rho_{E^* \otimes E^*}(g)h_0$. Hence it must be the same vector as h in $E^* \otimes E^*$. This shows that the $GL_n(\mathbb{R})$-orbit of h_0 in $E^* \otimes E^*$ is the entire set of Euclidean scalar products, or equivalently, the set of all positive definite symmetric matrices. This proves the following proposition.

PROPOSITION 6. *Let $E = \bigoplus_{j=1}^n \mathbb{R}e_j$ be a real n dimensional vector space with a chosen basis. The $GL_n(\mathbb{R})$-orbit through $h_0 = \sum_{j=1}^n e_j^* \otimes e_j^* \in E^* \otimes E^*$ can be identified with the moduli space of the Euclidean scalar products and we have the the following diagram.*

$$GL_n(\mathbb{R})/O(n) \xrightarrow{\cong} GL_n(\mathbb{R}) \cdot h_0 = \{\text{Euclidean scalar products in } E\} \subset E^* \otimes E^*$$

$$\updownarrow \cong$$

$$GL_n(\mathbb{R}) \ni M \mapsto {}^tM^{-1} \cdot M^{-1} \in \{\text{positive definite symmetric matrices }\} \subset M_n(\mathbb{R})$$

COROLLARY 7. *Any positive definite symmetric matrix in $M_n(\mathbb{R})$ is of the form $M \cdot {}^tM$ for some $M \in GL_n(\mathbb{R})$.*

(II)   The moduli space of complex structures as an orbit. Let $E = \bigoplus_{j=1}^{2N} \mathbb{R}e_j$ be a real $n = 2N$ dimensional vector space with a chosen basis. We consider the standard complex structure $I_0 \in E \otimes E^*$ given by

$$(7) \qquad I_0 = \sum_{j=1}^N e_{j+N} \otimes e_j^* - \sum_{j=1}^N e_j \otimes e_{j+N}^*.$$

Under the isomorphism $E \otimes E^* \cong \text{Hom}(E, E)$ in which $T : E \to E$ corresponds to $\sum_{j=1}^{2N} T(e_j) \otimes e_j^* \in E \otimes E^*$, the above element I_0 corresponds to a map such that $I_0(e_j) = e_{j+N}$, $I_0(e_{j+N}) = -e_j$, $1 \le j \le N$. Obviously $I_0^2 = -1$, so it defines a complex structure on E. Note also that we have an isomorphism $E \otimes E^*$ with $M_{2N}(\mathbb{R})$ using the chosen basis of E. The matrix associated to I_0 with respect to $\{e_1, \ldots, e_{2N}\}$ is $I_0 = \begin{pmatrix} 0 & -1_N \\ 1_N & 0 \end{pmatrix}$.

We consider the $G = GL_{2N}(\mathbb{R})$-orbit of I_0 in the representation space $E \otimes E^*$. Through the identification $E \otimes E^* \cong \text{Hom}(E, E)$, the matrix corresponding to a vector in the orbit $\rho_{E \otimes E^*}(g)(I_0)$ is given by $M_g \cdot I_0 \cdot M_g^{-1}$ by Corollary 3. We first look at the isotropy subgroup at I_0. An element $g \in GL_{2N}(\mathbb{R})$ is in the isotropy subgroup if and only if $M_g \cdot I_0 \cdot M_g^{-1} = I_0$. That is, if and only if the matrix M_g commutes with the matrix I_0, or $\rho_E(g)$ commutes with the complex structure I_0. This means that $M_g \in GL_N(\mathbb{C})$, where $GL_N(\mathbb{C})$ is the set of all the invertible complex linear maps on the complex vector space $E = $

$\bigoplus_{j=1}^{N} \mathbb{C}e_j$, where the complex structure comes from I_0 by $e_{j+N} = ie_j = I_0(e_j)$, $1 \leq j \leq N$. Thus, the orbit is a homogeneous space given by $GL_{2N}(\mathbb{R})/GL_N(\mathbb{C})$. Through the identification $E \otimes E^* \cong \text{Hom}(E, E) \cong M_{2N}(\mathbb{R})$, a point I in the orbit $I = \rho_{E \otimes E^*}(g)(I_0) = (\rho_E(g) \otimes \rho_{E^*}(g))(I_0)$ corresponds to a composition of the maps $\rho_E(g) \circ I_0 \circ \rho_E(g)^{-1} \in \text{Hom}(E, E)$ and to a product of matrices $M_g \cdot I_0 \cdot M_g^{-1} \in M_{2N}(\mathbb{R})$. We see that I, as an element in $\text{Hom}(E, E)$ or in $M_{2N}(\mathbb{R})$, satisfy $I^2 = -1$ since $I_0^2 = -1$, so it defines a complex structure in E. Thus, we have the following diagram.

$$GL_{2N}(\mathbb{R})/GL_N(\mathbb{C}) \xrightarrow{\cong} GL_{2N}(\mathbb{R}) \cdot I_0 \hookrightarrow \{\text{complex structures on } E\} \subset \text{Hom}(E, E)$$
$$\cap \qquad\qquad\qquad \updownarrow \cong$$
$$E \otimes E^* \qquad \{A \in M_{2N}(\mathbb{R}) \mid A^2 = -1 \} \subset M_{2N}(\mathbb{R})$$

We show that the middle inclusion map of the first line is actually surjective. To see this, let I be any complex structure on E. Let $\{v_1, \ldots, v_N\}$ be a complex basis of E. Then $\{v_1, \ldots, v_N, I(v_1), \ldots, I(v_N)\}$ is a basis of E over $\mathbb{R}$ and the map $\xi : (E, I_0) \to (E, I)$ defined by $\xi(e_j) = v_j$, $\xi(e_{N+j}) = \xi(I_0(e_j)) = I(v_j)$ for $1 \leq j \leq N$, is a complex linear map with respect to the respective complex structures. Let M_ξ be a matrix such that $(\xi(e_1), \ldots, \xi(e_{2N})) = (e_1, \ldots, e_{2N})M_\xi$. Let $g_\xi \in G = GL_{2N}(\mathbb{R})$ be an element given by M_ξ so that $\xi = \rho_E(g_\xi)$. Here recall that ρ_E is really an identity map. Since ξ commutes with complex structures I_0 and I, we have $I = \rho_E(g_\xi) \circ I_0 \circ \rho_E(g_\xi)^{-1}$. By definition, this is an element $\rho_{E \otimes E^*}(g_\xi)(I_0) \in E \otimes E^*$. So, I is in the G-orbit of I_0. Hence the orbit of $I_0 \in E \otimes E^*$ of the group $GL_{2N}(\mathbb{R})$ consists of all the complex structures in E^{2N} through the identification $E \otimes E^* \cong \text{Hom}(E, E)$.

PROPOSITION 8. *Let $E = \bigoplus_{j=1}^{2N} \mathbb{R}e_j$. The $GL_{2N}(\mathbb{R})$-orbit of $I_0 \in E \otimes E^*$ given in (7) under the action $\rho_{E \otimes E^*}$ can be identified with the moduli space of the complex structures in E.*

Since the orbit of the matrix I_0 in $M_{2N}(\mathbb{R})$ consists of those matrices whose squares are -1, we have the next corollary.

COROLLARY 9. *Let $A \in GL_{2N}(\mathbb{R})$ be such that $A^2 = -1$. Then there exists a matrix $M \in GL_{2N}(\mathbb{R})$ such that $A = M \cdot I_0 \cdot M^{-1}$, where $I_0 = \begin{pmatrix} 0 & -1_N \\ 1_N & 0 \end{pmatrix}$, i.e., A is conjugate to I_0 in $GL_{2N}(\mathbb{R})$.*

(III)   The moduli space of quaternionic structures as an orbit. Let the real dimension of E be a multiple of 4, say, $n = 2N = 4N'$. We consider the following two elements I_0 and J_0 in $E \otimes E^*$.

(8)
$$I_0 = \sum_{j=1}^{N} e_{N+j} \otimes e_j^* - \sum_{j=1}^{N} e_j \otimes e_{N+j}^*,$$

$$J_0 = \sum_{r=1}^{N'} (e_{r+N'} \otimes e_r^* + e_{N+r} \otimes e_{r+N+N'}^*) - \sum_{r=1}^{N'} (e_r \otimes e_{r+N'}^* + e_{r+N+N'} \otimes e_{r+N}^*).$$

Under the identification $E \otimes E^* \cong \mathrm{Hom}(E, E)$, these elements denote linear maps I_0, J_0 such that

(9)
$$I_0(e_j) = e_{j+N}, \qquad I_0(e_{j+N}) = -e_j, \qquad 1 \le j \le N,$$
$$J_0(e_r) = e_{r+N'}, \qquad J_0(e_{r+N'}) = -e_r,$$
$$J_0(e_{r+N}) = -e_{r+N+N'}, \qquad J_0(e_{r+N+N'}) = e_{r+N}, \qquad 1 \le r \le N'.$$

These maps satisfy $I_0^2 = J_0^2 = -1$, $I_0 J_0 = -J_0 I_0$. If we let $K_0 = J_0 I_0$, then I_0, J_0, K_0 together define the right $\mathbb{H}$-module structure on E, that is, the right multiplications by i, j, k are given by the maps I_0, J_0, K_0, which act on E from the left. The relations $i^2 = j^2 = -1$, $ij = -ji = k$ translate into the above relations. Since E has a chosen basis, we can further identify $E \otimes E^*$ with $\mathrm{M}_{4N'}(\mathbb{R})$. Then I_0, J_0 are represented by the following matrices.

(10)
$$I_0 = \begin{pmatrix} 0 & 0 & -1_{N'} & 0 \\ 0 & 0 & 0 & -1_{N'} \\ 1_{N'} & 0 & 0 & 0 \\ 0 & 1_{N'} & 0 & 0 \end{pmatrix}, \qquad J_0 = \begin{pmatrix} 0 & -1_{N'} & 0 & 0 \\ 1_{N'} & 0 & 0 & 0 \\ 0 & 0 & 0 & 1_{N'} \\ 0 & 0 & -1_{N'} & 0 \end{pmatrix}$$

As before, we make free use of the identification $E \otimes E^* \cong \mathrm{Hom}(E, E) \cong \mathrm{M}_{4N'}(\mathbb{R})$. The action of $G = \mathrm{GL}_{4N'}(\mathbb{R})$ on $E \otimes E^*$ described in Corollary 3 can be transferred to other two vector spaces. Suppose $T_v \in \mathrm{Hom}(E, E)$ and $M_v \in \mathrm{M}_{4N'}(\mathbb{R})$ are the elements corresponding to $v \in E \otimes E^*$. Then the results of the action of $g \in G$ on these elements are given by $\rho_E(g) \circ T_v \circ \rho_E(g)^{-1}$ and $M_g \cdot M_v \cdot M_g^{-1}$, respectively.

Now we consider the orbit of $(I_0, J_0) \in (E \otimes E^*) \oplus (E \otimes E^*)$ under the action of G. For convenience, we consider the orbit in the space of matrices $\mathrm{M}_{4N'}(\mathbb{R}) \oplus \mathrm{M}_{4N'}(\mathbb{R})$ through the above identification. Then, the action is given by

(11) $g \cdot (I_0, J_0) = (M_g \cdot I_0 \cdot M_g^{-1}, \, M_g \cdot J_0 \cdot M_g^{-1}), \quad g \in G = \mathrm{GL}_{4N'}(\mathbb{R}).$

From this formula, we immediately see that the isotropy subgroup of this action at (I_0, J_0) is given by those elements $g \in G$ such that the corresponding matrix M_g commutes with both I_0 and J_0. Thus, by definition, M_g is an element of $\mathrm{GL}_{N'}(\mathbb{H})$. Hence the $\mathrm{GL}_{4N'}(\mathbb{R})$-orbit of $(I_0, J_0) \in (E \otimes E^*) \oplus (E \otimes E^*)$ is a homogeneous space isomorphic to $\mathrm{GL}_{4N'}(\mathbb{R})/\mathrm{GL}_{N'}(\mathbb{H})$.

From the form of elements in this orbit given in (11), any point (I, J) in the orbit satisfy the relation $I^2 = J^2 = -1$, $IJ = -JI$. So, these points in the orbit define quaternionic structures in E. Different points on the orbit define different quaternionic structures. Thus, we have the following diagrams.

$$\mathrm{GL}_{4N'}(\mathbb{R}) \cdot (I_0, J_0) \quad \hookrightarrow \quad \{\text{ quaternionic structures in } E^{4N'}\}$$
$$\cap \qquad\qquad\qquad\qquad\qquad \updownarrow\cong$$
$$(E \otimes E^*) \oplus (E \otimes E^*) \quad \{I, J \in \mathrm{M}_{4N'}(\mathbb{R}) \mid I^2 = J^2 = -1, IJ = -JI\}$$

We show that the inclusion map in the first line above is in fact onto. To see this, let (I, J) be an arbitrary quaternionic structure in E and let $\{v_1, \ldots, v_{N'}\}$ be a right $\mathbb{H}$-basis of E with respect to the right $\mathbb{H}$-vector space structure on E induced from this quaternionic structure (I, J). Then, a basis of E over $\mathbb{R}$ is given by

$$(12) \quad \{v_1, \ldots, v_{N'}, J(v_1), \ldots, J(v_{N'}), I(v_1), \ldots, I(v_{N'}), IJ(v_1), \ldots, IJ(v_{N'})\}.$$

Let $\xi : (E; I_0, J_0) \to (E; I, J)$ be an $\mathbb{R}$-linear map sending the previously chosen and fixed basis $\{e_1, \ldots, e_{4N'}\}$ to the above basis in this order. Let M_ξ be the matrix representing this map with respect to the standard basis $\{e_1, \ldots, e_{4N'}\}$, i.e. $(\xi(e_1), \ldots, \xi(e_{4N'})) = (e_1, \ldots, e_{4N'})M_\xi$. By construction, the map ξ commutes with respective quaternionic structures and we have $I = \xi \cdot I_0 \cdot \xi^{-1}$, $J = \xi \cdot J_0 \cdot \xi^{-1}$. This means that the quaternionic structure (I, J) actually lie on the $\mathrm{GL}_{4N'}(\mathbb{R})$-orbit of $(I_0, J_0) \in \mathrm{Hom}(E, E) \oplus \mathrm{Hom}(E, E)$ in view of (11). In $\mathrm{M}_{4N'}(\mathbb{R}) \oplus \mathrm{M}_{4N'}(\mathbb{R})$, similar statement holds if we replace ξ by M_ξ. This proves the next proposition.

PROPOSITION 10. *Let E be a real $4N'$-dimensional vector space with a chosen $\mathbb{R}$-basis. Let I_0, J_0 be as in (8) or (9). Then, the moduli space of the quaternionic structures in E can be identified with the $\mathrm{GL}_{4N'}(\mathbb{R})$-orbit of (I_0, J_0) in the representation space $(E \otimes E^*) \oplus (E \otimes E^*)$, where the action is given by (11). This moduli space is isomorphic to a homogeneous space $\mathrm{GL}_{4N'}(\mathbb{R})/\mathrm{GL}_{N'}(\mathbb{H})$.*

Restating this proposition in terms of matrices using the chosen basis in E, we obtain the next Corollary.

COROLLARY 11. *Let $A, B \in \mathrm{M}_{4N'}(\mathbb{R})$ be two matrices such that $A^2 = B^2 = -1$, $AB = -BA$. Then, there exists a matrix $M \in \mathrm{GL}_{4N'}(\mathbb{R})$ such that $A = M \cdot I_0 \cdot M^{-1}$, $B = M \cdot J_0 \cdot M^{-1}$, i.e., the pair (A, B) is simultaneously conjugate to the pair (I_0, J_0), where matrices I_0, J_0 are given in (10).*

(IV) The moduli space of symplectic scalar product structures as an orbit. Let $(E; \langle \, , \, \rangle_0)$ be a Euclidean vector space of real dimension $n = 4N'$ with a chosen orthonormal basis $\{e_1, \ldots, e_{4N'}\}$. Let (I_0, J_0) be the standard quaternionic structure in E associated to this chosen basis defined as in (9). That is,

$$\begin{aligned}
I_0(e_j) &= e_{j+N}, & I_0(e_{j+N}) &= -e_j, & 1 \leq j \leq N, \\
J_0(e_r) &= e_{r+N'}, & J_0(e_{r+N'}) &= -e_r, & \\
J_0(e_{r+N}) &= -e_{r+N+N'}, & J_0(e_{r+N+N'}) &= e_{r+N}, & 1 \leq r \leq N'.
\end{aligned}$$

Note that these maps are isometries of E and satisfy $I_0^2 = J_0^2 = -1$, $I_0 J_0 = -J_0 I_0$. In (III), we considered the $\mathrm{GL}_{4N'}(\mathbb{R})$-orbit of (I_0, J_0) in the representation space $(E \otimes E^*) \oplus (E \otimes E^*)$. Here, we consider the $\mathrm{O}(4N')$-orbit of the same point in the same representation space. From the result of (III), isotroy subgroup of $\mathrm{GL}_{4N'}(\mathbb{R})$ action at (I_0, J_0) is given by $\mathrm{GL}_{N'}(\mathbb{H})$. So, the isotropy subgroup of $O(4N')$ action at the same point is given by $\mathrm{GL}_{N'}(\mathbb{H}) \cap O(4N') = \mathrm{Sp}(N')$. Thus, the $O(4N')$-orbit is a homogeneous space of the form $O(4N')/\mathrm{Sp}(N')$.

Now, through the identification $E \otimes E^* \cong \mathrm{M}_{4N'}(\mathbb{R})$, any element in the new orbit is of the form $(I, J) = (Q \cdot I_0 \cdot Q^{-1}, Q \cdot J_0 \cdot Q^{-1})$ for some $Q \in O(4N')$. Obviously, these are isometries of $(E; \langle\ ,\ \rangle_0)$, and satisfy quaternionic relations. Thus, every element in this orbit defines a symplectic scalar product structure in E. See Lemma 6 and the remark following it in §4.1. Conversely, we show that every symplectic scalar product structure in E is on this orbit, allowing us to identify this $O(4N')$-orbit with the moduli space of symplectic scalar product structures. To see this, let I, J be isometries and satisfy the quaternionic relations giving E the structure of a right module over $\mathbb{H}$. Let $\{v_1, \ldots, v_{N'}\}$ be a $\mathbb{H}$-basis of E which is also an $\mathbb{H}$-orthonormal basis with respect to the symplectic scalar product. Then, as in (12) of (III), we can construct a basis of E over $\mathbb{R}$ using I, J and we can construct a map $\xi : (E; I_0, J_0) \to (E; I, J)$ and a matrix M_ξ in the same way as before. But this time, since I, J are isometries, our basis of the form (12) is an $\mathbb{R}$-orthonormal basis of the Euclidean vector space $(E; \langle\ ,\ \rangle_0)$, and so ξ is also an isometry and $M_\xi \in O(4N')$. Thus, the above (I, J) is on the $O(4N')$-orbit. This proves the next proposition.

PROPOSITION 12. *Let $(E; \langle\ ,\ \rangle_0)$ be a real $4N'$-dimensional Euclidean vector space with a given $\mathbb{R}$-basis. The moduli space of symplectic scalar product structures in the Euclidean space $(E, \langle\ ,\ \rangle_0)$ can be identified with the $O(4N')$-orbit through (I_0, J_0) in the representation space $(E \otimes E^*) \oplus (E \otimes E^*)$. This orbit is a homogeneous space of the form $O(4N')/\mathrm{Sp}(N')$.*

We can restate the above proposition in terms of matrices using the chosen basis of E to obtain the following Corollary, which is the $O(n)$ version of the previous Corollary in (III).

COROLLARY 13. *Let $A, B \in O(4N')$ be such that $A^2 = B^2 = -1$, $AB = -BA$. Then, there exists a matrix $Q \in O(4N')$ such that $A = Q \cdot I_0 \cdot Q^{-1}$, $B = Q \cdot J_0 \cdot Q^{-1}$.*

(V) The moduli space of Hermitian structures as an orbit. Now, let E is a complex vector space with a chosen $\mathbb{C}$-basis $\{e_1, \ldots, e_N\}$ of E. We regard E as a representation ρ_E of $\mathrm{GL}_N(\mathbb{C})$ by

$$(13) \qquad (\rho_E(g)e_1, \ldots, \rho_E(g)e_N) = (e_1, \ldots, e_N)g, \qquad g \in \mathrm{GL}_N(\mathbb{C}).$$

Here, the right hand side is the matrix multiplication. Let h_0 be the standard Hermitian structure in E associated with this basis, i.e., $h_0(e_i, e_j) = \delta_{ij}$. When we regard h_0 as an element in $\overline{E}^* \otimes_{\mathbb{C}} E^*$, it can be expressed as

$$(14) \qquad h_0 = \sum_{j=1}^{N} \bar{e}_j^* \otimes e_j^*.$$

We consider the orbit of h_0 under the diagonal action $\rho_{\overline{E}^* \otimes E^*}$ of $G = \mathrm{GL}_N(\mathbb{C})$ on $\overline{E}^* \otimes E^*$ described in Lemma 5. As before, we identify any vector $v =$

$\sum_{ij} a_{ij} \bar{e}_i^* \otimes e_j^*$ in $\overline{E}^* \otimes E^*$ with a matrix $A = (a_{ij})$ in $M_N(\mathbb{C})$. Then, by Lemma 5 the matrix corresponding to a point in the orbit $\rho_{\overline{E}^* \otimes E^*}(g) h_0$ for $g \in G$ is given by $^t \overline{M}_g^{-1} M_g^{-1}$, which is a positive definite Hermitian symmetric matrix. Here, as before, M_g is the matrix representing the action $\rho_E(g)$ which is really the same as $g \in GL_N(\mathbb{C})$ itself in our present context, in view of (13). Since any positive definite Hermitian symmetric matrix $H = (h_{ij})$ defines a Hermitian scalar product h in E by letting $h(e_i, e_j) = h_{ij}$, we have an inclusion map from the orbit to the moduli space of Hermitian structures in E. The isotropy subgroup of this action at h_0 consists of those elements $g \in G$ such that $^t \overline{M}_g^{-1} M_g^{-1} = 1$, i.e., M_g is a unitary matrix. Thus, the isotropy subgroup is the unitary group $U(N)$ and the orbit is a homogeneous space of the form $GL_N(\mathbb{C})/U(N)$. We have the following diagram.

$$GL_N(\mathbb{C})/U(N) \xrightarrow{\cong} GL_N(\mathbb{C}) \cdot h_0 \qquad \hookrightarrow \{ \text{ Hermitian structures in } E \}$$

$$\cap \qquad\qquad\qquad \updownarrow \cong$$

$$(\overline{E}^* \otimes E^*) \oplus (\overline{E}^* \otimes E^*) \qquad \left\{ \begin{array}{c} \text{Positive definite Hermitian} \\ \text{symmetric matrices in } M_N(\mathbb{C}) \end{array} \right\}$$

We show that the inclusion map in the top line is actually a surjection. Let h be any Hermitian scalar product in E. We choose a Hermitian orthonormal basis $\{v_1, \ldots, v_N\}$ for E, for example, by the Gram-Schmidt process. Let $\xi : (E, h_0) \to (E, h)$ be an isometry defined by $\xi(e_j) = v_j$, $1 \leq j \leq N$. Let $g_\xi \in GL_N(\mathbb{C})$ be a matrix such that $(v_1, \ldots, v_N) = (e_1, \ldots, e_N) g_\xi$. That is, $\xi = \rho_E(g_\xi)$. For any $u, v \in E$, we then have $h(u \otimes v) = h_0(\xi^{-1}(u) \otimes \xi^{-1}(v)) = h_0 \left(\rho_E(g_\xi^{-1})(u) \otimes \rho_E(g_\xi^{-1})(v) \right) = \rho_{\overline{E}^* \otimes E^*}(g_\xi)(h_0)(u \otimes v)$. This means that $\rho_{\overline{E}^* \otimes E^*}(g_\xi)(h_0) = h$. Thus any Hermitian scalar product h in E is in the $GL_N(\mathbb{C})$-orbit of h_0.

PROPOSITION 14. *Let E be a complex N-dimensional vector space with a chosen $\mathbb{C}$-basis. The $GL_N(\mathbb{C})$-orbit of h_0 in $\overline{E}^* \otimes E^*$ can be identified with the moduli space of the Hermitian scalar products in E.*

Translating this fact into matrix terms, we have

COROLLARY 15. *Let $A \in M_N(\mathbb{C})$ be a positive definite Hermitian symmetric matrix, i.e., $^t \bar{A} = A$ and $^t \mathbf{x} A \mathbf{x} > 0$ for any $\mathbf{x} \neq 0 \in \mathbb{C}^N$. Then there exists a matrix $M \in GL_N(\mathbb{C})$ such that $A = {}^t \overline{M} \cdot M$. Such M is unique up to a left multiplication by an element in $U(N)$.*

(VI) The moduli space of special Hermitian structures as an orbit. We start from the definition of the special Hermitian structure in a complex vector space E. Let $(E; h)$ be a Hermitian vector space with a chosen Hermitian orthonormal basis $\{e_1, \ldots, e_N\}$. The Hermitian scalar product h on E determines a Hermitian scalar product on the exterior algebra $\bigwedge^* E$. In particular, the top exterior power $\bigwedge_\mathbb{C}^N E$ is a complex 1-dimensional Hermitian vector space generated by the unit vector $e_1 \wedge \cdots \wedge e_N$. Let $S_h(\bigwedge_{\mathbb{C}}^N E)$ be the set of unit length vectors. This is isomorphic to the unit circle $S^1 \subset \mathbb{C}$. Then, by definition, the special Hermitian structure on a complex vector space E is a pair (h, α), where $\alpha \in S_h(\bigwedge_{\mathbb{C}}^N E)$.

We consider the orbit of (h_0, α_0) in $(\overline{E}^* \otimes E^*) \oplus \bigwedge_{\mathbb{C}}^N E$, where $h_0 = \sum_{j=1}^N \bar{e}_j^* \otimes e_j^*$ is the canonical Hermitian structure associated to the above chosen basis and the vector α_0 is given by $\alpha_0 = e_1 \wedge \cdots \wedge e_N \in S_{h_0}(\bigwedge_{\mathbb{C}}^N E)$. Note that any element $\alpha \in S_{h_0}(\bigwedge^N E)$ is of the form $c \cdot \alpha_0$ with $|c| = 1$. Note also that if $g \cdot h_0 = \rho_{\overline{E}^* \otimes E^*}(h_0) = h$ for $g \in \mathrm{GL}_N(\mathbb{C})$, then, the basis $\{\rho(g)e_1, \ldots, \rho(g)e_N\}$ is a Hermitian orthonormal basis with respect to h. Now the action of $g \in \mathrm{GL}_N(\mathbb{C})$ on α_0 is given by $g \cdot \alpha_0 = \rho(g)e_1 \wedge \cdots \wedge \rho(g)e_N = (\det g)e_1 \wedge \cdots \wedge e_N$, which has unit length with respect to the new Hermitian pairing h. Thus, any element $g \cdot (h_0, \alpha_0)$ in the orbit gives a special Hermitian structure in E.

From (V), we know that $\mathrm{GL}_N(\mathbb{C})$ acts transitively on the space of the Hermitian structures in E and the isotropy group of the action is the unitary group. The unitary group acts on α_0 through the determinant, that is, $g \cdot \alpha_0 = (\det g)\alpha_0$. So, $\mathrm{U}(N)$ acts transitively on $S_h(\bigwedge_{\mathbb{C}}^N E)$. Hence, it acts transitively on the space of special Hermitian structures subordinate to the standard one h_0. The isotropy subgroup of the action of $\mathrm{GL}_N(\mathbb{C})$ on the element (h_0, α_0) is $\mathrm{SU}(N)$. Thus, our orbit is a homogeneous space of the form $\mathrm{GL}_N(\mathbb{C})/\mathrm{SU}(N)$.

Next, we show that $\mathrm{GL}_N(\mathbb{C})$ acts transitively on the moduli space of special Hermitian structures. To see this, let (h, α) be any special Hermitian structure in E. From the result of (V), the group $\mathrm{GL}_N(\mathbb{C})$ acts transitively on the moduli space of Hermitian structures. So, there exists an element $g \in \mathrm{GL}_N(\mathbb{C})$ such that $g \cdot h_0 = h$. Let $g \cdot (h_0, \alpha_0) = (h, \alpha')$. Here $\alpha' = \rho(g)e_1 \wedge \cdots \wedge \rho(g)e_N \in S_h(\bigwedge^N E)$. Since $|\alpha|_h = |\alpha'|_h = 1$ in the complex 1 dimensional space $\bigwedge_{\mathbb{C}}^N E$, we can write $\alpha' = c \cdot \alpha$ for some complex number $|c| = 1$. Let g' be a diagonal matrix whose $(1, 1)$ entry is c and the rest of the diagonal entries are all 1. Then, $g' \in \mathrm{U}(N)$ and $gg' \in \mathrm{GL}_N(\mathbb{C})$ and we have $(gg') \cdot (h_0, \alpha_0) = g \cdot (h_0, c\alpha_0) = (h, c\alpha') = (h, \alpha)$. This shows that any special Hermitian structure in E is in the $\mathrm{GL}_N(\mathbb{C})$-orbit.

Thus, the single $\mathrm{GL}_N(\mathbb{C})$-orbit can be identified with the moduli space of the special Hermitian structures.

PROPOSITION 16. *Let E be a complex N-dimensional vector space with a chosen $\mathbb{C}$-basis and let (h_0, α_0) be the canonically associated special Hermitian structure. The orbit $\mathrm{GL}_N(\mathbb{C}) \cdot (h_0, \alpha_0)$ in the representation space $(\overline{E}^* \otimes E^*) \oplus \bigwedge_{\mathbb{C}}^N E$ is a homogeneous space of the form $\mathrm{GL}_N(\mathbb{C})/\mathrm{SU}(N)$ and can be identified with the moduli space of the special Hermitian structures in E.*

The corresponding matrix version is that for any pair $(H, \alpha) \in \mathrm{M}_N(\mathbb{C}) \oplus \mathbb{C}$ of Hermitian symmetric matrix H and and a complex number α, there exists a matrix $M \in \mathrm{M}_N(\mathbb{C})$ such that $H = {}^t\overline{M} \cdot M$ and $\alpha = \det M$. Such M is unique up to the left multiplication by an element of $\mathrm{SU}(N)$.

(VII) The moduli space of pseudo-symplectic structures as an orbit. Let $(E; \langle \ , \ \rangle_0)$ be a Euclidean vector space of real dimension $n = 4N'$

with a chosen basis $\{e_1, \ldots, e_{4N'}\}$. A symplectic scalar product structure on this Euclidean space consists of three isometries I, J, K satisfying the relation $I^2 = J^2 = -1$, $IJ = -JI = -K$. Let (I_0, J_0, K_0) be the standard symplectic scalar product structure on E determined by the chosen basis as in (8), (9). We consider the three dimensional subspace $L = \mathbb{R}I \oplus \mathbb{R}J \oplus \mathbb{R}K$ of $\mathrm{End}(E)$ generated by symplectic scalar product structures (I, J, K). We call such a subspace a pseudo-symplectic structure in E. The purpose of (VII) is to realize the moduli space of pseudo-symplectic structures in E as an orbit of a suitable group G in a G-space. Before, orbits were considered in G-representations. Here, we consider an orbit in a certain G-manifold.

Two symplectic scalar product structures can give rise to the same pseudo-symplectic structure L. These symplectic scalar product structures are related by orthogonal transformations in L.

LEMMA 17. *Suppose we have* $\mathbb{R}I \oplus \mathbb{R}J \oplus \mathbb{R}K = \mathbb{R}I' \oplus \mathbb{R}J' \oplus \mathbb{R}K'$ *for two quaternionic structures* (I, J, K) *and* (I', J', K'). *Then, these are related by*

$$(I', J', K') = (I, J, K)Q, \qquad \text{for some } Q \in \mathrm{SO}(3).$$

Thus, if (I, J, K) *are isometries on* $(E; \langle \ , \ \rangle_0)$, *then* (I_0, J_0, K_0) *are also isometries.*

PROOF. Since I', J', K' are linear combinations of I, J, K, for some $Q \in M_3(\mathbb{R})$, we can write $(I', J', K') = (I, J, K)Q$. Using the quaternionic relations of (I, J, K), we can easily check that the relations $I'^2 = J'^2 = K'^2 = -1$ impose that the column vectors of Q, regarded as vectors in the standard Euclidean space $(\mathbb{R}^3, \langle \ , \ \rangle_0)$, must have unit length. Also, relations $I'J' = -J'I'$, $J'K' = -K'J'$, $K'I' = -I'K'$ imply that the column vectors of Q must be mutually orthogonal. Thus $Q \in \mathrm{O}(3)$. Furthermore, the relation $I'J' = -K'$ shows that $\det Q = 1$. Hence $Q \in \mathrm{SO}(3)$.

The last part of the statement follows by a simple calculation. $\square$

Since the moduli space of symplectic scalar product structures in E is isomorphic to a homogeneous space $\mathrm{O}(4N')/\mathrm{Sp}(N')$ from (IV), we get a next corollary which gives a realization of the moduli space as the base space of a principal bundle.

COROLLARY 18. *The moduli space of pseudo-symplectic structures in E is a base space of an* $\mathrm{SO}(3)$ *principal fibre bundle whose total space is* $\mathrm{O}(4N')/\mathrm{Sp}(N')$.

Next, we want to realize it as an orbit of $G = \mathrm{O}(4N')$ action on a G-manifold. We consider either a Grassmannian $G_3(E \otimes_\mathbb{R} E^*)$ of real three dimensional subspaces of $E \otimes_\mathbb{R} E^*$, or a projective space $\mathbb{P}(\bigwedge_\mathbb{R}^3 (E \otimes_\mathbb{R} E^*))$ of the third exterior power of $E \otimes_\mathbb{R} E^*$. Any pseudo-symplectic structure L determines an element in the above Grassmannian in an obvious way. It also determines a ray $\bigwedge_\mathbb{R}^3 L$ in $\bigwedge_\mathbb{R}^3 (E \otimes_\mathbb{R} E^*)$, so it determines a point of the above projective space. Thus, we have a map from the moduli space of pseudo-symplectic structures in E to the above Grassmannian or to the above projective space. The map to the Grassmannian is obviously an embedding. It turns out that the map to the projective space above is also an embedding, which follows from the following fact known as the Plücker embedding.

LEMMA 19. *Let V be a real n dimensional representation of a group G. Then, we have the following G-equivariant embedding*

$$G_r(V) \hookrightarrow \mathbb{P}\left(\bigwedge{}^r(V)\right) \subset \mathbb{P}\left(\bigwedge{}^*(V)\right), \qquad 0 \le r \le n.$$

PROOF. The above map is defined by associating the real r-th exterior power to a real r dimensional vector subspace of V. We show that this map is an embedding. Let W_1, W_2 be two r dimensional subspaces of V such that $\bigwedge^r W_1 = \bigwedge^r W_2$ as 1-dimensional subspaces of $\bigwedge^r V$. Let $W_1 = \bigoplus_{j=1}^r \mathbb{R}u_j$ and $W_2 = \bigoplus_{j=1}^r \mathbb{R}v_j$ for some vectors u_j and v_j. We show that $W_1 = W_2$. Suppose $W_1 \ne W_2$ in V. Then, there exists a vector $v \in W_2$ which doesn't belong to W_1. We may assume that $v = v_1$. Then we can choose a functional $f : V \to \mathbb{R}$ such that $f(W_1) = 0$, $f(v_1) \ne 0$, $f(v_j) = 0$, $2 \le j \le r$. We consider an interior product ι_f with f using the dual pairing. This map lowers the degree by 1, $\iota_f : \bigwedge^* V \to \bigwedge^{*-1} V$. We have $\iota_f(u_1 \wedge \cdots \wedge u_r) = 0$ using the derivation property of interior product, since $f(u_j) = 0$ for $1 \le j \le r$. Also, $\iota_f(v_1 \wedge \cdots \wedge v_r) = f(v_1) \cdot v_2 \wedge \cdots \wedge v_r \ne 0$ in $\bigwedge^* V$. Thus, $\iota_f(\bigwedge^r W_1) = 0$ and $\iota_f(\bigwedge^r W_2) \ne 0$ in $\bigwedge^* V$. This contradicts to our hypothesis $\bigwedge^r W_1 = \bigwedge^r W_2$ and hence we must have $W_1 = W_2$. This proves the injectivity of the above map.

The equivariance of the inclusion map follows from the way $g \in G$ acts on the exterior powers, that is, $g \cdot (v_1 \wedge \cdots \wedge v_r) = g(v_1) \wedge \cdots \wedge g(v_r)$, where $\{v_1, \ldots, v_r\}$ is any basis of a real r-dimensional subspace $W \in G_r(V)$. $\square$

Let $L_0 = \mathbb{R}I_0 \oplus \mathbb{R}J_0 \oplus \mathbb{R}K_0$ be the pseudo-symplectic structure canonically associated to the chosen $\mathbb{R}$-basis of E. We consider the $O(4N')$-orbit of the standard pseudo-symplectic structure L_0 of E in two $O(4N')$-manifolds, the Grassmannian $G_3(E \otimes E^*)$ and the projective space $\mathbb{P}\left(\bigwedge^3(E \otimes E^*)\right)$. Let Q_g be the matrix corresponding to the action $\rho_E(g)$ of $g \in O(4N')$ on E. We identify $E \otimes E^*$ with the set of linear maps or matrices, $\text{Hom}(E, E) \cong M_{4N'}(\mathbb{R})$ using the given basis of E. Then, the result of the action of an element $g \in O(4N')$ on L_0 in the Grassmannian is given by

$$(15) \quad g \cdot L_0 \equiv L_g = \mathbb{R}(Q_g \cdot I_0 \cdot Q_g^{-1}) \oplus \mathbb{R}(Q_g \cdot J_0 \cdot Q_g^{-1}) \oplus \mathbb{R}(Q_g \cdot K_0 \cdot Q_g^{-1}).$$

First, we determine the isotropy subgroup $G_0 \subset O(4N')$ of this action at L_0. Since the symplectic group $\text{Sp}(N')$ commutes with I_0, J_0, K_0, $\text{Sp}(N')$ is contained in G_0. However, the isotropy group G_0 is bigger than this. To see this, we recall that the conjugation action by any unit quaternions on $\mathbb{H}$ preserve the set of pure quaternions $\mathbb{R}i \oplus \mathbb{R}j \oplus \mathbb{R}k$ and this action induces the standard action of $SO(3)$ on $\mathbb{R}^3$. So, a subgroup $\text{Sp}(1) \subset O(4N')$ consisting of elements of the form $a1_E + bI_0 + cJ_0 + dK_0$, $a^2 + b^2 + c^2 + d^2 = 1$, preserves $L_0 = \mathbb{R}I_0 \oplus \mathbb{R}J_0 \oplus \mathbb{R}K_0$ under conjugation action and this action commutes with the action of $\text{Sp}(N')$, since elements in $\mathrm{Sp}(N')$ consists of those isometries commuting with I_0, J_0, K_0. The intersection of these two groups are given by $\{\pm 1_E\}$. Hence, $\mathrm{Sp}(N') \cdot \mathrm{Sp}(1) = \mathrm{Sp}(N') \times \mathrm{Sp}(1)/\pm 1$ is in the isotropy subgroup $\subset G_0 \subset \mathrm{O}(4N')$. Since it is known that the subgroup $\mathrm{Sp}(N') \cdot \mathrm{Sp}(1)$ is maximal in $\mathrm{O}(4N')$ and obviously G_0 is strictly smaller than $\mathrm{O}(4N')$, we conclude that $G_0 = \mathrm{Sp}(N') \cdot \mathrm{Sp}(1)$. See also §4.3 Lemma 9. Thus the $\mathrm{O}(4N')$-orbit through L_0 is a homogeneous space of the form $\mathrm{O}(4N')/(\mathrm{Sp}(N') \cdot \mathrm{Sp}(1))$.

Since any element in this $\mathrm{O}(4N')$-orbit is of the form (15) above for some $Q_g \in \mathrm{O}(4N')$, any element in this orbit gives a pseudo-symplectic structure. Since any pseudo-symplectic structure is spanned by a symplectic scalar product structure (I, J, K) and since $\mathrm{O}(4N')$ acts transitively on the set of symplectic scalar product structures, the moduli space of pseudo-symplectic scalar product structures is a single $\mathrm{O}(4N')$-orbit. Thus, we have the following proposition.

PROPOSITION 20. *Let $(E; \langle \ , \ \rangle_0)$ be a real $4N'$-dimensional Euclidean vector space with a chosen basis and let $L_0 = \mathbb{R}I_0 \oplus \mathbb{R}J_0 \oplus \mathbb{R}K_0$ be the pseudo-symplectic structure canonically associated to the chosen basis as in (8) or (9) with $K_0 = J_0 \cdot I_0$. The $\mathrm{O}(4N')$-orbit of L_0 in the $\mathrm{O}(4N')$-manifolds $G_3(E \otimes E^*) \subset \mathbb{P}(\bigwedge^3(E \otimes E^*))$ can be identified with the moduli space of pseudo-symplectic structures compatible with the Euclidean structure $\langle \ , \ \rangle_0$.*

Rephrasing this proposition in terms of matrices, we have

COROLLARY 21. *Let $L = \mathbb{R}I \oplus \mathbb{R}J \oplus \mathbb{R}K$ be a three dimensional subspace of $\mathrm{M}_{4N'}(\mathbb{R})$ spanned by orthogonal matrices I, J, K such that $I^2 = J^2 = -1$, $IJ = -IJ = -K$. Then, there exists an orthogonal matrix $Q \in \mathrm{O}(4N')$ such that $L = Q \cdot L_0 \cdot Q^{-1}$, where $L_0 = \mathbb{R}I_0 \oplus \mathbb{R}J_0 \oplus \mathbb{R}K_0$ with*

$$I_0 = \begin{pmatrix} 0 & 0 & -1_{N'} & 0 \\ 0 & 0 & 0 & -1_{N'} \\ 1_{N'} & 0 & 0 & 0 \\ 0 & 1_{N'} & 0 & 0 \end{pmatrix},$$

$$J_0 = \begin{pmatrix} 0 & -1_{N'} & 0 & 0 \\ 1_{N'} & 0 & 0 & 0 \\ 0 & 0 & 0 & 1_{N'} \\ 0 & 0 & -1_{N'} & 0 \end{pmatrix}, \quad K_0 = \begin{pmatrix} 0 & 0 & 0 & 1_{N'} \\ 0 & 0 & -1_{N'} & 0 \\ 0 & 1_{N'} & 0 & 0 \\ -1_{N'} & 0 & 0 & 0 \end{pmatrix}.$$

The matrix Q is unique up to right multiplication by an element in $\mathrm{Sp}(N') \cdot \mathrm{Sp}(1)$.

§4.3   A general theory of geometric structures in vector spaces and their compatibilities

In this section, we first discuss basic geometric structures in vector spaces and discuss their properties including (left and right) automorphism groups and moduli spaces. Here the word "basic" means that the corresponding moduli spaces are orbits of general linear groups $\mathrm{GL}_n(\mathbb{R})$ instead of their subgroups. We then introduce the notion of geometric types. The compatibilities of geometric types are defined and studied in the general context. We then describe moduli spaces of compatible basic geometric structures.

Basic geometric structures and their moduli spaces as G-orbits. In a real (or complex) vector space E, we can consider various geometric structures. We are particularly interested in the Euclidean structures, complex structures, symplectic structures, quaternionic structures, almost quaternionic structures and complex special linear structures. These are defined as follows.

DEFINITION 1 (BASIC GEOMETRIC STRUCTURES IN VECTOR SPACES).

(1) (Euclidean structure) Let E^n be a real n dimensional vector space. A Euclidean structure in E is a real valued positive definite symmetric bilinear pairing $h : E \otimes E \to \mathbb{R}$. Note that h is naturally a vector in $E^* \otimes E^*$.

(2) (Complex structures) Let E^{2N} be a real $2N$ dimensional vector space. A complex structure in E is a map $I : E \to E$ such that $I^2 = -1$. We have $I \in E \otimes E^*$.

(3) (Symplectic structures) Let E^{2N} be a real $2N$ dimensional vector space. A symplectic structure in E is a nondegenerate 2-form $\omega \in \bigwedge^2 E^*$ of the dual vector space. Note that the associated map $\overline{\omega} : E \to E^*$ defined by $(\overline{\omega}(u))(v) = \omega(u, v)$, $u, v \in E$, is an isomorphism.

(4) (Quaternionic structures) Let $E^{4N'}$ be a real $4N'$ dimensional real vector space. A quaternionic structure is a pair of complex structures (I, J) such that $IJ = -JI$. If we let $K = JI$, then, I, J, K satisfy quaternionic relations. We have $(I, J) \in (E \otimes E^*) \oplus (E \otimes E^*)$.

(5) (Almost quaternionic structures) Let $E^{4N'}$ be a real $4N'$ dimensional vector space. An almost quaternionic structure is a 3 dimensional subspace L of the vector space of endomorphisms $\mathrm{End}(E)$ spanned by a quaternionic structure, I, J, K, i.e., $L = \mathbb{R}I \oplus \mathbb{R}J \oplus \mathbb{R}K$. Note that an almost quaternionic structure L can be regarded as an element of the Grassmannian $G_3(E \otimes E^*)$ of real 3-dimensional vector subspaces of $E \otimes E^*$.

(6) (Special linear structures) Let E be a complex N dimensional vector space. A special linear structure α is a nonzero vector in the top exterior power $\bigwedge_{\mathbb{C}}^N E$.

REMARK. In (4), (5) above, the complex structures I, J, K satisfy relations which are a little different from the usual quaternion relations satisfied by i, j, k. This difference comes from the fact that the left module over the algebra $\mathbb{R}1_E \oplus \mathbb{R}I \oplus \mathbb{R}J \oplus \mathbb{R}K = \mathbb{R}1_E \oplus L$, where 1_E is an identity map of E, defines a right $\mathbb{H}$-module structure by letting $I(v) = v \cdot i$, $J(v) = v \cdot j$, $K(v) = v \cdot k$. The relations among I, J, K corresponding to the usual quaternion relations among i, j, k is the one mentioned above. In (5) above, if E is an Euclidean vector space and if we require in addition that I, J, K are isometries, then, the resulting structure is what we call pseudo-symplectic structure in E in §4.1 and §4.2.

Given a basis of E, there is a standard way to introduce the above type of geometric structures in E. If we use another basis, the structure we obtain may be different, but the expressions of these structures in terms of the bases are the same.

DEFINITION 2 (STANDARD GEOMETRIC STRUCTURES IN VECTOR SPACES ASSOCIATED TO A GIVEN BASIS).

(1) The Euclidean structure η in E associated to a basis $\{e_1, \ldots, e_n\}$ of E over $\mathbb{R}$ is given in terms of its dual basis $\{e_1^*, \ldots, e_n^*\}$ by

$$\eta = \sum_{j=1}^{n} e_j^* \otimes e_j^* \in E^* \otimes E^*.$$

(2) The complex structure I_0 in E^{2N} associated to a basis $\{e_1, \ldots, e_{2N}\}$ of E over $\mathbb{R}$ is given by

$$I = \sum_{j=1}^{N} e_{j+N} \otimes e_j^* - \sum_{j=1}^{N} e_j \otimes e_{j+N}^* \in E \otimes E^*.$$

(3) The symplectic structure ω in E associated to a basis $\{e_1, \ldots, e_{2N}\}$ of E over $\mathbb{R}$ is given by

$$\omega = \sum_{j=1}^{N} e_j^* \wedge e_{j+N}^* \in \bigwedge^2 E^*.$$

(4) The quaternionic structure (I, J) associated to a basis $\{e_1, \ldots, e_{4N'}\}$ of $E^{4N'}$ over $\mathbb{R}$ is given by I in (2) above and $J \in E \otimes E^*$ defined by

$$J = \sum_{r=1}^{N'} (e_{r+N'} \otimes e_r^* + e_{r+N} \otimes e_{r+N+N'}^*) - \sum_{r=1}^{N'} (e_r \otimes e_{r+N'}^* + e_{r+N+N'} \otimes e_{r+N'}^*).$$

(5) The almost quaternionic structure $L \subset \text{End}(E)$ in $E^{4N'}$ associated to a basis $\{e_1, \ldots, e_{4N'}\}$ of E over $\mathbb{R}$ is given by

$$L = \mathbb{R}I \oplus \mathbb{R}J \oplus \mathbb{R}K \in G_3(E \otimes E^*),$$

where $K = JI$ and I, J are given as in (4).

(6) The special linear structure α in a complex N dimensional vector space E associated to a $\mathbb{C}$-basis $\{e_1, \ldots, e_N\}$ of E is given by

$$\alpha = e_1 \wedge \cdots \wedge e_N \in \bigwedge_{\mathbb{C}}^{N} E.$$

The above formulae give a method to construct various geometric structures from an arbitrarily given basis. To introduce an interesting point of view concerning geometric structures, let $\mathcal{F}(E)$ (or $\mathcal{F}_{\mathbb{C}}(E)$) be the set of all real frames (complex frames, respectively) in a real (complex, respectively) vector space E. In the above, $\eta, I, \omega, (I, J), L, \alpha$ denote particular geometric structures. However, it is better to explicitly indicate the frame $\mathbf{e} = \{e_1, \ldots, e_n\}$ from which these structures come from, as in $\eta(\mathbf{e})$, $I(\mathbf{e})$, etc.

To exploit this point of view, let Σ denote a type of geometric structures, like Euclidean structures, complex structures, etc., as an abstract notion. Let $\mathfrak{M}(\Sigma) = \mathfrak{M}(\Sigma; E)$ denote the moduli space of geometric structures of the type Σ on E sitting inside of some vector spaces or some projective spaces.

NOTATION 3 : (MODULI SPACES OF GEOMETRIC STRUCTURES OF BASIC TYPES).

(1) $\mathfrak{M}(\eta) \subset E^* \otimes E^*$: Euclidean structures in E^n
(2) $\mathfrak{M}(I) \subset E \otimes E^*$: Complex structures in E^{2N}
(3) $\mathfrak{M}(\omega) \subset \bigwedge^2(E^*)$: Symplectic structures in E^{2N}
(4) $\mathfrak{M}((I, J)) \subset (E \otimes E^*) \oplus (E \otimes E^*)$: Quaternionic structures in $E^{4N'}$
(5) $\mathfrak{M}(L) \subset G_3(E \otimes E^*)$: Almost quaternionic structures in $E^{4N'}$
(6) $\mathfrak{M}(\alpha) \subset \bigwedge_{\mathbb{C}}^N E$: Special linear structures in $(E^{2N}; I)$

When Σ is regarded as a type of geometric structures rather than an individual geometric structure, it is natural to regard Σ as a map

(17) $$\Sigma : \mathcal{F}(E) \to \mathfrak{M}(\Sigma),$$

where for each frame $\mathbf{e} \in \mathcal{F}(E)$, $\Sigma(\mathbf{e})$ denotes a specific geometric structure constructed from a given frame $\mathbf{e}$. (We start our numbering of formulae from (17) because we want to reserve numbers up to (16) for geometric types in List 20 below.) For example, for any frame $\mathbf{e} = \{e_1, \ldots, e_n\}$, from Definition 2 we have

(18)
$$\eta(\mathbf{e}) = \sum_{j=1}^{n} e_j^* \otimes e_j^*,$$

$$I(\mathbf{e}) = \sum_{j=1}^{N} e_{j+N} \otimes e_j^* - \sum_{j=1}^{N} e_j \otimes e_{j+N}^*, \qquad n = 2N,$$

$$\omega(\mathbf{e}) = \sum_{j=1}^{N} e_j^* \wedge e_{j+N}^*, \qquad n = 2N,$$

$$J(\mathbf{e}) = \sum_{r=1}^{N'} (e_{r+N'} \otimes e_r^* + e_{r+N} \otimes e_{r+N+N'}^*)$$
$$- \sum_{r=1}^{N'} (e_r \otimes e_{r+N'}^* + e_{r+N+N'} \otimes e_{r+N'}^*), \qquad n = 4N',$$

$$L(\mathbf{e}) = \mathbb{R} I(\mathbf{e}) \oplus \mathbb{R} J(\mathbf{e}) \oplus \mathbb{R} K(\mathbf{e}), \qquad n = 4N',$$

$$\alpha(\mathbf{e}) = e_1 \wedge \cdots \wedge e_N, \qquad n = 2N.$$

Here, $K(\mathbf{e}) = J(\mathbf{e}) \cdot I(\mathbf{e})$ in $\text{Hom}(E, E)$.

LEMMA 4. *The map of geometric structures of type* $\Sigma = \eta, I, \omega, (I, J), L, \alpha$,

$$\Sigma : \mathcal{F}(E) \to \mathfrak{M}(\Sigma)$$

is surjective. That is, any geometric structure of the above type arises from a frame.

PROOF. The proofs are elementary. We use the previous numbering of geometric types for convenience. For (1), given a Euclidean structure η, choose an orthonormal basis by Gram-Schmidt process. Any such basis works.

For (2), given a complex structure I in E^{2N}, we regard E as a vector space over $\mathbb{C}$ where the multiplication by i is given by the map I, and choose any basis of E over $\mathbb{C}$, then extend it to a basis over $\mathbb{R}$ using I. Any such basis is the desired basis.

For (3), let ω be any nondegenerate symplectic structure in E^{2N}. We choose a basis as follows. Let e_1 be any nonzero vector. Let e_2 be a vector such that $\omega(e_1, e_2) = 1$. Such vector exists due to nondegeneracy of ω. Let $V_1 = \mathbb{R}e_1 \oplus \mathbb{R}e_2$. On V_1, ω restricts to $e_1^* \wedge e_2^*$. Let $W_1 \subset E$ be defined by $W_1 = \{v \in E \mid \omega(v, V_1) = 0\}$. We claim that $E = V_1 \oplus W_1$, a direct sum and ω restricted to W_1 is nondegenerate. To see this, let $v \in W_1 \cap V_1$ be a nonzero vector. Then, $v \in W_1$ implies that $\omega(V_1, v) = 0$. But ω is nondegenerate on V_1, so $\omega(V_1, v) \neq 0$ since $v \in V_1$. This contradiction shows that $V_1 \cap W_1 = 0$. To see that V_1 and W_1 together span E, let $u \in E$ be an arbitrary vector. Let $\omega(u, e_1) = a, \omega(u, e_2) = b$. Then, a vector $u' = u - ae_2 - be_1$ is such that $\omega(u', V_1) = 0$, so $u' \in W_1$. Thus, $E = V_1 \oplus W_1$. The nondegeneracy of ω restricted to W_1 is straightforward. If we have $\omega(u, W_1) = 0$ for some $u \in W_1$, then, $\omega(u, V) = 0$ since $\omega(u, V_1) = 0$ by the definition of vectors in W_1. But then, the nondegeneracy of the symplectic form ω on V imply that $u = 0$. Thus, ω is nondegenerate when restricted to W_1.

We repeat the above argument on W_1 to get V_2 and W_2, where $V_2 = \mathbb{R}e_3 \oplus \mathbb{R}e_4$ such that $\omega|_{V_2} = e_3^* \wedge e_4^*$ and the restriction of ω to W_2 is non-degenerate. We continue this process until W_N has dimension 0. Then, we have a decomposition $E = V_1 \oplus V_2 \oplus \cdots \oplus V_N$ and a basis $\{e_1, \ldots, e_{2N}\}$ such that $\omega(V_i, V_j) = 0$ for $i \neq j$ and ω restricted to V_j is of the form $e_{2j-1}^* \wedge e_{2j}^*$. By reindexing these basis vectors, we see that $\omega = \sum_{j=1}^{N} e_j^* \wedge e_{j+N}^*$. This proves the surjectivity of the symplectic structure map $\omega : \mathcal{F}(E) \to \mathfrak{M}(\omega)$.

For (4) and (5), we only have to choose any basis of E as a right $\mathbb{H}$ module defined by I, J, K. Then extend this to an $\mathbb{R}$-basis using I, J, K. (6) is obvious. $\square$

From now on, when we need to distinguish a geometric structure and its type, we use bold face to denote the type. For example, if a Euclidean structure η is associated to a frame $\mathbf{e}$, then we write $\eta = \boldsymbol{\eta}(\mathbf{e})$. For any geometric structure $\sigma \in \mathfrak{M}(\Sigma)$ of type Σ, any frame $\mathbf{e} \in \mathcal{F}(E)$ such that $\sigma = \Sigma(\mathbf{e})$ is called a σ-frame. Given a geometric structure σ, σ-frame is not unique. See Proposition 9 below.

We can rephrase the above lemma and its proof as follows.

Corollary and Terminology 5.

(1) $\eta = \boldsymbol{\eta}(\mathbf{e})$ if and only if $\mathbf{e} = \{e_1, \ldots, e_n\} \in \mathcal{F}(E)$ is an η-orthonormal frame, i.e., $\eta(e_j, e_k) = \delta_{jk}$.

(2) $I = \boldsymbol{I}(\mathbf{e})$ if and only if $\mathbf{e} \in \mathcal{F}(E)$ is an I-complex frame, i.e., a frame of the form $\mathbf{e} = \{e_1, \ldots, e_N, I(e_1), \ldots, I(e_N)\}$.

(3) $\omega = \boldsymbol{\omega}(\mathbf{e})$ if and only if $\mathbf{e} = \{e_1, \ldots, e_{2N}\}$ is ω-symplectic, namely,

$$\omega(e_j, e_{k+N}) = \delta_{jk}, \quad \omega(e_j, e_k) = 0, \quad \omega(e_{j+N}, e_{k+N}) = 0, \quad 1 \le j, k \le N.$$

(4) $(I, J) = (\boldsymbol{I}, \boldsymbol{J})(\mathbf{e})$ if and only if $\mathbf{e}$ is an (I, J)-quaternionic frame, that is,

$$\mathbf{e} = \{e_1, \ldots, e_{N'}, J(e_1), \ldots, J(e_{N'}), I(e_1), \ldots, I(e_{N'}), IJ(e_1), \ldots, IJ(e_{N'})\}.$$

(5) $L = \boldsymbol{L}(\mathbf{e})$ if and only if $\mathbf{e}$ is an (I, J)-quaternionic frame, where (I, J, K) is any quaternionic basis of L.

Now, on the space of frames $\mathcal{F}(E)$, the group $\mathrm{GL}(E)$ acts from the left and the group $\mathrm{GL}_n(\mathbb{R})$ acts from the right. Namely, for $\tau \in \mathrm{GL}(E)$, $\mathbf{e} = \{e_1, \ldots, e_n\} \in \mathcal{F}(E)$, $g \in \mathrm{GL}_n(\mathbb{R})$, we have $\tau \cdot \mathbf{e} = \{\tau(e_1), \ldots, \tau(e_n)\}$ and $\mathbf{e} \cdot g = (e_1, \ldots, e_n)g$, where the product in the right hand side of the second equation is a matrix multiplication.

The group $\mathrm{GL}(E)$ acts on the moduli spaces of geometric structures of the above types. This action is induced from the action of $\mathrm{GL}(E)$ on the ambient vector spaces $E^* \otimes E^*$, $E \otimes E^*$, or related spaces in which various geometric structures live.

Lemma 6 (Actions of $\mathrm{GL}(E)$ on the moduli spaces of geometric structures). *The action of $\mathrm{GL}(E)$ on the ambient spaces preserves the moduli spaces of geometric structures of the above types. The actions are given explicitly as follows:*

(1) $\tau \in \mathrm{GL}(E)$ acts on a pairing $\eta : E \otimes E \to \mathbb{R}$, in particular on a Euclidean structure, by

$$(\tau \cdot \eta)(u, v) = \eta(\tau^{-1}(u), \tau^{-1}(v)) \qquad \text{for any } u, v \in E.$$

(2) τ acts on a map $I : E \to E$, in particular on a complex structure I with $I^2 = -1$, by

$$\tau \cdot I = \tau \circ I \circ \tau^{-1}.$$

Here, elements in $E \otimes E^*$ are regarded as elements in $\mathrm{End}(E)$.

(3) τ acts on a symplectic structure $\omega \in \bigwedge^2 E^*$ by

$$(\tau \cdot \omega)(u, v) = \omega(\tau^{-1}(u), \tau^{-1}(v)), \qquad \text{for any } u, v \in E.$$

(4) τ acts on a quaternionic structure $(I, J) \in (E \otimes E^*) \oplus (E \otimes E^*)$ by

$$\tau \cdot (I, J) = (\tau \circ I \circ \tau^{-1}, \tau \circ J \circ \tau^{-1}).$$

(5) τ acts on an almost quaternionic structure $L = \mathbb{R}I \oplus \mathbb{R}J \oplus \mathbb{R}K \in G_3(E \otimes E^*)$ by

$$\tau \cdot L = \mathbb{R}(\tau \cdot I) \oplus \mathbb{R}(\tau \cdot J) \oplus \mathbb{R}(\tau \cdot K).$$

(6) τ acts on a complex special linear structure $\alpha \in \bigwedge_{\mathbb{C}}^{N} E$ by

$$\tau \cdot \alpha = (\det \tau)\alpha.$$

PROOF. Recall that an element $\tau \in \mathrm{GL}(E)$ acts on the dual vector space E^* by the dual of the inverse map $\tau \cdot = (\tau^{-1})^* : E^* \to E^*$.

For (1), we have $(\tau \cdot \eta)(u, v) = ((\tau^{-1})^* \otimes (\tau^{-1})^*)(\eta)(u, v) = \eta(\tau^{-1}(u), \tau^{-1}(v))$ for any $u, v \in E$.

Similarly, for (2), let $I(u)$ denote the pairing of $u \in E$ with the second tensor factor of $I \in E \otimes E^*$. Then, $(\tau \cdot I)(u) = ((\tau \otimes (\tau^{-1})^*) I)(u) = \tau (I(\tau^{-1}(u)))$.

(3) follows from (1) because one can regard $\bigwedge^2 E^* \subset E^* \otimes E^*$. The next two, (4), (5) immediately follows from (2). (6) is trivial. □

Now, for any frame $\mathbf{e} = \{e_1, \ldots, e_n\} \in \mathcal{F}(E)$, we can consider its dual frame $\mathbf{e}^* = \{e_1^*, \ldots, e_n^*\}$ in E^* defined by the property $e_i^*(e_j) = \delta_{ij}$. This gives us a map $* : \mathcal{F}(E) \to \mathcal{F}(E^*)$.

LEMMA 7. The dualizing map $* : \mathcal{F}(E) \to \mathcal{F}(E^*)$ is $\mathrm{GL}(E)$-equivariant. That is, the dual basis of $\{\tau e_1, \ldots, \tau e_n\}$ is $\{\tau \cdot e_1^*, \ldots, \tau \cdot e_n^*\}$. More concisely, $\tau \cdot e_j^* = (\tau e_j)^*$, $1 \le j \le n$.

PROOF. We have $\tau \cdot e_i^*(\tau e_j) = ((\tau^{-1})^* e_i^*)(\tau e_j) = e_i^*(\tau^{-1} \tau e_j) = e_i^*(e_j) = \delta_{ij}$. Hence $\tau \cdot e_i^* = (\tau e_i)^*$, $1 \le i \le n$. □

We are now in a position to prove,

LEMMA 8. Let $\Sigma = \boldsymbol{\eta}, \mathbf{I}, \boldsymbol{\omega}, (\mathbf{I}, \mathbf{J}), \mathbf{L}, \alpha$ be any of the above geometric types regarded as a map

$$\Sigma : \mathcal{F}(E) \to \mathfrak{M}(\Sigma; E) = \{\text{geometric structures of the type } \Sigma \text{ on } E\}.$$

Then, Σ is a $\mathrm{GL}(E)$-equivariant surjective map: $\Sigma(\tau \cdot \mathbf{e}) = \tau \cdot \Sigma(\mathbf{e})$ for any $\tau \in \mathrm{GL}(E)$ and $\mathbf{e} \in \mathcal{F}(E)$. Consequently, the moduli space $\mathfrak{M}(\Sigma)$ consists of a single $\mathrm{GL}(E)$-orbit.

PROOF. The surjectivity comes from Lemma 4. The proof of equivariance is straightforward. For example, for the case of complex structures, $\tau \cdot I(\mathbf{e}) = \tau(\sum e_{j+N} \otimes e_j^* - \sum e_j \otimes e_{j+N}^*) = \sum \tau e_{j+N} \otimes (\tau e_j)^* - \sum \tau e_j \otimes (\tau e_{j+N})^* = I(\tau \cdot \mathbf{e})$. Here we used Lemma 7 for the second equality. Other cases can be proved similarly. Since $\mathrm{GL}(E)$ acts transitively on $\mathcal{F}(E)$, $\mathrm{GL}(E)$ also acts transitively on its G-equivariant surjective image $\mathfrak{M}(\Sigma)$. □

For any geometric type Σ and a basis $\mathbf{e}$, the left automorphism group of the corresponding geometric structure $\Sigma(\mathbf{e})$ is defined by

(19) $\mathrm{Aut}_\ell(\Sigma(\mathbf{e})) = \{\tau \in \mathrm{GL}(E) \mid \tau \cdot \Sigma(\mathbf{e}) = \Sigma(\tau \cdot \mathbf{e}) = \Sigma(\mathbf{e})\}.$

For two different bases e and $e' = \tau' \cdot e$ with $\tau' \in GL(E)$, the corresponding automorphism groups are conjugate, $\mathrm{Aut}_\ell(\Sigma(e')) = \tau' \cdot \mathrm{Aut}_\ell(\Sigma(e)) \cdot \tau'^{-1}$. So, even if they are abstractly isomorphic, they sit in $GL(E)$ in different ways in general. However, there are better way of viewing this situation. Recall that the group $GL_n(\mathbb{R})$ acts on $\mathcal{F}(E)$ from the right. This action commutes with the right action of the group $GL(E)$.

We recall the definitions of various subgroups of $GL_n(\mathbb{R})$. Let
(20)
$$\Omega_N = \begin{pmatrix} 0 & 1_N \\ -1_N & 0 \end{pmatrix}, \qquad I_N = \begin{pmatrix} 0 & -1_N \\ 1_N & 0 \end{pmatrix}$$

$$J_{N'} = \begin{pmatrix} 0 & -1_{N'} & 0 & 0 \\ 1_{N'} & 0 & 0 & 0 \\ 0 & 0 & 0 & 1_{N'} \\ 0 & 0 & -1_{N'} & 0 \end{pmatrix}, \qquad K_{N'} = \begin{pmatrix} 0 & 0 & 0 & 1_{N'} \\ 0 & 0 & -1_{N'} & 0 \\ 0 & 1_{N'} & 0 & 0 \\ -1_{N'} & 0 & 0 & 0 \end{pmatrix}$$

These are the matrices corresponding to appropriate geometric structures. We use the following subgroups of $GL_n(\mathbb{R})$. Let $g \in GL_n(\mathbb{R})$. Letting $n = 2N$ for the second and the third cases, and $n = 4N'$ for the remaining cases, we have
(21)
$$g \in O(n) \iff {}^t g \cdot g = 1_n,$$
$$g \in GL_N(\mathbb{C}) \iff g \cdot I_N \cdot g^{-1} = I_N,$$
$$g \in Sp(n, \mathbb{R}) \iff {}^t g \cdot \Omega_N \cdot g = \Omega_N,$$
$$g \in GL_{N'}(\mathbb{H}) \iff g \cdot I_N \cdot g^{-1} = I_N, \qquad g \cdot J_{N'} \cdot g^{-1} = J_{N'},$$
$$g \in Sp(1) \iff g = a1_n + bI_N + cJ_{N'} + dK_{N'}, \quad a^2 + b^2 + c^2 + d^2 = 1.$$

Note that elements in $GL_{N'}(\mathbb{H})$ and elements in $Sp(1)$ always commute by definition. The right automorphism group for a geometric structure $\Sigma(e)$ turns out to be always the same independent of the frame $e \in \mathcal{F}(E)$ used to construct $\Sigma(e)$.

PROPOSITION 9 (AUTOMORPHISM GROUPS OF BASIC GEOMETRIC TYPES).

(1) *For any Euclidean structure $\eta(e)$ associated to any frame $e \in \mathcal{F}(E)$, we have $\eta(e \cdot g) = \eta(e)$ if and only if $g \in O(n)$.*
(2) *For any complex structure $I(e)$ associated to any frame $e \in \mathcal{F}(E)$, we have $I(e \cdot g) = I(e)$ if and only if $g \in GL_N(\mathbb{C})$.*
(3) *For any symplectic structure $\omega(e)$ associated to any frame $e \in \mathcal{F}(E)$, we have $\omega(e \cdot g) = \omega(e)$ if and only if $g \in Sp(n, \mathbb{R})$.*
(4) *For any quaternionic structure $(I, J)(e)$ associated to any frame $e \in \mathcal{F}(E)$, we have $(I, J)(e \cdot g) = (I, J)(e)$ if and only if $g \in GL_{N'}(\mathbb{H})$.*
(5) *For any almost quaternionic structure $L(e)$ associated to any frame $e \in \mathcal{F}(E)$, we have $L(e \cdot g) = L(e)$ if and only if $g \in GL_{N'}(\mathbb{H}) \cdot Sp(1)$.*
(6) *For any special linear structure $\alpha(e)$ associated to any frame $e \in \mathcal{F}_{\mathbb{C}}(E)$, we have $\alpha(e \cdot g) = \alpha(e)$ if and only if $g \in SL_N(\mathbb{C})$.*

PROOF. For notational convenience, we regard a basis $e = \{e_1, \ldots, e_n\}$ as a set of vectors arranged in a row. Similarly for its dual basis $e^* = \{e_1^*, \ldots, e_n^*\}$.

Their transposes are regarded in a similar fashion. We first prove

Claim :   Let $e' = e \cdot g$, $g \in \mathrm{GL}_n(\mathbb{R})$. Then, $e'^* = e^* \cdot {}^t g^{-1}$.

To see this, regarding the matrix multiplication as the dual pairing, we have ${}^t e'^* \cdot e' = 1_n$. Letting $e'^* = e^* \cdot A$, we have $1_n = ({}^t A \cdot {}^t e^*) \cdot e \cdot g = {}^t A \cdot g$. Hence $A = {}^t g^{-1}$. We can now begin the proof using the matrices in (20).

For (1), regarding the matrix multiplication as the tensor product, $\eta(e) = e^* \cdot {}^t e^*$. Thus, for $g \in \mathrm{GL}_n(\mathbb{R})$, $\eta(e \cdot g) = \eta(e') = e'^* \cdot {}^t e'^* = e^* \cdot {}^t g^{-1} \cdot g^{-1} \cdot {}^t e^*$. Hence, $\eta(e \cdot g) = \eta(e)$ holds if and only if ${}^t g^{-1} \cdot g^{-1} = 1_n$, i.e., $g \in \mathrm{O}(n)$.

For (2), as before, regarding the matrix multiplication as the tensor product, we can write $I(e) = e \cdot I_N \cdot {}^t e^*$. So, as before, for $g \in \mathrm{GL}_{2N}(\mathbb{R})$, $I(e \cdot g) = e \cdot g \cdot I_N \cdot g^{-1} \cdot {}^t e^*$. Hence $I(e \cdot g) = I(e)$ if and only if $g \cdot I_N \cdot g^{-1} = I_N$. This means that g commutes with the complex structure I_N, That is, $g \in \mathrm{GL}_N(\mathbb{C})$.

For (3), regarding the matrix multiplication as the wedge product, we can write $\omega(e) = \frac{1}{2} e^* \cdot \Omega_N \cdot {}^t e^*$. Then, for any $g \in \mathrm{GL}_{2N}(\mathbb{R})$, we have $\omega(e \cdot g) = \frac{1}{2} e^* \cdot {}^t g^{-1} \cdot \Omega_N \cdot g^{-1} \cdot {}^t e^*$. Thus, $\omega(e \cdot g) = \omega(e)$ if and only if ${}^t g^{-1} \cdot \Omega_N \cdot g^{-1} = \Omega_N$, which is the characterization of elements in the group $\mathrm{Sp}(n, \mathbb{R})$.

(4) $(I, J)(e \cdot g) = (I, J)(e)$ implies that $I(e \cdot g) = I(e)$, $J(e \cdot g) = J(e)$. From the proof of (2), this means that g commutes with the matrices corresponding to the quaternionic structure (I, J). This means that $g \in \mathrm{GL}_{N'}(\mathbb{H})$.

(5) Assume $g \in \mathrm{GL}_{4N'}(\mathbb{R})$ is such that $L(e \cdot g) = L(e)$ or $\mathbb{R}I(e \cdot g) \oplus \mathbb{R}J(e \cdot g) \oplus \mathbb{R}K(e \cdot g) = \mathbb{R}I(e) \oplus \mathbb{R}J(e) \oplus \mathbb{R}K(e)$. Then, there exists a linear relation among the two bases of L and we can write $(I(e \cdot g), J(e \cdot g), K(e \cdot g)) = (I(e), I(e), I(e))A$ for some $A \in M_3(\mathbb{R})$. The quaternion relations satisfied by these bases imply that $A \in \mathrm{SO}(3)$. For details, see §4.2 Lemma 17. Now, $\mathbb{R}1_n \oplus L = \mathbb{R}1_n \oplus \mathbb{R}I(e) \oplus \mathbb{R}J(e) \oplus \mathbb{R}K(e) \subset \mathrm{End}(E)$ is a subalgebra isomorphic to the skew field of quaternions $\mathbb{H}$. Let $\mathrm{Sp}_1(E; (I, J)(e))$ be a subgroup of $\mathrm{GL}(E)$ consisting of elements of the form $\{a + bI(e) + cJ(e) + dK(e) \in \mathrm{End}(E) \mid a^2 + b^2 + c^2 + d^2 = 1\}$. These correspond to unit quaternions. Note that for any $\tau \in \mathrm{Sp}_1(E; (I, J)(e))$, we have $\tau e = e \cdot h$ for some $h \in \mathrm{Sp}(1) \subset \mathrm{GL}_{4N'}(\mathbb{R})$. Now, recall that the double covering map $\mathrm{Sp}(1) \to \mathrm{SO}(3)$ is given by the adjoint action of $\mathrm{Sp}(1)$ on its Lie algebra $\mathbb{R}I_N \oplus \mathbb{R}J_{N'} \oplus \mathbb{R}K_{N'}$ with respect to this chosen basis. In our context, the Lie algebra of $\mathrm{Sp}_1(E; (I, J)(e))$ is given by $L = \mathbb{R}I(e) \oplus \mathbb{R}J(e) \oplus \mathbb{R}K(e)$. So, for some element $\tau \in \mathrm{Sp}_1(E; (I, J)(e))$, we have $(I(e \cdot g), J(e \cdot g), K(e \cdot g)) = (I(e), J(e), K(e))A = (\tau \cdot I(e) \cdot \tau^{-1}, \tau \cdot J(e) \cdot \tau^{-1}, \tau \cdot K(e) \cdot \tau^{-1})$. That is, $\tau^{-1} \cdot \Sigma(e \cdot g) \cdot \tau = \Sigma(e)$, for $\Sigma = I, J, K$. From the way $\mathrm{GL}(E)$ acts on I, J, K described above, this means that $\Sigma(\tau^{-1}(e \cdot g)) = \Sigma(e)$, for $\Sigma = I, J, K$. Since the left action of $\mathrm{GL}(E)$ and the right action of $\mathrm{GL}_{4N'}(\mathbb{R})$ on $\mathcal{F}(E)$ commutes, we have $\tau^{-1}(e \cdot g) = \tau^{-1}(e) \cdot g = e \cdot h^{-1}g$, where $\tau(e) = e \cdot h$. Thus, our conditions on g are $\Sigma(e \cdot h^{-1}g) = \Sigma(e)$ for $\Sigma = I, J, K$. From (4), this means that $h^{-1}g \in \mathrm{GL}_{N'}(\mathbb{H})$, or $g \in \mathrm{Sp}(1) \cdot \mathrm{GL}_{N'}(\mathbb{H})$. Since these two groups commute, we finally arrive at the conclusion $g \in \mathrm{GL}_{N'}(\mathbb{H}) \cdot \mathrm{Sp}(1)$. This proves (5).

For (6), just note that for any element $g \in \mathrm{GL}_N(\mathbb{C})$ and complex frame $e \in \mathcal{F}_{\mathbb{C}}(E)$, we have $\alpha(e \cdot g) = (\det g)\alpha(e)$. Thus, $\alpha(e \cdot g) = \alpha(e)$ if and only if $g \in \mathrm{SL}_N(\mathbb{C})$. $\square$

Previously, we defined the left automorphism group of a geometric structure $\Sigma(\mathbf{e})$ in (19). We can similarly define the right automorphism group $\mathrm{Aut}_r(\Sigma(\mathbf{e}))$ of a given geometric structure by

$$(22) \qquad \mathrm{Aut}_r(\Sigma(\mathbf{e})) = \{g \in \mathrm{GL}_n(\mathbb{R}) \mid \Sigma(\mathbf{e} \cdot g) = \Sigma(\mathbf{e})\}.$$

The above Lemma shows that for any frame $\mathbf{e} \in \mathcal{F}(E)$, the right automorphism group is independent of the particular geometric structure used to define the automorphism group and it depends only on the geometric type Σ.

$$(23) \qquad \begin{aligned} \mathrm{Aut}_r(\eta(\mathbf{e})) &= O(n), & \mathrm{Aut}_r(I(\mathbf{e})) &= \mathrm{GL}_N(\mathbb{C}), \\ \mathrm{Aut}_r(\omega(\mathbf{e})) &= \mathrm{Sp}(n, \mathbb{R}), & \mathrm{Aut}_r((I,J)(\mathbf{e})) &= \mathrm{GL}_{N'}(\mathbb{H}), \\ \mathrm{Aut}_r(L(\mathbf{e})) &= \mathrm{GL}_{N'}(\mathbb{H}) \cdot \mathrm{Sp}(1), & \mathrm{Aut}_r(\alpha(\mathbf{e})) &= \mathrm{SL}_N(\mathbb{C}). \end{aligned}$$

From the results we have so far, the next proposition is immediate.

PROPOSITION 10. *Let $\Sigma = \eta, I, \omega, (I, J), L, \alpha$ be a geometric type. Then, the moduli space $\mathfrak{M}(\Sigma)$ forms a single orbit of $\mathrm{GL}(E)$ with respect to the action given in Lemma 6. From (23), we obtain*

$$\{\text{Euclidean structures in } E^n\} = \mathrm{GL}(E) \cdot \eta \cong \mathcal{F}(E)/O(n),$$

$$\{\text{Complex structures in } E^{2N}\} = \mathrm{GL}(E) \cdot I \cong \mathcal{F}(E)/\mathrm{GL}_N(\mathbb{C}),$$

$$\{\text{Symplectic structures in } E^{2N}\} = \mathrm{GL}(E) \cdot \omega \cong \mathcal{F}(E)/\mathrm{Sp}(n, \mathbb{R}),$$

$$\{\text{Quaternionic structures in } E^{4N'}\} = \mathrm{GL}(E) \cdot (I, J) \cong \mathcal{F}(E)/\mathrm{GL}_{N'}(\mathbb{H}),$$

$$\{\text{Almost quaternionic structures in } E^{4N'}\} = \mathrm{GL}(E) \cdot L \cong \mathcal{F}(E)/\mathrm{GL}_{N'}(\mathbb{H}) \cdot \mathrm{Sp}(1),$$

$$\{\text{Special linear structures in } E^{2N} \cong \mathbb{C}^N\} = \mathrm{GL}_{\mathbb{C}}(E) \cdot \alpha \cong \mathcal{F}_{\mathbb{C}}(E)/\mathrm{SL}_N(\mathbb{C}).$$

General geometric structures and their compatibilities. So far, we have discussed six kinds of concrete but useful geometric structures separately. It is time to abstract basic concepts from these examples and obtain a more general concept from which we can view these examples in a more unified way. First, we define the types of geometric structures. This is an abstraction of (17) above.

DEFINITION 11 (TYPES OF GENERAL GEOMETRIC STRUCTURES AND THEIR MODULI SPACES). A type of a general geometric structure on a real vector space E is a $\mathrm{GL}(E)$-equivariant map

$$(24) \qquad \Sigma : \mathcal{F}(E) \to V,$$

where V is a representation of $\mathrm{GL}(E)$ or its projectivization. A geometric structure of type Σ is an element in the image of the map Σ, which is denoted by $\mathfrak{M}(\Sigma)$ and we call it the moduli space of geometric structures of type Σ. A definition of geometric structures in complex vector spaces can be formulated in exactly the same way.

The basic examples of V are tensor products of E's and E^*'s and their projectivizations.

Since $\mathcal{F}(E)$ is a left $\mathrm{GL}(E)$-space and right $\mathrm{GL}_n(\mathbb{R})$-space, we can define left and right automorphism groups of $\Sigma(\mathbf{e})$ as before by

(25)
$$\mathrm{Aut}_\ell(\Sigma(\mathbf{e})) = \{\tau \in \mathrm{GL}(E) \mid \tau \cdot \Sigma(\mathbf{e}) = \Sigma(\tau \cdot \mathbf{e}) = \Sigma(\mathbf{e})\},$$
$$\mathrm{Aut}_r(\Sigma(\mathbf{e})) = \{g \in \mathrm{GL}_n(\mathbb{R}) \mid \Sigma(\mathbf{e} \cdot g) = \Sigma(\mathbf{e})\}.$$

Previously, we proved some facts concerning these groups for the six basic geometric structures. See the paragraph right after (19) and Proposition 9. Similar identities hold in this abstract context.

LEMMA 12. *For any frame* $\mathbf{e} \in \mathcal{F}(E)$ *and any* $\tau \in \mathrm{GL}(E)$, *we have*

(26)
$$\mathrm{Aut}_\ell(\Sigma(\tau(\mathbf{e}))) = \tau \cdot \mathrm{Aut}_\ell(\Sigma(\mathbf{e})) \cdot \tau^{-1} \subset \mathrm{GL}(E).$$

Also, for any two frames $\mathbf{e}_1, \mathbf{e}_2 \in \mathcal{F}(E)$, *we have*

(27)
$$\mathrm{Aut}_r(\Sigma(\mathbf{e}_1)) = \mathrm{Aut}_r(\Sigma(\mathbf{e}_2)).$$

PROOF. The proof is formal. For (26), let $g \in \mathrm{Aut}_\ell(\Sigma(\mathbf{e}))$. Then, by the $\mathrm{GL}(E)$ equivariance of Σ and the definition of left automorphism group, $(\tau \cdot g \cdot \tau^{-1})(\Sigma(\tau\mathbf{e})) = \tau \cdot g(\Sigma(\mathbf{e})) = \tau(\Sigma(\mathbf{e})) = \Sigma(\tau\mathbf{e})$. This shows that $\tau \cdot g \cdot \tau^{-1}$ preserves the structure $\Sigma(\tau\mathbf{e})$, or $\tau \cdot \mathrm{Aut}_\ell(\Sigma(\mathbf{e})) \cdot \tau^{-1} \subset \mathrm{Aut}_\ell(\Sigma(\tau\mathbf{e}))$. The opposite inclusion relation can be proved in the same way.

For (27), let $\mathbf{e}_2 = \tau\mathbf{e}_1$ and let $g \in \mathrm{Aut}_r(\Sigma(\mathbf{e}_1))$. Then, we have $\Sigma(\mathbf{e}_2 \cdot g) = \Sigma(\tau(\mathbf{e}_1) \cdot g) = \Sigma(\tau(\mathbf{e}_1 \cdot g)) = \tau \cdot \Sigma(\mathbf{e}_1 \cdot g) = \tau \cdot \Sigma(\mathbf{e}_1) = \Sigma(\tau\mathbf{e}_1) = \Sigma(\mathbf{e}_2)$, where the second equality is due to the commutativity of $\mathrm{GL}(E)$ and $\mathrm{GL}_n(\mathbb{R})$, the third and the fifth equalities come from the $\mathrm{GL}(E)$ equivariance of Σ and the fourth one is due to the definition of g. Thus, the above g preserves $\Sigma(\mathbf{e}_2)$ and we have $\mathrm{Aut}_r(\Sigma(\mathbf{e}_1)) \subset \mathrm{Aut}_r(\Sigma(\mathbf{e}_2))$. By switching the role of $\mathbf{e}_1$ and $\mathbf{e}_2$, we get the opposite inclusion relation and we obtain Lemma 12. $\square$

In view of Lemma 12, the right automorphism group for $\Sigma(\mathbf{e})$ is independent of the frame chosen. Because of this fact, we use right automorphism groups more often than left automorphism groups and we give it a special name. We let

(28)
$$G(\Sigma) = \mathrm{Aut}_r(\Sigma(\mathbf{e})),$$

which is well defined and depends only on the geometric type Σ. As a simple corollary of the above lemma,

COROLLARY 13. *The moduli space of the geometric structures of the type* Σ *has the structure of a homogeneous space induced by the map* Σ,

(29)
$$\mathcal{F}(E)/G(\Sigma) \xrightarrow[\cong]{\Sigma} \mathfrak{M}(\Sigma).$$

These results in our current abstract context parallel those in the previous concrete examples. The advantage of this abstract concept is that we can define the notion of compatibility in a simple way which extends the obvious compatibility relations in our examples.

Compatible geometric structures. Let Σ_1, Σ_2 be two types of geometric structures, so we have $\Sigma_j : \mathcal{F}(E) \to V_j$ with moduli spaces $\mathfrak{M}(\Sigma)$ for $j = 1, 2$.

DEFINITION 14 (COMPATIBILITY). A pair of geometric structures of type Σ_1, Σ_2, $(\sigma_1, \sigma_2) \in \mathfrak{M}(\Sigma_1) \times \mathfrak{M}(\Sigma_2)$ is said to be compatible if it lies in the image of the diagonal map

$$(30) \quad \Delta : \mathcal{F}(E) \to (\mathcal{F}(E)/G(\Sigma_1)) \times (\mathcal{F}(E)/G(\Sigma_2)) \cong \mathfrak{M}(\Sigma_1) \times \mathfrak{M}(\Sigma_2).$$

In other words, (σ_1, σ_2) is a compatible pair if and only if there exists a frame $e \in \mathcal{F}(E)$ which is at the same time σ_1-frame and σ_2-frame.

At first glance, this definition looks unusual, but the above definition exposes the essence of the notion of compatibility. In fact, the above definition is equivalent to the usual notion of compatibilities, as the next Proposition shows.

PROPOSITION 15 (GEOMETRIC STRUCTURES COMPATIBLE WITH EUCLIDEAN STRUCTURES).

(i) *A Euclidean structure η and a complex structure I in E^{2N} is compatible if and only if I is an η-isometry, i.e., $\eta(I(u), I(v)) = \eta(u, v)$ for any $u, v \in E$.*

(ii) *A Euclidean structure η and a symplectic structure ω in E^{2N} are compatible if and only if there exists a complex structure I in E such that $\omega(u, v) = \eta(I(u), v)$ for any $u, v \in E$.*

(iii) *A Euclidean structure η and a quaternionic structure (I, J) are compatible if and only if I, J are η-isometries.*

(iv) *A Euclidean structure η and an almost quaternionic structure L are compatible if and only if L has a quaternionic basis I, J, K which are η-isometries.*

PROOF. For (i), if η and I are compatible, then there exists a frame $e = \{e_1, \ldots, e_{2N}\}$ such that $\eta = \boldsymbol{\eta}(e)$, $I = \boldsymbol{I}(e)$. Then, we have $\eta = \sum_{j=1}^{2N} e_j^* \otimes e_j^*$ and $I(e_j) = e_{j+N}$, $I(e_{j+N}) = -e_j$ for $1 \le j \le N$. Dualizing I, we get $I^*(e_j^*) = -e_{j+N}^*$, $I^*(e_{j+N}^*) = e_j^*$. Thus, $(I^* \otimes I^*)\eta = (I^* \otimes I^*)(\sum e_j^* \otimes e_j^*) = \sum_{j=1}^{N} e_{j+N}^* \otimes e_{j+N}^* + \sum_{j=1}^{N} e_j^* \otimes e_j^* = \eta$. Thus, I is an η-isometry.

Conversely, assume that I is an η-isometry. Then, we can define a Hermitian scalar product in E by $\langle \, , \, \rangle : E \otimes E \to \mathbb{C}$ by $\langle u, v \rangle = \eta(u, v) + i\eta(I(u), v)$. We then choose any Hermitian orthonormal basis of E, say, $\{e_1, \ldots, e_N\}$. Now, let e be a frame of E given by $\{e_1, \ldots, e_N, I(e_1), \ldots, I(e_N)\}$. This is an η-orthonormal basis and I-complex basis. Hence, $\eta = \boldsymbol{\eta}(e)$ and $I = \boldsymbol{I}(e)$ and η and I are compatible.

For (ii), first assume η and ω are compatible. Then, for some frame $e = \{e_1, \ldots, e_{2N}\}$, we have $\eta = \sum_{j=1}^{N} e_j^* \otimes e_j^*$ and $\omega = \sum_{j=1}^{N} e_j^* \wedge e_{j+N}^*$. We then define I as in (1). We then see that $(I^* \otimes 1)\eta = -\sum_{j=1}^{N} e_{j+N}^* \otimes e_j^* + \sum_{j=1}^{N} e_j^* \otimes e_{j+N}^* = \sum_{j=1}^{N} e_j^* \wedge e_{j+N}^* = \omega$. In other words, $\omega(u, v) = (I^* \otimes 1)\eta(u, v) = \eta(I(u), v)$ for any $u, v \in E$.

Conversely, suppose there exists a complex structure I such that $\omega(u, v) = \eta(I(u), v)$ for any $u, v \in E$. First note that any such I is necessarily an η-isometry because $\eta(I(u), I(v)) = \omega(u, I(v)) = -\omega(I(v), u) = -\eta(I^2(v), u) = \eta(v, u) =$

$\eta(u, v)$. We are in the situation of (i). We define a Hermitian pairing on E by $\langle u, v \rangle = \eta(u, v) + i\omega(u, v)$. As before, we consider Hermitian orthonormal basis $\{e_1, \ldots, e_N\}$ and its associated real frame $\mathbf{e} = \{e_1, \ldots, e_N, I(e_1), \ldots, I(e_N)\}$. This is an η-orthonormal frame, so $\eta = \boldsymbol{\eta}(\mathbf{e})$. By using our assumption $\omega(u, v) = \eta(I(u), v)$, we see that only nontrivial symplectic pairings are $\omega(e_j, e_{j+N}) = 1$, $1 \leq j \leq N$. Hence, $\omega = \sum_{j=1}^{N} e_j^* \wedge e_{j+N}^*$, which means $\omega = \boldsymbol{\omega}(\mathbf{e})$. This shows that η and I are compatible.

The proof of (iii) is similar to (i), and (iv) follows from (iii). $\square$

The above definition of compatible geometric structures has an immediate generalization, although we don't use it here.

DEFINITION 16 (COMPATIBILITY; GENERAL CASE). Let $\Sigma_1, \Sigma_2, \ldots, \Sigma_k$ be a collection of geometric types. A k-tuple of geometric structures $(\sigma_1, \sigma_2, \ldots, \sigma_k) \in \mathfrak{M}(\Sigma_1) \times \mathfrak{M}(\Sigma_2) \times \cdots \times \mathfrak{M}(\Sigma_k)$ of various types is said to be compatible if it is in the image of the diagonal map
(31)
$$\Delta : \mathcal{F}(E) \to (\mathcal{F}(E)/G(\Sigma_1)) \times \cdots \times (\mathcal{F}(E)/G(\Sigma_k)) \cong \mathfrak{M}(\Sigma_1) \times \cdots \times \mathfrak{M}(\Sigma_k).$$

For a pair of geometric structures (σ_1, σ_2) of geometric types Σ_1, Σ_2, we consider their left and right automorphism groups as before. The left automorphism group for the pair is defined by
(32)
$$\mathrm{Aut}_\ell(\sigma_1, \sigma_2) = \{\tau \in \mathrm{GL}(E) \mid \tau\sigma_i = \sigma_i, i = 1, 2\}.$$

For the right automorphism group, we start from frames $\mathbf{e}_1$ and $\mathbf{e}_2$ such that $\sigma_1 = \Sigma(\mathbf{e}_1)$, $\sigma_2 = \Sigma_2(\mathbf{e}_2)$. Then, the right automorphism group is defined by
(33)
$$\mathrm{Aut}_r(\sigma_1, \sigma_2) = \mathrm{Aut}_r\left(\Sigma_1(\mathbf{e}_1), \Sigma_2(\mathbf{e}_2)\right)$$
$$= \{g \in \mathrm{GL}_n(\mathbb{R}) \mid (\Sigma_1(\mathbf{e}_1 \cdot g), \Sigma_2(\mathbf{e}_2 \cdot g)) = (\Sigma_1(\mathbf{e}_1), \Sigma_2(\mathbf{e}_2))\}.$$

It is easy to check that this group is independent of the choices of σ_1-frame $\mathbf{e}_1$ and σ_2-frame $\mathbf{e}_2$ and depends only on σ_1 and σ_2. In fact, as in the single geometric structure case, it turns out that this right-automorphism group is completely independent of the choice of frames and depends only on geometric types.

LEMMA 17. *The automorphism groups of a pair of geometric structures are the intersections of individual automorphism groups. Namely,*

(34)
$$\mathrm{Aut}_\ell(\sigma_1, \sigma_2) = \mathrm{Aut}_\ell(\sigma_1) \cap \mathrm{Aut}_\ell(\sigma_2),$$
$$\mathrm{Aut}_r\left(\Sigma_1(\mathbf{e}_1), \Sigma_2(\mathbf{e}_2)\right) = \mathrm{Aut}_r\left(\Sigma_1(\mathbf{e}_1)\right) \cap \mathrm{Aut}_r\left(\Sigma_2(\mathbf{e}_2)\right)$$

Moreover, the right-automorphism group depends only on the geometric types and we have

(35)
$$\mathrm{Aut}_r(\Sigma_1, \Sigma_2) = G(\Sigma_1) \cap G(\Sigma_2) \equiv G(\Sigma_1, \Sigma_2).$$

PROOF. The proof is immediate from (27) in Lemma 12 and (28). $\square$

COROLLARY 18. *In general, for given geometric types* $\Sigma_1, \Sigma_2, \ldots, \Sigma_k$, *we have*

$$(36) \qquad \mathrm{Aut}_r(\Sigma_1, \ldots, \Sigma_k) = G(\Sigma_1) \cap \cdots \cap G(\Sigma_k) \equiv G(\Sigma_1, \ldots, \Sigma_k).$$

In the above identities (35) and (36), the symbol $\equiv$ is used to define notations. Next, we discuss interesting subsets of compatible geometric structures.

PROPOSITION 19 (MODULI OF GEOMETRIC STRUCTURES COMPATIBLE WITH A FIXED GEOMETRIC STRUCTURE). *Let* $\sigma_1 = \Sigma_1(\mathbf{e})$ *be a geometric structure of type* Σ_1 *associated to a frame* $\mathbf{e} \in \mathcal{F}(E)$. *Then, the set of geometric structures of type* Σ_2 *compatible with* σ_1, *denoted by* $\mathfrak{M}_{\sigma_1}(\Sigma_2)$, *is a single left* $\mathrm{Aut}_\ell(\sigma_1) \subset GL(E)$-*orbit in* $\mathfrak{M}(\Sigma_2)$ *given by*
$$(37)$$
$$(\mathbf{e} \cdot G(\Sigma_1))/(G(\Sigma_1) \cap G(\Sigma_2)) \xrightarrow[\Sigma_2]{\cong} \mathfrak{M}_{\sigma_1}(\Sigma_2) = \mathrm{Aut}_\ell(\sigma_1) \cdot \Sigma_2(\mathbf{e}) = \Sigma_2(\mathbf{e} \cdot G(\Sigma_1)) .$$

Here, $G(\Sigma_i)$ *is the right automorphism group of geometric structures of type* Σ_i, *which itself may be a collection of some geometric types.*

PROOF. In terms of the above frame $\mathbf{e}$, the totality of compatible geometric structures of type Σ_1 and Σ_2 is given by $\{(\Sigma_1(\mathbf{e} \cdot g), \Sigma_2(\mathbf{e} \cdot g)) \mid g \in GL_n(\mathbb{R})\}$. By definition, $\Sigma_1(\mathbf{e} \cdot g) = \Sigma_1(\mathbf{e})$ if and only if $g \in G(\Sigma_1)$. Thus, the set of geometric structures of type Σ_2 compatible with σ_1 is given by $\mathfrak{M}_{\sigma_1}(\Sigma_2) = \{\Sigma_2(\mathbf{e} \cdot g) \mid g \in G(\Sigma_1)\}$, which is a homogeneous space. Since the right automorphism group of any geometric structure $\Sigma_2(\mathbf{e}')$ of type Σ_2 is $G(\Sigma_2)$, the geometric structure $\Sigma_2(\mathbf{e} \cdot g)$ for $g \in G(\Sigma_1)$ depends only on the left coset of $G(\Sigma_1) \cap G(\Sigma_2)$ in $G(\Sigma_1)$. Hence, the moduli space of geometric structures of type Σ_2 compatible with a given geometric structure σ_1 is given by $\mathfrak{M}_{\sigma_1}(\Sigma_2) = (\mathbf{e} \cdot G(\Sigma_1))/(G(\Sigma_1) \cap G(\Sigma_2))$. $\square$

The following is the list of geometric structures in E which is of interest to us. The geometric types (1) to (6) are basic ones (see Definition 1) and others are compatible geometric types obtained by combining the basic geometric types.

LIST 20 : (GEOMETRIC TYPES).

(1) Euclidean structure in E^n.
(2) Complex structures in E^{2N}.
(3) Symplectic structures in E^{2N}.
(4) Quaternionic structures in $E^{4N'}$.
(5) Almost quaternionic structures in $E^{4N'}$.
(6) Special linear structures in a complex vector space $(E^{2N}; I)$.
(7) Hermitian structures in a complex vector space $(E^{2N}; I)$.
(8) Complex structures in a Euclidean space $(E^{2N}; \eta)$.
(9) Hermitian structures in a special complex vector space $(E^{2N}; I, \alpha)$.
(10) Special linear structures in a Hermitian vector space $(E^{2N}; I, h)$.
(11) Euclidean structures in a symplectic vector space $(E^{2N}; \omega)$.
(12) Symplectic structures in a Euclidean vector space $(E^{4N'}; \eta)$.

(13) Symplectic scalar products in a quaternionic vector space $(E^{4N'}; I, J, K)$.

(14) Quaternionic structures in a Euclidean vector space $(E^{4N'}; \eta)$.

(15) Euclidean structures in an almost quaternionic vector space $(E^{4N'}; L)$.

(16) Almost quaternionic structures in a Euclidean vector space $(E^{4N'}; \eta)$.

In the above, when there are two or more structures present, their compatibilities are defined as in Definition 14.

To each of these geometric structures in vector spaces, a Lie group is naturally associated, namely, the group of left and right automorphisms of the geometric structure. We recall that $O(n)$, $GL_N(\mathbb{C})$, $GL_{N'}(\mathbb{H})$, $GL_{N'}(\mathbb{H}) \cdot Sp(1)$, $Sp(n, \mathbb{R})$ are the right-automorphism groups of geometric types η, I, (I, J), L, ω, respectively. As we have already seen in Lemma 17, when there are two compatible geometric structures or more as in (7) to (16), then, the corresponding left and right automorphism groups are the intersections of the left and right automorphism groups of the individual geometric types. The inclusion relations among various automorphism groups are given in the diagram (39).

Let Σ_1, Σ_2, Σ_3 be some geometric types, each of which may be a collection of some geometric types. Then, the right automorphism groups $G(\Sigma_1, \Sigma_2)$, $G(\Sigma_1, \Sigma_3)$ are subgroups of $G(\Sigma_1)$ obtained as intersections with groups $G(\Sigma_2)$, $G(\Sigma_3)$, and their intersection preserves all the geometric types involved and it is the right-automorphism group for the union of all the geometric types involved, i.e., $G(\Sigma_1, \Sigma_2, \Sigma_3)$. The corresponding diagram is given by

(38)
$$
\begin{array}{ccc}
G(\Sigma_1, \Sigma_2) & \longrightarrow & G(\Sigma_1) \\
\uparrow & & \uparrow \\
G(\Sigma_1, \Sigma_2, \Sigma_3) & \longrightarrow & G(\Sigma_1, \Sigma_3),
\end{array}
$$

Here, $G(\Sigma_1, \Sigma_2, \Sigma_3) = G(\Sigma_1, \Sigma_2) \cap G(\Sigma_2, \Sigma_3)$. This inclusion relation is the general feature in the diagrams below, in which numbers on the arrows denote the relevant geometric structures listed in List 20 above.

(39)
$$
\begin{array}{cccccccccc}
GL_n(\mathbb{R}) & Sp(n, \mathbb{R}) & \xrightarrow{(3)} & GL_{2N}(\mathbb{R}) & \xleftarrow{(2)} & GL_N(\mathbb{C}) & \xrightarrow{(6)} & SL_N(\mathbb{C}) \\
{\scriptstyle(1)}\uparrow & {\scriptstyle(11)}\uparrow & & {\scriptstyle(1)}\uparrow & & {\scriptstyle(7)}\uparrow & & {\scriptstyle(9)}\uparrow \\
O(n)\,, & U(N) & \xrightarrow{(12)} & O(2N) & \xleftarrow{(8)} & U(N) & \xleftarrow{(10)} & SU(N) \\
\end{array}
$$

$$
\begin{array}{ccccccc}
GL_{N'}(\mathbb{H}) \cdot Sp(1) & \xrightarrow{(5)} & GL_{4N'}(\mathbb{R}) & \xleftarrow{(4)} & GL_{N'}(\mathbb{H}) \\
{\scriptstyle(15)}\uparrow & & {\scriptstyle(1)}\uparrow & & {\scriptstyle(13)}\uparrow \\
Sp(N') \cdot Sp(1) & \xrightarrow{(16)} & O(4N') & \xleftarrow{(14)} & Sp(N'),
\end{array}
$$

Note that the parallel arrows in the same row or columns have similar geometric contents. For example, arrows (1), (7), (9), (11), (13), (15) mean the reduction of the automorphism groups resulting from considering compatible Euclidean structures. Similarly, (2), (8) are relevant to considering compatible complex structures, (3), (12) considers compatible symplectic structures, reductions (4), (14) come from compatible quaternionic structures, (5), (16) have to do with compatible almost quaternionic structures and finally, (6), (10) deals with compatible special linear structures. Each square is an example of the diagram (38). This observation motivates us to think things in categorical terms, where an object is a pair of groups with an inclusion map between them and a morphism between our objects is a map of pairs. This point of view is in accordance with the moduli space point of view which is described in Proposition 22 below.

In the above diagrams, the unitary group $U(N)$ appears in two places. This is due to the following well known fact.

LEMMA 21. *The unitary group* $U(N)$ *is a maximal compact subgroup in* $Sp(n, \mathbb{R})$ *and in* $GL_N(\mathbb{C})$, *and we have*

$$(40) \qquad O(2N) \cap GL_N(\mathbb{C}) = U(N), \qquad O(2N) \cap Sp(n, \mathbb{R}) = U(N).$$

PROOF. We choose a basis of E and let η, I, ω be the corresponding compatible geometric structures. With respect to this chosen basis, we can identify elements of the above groups with linear transformations on E. This means that we are regarding the above groups as the *left* automorphism groups for the above structures.

For the first identity, from the compatible Euclidean structure η and the complex structure I in E, we can construct a Hermitian pairing in E given by $\langle u, v \rangle = \eta(u, v) + i\eta(I(u), v)$ by Proposition 15 (i). Let $\tau \in O(2N) \cap GL_n(\mathbb{C})$ be any element in the intersection. Then, τ preserves the pairing and commutes with I. So, $\langle \tau(u), \tau(v) \rangle = \eta(\tau(u), \tau(v)) + i\eta(I(\tau(u)), \tau(v)) = \eta(u, v) + i\eta(I(u), v) = \langle u, v \rangle$, which means that $\tau \in U(N)$.

Conversely, if $\tau \in U(N)$, then, it preserves the Hermitian pairing and by looking at the real part, we see that τ preserves the Euclidean pairing η. So, $\tau \in O(2N)$. As for the imaginary part, we have $\langle I(\tau(u)), \tau(v) \rangle = \langle I(u), v \rangle$ which is further equal to $\langle \tau(I(u)), \tau(v) \rangle$ since τ preserves the Hermitian pairing. Since this holds for any $u, v \in E$, we have $I \circ \tau = \tau \circ I$, or $\tau \in GL_N(\mathbb{C})$. Hence, $\tau \in O(2N) \cap GL_N(\mathbb{C})$. This proves the first identity.

For the second identity, given compatible η, ω, from Proposition 15 (ii), there exists a complex structure I in E such that $\omega(u, v) = \eta(I(u), v)$ for any $u, v \in E$. This shows that a $\mathbb{C}$-valued pairing defined by $\langle u, v \rangle = \eta(u, v) + i\omega(u, v)$ is in fact a Hermitian pairing. So, if $\tau \in O(2N) \cap Sp(n, \mathbb{R})$, then τ preserves η and ω, so it also preserves the Hermitian pairing $\langle \ , \ \rangle$, and $\tau \in U(N)$.

Conversely, if $\tau \in U(N)$, then it preserves the Hermitian pairing, so its real and imaginary parts are preserved. That is, τ preserves η and ω and $\tau \in O(2N) \cap Sp(n, \mathbb{R})$. This proves the second identity. $\square$

We are particularly interested in the geometric structures in vector spaces listed in List 20, because the associated right-automorphism groups of these geometric types, if they are compact, exhaust almost all the groups appearing in the classification theory of holonomy groups of Riemannian manifolds. We have already encountered $O(n)$, $Sp(N')$, $U(N)$, $SU(N)$, $Sp(N') \cdot Sp(1)$, where $n = 2N$ or $n = 4N'$. We are only missing $Spin(9)$, $Spin(7)$ and G_2 from this list. See the diagram (39) for the list of groups we are dealing with.

For each of the above geometric types in List 20, the collection of inequivalent structures of the given type form a moduli space which has the structure of a homogeneous space, as we saw before. Recall that given two geometric types Σ_1, Σ_2, we have an inclusion map of right-automorphism groups,

$$(41) \qquad\qquad G(\Sigma_1, \Sigma_2) \to G(\Sigma_1).$$

The moduli space of geometric structures of type Σ_2 compatible with a geometric structure $\Sigma_1(e)$, $e \in \mathcal{F}(E)$, is given by Proposition 19:

$$(42) \qquad\qquad \mathfrak{M}_{\Sigma_1(e)}(\Sigma_2) \xleftarrow[\Sigma_2]{\cong} e \cdot G(\Sigma_1)/G(\Sigma_1, \Sigma_2)$$

Thus, we have a (relative) moduli space for each object of the form $[G(\Sigma_1, \Sigma_2) \to G(\Sigma_1)]$. From a square diagram involving three geometric types as in (38), we have four moduli spaces, two associated to horizontal pairs and two associated to vertical pairs. These are $\mathfrak{M}_{\Sigma_1(e)}(\Sigma_2)$, $\mathfrak{M}_{(\Sigma_1(e),\Sigma_3(e))}(\Sigma_2)$ which are moduli spaces of geometric structures of type Σ_2 compatible with $\Sigma_1(e)$ or with a pair of geometric structures $(\Sigma_1(e), \Sigma_3(e))$, and $\mathfrak{M}_{\Sigma_1(e)}(\Sigma_3)$, $\mathfrak{M}_{(\Sigma_1(e),\Sigma_2(e))}(\Sigma_3)$ which are moduli spaces of geometric structures of type Σ_3 compatible with $\Sigma_1(e)$ or with a pair $(\Sigma_1(e), \Sigma_2(e))$. The square diagram (38) provides more. It provides morphism between pairs. In fact, this inclusion morphism between pairs induces an inclusion map of moduli spaces.

PROPOSITION 22. *Let Σ_1, Σ_2, Σ_3 be three geometric types. Then, for any frame $e \in \mathcal{F}(E)$, we have an inclusion map of moduli spaces,*
(43)

$$
\begin{array}{ccc}
\mathfrak{M}_{(\Sigma_1(e),\Sigma_3(e))}(\Sigma_2) & \longrightarrow & \mathfrak{M}_{\Sigma_1(e)}(\Sigma_2) \\
\Sigma_2 \uparrow \cong & & \Sigma_2 \uparrow \cong \\
e \cdot G(\Sigma_1) \cap G(\Sigma_3)/\,(G(\Sigma_1) \cap G(\Sigma_2) \cap G(\Sigma_3)) & \longrightarrow & e \cdot G(\Sigma_1)/\,(G(\Sigma_1) \cap G(\Sigma_2))
\end{array}
$$

PROOF. We only have to observe that the bottom arrow is an inclusion map. $\square$

Applying this result to the above list of (compatible) geometric types, we obtain various moduli spaces as homogeneous spaces.

COROLLARY 23. *Let $e \in \mathcal{F}(E)$ be a frame and let $\eta = \eta(e)$, $I = I(e)$, $\omega = \omega(e)$, $(I, J) = (I(e), J(e))$, $L = L(e)$, $\alpha = \alpha(e)$. We have the following diagrams of moduli spaces and inclusion relations among them:*

(i) *(Moduli spaces of Euclidean structures lying in $E^* \otimes E^*$ corresponding to (1), (7), (9), (11), (13), (15) for a real $2N$-dimensional vector space.) The moduli space $\mathfrak{M}(\eta)$ is a $GL(E)$-orbit of $\eta \in E^* \otimes E^*$, $\mathfrak{M}_I(\eta)$ is a $GL(E; I)$-orbit of η, $\mathfrak{M}_\omega(\eta)$ is a $Sp(E; \omega)$-orbit of η, $\mathfrak{M}_{(I,\alpha)}(\eta)$ is a $SL(E; I)$-orbit of η. By choosing a* *frame in* E^{2N}, *these groups can be identified with* $GL_{2N}(\mathbb{R})$, $GL_N(\mathbb{C})$, $Sp(n, \mathbb{R})$, $SL_N(\mathbb{C})$, *and we have the following diagrams, due to Proposition 22:*

$$
\begin{array}{cccc}
\mathfrak{M}_\omega(\eta) \xrightarrow{\hspace{2cm}} & \mathfrak{M}(\eta) \xleftarrow{\hspace{1cm}} & \mathfrak{M}_I(\eta) \xleftarrow{\hspace{0.6cm}} & \mathfrak{M}_{(I,\alpha)}(\eta) \\[2mm]
\eta \uparrow \cong & \eta \uparrow \cong & \eta \uparrow \cong & \eta \uparrow \cong \\[2mm]
Sp(n, \mathbb{R})/U(N) \longrightarrow & GL_{2N}(\mathbb{R})/O(2N) \longleftarrow & GL_N(\mathbb{C})/U(N) \longleftarrow & SL_N(\mathbb{C})/SU(N)
\end{array}
$$

The moduli space $\mathfrak{M}_{(I,J)}(\eta)$ *is a* $GL(E; I, J)$-*orbit of* η, $\mathfrak{M}_L(\eta)$ *is a* $GL(E; I, J) \cdot Sp_1(E; I, J)$-*orbit of* η. *A choice of a frame in* E^{2N} *identifies these groups with* $GL_{N'}(\mathbb{H})$, $GL_{N'}(\mathbb{H}) \cdot Sp(1)$ *and we have the next diagram:*

$$
\begin{array}{ccc}
\mathfrak{M}_L(\eta) \xrightarrow{\hspace{3cm}} & \mathfrak{M}(\eta) \xleftarrow{\hspace{1.5cm}} & \mathfrak{M}_{(I,J)}(\eta) \\[2mm]
\eta \uparrow \cong & \eta \uparrow \cong & \eta \uparrow \cong \\[2mm]
(GL_{N'}(\mathbb{H}) \cdot Sp(1))/(Sp(N') \cdot Sp(1)) \rightarrow & GL_{4N'}(\mathbb{R})/O(4N') \leftarrow & GL_{N'}(\mathbb{H})/Sp(N')
\end{array}
$$

(ii) (Moduli spaces of complex structures lying in $E \otimes E^*$ corresponding to (2) and (8) for a real $2N$-dimensional vector space.)   *The moduli space* $\mathfrak{M}(I)$ *is a* $GL(E)$-*orbit of* $I \in E \otimes E^*$ *and* $\mathfrak{M}_\eta(I)$ *is a* $O(\eta)$-*orbit of* $I \in E \otimes E^*$. *A frame in* E^{2N} *identifies these groups with* $GL_{2N}(\mathbb{R})$ *and* $O(2N)$ *and we obtain the following diagram:*

$$
\begin{array}{ccc}
\mathfrak{M}(I) & \xleftarrow[\quad I \quad]{\cong} & GL_{2N}(\mathbb{R})/GL_N(\mathbb{C}) \\[3mm]
\uparrow & & \uparrow \\[3mm]
\mathfrak{M}_\eta(I) & \xleftarrow[\quad I \quad]{\cong} & O(2N)/U(N)
\end{array}
$$

(iii) (Moduli spaces of symplectic structures lying in $\bigwedge^2 E^*$ corresponding to (3) and (12) for a real $2N$-dimensional vector space E.)   *The moduli space* $\mathfrak{M}(\omega)$ *is a* $GL(E)$-*orbit of* $\omega \in \bigwedge^2 E^*$ *and* $\mathfrak{M}_\eta(\omega)$ *is a* $O(\eta)$-*orbit of* ω. *By choosing a frame in* E^{2N}, *we obtain the next diagram:*

$$
\begin{array}{ccc}
\mathfrak{M}(\omega) & \xleftarrow[\quad \omega \quad]{\cong} & GL_{2N}(\mathbb{R})/Sp(n, \mathbb{R}) \\[3mm]
\uparrow & & \uparrow \\[3mm]
\mathfrak{M}_\eta(\omega) & \xleftarrow[\quad \omega \quad]{\cong} & O(2N)/U(N)
\end{array}
$$

(iv) (Moduli spaces of quaternionic structures lying in $(E \otimes E^*) \oplus (E \otimes E^*)$ corresponding to (4) and (14) for a real $4N'$-dimensional vector space.)   *The moduli space* $\mathfrak{M}((I, J))$ *is a* $GL(E)$-*orbit of* $(I, J) \in (E \otimes E^*) \oplus (E \otimes E^*)$ *and*

$\mathfrak{M}_\eta\left((I,J)\right)$ *is a* $O(\eta)$-*orbit of* (I,J). *Identifying these groups with* $GL_{4N'}(\mathbb{R})$, $O(4N')$ *by choosing a frame, we have the diagram below:*

$$\mathfrak{M}\left((I,J)\right) \xleftarrow[\;(I,J)\;]{\cong} GL_{4N'}(\mathbb{R})/GL_{N'}(\mathbb{H})$$

$$\uparrow \qquad\qquad\qquad\qquad \uparrow$$

$$\mathfrak{M}_\eta\left((I,J)\right) \xleftarrow[\;(I,J)\;]{\cong} O(4N')/Sp(N')$$

(v) (Moduli spaces of almost quaternionic structures lying in $G_3(E \otimes E^*)$ corresponding to (5) and (16) for a real $4N'$-dimensional vector space E.)   The *moduli space* $\mathfrak{M}(L)$ *is a* $GL(E)$-*orbit of* $L \in G_3(E \otimes E^*)$ *and* $\mathfrak{M}_\eta(L)$ *is a* $O(\eta)$-*orbit of* $L \in G_3(E \otimes E^*)$. *Identifying these groups with* $GL_{4N'}(\mathbb{R})$, $O(4N')$ *via a choice of a frame in* $E^{4N'}$, *we get the next diagram:*

$$\mathfrak{M}(L) \xleftarrow[\;L\;]{\cong} GL_{4N'}(\mathbb{R})/\left(GL_{N'}(\mathbb{H}) \cdot Sp(1)\right)$$

$$\uparrow \qquad\qquad\qquad\qquad\qquad \uparrow$$

$$\mathfrak{M}_\eta(L) \xleftarrow[\;L\;]{\cong} O(4N')/\left(Sp(N') \cdot Sp(1)\right)$$

(vi) (Moduli spaces of special linear structures lying in $\bigwedge_\mathbb{C}^N E$ corresponding to (6) and (10) when E is a complex N-dimensional vector space.)   *The moduli space* $\mathfrak{M}_I(\alpha)$ *is a* $GL(E;I)$-*orbit of* $\alpha \in \bigwedge_\mathbb{C}^N E$ *and* $\mathfrak{M}_{(I,\eta)}(\alpha)$ *is a* $U(E;I,\eta)$-*orbit of* $\alpha \in \bigwedge_\mathbb{C}^N E$. *These groups can be identified as* $GL_N(\mathbb{C})$ *and* $U(N)$ *through a choice of a frame in* $E^{4N'}$, *and we have the next diagram:*

$$\mathfrak{M}_I(\alpha) \xleftarrow[\;\alpha\;]{\cong} GL_N(\mathbb{C})/SL_N(\mathbb{C}) \cong \mathbb{C}^*$$

$$\uparrow \qquad\qquad\qquad\qquad \uparrow$$

$$\mathfrak{M}_{(I,\eta)}(\alpha) \xleftarrow[\;\alpha\;]{\cong} U(N)/SU(N) \cong S^1$$

§4.4   On connections and parallelism

In this section, we discuss the concept of connections in some detail. The concept of parallelism was first defined by Levi-Civita along a curve on a surface. This gives a nice geometric picture of the parallelism, and it is described as follows: One rolls a surface in $\mathbb{R}^3$ on a plane π without slipping. Let the points of contact move along a curve γ_S on S and along a curve γ_π on π. Then, the Euclidean parallelism in the plane π along γ_π corresponds the parallelism on S along γ_S with respect to the Riemannian connection of S in $\mathbb{R}^3$. The most general notion of connection can be defined on the total space of fibre bundles, following Ehresmann. One can refine this notion of Ehresmann connection on the principal fibre bundles to obtain the notion of connection 1-forms on the total space of principal bundles. If our fibre bundles are vector bundles, then one can refine the notion of Ehresmann connection to obtain the notion of covariant derivatives acting on sections of vector bundles. The connection 1-forms for principal bundles and covariant derivatives on vector bundles are two different aspects of the same object. We briefly describe these notions and discuss the relationships among these concepts. We also discuss the notion of parallelism associated with connections from various view points.

Associated to the concept of connections is the notion of curvatures. The simplest geometric way to understand the curvature is as the obstruction to the integrability of the horizontal distribution associated to connections. On principal bundles, one can picture the curvature geometrically in this way. However, it is on tangent bundles of Riemannian manifolds we usually start from by constructing covariant derivatives, namely the Levi-Civita connections, acting on vector fields, using metrics on manifolds. From here, we can construct connection 1-forms on orthonormal frame bundles and a horizontal distribution on the total space of this principal bundle. Thus, on Riemannian manifolds the covariant derivatives are more fundamental, although the meaning of curvature is not quite clear from its definition $\Omega = \nabla \circ \nabla$ in terms of the covariant derivative ∇.

In what follows, we describe various aspects of connections.

Why connections?. Let me try to motivate the notion of connections. Let M be a manifold and let $X \in TM$ be a tangent vector. For any function $f \in C^\infty(M)$, the differentiation of f by X is a well defined operation on the manifold, independent of the choice of local coordinates. We generalize the situation and consider a section s of a vector bundle $\pi : \mathcal{E} \to M$. We try to differentiate s in the direction of X. Recalling the definition of differentiation in calculus, the most natural thing to do would be to consider a curve $\gamma : [0, \ell] \to M$ such that $\dot{\gamma}(0) = X$ and write down the formula "$\lim_{t\to 0}\{s(\gamma(t)) - s(\gamma(0))\}/t$". But this formula doesn't quite make sense, because we can't take this difference $s(\gamma(t)) - s(\gamma(0))$ since these vectors live in different fibres $\mathcal{E}_{\gamma(t)}$ and $\mathcal{E}_{\gamma(0)}$. Even if the vector bundle is trivial, there are many ways to trivialize it. So, there is no canonical way to identify different fibres of the vector bundle $\mathcal{E}$. Unless we have a way to identify these vector spaces, at least along the curve γ, the above "definition" fails to make sense. This identification of fibres along any curves is what connections will do.

Let us try to define the differentiation of a section s by a tangent vector X from a different point of view. Since it makes sense to differentiate a vector valued *function* by a tangent vector, we try to write the section s of the vector bundle $\mathcal{E}$ as a vector valued function on a related space. This can be achieved through principal bundles. So, let $\pi : P \to M$ be a principal bundle with the group G. Let (E, ρ) be a representation of G. Then $\mathcal{E} = P \times_G E$ is a vector bundle and any section of this bundle can be written as a G-equivariant map $s : P \to E$, $s(pg) = \rho(g^{-1})s(p)$, $g \in G$, $p \in P$. This is a vector valued function on P. If $\tilde{X}$ is a tangent vector to P such that $\pi_*(\tilde{X}) = X$, then the differentiation of $s : P \to E$

by $\tilde{X}$ makes sense. However, there is an ambiguity as to the choice of a lift $\tilde{X}$ of the vector X. Different choices give rise to different results of differentiation, because the map s is not constant along the fibres, due to the equivariant nature of s along the fibres of P. In fact, if $\tilde{X}_1$, $\tilde{X}_2$ are two lifts of $X \in T_{\pi(p)}M$ to T_pP, then their difference is a vertical vector and can be written as a fundamental vector $\tilde{Y}_p$ for some $Y \in \mathfrak{g}$. Then $\tilde{X}_1 \cdot s - \tilde{X}_2 \cdot s = \tilde{Y} \cdot s = -\rho(Y)(s(p))$. The last identity is the infinitesimal form of the G equivariance of s. For any choice of lifts, this vanishes if the representation ρ is trivial $\rho = 0$ in which case s is a vector valued function on M, or the section vanishes at p, $s(p) = 0$. Otherwise, the derivative $\tilde{X} \cdot s$ depends on the choice of lifts. So, to define differentiation unambiguously, we need a systematic way to choose lifts of tangent vectors on M to tangent vectors on P. This is what connections will do.

The above two approaches are different aspects of the same device, connection. We now discuss various aspects of connections in detail.

Connections on Fibre Bundles. Let $\pi : Q \to M$ be a fibre bundle whose fibre is F. Here all spaces are smooth manifolds. The tangent bundle TQ of the total space has a subbundle $\mathcal{V} = \mathcal{V}Q$ consisting of tangent vectors to the fibres. We have the following exact sequence of bundles on Q:

$$(1) \qquad\qquad 0 \to \mathcal{V}Q \to TQ \to \pi^*TM \to 0.$$

DEFINITION 1. An (Ehresmann) connection on a fibre bundle $\pi : Q \to M^n$ is a splitting $\mathcal{H} : \pi^*(TM) \to TQ$ of the above exact sequence. Equivalently, a connection is a n-dimensional horizontal distribution $\mathcal{H}$ on Q such that $\mathcal{H} \oplus \mathcal{V} = TQ$.

The projections from TQ to subbundles $\mathcal{V}$ and $\mathcal{H}$ are also denoted by the same notation $\mathcal{V}$ and $\mathcal{H}$.

Since a fibre bundle $\pi : Q \to M$ is locally trivial, when it is equipped with a connection, we can consider horizontal lifts of curves in M. Let $\gamma : [0, \ell] \to M$ be a piecewise smooth curve in M to curves in Q. Then for any point $p \in Q_{\gamma(0)}$ in a fibre, there exists a unique horizontal lift $\tilde{\gamma} : [0, \ell] \to Q$ starting from p. This is a curve in Q such that its tangent vectors are horizontal, that is $\dot{\tilde{\gamma}}(t) \in \mathcal{H}$, $0 \leq t \leq \ell$, and $\pi_*(\dot{\tilde{\gamma}}(t)) = \dot{\gamma}(t)$. We define a map $\tau_\gamma : Q_{\gamma(0)} \to Q_{\gamma(\ell)}$ by $\tau_\gamma(p) = \tilde{\gamma}(\ell) \in Q_{\gamma(\ell)}$. Then τ_γ is a diffeomorphism between the fibres. This is due to the differentiable independence of the solutions to systems of linear first order ODE's on the initial conditions. The map τ_γ is the parallel translation along the curve γ. When the curve γ is a loop starting at $x \in M$, then τ_γ is a diffeomorphism of Q_x. The collection of τ_γ's for the loops γ based at x forms a subgroup $\mathrm{Hol}(x)$ of the diffeomorphism group of Q_x and it is called the holonomy group at x.

Whenever there is a distribution of subspaces of the tangent bundle of a manifold, we can consider its integrability in the sense of Frobenius. Given an Ehresmann connection on a fibre bundle, we have two distributions $\mathcal{V}$ and $\mathcal{H}$. Obviously the vertical distribution $\mathcal{V}$ is integrable because it is the tangent bundle to the fibres. For the horizontal distribution $\mathcal{H}$, the obstruction tensor field to the integrability is given by

(2)
$$\mathcal{O}_Q = \mathcal{O}^{\mathcal{H}(Q)} : \mathcal{H} \times \mathcal{H} \to \mathcal{V}, \quad \text{where} \quad \mathcal{O}^{\mathcal{H}(Q)}(X, Y) = -\mathcal{V}([X, Y]), \quad X, Y \in \mathcal{H}.$$

This is the negative of the vertical component of the bracket vector field $[X, Y]$. Note that $\mathcal{O}_Q(X, Y)_p$, $p \in Q$, depends only on the value of vector fields X, Y at $p \in Q$, since $\mathcal{O}_Q$ is a $C^\infty(Q)$ linear map. When this obstruction vanishes, then the distribution $\mathcal{H}$ is involutive and $\mathcal{H}$ is integrable, i.e., it comes from the tangent bundle of integral submanifolds.

The above quantity can be given another meaning when the fibre bundle $\pi : Q \to M$ is a Riemannian immersion. In this case, let D be the Levi-Civita covariant derivative on Q with respect to a metric on Q. This metric defines automatically the horizontal distribution $\mathcal{H}$ as the orthogonal complement of $\mathcal{V} \subset TQ$. Then for horizontal vectors $X, Y \in \mathcal{H}$, the vertical component of $D_X Y$ is given by $\frac{1}{2}\mathcal{V}([X, Y])$. For the case of principal bundles with a connection, the above obstruction $\mathcal{O}^{\mathcal{H}}$ to the integrability of $\mathcal{H}$ will be called the curvature of the connection. Also for the case in which Q is a vector bundle with a covariant derivative ∇, the above obstruction $\mathcal{O}^{\mathcal{H}}$ for the horizontal distribution associated with ∇ is essentially the curvature of the covariant derivative. But in this case, the relation is more subtle.

Principal Connections. Let $\pi : P \to M$ be a principal bundle with the structure group G. This means that on the total space P, G acts freely from the right with its orbit space P/G identified with M via the projection map π which is locally trivial. For $g \in G$, the right multiplication by g is denoted by R_g, i.e. $R_g(p) = pg$ for $p \in P$. Let $X \in \mathfrak{g}$ be an element of the Lie algebra of G. We can define a vertical vector field $\tilde{X}$, called a fundamental vector field associated to X on P, by $\tilde{X}_p = \frac{d}{dt}|_{t=0}(p \cdot e^{tX})$ for $p \in P$. These vector fields should be thought of as the analogue of left invariant vector fields on the Lie group G. Recall that on G, the left invariant vector fields transform by the adjoint action of G under the right translation. Similarly on P, we have

(3)
$$R_{g*}(\tilde{X}_p) = \frac{d}{dt}\bigg|_{t=0} p \cdot e^{tX} \cdot g = \frac{d}{dt}\bigg|_{t=0} pg \cdot e^{tg^{-1}Xg} = \widetilde{\mathrm{Ad}_{g^{-1}}(X)}_{pg}.$$

The set of vertical vectors forms a subbundle $\mathcal{V}$ of TP, the vertical distribution. Note that for a principal bundle P, $\mathcal{V}$ is trivial along each fibre.

Since a principal bundle is a fibre bundle, we can consider an Ehresmann connection, namely, a horizontal distribution $\mathcal{H}$ on P. Since P is a free G space, in the context of principal bundles, we require that the horizontal distribution be G-equivariant. If we denote the horizontal space at $p \in P$ by $\mathcal{H}_p$, the G-equivariance means that $R_{g*}(\mathcal{H}_p) = \mathcal{H}_{pg}$ for any $p \in P$ and $g \in G$. This is the principal connection on P.

This principal connection can be described by a connection 1-form $\omega : TP \to \mathfrak{g}$, which is a Lie algebra valued 1-form on P satisfying

(4)
$$R_g^* \omega = \mathrm{Ad}_{g^{-1}} \cdot \omega, \quad g \in G \quad \text{and} \quad \omega(\tilde{X}) = X, \quad X \in \mathfrak{g}.$$

Since the action of $g \in G$ on the differential form ω is given by $(g \cdot \omega)_p = R_g^*(\omega_{pg})$, the first condition above means that $\omega \in \left(\mathcal{A}^1(P) \otimes \mathfrak{g} \right)^G$. Also note that the infinitesimal form of the first condition above is given by

$$(5) \qquad \mathcal{L}_{\tilde{X}} \omega + [X, \omega] = 0, \qquad \text{for } X \in \mathfrak{g}.$$

Here $\mathcal{L}$ denotes the Lie derivative.

The notion of a connection 1-form ω on P and the notion of G-equivariant horizontal distribution $\mathcal{H}$ on P are equivalent. In fact, given ω, the kernel of the map $\omega : TP \to \mathfrak{g}$ is a G-equivariant horizontal distribution. Conversely, given $\mathcal{H}$, at each $p \in P$, we have a projection map $T_p P = \mathcal{H}_p \oplus \mathcal{V}_p \to \mathcal{V}_p \overset{\cong}{\longleftarrow} \mathfrak{g}$ where the last isomorphism is in terms of the G-action on P. This defines a Lie algebra valued 1-form on P, which can be easily shown to satisfy the above conditions and so it is a connection 1-form on P.

For $X \in TP$, let X^H be its horizontal component and let $\mathcal{V}(X)$ be its vertical component. The curvature of the connection ω is a Lie algebra valued 2-form on P defined by

$$(6) \quad \Omega^\omega(X, Y) = -\omega([X^H, Y^H]) = -\omega(\mathcal{V}([X^H, Y^H])) = \omega(\mathcal{O}^H(X^H, Y^H)) \in \mathfrak{g}$$

for any $X, Y \in TP$. We note that if one of X, Y is vertical, then Ω^ω is zero, i.e. Ω^ω is horizontal. This means that the above 2-form is really comes from a 2-form on the base manifold M. Recall that $\mathcal{O}^H(X^H, Y^H)$ measures the obstruction to the integrability of the horizontal distribution $\mathcal{H}$. We can rewrite this 2-form in terms of ω as in the next proposition, which is well known.

PROPOSITION 2. *If Ω^ω is the curvature of a principal connection ω on a principal bundle $\pi : P \to M$, then we have*

$$(7) \qquad \Omega^\omega = d\omega + \tfrac{1}{2}[\omega, \omega].$$

PROOF. First we show that the right hand side of the above identity denotes a horizontal differential form on P. Let $\tilde{X}$ be the fundamental vertical vector field corresponding to $X \in \mathfrak{g}$. Then an interior product with $\tilde{X}$ gives $\iota_{\tilde{X}} d\omega + \tfrac{1}{2} \iota_{\tilde{X}}[\omega, \omega] = \mathcal{L}_{\tilde{X}} \omega - d\iota_{\tilde{X}} \omega + [X, \omega]$, since $\mathcal{L}_Z = d \circ \iota_Z + \iota_Z \circ d$. Here, $\iota_{\tilde{X}} \omega = \omega(\tilde{X}) = X$ is a Lie algebra valued constant function on P, so its exterior derivative vanishes. On the other hand, by G-invariance of ω, $\mathcal{L}_{\tilde{X}} \omega + [X, \omega] = 0$ for any $X \in \mathfrak{g}$. Thus, the 2-form in right hand side is horizontal. We check values for two horizontal vector fields X and Y. From the definition of Ω^ω, $\Omega^\omega(X, Y) = -\omega([X, Y])$. On the other hand, $(d\omega + \tfrac{1}{2}[\omega, \omega])(X, Y) = \{X\omega(Y) - Y\omega(X) - \omega([X, Y])\} + [\omega(X), \omega(Y)]$. Since X, Y are horizontal $\omega(X) = \omega(Y) = 0$ as functions, so we are left with $-\omega([X, Y])$. This proves the identity. $\square$

For principal bundles, horizontal lifts and holonomy groups have extra properties due to G-invariance of the connection. Let $\gamma : [0, \ell] \to M$ be a piecewise smooth curve on M. Let $p \in \pi^{-1}(\gamma(0))$ and let $\tilde{\gamma}$ be the horizontal lift of γ to

P starting from p. We note that the horizontal lift starting from pg, $g \in G$, is given by $R_g(\tilde{\gamma})$. This is because $R_{g_*}(\dot{\tilde{\gamma}}) \in \mathcal{H}$ whenever $\dot{\tilde{\gamma}} \in \mathcal{H}$ since $R_{g_*}(\mathcal{H}) = \mathcal{H}$ by G-invariance. This implies that $\tau_\gamma(pg) = R_g(\tilde{\gamma})(\ell) = \tilde{\gamma}(\ell) \cdot g = \tau_\gamma(p) \cdot g$ for any $g \in G$. Hence the diffeomorphism $\tau_\gamma : P_{\gamma(0)} \to P_{\gamma(\ell)}$ is a G-equivariant map for any path γ on M. In particular, if the path γ is a loop based at $x \in M$, then given $p \in P_x$, the map τ_γ is completely determined by $\tau_\gamma(p)$ which is necessarily of the form $p \cdot h_\gamma$ for some $h_\gamma \in G$. Of course h_γ depends on our initial choice of $p \in P$. Since $\tau_{\gamma_1} \circ \tau_{\gamma_2}(p) = \tau_{\gamma_1}(p \cdot h_{\gamma_2}) = \tau_{\gamma_1}(p)h_{\gamma_2} = p \cdot h_{\gamma_1}h_{\gamma_2}$, the collection of elements $h_\gamma \in G$ for all possible loops starting at x forms a subgroup $\mathrm{Hol}(p)$ of G which is isomorphic to the holonomy group $\mathrm{Hol}(x) \subset \mathrm{Diff}(P_x)$. Note that $\mathrm{Hol}(x)$ depends only on x and acts on P_x from the left, whereas $\mathrm{Hol}(p) \subset G$ depends on $p \in P_x$ and acts on $p \in P_x$ from the right. We easily see that $\mathrm{Hol}(pg) = g^{-1} \cdot \mathrm{Hol}(p) \cdot g$.

Covariant Derivatives. Let $\pi : \mathcal{E} \to M$ be a vector bundle. This is a fibre bundle whose fibre is a vector space E and whose structure group is $\mathrm{GL}(E)$. As usual, let $\Gamma(\mathcal{E})$ be the set of smooth sections of $\mathcal{E}$. By definition, a covariant derivative ∇ in $\mathcal{E}$ is an $\mathbb{R}$-linear map

$$\nabla : \Gamma(\mathcal{E}) \to \Gamma(T^*M \otimes \mathcal{E}), \qquad \text{such that}$$

(8)

$$\nabla(fs) = df \otimes s + f \cdot \nabla s, \qquad \text{for any } f \in C^\infty(M), s \in \Gamma(\mathcal{E}).$$

In other words, for any tangent vector $X \in TM$, we have

(9)
$$\nabla_X(fs) = (Xf)s + f\nabla_X s.$$

Note that ∇_X is not a $C^\infty(M)$ linear map, and it is a differential operator rather than a tensor on M. The curvature Ω^∇ of the covariant derivative ∇ is defined by $\Omega^\nabla = \nabla^2$. That is, for any vector field X, Y on M,

(10)
$$\Omega^\nabla(X, Y) = \iota_Y \circ \iota_X \circ \nabla^2.$$

We have an alternate expression for the curvature given by

(11)
$$[\nabla_X, \nabla_Y] - \nabla_{[X,Y]}.$$

This identity can be shown as follows. Let $U \subset M$ be an open neighborhood of M and let $s_1, \ldots, s_m$ be sections of $\mathcal{E}$ on U which form a basis in each fibre. We can write $\nabla s = \sum_{j=1}^m \alpha_j \otimes s_j$ for some 1-forms α_j's on U. Then $\nabla^2 s = \sum_j d\alpha_j \otimes s_j - \sum_j \alpha_j \wedge \nabla s_j$. Now,

$$\nabla^2 s(X, Y) = \sum_{j=1}^m d\alpha_j(X, Y) \cdot s_j - \sum_{j=1}^m \alpha_j(X) \cdot \nabla_Y s_j + \sum_{j=1}^m \alpha_j(Y) \cdot \nabla_X s_j$$

Since $d\alpha_j(X, Y) = X\alpha_j(Y) - Y\alpha_j(X) - \alpha_j([X, Y])$, rearranging terms, we get

$$= \nabla_X \left(\sum_{j=1}^m \alpha_j(Y) \cdot s_j \right) - \nabla_Y \left(\sum_{j=1}^m \alpha_j(X) \cdot s_j \right) - \sum_{j=1}^m \alpha_j([X, Y]) \cdot s_j$$

$$= \nabla_X \nabla_Y s - \nabla_Y \nabla_X s - \nabla_{[X,Y]} s.$$

One can easily check that Ω^∇ is a $C^\infty(M)$ linear map and so it is a tensor field on M. That is, the value of $\Omega^\nabla(X, Y)_p$ for vector field X, Y depend only on their value X_p, Y_p at p.

Let $\gamma : [0, \ell] \to M$ be a path on M and and let $s : [0, \ell] \to \mathcal{E}$ be a section of $\mathcal{E}$ along γ, so that $\pi \circ s = \gamma$. It turns out that the covariant derivative $\nabla_{\dot\gamma(t)} s$ is well defined along γ. If this vanishes, then the section s is a horizontal lift of γ. For any vector $v \in \mathcal{E}_{\gamma(0)}$, there exists a unique horizontal lift of γ from v by solving $\nabla_{\dot\gamma(t)} s = 0, s(0) = v$. We define $\tau_\gamma : \mathcal{E}_{\gamma(0)} \to \mathcal{E}_{\gamma(\ell)}$ by $\tau_\gamma(v) = s(\gamma(\ell))$ where s is the horizontal lift of γ starting from $v \in \mathcal{E}_{\gamma(0)}$. This is a diffeomorphism due to the smooth dependence of the solutions of $\nabla_{\dot\gamma(t)} s = 0$ on the initial conditions. Since the equation $\nabla_{\dot\gamma(t)} s = 0$ is a linear equation in s, the map τ_γ is in fact an invertible linear map. We can define the horizontal distribution $\mathcal{H}$ on the total space E of the vector bundle by letting $\mathcal{H}_v$, $v \in \mathcal{E}$, consist of those vectors which are tangent to parallel(=horizontal) curves in $\mathcal{E}$ through v. Thus, given a covariant derivative on $\mathcal{E}$, by solving linear systems of ODE's, we can construct a horizontal distribution $\mathcal{H}$ on $\mathcal{E}$.

Conversely, given a horizontal distribution $\mathcal{H}$ on $\mathcal{E}$, we can construct a covariant derivative ∇ in $\mathcal{E}$ as follows. Let $X \in TM$ be any tangent vector on M and let $\gamma : [0, \ell] \to M$ be a curve such that $\dot\gamma(0) = X$. Let $\tau_{\gamma_t} : \mathcal{E}_{\gamma(0)} \to \mathcal{E}_{\gamma(t)}$ be the parallel transport for $0 \le t \le \ell$. Let s be a section of $\mathcal{E}$ along γ. Then, $\nabla_X s$ is defined by

$$(12) \qquad \nabla_X s = \lim_{t \to 0} \frac{1}{t} \left\{ \tau_{\gamma_t}^{-1}\big(s(\gamma(t))\big) - s\big(\gamma(0)\big) \right\}.$$

That is, we parallel transport $s(\gamma(t))$ from $\mathcal{E}_{\gamma(t)}$ back into $\mathcal{E}_{\gamma(0)}$ along γ, then we take the velocity vector at $t = 0$ of the curve $\tau_{\gamma_t}^{-1}\big(s(\gamma(t))\big)$ in $\mathcal{E}_{\gamma(0)}$ starting from $s(\gamma(0))$. This velocity vector is $\nabla_X s \in \mathcal{E}_{\gamma(0)}$ using the natural identification $T_{s(\gamma(0))} \mathcal{E} \cong \mathcal{E}_{\gamma(0)}$.

The fundamental theorem of Riemannian geometry states that for any metric on a given manifold, one can construct a unique torsion free, metric preserving covariant derivative, the Levi-Civita connection, acting on vector fields on the manifold. This covariant derivative gives rise to a connection 1-form on the associated principal orthonormal frame bundle on the manifold. This in turn allows us to construct covariant derivatives on arbitrary vector bundles associated to this principal bundle via representations of the structure group.

On the other hand, given covariant derivatives on vector bundles on $\mathcal{E}_1$ and $\mathcal{E}_2$, we can canonically construct a covariant derivative on their tensor product $\mathcal{E}_1 \otimes \mathcal{E}_2$ and their dual vector bundles using the derivation property and the contraction property of covariant derivatives. These new covariant derivatives are indeed those corresponding to connections on appropriate principal bundles which are naturally associated to original vector bundles. Given any vector bundle $\mathcal{E}$ equipped with a covariant derivative, the derivation property and the contraction property of covariant derivatives allow us to immediately construct a covariant derivative on the tensor algebra bundle of $\mathcal{E}$,

$$(13) \quad T\mathcal{E} = \bigoplus_{p,q \ge 0} T^{(p,q)}\mathcal{E}, \quad \text{where} \quad T^{(p,q)}\mathcal{E} = \underbrace{\mathcal{E} \otimes \cdots \otimes \mathcal{E}}_{p} \otimes \underbrace{\mathcal{E}^* \otimes \cdots \otimes \mathcal{E}^*}_{q}.$$

The tensor algebra bundles are important because various geometrically interesting structures on $\mathcal{E}$ can be expressed as covariantly invariant (=parallel) sections of tensor bundles. For example, the metric on TM is a parallel section of $T^*M \otimes T^*M$.

Although it is well known, we briefly discuss the derivation property and the contraction property of covariant derivatives in a general setting, not just on tensor bundles on TM, relating various concepts relevant to connections.

Let $(\mathcal{E}^{(i)}, \nabla^{(i)})$, $i = 1, 2$, be two vector bundles equipped with covariant derivatives. Let $\mathcal{H}^{(i)}$, $i = 1, 2$, be the horizontal distributions on $\mathcal{E}^{(i)}$ and let $\pi^{(i)} : \mathcal{E}^{(i)} \to M$ be the projection map. On the tensor product bundle $\mathcal{E}^{(1)} \otimes \mathcal{E}^{(2)}$, we consider the horizontal distribution $\mathcal{H}$ given by the fibre product

(14)
$$\mathcal{H}(\mathcal{E}^{(1)} \otimes \mathcal{E}^{(2)}) = \mathcal{H}^{(1)} \underset{TM}{\times} \mathcal{H}^{(2)}$$
$$= \{(X_1, X_2) \in \mathcal{H}^{(1)} \times \mathcal{H}^{(2)} \mid \pi^{(1)}{}_*(X_1) = \pi^{(2)}{}_*(X_2)\}.$$

If $(\mathcal{E}^{(i)}, \mathcal{H}^{(i)})$ comes from a principal bundle $P^{(i)}$ with a horizontal distribution $\mathcal{H}^{(i)}(P^{(i)})$, then we can consider a horizontal distribution $\mathcal{H}$ on $P^{(1)} \times P^{(2)}$ defined by the same formula as above. This principal horizontal distribution induces the above distribution on the tensor product of bundles, as will be discussed in the next subsection.

LEMMA 3. *The covariant derivative ∇ on $\mathcal{E}^{(1)} \otimes \mathcal{E}^{(2)}$ associated to the horizontal distribution $\mathcal{H}$ in (14) is given by*

(15)
$$\nabla = \nabla^{(1)} \otimes 1 + 1 \otimes \nabla^{(2)}.$$

That is, for sections $s_i \in \Gamma(\mathcal{E}^{(i)})$, $i = 1, 2$, and $X \in T_x M$, we have

$$\nabla_X(s_1 \otimes s_2) = \nabla_X s_1 \otimes s_2(x) + s_1(x) \otimes \nabla_X s_2.$$

PROOF. We use the description of covariant derivatives in terms of parallel transports (12). Let $\gamma : [0, \ell] \to M$ be a smooth path in M such that $\dot{\gamma}(0) = X$. Let $\tau_{\gamma_t}^{(i)} : \mathcal{E}_{\gamma(0)}^{(i)} \to \mathcal{E}_{\gamma(t)}^{(i)}$, $i = 1, 2$, be the parallel transports along γ for $0 \leq t \leq \ell$. Then the parallel transport in $(\mathcal{E}^{(1)} \otimes \mathcal{E}^{(2)}, \mathcal{H})$ is given by $\tau_{\gamma_t} = \tau_{\gamma_t}^{(1)} \otimes \tau_{\gamma_t}^{(2)}$, because the horizontal paths in the tensor product bundles are the tensor products of horizontal paths. Then, for any sections $s_i \in \Gamma(\mathcal{E}^{(i)})$, $i = 1, 2$, we have

$$\nabla_X(s_1 \otimes s_2) = \frac{d}{dt}\bigg|_{t=0} \left(\tau_{\gamma_t}^{-1}(s_1(\gamma(t)) \otimes s_2(\gamma(t)))\right)$$
$$= \frac{d}{dt}\bigg|_{t=0} \left\{\left(\tau_{\gamma_t}^{(1)}\right)^{-1} s_1(\gamma(t)) \otimes \left(\tau_{\gamma_t}^{(2)}\right)^{-1} s_2(\gamma(t))\right\} = \nabla_X s_1 \otimes s_2(x) + s_1(x) \otimes \nabla_X s_2.$$

This proves the derivation property of covariant derivatives. $\square$

Next, we discuss the contraction(=dual pairing) property of covariant derivatives. Given a vector bundle with covarint derivative $(\mathcal{E}, \nabla)$, we introduce a covariant derivative in the dual vector bundle $\mathcal{E}^*$ as follows. First, we introduce the parallel transport τ_γ in $\mathcal{E}^*$ along a path γ in M by

(16) $\qquad \tau_\gamma(r^*)(s) = r^*(\tau_\gamma^{-1}(s)),$ $\qquad$ for any $r^* \in \mathcal{E}_{\gamma(0)}^*,$ $s \in \mathcal{E}_{\gamma(\ell)}.$

This means that the dual pairing between $\mathcal{E}^*$ and $\mathcal{E}$ is preserved under parallel transport along any path γ, that is, $\tau_\gamma(r^*)(\tau_\gamma(s)) = r^*(s)$. This uniquely defines the horizontal distribution on $\mathcal{E}^*$. Later, we will deduce the above formula of parallel transport in $\mathcal{E}^*$ from the point of view of principal connections. For now, we prove the following formula of the covariant derivative in $\mathcal{E}^*$.

LEMMA 4. *For sections $r^* \in \Gamma(\mathcal{E}^*)$, $s \in \Gamma(\mathcal{E})$ and a vector $X \in TM$, the covariant derivative ∇_X in $\mathcal{E}^*$ satisfies*

(17) $\qquad\qquad X \cdot (r^*(s)) = (\nabla_X r^*)(s) + r^*(\nabla_X s).$

PROOF. Let $\gamma : [0, \ell] \to M$ be a path in M such that $\dot\gamma(0) = X$. we consider a function $\varphi : [0, \ell] \to \mathbb{R}$ defined by

$$\varphi(t) = \left(\tau_{\gamma_t}^{-1}(r^*(\gamma(t)))\right)\left(\tau_{\gamma_t}^{-1}(s(\gamma(t)))\right).$$

Differentiating at $t = 0$, we get $\varphi'(0) = (\nabla_X r^*)(s) + r^*(\nabla_X s)$. On the other hand, since the dual pairing remains invariant under parallel transport, $\varphi(t) = r^*(\gamma(t))(s(\gamma(t))) = (r^*(s))(\gamma(t))$. So, the derivative at $t = 0$ is given by $\varphi'(0) = X \cdot (r^*(s))$. This proves the lemma. $\square$

This Lemma shows that covariant derivatives commute with contractions. Namely, we have the following commutative diagram.

(18)
$$\begin{array}{ccc}
\mathcal{E}^* \otimes \mathcal{E} & \xrightarrow{\ \nabla_X\ } & \mathcal{E}^* \otimes \mathcal{E} \\
\downarrow{\scriptstyle C} & & \downarrow{\scriptstyle C} \\
C^\infty(M) & \xrightarrow{\ \nabla_X\ } & C^\infty(M).
\end{array}$$

Here C is the contraction map, that is, the dual pairing map.

The relations between principal connections and covariant derivatives. The notion of principal connections and that of covariant derivatives are essentially the same notions. One can construct vector bundles equipped with a connection from a principal bundle with a connection 1-form and vice versa. But there is a slight ambiguity as to the structure group of principal bundles in the correspondence between these two notions. Here is the detail.

From principal connections to covariant derivatives. We start from a principal bundle (P, ω) with the structure group G equipped with a principal connection ω. Let $\rho : G \to \mathrm{GL}(E)$ be a representation of G on E. We can then associate a vector bundle $\underline{E} = P \times_G E$. We can construct a covariant derivative acting on sections of $\underline{E}$ as follows. This will be an $\mathbb{R}$-linear map

$$(19) \qquad \nabla : \Gamma(\underline{E}) \to \Gamma(T^*M \otimes \underline{E}).$$

To construct this map, first we identify the above space of sections with objects on principal bundles. Let $\mathcal{A}^k(P)$ be the space of differential k-forms on P. The group G acts on this space by $g \cdot \alpha = R_{g^{-1}}{}^*(\alpha)$ for $\alpha \in \mathcal{A}^k(P)$, $g \in G$. On the vector space E, G acts by $g \cdot v = \rho(g)(v)$, for $g \in G$, $v \in E$. So we can let G act on $\mathcal{A}^k(P, E) = \mathcal{A}^k(P) \otimes E$ diagonally. We let

$$(20) \qquad \mathcal{A}^k(P, E)_{\text{basic}} = \{ \alpha \in \mathcal{A}^k(P, E) \mid g \cdot \alpha = \alpha, \ \iota_{\tilde{X}} \alpha = 0, X \in \mathfrak{g} \}.$$

Note that when $k = 0$, elements of $\mathcal{A}^0(P, E)$ is nothing but G-equivariant maps $P \to E$, hence they are sections of the vector bundle $\underline{E}$. Thus, basic differential forms are G-invariant horizontal differential forms. This notion is useful because it allows us to write objects on M in terms of G-invariant objects on P as in the following lemma, which can be easily checked.

LEMMA 5. *For $\alpha \in \mathcal{A}^k(P, E)_{\text{basic}}$, define $\alpha_M \in \mathcal{A}^k(M, \underline{E})$ by*

$$(21) \qquad \alpha_M(\pi_*(X_1), \dots, \pi_*(X_k))(x) = [p, \alpha(X_1, \dots, X_k)(p)] \in E_x.$$

Here, $\pi(p) = x$ and $X_1, \dots, X_k \in T_p P$. Then the correspondence $\alpha \to \alpha_M$ gives rise to an isomorphism

$$(22) \qquad \mathcal{A}^k(P, E)_{\text{basic}} \overset{\cong}{\to} \mathcal{A}^k(M, \underline{E}).$$

For example, $\Gamma(\underline{E}) \cong \mathcal{A}^0(P, E)^G$, and $\Gamma(T^*M \otimes \underline{E}) \cong \mathcal{A}^1(P, E)_{\text{basic}}$. Now, we consider the following map

$$(23) \qquad d + \rho(\omega) : \mathcal{A}^0(P, E)^G \to \mathcal{A}^1(P, E)_{\text{basic}},$$

For $s \in \mathcal{A}^0(P, E)^G$, $X_p \in T_p P$, we have

$$(24) \qquad ((d + \rho(\omega))s)(X_p) = X_p \cdot s + \rho(\omega(X_p)) \cdot (s(p)) = X_p^H \cdot s = ds(X_p^H).$$

Here, X_p^H is the horizontal part of X_p with respect to the connection ω. The middle identity in the above follows from the fact that G-invariance of s implies that $\tilde{X} \cdot s + \rho(X)s = 0$ for $X \in \mathfrak{g}$. From the last expression, it is clear that $(d + \rho(\omega))s$ is horizontal. Next, we show that $(d + \rho(\omega))s$ is G-invariant. Using G-invariance of ω and s, namely, $R_g^*\omega = \text{Ad}_{g^{-1}}(\omega)$, $R_g^*s = \rho(g^{-1}) \cdot s$, we have

$$R_g^* ((d + \rho(\omega))s) = dR_g^*s + \rho(R_g^*\omega)R_g^*s = d(\rho(g^{-1}) \cdot s) + \rho(\text{Ad}_{g^{-1}}(\omega))\rho(g^{-1}) \cdot s$$

Since $\rho(\text{Ad}_{g^{-1}}(\omega)) = \rho(g^{-1})\rho(\omega)\rho(g)$, this is equal to

$$= \rho(g^{-1})ds + \rho(g^{-1})\rho(\omega)s = \rho(g^{-1})((d + \rho(\omega))s).$$

This shows that E-valued 1-form $(d + \rho(\omega))s$ is G-invariant. Hence $(d + \rho(\omega))s$ belongs to $\mathcal{A}^1(P, E)_{\text{basic}}$. Now the covariant derivative $\nabla = \nabla^\omega$ is defined by the following commutative diagram.

(25)
$$
\begin{array}{ccc}
\mathcal{A}^0(P, E)^G & \xrightarrow{\ d+\rho(\omega)\ } & \mathcal{A}^1(P, E)_{\text{basic}} \\
\Big\downarrow{\scriptstyle\simeq} & & {\scriptstyle\simeq}\Big\downarrow \\
\Gamma(\underline{E}) & \xrightarrow{\ \ \nabla^\omega\ \ } & \Gamma(T^*M \otimes \underline{E}).
\end{array}
$$

We record this result here, which will be very useful later.

LEMMA 6. *Let $(P, G, M; \omega)$ be a principal bundle with a connection and let $\underline{E} = P \times_G E$ be the vector bundle associated to a representation $\rho : G \to \mathrm{GL}(E)$. Then, the covariant derivative ∇ acting on sections of $\underline{E}$ is given by*

(26)
$$
\widetilde{\nabla_X s} = X^H \cdot \tilde{s},
$$

where $X \in TM$, $s \in \Gamma(\underline{E})$, X^H is the horizontal lift of X on P, and $\tilde{s} : P \to E$ is the G-equivariant map associated to s.

Let $\gamma : [0, \ell] \to M$ be a piecewise smooth curve. The parallel transport of vectors in $\underline{E}_{\gamma(0)}$ to vectors in $\underline{E}_{\gamma(\ell)}$ has a simple description in terms of the parallel transport in P. Let $v \in \underline{E}_{\gamma(0)}$. We must solve the equation $\nabla_{\dot\gamma(t)} s(t) = 0$, $s(0) = v$, $0 \le t \le \ell$, for a curve $s(t)$ in $\underline{E}$ lying above γ. The vector $v \in P \times_G E$ has many representatives. Take one representative, say, $v = [p, e]$. Let $\tilde\gamma(t)$ be the lift of γ to P such that $\tilde\gamma(0) = p$. Then the path $r(t)$ in $\underline{E}$ given by $r(t) = [\tilde\gamma(t), e]$ satisfy the above equation. To see this, let $\tilde{r}$ be the associated G-equivariant map $\tilde{r} : \pi_P^{-1}(\gamma) \to E$. Then, we note that along $\tilde\gamma(t)$ this is a constant map with constant value $e \in E$. To calculate $\nabla_{\dot\gamma(t)} r(t)$, we work on P. By the above lemma, this is equal to $\dot{\tilde\gamma}(t)_{\tilde\gamma(t)} \cdot \tilde{r}(\tilde\gamma(t)) = 0$, because the function $\tilde{r}(\tilde\gamma(t))$ is a constant function. This shows that $r(t) = [\tilde\gamma(t), e]$ is the parallel transport of $[\tilde\gamma(0), e]$ along $\gamma(t)$. We record this observation in the next lemma.

LEMMA 7. *With the above notation, the parallel transport of $[p, e] \in \underline{E}_{\gamma(0)}$ along γ is given by $[\tilde\gamma(t), e] \in \underline{E}_{\gamma(t)}$ where $\tilde\gamma(t)$ is the horizontal lift of $\gamma(t)$ starting from $p \in P$.*

By considering tangent vectors to the horizontal paths in $\underline{E}$ constructed above, we can construct the horizontal distribution $\mathcal{H}^\nabla \subset T\underline{E}$ with respect to the covariant derivative $\nabla = \nabla^\omega$. But there is a more direct and simple way to construct $\mathcal{H}^\nabla$. Taking the differential of the quotienting map $q : P \times E \to P \times_G E = \underline{E}$, we have $q_* : TP \times TE \to T\underline{E}$. Here, note that the horizontal distribution $\mathcal{H}^\omega \subset TP$ can be regarded as a subbundle $\mathcal{H}^\omega \times \{0\}$ of $TP \times TE$.

LEMMA 8. *With the above notation, $q_*(\mathcal{H}^\omega \times \{0\}) \subset T\underline{E}$ is the horizontal distribution $\mathcal{H}^\nabla$ of the covariant derivative ∇^ω.*

PROOF. This is clear from the previous picture of horizontal paths in $\underline{E}$ in terms of horizontal paths in P. $\quad\square$

We give another proof of the derivation property and the contraction property of covariant derivatives from the point of view of principal connections. For this, we start from a principal bundle $(P, G, M; \omega)$ equipped with a connection ω. Let bundles $\mathcal{E}^{(i)}$ be obtained through representations $\rho_i : G \to \mathrm{GL}(E_i)$, $i = 1, 2$. For any sections $s_i \in \Gamma(\mathcal{E}^{(i)})$, let $\tilde{s}_i : P \to E_i$ be associated G-equivariant maps. Also, for $X \in T_x M$, let X^H be the horizontal lift to $P_x = \pi^{-1}(x)$. Then, (26) gives

$$
(27) \quad
\begin{aligned}
\nabla_X (s_1 \otimes s_2) = X^H (\tilde{s}_1 \otimes \tilde{s}_2) &= (X^H \tilde{s}_1) \otimes \tilde{s}_2 + \tilde{s}_1 \otimes X^H \tilde{s}_2 \\
&= \nabla_X s_1 \otimes s_2 + s_1 \otimes \nabla_X s_2.
\end{aligned}
$$

This proves the derivation formula.

Now we discuss parallel transports and covariant derivatives in the dual bundle $\mathcal{E}^*$ from the point of view of principal connections. Assume that $(\mathcal{E}, \nabla)$ is a vector bundle associated to a principal bundle $(P, G, M; \omega)$ with a connection, through a representation $\rho : G \to \mathrm{GL}(E)$. Then, the dual bundle $\mathcal{E}^*$ is constructed using the dual representation $\rho^* : G \to \mathrm{GL}(E^*)$, where $\rho^*(g)(r^*) = r^* \circ \rho(g)$ for $g \in G$, $r^* \in E^*$. Note that the dual pairing map $C : E^* \otimes E \to \mathbb{R}$ is G-equivariant inducing a pairing map $\mathcal{E}^* \otimes \mathcal{E} \to \mathbb{R}$. We examine the parallel transport in $\mathcal{E}^*$ induced from the connection on P. Let $\gamma : [0, \ell] \to M$ be a path in M and let $\tilde{\gamma}$ be a lift to P starting from $p \in P_{\gamma(0)}$. If $r^* \in \mathcal{E}^*$, $s \in \mathcal{E}$ have representatives $r^* = [p, \tilde{r}^*]$, $s = [p, \tilde{s}]$, the parallel transports along γ of these vectors are given by $\tau_\gamma(r^*) = [\tilde{\gamma}(\ell), \tilde{r}^*]$ and $\tau_\gamma(s) = [\tilde{\gamma}(\ell), \tilde{s}]$, by Lemma 7. Then, taking pairings, we see that $(\tau_\gamma(r^*))(\tau_\gamma(s)) = \tilde{r}^*(\tilde{s})$. On the other hand we also have $r^*(s) = \tilde{r}^*(\tilde{s})$. Thus we have shown that the pairing remains invariant under parallel transports proving (16), $\tau_\gamma(r^*)(s) = r^*(\tau_\gamma^{-1}(s))$, which was used as a definition there.

We can also prove a formula for the covariant derivative in the dual bundle $\mathcal{E}^*$ using the connection ω of the principal bundle P. For this, let $\tilde{r}^* : P \to E^*$, $\tilde{s} : P \to E$ be the G-equivariant maps corresponding to sections $r^* \in \Gamma(\mathcal{E}^*)$, $s \in \Gamma(\mathcal{E})$. Then the pairing $\tilde{r}^*(\tilde{s}) : P \to \mathbb{R}$ is G-invariant because $\tilde{r}^*_{pg}(\tilde{s}_{pg}) = (\tilde{r}^*_p \circ \rho(g))(\rho(g^{-1})(\tilde{s}_p)) = \tilde{r}^*_p(\tilde{s}_p)$, and hence factors through M, resulting in a map $r^*(s) : M \to \mathbb{R}$. Let X^H be the horizontal lift of $X \in TM$. Then, (26) implies

$$
(28)
$$
$$
X \cdot (r^*(s)) = X^H \cdot (\tilde{r}^*(\tilde{s})) = (X^H \cdot \tilde{r}^*)(\tilde{s}) + \tilde{r}^*(X^H \cdot \tilde{s}) = (\nabla_X r^*)(s) + r^*(\nabla_X s).
$$

From covariant derivatives to principal connections. Conversely, given a real n dimensional vector bundle $\pi : \mathcal{E} \to M$ with a covariant derivative ∇, we consider the frame bundle $\mathcal{F}(\mathcal{E})$ of $\mathcal{E}$ with the structure group $G = \mathrm{GL}_n(\mathbb{R})$. We recover the vector bundle by $\mathcal{E} = \mathcal{F}(\mathcal{E}) \times_G E$, where $E = \mathbb{R}^n$. The covariant derivative allows us to construct a horizontal distribution $\mathcal{H}$ on the total space of $\mathcal{F}(\mathcal{E})$. To do this, we consider parallel transport of a basis of a fibre of $\mathcal{E}$ along a curve in M. This is just parallel transport of individual basis vectors. This gives us a horizontal lift of a path in M to a path in $\mathcal{F}(E)$ starting from a given basis. By definition, the horizontal space $\mathcal{H}_p \subset T_p \mathcal{F}(\mathcal{E})$, $p \in \mathcal{F}(E)$, consists of tangent vectors at p of horizontal path in $\mathcal{F}(E)$ passing through p. Note that this distribution is automatically $G = \mathrm{GL}_n(\mathbb{R})$-equivariant. This is because $(v_1(t), \ldots, v_n(t))A$ is a horizontal path in $\mathcal{F}(E)$ for any $A \in \mathrm{GL}_n(\mathbb{R})$ if $(v_1(t), \ldots, v_n(t))$ is a horizontal path in $\mathcal{F}(\mathcal{E})$ due to the linearity of the equation for the parallel transport. From this horizontal distribution on $\mathcal{F}(\mathcal{E})$, one can construct a connection 1-form ω as before.

However, we can construct a connection 1-form directly from the covariant derivative ∇. The idea is to use the formula discussed above, "$\nabla = d + \rho(\omega)$" in the other direction. To define $\omega \in \mathcal{A}^1(\mathcal{F}(\mathcal{E}), \mathfrak{gl}_n(\mathbb{R}))^G$, we have to assign an endomorphism $\omega_p(X)$ of $\mathbb{R}^n$ for any $p \in \mathcal{F}(\mathcal{E})$ and $X \in T_p\mathcal{F}(\mathcal{E})$. Let $\pi(p) = x \in M$. For any $v \in \mathbb{R}^n$, we let this endomorphism defined by

$$(29) \qquad \omega_p(X)(v) = (\widetilde{\nabla_{(\pi_* X)} s_M})(p) - (X \cdot s)(p) \in \mathbb{R}^n.$$

Here, $s : \mathcal{F}(\mathcal{E}) \to \mathbb{R}^n$ is any G-equivariant map such that $s(p) = v$ and s_M is the corresponding section of the vector bundle $\mathcal{E}$. Also, $(\widetilde{\nabla_{(\pi_* X)} s_M}) : \mathcal{F}(\mathcal{E})_x \to \mathbb{R}^n$ is the G-equivariant map corresponding to a vector $\nabla_{(\pi_* X)} s_M \in \mathcal{E}_x$. We first check that this is well defined, that is, it is independent of the choice of s such that $s(p) = v$. This follows from the next claim:

$$(30) \qquad (\widetilde{\nabla_{\pi_* X} s_M})(p) - (X \cdot s)(p) = 0, \qquad \text{if} \quad s(p) = 0.$$

Note that if the vector X is a horizontal vector in $\mathcal{H}$, then this identity holds without the condition $s(p) = 0$ by (26). To show this claim, note that the condition $s(p) = 0$ implies that $s(\mathcal{F}(\mathcal{E})_x) = 0$ by G-equivariance. So, the section s_M of $\mathcal{E}$ vanishes at $x \in M$. Then we can write s_M as $s_M(m) = f(m)r(m)$ for some $r \in \Gamma(\mathcal{E})$ and $f \in C^\infty(M)$ such that $f(x) = 0$. Let $\tilde{f} : \mathcal{F}(\mathcal{E}) \to \mathbb{R}$, $\tilde{r} : \mathcal{F}(E) \to \mathbb{R}^n$ be the $\mathrm{GL}_n(\mathbb{R})$-invariant liftings of f and r. Then, $s = \tilde{f} \cdot \tilde{r}$ and $\tilde{f}(p) = 0$. Now, $\nabla_{\pi_* X}(f \cdot r) = (\pi_* X)_x f \cdot r(x) + f(x) \cdot \nabla_{\pi_* X} r$. Since the second term vanishes because $f(x) = 0$, lifting the first term to $\mathcal{F}(\mathcal{E})$, we have $(\widetilde{\nabla_{\pi_* X} s_M})(p) = (X_p \tilde{f}) \cdot \tilde{r}(p)$. On the other hand, $X_p \cdot s = X_p \cdot (\tilde{f}\tilde{r}) = (X_p \tilde{f}) \cdot \tilde{r}(p) + \tilde{f}(p) \cdot (X_p \tilde{r}) = (X_p \tilde{f}) \cdot \tilde{r}(p)$ since $\tilde{f}(p) = 0$. This is the same as before. Hence we have the claim and the above definition (29) of ω is well defined.

Next, we show that ω defined above is G invariant. Recall that $G = \mathrm{GL}_n(\mathbb{R})$ acts on $\mathrm{Lie}(G) = \mathfrak{gl}_n(\mathbb{R}) = \mathrm{End}(\mathbb{R}^n)$ by the adjoint representation, that is, $g \cdot \varphi = \rho(g) \circ \varphi \circ \rho(g^{-1})$ for any $\varphi \in \mathfrak{gl}_n(\mathbb{R})$. Here $\rho : \mathrm{GL}_n(\mathbb{R}) \to \mathrm{GL}(\mathbb{R}^n)$ is the identity map. G invariance of $\omega \in \mathcal{A}^1(\mathcal{F}(\mathcal{E})) \otimes \mathfrak{gl}_n(\mathbb{R})$ is stated as

$$(31) \qquad \begin{aligned} R_g^* \omega &= \mathrm{Ad}_{\rho(g^{-1})}(\omega), \qquad \text{or} \\ \rho(g) \circ (R_g^* \omega)(X) \circ \rho(g^{-1}) &= \omega(X) \in \mathfrak{gl}_n(\mathbb{R}). \end{aligned}$$

for any $g \in G, X \in T\mathcal{F}(\mathcal{E})$. In the above, R_g^* acts on differential forms on $\mathcal{F}(\mathcal{E})$ and Ad_g acts on $\mathfrak{gl}_n(\mathbb{R})$. We apply the left hand side of the second identity above on the vector $s(p) \in \mathbb{R}^n$ for any G-equivariant map $s : \mathcal{F}(\mathcal{E}) \to \mathbb{R}^n$ satisfying $s(pg) = \rho(g^{-1})s(p)$. We have

$$\rho(g) \circ (R_g^* \omega_{pg})(X_p) \circ \rho(g^{-1})s(p) = \rho(g) \circ \omega_{pg}(R_{g_*}X)_{pg}s(pg)$$
$$= \rho(g) \left\{ \left(\widetilde{\nabla_{\pi_*(R_{g_*}X)}s} \right)(pg) - \left((R_{g_*}X) \cdot s \right)(pg) \right\}.$$

Here, $\pi_*(R_{g_*}X) = \pi_* X$ and $\left(\widetilde{\nabla_{(\pi_* X)}s} \right)(pg) = \rho(g^{-1}) \left(\widetilde{\nabla_{(\pi_* X)}s} \right)(p)$. For the second term in $\{\ \}$, $(R_{g_*}X) \cdot s = ds(R_{g_*}X) = R_g^*(ds)(X) = \rho(g^{-1}) \cdot ds(X)$. Thus the above is equal to

$$\rho(g) \left\{ \rho(g^{-1}) \left(\widetilde{\nabla_{(\pi_* X)}s} \right)(p) - \rho(g^{-1}) \cdot (Xs)(p) \right\} = \omega_p(X)(s(p)),$$

from the definition (29) of ω. This proves G-invariance of ω. To show that ω is a connection 1-form, we must show that $\omega(\tilde{X}) = X$ for any $X \in \mathfrak{gl}_n(\mathbb{R})$, where $\tilde{X}$ is the fundamental vector field on $\mathcal{F}(\mathcal{E})$ associated to X. Since $\pi_* \tilde{X} = 0$ because $\tilde{X}$ is vertical, from (29) we have $\omega_p(\tilde{X})(s(p)) = -(\tilde{X}s)(p) = \rho(X)(s(p))$, using the infinitesimal equivariance of s, $\tilde{X}_p s + \rho(X)(s(p)) = 0$. Since ρ is an identity map, $\rho(X) = X$ and we have $\omega(\tilde{X}) = X$ for any $X \in \mathfrak{gl}_n(\mathbb{R})$. This completes the proof that ω is a connection 1-form on the principal frame bundle $\mathcal{F}(\mathcal{E})$ of $\mathcal{E}$ with the structure group $G = \mathrm{GL}_n(\mathbb{R})$.

Correspondence of Curvatures of principal connections and of covariant derivatives. Let $(P, G, M; \omega)$ be a principal bundle over M with the group G equipped with a connection 1-form ω. For $X \in T_x M$, let $\tilde{X} \in T_p P$ be any vector at $p \in \pi^{-1}(x)$ such that $\pi_*(\tilde{X}) = X$. Let $\tilde{X}^H$ denote the horizontal part of the vector $\tilde{X}$ with respect to the connection ω. In general, we use tilde notation to denote those objects on P which correspond to objects on M.

Let $\rho : G \to \mathrm{GL}(E)$ be a representation of G and let $\underline{E} = P \times_G E$ be the associated vector bundle. Let $\nabla = \nabla^\omega$ be the covariant derivative associated to ω acting on sections of $\underline{E}$. Recall that we have the notion of curvature forms

$$(32) \qquad \Omega^\omega \in \mathcal{A}^2(P, \mathfrak{g})_{\mathrm{basic}}, \qquad \Omega^\nabla \in \mathcal{A}^2(M, \mathrm{End}(\underline{E})),$$

associated to the connection 1-form ω and the covariant derivative ∇. Note that by definition, Ω^ω lives on P and Ω^∇ lives on M. But these objects are in fact essentially the same as we now show. First we recall the definitions of these forms.

(33)
$$\Omega_p^\omega(\tilde{X}, \tilde{Y}) = (d\omega + \tfrac{1}{2}[\omega, \omega])_p(\tilde{X}, \tilde{Y}) = -\omega_p([\tilde{X}^H, \tilde{Y}^H]) \in \mathfrak{g}, \qquad \tilde{X}, \tilde{Y} \in \mathcal{X}(P)$$
$$\Omega_x^\nabla(X, Y) = [\nabla_X, \nabla_Y] - \nabla_{[X, Y]} \in \mathrm{End}(\underline{E}_x), \qquad X, Y \in \mathcal{X}(M).$$

Here, $\mathcal{X}(P)$, $\mathcal{X}(M)$ denote the Lie algebras of smooth vector fields on P, M. From our previous discussions, we know how to calculate covariant derivatives in terms of connection 1-form ω or in terms of the horizontal distribution $\mathcal{H}^\omega$ on

P. Namely, $\nabla_X s = \tilde{X}^H \tilde{s} : P_x \to E$ for any section s of $\underline{E}$ and its lift $\tilde{s}$. Using this, we have,

$$(\widetilde{\Omega_x^\nabla(X,Y)}s)_p = \tilde{X}_p^H \cdot \tilde{Y}^H\tilde{s} - \tilde{Y}_p^H \cdot \tilde{X}^H\tilde{s} - \widetilde{[X,Y]}_p^H \tilde{s}$$
$$= \left([\tilde{X}^H, \tilde{Y}^H]_p - [\tilde{X}^H, \tilde{Y}^H]_p^H\right)\tilde{s}$$

Since the inside of the parenthesis is a vertical vector $\omega_p([\tilde{X}^H, \tilde{Y}^H])$, the $\mathfrak{g}$-invariance of ω and the identity $\tilde{Z} \cdot \tilde{s} + \rho(Z)\tilde{s} = 0$, $Z \in \mathfrak{g}$, tells us that this is equal to

$$= -\rho\left(\omega_p([\tilde{X}^H, \tilde{Y}^H])\right)(\tilde{s}(p)) = \rho\left(\Omega_p^\omega(\tilde{X}, \tilde{Y})\right)(\tilde{s}(p)).$$

This proves the next lemma.

PROPOSITION 9. *Two curvatures Ω^ω and Ω^∇ in (32) are related by*

$$(34) \qquad \rho\left(\Omega_p^\omega(\tilde{X}, \tilde{Y})\right) = \widetilde{\Omega_x^\nabla(X,Y)}(p) \in \text{End}(E).$$

In other words,

$$(35) \qquad [p, \rho\left(\Omega_p^\omega(\tilde{X}, \tilde{Y})\right)e] = \Omega_x^\nabla(X,Y)v, \qquad v = [p,e] \in \underline{E}.$$

On the relation between the integrability obstruction $\mathcal{O}_\mathcal{E}$ and the curvature Ω^∇ of vector bundles $(\mathcal{E}, \nabla, \mathcal{H})$. For any fibre bundle with a horizontal distribution $\mathcal{H}$, we can associate the obstruction $\mathcal{O}^\mathcal{H}$ to the integrability of this distribution given by $\mathcal{O}^\mathcal{H} : \mathcal{H} \times \mathcal{H} \to \mathcal{V}$, where $\mathcal{V}$ is the vertical distribution. On a principal bundle with a connection ω, this obstruction is precisely the curvature Ω^ω of the connection. For a vector bundle with a covariant derivative ∇, the above obstruction $\mathcal{O}^\mathcal{H}$ is not quite the same as the curvature $\Omega^\nabla \in \mathcal{A}^2(M, \text{End}(\mathcal{E}))$ of ∇. Note that for the case of vector bundles, the vertical distribution $\mathcal{V}$ of $T\mathcal{E}$ can be canonically identified with $\mathcal{E}$, so that our obstruction can be regarded as a map $\mathcal{O}^{\mathcal{H}(\mathcal{E})} : \mathcal{H} \times \mathcal{H} \to \mathcal{E}$. At this point, one notable difference between Ω^∇ and $\mathcal{O}^{\mathcal{H}(\mathcal{E})}$ is clear, Ω^∇ is $\text{End}(\mathcal{E})$-valued, whereas $\mathcal{O}^{\mathcal{H}(\mathcal{E})}$ is $\mathcal{E}$-valued. We give a precise relationship between Ω^∇ and $\mathcal{O}^{\mathcal{H}(\mathcal{E})}$ in the next Theorem, which doesn't seem to be found in usual literatures.

THEOREM 10. *Let $\pi : \mathcal{E} \to M$ be a vector bundle equipped with a covariant derivative ∇. Let $\mathcal{H} \subset T\mathcal{E}$ be the horizontal distribution. For any $v \in \mathcal{E}$, we can identify the vertical space $\mathcal{V}_v \subset T_v\mathcal{E}$ with $\mathcal{E}_x$ where $x = \pi(v)$. With this identification, we have*

$$(36) \qquad \Omega_x^\nabla(X,Y)v = \mathcal{O}_v^\mathcal{H}(X^H, Y^H) = [X,Y]_v^H - [X^H, Y^H]_v,$$

for any vector field $X, Y \in \mathcal{X}(M)$. Here, Z^H is the horizontal vector field on $\mathcal{E}$ lifting a vector field Z on M.

We prove the above Theorem by working with principal bundles where the geometry is simpler. So, for a given vector bundle with a covariant derivative, we choose a principal bundle with a connection $(P, G, M; \omega)$ and a representation $\rho : G \to \mathrm{GL}(E)$ so that $\mathcal{E} = P \times_G E$. Let $q : P \times E \to \mathcal{E}$ be the quotienting map amd $q_* : T_{(p,e)}(P \times E) \to T_v \mathcal{E}$ be the differential of q, where $v = [p, e] = q((p, e)) \in \mathcal{E}_x$, $\pi(v) = \pi(p) = x$. Using the connections on P and $\mathcal{E}$, we have a decomposition of tangent spaces as

$$(37) \qquad T_{(p,e)}(P \times E) = \mathcal{H}_p^\omega \oplus \mathcal{V}_p \oplus T_e E, \qquad T_v \mathcal{E} = \mathcal{H}_v^\nabla \oplus \mathcal{V}_v.$$

From our previous observations in Lemma 7 and Lemma 8, q_* is an isomorphism between the horizontal vector spaces: $q_* : \mathcal{H}_p^\omega \xrightarrow{\cong} \mathcal{H}_v^\nabla$. As for the remaining part of the tangent spaces, since $\mathcal{E}_x = P_x \times_G E = q(P_x \times E)$, we have $q_*(\mathcal{V}_p \oplus T_e E) = \mathcal{V}_v = T_v \mathcal{E}_x$. This map is described as follows.

LEMMA 11. *For any $(p, e) \in P \times E$, we have the following exact sequence.*

$$(38) \qquad\qquad 0 \to \mathfrak{g} \xrightarrow{\iota_*} \mathcal{V}_p \oplus T_e E \xrightarrow{q_*} T_v \mathcal{E}_x \to 0,$$

where $\iota_*(X) = (\tilde{X}_p, -\rho_*(X)e)$ for $X \in \mathfrak{g}$, and $\tilde{X}$ is the fundamental vector field on the total space of the principal bundle P.

PROOF. The surjectivity of q_* is obvious. Now, the G orbit through $(p, e) \in P \times E$ is given by $\iota : G \to P \times E$ where $\iota(g) = (pg, \rho(g^{-1})e)$. Taking the differential, we get $\iota_*(X) = (\tilde{X}_p, -\rho_*(X)e)$. ι_* is injective because ι_* followed by the projection onto $\mathcal{V}_p$ is an isomorphism. Its inverse is given by the connection 1-form ω. Since q_* collapses the tangent spaces of G orbits, $q_* \circ \iota_* = 0$. By dimension counting of the tangent spaces, we see that $\mathrm{Ker}\, q_* = \mathrm{Im}\, \iota_*$ and the above sequence is exact. $\square$

In this exact sequence, we can naturally identify $T_v \mathcal{E}_x$ with $\mathcal{E}_x$. Also, $T_e E$ can be identified with E for any $e \in E$.

LEMMA 12. *The restriction of q_* to $\{0\}_p \times T_e E \in T_{(p,e)}(P \times E)$ is an isomorphism to $\mathcal{E}_x$:*

$$(39) \qquad\qquad q_* : \{0\}_p \times T_e E \xrightarrow{\cong} \mathcal{E}_x,$$

which is given by $q_*(w) = [p, w]$ for any $w \in E \cong T_e E$.

PROOF. Let $\gamma : [0, \ell] \to P \times E$ be a curve given by $\gamma(t) = (p, e + wt)$ for $w \in E$. Then, $\gamma(0) = (p, e)$ and $\dot\gamma(0) = (0, w)$, identifying $T_e E$ with E. Its image curve $q(\gamma(t)) = [p, e + wt]$ is contained in the fibre $\mathcal{E}_x$ and $q_*(w) = q_*(\dot\gamma(0)) = [p, w] \in \mathcal{E}_x$. $\square$

We can now describe the map q_* in Lemma 11 which is the key lemma for our purpose.

PROPOSITION 13. *The map* $q_* : \mathcal{V}_p \oplus T_e E \to \mathcal{E}_x$ *is given by*

(40) $$q_*(\tilde{X}_p, w) = [p, \rho(X)e + w] \in \mathcal{E}_x$$

for any $X \in \mathfrak{g}$ *and* $w \in E \cong T_e E$.

PROOF. Since $q_* \circ \iota_* = 0$ and $\iota_*(X) = (\tilde{X}_p, -\rho_*(X)e)$, we have $q_*(\tilde{X}_p) = q_*(\rho(X)e)$. Thus, $q_*(\tilde{X}_p + w) = q_*(\rho(X)e + w) = [p, \rho(X)e + w]$ by Lemma 12. □

Next, recall that the horizontal distributions $\mathcal{H}^\omega \subset TP$ and $\mathcal{H}^\nabla \subset T\mathcal{E}$ are related by the map q_*, $q_*(\mathcal{H}^\omega) = \mathcal{H}^\nabla$. We compare the obstructions to the integrability of these distributions on P and on $\mathcal{E}$. These are

$$\mathcal{O}_P(X, Y) = [X, Y]^H - [X, Y] \in \mathcal{V}(P), \qquad X, Y \in \Gamma(\mathcal{H}^\omega)$$

(41)

$$\mathcal{O}_\mathcal{E}(X', Y') = [X', Y']^H - [X', Y'] \in \mathcal{V}(\mathcal{E}), \qquad X', Y' \in \Gamma(\mathcal{H}^\nabla).$$

It is easy to check that $\mathcal{O}_P$ is $C^\infty(P)$ linear and $\mathcal{O}_\mathcal{E}$ is $C^\infty(\mathcal{E})$ linear. So, $\mathcal{O}_P(X, Y)_p$, $\mathcal{O}_\mathcal{E}(X'Y')_v$ actually depend only on the values of vector fields at $p \in P$ or at $v \in \mathcal{E}$. Note that $\mathcal{O}_P$ has values in $\mathcal{V}(P)$ and $\mathcal{O}_\mathcal{E}$ has values in $\mathcal{V}(\mathcal{E}) \cong \mathcal{E}$ and $q_*(\mathcal{V}(P)) \subset \mathcal{V}(\mathcal{E})$. The above two obstructions are related in the following way.

PROPOSITION 14. *For any* $X, Y \in \mathcal{H}_p^\omega$, $p \in P$, *regarding* $\mathcal{O}_P(X, Y)_p \in \mathcal{V}_p \subset T_{(p,e)}(P \times E)$ *for* $e \in E$, *we have*

(42) $$q_* \left(\mathcal{O}_P(X, Y)_{(p,e)} \right) = \mathcal{O}_\mathcal{E} \left(q_*(X), q_*(Y) \right)_v,$$

where $v = [p, e] \in \mathcal{E}$.

PROOF. Since $q_*(\mathcal{H}^\omega) = \mathcal{H}^\nabla$, we note that $q_*(X), q_*(Y) \in \mathcal{H}^\nabla$ for $X, Y \in \mathcal{H}^\omega$ and the right hand side of the above identity makes sense. Now, let $\pi_*(X)^\sim$, $\pi_*(Y)^\sim$ be vector fields on M extending $\pi_*(X)$ and $\pi_*(Y)$ in an arbitrary way. Let $\tilde{X}, \tilde{Y}$ be vector fields on P which are horizontal lifts of $\pi_*(X)^\sim$, $\pi_*(Y)^\sim$. Similarly, we let $\tilde{X}'$ and $\tilde{Y}'$ be horizontal lifts to $\mathcal{E}$ of $\pi_*(X)^\sim$, $\pi_*(Y)^\sim$. We then have $q_*(\tilde{X}) = \tilde{X}'$ and $q_*(\tilde{Y}) = \tilde{Y}'$. Since q_* commutes with bracketing, we have $q_*([\tilde{X}, \tilde{Y}]) = [q_*(\tilde{X}), q_*(\tilde{Y})] = [\tilde{X}', \tilde{Y}']$. Also, due to Lemma 8, we have $q_*([\tilde{X}, \tilde{Y}]^H) = [\tilde{X}', \tilde{Y}']^H$ since these vectors are horizontal with the same projections $[\pi_*(X), \pi_*(Y)]$ on M. Thus,

$$q_* \left(\mathcal{O}_P(\tilde{X}, \tilde{Y}) \right) = q_* \left([\tilde{X}, \tilde{Y}]^H - [\tilde{X}, \tilde{Y}] \right) = [\tilde{X}', \tilde{Y}']^H - [\tilde{X}', \tilde{Y}']$$

$$= \mathcal{O}_\mathcal{E}(\tilde{X}', \tilde{Y}') = \mathcal{O}_\mathcal{E}(q_*(\tilde{X}), q_*(\tilde{Y})).$$

Evaluating at $(p, e) \in P \times E$ and at $v = [p, e] \in \mathcal{E}$, we get our result. □

(PROOF OF THEOREM 10). For $X, Y \in T_x M$, let $X^H, Y^H \in \mathcal{H}^\omega \subset T_p P$, $X^{H'}, Y^{H'} \in \mathcal{H}^\nabla \subset T_v \mathcal{E}$ be horizontal lifts of X, Y, where $\pi(p) = \pi(v) = x$.

On P, from (6) the obstruction $\mathcal{O}_P^\omega(X^H, Y^H) \in \mathcal{V}_p$ to the integrability of $\mathcal{H}^\omega$ is related to the curvature by

$$\Omega_p^\omega(X^H, Y^H) = \omega\left(\mathcal{O}_P(X^H, Y^H)_p\right) \in \mathfrak{g}.$$

In other words, $\mathcal{O}_P(X^H, Y^H)_p$ is the fundamental vector corresponding to a Lie algebra element $\Omega_p^\omega(X^H, Y^H)$. We calculate $q_*\left(\mathcal{O}_P(X^H, Y^H)_p\right)$ in two different ways. First from Lemmas 12 and Proposition 13,

$$q_*\left(\mathcal{O}_P(X^H, Y^H)_p\right) = q_*\left(\rho(\Omega_p^\omega(X^H, Y^H))e\right) = [p, \rho\left(\Omega_p^\omega(X^H, Y^H)\right)e]$$
$$= \Omega_x^\nabla(X, Y)v.$$

Here we used our our previous result on the relationship of two curvatures in (P, ω) and $(\mathcal{E}, \nabla)$ given in (35) of Proposition 9. On the other hand, Proposition 14 gives that $q_*\left(\mathcal{O}_P(X^H, Y^H)_{(p,e)}\right) = \mathcal{O}_{\mathcal{E}}(X^{H'}, Y^{H'})_v$. Thus, we have $\Omega_x^\nabla(X, Y)v = \mathcal{O}_{\mathcal{E}}(X^{H'}, Y^{H'})_v$. This completes the proof of Theorem 10. $\square$

§4.5   The reduction of structure groups on manifolds and parallel sections

Locally, manifolds are all alike, they are all locally Euclidean. But globally, manifolds can be very different. Some of the global topological or differential geometric properties are reflected in the structure groups of these manifolds. We discuss various ways the reductions of structure groups occur for manifolds.

Given a smooth manifold M^n, the most natural principal bundle over M is the frame bundle $\mathcal{F}(M)$ with the structure group $\mathrm{GL}_n(\mathbb{R})$ consisting of all the bases in the tangent bundle. Let $G \subset \mathrm{GL}_n(\mathbb{R})$ be a subgroup. If there exists a principal subbundle $P \subset \mathcal{F}(M)$ over M with the structure group G, then we say that the structure group of M reduces to G. The structure group of M reduces to G if there exists an open covering of M by coordinate neighborhoods such that the induced transition functions of the frame bundle $\mathcal{F}(M)$ have values in G. The existence of a reduction of the structure group of M to G means that globally on M, we can systematically choose a certain "equivalence class" of frames on which G acts transitively and effectively in each tangent space. Here, roughly, two frames are equivalent if they define the same geometric structures associated to G in the tangent spaces. The geometric structure we are referring to is the one whose moduli space is given by $\mathrm{GL}_n(\mathbb{R})/G$ and whose automorphism group is given by G. For example, if $G = \mathrm{O}(n)$, the relevant geometric structure is a Euclidean scalar product and the homogeneous space $\mathrm{GL}_n(\mathbb{R})/\mathrm{O}(n)$ is the totality of Euclidean structures in $\mathbb{R}^n$. Thus, the existence of a reduction of the structure group means the global existence of a certain interesting geometric structure on the manifold.

In general, there are topological obstructions to the existence of reductions of the structure group of M. The more the manifold is structured and special, the more the structure group reduces. Once one knows that a reduction to a group $G \subset GL_n(\mathbb{R})$ exists, then in general there are many ways the structure group reduces, i.e., there are many choices of the principal G-ubbundle of $\mathcal{F}(M)$. For example, when the structure group of M reduces to a trivial group consisting only of a single element $\{e\}$, then the tangent bundle must be trivial and the $\{e\}$-structure is a choice of a trivialization of the tangent bundle. Obviously this doesn't happen unless obstructions like characteristic classes of the tangent bundle vanish. But when the tangent bundle is trivial, there are many trivializations and the totality of the trivializations is parametrized by $\mathrm{Map}(M, GL_N(\mathbb{R}))$ after choosing a single trivialization.

The next well-known proposition is basic concerning the reduction of the structure groups.

PROPOSITION 1 (REDUCTION OF STRUCTURE GROUPS). *Let $\pi : P = P_G \to M$ be a principal fibre bundle with the structure group G. The structure group reduces to a subgroup $H \subset G$ if and only if the associated fibre bundle $P \times_G (G/H) \to M$ has a global section. There is the following $1 : 1$ correspondence between H-subbundles and sections of $P/H = P \times_G (G/H)$:*

$$\{ H\text{--subbundles of } P \} \longleftrightarrow \Gamma\left(P \times_G (G/H)\right) = \Gamma(P/H) = \mathrm{Hom}_G(P, G/H).$$

Given a G-equivariant map $s : P_G \to G/H$, the associated principal H-subbundle is given by $s^{-1}([H]) \subset P_G$, where $[H] \in G/H$ is the identity left coset in G/H.

To see why this is true, suppose that we are given a principal H-subbundle P_H of P_G. At each point $x \in M$, the fibre $(P_H)_x$ over x is a left H coset of $(P_G)_x$. So, it determines a point in $(P_G)_x/H$. Letting this point be $s(x)$, we have a section $s : M \to P_G/H$.

Conversely, given a section $s : M \to P_G/H$, $s(x)$ is a left H coset in $(P_G)_x$ for each $x \in M$. Collecting these H cosets in P_G, we get a principal H-bundle P_H. Equivalently, the G-equivariant map $\hat{s} : P_G \to G/H$ corresponding to a section $s : M \to P_G/H$ is given by $\hat{s}(ug) = g^{-1}[H] \in G/H$, $g \in G$, where $s(x) = [uH] \in P_G/H$, $x = \pi(u) \in M$. This definition is well defined, since if $s(x) = [u'H]$, $u' = uh$, $h \in H$, then, $\hat{s}(ug) = \hat{s}(u'h^{-1}g) = g^{-1}h[H] = g^{-1}[H]$. So the value of $\hat{s}$ at ug is independent of the representatives of $s(x)$. Note that by letting $g \in H$, we see that $\hat{s}(uH) = [H] \in G/H$. So, $\hat{s}^{-1}([H]) \subset P_G$ is a H principal subbundle and coincides with P_H constructed from the section s, since $\hat{s}^{-1}([H]) \cap (P_G)_x = uH$ and $s(x) = [uH]$.

We remark that in the above statement, the homogeneous space G/H can be thought of as the moduli space of H-structures compatible with the given G-structure.

There are various ways the reductions of structure groups occur. We discuss the following three ways. We are especially interested in (III) for our later purpose.

(I) Smooth assignments of geometric structures in tangent spaces of manifolds.
(II) Holonomy subbundles of principal bundles with connections.

(III) Parallel assignments of geometric structures in tangent spaces with respect to connections on manifolds.

Note that (III) is the combination of (I) and (II). We discuss each cases separately.

Case (I): Smooth assignments of Geometric Structures in Tangent Bundle. On a manifold, a smooth assignment of geometric structures of a given geometric type in the tangent spaces corresponds to a section of (sums of) the associated tensor bundle or an associated fibre bundle. For example, any Riemannian structure on a manifold is a section of the vector bundle $T^*M \otimes T^*M$. However, it is obviously not the case that any section of this bundle is a Riemannian structure, we merely get a pairing on the tangent bundle without positive definiteness nor symmetry of the pairing. Similar remarks apply to other types of geometric structures on manifolds. We must have a way to control the values of these sections in tensor bundles or in related fibre bundles. To clarify the situation, we define the notion "geometric structures" on manifolds in a more general, yet in a sense restricted way.

To explain our notion of geometric structures on manifolds, let M^n be a manifold with a G-structure given by a principal G-bundle $\pi = \pi_G : P \to M$. To any G-space F, we can associate a fibre bundle $\pi_F : P \times_G F \to M$. The fibre F may be a vector space, or more generally may be a G-manifold. For example, F can be a tensor algebra of $\mathbb{R}^n$ or one of the homogeneous spaces encountered as moduli spaces of various geometric structures in previous sections. Any section T of the fibre bundle π_F can be thought of as a G-equivariant map $T : P \to F$.

DEFINITION 2 (GENERALIZED GEOMETRIC STRUCTURE). A geometric structure on M associated to a given G-structure $\pi_G : P \to M$ is a section $T \in \Gamma(P \times_G F)$ such that, when regarded as a G-equivariant map $T : P \to F$, the image $(\operatorname{Im} T) \subset F$ is contained in a *single* G-orbit $\mathcal{O}$ in F. Here, F is a G-space.

If the above G-space F has a non-empty fixed point set F^G, then for any G-bundle P, we have a geometric structure T by mapping the whole space P to any of the fixed point of F. Such geometric structure doesn't say anything about P. The bigger the G-orbit $\mathcal{O}$ in F is, the more interesting the geometric structure T becomes. The next lemma is straightforward from this definition.

LEMMA 3. *Suppose that a G-equivariant map $T : P_G \to F$ is a geometric structure and suppose $(\operatorname{Im} T) \subset \mathcal{O}$ for some G-orbit $\mathcal{O}$ in F. Then, $(\operatorname{Im} T)$ is in fact the entire orbit, that is, $\operatorname{Im} T = \mathcal{O}$.*

PROOF. This is clear since the image set $(\operatorname{Im} T)$ must be G invariant. $\square$

Let $T : P_G \twoheadrightarrow \mathcal{O} \subset F$ be a geometric structure in the above sense. For any point $t \in \mathcal{O}$, let H_t be the isotropy subgroup of G at t, so that the orbit can now be identified with G/H_t. Thus, the above geometric structure T is now a G-equivariant map $T : P_G \to G/H_t$. From the previous Proposition 1, this map defines an H_t principal subbundle of P_G given by $T^{-1}(t)$. Note that if we use different point $g \cdot t$ in the orbit $\mathcal{O}$, then the isotropy subgroup is replaced by

$H_{g\cdot t} = g \cdot H_t \cdot g^{-1}$ and we get $H_{g\cdot t}$ principal bundle as $T^{-1}(g \cdot t)$. We record this observation for future reference.

PROPOSITION 4. *Let* $T : P_G \twoheadrightarrow \mathcal{O} \subset F$ *be a geometric structure associated to a G-structure on* M. *Then, for any* $t \in \mathcal{O}$, $T^{-1}(t)$ *is a principal* H_t-*bundle and the structure group of* M *reduces from* G *to* H_t, *where* H_t *is the isotropy subgroup of* G *at* $t \in \mathcal{O}$.

We discuss geometric structures on manifolds in detail in connection with the moduli spaces of geometric structures in vector spaces. Recall that on a manifold M^n, $\mathcal{F}(M)$ denotes the bundle of frames in tangent spaces to M. At each point $x \in M$, we can consider the group of linear isomorphisms of $T_x M$, denoted by $\mathrm{GL}(T_x M)$. Let $\mathrm{GL}(M)$ be the bundle of groups whose fibre over $x \in M$ is $\mathrm{GL}(T_x M)$.

Recall that for a vector space E, the group $\mathrm{GL}(E)$ acts on the set of all frames $\mathcal{F}(E)$ from the left, and the group $\mathrm{GL}_n(\mathbb{R})$ acts on the same space from the right. Similarly on a manifold M, the bundle of groups $\mathrm{GL}(M)$ acts on $\mathcal{F}(M)$ from the left, and the group $\mathrm{GL}_n(\mathbb{R})$ acts on $\mathcal{F}(E)$ from the right. Note that an element $g \in \mathrm{GL}_n(\mathbb{R})$ acts on the whole space $\mathcal{F}(M)$, whereas an element $\tau \in \mathrm{GL}(T_x M)$ acts on a single fibre of $\mathcal{F}(M)_x$. Also note that there is no way to identify two fibres of the bundle of groups $\mathrm{GL}(M)$. So, there is no canonical way to let $\tau \in \mathrm{GL}(M)$ act on the whole space $\mathcal{F}(M)$. Thus, it is more convenient to deal with the right action of $\mathrm{GL}_n(\mathbb{R})$ on $\mathcal{F}(M)$.

Now, a vector space E can be recovered from its set of frames $\mathcal{F}(E)$ through $E = \mathcal{F}(E) \times_{\mathrm{GL}_n(\mathbb{R})} \mathbb{R}^n = \mathrm{Hom}_{\mathrm{GL}_n(\mathbb{R})}(\mathcal{F}(E), \mathbb{R}^n)$, where $\mathbb{R}^n$ is the vector space of column vectors of n-tuple of real numbers. Given a vector $v \in E$ and a frame $\mathbf{e} = \{e_1, \ldots, e_n\} \in \mathcal{F}(E)$, we can express the vector v as $v = x_1 e_1 + \cdots + x_n e_n$ for some coefficient column vector $\vec{x} \in \mathbb{R}^n$ such that ${}^t\vec{x} = (x_1, \ldots, x_n)$. The $\mathrm{GL}_n(\mathbb{R})$-equivariant map $\bar{v} \in \mathrm{Hom}_{\mathrm{GL}_n(\mathbb{R})}(\mathcal{F}(M), \mathbb{R}^n)$ corresponding to v is defined by $\bar{v}(\mathbf{e}) = \vec{x}$. If we choose different frame $\mathbf{e}'$, then we get different coefficient vector $\vec{x}'$ with the property $v = \mathbf{e}' \cdot \vec{x}'$. If $\mathbf{e}' = \mathbf{e} \cdot g$ for some $g \in \mathrm{GL}_n(\mathbb{R})$, then, the relation $\mathbf{e} \cdot \vec{x} = \mathbf{e}' \cdot \vec{x}'$ implies that $\vec{x}' = g^{-1} \cdot \vec{x}$. Thus, $(\mathbf{e}, \vec{x})$ and $(\mathbf{e} \cdot g, g^{-1} \cdot \vec{x}) \in \mathcal{F}(M) \times \mathbb{R}^n$ represent the same vector in E. This is where the action of the group $\mathrm{GL}_n(\mathbb{R}^n)$ comes in to identify these two pairs. Similarly, on a manifold M, the tangent bundle TM can be recovered from the frame bundle $\mathcal{F}(M)$ by $TM = \mathcal{F}(M) \times_{\mathrm{GL}_n(\mathbb{R})} \mathbb{R}^n$.

Let Σ be one type of geometric structures in real n dimensional vector spaces. In each tangent space $T_x M$, $x \in M$, we can consider the moduli space of geometric structures of type Σ, which we denote by $\mathfrak{M}^{T_x M}(\Sigma)$. The collection of these moduli spaces over $x \in M$, $\coprod_{x \in M} \mathfrak{M}^{T_x M}(\Sigma)$, forms a fibre bundle over M which we denote by $\mathcal{M}^{TM}(\Sigma)$. A section of this fibre bundle is a geometric structure on M of type Σ. The totality of geometric structures on M of type Σ is denoted by $\mathfrak{M}^{TM}(\Sigma)$. Thus, this is the totality of the smooth sections of the fibre bundle $\mathcal{M}^{TM}(\Sigma)$. We reformulate this space from the point of view of principal bundles and equivariant "classifying" spaces and show that these geometric structures indeed satisfy the requirements of the Definition 2.

"Classifying" spaces for geometric structures. In algebraic topology, there is a notion of classifying space for a group G. This is a space BG such that the set of homotopy classes of continuous maps to BG from a "good" space X is in 1:1 correspondence with the isomorphism classes of G principal bundles over X. We use the term "classifying" spaces for geometric structures in analogy to this. Since we are concerned with geometric structures in tangent spaces of manifolds, the relevant space for a manifold M is not the manifold itself, rather its frame bundle $\mathcal{F}(M)$.

DEFINITION 5 (THE CLASSIFYING SPACES FOR GEOMETRIC STRUCTURES ON MANIFOLDS). For a given type Σ of geometric structure in dimension n, our "classifying" space is a pointed homogeneous $\mathrm{GL}_n(\mathbb{R})$-space $(B\Sigma, \Sigma_0)$ such that the identity

$$\mathfrak{M}^{TM}(\Sigma) = \mathrm{Hom}_{\mathrm{GL}_n(\mathbb{R})}(\mathcal{F}(M^n), B\Sigma)$$

holds for any real n dimensional manifold M.

Note that any $\mathrm{GL}_n(\mathbb{R})$-equivariant map σ from $\mathcal{F}(M)$ to $B\Sigma$ is a surjective map since $B\Sigma$ is a $\mathrm{GL}_n(\mathbb{R})$-homogeneous space. Let $G(\Sigma)$ be the isotropy subgroup of $\mathrm{GL}_n(\mathbb{R})$ at the base point $\Sigma_0 \in B\Sigma$. Then, we have an isomorphism

$$\mathrm{GL}_n(\mathbb{R})/G(\Sigma) \xrightarrow[\cong]{\cdot\Sigma_0} B\Sigma.$$

Note the similarity of the above Definition 5 to the definition of geometric structures of type Σ in a vector space E. We have $\mathfrak{M}^E(\Sigma) = \mathrm{Hom}_{\mathrm{GL}_n(\mathbb{R})}(\mathcal{F}(E), B\Sigma)$. In this context, the map $\mathcal{F}(E) \to \mathfrak{M}^E(\Sigma)$ sends a frame e to a $\mathrm{GL}_n(\mathbb{R})$-equivariant map which sends the frame e to $\Sigma_0 \in B\Sigma$.

The next proposition is immediate from Proposition 4.

LEMMA 6. *Let* $\sigma : \mathcal{F}(M) \to B\Sigma$ *be an* $\mathrm{GL}_n(\mathbb{R})$-*equivariant map, that is, a type* Σ *geometric structure on* M. *Then,* $\mathcal{F}_\sigma(M) \equiv \sigma^{-1}(\Sigma_0)$ *is a principal* $G(\Sigma)$-*bundle over* M.

We call the subbundle $\mathcal{F}_\sigma(M)$ the bundle of frames *adapted* to the geometric structure $\sigma \in \mathfrak{M}^{TM}(\Sigma)$. This notion is useful when we consider compatible geometric structures later.

EXAMPLES 7. We list some classifying spaces $B\Sigma$ together with their reference points Σ_0's. We use bold faces to denote types of geometric structures and ordinary faces to denote individual geometric structures when distinction between these two needs to be made.

(i) The classifying space for the Euclidean structure:

$B(\boldsymbol{\eta}) = \{$ positive definite symmetric bilinear forms over $\mathbb{R}$ in n variables $\}$

$\quad\quad = \{$ positive definite symmetric matrices $\} \subset \mathrm{M}_n(\mathbb{R})$,

$$\eta_0(\vec{x}, \vec{y}) = \sum_{j=1}^n x_j y_j, \qquad \vec{x}, \vec{y} \in \mathbb{R}^n.$$

The first set above is contained in $\subset S^2(\mathbb{R}^n)^*$. The action of $g \in GL_n(\mathbb{R})$ on an element $\eta \in B(\eta)$ is given by $g \cdot \eta(\vec{x}, \vec{y}) = \eta(g^{-1} \cdot \vec{x}, g^{-1} \cdot \vec{y})$, where the dots on the right hand side denote the matrix multiplication.

(ii) The classifying space for the complex structures:

$$B(I) = \{ 2N \times 2N \text{ real matrices } I \text{ such that } I^2 = -1_{2N} \} \subset M_{2N}(\mathbb{R}),$$

$$I_0 = \begin{pmatrix} 0 & -1_N \\ 1_N & 0 \end{pmatrix}.$$

The action of $g \in GL_{2N}(\mathbb{R})$ on $I \in B(I)$ is given by $g \cdot I = g \cdot I \cdot g^{-1}$, where the dots in the right hand side denote the matrix multiplications.

(iii) The classifying space for the symplectic structures:

$$B(\omega) = \{ \text{ nondegenerate alternate bilinear forms on } \mathbb{R}^{2N} \} \subset \bigwedge^2(\mathbb{R}^n)^*,$$

$$\omega_0(\vec{x}, \vec{y}) = \sum_{j=1}^{N}(x_j y_{j+N} - x_{j+N} y_j), \qquad \vec{x}, \vec{y} \in \mathbb{R}^{2N}.$$

The action of $g \in GL_n(\mathbb{R})$ on $\omega \in B(\omega)$ is given by $g \cdot \omega(\vec{x}, \vec{y}) = \omega(g^{-1} \cdot \vec{x}, g^{-1} \cdot \vec{y})$.

(iv) The classifying space for quaternionic structures:

$$B((I, J)) = \left\{ \begin{array}{l} \text{pairs of matrices } I, J \in M_{4N'}(\mathbb{R}) \text{ such that} \\ I^2 = J^2 = -1_{4N'}, IJ = -JI \end{array} \right\},$$

$$I_0 = \begin{pmatrix} 0 & -1_N \\ 1_N & 0 \end{pmatrix}, \qquad J_o = \begin{pmatrix} 0 & -1_{N'} & 0 & 0 \\ 1_{N'} & 0 & 0 & 0 \\ 0 & 0 & 0 & 1_{N'} \\ 0 & 0 & -1_{N'} & 0 \end{pmatrix}.$$

The action of $g \in GL_{4N'}(\mathbb{R})$ on (I, J) is as in (ii) by conjugation.

(v) The classifying space for almost quaternionic structures:

$$B(L) = \left\{ \begin{array}{l} \text{3-dimensional subspace } L \text{ of } M_{4N'}(\mathbb{R}) \text{ spanned by matrices} \\ I, J, K \text{ such that } I^2 = J^2 = -1_{4N'}, IJ = -JI = -K \end{array} \right\},$$

$$L_0 = \mathbb{R}I_0 \oplus \mathbb{R}J_0 \oplus \mathbb{R}K_0 = \left\{ \begin{pmatrix} 0 & -b & -a & c \\ b & 0 & -c & -a \\ a & c & 0 & b \\ -c & a & -b & 0 \end{pmatrix} \middle| a, b, c \in \mathbb{R} \right\} \subset M_{4N'}(\mathbb{R}).$$

The action of $g \in GL_{4N'}(\mathbb{R})$ on $L = \mathbb{R}I \oplus \mathbb{R}J \oplus \mathbb{R}K$ is given by the action in (ii). Namely, $g \cdot L = \mathbb{R}(g \cdot I \cdot g^{-1}) \oplus \mathbb{R}(g \cdot J \cdot g^{-1}) \oplus \mathbb{R}(g \cdot K \cdot g^{-1})$.

(vi) The classifying space of complex special linear structures:

$$B(\alpha) = \{ \text{ determinants on } M_N(\mathbb{C}) \} \cong \mathbb{C}^*,$$

$$\alpha_0(1_N) = 1.$$

The action of $g \in \mathrm{GL}_N(\mathbb{C})$ on α is given by $g \cdot \alpha = \det(g)\,\alpha$.

In the above (ii) and (iii), the matrix I_0 and the alternate form ω_0 correspond to each other in a certain sense, although the actions of the group $\mathrm{GL}_n(\mathbb{R})$ are different. This relation becomes more clear and in fact the actions become identical when we consider these structures under a Euclidean structure. See Lemma 11 (i), (ii) below and §4.3 Lemma 21.

For a real n dimensional vector space E, the moduli space $\mathfrak{M}^E(\varSigma)$ of geometric structure of type $\varSigma$ can be identified with $B(\varSigma)$, but not canonically. The identifications depend on the choices of the basis of E. So, $B(\varSigma)$ is given by $\mathfrak{M}^{\mathbb{R}^n}(\varSigma)$, the moduli space of geometric structures of type $\varSigma$ in the standard n-dimensional vector space $\mathbb{R}^n$ which is equipped with the standard basis.

From the definition, these classifying spaces are homogeneous spaces. The precise form of these spaces as homogeneous spaces are given as follows, whose proof is essentially given in §4.3.

LEMMA 8 (CLASSIFYING SPACES AS HOMOGENEOUS SPACES). *The action of the group* $\mathrm{GL}_n(\mathbb{R})$ *on the classifying spaces* $B\varSigma$ *in Example 7, (i), ...,(v) is transitive. With respect to the reference point* $\varSigma_0$, *we have the following isomorphisms as homogeneous spaces*:

(i) $B(\boldsymbol{\eta}) \cong \mathrm{GL}_n(\mathbb{R})/O(n)$, (ii) $B(\boldsymbol{I}) \cong \mathrm{GL}_{2N}(\mathbb{R})/\mathrm{GL}_N(\mathbb{C})$,

(iii) $B(\boldsymbol{\omega}) \cong \mathrm{GL}_{2N}(\mathbb{R})/\mathrm{Sp}(N,\mathbb{R})$, (iv) $B((\boldsymbol{I},\boldsymbol{J})) \cong \mathrm{GL}_{4N'}(\mathbb{R})/\mathrm{GL}_{N'}(\mathbb{H})$,

(v) $B(\boldsymbol{L}) \cong \mathrm{GL}_{4N'}(\mathbb{R})/\mathrm{GL}_{N'}(\mathbb{H}) \cdot \mathrm{Sp}(1)$.

There is a natural map from $B((\boldsymbol{I},\boldsymbol{J}))$ to $B(\boldsymbol{L})$ which sends (I, J) to the vector space $\mathbb{R}I \oplus \mathbb{R}J \oplus \mathbb{R}K$, where $K = -IJ$. This map has a structure of a principal bundle with the structure group $\mathrm{SO}(3)$, that is $B(\boldsymbol{L}) = B((\boldsymbol{I},\boldsymbol{J}))/\mathrm{SO}(3)$. This is essentially because the group $\mathrm{Sp}(1)$ acts on the real 3-dimensional vector space of endomorphisms $\mathbb{R}I \oplus \mathbb{R}J \oplus \mathbb{R}K$ by conjugation and its effect is the same as the standard action of $\mathrm{SO}(3) = \mathrm{Sp}(1)/ \pm 1$. This was proved in §4.2 Lemma 17 and in §4.3 Proposition 9.

The spaces introcuced above in Example 7 in terms of matrices and bilinear forms on $\mathbb{R}^n$ are indeed classifying spaces in the sense of Definition 5.

LEMMA 9. *The classifying spaces in Example 7 give rise to the following isomorphisms.*
(i) *Riemannian structures on* M^n *are classified by* $B(\boldsymbol{\eta})$:

$$\mathfrak{M}^{TM}(\boldsymbol{\eta}) = \mathrm{Hom}_{\mathrm{GL}_n(\mathbb{R})}(\mathcal{F}(M), B(\boldsymbol{\eta})).$$

(ii) *Almost complex structures on* M^{2N} *are classified by* $B(\boldsymbol{I})$:

$$\mathfrak{M}^{TM}(\boldsymbol{I}) = \mathrm{Hom}_{\mathrm{GL}_{2N}(\mathbb{R})}(\mathcal{F}(M), B(\boldsymbol{I})).$$

(iii) *Symplectic structures on* M^{2N} *are classified by* $B(\boldsymbol{\omega})$:

$$\mathfrak{M}^{TM}(\boldsymbol{\omega}) = \mathrm{Hom}_{\mathrm{GL}_{2N}(\mathbb{R})}(\mathcal{F}(M), B(\boldsymbol{\omega})).$$

(iv) $GL_{N'}(\mathbb{H})$-*structures on* $M^{4N'}$ *are classified by* $B((I,J))$:

$$\mathfrak{M}^{TM}((I,J)) = \mathrm{Hom}_{GL_{4N'}(\mathbb{R})}(\mathcal{F}(M), B((I,J))).$$

(v) *Almost quaternionic structures on* $M^{4N'}$ *are classified by* $B(L)$:

$$\mathfrak{M}^{TM}(L) = \mathrm{Hom}_{GL_{4N'}(\mathbb{R})}(\mathcal{F}(M), B(L)).$$

Here, $GL_n(\mathbb{R})$ *actions on classifying spaces* $B\Sigma$ *are the ones in Example* 7.

PROOF. This is immediate, since for each $x \in M$, $\mathrm{Hom}_{GL_n(\mathbb{R})}(\mathcal{F}(T_xM), B\Sigma)$ is the set of all geometric structures in T_xM of type Σ, that is, $\mathfrak{M}^{T_xM}(\Sigma)$. So, any smooth $GL_n(\mathbb{R})$-equivariant map $\sigma : \mathcal{F}(M) \to B(\Sigma)$ selects a geometric structure of type Σ at each $x \in M$ in a smooth way giving rise to a globally defined geometric structure of type Σ on a manifold M^n, $n = 2N$ or $n = 4N'$. □

NOTATION 10. We use the following notations for various principal bundles associated to various geometric structures on the manifold M.

$\mathcal{F}(M) =$ The principal $GL_n(\mathbb{R})$-bundle of linear frames in the tangent bundle TM^n.

$\mathcal{Q}(M) =$ The principal $O(n)$-bundle of orthonormal frames on a Riemannian manifold (M^n, η).

$\mathcal{W}(M) =$ The principal $GL_N(\mathbb{C})$-bundle of complex frames on an almost complex manifold (M^{2N}, I).

$\mathcal{S}(M) =$ The principal $Sp(N, \mathbb{R})$-bundle of symplectic frames on a symplectic manifold (M^{2N}, ω).

$\mathcal{U}(M) =$ The principal $U(N)$-bundle of Hermitian orthonormal frames for a manifold M^{2N} with a Hermitian structure defined by compatible geometric structures (η, I).

$\mathcal{H}(M) =$ The principal $GL_{N'}(\mathbb{H})$-bundle of quaternionic frames for a manifold $M^{4N'}$ with right $\mathbb{H}$ module structure on its tangent bundle induced from (I, J).

$\mathcal{L}(M) =$ The principal $GL_{N'}(\mathbb{H}) \cdot Sp(1)$-bundle of an almost quaternionic manifold $M^{4N'}$ defined by L.

The above principal bundles are all subbundles of the frame bundle $\mathcal{F}(E)$. Using our previous notation $\mathcal{F}_\sigma(M)$ in Lemma 6 for the bundle of frames adapted to the geometric structure σ, we see that $\mathcal{Q}(M) = \mathcal{F}_\eta(M)$, $\mathcal{W}(M) = \mathcal{F}_I(M)$, $\mathcal{S}(M) = \mathcal{F}_\omega(M)$, $\mathcal{U}(M) = \mathcal{F}_{(\eta,I)}(M)$, $\mathcal{H}(M) = \mathcal{F}_{(I,J)}(M)$, $\mathcal{L}(M) = \mathcal{F}_L(M)$. Equivariant maps from these principal bundles to suitable spaces classify geometric structures compatible with the given geometric structure. Here, two geometric structures σ_1 and σ_2 are compatible if they are compatible in each tangent space in the sense of Definition 14 in §4.3.

Let $\sigma \in \mathfrak{M}^{TM}(\Sigma)$ be a geometric structure on M of type Σ. Let Σ' be another geometric type. We let

$$\mathfrak{M}_\sigma^{TM}(\Sigma') = \{ \text{ geometric structure of type } \Sigma' \text{ on } M \text{ compatible with } \sigma \}.$$

We discuss the classifying spaces of geometric structures of a given type compatible (a) with a given Riemannian structure, (b) with a given almost complex structure, (c) with other structures.

(a) *Geometric structures compatible with a given Riemannian structure.* Let $\eta \in \mathfrak{M}^{TM}(\eta)$ be a Riemannian structure on M. From η, we obtain a principal $O(n)$-bundle $\mathcal{Q}(M)$ of orthonormal frames. We can consider $O(n)$-equivariant maps to various classifying spaces $B(I)$, $B(\omega)$, etc. However, the set of $O(n)$-equivariant maps to these spaces do not actually give rise to the set of compatible geometric structures. These spaces are too large. The correct classifying spaces for geometric structures compatible with the given Riemannian structure is the $O(n)$-orbits of the reference points of these classifying spaces.

LEMMA 11 (CLASSIFYING SPACES COMPATIBLE WITH RIEMANNIAN STRUCTURES).

(i) *Let $B_\eta(I)$ be the $O(2N)$-orbit of $I_0 \in B(I)$. Then,*

$$O(2N)/U(N) \xrightarrow{\cdot I_0}_{\cong} B_\eta(I) = \left\{ \begin{array}{l} \text{orthogonal matrices } I \in O(2N) \\ \text{such that } I^2 = -1_{2N} \end{array} \right\}.$$

(ii) *Let $B_\eta(\omega)$ be the $O(2N)$-orbit of $\omega_0 \in B(\omega)$. Then,*

$$O(2N)/U(N) \xrightarrow{\cdot \omega_0}_{\cong} B_\eta(\omega) = \left\{ \begin{array}{l} \text{nondegenerate alternate forms } \omega \text{ on } \mathbb{R}^n \\ \text{of the form } \omega(\vec{x}, \vec{y}) = {}^t\vec{x} \cdot I \cdot \vec{y}, \\ \text{for some } I \in O(2N) \text{ with } I^2 = -1_{2N} \end{array} \right\}.$$

(iii) *Let $B_\eta((I, J))$ be the $O(4N')$-orbit of $(I_0, J_0) \in B((I, J))$. Then,*

$$O(4N')/Sp(N') \xrightarrow{\cdot (I_0, J_0)}_{\cong} B_\eta((I, J)) = \left\{ \begin{array}{l} \text{pairs of anticommuting orthogonal} \\ \text{matrices } I, J \text{ such that } I^2 = J^2 = -1 \end{array} \right\}.$$

(iv) *Let $B_\eta(L)$ be the $O(4N')$-orbit of $L_0 \in B(L)$. Then,*

$$O(4N')/(Sp(N') \cdot Sp(1)) \xrightarrow{\cdot L_0}_{\cong} B_\eta(L) = \left\{ \begin{array}{l} \text{3-dimensional subspace} \\ \mathbb{R}I \oplus \mathbb{R}J \oplus \mathbb{R}IJ \text{ in } M_{4N'}(\mathbb{R}), \\ \text{where } (I, J) \in B_\eta((I, J)) \end{array} \right\}.$$

In the above, note that $B_\eta(I) \cong B_\eta(\omega)$. The set of geometric structures on M compatible with a given Riemannian structure are given by equivariant maps to these spaces as stated in Proposition 12 below. Thus, these spaces are indeed classifying spaces for the compatible geometric structures.

PROPOSITION 12 (COMPATIBLE GEOMETRIC STRUCTURES). *Let η be a Riemannian structure on M.*

(i) *The moduli space of Hermitian structures on M^{2N} compatible with the Riemannian structure η is given by*

$$\mathfrak{M}_\eta^{TM}(\boldsymbol{I}) = \operatorname{Hom}_{O(2N)}\left(\mathcal{Q}(M), B_\eta(\boldsymbol{I})\right).$$

For any isometric almost complex structure $I : Q(M) \to B_\eta(\boldsymbol{I}) \ni I_0$, $I^{-1}(I_0) = \mathcal{Q}_I(M) = \mathcal{U}(M)$ is a principal $U(N)$-bundle over M^{2N}.

(ii) *The moduli space of symplectic structures on M^{2N} compatible with the Riemannian structure η is given by*

$$\mathfrak{M}_\eta^{TM}(\boldsymbol{\omega}) = \operatorname{Hom}_{O(2N)}\left(\mathcal{Q}(M), B_\eta(\boldsymbol{\omega})\right).$$

For any compatible symplectic structure $\omega : Q(M) \to B_\eta(\boldsymbol{\omega}) \ni \omega_0$, $\omega^{-1}(\omega_0) = \mathcal{Q}_\omega(M) = \mathcal{U}(M)$ is a principal $U(N)$-bundle over M.

(iii) *The moduli space of symplectic scalar products on $M^{4N'}$ compatible with the Riemannian structure η is given by*

$$\mathfrak{M}_\eta^{TM}\left((\boldsymbol{I}, \boldsymbol{J})\right) = \operatorname{Hom}_{O(4N')}\left(\mathcal{Q}(M), B_\eta\left((\boldsymbol{I}, \boldsymbol{J})\right)\right).$$

For any symplectic scalar product structure $(I, J) : Q(M) \to B_\eta\left((\boldsymbol{I}, \boldsymbol{J})\right)$, the inverse image of the base point $(I, J)^{-1}(I_0, J_0) = \mathcal{Q}_{(I,J)}(M)$ is a principal $Sp(N')$-bundle over $M^{4N'}$.

(iv) *The moduli space of almost quaternionic structures on $M^{4N'}$ compatible with the Riemannian structure η is given by*

$$\mathfrak{M}_\eta^{TM}(\boldsymbol{L}) = \operatorname{Hom}_{O(4N')}\left(\mathcal{Q}(M), B_\eta(\boldsymbol{L})\right).$$

For any compatible almost quaternionic structure $L : Q(M) \to B_\eta(\boldsymbol{L}) \ni L_0$, $L^{-1}(L_0) = \mathcal{Q}_L(M)$ is a principal $Sp(N') \cdot Sp(1)$-bundle over $M^{4N'}$.

Here, $O(n)$ actions on the classifying spaces are the restrictions of the $GL_n(\mathbb{R})$ action given in Example 7.

(b) *Geometric structures compatible with a given almost complex structure.* Let $I \in \mathfrak{M}^{TM}(\boldsymbol{I})$ be an almost complex structure on M^{2N}. As in Notation 10, we let $\mathcal{W}(M) = \mathcal{F}_I(M)$ be the principal $GL_N(\mathbb{C})$-bundle of complex frames on M with respect to the almost complex structure I. We consider two compatible geometric structures, special linear structures and hermitian structures.

LEMMA 13.

(i) *Special linear structures on an almost complex manifold M^{2N} are classified by $GL_N(\mathbb{C})$-equivariant maps to the space of determinants $B(\alpha) \cong \mathbb{C}^*$. That is,*

$$\mathfrak{M}_I^{TM}(\alpha) = \operatorname{Hom}_{GL_N(\mathbb{C})}\left(\mathcal{W}(M), B(\alpha)\right).$$

For any compatible special linear structure $\alpha : \mathcal{W}(M) \to B(\alpha)$, $\alpha^{-1}(\alpha_0) = \mathcal{W}_\alpha(M)$ is a principal $SL_N(\mathbb{C})$-bundle over M^{2N}.

(ii) *Hermitian structures on an almost complex manifold M^{2N} are classified by $\mathrm{GL}_N(\mathbb{C})$-equivariant maps to $B_I(\eta)$:*

$$\mathfrak{M}_I^{TM}(\eta) = \mathrm{Hom}_{\mathrm{GL}_N(\mathbb{C})}\left(\mathcal{W}(M), B_I(\eta)\right),$$

where $B_I(\eta)$ is the $\mathrm{GL}_N(\mathbb{C})$-orbit of $\eta_0 \in B(\eta)$ described by

$$\mathrm{GL}_N(\mathbb{C})/\mathrm{U}(N) \xrightarrow[\cong]{\cdot \eta_0} B_I(\eta) = \{\text{positive definite Hermitian pairings on } \mathbb{C}^N \ \}.$$

For any compatible Riemannian structure η, $\eta^{-1}(\eta_0) = \mathcal{W}_\eta(M) = \mathcal{U}(M)$ is a principal $\mathrm{U}(N)$-bundle over M^{2N}.

(c) *Other compatible geometric structures.* We list other compatible geometric structures which are relevant to the commutative diagram of the structure groups described in the previous section. We use notations appearing in Notation 10.

(i) Given a Hermitian structure (η, I), the set of compatible special linear structures is given by

$$\mathfrak{M}_{(\eta,I)}^{TM}(\alpha) = \mathrm{Hom}_{\mathrm{U}(N)}\left(\mathcal{U}(M), B(\alpha)\right).$$

For any such compatible special linear structure σ, $\sigma^{-1}(\alpha_0) \subset \mathcal{U}(M)$ is an principal $\mathrm{SU}(N)$-bundle on M.

(ii) Given a symplectic structure ω on M^{2N}, the moduli space of compatible Riemannian structures is given by

$$\mathfrak{M}_\omega^{TM}(\eta) = \mathrm{Hom}_{\mathrm{Sp}(N,\mathbb{R})}\left(\mathcal{S}(M), B_\omega(\eta)\right).$$

Here, $B_\omega(\eta)$ is the $\mathrm{Sp}(N,\mathbb{R})$-orbit of $\eta_0 \in B(\eta)$ and is isomorphic to a homogeneous space $\mathrm{Sp}(N,\mathbb{R})/\mathrm{U}(N)$. For any such compatible Riemannian structure σ, $\sigma^{-1}(\eta_0) \subset \mathcal{S}(M)$ is a principal $\mathrm{U}(N)$-bundle over M.

(iii) Given a complex special linear structure (I,α) on M^{2N}, the moduli space of compatible Riemannian structures on M is given by

$$\mathfrak{M}_{(I,\alpha)}^{TM}(\eta) = \mathrm{Hom}_{\mathrm{SL}_N(\mathbb{C})}\left(\mathcal{W}_\alpha(M), B_{(I,\alpha)}(\eta)\right).$$

Here, $B_{(I,\alpha)}(\eta)$ is the $\mathrm{SL}_N(\mathbb{C})$-orbit of $\eta_0 \in B(\eta)$ and has a structure of a homogeneous space $\mathrm{SL}_N(\mathbb{C})/\mathrm{SU}(N)$. For any such compatible Riemannian structure σ, $\sigma^{-1}(\eta_0) \subset \mathcal{W}_\alpha(M)$ is a principal $\mathrm{SU}(N)$-bundle over M.

(iv) Given a right $\mathbb{H}$-module structure (I,J,K), $K = JI$, on the tangent bundle of $M^{4N'}$, the moduli space of compatible Riemannian structures on M is given by

$$\mathfrak{M}_{(I,J)}^{TM}(\eta) = \mathrm{Hom}_{\mathrm{GL}_{N'}(\mathbb{H})}\left(\mathcal{H}(M), B_{(I,J)}(\eta)\right).$$

Here, $B_{(I,J)}(\eta)$ is the $\mathrm{GL}_{N'}(\mathbb{H})$-orbit of $\eta_0 \in B(\eta)$. This classifying space is isomorphic to $\mathrm{GL}_{N'}(\mathbb{H})/\mathrm{Sp}(N')$. If σ is any such compatible Riemannian structure on $M^{4N'}$, then $\sigma^{-1}(\eta_0) \subset \mathcal{H}(M)$ is a principal $\mathrm{Sp}(N')$-bundle over M.

(v)   Finally, given an almost quaternionic structure L on $M^{4N'}$, the moduli space of compatible Riemannian structure on M is given by

$$\mathfrak{M}_L^{TM}(\eta) = \mathrm{Hom}_{\mathrm{GL}_{N'}(\mathbb{H}) \cdot \mathrm{Sp}(1)}(\mathcal{L}(M), B_L(\eta)).$$

Here, $B_L(\eta)$ is the $\mathrm{GL}_{N'}(\mathbb{H}) \cdot \mathrm{Sp}(1)$-orbit of $\eta_0 \in B(\eta)$ and is isomorphic to a homogeneous space $(\mathrm{GL}_{N'}(\mathbb{H}) \cdot \mathrm{Sp}(1)) / (\mathrm{Sp}(N') \cdot \mathrm{Sp}(1))$. Let σ be any Riemannian structure on M compatible with the given almost quaternionic structure L, then, $\sigma^{-1}(\eta_0) \subset \mathcal{L}(M)$ is a principal $\mathrm{Sp}(N') \cdot \mathrm{Sp}(1)$-bundle over $M^{4N'}$.

Case (II): Holonomy subbundles of principal bundles with connections. Let $\pi : P \to M^n$ be a principal G-bundle with a connection. Recall that one way to describe a connection on P is in terms of a G-equivariant horizontal distribution $\mathcal{H}$ on P. For details, see §4.4. This is an n dimensional distribution on P such that for any $p \in P$, π_* maps $\mathcal{H}_p$ isomorphically onto $T_{\pi(p)}M$ and $R_{g_*}(\mathcal{H}_p) = \mathcal{H}_{pg}$ for any $g \in G$. For any path $\gamma : [0, \ell] \to M$ in M, there is a unique lift of $\gamma(t)$ to a horizontal path in P starting from any point in the fibre $P_{\gamma(0)}$. Evaluating this horizontal lift at $t = \ell$, we have an element in $P_{\gamma(\ell)}$. Thus, we have a map $\tau_\gamma : P_{\gamma(0)} \to P_{\gamma(\ell)}$ which is G-equivariant thanks to the G-invariance of $\mathcal{H}$. For $x \in M$, the collection of all maps τ_γ's for all possible piecewise smooth loops γ in M based at x forms a group of G-equivariant maps from $P_{\gamma(0)}$ to itself and is called the holonomy group at $x \in M$, denoted by $\mathrm{Hol}(x) \subset \mathrm{Aut}_G(P_x)$. Since, for $p \in P_x$, $\tau_\gamma(p)$ is again an element in P_x, so it is of the form $p \cdot g_{\gamma,p}$ for some element $g_{\gamma,p} \in G$. Collection of elements of the form $g_{\gamma,p}$ for a fixed $p \in P$ is a subgroup of G denoted by $\mathrm{Hol}(p) \subset G$ called the holonomy group at p. The relation between these two holonomy groups is given by $\mathrm{Hol}(x) \cdot p = p \cdot \mathrm{Hol}(p)$. Note that $\mathrm{Hol}(x)$ depends only on $x \in M$, where $\mathrm{Hol}(p)$ depends on elements in the fibre $p \in P_x$ over x. Different points p in the fibre P_x correspond to conjugate subgroups in G.

Suppose that the base manifold M is connected. For $u \in P$, let $P(u)$ be the set of all elements in P which can be connected with the point u by some horizontal path γ. Obviously, $P(u) \cap P_x = u \cdot H$, where $H = \mathrm{Hol}(u)$, $x = \pi(u)$, from the definition of the holonomy group. Since τ_γ is G-equivariant, we see that H acts transitively on $P(u)_y$ for any $y \in M$. Thus, we see that $P(u)$ has the structure of a principal bundle with the structure group $H = \mathrm{Hol}(u)$. Since horizontal paths in P starting from a point in $P(u)$ stays in $P(u)$, by definition, the restriction of the horizontal distribution $\mathcal{H}$ to $P(u)$ is tangent to $P(u)$ and automatically gives a horizontal H-invariant distribution on $P(u)$. Thus, the connection restricts from the principal G-bundle P to the principal $\mathrm{Hol}(u)$-bundle $P(u)$. Note that we can recover the original connection on P by extending the horizontal distribution on $P(u)$ using G-invariance of the horizontal distribution. We record this observation in the next lemma.

LEMMA 14. *Let $\pi_G : P_G \to M$ be a principal G-bundle with a connection. For any $u \in P_G$, the holonomy bundle $P(u)$ defines a principal $\mathrm{Hol}(u)(\subset G)$-bundle and the restriction of the horizontal distribution $\mathcal{H}$ from P_G to $P(u)$ defines a connection on $P(u)$.*

Case (III): Parallel geometric structures. This case combines the previous cases (I) and (II). Thus, we are given both a geometric structure associated to a G-structure on a manifold, that is, a G-equivariant map $T : P_G \to F$, and a connection on P_G in a compatible way, i.e., the given geometric structure T is *parallel* with respect to this connection. By definition, this means that T is constant along any horizontal paths in P_G. In fact, if a G-equivariant map T is parallel, then it turns out that it is a geometric structure in the sense of our Definition 2.

LEMMA 15. *Let $\pi_G : P_G \to M$ be a G-structure with a connection on a connected manifold M. If a G-equivariant map $T : P_G \to F$ is parallel for a G-manifold F, then it must be a geometric structure, i.e., its image must be a single G-orbit.*

PROOF. Let $u \in P$ and let $P(u)$ be the associated holonomy bundle. Also, let $T(u) = t \in F$ be the image of u under T. Let $u' \in P(u)$ be an arbitrary element. Then, we can connect u and u' by an horizontal path in P. Since T is parallel, T is constant along this path. So $T(u) = T(u')$ and T is constant on all of $P(u)$, i.e. $T(P(u)) = t \in F$. Now any element in P_G is of the form $p = u' \cdot g$ for $u' \in P(u)$ and $g \in G$. Then, the G-equivariance of T gives that $T(p) = g^{-1}T(u') = g^{-1} \cdot t$, which is in the G-orbit through $t \in F$. Thus, the image of the map T is in a single G-orbit and T is a geometric structure in the sense of Definition 2. $\square$

As a corollary of the above proof, we have

COROLLARY 16. *Let $\pi_G : P_G \to M$ be a principal G-bundle with a connection on a connected manifold M, and let F be a G-manifold. If $T : P_G \to F$ is a parallel G-equivariant map, then T is constant on any holonomy subbundle $P(u)$ of P_G and the connection on P_G reduces to a connection on $P(u)$.*

In Case (I), we have seen that whenever there is a geometric structure associated to a G-structure, we can reduce the structure group from G to its subgroup, namely, the automorphism group of the geometric structure compatible with the G-structure in each tangent space. When the geometric structure is compatible with the connection, that is, when it parallel, more is true.

PROPOSITION 17. *Suppose a G-equivariant map $T : P_G \to F$ is parallel with respect to a connection on P_G. Then, for any $t \in (\mathrm{Im}\,T)$, $T^{-1}(t)$ is a principal H_t-bundle and the connection reduces to this principal subbundle of P_G. Here, H_t is the isotropy subgroup of G at $t \in F$.*

PROOF. We only have to show that the connection reduces to $T^{-1}(t)$. To show this, it is enough to show that any horizontal path starting from a point in $T^{-1}(t)$ stays in $T^{-1}(t)$. Let $\gamma : [0, \ell] \to P_G$ be such a horizontal path with $\gamma(0) \in T^{-1}(t)$. Now, since T is horizontal, T is constant along any horizontal path. So, $T(\gamma) = T(\gamma(0)) = t$, which means that the entire path lies inside of $T^{-1}(t)$. This is what we wanted to show. $\square$

In the above proof, we could use Lemma 14. We note that $T^{-1}(t)$ is the union of holonomy subbundles $P(u)$ for all $u \in T^{-1}(t)$. From Lemma 14, any horizontal path starting from a point in $P(u)$ stays in $P(u)$. Thus, it is clear that any horizontal path starting from a point in $T^{-1}(t)$ stays in $T^{-1}(t)$.

NOTATION 18. In view of the usefulness of Proposition 17, we let

$$\text{P–Hom}_G(P_G, F) = \{ \text{ parallel } G\text{-equivariant maps from } P_G \text{ to } F \}.$$

We give a list of parallel geometric structures which are of interest to us.

Example 19 : *Some parallel geometric structures on Riemannian manifolds.*

(i) KÄHLERIAN MANIFOLDS. A Riemannian manifold M^{2N} is called Kählerian if it admits a parallel almost complex structure I which is an isometry in each tangent space. Thus,

$$\left\{ \begin{array}{l} \text{Kählerian structures on a} \\ \text{Riemannian manifold } (M^{2N}, \eta) \end{array} \right\} = \text{P–Hom}_{O(2N)} \left(\mathcal{Q}(M), B_\eta(\boldsymbol{I}) \right).$$

For a Kählerian manifold M^{2N}, the structure group reduces to a subgroup of $U(N)$. See Proposition 12 (i). It is well known that a parallel almost complex structure is always integrable and any Kählerian manifold is in fact a complex Kähler manifold.

(ii) SPECIAL KÄHLER MANIFOLDS. A Kählerian manifold M^{2N} is called special if it possesses a complex volume form, i.e., a nontrivial parallel form α of type $(N, 0)$. Thus,

$$\left\{ \begin{array}{l} \text{Special Kähler structure on a} \\ \text{Kählerian manifold } (M^{2N}; \eta, I) \end{array} \right\} = \text{P–Hom}_{U(N)} \left(\mathcal{U}(M), B(\alpha) \right).$$

Here, $B(\alpha)$ is regarded as the set of nonzero vectors in the top exterior power $\Lambda_{\mathbb{C}}^N(\mathbb{C}^N)$. For a special Kähler manifold, the structure group reduces to a subgroup of $SU(N)$. See (c) (i) above.

(iii) HYPERKÄHLER MANIFOLDS. A Riemannian manifold $M^{4N'}$ is called hyperkähler if there exists three parallel almost complex structures I, J, K which are isometries in the tangent spaces and which satisfy quaternionic relations $I^2 = J^2 = -1_{4N'}$, $IJ = -JI = -K$. Thus,

$$\left\{ \begin{array}{l} \text{Hyperkähler structures on a} \\ \text{Riemannian manifold } (M^{4N'}, \eta) \end{array} \right\} = \text{P–Hom}_{O(4N')} \left(\mathcal{Q}(M), B_\eta((\boldsymbol{I}, \boldsymbol{J})) \right).$$

The structure group of a hyperkähler manifold $M^{4N'}$ reduces to a subgroup of $Sp(N')$. See Proposition 12 (iii).

(iv) QUATERNION-KÄHLER MANIFOLDS. A Riemannian manifold $M^{4N'}$ is called quaternion-Kähler if there exists a parallel almost quaternionic structure L on $M^{4N'}$ which is locally spanned by three isometric almost complex structures

I, J, K such that $I^2 = J^2 = -1_{4N'}$, $IJ = -JI = -K$. Here, L being parallel means that $\nabla_X L \subset L$ for any $X \in TM$. In this case, the quaternion relations among I, J, K imposes that the covariant derivatives of I, J, K must be of the following form for some 1-forms α, β, γ.

$$
\begin{cases}
\nabla_X I = \qquad\quad \alpha(X)J - \beta(X)K \\
\nabla_X J = -\alpha(X)I \qquad\quad + \gamma(X)K \\
\nabla_X K = \beta(X)I - \gamma(X)K
\end{cases}
$$

In terms of parallel geometric structures, we have

$$
\left\{
\begin{array}{l}
\text{Quaternion-Kähler structures on} \\
\text{a Riemannian manifold } (M^{4N'}, \eta)
\end{array}
\right\} = \text{P-Hom}_{O(4N')}\left(\mathcal{Q}(M), B_\eta(L) \right).
$$

The structure group, that is, the holonomy group of a quaternion-Kähler manifold reduces to a subgroup of $\text{Sp}(N') \cdot \text{Sp}(1)$. See Proposition 12 (iv).

On the existence of parallel sections and the structure of the totality of parallel sections. So far, given a G-structure P_G on M with a connection, we assumed the existence of a parallel section $T : P_G \to F$ for some G-manifold F. Now, we ask the existence question. Given P_G, for which G-manifold F does a parallel section exist?

First we deduce a necessary condition. Let M be connected. Suppose $T : P_G \to F$ is a parallel G-equivariant map. Then, due to the Corollary 16, T is constant on any holonomy subbundle $P(u)$, $u \in P_G$, of P_G, say, $T(P(u)) = t \in F$. Let $H = \text{Hol}(u) \subset G$ be the holonomy group with respect to $u \in P$. This is the structure group of $P(u)$. By H-equivariance of T restricted to $P(u)$, we see that $t \in F$ must be fixed by the action of the holonomy group H. This means in particular that if the set of H-fixed points in F is empty, then, there are no parallel G-equivariant maps to F. When there is H-fixed points in F, we have a following correspondence for each choice of $u \in P_G$.

$$
\{\text{parallel } G\text{-equivariant maps } T : P_G \to F\} \longrightarrow F^{\text{Hol}(u)}.
$$

We show that this map is a bijection once $u \in P_g$ is chosen. To see that it is injective, let T_1, T_2 be two parallel G-equivariant maps from P_G to F such that $T_1(P(u)) = T_2(P(u)) = t \in F^{\text{Hol}(u)}$. Let $v \in P_G$ be any point in P_G. We can write $v = u' \cdot g$ for some $u' \in P(u)$ and $g \in G$. Note that $T_1(u') = T_2(u') = t$. Then, $T_1(v) = T_1(u' \cdot g) = g^{-1}T_1(u') = g^{-1} \cdot t$. Similarly, $T_2(v) = T_2(u' \cdot g) = g^{-1} \cdot t$. Thus, $T_1(v) = T_2(v)$ for any $v \in P_G$. Hence, T_1 and T_2 agrees everywhere on P_G and the above map is injective.

To see the surjectivity of the above correspondence, let $t \in F^{\text{Hol}(u)}$. Then, a constant map $\tilde{T} : P(u) \to t \in F$ is a parallel $H = \text{Hol}(u)$-equivariant map. We extend this map to a map $T : P_G \to F$ so that it is G-equivariant. There is a unique such extension. We have to show that this extended map is automatically parallel. To see this, let $\gamma : [0, \ell] \to P_G$ be a horizontal path in P_G. We choose

$g \in G$ so that $\gamma(0) \cdot g \in P(u)$. Since $P(u)$ is a holonomy bundle, it follows by definition that the entire horizontal path $\gamma \cdot g$ must lie inside of $P(u)$. Thus, $T(\gamma) = g \cdot T(\gamma \cdot g) = g \cdot t \in F$ and T is constant along any arbitrarily chosen horizontal path γ. This shows that T is in fact parallel. Thus, we have proved the next Proposition.

PROPOSITION 20. *Let* $\pi_G : P_G \to M$ *be a G-structure on M with a connection. For* $u \in P_G$, *let* $\mathrm{Hol}(u) \subset G$ *be the holonomy group with respect to u. Let F be a G-manifold. Then, a parallel section exists in the fibre bundle* $\underline{F} = P_G \times_G F$ *if and only if* $F^{\mathrm{Hol}(u)} \neq \emptyset$. *In fact, there is a bijection*

$$\{\text{parallel sections in the fibre bundle } \underline{F} \} \xrightarrow{1:1} F^{\mathrm{Hol}(u)} = \{x \in F \mid \mathrm{Hol}(u) \subset G_x\},$$

where G_x *is the isotropy subgroup of G at* $x \in F$.

REMARK. If we choose another point $v \in P_G$, then the associated holonomy group $\mathrm{Hol}(v)$ is conjugate to $\mathrm{Hol}(u)$ in G, say, $\mathrm{Hol}(v) = g \cdot \mathrm{Hol}(u) \cdot g^{-1}$ for some $g \in G$. Then, the fixed point sets of these two holonomy groups are isomorphic via the action of the element $g \in G$, $F^{\mathrm{Hol}(v)} = g \cdot F^{\mathrm{Hol}(u)}$. So, the above non-emptiness condition of fixed points is independent of the point $u \in P_G$ chosen.

Parallel sections in vector bundles and G-invariants in a fibre. In this subsection, we restrict ourselves to vector bundles, rather than general fibre bundles. As before, let $\pi_G : (P_G, \omega) \to M$ be a G-structure on M with a connection ω and let E be a representation of G. We can form a vector bundle $\mathcal{E} = P_G \times_G E$ equipped with a covariant derivative $\nabla = \nabla^\omega : \Gamma(\mathcal{E}) \to \Gamma(T^* M \otimes \mathcal{E})$. For details of the construction of the covariant derivative ∇ from the principal connection ω, see §4.4. Recall that a section s of this bundle is called parallel if it is co-variantly constant, $\nabla s = 0$. If we regard s as a G-equivariant map $s : P_G \to E$, then s is parallel if and only if it is constant along any horizontal path γ in P_G. Thus, s is a geometric structure in the sense of Definition 2 due to Lemma 15. First we establish a notation for an object of our interest.

NOTATION 21 (PARALLEL SECTIONS). Let $\mathcal{E} = P_G \times_G E$ be as above. Let $\mathcal{P}(\mathcal{E}) = \mathcal{P}(\mathcal{E}; \nabla)$ denote the set of parallel sections in the vector bundle $\mathcal{E}$ equipped with a covariant derivative ∇. That is, $\mathcal{P}(\mathcal{E}) = \{s \in \Gamma(\mathcal{E}) \mid \nabla s = 0\}$.

Obviously, this space $\mathcal{P}(\mathcal{E})$ depends on the G-representation space E chosen. Our objective is to identify the property of the representation E which is relavant to $\mathcal{P}(\mathcal{E})$. Although $\mathcal{P}(\mathcal{E})$ is a global object on the manifold M, it is very rigid in the sense that (quasi-)local data completely determine this set in the following sense.

LEMMA 22. *The evaluation (or restriction) map* $\mathrm{Res}_x : \mathcal{P}(\mathcal{E}) \to \mathcal{E}_x$ *at a point* $x \in M$ *is an injective map and its image can be characterized by the invariance under the holonomy goup, i.e.,* $\mathrm{Im}(\mathrm{Res}_x) = \mathcal{E}_x^{\mathrm{Hol}(x)}$, *where* $\mathrm{Hol}(x) \subset \mathrm{GL}(\mathcal{E}_x)$ *is the holonomy group at* $x \in M$.

PROOF. Let $s \in \mathcal{P}(\mathcal{E})$ be a parallel section. Let γ be any piecewise smooth loop in M based at $x \in M$. The parallel transport of the vector $s_x \in \mathcal{E}_x$ along γ

gives the same vector s_x since s is parallel. So, $\tau_\gamma(s_x) = s_x$ for any loop γ based at $x \in M$. This means that the holonomy group $\text{Hol}(x)$ at $x \in M$ acts trivially on the vector s_x, that is, $s_x \in \mathcal{E}^{\text{Hol}(x)}$.

Conversely, let $r_x \in \mathcal{E}_x^{\text{Hol}(x)}$ be any vector in $\mathcal{E}_x$ fixed by the holonomy group $\text{Hol}(x)$ at x. We show that there is a unique extension of r_x to a global parallel section $r \in \mathcal{P}(\mathcal{E})$. To see this, let $y \in M$ be any point of the manifold M and let γ be an arbitrary path on M from x to y. We define a vector $r_y \in \mathcal{E}_y$ by $r_y = \tau_\gamma(r_x)$. We claim that this vector is independent of the path γ chosen. To show this, let γ' be another path from x to y. Then, the composed path $\gamma'^{-1} \circ \gamma$ is a loop based at x, so the corresponding parallel transport $\tau_{\gamma'}^{-1} \circ \tau_\gamma$ fixes r_x because it is $\text{Hol}(x)$-invariant. Hence $r_y = \tau_\gamma(r_x) = \tau_{\gamma'}(r_x)$ and the vector r_y is independent of the path chosen for its definition. But then, the collection of these vectors $r = \{r_y\}_{y \in M}$ forms a smooth parallel section of $\mathcal{E}$ and this is the only parallel section whose restriction at $x \in M$ is $r_x \in \mathcal{E}_x$. This proves Lemma 22. □

Next, we characterize the vector space $\mathcal{P}(\mathcal{E})$ in terms of a property of the representation E of G.

PROPOSITION 23 (PARALLEL SECTIONS AND G-INVARIANTS). *Let $u \in P_G$ be any point and let $\pi_G(u) = x \in M$. Let $P(u)$ be the holonomy subbundle through u with the holonomy group $\text{Hol}(u) = G(u) \subset G$ as its structure group. Then,*

$$\mathcal{P}(\mathcal{E}) \xrightarrow[\cong]{\text{Res}_x} \mathcal{E}_x^{\text{Hol}(x)} = \{[u, e] \in \mathcal{E}_x \mid e \in E^{G(u)}\} \xleftarrow{\cong} E^{G(u)}.$$

PROOF. We only have to show the equality in the middle. First, note that any vector in $\mathcal{E}_x$ can be represented as $[u, e]$ for some vector $e \in E$. The action of an element τ in the holonomy group $\text{Hol}(x)$ on a vector $[u, e] \in \mathcal{E}_x$ is given by $\tau([u, e]) = [\tau(u), e]$, where τ on the L.H.S. is regarded as an element in $GL(\mathcal{E}_x)$ and τ on the R.H.S. is regarded as a G-equivariant map $\tau : P_x \to P_x$, both acting from the left. Since the two holonomy groups $\text{Hol}(x)$ and $G(u)$ are related by $\text{Hol}(x) \cdot u = u \cdot G(u)$, for a given $\tau \in \text{Hol}(x)$, there exists an element $g = g_{\tau,u} \in G(u)$ such that $\tau(u) = u \cdot g$. Then, $[\tau(u), e] = [u \cdot g, e] = [u, g \cdot e]$. Thus, $[u, e] \in \mathcal{E}_x^{\text{Hol}(x)}$ if and only if $\tau([u, e]) = [u, e]$ for any $\tau \in \text{Hol}(x)$ if and only if $[u, g \cdot e] = [u, e]$ for any $g \in G(u)$, or $e \in E^{G(u)}$. This proves Proposition 23. □

REMARK. In Proposition 23, we made a choice of a point u in the fibre P_x over $x \in M$. Choosing another point u' in the same fibre doesn't make any difference in our conclusion. But, we clarify the situation. Let $u' = u \cdot g^{-1}$ for $g \in G$. Then, the corresponding holonomy bundles and the holonomy groups are given by $P(u') = P(u) \cdot g^{-1}$, $G(u') = g \cdot G(u) \cdot g^{-1}$. So, the invariant subspaces $E^{G(u)}$ and $E^{G(u')}$ under these groups are different in general. However, they are isomorphic via the action of $g \cdot$ and we have the following commutative diagram:

$$
\begin{array}{ccc}
\mathcal{E}_x^{\mathrm{Hol}(x)} & \overset{\cong}{\underset{[u,\,]}{\longleftarrow}} & E^{G(u)} \\[2ex]
\Big\| & & \cong \Big\downarrow g\cdot \\[2ex]
\mathcal{E}_x^{\mathrm{Hol}(x)} & \overset{\cong}{\underset{[u',\,]}{\longleftarrow}} & E^{G(u')}
\end{array}
$$

Here, the top arrow maps $e \in E^{G(u)}$ to $[u, e] \in \mathcal{E}_x^{\mathrm{Hol}(x)}$ and similarly the bottom arrow maps $e' \in E^{G(u')}$ to $[u', e'] \in \mathcal{E}_x^{\mathrm{Hol}(x)}$. The commutativity can be checked by a simple calculation, $[u', g \cdot e] = [u \cdot g^{-1}, g \cdot e] = [u, e]$.

CHAPTER V

INFINITE DIMENSIONAL SYMMETRIES IN ELLIPTIC GENERA FOR KÄHLER MANIFOLDS

§5.1 Kählerian manifolds

Usually, a Kähler manifold M is defined as a complex manifold equipped with a Hermitian metric h whose associated fundamental 2-form Ω, which is the imaginary part of the Hermitian metric h up to a constant multiple, is closed, i.e., $d\Omega = 0$. In the previous section, we discussed a Kähler manifold M from Riemannian geometry point of view, that is, from the point of view of holonomy groups and parallel geometric structures on the manifold M. We discuss Kähler geometry in more detail in this section from the Riemannian geometry point of view. In particular, we discuss Kählerian manifolds, special Kähler manifolds, hyperkähler manifolds and quaternion-Kähler manifolds in much more detail in connection with their holonomy groups. The importance of these types of manifolds comes from their special form of holonomy groups.

First, we review the classification theory of holonomy groups of Riemannian manifolds. The next theorem says that if the holonomy representation is reducible, then, the manifold is locally a product.

THEOREM 1 (DE RHAM DECOMPOSITION). *Let M be a Riemannian manifold. If the holonomy representation is reducible, then locally it is a Riemannian product.*

The point of this Theorem is that if the representation reduces, the holonomy group itself becomes a product and no diagonal representations appear. Thus, to classify holonomy representations, we can restrict ourselves to irreducible holonomy representations. It is well known that for non-symmetric spaces, possible holonomy representations are very restricted. Recall that the restricted holonomy group Hol^0 is the identity component of the holonomy group Hol.

THEOREM 2 (HOLONOMY CLASSIFICATION). *Let M^n be a Riemannian manifold which is not locally symmetric and whose holonomy representation of the restricted holonomy group Hol^0 is irreducible. Then, the holonomy representation of Hol^0 is one of the followings:*

(1) $\mathrm{Hol}^0 = \mathrm{SO}(n)$ *acting on* $\mathbb{R}^n$, $n \geq 2$.
(2) $\mathrm{Hol}^0 = \mathrm{U}(N)$ *acting on* $\mathbb{R}^{2N} = \mathbb{C}^N$, $N \geq 2$.
(3) $\mathrm{Hol}^0 = \mathrm{SU}(N)$ *acting on* $\mathbb{R}^{2N} = \mathbb{C}^N$, $N \geq 2$.
(4) $\mathrm{Hol}^0 = \mathrm{Sp}(N') \cdot \mathrm{Sp}(1)$ *acting on* $\mathbb{R}^{4N'} = \mathbb{H}^{N'}$, $N' \geq 2$.

(5) $\mathrm{Hol}^0 = \mathrm{Sp}(N')$ *acting on* $\mathbb{R}^{4N'} = \mathbb{H}^{N'}$, $N' \geq 2$.
(6) $\mathrm{Hol}^0 = \mathrm{Spin}(9)$ *acting on* $\mathbb{R}^{16}$.
(7) $\mathrm{Hol}^0 = \mathrm{Spin}(7)$ *acing on* $\mathbb{R}^8$.
(8) $\mathrm{Hol}^0 = G_2$ *acting on* $\mathbb{R}^7$.

The restrictions on N, N' in the above list comes from the fact that for (5) $\mathrm{Sp}(1) = \mathrm{SU}(2)$, for (2) $\mathrm{U}(1) = \mathrm{SO}(2)$, and for (4) $\mathrm{SO}(4) = \mathrm{Sp}(1) \cdot \mathrm{Sp}(1) = \mathrm{Sp}(1) \times \mathrm{Sp}(1)/\{\pm 1\}$. To represent the holonomy group, we choose a basis of the tangent space of a point $x \in M$. Thus, the holonomy group we are dealing with is a subgroup of $\mathrm{O}(n)$ which acts on the principal bundle of orthonormal frames from the right. This holonomy group depends on the choice of the frame. Different bases give rise to conjugate holonomy groups. So, it is understood that when we say the holonomy group is G, it means that the holonomy group with respect to a given basis in a tangent space is conjugate to G in $\mathrm{GL}_n(\mathbb{R})$.

Now, let M^n be a Riemannian manifold with an almost complex structure J. Let its Riemannian metric be denoted by $g(\,,\,) = \langle\,,\,\rangle$.

DEFINITION 3. On a Riemannian manifold M^{2N} with an almost complex structure J, the metric $\langle\,,\,\rangle$ is called *Hermitian* if the almost complex structure J is an isometry, i.e.,

$$\langle JX, JY \rangle = \langle X, Y \rangle, \qquad \text{for any } X, Y \in TM.$$

In this case, we can construct a $\mathbb{C}$-valued Hermitian metric in two different ways.

$$(X, Y)_1 = \langle X, Y \rangle - i\langle J(X), Y \rangle,$$
$$(X, Y)_2 = \langle X, Y \rangle + i\langle J(X), Y \rangle.$$

Note that the first one is sesqui-linear in the second variable and the second Hermitian form is sesqui-linear in the first variable. These two Hermitian metrics are conjugate to each other, $(X, Y)_1 = \overline{(X, Y)_2}$. Since there is no canonical choice among these two Hermitian metrics, by Hermitian metric we mean a Riemannian metric g together with an isometric almost complex structure J. A manifold with a Hermitian metric is called a *Hermitian manifold*.

Kählerian manifolds. We start from a basic definition.

DEFINITION 4. Let $(M^{2N}; g, J)$ be a Hermitian manifold. Then, the metric g is called *Kähler* if the almost complex structure J is parallel with respect to the Levi-Civita connection ∇ of the metric, i.e., $\nabla J = 0$.

DEFINITION 5. A Hermitian manifold $(M^{2N}; g, J)$ is called *Kählerian* if the metric g is Kähler.

To understand the meaning of the condition $\nabla J = 0$, we consider the Kähler form $\Omega \in A^2(M)$ defined by

$$\Omega(X, Y) = \langle J(X), Y \rangle, \qquad X, Y \in TM.$$

Note that this is essentially the imaginary part of the Hermitian metric. In the usual literature, Ω is defined by the above formula up to a constant multiple. Since this constant is not important for our purpose, we use 2-form Ω defined by the above formula. This real 2-form can be defined on any Hermitian manifold. But when J is parallel, Ω is actually closed. See Proposition 7 below.

In general, on a given almost complex Riemannian manifold (M, g, I) and given two tensor fields A, B of type $(1,1)$, we can consider their *torsion* which is a tensor field of type $(1,2)$. In our context, if $A = B = J$, the corresponding torsion tensor field is given by (up to a constant multiple)

$$N(X,Y) = [J(X), J(Y)] - [X, Y] - J([X, J(Y)]) - J([J(X), Y]) \in \mathcal{X}(M),$$

where $X, Y \in \mathcal{X}(M)$. Here, $\mathcal{X}(M)$ is the Lie algebra of vector fields on M. This tensor field is important in view of the next well known theorem.

THEOREM 6 (NEWLANDER-NIRENBERG). *An almost complex structure J on a Riemannian manifold is a complex structure if and only if it has no torsion, i.e., $N \equiv 0$.*

Now, the condition $\nabla J = 0$ has the following consequence:

PROPOSITION 7. *Let $(M^{2N}; g, J)$ be a Hermitian manifold and let ∇ be the associated Levi-Civita connection. Then,*

$$\nabla J = 0 \iff \begin{cases} d\Omega = 0, \\ N = 0. \end{cases}$$

Thus, if $\nabla J = 0$, then the Hermitian manifold is in fact complex Kähler.

PROOF. See [K-N] page 148,   □

Kählerian manifolds defined above can be characterized in terms of holonomy groups.

PROPOSITION 8. *Let $(M^{2N}; g)$ be a Riemannian manifold. Then, there exists an almost complex structure J on M for which M is Kählerian if and only if the holonomy group of M^{2N} is contained in the unitary group $U(N)$ (after a conjugation in $GL_{2N}(\mathbb{R})$).*

PROOF. If M is Kählerian, then the almost complex structure J is parallel, which means that the operator J commutes with parallel transports along any curves on M. Let $x \in M$ be any point. A choice of a Hermitian orthonormal basis in $T_x M$ identifies $T_x M$ with $\mathbb{C}^N$. Under this identification, the operator J corresponds to the usual multiplication by i on $\mathbb{C}^N$. Now, to any piecewise smooth loop γ based at x, the corresponding parallel transport τ_γ is an element in $O(2N)$ since parallel transports are isometries between tangent spaces. Since this element in $O(2N)$ must commute with the operator J, or with multiplication by i, it must belong to $U(N)$. Thus, the holonomy group is a subgroup of $U(N)$. Conversely, suppose that the holonomy group at $x \in M$ is a subgroup of $U(N)$ with respect to an orthonormal basis and an isometric complex structure J_x in

$T_x M$. At this point, we have chosen an arbitrary isometric complex structure in the tangent space *only* at $x \in M$. We must show that there exists *globally* an almost complex structure J on M which is parallel, that is, $J \circ \tau_\gamma = \tau_\gamma \circ J$ for any piecewise smooth curve γ in M. Let γ_i, $i = 1, 2$, be a curve in M from x to y. Let $\tau_{\gamma_1}, \tau_{\gamma_2}$ be the parallel transports along these curves and let linear isomorphisms $J', J'' : T_y M \to T_y M$ be defined by

$$\tau_{\gamma_1}\left(J_x(v)\right) = J'\left(\tau_{\gamma_1}(v)\right), \qquad \tau_{\gamma_2}\left(J_x(v)\right) = J''\left(\tau_{\gamma_2}(v)\right),$$

for any $v \in T_x M$. The curve $\gamma_2^{-1} \circ \gamma_1$ is a loop in M based at x. Thus, $\tau_{\gamma_2^{-1} \circ \gamma_1} = \tau_{\gamma_2}^{-1} \circ \tau_{\gamma_1}$ commutes with J_x. So, for any $v \in T_x M$, we have

$$\tau_{\gamma_2}\left(J_x \circ \tau_{\gamma_2}^{-1} \circ \tau_{\gamma_1}(v)\right) = \tau_{\gamma_2}\left(\tau_{\gamma_2}^{-1} \circ \tau_{\gamma_1} \circ J_x(v)\right) = \tau_{\gamma_1}\left(J_x(v)\right).$$

On the other hand, using J'', we have that $\tau_{\gamma_2}\left(J_x \circ \tau_{\gamma_2}^{-1} \circ \tau_{\gamma_1}(v)\right)$ is equal to $J''\left(\tau_{\gamma_2}(\tau_{\gamma_2}^{-1} \circ \tau_{\gamma_1}(v))\right) = J''\left(\tau_{\gamma_1}(v)\right)$. Thus, we have $\tau_{\gamma_1}\left(J_x(v)\right) = J''\left(\tau_{\gamma_1}(v)\right)$. Comparing this formula with the formula for J' above, we see that $J' = J''$ as isomorphisms from $T_y M$ to $T_y M$, since τ_γ's are isomorphisms. This shows that we can define a complex structure $J_y = J' = J''$ in $T_y M$ independent of the choice of the path from x to y. Since $y \in M$ is arbitrary, this defines an almost complex structure on M which is parallel by construction. Thus, the manifold M has the structure of a Kählerian manifold. This proves Proposition 8.   □

REMARK. We note that the condition that an almost complex structure I is parallel on a Riemannian manifold (M, g) means that for any curve γ in M from a point x to a point y, the induced parallel transform τ_γ makes the following diagram commutative:

$$
\begin{array}{ccc}
T_x M & \xrightarrow{\ I_x\ } & T_x M \\
{\scriptstyle \tau_\gamma} \downarrow & & \downarrow {\scriptstyle \tau_\gamma} \\
T_y M & \xrightarrow{\tau_\gamma(I_x) = I_y} & T_y M
\end{array}
$$

Special Kähler manifolds. Recall that the curvature tensor R is a tensor field of type $(1,3)$ defined by

$$R(X, Y)Z = [\nabla_X, \nabla_Y]Z - \nabla_{[X,Y]}Z \in \mathcal{X}(M), \qquad X, Y, Z \in \mathcal{X}(M).$$

Although one uses vector fields to define the Riemannian curvature tensor R, the value of the resulting vector field at $x \in M$ depends only on the values of vector fields X, Y, Z at $x \in M$. Since this is a tensor field of type $(1,3)$, we can obtain new tensor fields by contractions. One such tensor field is the Ricci tensor field. The Ricci tensor r is a 2 tensor field defined by

$$r(X, Y) = \text{Tr}\left(Z \to R(X, Z)Y\right), \qquad \text{for } X, Y \in TM.$$

A Riemannian manifold is called *Ricci-flat* if $r \equiv 0$ on M.

DEFINITION 9. A Ricci-flat Kähler manifold is called a *special Kähler* manifold.

To study special Kähler manifolds, the following notion is useful.

DEFINITION 10. A *complex volume form* θ on a Kähler manifold M^{2N} is a nowhere vanishing parallel differential form of type $(N, 0)$.

The special Kähler manifolds are characterized among Kähler manifolds by vanishing of their Ricci tensors or by their holonomy groups.

PROPOSITION 11. (i) *A Kähler manifold M^{2N} has a complex volume form if and only if its entire holonomy group* Hol *is contained in* SU(N).

(ii) *A Kähler manifold M^{2N} is special if and only if its restricted holonomy group* Hol^0 *is contained in* SU(N).

PROOF. We choose a Hermitian orthonormal basis at $x \in M$. Since our manifold is Kähler, any holonomy transformation τ_γ along a loop based at x can be represented by a unitary matrix with respect to the above basis, in view of Proposition 8. This τ_γ acts on the fibre over $x \in M$ of the canonical line bundle $\bigwedge_{\mathbb{C}}^N (T^{(1,0)}M)$ by its determinant.

For (i), suppose that the manifold M^{2N} possesses a complex volume form. The existence of the complex volume form θ means that this canonical line bundle is trivialized by the parallel section θ. Thus, the action of τ_γ on θ_x is trivial, which means that τ_γ is represented by an element in U(N) with the determinant 1, i.e., by an element in SU(N). Conversely, if the holonomy group is contained in SU(N), then the action of τ_γ on the fibre at x of the canonical line bundle is trivial for any path γ in M. Thus, by parallel transporting any nonzero element $\theta_x \in \bigwedge_{\mathbb{C}}^N T_x^{(1,0)} M$ along any path from x to y we can define an element θ_y independent of the path. The collection of these vectors $\{\theta_y\}_{y \in M}$ gives a parallel complex volume form θ on M.

For (ii), we recall that the curvature of the canonical line bundle $\bigwedge_{\mathbb{C}}^N T^{(1,0)} M$ with respect to the connection induced from the Levi-Civita connection on TM is, up to a constant multiple, given by the Ricci *form* ρ defined by $\rho(X, Y) = r(J(X), Y)$ in terms of the Ricci tensor r. When Ricci tensor vanishes on a Kähler manifold M^{2N}, that is, when the manifold is special Kähler, the associated Ricci form also vanishes and the canonical line bundle is flat, which means that along arbitrary contractible loops γ, the corresponding parallel transports τ_γ act as identity operators on the complex canonical line bundle. So, the restricted holonomy group is contained in SU(N). Conversely, if the restricted holonomy group is contained in SU(N), then any parallel transport along any contractible loops leave canonical line bundle fixed, which is the same as saying that the curvature of the line bundle vanishes, which in turn is equivalent to vanishing of the Ricci tensor. So, the manifold is special Kähler. This proves Proposition 11. □

Note that the existence of the complex volume form θ is a stronger condition than the vanishing of the Ricci tensor. The former implies the later. These two conditions are equivalent for simply connected manifolds.

As we have seen in Proposition 11, it is simpler to characterize various subclasses of Riemannian manifolds in terms of their holonomy groups. We will do so for the remaining two subclasses of Riemannian manifolds.

Hyperkählerian manifolds.

DEFINITION 12. A Riemannian manifold $M^{4N'}$ is called *hyperkählerian* if its holonomy group is contained in $\mathrm{Sp}(N') \subset \mathrm{O}(4N')$.

We examine geometric consequences of this definition.

PROPOSITION 13. *A Riemannian manifold $M^{4N'}$ is hyperkählerian if and only if there exist three almost complex structures I, J, K such that*

(1) $I, J, K : TM \to TM$ *are isometries on M.*
(2) $IJ = -JI = -K$.
(3) I, J, K *are parallel with respect to the Levi-Civita connection.*

REMARK. With the existence of the above I, J, K, we note that any map of the form $aI + bJ + cK$ with $a^2 + b^2 + c^2 = 1$, $a, b, c \in \mathbb{R}$, is also a parallel isometric almost complex structure of M.

PROOF. Suppose that there exist almost complex structures I, J, K on M satisfying the above conditions (1), (2), (3). Let γ be any loop based at $x \in M$ and we choose an orthonormal basis ξ in $T_x M$ such that the canonical quaternion structure (I_0, J_0, K_0) in $\mathbb{R}^{4N'}$ corresponds to (I_x, J_x, K_x) under the isomorphism $\xi : \mathbb{R}^{4N'} \to T_x M$. Since M is Riemannian, the parallel transport τ_γ can be represented by an element in $\mathrm{O}(4N')$. Since I, J, K are parallel, this element τ_γ commutes with I_x, J_x, K_x which satisfy quaternion relations. Hence, the element τ_γ is in fact belongs to $\mathrm{GL}_{N'}(\mathbb{H})$. Since $\mathrm{GL}_{N'}(\mathbb{H}) \cap \mathrm{O}(4N') = \mathrm{Sp}(N')$, we have $\tau_\gamma \in \mathrm{Sp}(N')$. Since γ is any loop based at x, the holonomy group of $M^{4N'}$ at x with respect to the above chosen basis is a subgroup of $\mathrm{Sp}(N')$.

Conversely, suppose the holonomy group of $M^{4N'}$ at $x \in M$ is a subgroup of $\mathrm{Sp}(N')$ with respect to an orthonormal basis ξ in $T_x M$. We transport the standard quaternionic structure (I_0, J_0, K_0) in $\mathbb{R}^{4N'}$ to $T_x M$ and let it be denoted by (I_x, J_x, K_x). Note that these are isometries of $T_x M$. We show that the parallel transport of this quaternionic structure at $x \in M$ to all over M is well defined. Let γ be a loop based at $x \in M$ and let τ_γ be the induced parallel transport. The parallel transport of I_x along γ is given by $\tau_\gamma(I_x) = \tau_\gamma \circ I_x \circ \tau_\gamma^{-1}$. By the assumption $\tau_\gamma \in \mathrm{Sp}(N')$, τ_γ commutes with I_x. So, $\tau_\gamma(I_x) = I_x$. Similarly we have $\tau_\gamma(J_x) = J_x$ and $\tau_\gamma(K_x) = K_x$ for any loop γ based at x. This means that the parallel transport of I_x, J_x, K_x define well defined tensor fields I, J, K on M. Namely, the values of these tensor fields at $y \in M$ are given by $I_y = \tau_\gamma \circ I_x \circ \tau_\gamma^{-1}$, $J_y = \tau_\gamma \circ J_x \circ \tau_\gamma^{-1}$, $K_y = \tau_\gamma \circ K_x \circ \tau_\gamma^{-1}$ where γ is any path from x to y. Since Riemannian connection preserves Riemannian metric, τ_γ is an isometry. Since I_x, J_x, K_x are isometries of $T_x M$ by their construction, I_y, J_y, K_y are isometries of $T_y M$. Also, since I_x, J_x, K_x satisfy quaternion relations, it follows that I_y, J_y, K_y also satisfy quaternion relations for any $y \in M$. Thus, tensor fields I, J, K are isometries and satisfy quaternion relations. They are also parallel by construction. This shows that I, J, K satisfy the condition (1), (2), (3) in Proposition 13. □

There is another characterization of hyperkähler manifolds.

DEFINITION 14. A closed holomorphic 2-form of maximal rank on a Kähler manifold is called a *complex symplectic structure*.

THEOREM 15 (BEAUVILLE[BEA]). *Let (M^{2N}, I) be a compact Kähler manifold. The followings are equivalent:*

(1) *(M, I) has a complex symplectic structure.*
(2) *There is a hyperkähler metric g on M such that I is g-parallel and the Kähler class of (g, I) is the given Kähler class of M.*

Quaternion-Kähler manifolds.

DEFINITION 16. A Riemannian manifold $M^{4N'}$ is called *quaternion-Kähler* if its holonomy group is contained in $\mathrm{Sp}(N') \cdot \mathrm{Sp}(1) \subset \mathrm{O}(4N')$.

Here, the group $\mathrm{Sp}(N') \cdot \mathrm{Sp}(1)$ is contained in $\mathrm{O}(4N')$ as follows. First, we identify $\mathbb{R}^{4N'}$ with $\mathbb{H}^{N'}$. The group $\mathrm{Sp}(1)$ corresponds to transformations of $\mathbb{H}^{N'}$ by the multiplications of elements of the form $a+bi+cj+dk$, $a^2+b^2+c^2+d^2 = 1$, $a, b, c, d \in \mathbb{R}$, from the right. The group $\mathrm{Sp}(N')$ corresponds to those orthogonal transformations of $\mathbb{R}^{4N'} = \mathbb{H}^{N'}$ commuting with the multiplications of i, j, k from the right.

We remark that a quaternion-Kähler manifold $M^{4N'}$ is not necessarily a Kähler manifold, since the holonomy group $\mathrm{Sp}(N') \cdot \mathrm{Sp}(1)$ is not contained in $\mathrm{U}(N)$, $N = 2N'$. It is rather unfortunate that the term "Kähler" is used for some non-Kähler manifolds.

Now we deal with some geometric implications of $\mathrm{Sp}(N') \cdot \mathrm{Sp}(1)$ holonomy.

PROPOSITION 17. *A Riemannian manifold $M^{4N'}$ is quaternion-Kähler if and only if there exists a real 3-dimensional subbundle $L \subset \mathrm{End}(TM)$ with the following properties:*

(1) *(Holonomy Invariance of L) $\nabla_X L \subset L$ for any $X \in TM$, i.e., L is invariant under parallel transport.*
(2) *(Local quaternionic basis of L) For any point $y \in M$, there exists an open neighborhood U such that L is spanned by isometric almost complex structures I, J, K satisfying quaternion relations $IJ = -JI = -K$ on U.*

REMARK. The local basis $\{I, J, K\}$ in (2) above does not necessarily consist of parallel endomorphisms.

PROOF. Suppose that the given manifold is quaternion-Kähler, that is, suppose that the holonomy group of M is contained in $\mathrm{Sp}(N') \cdot \mathrm{Sp}(1)$. This means that at any point $x \in M$, there exists an orthonormal basis ξ in $T_x M$ such that after the identification $\xi : \mathbb{R}^n \to T_x M$, any parallel transport τ_γ is an element in $\mathrm{Sp}(N') \cdot \mathrm{Sp}(1) \subset \mathrm{GL}(\mathbb{R}^{4N'})$ for any loop γ based at x. Here, the quaternion structure (I_x, J_x, K_x) in $T_x M$ is the one induced from the standard quaternionic structure (I_0, J_0, K_0) in $\mathbb{R}^{4N'}$ via ξ. Since the action of $\mathrm{Sp}(N')$ on $\mathrm{End}(T_x M)$ fixes almost complex structures I_x, J_x, K_x and $\mathrm{Sp}(1)$ acts by on the real 3-dimensional vector space L_x spanned by these endomorphisms, it follows that $\mathrm{Sp}(N') \cdot \mathrm{Sp}(1)$ preserves the real 3-dimensional subspace $L_x \subset \mathrm{End}(T_x M)$. This means that for any point $y \in M$ and for any curve γ from x to y, the vector space $L_y = \tau_\gamma(L_x) = \tau_\gamma \circ L_x \circ \tau_\gamma^{-1}$ is a well defined real 3-dimensional subspace of $\mathrm{End}(T_y M)$ independent of the path γ chosen. The collection $\{L_y\}_{y \in M}$ defines a real 3-dimensional subbundle $L \subset \mathrm{End}(TM)$ which is invariant under parallel transport, and we have $\nabla_X L \subset L$ for any $X \in TM$.

For the existence of the local basis of L around an arbitrary point $y \in M$, we choose a path from x to y and let (I_y, J_y, K_y) be the result of the parallel transport of (I_x, J_x, K_x) along the path γ. We have $I_y = \tau_\gamma \circ I_x \circ \tau_\gamma^{-1}$, etc. Since τ_γ is an isometry, we see that I_y, J_y, K_y are isometries and satisfy quaternion relations. Around $y \in M$, we consider a "star-like" neighborhood U such that for any point $z \in U$, there exists a unique geodesic lying in U connecting y and z. For any $z \in U$, we parallel transport (I_y, J_y, K_y) along the unique geodesic from y to z in U to obtain a quaternionic structure (I_z, J_z, K_z) in $T_z M$. This gives a smooth tensor field I, J, K on U satisfying quaternionic relations. Although these tensor fields are not parallel in general, they form an $\mathbb{R}$-basis of L on an open set U. This shows the existence of the local quaternionic basis of L described in Proposition 17.

Conversely, suppose there exists a 3-dimensional subbundle $L \subset \mathrm{End}(TM)$ satisfying the properties specified in the above proposition. We show that the holonomy group of $M^{4n'}$ is contained in $\mathrm{Sp}(N') \cdot \mathrm{Sp}(1)$. Let $x \in M$ and let (I_x, J_x, K_x) be a quaternionic structure in $T_x M$ consisting of isometries which also form the basis of L at $x \in M$. We choose an orthonormal basis ξ of $T_x M$ such that the standard quaternionic structure of $\mathbb{R}^{4N'}$, (I_0, J_0, K_0), corresponds to (I_x, J_x, K_x) under the isomorphism $\xi : \mathbb{R}^{4N'} \to T_x M$. This is possible because all the isometric quaternionic structures in the real $4N'$ dimensional Euclidean vector space is an $O(4N')$-homogeneous space $O(4N')/\mathrm{Sp}(N')$. Let $G \subset O(4N')$ be the holonomy group of M at x with respect to the basis ξ. Our assumption that L is invariant under parallel transports implies that G preserves the vector space $L_x = \mathbb{R} I_x \oplus \mathbb{R} J_x \oplus \mathbb{R} K_x \subset \mathrm{End}(T_x M)$. On the other hand, $\mathrm{Sp}(N') \cdot \mathrm{Sp}(1)$ is the maximal subgroup of $O(4N')$ preserving the 3-dimensional subspace $L_0 = \mathbb{R} I_0 \oplus \mathbb{R} J_0 \oplus \mathbb{R} K_0$ of $\mathrm{End}(\mathbb{R}^{4N'})$. Hence $G \subset \mathrm{Sp}(N') \cdot \mathrm{Sp}(1)$ and Proposition 17 is proved. $\square$

A quaternion-Kähler manifold may not admit $\mathbb{H}$-bundle structure on its tangent bundle globally. It may not even admit a single almost complex structure at all on its tangent bundle. For example, the quaternion projective space $\mathbb{H}P^{N'}$ is quaternion-Kähler, but it doesn't admit any almost complex structures, for a topological reason. So, it doesn't make sense to talk about Kähler forms on quaternion-Kähler manifolds $M^{4N'}$. However, we have a differential 4-forms canonically associated to the quaternion-Kähler structure $L \subset \mathrm{End}(TM)$ of $M^{4N'}$.

From Proposition 17, on a quaternion-Kähler manifold $M^{4N'}$ we can locally choose a quaternionic basis I, J, K of the structure bundle $L \subset \text{End}(TM)$ on an open set $U \subset M$ which acts as isometries on TM. This amounts to choosing an $\mathbb{H}$-bundle structure of TM over U compatible with the Riemannian metric g. For each of these almost complex structures I, J, K on U, we can consider the associated 2-forms κ_I, κ_J, κ_K on U defined by

$$\kappa_I(X, Y) = g(I(X), Y), \quad \kappa_J(X, Y) = g(J(X), Y), \quad \kappa_K(X, Y) = g(K(X), Y).$$

Since these almost complex structures I, J, K need not be parallel on U, the above 2-forms $\kappa_I, \kappa_J, \kappa_K$ may not be parallel nor closed. However, the associated 4-form $\kappa_I^2 + \kappa_J^2 + \kappa_K^2$ turns out to be parallel and closed on U. This enables us to construct a globally defined parallel closed 4-form κ_{Q-K} on a quaternion-Kähler manifold.

PROPOSITION 18 (QUATERNION-KÄHLER FORMS). *Let M^{4N} be a quaternion-Kähler manifold with the structure bundle $L \subset \text{End}(TM)$. Then, there exists a globally defined parallel closed 4-form κ_{Q-K}, the quaternion-Kähler form, such that if $L|_U = \mathbb{R}I \oplus \mathbb{R}J \oplus \mathbb{R}K$ for some quaternionic isometric almost complex structures I, J, K on an open set $U \subset M$, then,*

$$\kappa_{Q-K}|_U = \kappa_I^2 + \kappa_J^2 + \kappa_K^2,$$

where $\kappa_I, \kappa_J, \kappa_K$ are 2-forms associated to I, J, K, as above.

PROOF. First, we show that on U, the 4-form $\kappa_I^2 + \kappa_J^2 + \kappa_K^2$ is independent of the choice of isometric quaternionic basis of L over the open set U. So, let (I', J', K') be another basis of L_x, $x \in U$, so that there exists an element $g \in \text{GL}_3(\mathbb{R})$ such that $(I', J', K') = (I, J, K) \cdot g$. The quaternionic relations among (I, J, K) and (I', J', K') implies that $g \in \text{SO}(3)$. See §4.2 Lemma 17. It then automatically follows that if I, J, K are isometries, then, I', J', K' are also isometries. Since the dependence of 2-form κ_J on an almost complex structure $\mathcal{J}$ is linear, that is, $\kappa_{aI+bJ+cK} = a\kappa_I + b\kappa_J + c\kappa_K$ due to §3.9 Theorem 15, it follows that $(\kappa_{I'}, \kappa_{J'}, \kappa_{K'}) = (\kappa_I, \kappa_J, \kappa_K) \cdot g$ with the same $g \in \text{SO}(3)$. Hence,

$$\kappa_{I'}^2 + \kappa_{J'}^2 + \kappa_{K'}^2 = (\kappa_{I'} \quad \kappa_{J'} \quad \kappa_{K'}) \begin{pmatrix} \kappa_{I'} \\ \kappa_{J'} \\ \kappa_{K'} \end{pmatrix} = (\kappa_I \quad \kappa_J \quad \kappa_K) g \cdot {}^t g \begin{pmatrix} \kappa_I \\ \kappa_J \\ \kappa_K \end{pmatrix}$$
$$= \kappa_I^2 + \kappa_J^2 + \kappa_K^2.$$

Thus, the form $\kappa_U = \kappa_I^2 + \kappa_J^2 + \kappa_K^2$ depends only on the structure bundle $L|_U$ on U and independent of the choice of local quaternionic basis of L. This shows the uniqueness of the quaternion-Kähler form κ_{Q-K} on U. Let $V \subset M$ be another open set such that $U \cap V \neq \emptyset$. We consider the quaternion-Kähler form κ_V on V using a local quaternionic basis of L on V. Then, on $U \cap V$, the restrictions $\kappa_U|_{U \cap V}$ and $\kappa_V|_{U \cap V}$ must coincide due to the uniqueness of such forms on $U \cap V$. Thus, we can patch together κ_U and κ_V to obtain a form on $U \cup V$. Thus, from a covering of M by open sets over which L is trivialized by quaterninic basis, we can obtain a globally defined 4-form κ_{Q-K} with the property that

$$\kappa_{Q-K}|_U = \kappa_I^2 + \kappa_J^2 + \kappa_K^2,$$

for any open set U over which L has a quaternionic local basis

Next, we show that the quaternion-Kähler form κ_{Q-K} is parallel, $\nabla_X \kappa_{Q-K} = 0$, for any $X \in TM$. Here, ∇ is the torsion free covariant derivative associated to the Riemannian metric g. Since this is a local property, we can work in an open set U over which we have a quaternionic basis I, J, K of L, although they may not be parallel. Let $\gamma : [0,1] \to U \subset M$ be a smooth curve in U such that $\gamma(0) = x$ and $\dot\gamma(0) = X \in T_x M$. Let $\gamma_t = \gamma|_{[0,t]}$ be its restriction. Let τ_{γ_t} be the parallel transport of tensors at x to tensors at $\gamma(t)$. We consider parallel transport in the bundle $\text{End}(TM)$. Since $M^{4N'}$ is quaternion-Kähler, by Proposition 17, the structure bundle L is preserved by the parallel transport. In particular, for any vector $v \in L_{\gamma(t)}$, $\tau_{\gamma_t}^{-1}(v) \in L_{\gamma(0)} = L_x$ is a linear combination of I_x, J_x, K_x. Thus, we can write

$$\begin{cases} \tau_{\gamma_t}^{-1}\left(I_{\gamma(t)}\right) = c_{II}(t)I_x + c_{IJ}(t)J_x + c_{IK}(t)K_x, \\ \tau_{\gamma_t}^{-1}\left(J_{\gamma(t)}\right) = c_{JI}(t)I_x + c_{JJ}(t)J_x + c_{JK}(t)K_x, \\ \tau_{\gamma_t}^{-1}\left(K_{\gamma(t)}\right) = c_{KI}(t)I_x + c_{KJ}(t)J_x + c_{KK}(t)K_x, \end{cases}$$

for some smooth coefficient functions $c_{II}, c_{IJ}, \ldots$. By definition, the covariant derivatives of I, J, K at x by X are exactly the derivatives of the above functions in t at $t = 0$. Thus,

$$\begin{cases} \nabla_X I = \dot c_{II}(0)I_x + \dot c_{IJ}(0)J_x + \dot c_{IK}(0)K_x, \\ \nabla_X J = \dot c_{JI}(0)I_x + \dot c_{JJ}(0)J_x + \dot c_{KK}(0)K_x, \\ \nabla_X K = \dot c_{KI}(0)I_x + \dot c_{KJ}(0)J_x + \dot c_{KK}(0)K_x. \end{cases}$$

We can deduce relations among these coefficients from the quaternionic relations of I, J, K. Since ∇ is a derivation, the relation $I^2 = -1$ implies that $\nabla_X I \cdot I + I \cdot \nabla_X I = 0$, which in turn implies that $\dot c_{II}(0) = 0$. Similarly, the relations $J^2 = -1, K^2 = -1$ imply that $\dot c_{JJ}(0) = \dot c_{KK}(0) = 0$. Next, the relation $IJ = -K$ implies that $\nabla_X I \cdot J + I \cdot \nabla_X J = -\nabla_X K$ which imposes that $\dot c_{IJ}(0) + \dot c_{JI}(0) = 0$, $\dot c_{JK}(0) + \dot c_{KJ}(0) = 0$, $\dot c_{KI}(0) + \dot c_{IK}(0) = 0$. Since these coefficients c_{IJ}'s depend linearly on X, we can write the result of our calculation in terms of certain 1-forms α, β, γ on M as follows:

$$\begin{cases} \nabla_X I = \qquad\qquad \gamma(X)J - \beta(X)K \\ \nabla_X J = -\gamma(X)I \qquad\qquad + \alpha(X)K \\ \nabla_X K = \beta(X)I - \alpha(X)J \end{cases}$$

Since the 2-forms κ_J depends linearly on almost complex structures J by Theorem 15 in §3.9, we have similar formulae for $\kappa_I, \kappa_J, \kappa_K$ concerning the parallel transports along γ_t. Namely,

$$
\begin{cases}
\tau_{\gamma_t}^{-1}\left(\kappa_{I,\gamma(t)}\right) = c_{II}(t)\kappa_{I,x} + c_{IJ}(t)\kappa_{J,x} + c_{IK}(t)\kappa_{K,x}, \\
\tau_{\gamma_t}^{-1}\left(\kappa_{J,\gamma(t)}\right) = c_{JI}(t)\kappa_{I,x} + c_{JJ}(t)\kappa_{J,x} + c_{JK}(t)\kappa_{K,x}, \\
\tau_{\gamma_t}^{-1}\left(\kappa_{K,\gamma(t)}\right) = c_{KI}(t)\kappa_{I,x} + c_{KJ}(t)\kappa_{J,x} + c_{KK}(t)\kappa_{K,x}.
\end{cases}
$$

Since the coefficients are the same as in the almost complex structure case, differentiating the above formulae at $t = 0$, we obtain,

$$
\begin{cases}
\nabla_X \kappa_I = \qquad\qquad \gamma(X)\kappa_J - \beta(X)K \\
\nabla_X \kappa_J = -\gamma(X)\kappa_I \qquad\quad + \alpha(X)\kappa_K \\
\nabla_X \kappa_K = \beta(X)\kappa_I - \alpha(X)\kappa_J
\end{cases}
$$

Now, we are ready to show that the quaternion-Kähler form κ_{Q-K} is parallel. The covariant derivative $\nabla_X(\kappa_{Q-K})$ for $X \in TM$ is given by

$$
\begin{aligned}
\nabla_X(\kappa_I^2 + \kappa_J^2 + \kappa_K^2) &= 2\kappa_I \cdot \nabla_X \kappa_I + 2\kappa_J \cdot \nabla_X \kappa_J + 2\kappa_K \cdot \nabla_X \kappa_K \\
&= 2\gamma(X)\kappa_I \cdot \kappa_J - 2\beta(X)\kappa_I \cdot \kappa_K - 2\gamma(X)\kappa_J \cdot \kappa_I + 2\alpha(X)\kappa_J \cdot \kappa_K \\
&\qquad + 2\beta(X)\kappa_K \cdot \kappa_I - 2\alpha(X)\kappa_K \cdot \kappa_J = 0
\end{aligned}
$$

Hence, the canonical 4-form κ_{Q-K} is parallel and consequently, it is closed because the Levi-Civita connection is torsion free. □

§5.2 Generalizations of Riemannian metrics, Riemannian volume forms, Kähler forms, complex volume forms, and quaternion-Kähler forms

We study certain generalizations of basic tensor fields on Riemannian manifolds such as metric tensors and volume forms on oriented manifolds, Kähler forms, Hermitian pairings, and complex volume forms on complex Kähler manifolds, and quaternion-Kähler forms on quaternion-Kähler manifolds. Our generalization takes place in an infinite dimensional vector bundle $\underline{V}$ which is the spin bundle on the free loop space $\mathcal{L}M^{2N}$ restricted to $M \subset \mathcal{L}M$. This graded vector bundle contains the total exterior bundle $\bigwedge^* T_{\mathbb{C}}^* M$ of the cotangent bundle of M. The above generalized tensor fields of interest to us are nowhere vanishing parallel tensor fields on M with respect to the given Riemannian connection. Later, we will see that these parallel tensor fields give rise to a vertex operator super algebra acting on the elliptic genus of the Spin manifold M^{2N}.

The algebra of tensor fields on manifolds and geometric structures. Let TM be the (real) tangent bundle of a manifold M. The associated tensor algebra bundle of M is given by

$$
(1) \qquad \mathcal{T}M = \bigoplus_{p,q \geq 0} \mathcal{T}^{(p,q)}(M), \quad \text{where} \quad \mathcal{T}^{(p,q)}(M) = \bigotimes^p TM \otimes \bigotimes^q T^*M.
$$

A tensor field of type (p,q) on M is a section of the bundle $\mathcal{T}^{(p,q)}(M)$. If there exists a nowhere vanishing tensor field of type (p,q) on M, then it splits off a trivial real line bundle from $\mathcal{T}^{(p,q)}(M)$. Thus, the existence of nowhere vanishing tensor field depends on the global topological and geometric properties of the manifold.

We list several nowhere vanishing tensor fields whose existence follows from the existence of certain geometric structures.

RIEMANNIAN METRIC: At each point x of a manifold M, a Riemannian metric $g = \{g_x\}_{x \in M}$ consists of positive definite symmetric bilinear pairings g_x on each tangent space $T_x M$. A Riemannian metric defines a nowhere vanishing tensor field

$$(2) \qquad\qquad g \in \Gamma(T^*M \otimes T^*M).$$

By the very definition of the Riemannian connection ∇, the metric tensor g is a parallel tensor field, i.e., $\nabla g = 0$.

RIEMANNIAN VOLUME FORMS: On an oriented manifold M^n, the highest exterior power $\bigwedge^n T^*M^n$ is a trivial line bundle. When M^n is also Riemannian, we can introduce a metric on the total exterior bundle $\bigwedge^* T^*M$, and hence $\bigwedge^n T^*M^n$ also carries a metric. Since it is trivial, there exists a unique section ξ such that $|\xi| = 1$ belonging to the orientation class of M^n. This section is the Riemannian volume form. Thus, $\xi \in \Gamma(\bigwedge^n T^*M^n)$. Since parallel transport preserves the length of vectors and $\bigwedge^n T^*M^n$ is real 1-dimensional, we see that the Riemannian volume form ξ must be parallel. Consequently, $-\xi$ is also parallel.

ALMOST COMPLEX STRUCTURES: An almost complex structure on an even dimensional manifold M^{2N} assigns a complex structure I_x, $I_x^2 = -1$, to each tangent space $T_x M$ at $x \in M$. Thus, I defines a nowhere vanishing tensor field of type $(1,1)$ in the sense of (1) above and $I \in \Gamma(T^*M \otimes TM)$. Let the contraction map or the dual pairing map be denoted by C, so $C : TM \otimes T^*M \to \mathbb{R}$. The following map induced by C,

$$C = 1 \otimes C \otimes 1 : (T^*M \otimes TM) \otimes (T^*M \otimes TM) \to T^*M \otimes \mathbb{R} \otimes TM,$$

is such that $C(I \otimes I) = -\sum_{j=1}^{2N} e_j^* \otimes e_j$ for any basis $\{e_1, \ldots, e_{2N}\}$ of $T_x M$. Here, $\{e_1^*, \ldots, e_{2N}^*\}$ is the dual basis. This is another way of saying that $I^2 = -1$ when I is regarded as an element in $\mathrm{End}(TM) \cong T^*M \otimes T^*M$.

An almost complex structure canonically splits the complexified tangent bundle into eigenbundles of I, $TM \otimes \mathbb{C} = \mathcal{A}_M^+ \oplus \mathcal{A}_M^-$, where I acts on the subbundle $\mathcal{A}_M^+$ by multiplication by i and on $\mathcal{A}_M^-$ by multiplication by $-i$. These subbundles are also denoted by $T_{\mathbb{C}}^{(1,0)} M$ and $T_{\mathbb{C}}^{(0,1)} M$, respectively. But we use the above notation $\mathcal{A}_M^{\pm}$ when we have generalized tensor bundles and vertex operator super algebras in mind.

KÄHLER FORMS: If an almost complex structure I on an even dimensional Riemannian manifold (M^{2N}, g) is is an isometry on TM, then we can define a differential 2-form κ on M by $\kappa(X,Y) = g(I(X),Y)$ for $X, Y \in TM$.

Since I is an isometry with respect to the symmetric tensor g, it is immediate to check that κ is an alternating tensor field, i.e. a differential 2-form $\kappa \in \Gamma\left((A_M^+ \otimes A_M^-) \oplus (A_M^- \otimes A_M^+)\right)$. The 2-form κ is a nowhere vanishing differential form on M since g is nondegenerate and I is an isomorphism.

In general, this differential form κ is *not* closed. But if the almost complex structure I is parallel with respect to the Riemannian connection, then it is well known that the almost complex structure I is integrable and that the manifold M^{2N} itself has a structure of a complex Kähler manifold and the form κ is parallel and closed [Proposition 7 §4.1]. In this case, κ is the Kähler form for the Kähler metric g. If furthermore M is compact, the cohomology classes $[\kappa]$, $[\kappa^2]$, ..., $[\kappa^N]$ are all nontrivial. If I is not parallel, then κ is merely the imaginary part of the Hermitian pairing associated to the metric g and the isometric almost complex structure I.

HERMITIAN PAIRINGS: Let M^{2N} be a Riemannian manifold with a metric g. Suppose we have an almost complex structure I on M acting on TM as an isometry. Here the tensor field I is not necessarily assumed to be parallel with respect to the Levi-Civita connection associated to g. In this context, we can consider two Hermitian pairings $h^\pm : TM \otimes TM \to \mathbb{C}$, conjugate to each other, defined by

$$(3) \qquad h^\pm(X,Y) = g(X,Y) \pm ig(I(X),Y) = g(X,Y) \pm i\kappa(X,Y),$$

where $X, Y \in TM$. h^+ is conjugate linear in the first entry and complex linear in the second entry. For h^- this is reversed, since $\overline{h^+(X,Y)} = h^-(X,Y)$. These Hermitian pairings define nowhere vanishing tensor fields $h^+ \in \Gamma(A_M^+ \otimes A_M^-)$, $h^- \in \Gamma(A_M^- \otimes A_M^+)$. Also note that $\frac{1}{2}(h^+ + h^-) = g$, $\frac{1}{2i}(h^+ - h^-) = \kappa$.

When the almost complex structure I is parallel, κ is parallel and so Hermitian tensors $h^\pm = g \pm i\kappa$ are also parallel.

COMPLEX VOLUME FORMS: On an almost complex Riemannian manifold M^{2N}, the top exterior bundle of the $(1,0)$ part of $T_\mathbb{C}M$, $\bigwedge_\mathbb{C}^N T_\mathbb{C}^{(1,0)}M$, is a complex line bundle on M whose first Chern class if the same as the first Chern class of the bundle $T_\mathbb{C}^{(1,0)}M = A_M^+$. If the first Chern class of $T_\mathbb{C}^{(1,0)}M$ vanishes, then the above line bundle is topologically trivial and we have a nowhere vanishing section of the line bundle $\bigwedge_\mathbb{C}^N T_\mathbb{C}^{(1,0)}M$. If the almost complex structure I is parallel with respect to the Riemannian connection, then the manifold is in fact a complex Kähler manifold. In this case, if the manifold is compact, then by the Calabi conjecture the vanishing of the first Chern class implies that there exists a Kähler metric on M with the same Kähler class as the original metric and with vanishing Ricci tensor, which implies that the restricted holonomy group of the manifold with respect to the new metric is contained in $SU(N)$.

If the entire holonomy group of M is contained in $SU(N)$, then there exists nowhere vanishing parallel sections in $\bigwedge_\mathbb{C}^N A_M^+$:

$$(4) \qquad \xi^+ \in \Gamma_{\text{parallel}}\left(\bigwedge_\mathbb{C}^N T_\mathbb{C}^{(1,0)}M\right) = \Gamma_{\text{parallel}}\left(\bigwedge_\mathbb{C}^N A_M^+\right) \cong \mathbb{C}.$$

A nonzero parallel section of this holomorphic line bundle is a complex volume form [Proposition 11, §5.1]. Their complex conjugates $\xi^- \in \Gamma(\wedge_{\mathbb{C}}^N \mathcal{A}_M^-)$ are also parallel sections.

QUATERNION-KÄHLER FORMS: On a $4N'$-dimensional Riemannian manifold suppose that the endomorphism bundle $\text{End}(TM)$ of the tangent bundle has a real 3-dimensional subbundle L such that for any $x \in M$, there exists an open neighborhood U of x over which L is spanned by isometric endomorphism fields I, J, K satisfying $I^2 = J^2 = -1$, $IJ = -JI = -K$. Although the corresponding 2-forms κ_I, κ_J, κ_K depend on the choice of I, J, K, their sum of squares

$$(5) \qquad \kappa_{Q-K} = \kappa_I^2 + \kappa_J^2 + \kappa_K^2$$

is independent of the choice of I, J, K and depends only on L. It is also uniquely determined by L. If furthermore L is invariant under the parallel transport, then the canonically defined 4-form κ_{Q-K} is parallel and this form is a quaternion-Kähler form on $M^{4N'}$ [§5.1 Proposition 18].

LSpin representations and generalized tensor bundles. Let E^{2N} be a real $2N$ dimensional vector space with an inner product $\langle \ , \ \rangle$, and let $A = E \otimes \mathbb{C}$ be its complexification. Let $A(-n - \frac{1}{2})$ be a copy of A with weight $n + \frac{1}{2}$. The vector space formed by an infinite tensor product

$$(6) \qquad V = \bigotimes_{n \geq 0} \wedge_{q^{n+\frac{1}{2}}}^* A(-n - \tfrac{1}{2}) \cdot \Omega$$

has the structure of a direct sum of two spin representations of the orthogonal affine Lie algebra $\hat{o}(2N)$. When the vector space E^{2N} has a complex structure $I : E^{2N} \to E^{2N}$, $I^2 = -1$, the complexified vector space A splits into a sum of two eigenspaces of I, $A = A^+ \oplus A^-$, where I acts by multiplication by i on A^+ and by multiplication by $-i$ on A^-. A choice of a Hermitian orthonormal basis $\{e_1, \ldots, e_{2N}\}$ of E^{2N} such that $I(e_j) = e_{j+N}$ give rise to a basis $\{a_1, \ldots, a_N\}$, $\{a_1^*, \ldots, s_N^*\}$ of A^+, A^- such that with respect to the pairing extended $\mathbb{C}$-linearly, we have $\langle a_j, a_k^* \rangle = \langle a_j^*, a_k \rangle = \delta_{jk}$, $\langle a_j, a_k \rangle = \langle a_j^*, a_k^* \rangle = 0$ for $1 \leq j, k \leq N$. In fact, in terms of e_j's, these basis vectors are given by

$$(7) \qquad a_j = \frac{e_j - iI(e_j)}{\sqrt{2}}, \qquad a_j^* = \frac{e_j + iI(e_j)}{\sqrt{2}}, \qquad 1 \leq j \leq N.$$

We recall that an exterior algebra $\mathcal{A}$ is defined by

$$(8) \qquad \mathcal{A} = \bigotimes_{n \geq 0} \wedge^* A(-n - \tfrac{1}{2}).$$

As a vector space, this is isomorphic to V via $[\ ] : \mathcal{A} \to V$ given by $[\alpha] = \alpha \cdot \Omega$ for $\alpha \in \mathcal{A}$, where Ω is the vacuum vector. We regard $\mathcal{A}$ as an algebra of operators and V as a vector space generated by operators in $\mathcal{A}$. The following elements in $\mathcal{A}$ are of interest to us:

$$\omega = -\tfrac{1}{2}\sum_{j=1}^{N}\left\{a_j(-\tfrac{3}{2})a_j^*(-\tfrac{1}{2}) + a_j^*(-\tfrac{3}{2})a_j(-\tfrac{1}{2})\right\},$$

$$\omega_+ = -\sum_{j=1}^{N}a_j(-\tfrac{3}{2})a_j^*(-\tfrac{1}{2}), \qquad\qquad \omega_- = -\sum_{j=1}^{N}a_j^*(-\tfrac{3}{2})a_j(-\tfrac{1}{2}),$$

(9)
$$h = \sum_{j=1}^{N}a_j(-\tfrac{1}{2})a_j^*(-\tfrac{1}{2}),$$

$$\rho(-\tfrac{1}{2}) = a_1(-\tfrac{1}{2})a_1^*(-\tfrac{1}{2})\cdots a_N(-\tfrac{1}{2})a_N^*(-\tfrac{1}{2}),$$

$$\rho^+(-\tfrac{1}{2}) = a_1(-\tfrac{1}{2})\cdots a_N(-\tfrac{1}{2}), \qquad \rho^-(-\tfrac{1}{2}) = a_1^*(-\tfrac{1}{2})\cdots a_N^*(-\tfrac{1}{2}).$$

Let M^{2N} be a Riemannian manifold and let $T_{\mathbb{C}}M(-n-\tfrac{1}{2})$ be a copy of the complexified tangent bundle $T_{\mathbb{C}}M$ carrying the weight $n+\tfrac{1}{2}$ for $n \in \mathbb{Z}$. We apply the construction similar to (6) or (8) to the tangent bundle TM. So, we let

$$(10) \qquad \underline{V} = \bigotimes_{n \geq 0}\bigwedge\nolimits_{q^{n+\frac{1}{2}}}^{*}T_{\mathbb{C}}M(-n-\tfrac{1}{2})\cdot 1_M.$$

Here, 1_M is a constant function on M. For each $x \in M$, we can construct an infinite dimensional spin representation V_x as in (6) using T_xM as E. The above bundle $\underline{V}$ is nothing but the collection of these representations V_x, one for each point $x \in M$. This fibrewise point of view is often useful.

Now, assume that the Riemannian manifold M^{2N} carries an almost complex structure I which acts as an isometry with respect to the metric g. So, for each $x \in M$, the tangent space T_xM is a Euclidean vector space with a complex structure I_x which acts as an isometry. So, its complexification splits into a direct sum of eigenspaces of I_x, $T_xM \otimes \mathbb{C} = A_x^+ \oplus A_x^-$. We can choose a basis of $A_x^{\pm}$ in a way similar to the formulae in (7). That is, $A_x^+ = \bigoplus_{j=1}^{N}\mathbb{C}a_{j,x}$, $A_x^- = \bigoplus_{j=1}^{N}\mathbb{C}a_{j,x}^*$ where basis vectors satisfy $g_x(a_{j,x}, a_{k,x}^*) = g_x(a_{j,x}^*, a_{k,x}) = \delta_{jk}$, $g_x(a_{j,x}, a_{k,x}) = g_x(a_{j,x}^*, a_{k,x}^*) = 0$. This is done by first choosing any Hermitian orthonormal basis $\{e_{1,x}, \ldots, e_{2N,x}\}$ of T_xM with the property $I(e_{j,x}) = e_{j+N,x}$ for $1 \leq j \leq N$. We then let

$$a_{j,x} = \frac{e_{j,x} - iI(e_{j,x})}{\sqrt{2}}, \qquad a_{j,x}^* = \frac{e_{j,x} + iI(e_{j,x})}{\sqrt{2}}, \qquad 1 \leq j \leq N.$$

We can do the same for the weight $n + \tfrac{1}{2}$ space $T_xM(-n-\tfrac{1}{2})\otimes\mathbb{C}$:

(11)
$$T_xM(-n-\tfrac{1}{2})\otimes\mathbb{C} = A^+(-n-\tfrac{1}{2})_x \oplus A^-(-n-\tfrac{1}{2})_x, \qquad \text{where}$$

$$A^+(-n-\tfrac{1}{2})_x = \bigoplus_{j=1}^{N}\mathbb{C}a_j(-n-\tfrac{1}{2})_x, \quad A^-(-n-\tfrac{1}{2})_x = \bigoplus_{j=1}^{N}\mathbb{C}a_j^*(-n-\tfrac{1}{2})_x.$$

These basis vectors have similar properties with respect to a generalized metric tensor $g_{(n,n)}$ in the bundle $TM(-n-\tfrac{1}{2})$ discussed below. In $(\underline{V})_x$, we can consider elements corresponding to ω, $\omega_\pm$, $\rho^\pm$, etc in $\mathcal{A}$. These elements are then denoted with subscript x, and with a bracket, like $[\omega_x]$, $[\omega_{\pm,x}]$, $[\rho_x^\pm]$:

(12)
$$[\omega_x] = -\tfrac{1}{2}\sum_{j=1}^{N}\left\{a_j(-\tfrac{3}{2})_x a_j^*(-\tfrac{1}{2})_x + a_j^*(-\tfrac{3}{2})_x a_j(-\tfrac{1}{2})_x\right\} \cdot 1_M,$$

$$[\omega_{+,x}] = -\sum_{j=1}^{N}a_j(-\tfrac{3}{2})_x a_j^*(-\tfrac{1}{2})_x \cdot 1_M, \qquad [\omega_{-,x}] = -\sum_{j=1}^{N}a_j^*(-\tfrac{3}{2})_x a_j(-\tfrac{1}{2})_x \cdot 1_M,$$

$$[\rho(-\tfrac{1}{2})_x] = a_1(-\tfrac{1}{2})_x a_1^*(-\tfrac{1}{2})_x \cdots a_N(-\tfrac{1}{2})_x a_N^*(-\tfrac{1}{2})_x \cdot 1_M,$$

$$[h_x] = \sum_{j=1}^{N}a_j(-\tfrac{1}{2})_x a_j^*(-\tfrac{1}{2})_x \cdot 1_M,$$

$$[\rho^+(-\tfrac{1}{2})_x] = a_1(-\tfrac{1}{2})_x \ldots a_N(-\tfrac{1}{2})_x \cdot 1_M,$$

$$[\rho^-(-\tfrac{1}{2})_x] = a_1^*(-\tfrac{1}{2})_x \ldots a_N^*(-\tfrac{1}{2})_x \cdot 1_M.$$

We often omit the subscript x from $a_{j,x}$, $a_{j,x}^*$ etc. and simply write a_j or a_j^* etc. when it is clear at which point we are considering these elements.

Generalized Riemannian tensors. We generalize Riemannian metric tensor g to "symmetric" tensor fields in $\underline{V}$.

DEFINITION 1. For integers $n_1, n_2 \geq 0$, a generaized Riemannian tensor

$$g_{(n_1,n_2)} : T_\mathbb{C}M(n_1 + \tfrac{1}{2}) \otimes T_\mathbb{C}M(n_2 + \tfrac{1}{2}) \to \mathbb{C},$$

is a pairing defined by $g_{(n_1,n_2)}(v_1(n_1 + \tfrac{1}{2}), v_2(n_2 + \tfrac{1}{2})) = g(v_1, v_2)$ for $v_1, v_2 \in T_\mathbb{C}M$. So, the pairing $g_{(n_1,n_2)}$ is the most obvious one induced from the Riemannian metric $g : TM \otimes TM \to \mathbb{R}$ extended $\mathbb{C}$-linearly.

Since $g_{(n_1,n_2)}$ pairs vectors from different vector spaces when $n_1 \neq n_2$, we can't talk about the symmetry of this tensor field. However, the symmetry of the Riemannian tensor field g implies :

LEMMA 2 (SYMMETRY OF GENERALIZED RIEMANNIAN TENSORS). *The generalized Riemannian tensor field is "symmetric" in the sense that*

$$g_{(n_1,n_2)}(v_1(n_1 + \tfrac{1}{2}), v_2(n_2 + \tfrac{1}{2})) = g_{(n_1,n_2)}(v_2(n_1 + \tfrac{1}{2}), v_1(n_2 + \tfrac{1}{2})),$$

for $v_1, v_2 \in T_\mathbb{C}M$, and for $n_1, n_2 \geq 0$.

We recall that among the vector spaces $A(m)$ for $m \in \mathbb{Z} + \tfrac{1}{2}$, a pairing can be defined by $\langle a(m), b(n)\rangle = \langle a, b\rangle \delta_{m,-n}$. In our present context this means that the vector bundle dual to $T_\mathbb{C}M(n + \tfrac{1}{2})$ is $T_\mathbb{C}M(-n - \tfrac{1}{2})$, with respect to the fibrewise pairing. Note that we are using two different types of pairings here. One type of pairings consists of generalized Riemannian tensors given in Definition 1. The other type of pairings consists of the Clifford pairings which come from the grading on $\bigoplus_{n\in\mathbb{Z}} A(n + \tfrac{1}{2})$ and pairs nontrivially only for vectors of complementary weights. In view of the Clifford pairings, the generalized Riemannian tensor $g_{(n_1,n_2)}$ can be thought of as a section in the bundle $T_{\mathbb{C}}M(-n_1 - \frac{1}{2}) \otimes T_{\mathbb{C}}M(-n_2 - \frac{1}{2})$, $n_1, n_2 \geq 0$. Furthermore, since the Riemannian tensor $g : TM \otimes TM \to \mathbb{R}$ is a parallel section in the vector bundle $TM \otimes TM$ with respect to the induced Levi-Civita connection, it follows that the generalized Riemannian tensor $g_{(n_1,n_2)}$ is also a parallel section.

The generalized Riemannian tensors can be defined for any pair of integers $n_1, n_2 \in \mathbb{Z}$. But those tensors with $n_1 < 0$ or $n_2 < 0$ do not belong to $\underline{V}$. Furthermore, when $n_1 = n_2 = n$, $g_{(n,n)}$ is no longer a section of the exterior bundle $\underline{V}$. It is a section of the symmetric power bundle $S^2(T_{\mathbb{C}}M(n + \frac{1}{2}))$. Thus, from now on, we restrict ourselves to the case $n_1 \neq n_2$ and $n_1, n_2 \geq 0$.

When we regard generalized Riemannian tensor $g_{(n_1,n_2)}$ as a section of the vector bundle $\underline{V}$, then we use the notation $[g_{(n_1,n_2)}]$ to distinguish it from the operator acting on $\underline{V}$ by the Clifford multiplication. In our notations, vectors in representation space $\underline{V}$ have brackets and operators acting on $\underline{V}$ by Clifford multiplications don't have brackets. This rule applies throughout this paper.

Note that for a parallel section s in any bundle, if it doesn't vanish at some point $x \in M$, then it is nowhere vanishing, since parallelism implies that the norm $\|s\|$ is always constant on M. Collecting our observations, we obtain :

PROPOSITION 3. *The generalized Riemannian tensors*

$$[g_{(n_1,n_2)}] \in \Gamma\left(T_{\mathbb{C}}M(-n_1 - \tfrac{1}{2}) \otimes T_{\mathbb{C}}M(-n_2 - \tfrac{1}{2})\right) \in \Gamma(\underline{V})_{n_1+n_2+1},$$

are nowhere vanishing parallel sections of $\underline{V}$ of weight $n_1 + n_2 + 1$ for $n_1, n_2 \geq 0$ with $n_1 \neq n_2$. In other words, $[g_{(n_1,n_2)}] \in \mathcal{P}_M(\underline{V})_{n_1+n_2+1}$. Thus, the algebra generated by generalized Riemannian metric tensors is a subalgebra of $\mathcal{P}_M(\underline{V})$, the algebra of parallel sections in $\underline{V}$. The lowest weight generalized Riemannian tensor is $[g_{(1,0)}] \in \mathcal{P}_M(\underline{V})_2$.

Now we look at $[g_{(n_1,n_2)}]$ fibre-wise and identify its local expression in $(\underline{V})_x$.

PROPOSITION 4 (LOCAL EXPRESSIONS OF GENERALIZED RIEMANNIAN TENSORS). *For any $n_1, n_2 \in \mathbb{Z}_+$, $n_1 \neq n_2$, the generalized Riemannian tensor $[g_{(n_1,n_2)}]$ has the local expression at $x \in M$ given by*

$$[g_{(n_1,n_2)}]_x = \sum_{j=1}^{N}[a_j(-n_1 - \tfrac{1}{2}) \otimes a_j^*(-n_2 - \tfrac{1}{2})] + \sum_{j=1}^{N}[a_j^*(-n_1 - \tfrac{1}{2}) \otimes a_j(-n_2 - \tfrac{1}{2})]$$
$$= -[\omega_+(n_1, n_2)_x] - [\omega_-(n_1, n_2)_x] = -2[\omega(n_1, n_2)_x].$$

In particular at $x \in M$, the fundamental Riemannian tensor $[g_{(1,0)}]$ is given by $[g_{(1,0)}]_x = -2[\omega_x]$ where $[\omega_x]$ is the canonical Virasoro element in V_x.

PROOF. Since $g_{(n_1,n_2)}$ is essentially the Riemannian metric g, we have

$$g_{(n_1,n_2)}\left(a_i(n_1 + \tfrac{1}{2}), a_j^*(n_2 + \tfrac{1}{2})\right) = \delta_{ij} = g_{(n_1,n_2)}\left(a_i^*(n_1 + \tfrac{1}{2}), a_j(n_2 + \tfrac{1}{2})\right),$$
$$g_{(n_1,n_2)}\left(a_i(n_1 + \tfrac{1}{2}), a_j(n_2 + \tfrac{1}{2})\right) = 0 = g_{(n_1,n_2)}\left(a_i^*(n_1 + \tfrac{1}{2}), a_j^*(n_2 + \tfrac{1}{2})\right).$$

Now from this table of pairings, we observe that the following two bases, one of $T_{\mathbb{C}}M(n + \frac{1}{2})$ and the other of $T_{\mathbb{C}}M(-n - \frac{1}{2})$, are dual to each other.

$$\{a_1(n + \tfrac{1}{2}), \ldots, a_N(n + \tfrac{1}{2})\} \cup \{a_1^*(n + \tfrac{1}{2}), \ldots, a_N^*(n + \tfrac{1}{2})\},$$
$$\{a_1^*(-n - \tfrac{1}{2}), \ldots, a_N^*(-n - \tfrac{1}{2})\} \cup \{a_1(-n - \tfrac{1}{2}), \ldots, a_N(-n - \tfrac{1}{2})\}.$$

Lemma now follows from these formulae. We use brackets in our expressions above, since we are dealing with vectors in $\underline{V}$ rather than operators. $\square$

The Riemannian tensor $g = g_{(0,0)}$ is a symmetric tensor field and can be identified with the original Riemannian metric g, but doesn't belong to the algebra $\mathcal{P}_M(\underline{V})$.

Generalized Riemannian volume forms. Let M^n be an oriented Riemannian manifold. The top exterior bundle $\bigwedge^n_{\mathbb{R}} T^*M$ is a trivial line bundle with metric trivialized by the Riemannian volume form ξ such that $|\xi_x| = 1$ for any $x \in M$. We consider generalizations of ξ in the LSpin bundle $\underline{V}$ of (19).

DEFINITION 5 (GENERALIZED RIEMANNIAN VOLUME FORMS). Let M^n be an oriented Riemannian manifold. For $\ell \in \mathbb{Z}$, a generalized Riemannian volume form $\xi_{(\ell)}$ is a section

$$\xi_{(\ell)} \in \Gamma\big(\bigwedge^n_{\mathbb{R}} TM(-\ell - \tfrac{1}{2})\big)$$

in the orientation class such that $|\xi_{(\ell),x}| = 1$ for all $x \in M$.

As we have seen before, the above unit length condition actually implies that $\xi_{(\ell)}$ is parallel.

PROPOSITION 6 (PARALLEL GENERALIZED RIEMANNIAN VOLUME FORMS). *On an oriented Riemannian manifold M^n, generalized Riemannian volume forms $\xi_{(\ell)}$ for $\ell \in \mathbb{Z}$ are parallel and spans the space of all parallel sections in the top exterior power $\bigwedge^n_{\mathbb{R}} TM(-\ell - \frac{1}{2})$. That is,*

$$\mathbb{R} \cdot \xi_{(\ell)} = \Gamma_{\text{parallel}}\big(\bigwedge^n_{\mathbb{R}} TM(-\ell - \tfrac{1}{2})\big).$$

When $\ell \geq 0$, $\xi_{(\ell)}$ can be regarded as a section of weight $n(\ell + \frac{1}{2})$ of the complex vector bundle $\underline{V}$. Thus, $\xi_{(\ell)} \in \mathcal{P}_M(\underline{V})_{n(\ell + \frac{1}{2})}$.

Generalized Kähler forms. Let M^{2N} be a Hermitian manifold, that is, M^{2N} is a Riemannian manifold with an almost complex structure I which acts as an isometry on TM. Let $T_{\mathbb{C}}M(n + \frac{1}{2})$ be a copy of the complexified tangent bundle. As before, we have the Clifford pairing $\langle\ ,\ \rangle$ on the graded vector space $\bigoplus_{n \in \mathbb{Z}} T_{\mathbb{C}}M(n + \frac{1}{2})$ defined by

(13) $$\langle v_1(n_1 + \tfrac{1}{2}), v_2(-n_2 - \tfrac{1}{2})\rangle = \langle v_1, v_2\rangle \delta_{n_1, n_2},$$

for any $v_1, v_2 \in T_{\mathbb{C}}M$ and any $n_1, n_2 \in \mathbb{Z}$. With respect to this pairing, the dual vector bundle of $T_{\mathbb{C}}M(n + \frac{1}{2})$ is $T_{\mathbb{C}}M(-n - \frac{1}{2})$. We can also extend the definition of the almost complex structure I to the graded vector space $\bigoplus_{n \in \mathbb{Z}} T_{\mathbb{C}}M(n + \frac{1}{2})$ as a weight preserving map $I(v(n + \frac{1}{2})) = (I(v))(n + \frac{1}{2})$ for $v \in T_{\mathbb{C}}M, n \in \mathbb{Z}$

DEFINITION 7 (GENERALIZED KÄHLER FORMS). Let $(M^{2N}; g, I)$ be a Hermitian manifold. For any $n_1, n_2 \in \mathbb{Z}$, a generalized Kähler form

$$\kappa_{(n_1,n_2)} = \kappa_{I,(n_1,n_2)} : T_{\mathbb{C}}M(n_1 + \tfrac{1}{2}) \otimes T_{\mathbb{C}}M(n_2 + \tfrac{1}{2}) \to \mathbb{C}$$

is defined by $\kappa_{(n_1,n_2)}\left(v_1(n_1 + \tfrac{1}{2}), v_2(n_2 + \tfrac{1}{2})\right) = g_{\mathbb{C}}(I(v_1), v_2)$ for any $v_1, v_2 \in T_{\mathbb{C}}M$, where $g_{\mathbb{C}}$ denotes the $\mathbb{C}$-linear extension of the Riemannian metric g.

When $n_1 = n_2 = n$, the above tensor is actually a differential form, that is, an alternating form on $TM(n + \tfrac{1}{2})$. However, when $n_1 \neq n_2$, the generalized Kähler form defined above is not a form. But they are "alternating" in the following sense.

$$(14) \quad \kappa_{(n_1,n_2)}\left(v_1(n_1 + \tfrac{1}{2}), v_2(n_2 + \tfrac{1}{2})\right) = -\kappa_{(n_1,n_2)}\left(v_2(n_1 + \tfrac{1}{2}), v_1(n_2 + \tfrac{1}{2})\right),$$

for any $v_1, v_2 \in TM$. This tensor field $\kappa_{(n_1,n_2)}$ can be thought of as a nowhere vanishing tensor field of a bundle

$$(15) \quad \begin{cases} \kappa_{I,(n_1,n_2)} \in \Gamma\left(T_{\mathbb{C}}M(-n_1 - \tfrac{1}{2}) \otimes T_{\mathbb{C}}M(-n_2 - \tfrac{1}{2})\right), & \text{if } n_1 \neq n_2, \\ \kappa_{I,(n)} = \kappa_{I,(n,n)} \in \Gamma(\bigwedge^2 T_{\mathbb{C}}M(-n - \tfrac{1}{2})), & \text{if } n_1 = n_2 = n. \end{cases}$$

in view of the duality induced by the Clifford pairing.

As with the generalized Riemannian tensors, we use the notation $[\kappa_{(n_1,n_2)}]$ with brackets when the generalized Kähler forms are regarded as sections in $\underline{V}$. When we do regard them as operators acting on the Clifford module $\underline{V}$, we don't use brackets.

Note that the above $\kappa_{I,(n_1,n_2)}$ is a generalization of the usual Kähler form κ on M defined by $\kappa(v_1, v_2) = \langle I(v_1), v_2 \rangle$ for $v_1, v_2 \in T_{\mathbb{C}}M$. As with the usual κ, generalized Kähler forms $\kappa_{(n_1,n_2)}$ are the imaginary parts of generalized Hermitian pairings discussed next. We can identify the Kähler form κ with $[\kappa_{(0,0)}]$ through the identification of $T_{\mathbb{C}}^*M$ with $T_{\mathbb{C}}M(-\tfrac{1}{2})$.

Note that unlike the generalized Riemannian tensor $[g_{(n_1,n_2)}]$ which didn't belong to $\mathcal{P}_M(\underline{V})_*$ when $n_1 = n_2$, the generalized Kähler forms $[\kappa_{(n_1,n_2)}]$ always belong to $\mathcal{P}_M(\underline{V})_*$ for all $n_1, n_2 \geq 0$.

At $x \in M$, we choose an orthonormal basis $\{e_1, \ldots, e_N, I(e_1), \ldots, I(e_N)\}$ of $T_x M$ adapted to the complex structure I and let $\{a_1, \ldots, a_N\} \cup \{a_1^*, \ldots, a_N^*\}$ be the complex basis of the decomposition $T_x M \otimes \mathbb{C} = A_x^+ \oplus A_x^-$ into maximal isotropic subspaces with respect to I. See (12) for notations.

PROPOSITION 8 (LOCAL EXPRESSIONS OF GENERALIZED KÄHLER FORMS). Let $(M^{2N}; g, I)$ be a Hermitian Manifold. With respect to the above basis, the local expression of the generalized Kähler form $[\kappa_{(n_1,n_2)}] = [\kappa_{I,(n_1,n_2)}]$ at $x \in M$ for $n_1 > n_2 \geq 0$ with respect to a complex structure I is given by

$$[\kappa_{(n_1,n_2)}]_x = -i \sum_{j=1}^{N} [a_j(-n_1 - \tfrac{1}{2})_x \otimes a_j^*(-n_2 - \tfrac{1}{2})_x]$$

$$+ i \sum_{j=1}^{N} [a_j^*(-n_1 - \tfrac{1}{2})_x \otimes a_j(-n_2 - \tfrac{1}{2})_x]$$

$$= i[\omega_+(n_1, n_2)_x] - i[\omega_-(n_1, n_2)_x].$$

In particular, the weight 2 form $[\kappa_{(1,0)}]$ is such that $[\kappa_{(1,0)}]_x = i([\omega_{+,x}] - [\omega_{-,x}])$. When $n_1 = n_2 = n \geq 0$, we have

$$[\kappa_{(n,n)}]_x = [\kappa_{I,(n,n)}]_x = -i \sum_{j=1}^{N} [a_j(-n-\tfrac{1}{2})_x \wedge a_j^*(-n-\tfrac{1}{2})_x] = -i \cdot [h(-n-\tfrac{1}{2})_x].$$

For the ordinary Kähler form $[\kappa] = [\kappa_{(0,0)}]$, we have $[\kappa]_x = -i[h_x]$.

PROOF. Recall that the basis

$$\{a_1(n+\tfrac{1}{2}), \ldots, a_N(n+\tfrac{1}{2}), a_1^*(n+\tfrac{1}{2}), \ldots, a_N^*(n+\tfrac{1}{2})\}$$

of $T_x M(n+\tfrac{1}{2}) \otimes \mathbb{C}$ is dual to the basis

$$\{a_1^*(-n-\tfrac{1}{2}), \ldots, a_N^*(-n-\tfrac{1}{2}), a_1(-n-\tfrac{1}{2}), \ldots, a_N(-n-\tfrac{1}{2})\}$$

of $T_x M(-n-\tfrac{1}{2}) \otimes \mathbb{C}$ by the Clifford pairing. Since I acts on A^+ by i, and on A^- by $-i$, a simple calculation shows the first part of the above result.
When $n_1 = n_2$, we observe that

$$a_j(-n-\tfrac{1}{2}) \wedge a_j^*(-n-\tfrac{1}{2}) = a_j(-n-\tfrac{1}{2}) \otimes a_j^*(-n-\tfrac{1}{2}) - a_j^*(-n-\tfrac{1}{2}) \otimes a_j(-n-\tfrac{1}{2})$$

in $T_{\mathbb{C}}M(-n-\tfrac{1}{2}) \wedge T_{\mathbb{C}}M(-n-\tfrac{1}{2}) \subset T_{\mathbb{C}}M(-n-\tfrac{1}{2}) \otimes T_{\mathbb{C}}M(-n-\tfrac{1}{2})$. So, the first formula reduces to the second. □

Note that signs in front of these summations in the first formula Proposition 6 are different. This is in contrast with the local expression of generalized Riemannian tensors $g_{(n_1,n_2)}$ in Proposition 4.

We have seen that the generalized Riemannian tensors $g_{(n_1,n_2)}$ are parallel on Riemannian manifolds. A similar statement holds for generalized Kähler forms.

PROPOSITION 9 (PARALLEL GENERALIZED KÄHLER FORMS). *Let $(M^{2N}; g, I)$ be a Kählerian manifold with an almost complex structure I. That is, M^{2N} is a Hermitian manifold for which I is parallel. For $n_1, n_2 \geq 0$, the generalized Kähler form $[\kappa_{I,(n_1,n_2)}]$ is a parallel section of $\underline{V}$ of weight $n_1 + n_2 + 1$. Namely,*

$$[\kappa_{I,(n_1,n_2)}] \in \mathcal{P}_M(\underline{V})_{n_1+n_2+1}.$$

Thus, the algebra of generalized Kähler forms are contained in the vertex operator super algebra $\mathcal{P}_M(\underline{V})$. In particular, $[\kappa] = [\kappa_{(0,0)}] \in \mathcal{P}_M(\underline{V})_1$, $[\kappa_{(1,0)}] \in \mathcal{P}_M(\underline{V})_2$.

PROOF. First we note that the generalized Kähler form $[\kappa_{I,(n_1,n_2)}]$ is obtained by applying the contraction map C to the tensor product of generalized Riemannian tensor $[g_{(n_1,n_2)}]$ and the generalized complex structure I. That is, $[\kappa_{(n_1,n_2)}]$ is the composition of the following maps.

$$M \xrightarrow{I \otimes g_{(n_1,n_2)}} T_{\mathbb{C}}M(-n_1-\tfrac{1}{2}) \otimes T_{\mathbb{C}}M(-n_1-\tfrac{1}{2})^* \otimes T_{\mathbb{C}}M(-n_1-\tfrac{1}{2}) \otimes T_{\mathbb{C}}M(-n_2-\tfrac{1}{2})$$

$$\xrightarrow{1 \otimes C \otimes 1} T_{\mathbb{C}}M(-n_1-\tfrac{1}{2}) \otimes T_{\mathbb{C}}M(-n_2-\tfrac{1}{2}).$$

Then, since the covariant derivative ∇ is a derivation commuting with the contraction operator C,

$$\nabla[\kappa_{(n_1,n_2)}] = \nabla\left(C(I \otimes [g_{(n_1,n_2)}])\right) = C\left(\nabla I \otimes [g_{(n_1,n_2)}] + I \otimes \nabla[g_{(n_1,n_2)}]\right).$$

Since the generalized Riemannian tensors are parallel, we have $\nabla[g_{(n_1,n_2)}] = 0$. Since M is Kählerian, the complex structure is parallel, $\nabla I = 0$. Thus, we see that $\nabla[\kappa_{I,(n_1,n_2)}] = 0$ and the generalized Kähler forms on Kählerian manifolds are parallel. $\square$

Generalized Hermitian Tensors. Let M^{2N} be a Hermitian manifold, that is, a Riemannian manifold with a compatible almost complex structure I. We can then define generalized Hermitian tensors.

DEFINITION 10 (GENERALIZED HERMITIAN TENSORS). On a Hermitian manifold $(M^{2N}; g, I)$, generalized Hermitian tensors $\eta^{\pm}_{I,(n_1,n_2)}$ are defined by

$$[\eta^{\pm}_{I,(n_1,n_2)}] = [g_{(n_1,n_2)}] \pm i[\kappa_{I,(n_1,n_2)}] : T_{\mathbb{C}}M(n_1 + \tfrac{1}{2}) \otimes T_{\mathbb{C}}M(n_2 + \tfrac{1}{2}) \to \mathbb{C}$$

where $g_{(n_1,n_2)}$ and $\kappa_{(n_1,n_2)}$ are generalized Riemannian tensors and generalized Kähler forms defined before for $n_1, n_2 \in \mathbb{Z}$.

The weight 2 Hermitian tensors $\eta^{\pm}_{I,(1,0)} \in T_{\mathbb{C}}M(-\tfrac{3}{2}) \otimes T_{\mathbb{C}}M(-\tfrac{1}{2})$ are denoted by $\eta^{\pm}_I$. If we deal with only one complex structure I, then I is often omitted from the notation and the above elements are simply written as $\eta^{\pm}$.

When the manifold is just Hermitian, the above Hermitian tensors are not necessarily parallel. But they are so for Kählerian manifolds.

PROPOSITION 11 (PARALLEL GENERALIZED HERMITIAN TENSORS). *Suppose a Hermitian manifold $(M^{2N}; g, L)$ is Kähler with respect to an almost complex structure I. Then, for any $n_1, n_2 \in \mathbb{Z}$, the generalized Hermitian tensors $\eta^{\pm}_{I,(n_1,n_2)}$ are parallel tensors of weight $n_1 + n_2 + 1$ for $n_1, n_2 \in \mathbb{Z}$. When $n_1 \neq n_2$, $n_1, n_2 \in \mathbb{Z}_+$, these sections belong to $\Gamma(\underline{V})$ and we have $[\eta^{\pm}_{I,(n_1,n_2)}] \in \mathcal{P}_M(\underline{V})_{n_1+n_2+1}$. We can recover generalized Riemannian tensors and generalized Kähler forms by*

$$[g_{(n_1,n_2)}] = \frac{1}{2}([\eta^+_{(n_1,n_2)}] + [\eta^-_{(n_1,n_2)}]), \qquad [\kappa_{(n_1,n_2)}] = \frac{1}{2i}([\eta^+_{(n_1,n_2)}] - [\eta^-_{(n_1,n_2)}]).$$

For weight 2 parallel tensors, $[g_{(1,0)}] = \frac{1}{2}([\eta^+] + [\eta^-])$, $[\kappa_{(1,0)}] = \frac{1}{2i}([\eta^+] - [\eta^-])$. The local expressions at $x \in M$ of generalized Hermitian tensors are given by

$$[\eta^+_{I,(n_1,n_2)}]_x = -2[\omega_+(n_1, n_2)_x], \qquad [\eta^-_{I,(n_1,n_2)}]_x = -2[\omega_-(n_1, n_2)_x].$$

For the weight 2 Hermitian tensors $[\eta^{\pm}]$, $[\eta^+]_x = -2[\omega_{+,x}]$, $[\eta^-]_x = -2[\omega_{-,x}]$.

The proof follows from the corresponding properties of Generalized Riemannian tensors and generalized Kähler forms.

Generalized complex volume forms. Let M^{2N} be a Riemannian manifold with an almost complex structure I such that

(i) I is isometric, that is, $g(I(X), I(Y)) = g(X, Y)$ for $X, Y \in TM$,

(ii) I is parallel with respect to the Levi-Civita connection, $\nabla I = 0$.

Under these conditions, M^{2N} is in fact a complex Kähler manifold. If furthermore,

(iii) the entire holonomy group is contained in $\mathrm{SU}(N)$,

then the top exterior power bundle of the holomorphic tangent bundle $\bigwedge_{\mathbb{C}}^N \mathcal{A}_M^+$ is flat, in fact, holomorphically trivial. Thus, it has one dimensional family of parallel sections, nonzero members of which are called complex volume forms.

We consider complex volume forms in our general setting of generalized tensor bundles $\underline{V}$. So, we assume that the Riemannian manifold M^{2N} has an almost complex structure I and let $\mathcal{A}_M^{\pm}(-n - \frac{1}{2})$ be a copy of the bundle $\mathcal{A}_M^{\pm}$, carrying the weight $n + \frac{1}{2}$ for $n \in \mathbb{Z}$. The top exterior bundles $\bigwedge_{\mathbb{C}}^N \mathcal{A}_M^{\pm}(-n - \frac{1}{2})$ for $n \in \mathbb{Z}$ are complex line bundles with an induced Hermitian metric and an induced Riemannian connection.

DEFINITION 12 (GENERALIZED COMPLEX VOLUME FORMS). *Suppose M^{2N} is equipped with an almost complex structure. For $n \in \mathbb{Z}$, a nowhere vanishing section $[\xi_{(n)}^+] \in \Gamma(\bigwedge_{\mathbb{C}}^N \mathcal{A}_M^+(-n - \frac{1}{2})) \subset \Gamma(\underline{V})$ is called a generalized complex volume form. Its conjugate $[\xi_{(n)}^-]$ can be defined in a similar way. The elements corresponding to the usual complex volume forms $\xi^{\pm}$ through the identification of $T_{\mathbb{C}}M$ with $\mathcal{A}_M(-\frac{1}{2})$ are $\xi_{(0)}^{\pm}$. These elements are denoted by $\xi^{\pm}$.*

Here again, we use the bracket notation to distinguish vectors in $\underline{V}$ and operators acting on $\underline{V}$. Generalized complex volume forms exist only for those manifolds M^{2N} which are special Kähler and whose holonomy group is contained in $\mathrm{SU}(N)$, i.e., those manifolds satisfying those conditions (i), (ii), (iii) above.

Since the covariant derivative ∇ is a real differential operator, it commutes with the complex conjugation.

PROPOSITION 13 (PARALLEL GENERALIZED COMPLEX VOLUME FORMS FOR SPECIAL KÄHLER MANIFOLDS). *Let $(M^{2N}; g, I)$ be a Hermitian manifold whose holonomy group is contained in $\mathrm{SU}(N)$. Then, for any $n \in \mathbb{Z}$, the vector space of parallel sections $\Gamma_{\mathrm{parallel}}(\bigwedge_{\mathbb{C}}^N \mathcal{A}_M^+(-n - \frac{1}{2}))$ is one dimensional generated by the generalized complex volume form $[\xi_{(n)}^+] \in \mathcal{P}_M(\underline{V})_{N(n+\frac{1}{2})}$. A similar statement holds with the super script $+$ replaced by $-$.*

PROOF. Let $G \subset \mathrm{SU}(N)$ be the holonomy group of the manifold. By the general theory, the restriction map to the fibre over $x \in M$, $\Gamma_{\mathrm{parallel}}(\bigwedge_{\mathbb{C}}^N \mathcal{A}_M^{\pm}) \xrightarrow{\cong} (\bigwedge_{\mathbb{C}}^N \mathcal{A}_x^{\pm})^G$, is an isomorphism of vector spaces [§4.5 Proposition 23]. Since the vector space $(\bigwedge_{\mathbb{C}}^N \mathcal{A}_x^{\pm})$ is fixed by the action of $\mathrm{SU}(N)$, the right hand side of the above isomorphism is the entire vector space itself, $(\bigwedge_{\mathbb{C}}^N \mathcal{A}_x^{\pm})$, which is complex one dimensional generated by the restrictions of complex volume forms and their conjugates. Hence, all the parallel sections in the line bundles $(\bigwedge_{\mathbb{C}}^{N} \mathcal{A}_M^{\pm})$ form one dimensional vector spaces generated by complex volume forms and their conjugates. $\square$

We look at local expressions for generalized complex volume forms and their conjugates.

PROPOSITION 14. *Let* $x \in M$. *Then, for some* $\alpha, \beta \in \mathbb{C}^*$, *the local expressions of complex volume forms and their conjugates are given by*

$$[\xi_{(n)}^+]_x = \alpha \cdot a_1(-n - \tfrac{1}{2})_x a_2(-n - \tfrac{1}{2})_x \dots a_N(-n - \tfrac{1}{2})_x \cdot \Omega = \alpha[\rho_x^+],$$

$$[\xi_{(n)}^-]_x = \beta \cdot a_1^*(-n - \tfrac{1}{2})_x a_2^*(-n - \tfrac{1}{2})_x \dots a_N^*(-n - \tfrac{1}{2})_x \cdot \Omega = \beta[\rho_x^-].$$

In particular, we can adjust the constants so that $[\xi^{\pm}]_x = [\rho_x^{\pm}]$.

PROOF. This is clear because the fibre $(\bigwedge_{\mathbb{C}}^{N} \mathcal{A}_M^{\pm})$ at $x \in M$ is one dimensional. $\square$

Generalized quaternion-Kähler forms. Let M be a $4N'$-dimensional Riemannian manifold. Let $\mathrm{End}(TM)$ be its endomorphism bundle.

DEFINITION 15 (ALMOST QUATERNION-KÄHLER STRUCTURE). A subbundle L of real dimension 3 of $\mathrm{End}(TM)$ is called an *almost quaternion-Kähler structure* on a Riemannian manifold $M^{4N'}$ if at any point $x \in M$, there exists an open neighborhood U of x such that $L|_U = \mathbb{R}I \oplus \mathbb{R}J \oplus \mathbb{R}K$ where I, J, K are isometric almost complex structures on U satisfying the quaternionic relations $I^2 = J^2 = -1$, $IJ = -JI = -K$. A Riemannian manifold with an almost quaternion-Kähler structure is called an *almost quaternion-Kähler manifold*. If the almost quaternion-Kähler structure L is parallel, that is, invariant under the parallel transport, then L is called a quaternion-Kähler structure and a manifold with a quaternion-Kähler structure is called a quaternion-Kähler manifold.

On an almost quaternion-Kähler manifold, we can consider the quaternion-Kähler form κ_{Q-K} which is canonically associated to the almost quaternion-Kähler structure L by $\kappa_{Q-K}|_U = \kappa_I^2 + \kappa_J^2 + \kappa_K^2$, where κ_I, κ_J, κ_K are Kähler forms defined on an open set $U \subset M$ on which I, J, K are defined.

We consider a generalization of quaternion-Kähler form in the bundle $\underline{V}$. On an open set $U \subset M$ as in Definition 15, we can consider generalized Kähler form

$$(16) \qquad \kappa_{\mathcal{J},(n)} = \kappa_{\mathcal{J},(n,n)} \in \bigwedge^2 T_{\mathbb{C}}M(-n - \tfrac{1}{2}),$$

for $\mathcal{J} = I, J, K$ as in Definition 7.

DEFINITION 16 (GENERALIZED QUATERNION-KÄHLER FORMS). Let $M^{4N'}$ be an almost quaternion-Kähler manifold with an almost quaternion-Kähler structure L. Then, for any $n \in \mathbb{Z}$, a generalized quaternion-Kähler form $\kappa_{Q-K,(n)}$ is defined by

$$[\kappa_{Q-K,(n)}]\big|_U = [\kappa_{I,(n)}^2] + [\kappa_{J,(n)}^2] + [\kappa_{K,(n)}^2] \in \Gamma(\bigwedge^4 TM(-n - \tfrac{1}{2})).$$

Here $U \subset M$ is an open set such that $L|_U = \mathbb{C}I \oplus \mathbb{C}J \oplus \mathbb{C}K$. The above expression is independent of the choice of quaternionic basis (I, J, K) of $L|_U$ and completely determined by L only [§5.1 Proposition 18].

As for the usual quaternion-Kähler forms κ_{Q-K}, the generalized quaternion-Kähler forms are parallel on quaternion-Kähler manifolds.

PROPOSITION 16 (PARALLEL GENERALIZED QUATERNION-KÄHLER FORMS IN $\underline{V}$). *Let* $(M^{4N'}; g, L)$ *be a quaternion-Kähler manifold. For* $n \in \mathbb{Z}$, *a generalized quaternion-Kähler form* $\kappa_{Q-K,(n)}$ *is a parallel closed form. For* $n \geq 0$, *it is a parallel section of the LSpin bundle* $\underline{V}$ *of weight* $4(n + \frac{1}{2})$, *that is,* $[\kappa_{Q-K,(n)}] \in \mathcal{P}_M(\underline{V})_{4(n+\frac{1}{2})}$.

At $x \in M$, *restriction of the generalized quaternion-Kähler form to* $\underline{V}_x$ *is given by*

$$[\kappa_{Q-K,(n)}]_x = [\omega_I(-n - \tfrac{1}{2})_x^2] + [\omega_J(-n - \tfrac{1}{2})_x^2] + [\omega_K(-n - \tfrac{1}{2})_x^2]$$
$$= -h(-n - \tfrac{1}{2})^2 + 4x(-n - \tfrac{1}{2})x^*(-n - \tfrac{1}{2}).$$

Here elements on the right hand side are given in (35), (36) *of* §3.8.

PROOF. This follows from the corresponding statements on usual quaternion-Kähler forms. □

For later reference, we collect some local identities here induced by the injective evaluation map at $x \in M$.

$$\Gamma_{\text{parallel}}(\underline{V}) \to (\underline{V})_x \xleftarrow{\cong} V.$$
$$[g_{(1,0)}]_x = -2[\omega_x],$$
$$[\kappa]_x = [\kappa_I]_x = -i[h_x],$$
$$[\kappa_J]_x = [(x + x^*)_x], \qquad [\kappa_K]_x = [-i(x - x^*)_x],$$
(17)
$$[\kappa_{(1,0)}]_x = i\left([\omega_{+,x}] - [\omega_{-,x}]\right),$$
$$[\eta^+]_x = -2[\omega_{+,x}], \qquad [\eta^-]_x = -2[\omega_{-,x}],$$
$$[\xi^+]_x = [\rho_x^+], \qquad [\xi^-]_x = [\rho_x^-],$$
$$[\kappa_{Q-K}]_x = [\omega_{I,x}^2] + [\omega_{J,x}^2] + [\omega_{K,x}^2]$$
$$= -[h_x^2] + 4[(xx^*)_x].$$

§5.3   Twisted G-Dirac operators

To study the elliptic genus $\Phi_{\text{ell}}^*(M)$ of a Spin manifold from a representation theoretic point of view, we need to do some fine tuning of the geometric setting. Although the (numerical) index of an elliptic operator acting on sections of a vector bundle over a manifold is a topological invariant in general, dimension of the actual kernel space and the cokernel space is not topological invariant and they do depend on geometric data. Although the standard twisted Dirac operators coming from the Levi-Civita connection of a Riemannian metric are the operators we will use later, we can consider twisted Dirac operators in more general setting. This is the notion of G-Dirac operators, where G is a subgroup of a Spin group. Essentially, this is a twisted Dirac operator constructed from any connection on a G-principal subbundle of a Spin principal bundle over M. We give a detailed exposition on twisted Dirac operators and their properties, including self-adjointness and multiplicative property.

General framework. Let M^{2N} be a oriented Riemannian Spin manifold. Let $\mathbb{R}^{2N} = \bigoplus_{i=1}^{2N} \mathbb{R} e_i$ be the standard real $2N$ dimensional inner product space equipped with an orthonormal basis $\{e_1, \ldots, e_N\}$. To M, we can associate the orthonormal frame bundle $\pi : F_{\mathrm{SO}(2N)} \to M$ with structure group $\mathrm{SO}(2N)$ whose fibre over $m \in M$ denoted by $F_{\mathrm{SO}(2N),m}$ is given by

(1) $F_{\mathrm{SO}(2N),m} = \{p : \mathbb{R}^{2N} \to T_m M \mid p$ is an orientation preserving isometry.$\}$

The group $\mathrm{SO}(2N)$ acts on this frame bundle from the right. Any tangent vector $v \in T_m M$ defines an $\mathrm{SO}(2N)$-equivariant map $\overline{v} : F_{\mathrm{SO}(2N),m} \to \mathbb{R}^{2N}$ by $\overline{v}(p) = p^{-1}(v)$. This map gives the coefficient vector for $v \in T_m M$ when a frame p is used to express the tangent vector v. This is $\mathrm{SO}(2N)$ equivariant, since for any $g \in \mathrm{SO}(2N)$ we have $\overline{v}(pg) = (pg)^{-1}(v) = g^{-1}(p^{-1}(v)) = g^{-1}\overline{v}(p)$.

The Spin group $\mathrm{Spin}(2N)$ is a subgroup of invertible elements in the Clifford algebra $\mathrm{Cliff}(\mathbb{R}^{2N})$ preserving $\mathbb{R}^{2N}$ under conjugation, $g \cdot \mathbb{R}^{2N} \cdot g^{-1} \subset \mathbb{R}^{2N}$, with respect to the Clifford multiplication. This gives us a ready made double covering map $\rho : \mathrm{Spin}(2N) \to \mathrm{SO}(2N)$ with the property

(2) $g \cdot v \cdot g^{-1} = \rho(g)(v),$ for any $v \in \mathbb{R}^{2N}$, $g \in \mathrm{Spin}(2N)$.

Since M is a Spin manifold, the oriented orthogonal frame bundle is doubly covered by a principal $\mathrm{Spin}(2N)$ bundle.

(3) $\pi : P_{\mathrm{Spin}(2N)} \xrightarrow{\rho} F_{\mathrm{SO}(2N)} \xrightarrow{\pi} M$

Here, ρ is the covering map and we denote the above composition $\pi \circ \rho$ by π also. Our object of interest is a principal subbundle $\pi = \pi_G : P_G \to M$ of $P_{\mathrm{Spin}(2N)}$ with the structure group $G \subset \mathrm{Spin}(2N)$. We also use ρ to denote the following composition map.

(4) $\rho : P_G \hookrightarrow P_{\mathrm{Spin}(2N)} \xrightarrow{\rho} F_{\mathrm{SO}(2N)}.$

The composition of maps between structure groups $G \hookrightarrow \mathrm{Spin}(2N) \twoheadrightarrow \mathrm{SO}(2N)$ is also denoted by ρ. Any element $p \in P_G$ is a Spin frame compatible with the G-structure which in turn determine a preferred orthonormal frame $\rho(p)$ in the tangent space at $\pi(p) \in M$. For $v \in T_m M$, let $\overline{v}_G$ denote the G-equivariant map

(5) $\overline{v}_G : P_{G,m} \xrightarrow{\rho} F_{\mathrm{SO}(2N)} \xrightarrow{\overline{v}} \mathbb{R}^{2N},$ $\overline{v}_G(pg) = \rho(g)^{-1}(\overline{v}_G(p)) = g^{-1} \cdot \overline{v}_G(p) \cdot g,$

for $g \in G$ and $p \in P_G$. We simply write $\bar{v}$ for $\bar{v}_G$. Let F be a finite dimensional G-representation space and we let $\underline{F} = P_G \times_G F$. This is a vector bundle and its section $s \in \Gamma(\underline{F})$ can be identified with a G equivariant map $\bar{s} : P_G \to F$ such that $\bar{s}(pg) = \bar{s} \circ R_g(p) = g^{-1} \cdot \bar{s}(p)$ for $g \in G$ and $p \in P_G$ so that the equivalence class $[p, \bar{s}(g)]$ is a well defined vector in $\underline{F}_{\pi(p)}$. Here, $R_g : P_G \to P_G$ is right multiplication by $g \in G$ on P_G. Taking the differential of this map, we get

$$(6) \qquad d\bar{s}_{pg}\left(R_{g_*}(v)\right) = g^{-1}\left(d\bar{s}_p(v)\right) \quad \text{for } v \in T_p P_G.$$

To define a G-Dirac opertaor, we need to choose a connection θ on the G-bundle P_G. This is a G-equivariant real $2N$ dimensional horizontal distribution H^θ of P_G. This distribution has the property that (i) at each point $p \in P_G$, we have a $2N$ dimensional vector space $H_p^\theta \subset T_p P_G$ such that the restriction of the map $\pi_* : T_p P_G \to T_{\pi(p)} M$ to the subspace $H_p^\theta \subset T_p P_G$ is an isomorphism onto, and (ii) for each $g \in G$, the differential R_{g_*} of the right translation R_g by g is such that $H_{pg}^\theta = R_{g_*}(H_p^\theta)$. For our purpose, it doesn't matter whichever connection we choose on P, so we omit the notation θ from now on. Since $P_G \subset P_{\mathrm{Spin}(2N)}$, we can extend the connection from P_G to $P_{\mathrm{Spin}(2N)}$. This connection and the Levi-Civita connection coming from the metric on M may not agree. An interesting and important situation arises when the holonomy group of the Riemannian Spin manifold reduces to $G \subset \mathrm{Spin}(2N)$. In this case, we can restrict the Levi-Civita connection on $P_{\mathrm{Spin}(2N)}$ to P_G. Then, the principal bundle P_G is preserved under the parallel transport in $P_{\mathrm{Spin}(2N)}$ and we have a torsion free connection on P_G. For example, if M^{2N} has the structure of a Kähler manifold, then the holonomy group reduces to a subgroup of $U(N)$, and if it is Ricci flat, then the restricted holonomy group reduces to a subgroup of $SU(N)$, which is naturally a subgroup of $\mathrm{Spin}(2N)$. Furthermore, if the manifold $M^{4N'}$ is hyperkähler, then the holonomy group reduces to a subgroup of $Sp(N')$. These situations in which the connection arises from the restriction of Levi-Civita connection are particularly important because we can then consider torsion free connections on the principal G-bundle and with respect to torsion free connections parallel differential forms on M are closed.

Twisted G-Dirac operators. Now, we define twisted G-Dirac operators. Let (P_G, θ) be a principal G-bundle with a connection θ. Let F be a finite dimensional complex representation of $\mathrm{Spin}(2N)$ as before. Let $\Delta^\pm$ denote half spin representations of complex dimension 2^{N-1}. We consider the associated bundle on M.

$$(7) \qquad P_G \times_G (\Delta^\pm \otimes F) = \underline{\Delta}^\pm \otimes \underline{F}$$

Here, G acts diagonally on the tensor product representations $\Delta^\pm \otimes F$. Any section $s \in \Gamma(\underline{\Delta}^+ \otimes \underline{F})$ of this bundle can be identified with a G-equivariant map

$$(8) \qquad \bar{s} : P_G \to \Delta^\pm \otimes F, \qquad \bar{s}(pg) = g^{-1}\left(\bar{s}(p)\right).$$

Let $d\bar{s}^H : H \to \Delta^\pm \otimes F$ be the restriction of the differential $d\bar{s} : TP_G \to \Delta^\pm \otimes F$ to the real $2N$-dimensional subbundle H of P_G given by the connection θ.

DEFINITION 1 (TWISTED G-DIRAC OPERATORS). Let a Spin manifold M^{2N} has a G-structure P_G with $G \subset \mathrm{Spin}(2N)$ and let F be a complex representation of $\mathrm{Spin}(2N)$. The twisted G-Dirac operator $d_F : \Gamma(\underline{\Delta}^+ \otimes \underline{F}) \to \Gamma(\underline{\Delta}^- \otimes \underline{F})$ is defined as follows: For any smooth section $s \in \Gamma(\underline{\Delta}^+ \otimes \underline{F})$, the section $d_F s$ in $\Gamma(\underline{\Delta}^- \otimes \underline{F})$ is given by a G-equivariant map $d_F \bar{s} : P_G \to \Delta^- \otimes F$ defined by

$$(9) \qquad d_F\bar{s}(p) = \sum_{i=1}^{2N} \bar{v}_i(p) \cdot d\bar{s}_p^H(\tilde{v}_i), \qquad \text{for} \quad p \in P_G$$

where $(v_1, v_2, \ldots, v_{2N})$ is any orthonormal frame in $T_{\pi(p)}M$ and $(\tilde{v}_1, \ldots, \tilde{v}_{2N})$ is the lift to $H_p \subset T_p P_G$. The dot "$\cdot$" denotes Clifford multiplication of $\bar{v}_i(p) \in \mathbb{R}^{2N}$ on the first factor Δ^+, where $\bar{v}_i : P_G \to \mathbb{R}^{2N}$ is as in (6).

Note that Clifford multiplication by $\bar{v}_i(p)$ on Δ is such that

$$\bar{v}_i(p) \cdot \bar{v}_j(p) + \bar{v}_j(p) \cdot \bar{v}_i(p) = -\delta_{ij}.$$

We check that the above procedure is well defined. First, we show the independence of choices of an orthonormal basis used. Let $(u_1, u_2, \ldots, u_{2N})$ be another orthonormal basis of $T_{\pi(p)}M$. Then, $(v_1, v_2, \ldots, v_{2N}) = (u_1, u_2, \ldots, u_{2N})h$ for some $h \in \mathrm{SO}(2N)$. The lifted bases in H_p also have the same relation because $\pi_* : H_p \to T_{\pi(p)}M$ is an isomorphism. Then, regarding $\bar{v}(p)$ as a row vector and $d\bar{s}_p^H(\tilde{v})$ as a column vector, we have $\bar{v}(p) \cdot d\bar{s}_p^H(\tilde{v}) = \bar{u}(p)h \cdot h^t d\bar{s}_p^H(\tilde{u}) = \bar{u}(p) \cdot d\bar{s}_p^H(\tilde{u})$, which shows that (9) is independent of the orthonormal basis chosen.

Next, we must show that the map $d_F\bar{s} : P_G \to \Delta^- \otimes F$ is G-equivariant. That is $d_F\bar{s}(pg) = g^{-1} \cdot d_F\bar{s}(p)$, for $g \in G \subset \mathrm{Spin}(2N) \subset \mathrm{Cliff}(\mathbb{R}^{2N})$. Now,

$$d_F\bar{s}(pg) = \sum_{i=1}^{2N} \bar{v}_i(pg) \cdot d\bar{s}_{pg}^H(\tilde{v}_i) = \sum_{i=1}^{2N} \left(\rho(g)^{-1}(\bar{v}_i(p))\right) \cdot g^{-1}\left(d\bar{s}_p^H(\tilde{v}_i)\right)$$

$$= \sum_{i=1}^{2N} \left(g^{-1} \cdot \bar{v}_i \cdot g\right) \cdot g^{-1}\left(d\bar{s}_p^H(\tilde{v})\right) = \sum_{i=1}^{2N} g^{-1}\left(\bar{v}_i(p) \cdot d\bar{s}_p^H(\tilde{v}_i)\right)$$

$$= g^{-1} d_F\bar{s}(p).$$

Here, the second equality is due to (5) and (6). Note that $\tilde{v}_i$ is a R_{g_*}-invariant vector field on the fibre $(P_G)_m$ over $m \in M$ for $g \in G$. The third equality is by (2). For the fourth equality, note that in the fourth formula, g^{-1} in front of the second parenthesis acts diagonally on the tensor product and the objects in the first parenthesis act on Δ^+ by Clifford multiplication to produce a vector in Δ^-. So the cancellation of $g \cdot g^{-1}$ occurs on the first factor Δ^+, and the action of g^{-1} on the second factor F remains intact. So, g^{-1} acts on $\Delta^- \otimes F$ diagonaly in the final formula.

REMARK. Under the map $\rho : P_G \to F_{\mathrm{SO}(2N)}$, for any $p \in P_G$, $\rho(p) : \mathbb{R}^{2N} \to T_{\pi(p)}M$ gives a natural frame $v_i = \rho(p)(e_i)$, $1 \leq i \leq 2N$, in the tangent space

$T_{\pi(p)}M$. Here, $\{e_1, e_2, \ldots, e_{2N}\}$ is the standard orthonormal basis of $\mathbb{R}^{2N}$. When these vectors $v_1, \ldots, v_{2N}$ are regarded as G-equivariant maps on the fibre, these maps have the property $\bar{v}_i(p) = e_i$ for $1 \le i \le 2N$. So, our twisted G-Dirac operator can be written as

$$(10) \qquad d_F \bar{s}(p) = \sum_{i=1}^{2N} e_i \cdot \widetilde{d\bar{s}_p^H (\rho(p)(e_i))}.$$

It is often convenient to have intrinsic expressions of twisted G-Dirac operators in terms of the covariant derivative ∇ associated to the connection on P_G. We briefly recall the definition of ∇ in terms of the parallel transports defined by the connection. Let R be a complex linear representation of G and let $\underline{R}$ be the associated vector bundle. The covariant derivative $\nabla_X s$ of a section $s \in \Gamma(\underline{R})$ by a vector $X \in T_m M$ is defined as follows. Let $\gamma : (-\varepsilon, \varepsilon) \to M$ be a smooth curve in M such that $\gamma(0) = m$ and $\dot{\gamma}(0) = X \in T_m M$. Let γ_t be its restriction on $[0, t]$ and let $\tau_{\gamma_t} : T_m M \to T_{\gamma(t)} M$ be the parallel transport along γ_t. Then, the covariant derivative $\nabla_X s$ is defined by

$$\nabla_X s = \frac{d}{dt} \tau_{\gamma_t}^{-1} (s(\gamma(t))) \Big|_{t=0} \in (\underline{R})_m.$$

We explicitly calculate this object in $\underline{R}$. Let $\tilde{\gamma}(t)$ be a path which is the horizontal lift to P_G of $\gamma(t)$ starting from $p \in P_G$. So, $\dot{\tilde{\gamma}}(0) = \tilde{X} \in H_p$. We may write $s(\gamma(t)) = [\tilde{\gamma}(t), v(t)] \in (P_G)_{\gamma(t)}$ with $v(t) \in R$. Then, from the definition of parallel transport, $\tau_{\gamma_t}^{-1} (s(\gamma(t))) = [p, v(t)] \in (\underline{R})_m$. So, $\frac{d}{dt} \tau_{\gamma_t}^{-1} (s(\gamma(t))) |_{t=0} = [p, \dot{v}(0)] \in (\underline{R})_m$. Hence, $(\nabla_X s)_m = [p, \dot{v}(0)]$. On the other hand, when $s \in \Gamma(\underline{R})$ is regarded as G-equivariant map, $d\bar{s}_p^H (\tilde{X}) = \frac{d}{dt} \bar{s}(\tilde{\gamma}(t))|_{t=0} = \frac{d}{dt} v(t)|_{t=0} = \dot{v}(0)$. Thus,

$$(11) \qquad (\nabla_X s)_m = [p, d\bar{s}_p^H (\tilde{X})], \qquad \text{for any } p \in (P_G)_m.$$

Here, the equivalence class in the right hand side is independent of the choice of p due to (6).

The Clifford multiplication of a tangent vector $v = [p, \bar{v}(p)] \in T_m M$ on $s = [p, \bar{s}(p)] \in \Gamma(\underline{\Delta^+})$ is given by $(v \cdot s)_m = [p, \bar{v}(p) \cdot \bar{s}(p)]$ for $p \in (P_G)_m$, where $\bar{v}(p) \in \mathbb{R}^{2N}$ and $\bar{s}(p) \in \Delta^+$. This definition is independent of p used. To see this, from (5) and (8),

$$[pg, \bar{v}(pg) \cdot \bar{s}(pg)] = [pg, g^{-1} \cdot \bar{v}(p) \cdot g \cdot g^{-1} \cdot \bar{s}(p)] = [pg, g^{-1} \cdot \bar{v}(p) \cdot \bar{s}(p)] = [p, \bar{v}(p) \cdot \bar{s}(p)].$$

Hence, the above definition of $v \cdot s$ is well defined in the bundle $\underline{\Delta}$. Applying these calculations for the case $R = \Delta^+ \otimes F$ for some complex G-representation F, we see that

$$\sum_{i=1}^{2N} (v_i \cdot \nabla_{v_i} s)_m = \sum_{i=1}^{2N} [p, \bar{v}_i(p) \cdot \overline{(\nabla_{v_i} s)}(p)] = \sum_{i=1}^{2N} [p, \bar{v}_i(p) \cdot d\bar{s}_p^H (\tilde{v}_i)]$$
$$= [p, d_F \bar{s}(p)].$$

In the second equality, we used (11). The third equality is due to (9). Thus, we can give an alternate definition of twisted G-Dirac operators which is equivalent to Definition 1. We use the same notation as above.

DEFINITION 1' (TWISTED G-DIRAC OPERATORS). Let M has a G-structure P_G. Then, the result of the G-Dirac operator $d_E : \Gamma(\underline{\Delta}^+ \otimes \underline{F}) \to \Gamma(\underline{\Delta}^- \otimes \underline{F})$ applied to $s \in \Gamma(\underline{\Delta}^+ \otimes \underline{F})$ at $m \in M$ is defined by

$$(12) \qquad (d_F s)_m = \sum_{i=1}^{2N} v_i \cdot (\nabla_{v_i} s)_m,$$

where $\{v_1, \ldots, v_{2N}\}$ is an orthonormal basis in $T_m M$.

Note that in the above definition (12), G-structure is not used and (12) makes sense for any Spin bundles with a connection.

Some Properties of twisted G-Dirac operators. We prove two propositions in this subsection concerning properties of twisted G-Dirac operators. We also give an application to elliptic genera.

PROPOSITION 2. *Let M^{2N} be a Spin manifold, and let $\pi : P_{\mathrm{Spin}(2N)} \to M$ be the structure bundle with a principal connection. Suppose that M has a G structure $P_G \subset P_{\mathrm{Spin}(2N)}$ for some group $G \subset \mathrm{Spin}(2N)$. Let E be a linear complex representation of $\mathrm{Spin}(2N)$ and let*

$$\mathrm{End}_G(F) = \{f \in \mathrm{End}(F) \mid f \text{ is a } \mathbb{C}\text{-linear } G\text{-equivariant map}\}.$$

Then, for any $f \in \mathrm{End}_G(F)$, it induces a bundle map $\underline{f}$ commuting with the differential operator d_F, making the following diagram commutative whenever operators are defined.

$$(13) \qquad \begin{array}{ccc} \Gamma(\underline{\Delta}^+ \otimes \underline{F}) & \xrightarrow{\;d_F\;} & \Gamma(\underline{\Delta}^- \otimes \underline{F}) \\ {\scriptstyle \Gamma(1 \otimes \underline{f})} \downarrow & & \downarrow {\scriptstyle \Gamma(1 \otimes \underline{f})} \\ \Gamma(\underline{\Delta}^+ \otimes \underline{F}) & \xrightarrow{\;d_F\;} & \Gamma(\underline{\Delta}^- \otimes \underline{F}) \end{array}$$

Thus, the vector spaces $\mathrm{Ker}\, d_F$ and $\mathrm{Coker}\, d_F$ are representations of the algebra $\mathrm{End}_G(F)$.

PROOF. Since $f \in \mathrm{End}_G(F)$ is a G-equivariant map, it induces a bundle map $\underline{f} : \underline{F} \to \underline{F}$, where $\underline{F} = P_G \times_G F$, by $\underline{f}([p, v]) = [p, f(v)]$, where square bracket denotes the equivalence class. This is well defined because for another representative $[pg, g^{-1}v]$, $g \in G$, we have $\underline{f}([pg, g^{-1}v]) = [pg, f(g^{-1}v)] = [pg, g^{-1}f(v)] = [p, v]$. Next, we have to show that the bundle map $1 \otimes \underline{f}$ and d_F commutes. We only have to show this for sections $s \in \Gamma(\underline{\Delta}^+ \otimes \underline{F})$ of the form $s = s_1 \otimes s_2$ where $\bar{s}_1 : P_G \to \Delta^+$, $\bar{s}_2 : P_G \to F$ are G-equivariant maps. Since $\bar{s} = \bar{s}_1 \otimes \bar{s}_2 : P_G \to \Delta^+ \otimes F$ and $d\bar{s} = d\bar{s}_1 \otimes s_2 + \bar{s}_1 \otimes d\bar{s}_2$,

$$(1 \otimes f) \circ d_F \bar{s}(p) = (1 \otimes f) \circ \sum_{i=1}^{2N} \{\bar{v}_i(p) \cdot d\bar{s}_{1,p}^H(\tilde{v}_1) \otimes \bar{s}_2 + \bar{v}_i(p) \cdot \bar{s}_1 \otimes d\bar{s}_{2,p}^H(\tilde{v}_i)\}$$

$$= \sum_{i=1}^{2N} \{\bar{v}_i(p) \cdot d\bar{s}_{1,p}^H(\tilde{v}_1) \otimes (f \circ \bar{s}_2) + \bar{v}_i(p) \cdot \bar{s}_1 \otimes f \circ d\bar{s}_{2,p}^H(\tilde{v}_i)\}$$

Since f is a linear map between vector spaces, we have $df = f$ and $d(f \circ \bar{s}_2)^H = f \circ d\bar{s}_2^H$. So,

$$= d_F(\bar{s}_1 \otimes f \circ \bar{s}_2)(p) = d_F((1 \otimes f) \circ \bar{s})(p).$$

This proves that f commutes with d_F. $\square$

REMARK. In the above proposition, we can take the vector space F to be a graded vector space F_* and let $\text{End}_G(F_*)$ be the set of all G-equivariant maps which do not necessarily preserve the grading of F_*. Then, the same conclusion holds.

PROPOSITION 3. *Let M^{2N} be a Spin manifold with a principal connection on its $\text{Spin}(2N)$ structure bundle. Let F be a linear complex representation of $\text{Spin}(2N)$. Suppose an operator $\chi : \Gamma(\underline{F}) \to \Gamma(\underline{F})$ acting on sections of $\underline{F}$ commutes with the covariant derivative in $\underline{F}$, i.e., $[\nabla, \chi] = 0$. Then, the operator $1\hat{\otimes}\chi$ on $\Gamma(\underline{\Delta}^{\pm} \otimes \underline{F})$ commutes with the twisted Dirac operator d_F making the following diagram commute whenever the operators are defined:*

(14)
$$
\begin{array}{ccc}
\Gamma(\underline{\Delta}^+ \otimes \underline{F}) & \xrightarrow{\ d_F\ } & \Gamma(\underline{\Delta}^- \otimes \underline{F}) \\
{\scriptstyle 1\hat{\otimes}\chi}\big\downarrow & & \big\downarrow{\scriptstyle 1\hat{\otimes}\chi} \\
\Gamma(\underline{\Delta}^+ \otimes \underline{F}) & \xrightarrow{\ d_F\ } & \Gamma(\underline{\Delta}^- \otimes \underline{F})
\end{array}
$$

PROOF. It is sufficient to prove the commutativity of the above diagram when a section $s \in \Gamma(\underline{\Delta}^+ \otimes \underline{F})$ is of the form $s = s_1 \otimes s_2$ where $s_1 \in \Gamma(\underline{\Delta}^+)$ and $s_2 \in \Gamma(\underline{F})$. Now, using the description of the twisted Dirac operator d_F given in Definition 1′, for any orthonormal basis $\{v_1, \ldots, v_{2N}\}$ in $T_m M$, using the commutativity of ∇ and χ, we have

$$d_F \circ (1\hat{\otimes}\chi)(s) = d_F(s_1 \otimes \chi(s_2)) = \sum_{i=1}^{2N} v_i \cdot \nabla_{v_i}(s_1 \otimes \chi(s_2))$$

$$= \sum_{i=1}^{2N} v_i \cdot ((\nabla_{v_i} s_1) \otimes \chi(s_2)) + \sum_{i=1}^{2N} v_i \cdot (s_1 \otimes \nabla_{v_i}\chi(s_2))$$

$$= \sum_{i=1}^{2N} (v_i \cdot \nabla_{v_i} s_1) \otimes \chi(s_2) + \sum_{i=1}^{2N} (v_i \cdot s_1) \otimes \chi(\nabla_{v_i} s_2)$$

$$= (1 \otimes \chi)\Big(\sum_{i=1}^{2N}(v_i \cdot \nabla_{v_i} s_1) \otimes s_2 + \sum_{i=1}^{2N}(v_i \cdot s_1) \otimes \nabla_{v_i} s_2\Big)$$

$$= (1\hat{\otimes}\chi)d_F s.$$

This proves the commutativity of the above diagram for sections of the form $s = s_1 \otimes s_2$. Since the subspace of $\Gamma(\underline{\Delta}^+ \otimes \underline{F})$ consisting of finite sums of sections of this form is dense in $\Gamma(\underline{\Delta}^+ \otimes \underline{F})$, Proposition 3 is proved. $\square$

Previously, we defined super-pairs of graded vector spaces $\hat{A}(M)$, $\Phi_\theta(M)$, $\Phi_{\text{ell}}(M)$ in §1.1 and in §1.6. These are the geometric $\hat{A}$-genus, the geometric Witten genus, and the geometric elliptic genus. These graded vector spaces are defined in terms of twisted Dirac operators. These super-spaces have the following inclusion relations.

PROPOSITION 4. *Let M^{2N} be a closed Riemannian Spin manifold. Then, there exist the following inclusion relations:*

$$(15) \qquad \hat{A}(M) \subset \Phi_\theta(M) \subset \Phi^*_{\text{ell}}(M), \qquad for * = Q, R.$$

PROOF. The first inclusion relation follows because $\hat{A}(M)$ is the lowest weight super-pair in the super-pair of graded vector spaces $\Phi_\theta(M)$, as we observed before. For the second inclusion, we first note that there is an inclusion of bundles

$$\iota : \underline{\Delta}^\pm \otimes \underline{S}_q \otimes 1 \longrightarrow \underline{\Delta}^\pm \otimes \underline{S}_q \otimes (\underline{V}_q \text{ or } \underline{W}_q)$$

inducing the inclusions in the spaces of smooth sections. We show that the twisted graded Dirac operator d_* commute with this inclusion map making the following diagram commutative:

$$
\begin{array}{ccc}
\Gamma(\underline{\Delta}^+ \otimes \underline{S}_q) & \overset{\iota}{\longrightarrow} & \Gamma(\underline{\Delta}^+ \otimes \underline{S}_q \otimes \underline{V}_q) \\
\downarrow{\scriptstyle d_*} & & \downarrow{\scriptstyle d'_*} \\
\Gamma(\underline{\Delta}^- \otimes \underline{S}_q) & \overset{\iota}{\longrightarrow} & \Gamma(\underline{\Delta}^- \otimes \underline{S}_q \otimes \underline{V}_q)
\end{array}
$$

Tosee this, let a section $s \in \Gamma(\underline{\Delta}^+ \otimes \underline{S}_q)$ be of the form $s = s_1 \otimes s_2$ where $s_1 \in \Gamma(\underline{\Delta}^+)$ and $s_2 \in \Gamma(\underline{S}_q)$. We choose an orthonormal basis $\{v_1, \ldots, v_{2N}\}$ at $x \in M$. Then,

$$\left(\iota \circ d_*(s)\right)_x = \iota\left(\sum_{i=1}^{2N} v_i \cdot \nabla_{v_i}(s_1 \otimes s_2)\right) = \sum_{i=1}^{2N}(v_i \cdot \nabla_{v_i} s_1) \otimes s_2 \otimes 1 + \sum_{i=1}^{2N} v_i \cdot s_1 \otimes \nabla_{v_i} s_2 \otimes 1.$$

On the other hand,

$$(d'_* \circ \iota(s))_x = d'_*(s_1 \otimes s_2 \otimes 1) = \sum_{i=1}^{2N} v_i \cdot \nabla_{v_i}(s_1 \otimes s_2 \otimes 1)$$

$$= \sum_{i=1}^{2N}(v_i \cdot \nabla_{v_i} s_1) \otimes s_1 \otimes 1 + \sum_{i=1}^{2N}(v_i \cdot s_1) \otimes \nabla_{v_i} s_2 \otimes 1.$$

Hence, the above diagram commutes for the sections of the form $s = s_1 \otimes s_2$. Consequently, the above diagram commutes on the dense subspace consisting of sections of the form $\sum_i s_{1,i} \otimes s_{2,i}$ where $s_{1,i} \in \Gamma(\underline{\Delta}^+)$ and $s_{2,i} \in \Gamma(\underline{S}_q)$. In general, if $s \in \Gamma(\underline{\Delta}^+ \otimes \underline{S}_q)$ is a section on which d_* is defined, then, $\iota \circ d_*(s) = d_*(s) \otimes 1$ and $d'_* \circ \iota(s) = d'_*(s \otimes 1) = d_*(s) \otimes 1$ and the above diagram commutes.

Hence, by looking at the kernels and the cokernels, we have $\text{Ker}\, d_* \subset \text{Ker}\, d'_*$, $\text{Coker}\, d_* \subset \text{Coker}\, d'_*$, and it follows that $\Phi_\theta(M) \subset \Phi^*_{\text{ell}}(M)$. $\square$

Self-adjointness of twisted Dirac operators. We show that the twisted Dirac operators we are considering are formally self-adjoint. To show this, we need to recall the construction of the Spin representations of the group $\mathrm{Spin}(2N)$.

Let $(E^{2N}; \langle\ ,\ \rangle)$ be a real $2N$-dimensional Euclidean vector space. We can construct a Spin representation by choosing an isometric complex structure in E. Let I be such a complex structure. Let $\{e_1, \ldots, e_N, e'_1, \ldots, e'_N\}$ be an orthonormal basis of E such that $I(e_j) = e'_j$ for $1 \le j \le N$. Let $a_j = (e_j - ie'_j)/\sqrt{2}$, $a_j^* = (e_j + ie'_j)/\sqrt{2}$. Then, $A^+ = \bigoplus_{j=1}^N \mathbb{C}a_j$, $A^- = \bigoplus_{j=1}^N \mathbb{C}a_j^*$ are $\pm i$-eigenspaces of I of the complexification $A = E \otimes \mathbb{C}$. The sets $\{a_1, \ldots, a_N\}$, $\{a_1^*, \ldots, a_N^*\}$ are Hermitian orthonormal basis for A^+ and A^- with respect to the Hermitian pairing $(\ ,\ )$ defined by $(a, b) = \langle \bar{a}, b \rangle$ where $\langle\ ,\ \rangle$ is the $\mathbb{C}$-linear extension of $\langle\ ,\ \rangle$ to A. With respect to $\langle\ ,\ \rangle$, $A^\pm$ are maximal isotropic subspaces of A and $\langle a_j, a_k^* \rangle = \delta_{jk}$.

As the Spin representation Δ, we take the total exterior power of A^-. That is, $\Delta = \bigwedge^* A^-$. On Δ, vectors in A^- act by exterior multiplication, and vectors in A^+ act by derivation with respect to the $\mathbb{C}$-linear pairing $\langle\ ,\ \rangle$. Let $J = (j_1, j_2, \ldots, j_r)$ be a sequence of distinct integers between 1 and N, where $0 \le r \le N$. When $r = 0$, we set $J = \emptyset$. To each such J, we associate a vector a_J^* in Δ given by $a_J^* = a_{j_1}^* \cdot a_{j_2}^* \cdots a_{j_r}^*$. We let $a_\emptyset^* = 1$. When $j \notin J$, we let $a_{\{j\}\cup J}^* = a_j^* \cdot a_J^*$. When $j \in J$, we let $a_{J-\{j\}}^* = a_{j_1}^* \cdots \widehat{a_j^*} \cdots a_{j_r}^*$. The total Spin representation has an induced Hermitian pairing given by $(a_I^*, a_J^*) = \mathrm{sgn}(I, J)\delta_{\{I\},\{J\}}$. Here, $\mathrm{sgn}(I, J) = 0$ if $\{I\} \ne \{J\}$ as sets and $\mathrm{sgn}(I, J) = \mathrm{sgn}(\sigma)$ if I is a permutation of J induced by $\sigma \in \mathfrak{S}_r$. Also, $\delta_{\{I\},\{J\}} = 1$ only when we have $\{I\} = \{J\}$ as sets. The set $\{a_J^*\}$ where J runs through all the subsequences of $0 < 1 < 2 < \cdots < N$ form a Hermitian orthonormal basis of Δ. The even exterior part Δ^+ and odd exterior part Δ^- are preserved by the action of the group $\mathrm{Spin}(2N)$ and they are the irreducible half Spin representations.

First, we prove a formula that describes the effect of Clifford multiplication by a vector $v \in A = E \otimes \mathbb{C}$ on vectors in Δ in terms of Hermitian pairings.

Recall that for $v_1, v_2 \in A$, the Clifford multiplication has the property that $v_1 \cdot v_2 + v_2 \cdot v_1 = \langle v_1, v_2 \rangle$ in the Clifford algebra $\mathrm{Cliff}\, A$. In particular, $v \cdot v = \frac{1}{2}\langle v, v \rangle$ for $v \in A$. For example, if $e \in E$ is a unit vector $|e| = 1$, then $e \cdot e = \frac{1}{2}$.

The complex vector space $A = E \otimes \mathbb{C}$ contains not only E but also $iE = E \otimes i$. We denote this real vector subspace by iE. Note that the $\mathbb{C}$-linear pairing $\langle\ ,\ \rangle$ on A is positive definite on E and it is negative definite on iE.

LEMMA 5. *Let $v \in E$ or $v \in iE$. Then, for any $\sigma_1, \sigma_2 \in \Delta$, the Clifford multiplication by v has the property*

$$(16) \qquad (v \cdot \sigma_1, v \cdot \sigma_2) = |v \cdot v|(\sigma_1, \sigma_2) = \frac{1}{2}|\langle v, v \rangle|(\sigma_1, \sigma_2) = \frac{1}{2}\langle v, v \rangle(\sigma_1, \sigma_2).$$

Here $v \cdot v \ge 0$ if $v \in E$, and $v \cdot v \le 0$ if $v \in iE$. In particular,

$$(17) \qquad (v \cdot \sigma_1, \sigma_2) = \begin{cases} (\sigma_1, v \cdot \sigma_2), & \text{if } v \in E, \\ -(\sigma_1, v \cdot \sigma_2), & \text{if } v \in iE. \end{cases}$$

PROOF. Since the Hermitian pairing is sesquilinear, we only have to check the above identities when σ_1, σ_2 are basis vectors of Δ described above. First, we examine the effect of Clifford multiplication by elements of A on Δ. Vectors in A^- acts on Δ by exterior multiplication. So, if $j \in J$, then $a_j^* \cdot a_J^* = 0$. If $j \notin J$, then $a_j^* \cdot a_J^* = a_{\{j\} \cup J}^*$. Next, vectors in A^+ acts on Δ by derivation. If $j \in J$, then $a_j \cdot a_J^* = (-1)^t a_{J-\{j\}}^*$ if $J = (j_1, \ldots, j_{t+1} = j, \ldots, j_r)$. If $j \notin J$, then $a_j a_J^* = 0$. Since $e_j = (a_j + a_j^*)/\sqrt{2}$ for $1 \leq j \leq N$, from the above calculations,

$$\sqrt{2} e_j \cdot a_J^* = \begin{cases} (-1)^t a_{J-\{j\}}^*, & \text{if } j \in J, \\ a_{\{j\} \cup J}^*, & \text{if } j \notin J. \end{cases}$$

Thus, by Clifford multiplication by $\sqrt{2} e_j$, the Hermitian orthonormal basis vectors a_J^* in Δ are mapped into another Hermitian orthonormal basis vectors. Similarly, since $\sqrt{2} e_j' = i(a_j - a_j^*)$ for $1 \leq j \leq N$, letting $e_{j+N} = e_j'$, we have

$$\sqrt{2} e_{j+N} \cdot a_J^* = \begin{cases} i(-1)^t a_{J-\{j\}}^*, & \text{if } j \in J, \\ -i a_{\{j\} \cup J}^*, & \text{if } j \notin J. \end{cases}$$

Thus, for basis vectors e_j, we have $(e_j \cdot \sigma_1, e_j \cdot \sigma_2) = \frac{1}{2}(\sigma_1, \sigma_2)$ for $1 \leq j \leq 2N$. Using this relation, for $1 \leq j, k \leq 2N$, we have

$$(e_j \cdot \sigma_1, e_k \cdot \sigma_2) + (e_k \cdot \sigma_1, e_j \cdot \sigma_2) = 2(e_j \cdot e_j \cdot \sigma_1, e_j \cdot e_k \cdot \sigma_2) + 2(e_k \cdot e_k \cdot \sigma_1, e_k \cdot e_j \sigma_2)$$
$$= (\sigma_1, e_j \cdot e_k \cdot \sigma_2) + (\sigma_1, e_k \cdot e_j \cdot \sigma_2) = (\sigma_1, \langle e_j, e_k \rangle \sigma_2) = \delta_{jk}(\sigma_1, \sigma_2).$$

In the above, we used $e_j \cdot e_j = \frac{1}{2}$, $e_k \cdot e_k = \frac{1}{2}$, and $e_j e_k + e_k e_j = \langle e_j, e_k \rangle$. Note that this formula contains the above formula by letting $j = k$. Now, we are ready to prove (16). If $v \in E$, we can write it as $v = \sum_{j=1}^{2N} r_j e_j$ for some $r_j \in \mathbb{R}$, $1 \leq j \leq 2N$. Then,

$$(v \cdot \sigma_1, v \cdot \sigma_2) = \sum_{j=1}^{2N} r_j^2 (e_j \cdot \sigma_1, e_j \cdot \sigma_2) + \sum_{j > k} r_j r_k \{(e_j \cdot \sigma_1, e_k \cdot \sigma_2) + (e_k \cdot \sigma_1, e_j \cdot \sigma_2)\}$$

$$= \frac{1}{2} \Big(\sum_{j=1}^{2N} r_j^2 \Big)(\sigma_1, \sigma_2) = \frac{1}{2} \langle v, v \rangle (\sigma_1, \sigma_2) = \frac{1}{2}(v, v)(\sigma_1, \sigma_2).$$

If $v' \in iE$, then we may write $v = i \sum_j r_j e_j$. A similar calculation yields that

$$(v' \cdot \sigma_1, v' \cdot \sigma_2) = \frac{1}{2} \Big(\sum r_j^2 \Big)(\sigma_1, \sigma_2) = -\frac{1}{2} \langle v', v' \rangle (\sigma_1, \sigma_2) = \frac{1}{2}(v', v')(\sigma_1, \sigma_2).$$

Since $v \cdot v = \frac{1}{2} \langle v, v \rangle$, this proves (16). To prove (17), note that (16) implies that for $v \in E$ or $v \in iE$,

$$(v \cdot \sigma_1, \sigma_2) = \frac{1}{|v \cdot v|}(v \cdot v \cdot \sigma_1, v \cdot \sigma_2) = \frac{v \cdot v}{|v \cdot v|}(\sigma_1, v \cdot \sigma_2).$$

Since $v \in E$ or $v \in iE$, $v \cdot v \in \mathbb{R}$. So, in the second equality above, $v \cdot v$ comes out of the Hermitian pairing which is sesquilinear in the first variable. Since $v \cdot v \geq 0$ if $v \in E$ and $v \cdot v \leq 0$ if $v \cdot v \in iE$, (17) follows. $\square$

A slight extension of Lemma 5 is given as follows.

COROLLARY 6. *Let F be any Hermitian vector space. Let the vector space $A = E \otimes \mathbb{C}$ act on the Hermitian vector space $\Delta \otimes F$ on the first factor by Clifford multiplication, where Δ is the total Spin representation for $(E^{2N}, \langle \ , \ \rangle)$. Then, for any $v \in E$ or $v \in iE$, and for any $\sigma_1, \sigma_2 \in \Delta \otimes F$, we have $(v \cdot \sigma_1, v \cdot \sigma_2) = |v \cdot v|(\sigma_1, \sigma_2)$, where $v \cdot v \in \mathbb{R}$.*

Recall that in a Hermitian vector bundle $\underline{F}$, a compatible connection ∇ is the one with the property that $X(s_1, s_2) = (\nabla_X s_1, s_2) + (s_1, \nabla_X s_2)$ for any section s_1, s_2 of $\underline{F}$ and any tangent vector $X \in TM$.

On a Riemannian manifold M^{2N}, the tangent bundle TM is identified with $i\underline{E}$ in the Clifford algebra bundle Cliff $\underline{A}$ so that the pairing on $TM \subset$ Cliff $(\underline{A})$ is negative definite. So, if $v \in TM$, (17) implies that $(v \cdot \sigma_1, \sigma_2) = -(\sigma_1, v \cdot \sigma_2)$ for $\sigma_1, \sigma_2 \in \underline{\Delta}$.

PROPOSITION 7 (SELF-ADJOINTNESS OF TWISTED DIRAC OPERATORS). *Let M^{2N} be a closed Riemannian Spin manifold. Let $\underline{F}$ be a Hermitian vector bundle with a compatible connection. Then, the twisted Dirac operator*

$$(18) \qquad d_{\underline{F}} : \Gamma(\underline{\Delta} \otimes \underline{F}) \to \Gamma(\underline{\Delta} \otimes \underline{F})$$

is formally self-adjoint. That is, $(d_{\underline{F}}\sigma_1, \sigma_2) = (\sigma_1, d_{\underline{F}}\sigma_2)$ for any $\sigma_1 \in \Gamma(\underline{\Delta}^+ \otimes \underline{F})$ and $\sigma_2 \in \Gamma(\underline{\Delta}^- \otimes \underline{F})$.

PROOF. Let $v_1, \ldots, v_{2N}$ be orthonormal vector fields on an open set $U \subset M$. We use "·" to denote the Clifford multiplication. Then, for any sections $\sigma_1 \in \Gamma(\underline{\Delta}^+ \otimes \underline{F})$ and $\sigma_2 \in \Gamma(\underline{\Delta}^- \otimes \underline{F})$, we have

$$(d_{\underline{F}}\sigma_1, \sigma_2) = \sum_{j=1}^{2N}(v_j \cdot \nabla_{v_j}\sigma_1, \sigma_2) = -\sum_{j=1}^{2N}(\nabla_{v_j}\sigma_1), v_j \cdot \sigma_2)$$

$$= \sum_{j=1}^{2N}\left(\sigma_1, \nabla_{v_j}(v_j \cdot \sigma_2)\right) - \sum_{j=1}^{2N}v_j(\sigma_1, v_j \cdot \sigma_2)$$

$$= \sum_{j=1}^{2N}(\sigma_1, v_j \cdot \nabla_{v_j}\sigma_2) + \sum_{j=1}^{2N}\left(\sigma_1, (\nabla_{v_j}v_j) \cdot \sigma_2\right) - \sum_{j=1}^{2N}v_j(\sigma_1, v_j \cdot \sigma_2)$$

as functions on U. Here, we used a fact that the covariant derivative act on Clifford multiplications by derivation. This is because Clifford multiplication is a combination of exterior multiplications and pairings, on both of which the covariant derivative acts by derivation. In the above, $v_j\cdot$ denotes the Clifford multiplication by v_j, and v_j in front of $(\sigma_1, v_j \cdot \sigma_2)$ denotes the derivation of the function $(\sigma_1, v_j \cdot \sigma_2)$ in the direction of v_j.

Let α be a 1-form on M defined by $\alpha(X) = (\sigma_1, X \cdot \sigma_2)$. Here, the second entry is the Clifford multiplication of X regarded as a vector in $i\underline{E}$ on σ_2, and $(\ , \ )$ on the right hand side is the Hermitian pairing in Δ. Then, the last two summations above can be written as

$$\sum_{j=1}^{2N}\alpha(\nabla_{v_j}v_j) - \sum_{j=1}^{2N}v_j\left(\alpha(v_j)\right) = -\sum_{j=1}^{2N}(\nabla_{v_j}\alpha)(v_j) = -\mathrm{Tr}\,(\nabla\alpha) = d^*\alpha,$$

where d^* is the adjoint of the exterior differentiation $d : \Omega^0(M) \to \Omega^1(M)$. Thus, as functions on $U \subset M$, we have

$$(d_{\underline{F}}\sigma_1, \sigma_2) = (\sigma_1, d_{\underline{F}}\sigma_2) + d^*\alpha.$$

Since this expression is independent of U and each objects are globally defined, the above identity holds as functions on M. Since the integral of the divergence $d^*\alpha$ over M vanishes, we have

$$\int_M (d_{\underline{F}}\sigma_1, \sigma_2)\text{vol} = \int_M (\sigma_1, d_{\underline{F}}\sigma_2)\text{vol}$$

for any sections σ_1, σ_2 as above. Here, "vol" is the Riemannian volume form. This shows that the twisted Dirac operator $d_{\underline{F}}$ is formally self-adjoint. $\quad\square$

§5.4   Elliptic genera as modules over vertex operator super algebras of parallel sections

Let $(M^{2N}; g)$ be a closed oriented Riemannian Spin manifold and let $Q(M^{2N})$ be the principal $SO(2N)$-bundle of orthonormal frames on M with the principal Levi-Civita connection. The major emphasis of this paper is that we should view the elliptic genus as a geometric device which associates, to each closed Riemannian Spin manifold M^{2N}, a super-pair of modules $\Phi_{\text{ell}}(M)$ over a vertex operator super algebra $\mathcal{P}_M(\underline{V})_*$ of parallel sections in the graded vector bundle $\underline{V} = Q(M^{2N}) \times_{SO(2N)} V$ equipped with the covariant derivative. See Theorem 7 below. Here, V is a Spin representation of $\hat{o}(2N)$. Schematically,

$$(M^{2N}; g) \xrightarrow{\text{elliptic genera}} \left(\mathcal{P}_M(\underline{V}), \Phi_{\text{ell}}(M^{2N})\right).$$

Let G be the structure group of a holonomy bundle $P_G \subset Q(M)$ on M^{2N}. We will show that the vertex operator super algebra $\mathcal{P}_M(\underline{V})$ is canonically isomorphic to the vertex operator super algebra V^G of G-invariant vectors in V (Theorem 9). This isomorphism depends only on the choice of the holonomy bundle P_G. Different choices of holonomy bundles give rise to isomorphic invariant subspaces of V.

To see why elliptic genus of M^{2N} can possibly have this property, we recall its construction below and note that all of the level 1 Spin representations of the orthogonal affine Lie algebra $\hat{o}(2N)$ are present in the construction of the elliptic genus $\Phi_{\text{ell}}(M^{2N})$, and these Spin representations of $\hat{o}(2N)$ have the structure of a vertex operator super algebra and its modules. After all, elliptic genus is the S^1-equivariant signature of the loop space on the manifold.

The vertex operator super algebra $\mathcal{P}_M(\underline{V})$ of parallel sections in $\underline{V}$. Let M^{2N} be a closed Riemannian Spin manifold. At each point $x \in M$, we have a Euclidean vector space (T_xM, g_x) with respect to the Riemannian metric g_x. The negative definite inner product $-g_x$ can be $\mathbb{C}$-linearly extended to a pairing $\langle \ , \ \rangle_x$ on its complexification $T_x M \otimes \mathbb{C}$. It can be further extended to a nondegenerate pairing on the graded vector space

$$A_x(\mathbb{Z} + \tfrac{1}{2}) = \bigoplus_{n \in \mathbb{Z}} T_x M(-n - \tfrac{1}{2}) \otimes \mathbb{C}$$

by letting

$$\langle v_1(-n - \tfrac{1}{2}), v_2(m + \tfrac{1}{2}) \rangle_x = \langle v_1, v_2 \rangle_x \delta_{m,n},$$

where $v_1, v_2 \in T_x M \otimes \mathbb{C}$. From $(A_x(\mathbb{Z} + \tfrac{1}{2}), \langle \ , \ \rangle_x)$, we can construct an infinite dimensional Clifford algebra $\mathrm{Cliff}_x(\mathbb{Z} + \tfrac{1}{2})$. Its Clifford module V_x is given by

$$(1) \qquad V_x = \bigoplus_{0 \le \ell \in \frac{1}{2}\mathbb{Z}} (V_x)_\ell = \bigotimes_{0 \le n \in \mathbb{Z}} \bigwedge{}^* T_x M(-n - \tfrac{1}{2}) \cdot 1_x.$$

Here 1 is the constant function on M with value 1 and is the "vacuum" vector such that $(V_x)_0 = \mathbb{C} \cdot 1_x$. The vectors in $T_x M(-n - \tfrac{1}{2})$ can be regarded as elements in the Clifford algebra and hence as operators. The graded vector space V_x has a structure of a vertex operator super algebra for each $x \in M$. It also has the structure of a graded Hermitian vector space. The Hermitian structure $(\ , \ )$ on V_x is the one considered in §2.4.

The Clifford algebra $\mathrm{Cliff}_x(\mathbb{Z} + \tfrac{1}{2})$ is a proper subalgebra of the graded algebra of endomorphisms

$$\mathrm{End}(V_x) = \bigoplus_{\ell \in \frac{1}{2}\mathbb{Z}} \mathrm{End}_\ell(V_x) \cong V_x \hat{\otimes} (V_x)^*,$$

where $(V_x)^*$ is the graded dual. In this isomorphism for $\mathrm{End}(V_x)$, we have to use the completed tensor product because a certain kind of infinite sum must be allowed. As a module over the Clifford algebra, vectors in $T_x M(-n - \tfrac{1}{2})$, $n \ge 0$, act on V_x by exterior multiplication, and vectors in $T_x M(m + \tfrac{1}{2})$, $m \ge 0$, act on V_x by derivation. In terms of Clifford multiplication and a completed tensor product, we have

$$(2) \qquad \mathrm{End}(V_x) \cong \bigotimes_{0 \le n \in \mathbb{Z}} \bigwedge{}^* T_{\mathbb{C}} M(-n - \tfrac{1}{2})_x \hat{\otimes} \bigotimes_{0 \le m \in \mathbb{Z}} \bigwedge{}^* T_{\mathbb{C}} M(m + \tfrac{1}{2})_x.$$

Note that for a given homogeneous vector $v \in V_x$, only finitely many summands of equal homogeneous degree in (2) act nontrivially on v via Clifford multiplication. So, the above completed tensor product makes sense as operators on V_x. To see the above isomorphism, we define the following object V_x^* which is a subalgebra of the Clifford algebra $\mathrm{Cliff}_x(\mathbb{Z} + \tfrac{1}{2})$.

$$V_x^* = \bigoplus_{0 \le r \in \frac{1}{2}\mathbb{Z}} (V_x^*)_{(-r)} = \bigotimes_{0 \le m \in \mathbb{Z}} \bigwedge{}^* T_x M(m + \tfrac{1}{2}) \otimes \mathbb{C}.$$

We observe that when the action of a subspace $(V_x)_{r+\ell} \otimes (V_x^*)_{(-r)}$ of the Clifford algebra $\mathrm{Cliff}_x(\mathbb{Z} + \frac{1}{2})$ is restricted to $(V_x)_r \subset V_x$, Clifford multiplication gives an isomorphism from $(V_x)_{r+\ell} \otimes (V_x^*)_{(-r)}$ to $\mathrm{Hom}\left((V_x)_r, (V_x)_{r+\ell}\right)$. From this, by induction using Five Lemmas, we can easily show that for any $0 \le q \in \frac{1}{2}\mathbb{Z}$,

$$\bigoplus_{0 \le r \le q} (V_x)_{r+\ell} \otimes (V_x^*)_{(-r)} \cong \mathrm{Hom}\left(\bigoplus_{0 \le r \le q} (V_x)_r, \bigoplus_{0 \le r \le q} (V_x)_{r+\ell}\right).$$

Taking the limit in q, we see that $\bigoplus_{0 \le r \in \frac{1}{2}\mathbb{Z}}(V_x)_{r+\ell} \otimes (V_x^*)_{(-r)} \cong \mathrm{End}_\ell(V_x, V_x)$. Now, by taking the direct sum over ℓ, (2) follows.

Let $e_1, \ldots, e_{2N}$ be an orthonormal basis of T_xM. Then, in $T_xM(-n - \frac{1}{2}) \otimes \mathbb{C}$, vectors $\{e_1(-n - \frac{1}{2}), \ldots, e_{2N}(-n - \frac{1}{2})\}$ form a Hermitian orthonormal basis for $0 \le n \in \mathbb{Z}$. At each $x \in M$, the graded vector space V_x is spanned by exterior products of vectors of the form $e_{j_1}(-n_1 - \frac{1}{2}) \cdots e_{j_r}(-n_r - \frac{1}{2}) \cdot 1$ for $0 \le n_1, \ldots, n_r \in \mathbb{Z}$ and $1 \le j_1, \ldots, j_r \le 2N$. The collection of these algebras V_x for $x \in M$ forms a *vector bundle of vertex operator super algebras* $\underline{V}$ given by

$$(3) \qquad \underline{V} = \bigoplus_{0 \le \ell \in \frac{1}{2}\mathbb{Z}} (\underline{V})_\ell = \bigotimes_{0 \le n \in \mathbb{Z}} \bigwedge^* T_{\mathbb{C}}M(-n - \tfrac{1}{2}).$$

Its fibre over $x \in M$, $(\underline{V})_x$, is V_x in (1). $(\underline{V})_0$ is a 1-dimensional trivial line bundle generated by the constant function 1_M. If M is connected, then the space of constant functions is 1-dimensional. If M is not connected and has r connected components, then the space of constant sections is an r-dimensional vector space.

Similarly, the collection of Clifford algebras $\mathrm{Cliff}_x(\mathbb{Z} + \frac{1}{2})$ for $x \in M$ gives rise to a *bundle of Clifford algebras* $\mathrm{Cliff}_M(\mathbb{Z} + \frac{1}{2})$. As graded vector bundles, but not as vector bundles of algebras, it is isomorphic to a (not completed) tensor product:

$$(4) \qquad \mathrm{Cliff}_M(\mathbb{Z} + \tfrac{1}{2}) \cong \bigotimes_{0 \le n \in \mathbb{Z}} \bigwedge^* T_{\mathbb{C}}M(-n - \tfrac{1}{2}) \otimes \bigotimes_{0 \le m \in \mathbb{Z}} \bigwedge^* T_{\mathbb{C}}M(m + \tfrac{1}{2}).$$

The bundle of endomorphism algebras $\mathrm{End}(\underline{V}) \cong \underline{V} \hat{\otimes} (\underline{V})^*$, *where* $(\underline{V})^*$ *is the graded dual of* $\underline{V}$, *properly contains the bundle of Clifford algebras* (4). From (2), the endomorphism bundle is given by the completed tensor product

$$(5) \qquad \mathrm{End}(\underline{V}) \cong \bigotimes_{0 \le n \in \mathbb{Z}} \bigwedge^* T_{\mathbb{C}}M(-n - \tfrac{1}{2}) \hat{\otimes} \bigotimes_{0 \le m \in \mathbb{Z}} \bigwedge^* T_{\mathbb{C}}M(m + \tfrac{1}{2}).$$

The Levi-Civita connection on the tangent bundle TM induces a connection on $Q(M)$, and then on the following vector bundles

$$\underline{V} = Q(M) \times_{\mathrm{SO}(2N)} V, \qquad \mathrm{End}(\underline{V}) = Q(M) \times_{\mathrm{SO}(2N)} \mathrm{End}(V).$$

The associated parallel transports in $\underline{V}$ and in $\mathrm{Cliff}_M(\mathbb{Z} + \frac{1}{2}) \subset \mathrm{End}(\underline{V})$ can be described in the following way. Let $\gamma : [0, 1] \to M$ be a piecewise smooth curve in M such that $\gamma(0) = x$. Let $\gamma_t : [0, t] \to M$ be its restriction on $[0, t]$. Let the parallel transport induced by γ_t be denoted by $\tau_{\gamma_t} : T_x M \to T_{\gamma(t)} M$. This is a linear isomorphism. The parallel transport in the bundle $TM(-n - \frac{1}{2})$ of weight $n + \frac{1}{2}$, $\tau_{\gamma_t} : T_x M(-n - \frac{1}{2}) \to T_{\gamma(t)} M(-n - \frac{1}{2})$, is essentially the same as the case of TM and is given by the formula $\tau_{\gamma_t}\left(e_j(-n - \frac{1}{2})\right) = \tau_{\gamma_t}(e_j)(-n - \frac{1}{2})$ for $1 \le j \le 2N, n \in \mathbb{Z}$. Since parallel transports commute with wedge products and tensor products of vectors, we obtain parallel transports $\tau_{\gamma_t} : (\underline{V})_x \to (\underline{V})_{\gamma(t)}$ and $\tau_{\gamma_t} : \mathrm{End}(\underline{V})_x \to \mathrm{End}(\underline{V})_{\gamma(t)}$ in view of (3), (4), (5). More precisely, on a basis vector of the form

$$v = e_{j_1}(-n_1 - \tfrac{1}{2}) \cdots e_{j_r}(-n_r - \tfrac{1}{2}) e_{k_1}(m_1 + \tfrac{1}{2}) \cdots e_{k_s}(m_s + \tfrac{1}{2}),$$

with $n_1, \ldots, n_r, m_1, \ldots, m_s \ge 0$, τ_{γ_t} is described by

(6)
$$\begin{aligned}
\tau_{\gamma_t}(v) &= \tau_{\gamma_t}\left(e_{j_1}(-n_1 - \tfrac{1}{2})\right) \cdots \tau_{\gamma_t}\left(e_{k_s}(m_s + \tfrac{1}{2})\right) \\
&= \tau_{\gamma_t}(e_{j_1})(-n_1 - \tfrac{1}{2}) \cdots \tau_{\gamma_t}(e_{k_r})(m_r + \tfrac{1}{2}).
\end{aligned}$$

Note that if vectors from $T_{\mathbb{C}} M(m + \frac{1}{2})$ for $m > 0$ are absent, then $v \in \underline{V}$. Otherwise, the above v is regarded as a vector in $\mathrm{Cliff}_M(\mathbb{Z} + \frac{1}{2}) \subset \mathrm{End}(\underline{V})$.

Now, we consider a relation between parallel transport and vertex operators $Y(v, \zeta)$ acting on $\underline{V}$ for $v \in \underline{V}$. Strictly speaking, for each $n \in \mathbb{Z}$, the vertex operator $\{v\}_n$ is an infinite sum of vectors in the Clifford algebra bundle $\mathrm{Cliff}_M(\mathbb{Z} + \frac{1}{2})$. So operators $\{v\}_n$ are not vectors in the bundle $\mathrm{Cliff}_M(\underline{V})$, but instead they are in $\mathrm{End}(\underline{V})$. For the purpose of the parallel transport, it is enough to consider it in each summand of (5). From the above calculation (6), the vertex operators corresponding to $\tau_{\gamma_t}(v)$ is given by

$$Y\left(\tau_{\gamma_t}(v), \zeta\right) =: \tau_{\gamma_t}(e_{j_1})^{(n_1)}(\zeta) \cdots \tau_{\gamma_t}(e_{j_k})^{(n_k)}(\zeta) : .$$

From the definition of operators $e^{(n)}(\zeta)$, we have

$$\begin{aligned}
\left(\tau_{\gamma_t}(e_j)\right)^{(n)}(\zeta) &= \frac{1}{n!} \frac{d^n}{d\zeta^n}\left(\sum_{m \in \mathbb{Z}} \tau_{\gamma_t}(e_j)(m + \tfrac{1}{2}) \zeta^{-m-1}\right) \\
&= \tau_{\gamma_t}\left\{\frac{1}{n!} \frac{d^n}{dz\zeta^n}\left(\sum_{m \in \mathbb{Z}} e_j(m + \tfrac{1}{2}) \zeta^{-m-1}\right)\right\} = \tau_{\gamma_t}\left(e_j^{(n)}(\zeta)\right).
\end{aligned}$$

Thus, the above vertex operator acting on the fibre $(\underline{V})_{\gamma(t)}$ then becomes

(7) $$Y\left(\tau_{\gamma_t}(v), \zeta\right) = \tau_{\gamma_t}\left(: e_{j_1}^{(n_1)}(\zeta) \cdots e_{j_k}^{(n_k)}(\zeta) :\right) = \tau_{\gamma_t}\left(Y(v, \zeta)\right).$$

This shows that parallel transport τ_{γ_t} and vertex operator map commute on basis vectors of $\underline{V}$. Since the vertex operator map $Y(\ , \zeta)$ acting on $\underline{V}$ and the parallel transport map τ_{γ_t} are linear maps on vectors in $\underline{V}$, the formula (7) holds for any vector in $(\underline{V})_x$.

Similarly, we can consider a *bundle of modules* over the bundle of vertex operator super algebras $\underline{V}$ given by

$$(8) \qquad \underline{W} = \Delta(TM) \otimes \bigotimes_{0 < m \in \mathbb{Z}} \bigwedge{}^* T_{\mathbb{C}}M(-m).$$

Here, $\Delta(TM)$ is the complex 2^N-dimensional total Spin bundle over M. The above graded vector bundle $\underline{W}$ is a module over the bundle of Clifford algebras

$$(9) \qquad \text{Cliff}_M(\mathbb{Z}) = \text{Cliff}(TM) \otimes \bigotimes_{0 < m \in \mathbb{Z}} \bigwedge{}^* T_{\mathbb{C}}M(-m) \otimes \bigotimes_{0 < m \in \mathbb{Z}} \bigwedge{}^* T_{\mathbb{C}}M(m).$$

As before, in terms of Clifford multiplication, the bundle of endomorphisms of $\underline{W}$, $\text{End}(\underline{W}) = \bigoplus_{\ell \in \mathbb{Z}} \text{End}_\ell(\underline{W})$ is isomorphic to a completed tensor product

$$(10) \qquad \text{End}(\underline{W}) \cong \text{Cliff}(TM) \hat{\otimes} \bigotimes_{0 < m \in \mathbb{Z}} \bigwedge{}^* T_{\mathbb{C}}M(-m) \hat{\otimes} \bigotimes_{0 < m \in \mathbb{Z}} \bigwedge{}^* T_{\mathbb{C}}M(m).$$

These bundles $\underline{W}$, $\text{Cliff}_M(\mathbb{Z}) \subset \text{End}(\underline{W})$ have connections induced from the Riemannian metric via

$$\underline{W} = \tilde{Q}(M) \times_{\text{SO}(2N)} W, \qquad \text{End}(\underline{W}) = \tilde{Q}(M) \times_{\text{SO}(2N)} \text{End}(W),$$

where $\tilde{Q}(M)$ is the principal $\text{Spin}(2N)$ bundle with the Levi-Civita connection. For vertex operators $Y(v, \zeta)$, $v \in \underline{V}$, acting on $\underline{W}$, similar argument as above using the induced Levi-Civita connection shows that the formula (7) above also holds in $\text{End}(\underline{W})$.

PROPOSITION 1 (PARALLEL TRANSPORTS OF VERTEX OPERATORS). *On a Riemannian manifold* M^{2N}, *let* $\gamma : [0,1] \to M$ *be a piecewise smooth path. Then, parallel transports of vertex operators in* $\text{End}(\underline{V})$ *or in* $\text{End}(\underline{W})$ *are vertex operators of parallel transports in* $\underline{V}$. *That is, for any vector* $v \in (\underline{V})_x$, *we have*

$$(11) \qquad Y\left(\tau_\gamma(v), \zeta\right) = \tau_\gamma\left(Y(v, \zeta)\right).$$

In other words, $\{\tau_\gamma(v)\}_n = \tau_\gamma(\{v\}_n)$ *for* $n \in \mathbb{Z}$. *Here, the vertex operator* $Y(v, \zeta)$ *acts on* $\underline{V}$ *and on* $\underline{W}$.

Vector bundles $\underline{V}$, $\text{End}(\underline{V})$, $\text{End}(\underline{W})$ have covariant derivatives ∇ induced from the Levi-Civita connection on the tangent bundle TM. The infinitesimal form of Proposition 1 in terms of ∇ is given as follows.

COROLLARY 2 (COVARIANT DERIVATIVES OF VERTEX OPERATORS). *Let* M *be a Riemannian manifold and let* ∇ *be the associated covariant derivative extended to* $\underline{V}$, $\text{End}(\underline{V})$, $\text{End}(\underline{W})$. *Then, for any smooth section* $\sigma \in \Gamma(\underline{V})$ *and for any vector field* X *on* M, *we have*

$$(12) \qquad Y\left(\nabla_X(\sigma), \zeta\right) = \nabla_X\left(Y(\sigma, \zeta)\right).$$

In other words, $\{\nabla_X(\sigma)\}_n = \nabla_X(\{\sigma\}_n)$ *for* $n \in \mathbb{Z}$.

PROOF. We consider a smooth curve $\gamma(t)$ starting from $x \in M$ with $\dot{\gamma}(0) = X_x$. From (11), we have $Y(\tau_{\gamma_t}^{-1}(\sigma(\gamma_t(x))), \zeta) = \tau_{\gamma_t}^{-1}(Y(\sigma(\gamma_t(x)), \zeta))$ over $x \in M$. We then take the derivative at $t = 0$. $\square$

COROLLARY 3. *Let* $\sigma \in \mathcal{P}_M(\underline{V})$ *and let* $Y(\sigma, \zeta) = \sum_{n \in \mathbb{Z}} \{\sigma\}_n \zeta^{-n-1}$ *be the corresponding vertex operator. Then, as sections of vector bundles* $\mathrm{End}(\underline{V})$, $\mathrm{End}(\underline{W})$, *the vertex operators* $\{\sigma\}_n$ *are also parallel, that is,* $\nabla(\{\sigma\}_n) = 0$ *for* $n \in \mathbb{Z}$ *in* $\mathrm{End}(\underline{V})$, $\mathrm{End}(\underline{W})$.

PROOF. We let $\nabla(\sigma) = 0$ in (12) above. Then, we get $\nabla Y(\sigma, \zeta) = 0$ and the coefficient of ζ^{-n-1} gives $\nabla(\{\sigma\}_n) = 0$. $\square$

The graded vector spaces of parallel sections $\mathcal{P}_M(\mathrm{End}(\underline{V}))$, $\mathcal{P}_M(\mathrm{End}(\underline{V}))$ in the endomorphism bundle $\mathrm{End}(\underline{V})$, $\mathrm{End}(\underline{W})$ are determined by their values at any point $x \in M$ as for the case of $\mathcal{P}_M(\underline{V})$.

PROPOSITION 4 (PARALLEL BUNDLE MAPS). *Let* M^{2N} *be a connected Riemannian Spin manifold. The evaluation of parallel sections and parallel endomorphism at any point* $x \in M$ *induces isomorphisms of graded vector spaces*

$$\mathcal{P}_M(\underline{V}) \xrightarrow{\cong} (V_x)^{G_x} \xleftarrow{\cong} (V)^{G_u},$$

(13)
$$\mathcal{P}_M(\mathrm{End}(\underline{V})) \xrightarrow{\cong} \mathrm{End}_{G_x}(\underline{V})_x \xleftarrow{\cong} \mathrm{End}_{G_u}(V),$$

$$\mathcal{P}_M(\mathrm{End}(\underline{W})) \xrightarrow{\cong} \mathrm{End}_{G_x}(\underline{W})_x \xleftarrow{\cong} \mathrm{End}_{G_u}(W),$$

where G_x *is the holonomy group at* $x \in M$ *and* G_u *is the holonomy group with respect to a frame* u *such that* $\pi(u) = x$.

PROOF. This follows by applying Proposition 23 in §4.5 to each finite dimensional homogeneous subbundle which is preserved by parallel transports. Note that for any G-representation E, $\left(\mathrm{End}(E)\right)^G = \mathrm{End}_G(E)$. Note also that since M is connected, any point in M can be connected to $x \in M$ by a piecewise smooth path. $\square$

Both covariant derivative ∇ and vertex operators $\{\sigma\}_n$ for $\sigma \in \Gamma(\underline{V})$ are operators acting on the spaces of sections of the bundles $\underline{V}$, $\underline{W}$. When σ is parallel, these operators actually commute.

PROPOSITION 5 (PARALLEL VERTEX OPERATORS). *Let the vertex operator corresponding to a parallel section* $\sigma \in \mathcal{P}_M(\underline{V})$ *be* $Y(\sigma, \zeta) = \sum_{n \in \mathbb{Z}} \{\sigma\}_n \zeta^{-n-1}$. *Then, as operators acting on* $\Gamma(\underline{V})$, *the covariant derivative* ∇ *and the vertex operators* $\{\sigma\}_*$ *commute, i.e.* $[\nabla, \{\sigma\}_n] = 0$ *for any* $n \in \mathbb{Z}$. *In particular,* $\{\sigma\}_n$ *induces a map* $\{\sigma\}_n : \mathcal{P}_M(\underline{V}) \to \mathcal{P}_M(\underline{V})$ *between the graded vector space of parallel sections. Thus, we have an injective map*

(14)
$$Y(\,,\zeta) : \mathcal{P}_M(\underline{V}) \to \mathrm{End}\,(\mathcal{P}_M(\underline{V}))\,[[\zeta, \zeta^{-1}]].$$

Furthermore, if $\sigma_1, \sigma_2 \in \mathcal{P}_M(\underline{V})$, *then the associated vertex operators are parallel endomorphisms in* $\underline{V}$ *and they satisfy Jacobi identity:*

(15)
$$\sum_{0 \le i \in \mathbb{Z}} (-1)^i \binom{r}{i} \left(\{\sigma_1\}_{m+r-i} \{\sigma_2\}_{n+i} - (-1)^{|\sigma_1||\sigma_2|+r} \{\sigma_2\}_{n+r-i} \{\sigma_1\}_{m+i} \right)$$
$$= \sum_{0 \le k \in \mathbb{Z}} \binom{m}{k} \{\{\sigma_1\}_{r+k} \sigma_2\}_{m+n-k},$$

where $|\sigma_i| = 2\mathrm{wt}(\sigma_i) \mod 2$ *for* $i = 1, 2$. *The action of vertex operators* $\{\sigma\}_n$ *on the graded vector space* $\mathcal{P}_M(\underline{V})$ *is fibrewise and for* $\eta \in \mathcal{P}_M(\underline{V})$, *it is given by* $(\{\sigma\}_n \eta)_x = \{\sigma_x\}_n \eta_x = [u, \{\overline{\sigma}\}_n \overline{\eta}]$, *where* $\sigma = [u, \overline{\sigma}], \eta = [u, \overline{\eta}]$ *in* $P_G \times_G V$ *for the same* $u \in (P_G)_x$.

Similar statements hold when vertex operator $Y(\sigma, \zeta)$ *for* $\sigma \in \mathcal{P}_M(\underline{V})$ *acts on* $\Gamma(\underline{W})$ *and we have a map*

$$(16) \qquad Y(\ , \zeta) : \mathcal{P}_M(\underline{V}) \to \mathrm{End}(\mathcal{P}_M(\underline{W}))[[\zeta^{\frac{1}{2}}, \zeta^{-\frac{1}{2}}]]$$

satisfying the Jacobi identity above, although this map is not necessarily injective.

PROOF. We recall that the covariant differentiation ∇ and the contractions (=dual pairings and Riemannian pairings) commute. And ∇ acts by derivation on exterior products. Since the action of a vertex operator $Y(v_1, \zeta)$ for $v_1 \in \Gamma(\underline{V})$ on $v_2 \in \Gamma(\underline{V}), \Gamma(\underline{W})$ is a fibrewise operation and it can be expressed in terms of exterior multiplications and dual pairings in $\underline{V}$ or in $\underline{W}$, we see that the covariant derivative acts as a derivation on the action of vertex operators. That is, $\nabla(\{v_1\}_n(v_2)) = (\nabla\{v_1\}_n)(v_2) + \{v_1\}_n(\nabla(v_2))$. If $v_1 = \sigma$ is a parallel section, then $\nabla\{\sigma\}_n = 0$ due to Corollary 3 and the first term vanishes, and ∇ and $\{\sigma\}_n$ commute. So, if $\sigma_1, \sigma_2 \in \mathcal{P}_M(\underline{V})$, we have $\nabla(\{\sigma_1\}_n \sigma_2) = \{\sigma_1\}_n(\nabla\sigma_2) = 0$ and $\{\sigma_1\}_n \sigma_2 \in \mathcal{P}_M(\underline{V})$ for any $n \in \mathbb{Z}$. This shows that vertex operators $\{\sigma\}_n$ preserve $\mathcal{P}_M(\underline{V})$. For the injectivity of the vertex operator map, if $\{\sigma\}_n = 0$ for all $n \in \mathbb{Z}$ and for some $\sigma \in \mathcal{P}_M(\underline{V})$, then, $\sigma = \{\sigma\}_{-1} \cdot 1_M = 0$ from the reproducing property of vertex operators.

For the Jacobi identity, first note that all objects in this formula are parallel endomorphisms. So, we can restrict them to any point $x \in M$. Then, the above commutation relation follows from that of the vertex operator super algebra V_x.

The local formula for the action of the vertex operator $\{\sigma\}_n$ on $\mathcal{P}_M(\underline{V})$ is given through the identity $V_x = (P_G)_x \times_G V$.

The similar argument applies for $\underline{W}$ making $\mathcal{P}_M(\underline{W})$ into a module over the vertex operator super algebra $\mathcal{P}_M(\underline{V})$. $\square$

Recall that the vector $\omega \in (V)_2$ given in (9) of §2.3 generates a rank N Virasoro algebra. From §5.2, at each point $x \in M$ we have $[g_{(1,0)}]_x = -2\omega_x \in (V_x)_2$. Thus, the generalized Riemannian tensor $[g_{(1,0)}]$ is of central importance in $\mathcal{P}_M(\underline{V})$, as the next corollary shows.

COROLLARY 6. *Let* M^{2N} *be a Riemannian Spin manifold. The vertex operator* $Y(-[g_{(1,0)}]/2, \zeta) = \sum_{n \in \mathbb{Z}} D(n)\zeta^{-n-2}$ *corresponding to the generalized Riemannian tensor* $-\frac{1}{2}[g_{(1,0)}] \in \mathcal{P}_M(\underline{V})$ *commutes with* ∇ *as operators on* $\Gamma(\underline{V}), \Gamma(\underline{W})$ *and induces maps*

$$(17) \qquad \begin{aligned} D(n) &: \mathcal{P}_M(\underline{V}) \to \mathcal{P}_M(\underline{V}), \\ D(n) &: \mathcal{P}_M(\underline{W}) \to \mathcal{P}_M(\underline{W}), \end{aligned}$$

for any $n \in \mathbb{Z}$ *satisfying the rank* N *Virasoro relations:*

$$(18) \qquad [D(m), D(n)] = (n - m)D(m + n) + \frac{m(m^2 - 1)}{12}\delta_{m, -n} N \cdot \mathrm{Id}_V.$$

PROOF. (17) is a consequence of Proposition 5. The action of vertex operators is fibrewise, and in $(\underline{V})_x = V_x$ for $x \in M$ the vector $-\frac{1}{2}[g_{(1,0)}]_x \in (V_x)_2$ generates a rank N Virasoro algebra acting on V_x and on W_x satisfying the above commutation relations. Proposition 4 globalizes this commutation relation to obtain (18). $\square$

THEOREM 7. *Let* M^{2N} *be a Riemannian Spin manifold. The graded vector space* $\mathcal{P}_M(\underline{V})$ *of parallel sections in the bundle* $\underline{V}$ *has a structure of a vertex operator super algebra of rank* N. *The graded vector space* $\mathcal{P}_M(\underline{W})$ *is the* ($\mathbb{Z}_2$-*twisted*) *module of* $\mathcal{P}_M(\underline{V})$.

PROOF. The existence of vertex operators $Y(\sigma, \zeta)$ for $\sigma \in \mathcal{P}_M(\underline{V})$ acting on $\mathcal{P}_M(\underline{V})$ and on $\mathcal{P}_M(\underline{W})$ satisfying the Jacobi identity is shown in Proposition 5. The canonical Virasoro element in $\mathcal{P}_M(\underline{V})$ is given by $-\frac{1}{2}[g_{(1,0)}] \in \mathcal{P}_M(\underline{V})_2$. The Corollary 6 shows that this vector generates rank N Virasoro algebra (18) above acting on $\mathcal{P}_M(\underline{V})$, $\mathcal{P}_M(\underline{W})$. Other requirements to be a vertex operator super algebra follow by globalizing the corresponding local versions in V_x using Proposition 4. This proves Theorem 7. $\square$

The structure of the graded vector space $\mathcal{P}_M(\underline{V})$ depends only on the holonomy group of the Riemannian Spin manifold M^{2N}. Let $G \subset \mathrm{Spin}(2N)$ be the structure group of a holonomy bundle $P_G \subset \tilde{Q}(M)$ on M, where $\tilde{Q}(M)$ is the bundle of Spin frames on M with a connection induced from the Levi-Civita connection. The G-principal bundle P_G inherits a connection from $\tilde{Q}(M)$ and provides a set of preferred frames in each tangent space $T_x M$.

PROPOSITION 8. *Let* M *be a Riemannian Spin manifold. Let* $G \subset \mathrm{Spin}(2N)$ *be the structure group of a holonomy bundle* P_G. *Then, for any* G-*invariant vector* $v \in V^G$, *the corresponding vertex operator* $Y(v, z) = \sum_{n \in \mathbb{Z}} \{v\}_n z^{-n-1}$ *induces smooth bundle maps*

$$\{v\}_n : \underline{V} \to \underline{V}, \qquad \{v\}_n : \underline{W} \to \underline{W}, \quad for\ n \in \mathbb{Z}$$

given for $\eta \in (\underline{V})_x$, (*or for* $\eta \in (\underline{W})_x$) *by* $\{v\}_n(\eta) = [u, \{v\}_n(\overline{\eta})]$, *where* $\eta = [u, \overline{\eta}] \in \underline{V} = P_G \times_G V$ (*or* $\in \underline{W} = P_G \times_G W$), $u \in (P_G)_x$, $\overline{\eta} \in V$ (*or* $\in W$, *respectively*). *As a consequence, it also induces maps between the spaces of smooth sections*

(19) $\{v\}_n : \Gamma(\underline{V}) \to \Gamma(\underline{V})$, $\qquad \{v\}_n : \Gamma(\underline{W}) \to \Gamma(\underline{W})$, *for* $n \in \mathbb{Z}$,

preserving the subspaces of parallel sections, giving rise to a module structure in $\mathcal{P}_M(\underline{V})$, $\mathcal{P}_M(\underline{W})$ *over a vertex operator super algebra* V^G.

PROOF. The existence of the holonomy bundle $P_G \subset \tilde{Q}(M)$ dictates which frames to use at each point $x \in M$ in a coherent way over M. We first show that the above definition of the map $\{v\}_n$, $n \in \mathbb{Z}$, is well defined and depends only on the choice of the holonomy bundle P_G, that is, the definition is independent of the choice of the representatives $[u, \overline{\eta}]$ of η such that $u \in P_G$. Let $u \cdot g \in (P_G)_x$

be another G-frame at $x \in M$ for $g \in G$. Then, the vector $\eta \in \underline{V}$ can also be written as $\eta = [ug, g^{-1}\overline{\eta}]$. Since the vertex operator $\{v\}_n$, $n \in \mathbb{Z}$, acting on V commutes with the action of $g \in G$ due to Lemma 1 in §3.1, in terms of this new representative, the action can be described by

$$\{v\}_n(\eta) = [ug, \{v\}_n(g^{-1} \cdot \overline{\eta})] = [ug, g^{-1} \cdot \{v\}_n(\overline{\eta})] = [u, \{v\}_n(\overline{\eta})].$$

This shows that the definition of bundle maps $\{v\}_n$ on $\underline{V}$ or on $\underline{W}$ is independent of the choice of the representative and depends only on the holonomy bundle P_G. This defines the bundle map fiberwise. The vertex operator $\{v\}_n$ defines a smooth bundle map mapping smooth sections to smooth sections since the holonomy bundle P_G is a smooth fibre bundle and we can choose local smooth sections of P_G.

Next we show that $\{v\}_n$ maps parallel sections into parallel sections. Let $\eta \in \mathcal{P}_M(\underline{V})$ and let $\gamma : [0,1] \to M$ be a smooth curve such that $\gamma(0) = x$, $\dot{\gamma}(0) = X \in T_x M$. Let $\tilde{\gamma}$ be a horizontal lift of γ to P_G. We let $\tilde{\gamma}(0) = u \in (P_G)_x$. Along γ, η is given by $\eta(t) = [\tilde{\gamma}(t), \overline{\eta}]$ where $\eta_x = [u, \overline{\eta}] \in (\underline{V})_x = (P_G)_x \times_G V$ for some $\overline{\eta} \in V$. Then, along γ, $\{v\}_n\eta = [\tilde{\gamma}(t), \{v\}_n\overline{\eta}]$. From this, we have

$$\nabla_X\left(\{v\}_n\eta\right) = \frac{d}{dt}\tau_{\gamma_t}^{-1}\left(\{v\}_n\eta\right)_x\bigg|_{t=0} = \frac{d}{dt}[u, \{v\}_n\overline{\eta}] = 0.$$

Hence, $\{v\}_n\eta$ is a parallel section and so it belongs to $\mathcal{P}_M(\underline{V})$. Other requirements such as Jacobi identities follow from that for V or for W by working with the representatives of vectors in $\underline{V}$ and in $\underline{W}$. $\square$

THEOREM 9. *Let $G \subset \mathrm{Spin}(2N)$ be the structure group of a holonomy bundle on a Riemannian Spin manifold M^{2N}. Then, the graded vector space of parallel sections $\mathcal{P}_M(\underline{V})$ in the graded vector bundle $\underline{V}$ is a cyclic module over V^G. In fact, there exists a canonical isomorphism*

$$(20) \qquad \{\ \}_{(-1)} : V^G \to \mathcal{P}_M(\underline{V})$$

of vertex operator super algebras given by $\{v\}_{(-1)} = \{v\}_{(-1)} \cdot 1_M$ for $v \in V^G$. Here, 1_M is a constant function on M with value 1 and the operator $\{v\}_{(-1)}$ is as in (19). The above map is essentially the inverse of the restriction map to a fibre.

Through this isomorphism, if $\sigma = \{v\}_{(-1)} \cdot 1_M \in \mathcal{P}_M(\underline{V})$ for $v \in V^G$, the action of $Y(v, \zeta)$ on $\mathcal{P}_M(\underline{V})$ as a module over V^G is the same as the action of $Y(\sigma, \zeta)$ on $\mathcal{P}_M(\underline{V})$ as a vertex operator super algebra.

PROOF. We first show that for any $v \in V^G$, $\{v\}_{(-1)} \cdot 1_M \in \Gamma(\underline{V})$ is a parallel section of $\underline{V}$. Although this follows from Proposition 8 where the vanishing of covariant derivatives are shown, here we prove it using parallel transports. At $x \in M$, as a representative of $1_x \in (\underline{V})_0$, we may take $1_x = [u, 1]$ where u is an arbitrary element in $(P_G)_x$ and $1 \in (V)_0$ is the vacuum vector. Then, from our definition of the action of the vertex operator $\{v\}_n$ on $\underline{V}$ given in Proposition 8,

$$\{v\}_{(-1)} \cdot 1_x = [u, \{v\}_{(-1)}1] = [u, v] \in (\underline{V})_x.$$

Here, we used the reproducing property of vertex operators in V. Note that this is independent of the choice of the representative of the constant function 1_x, i.e., independent of the choice of $u \in (P_G)_x$, since $v \in V^G$ is G-invariant. Let γ be a piecewise smooth curve in M connecting points x and y of M and let τ_γ be the associated parallel transport in the graded vector bundle $\underline{V}$. Since P_G is a holonomy bundle and invariant under parallel transport τ_γ, $\tau_\gamma(u) \in (P_G)_y$ and we have

$$\tau_\gamma\left(\{v\}_{(-1)} \cdot 1_x\right) = \tau_\gamma([u,v]) = [\tau_\gamma(u), v] = [\tau_\gamma(u), \{v\}_{(-1)} \cdot 1] = \{v\}_{(-1)} \cdot 1_y.$$

Here, $1_y = [\tau_\gamma(u), 1] \in (P_G)_y \times_G \mathbb{C}$ is independent of the frame in $(P_G)_y$. Hence, $\{v\}_{(-1)} \cdot 1_M$ is a parallel section and thus it belongs to $\mathcal{P}_M(\underline{V})$. The map $\{\ \}_{(-1)}$ is an isomorphism of graded vector spaces because the composition

$$V^G \xrightarrow{\ \{\ \}_{(-1)}\ } \mathcal{P}_M(\underline{V}) \xrightarrow[\text{Res}]{\cong} (\underline{V})_x^{\text{Hol}(x)} \xleftarrow{\cong} V^G$$

of $\{\ \}_{(-1)}$ with the isomorphisms of Proposition 23 of §4.5 is an identity map of $V^G = V^{G_u}$, which can be easily checked. Here, G_u is the holonomy group of M with respect to a Spin frame $u \in \tilde{Q}(M)$.

We have to show that the map $\{\ \}_{(-1)} : V^G \to \mathcal{P}_M(\underline{V})$ preserves the vertex operator super algebra structure in these graded vector spaces. That is, if $\sigma = \{v\}_{(-1)} \in \mathcal{P}_M(\underline{V})$ for $v \in V^G$, we must show the commutativity of the following diagram.

$$
\begin{array}{ccc}
V^G & \xrightarrow{\ \{v\}_n\ } & V^G \\
{\scriptstyle \{\ \}_{(-1)}}\downarrow & & \downarrow{\scriptstyle \{\ \}_{(-1)}} \\
\mathcal{P}_M(\underline{V}) & \xrightarrow{\ \{\sigma\}_n\ } & \mathcal{P}_M(\underline{V})
\end{array}
$$

Let $v_1 \in V^G$ be an arbitrary vector in V^G. We let $(\sigma)_x = (\{v\}_{(-1)})_x = [u, \overline{\sigma}]$, $(\sigma_1)_x = (\{v_1\}_{(-1)})_x = [u, \overline{\sigma}_1] \in \underline{V}_x$ for some $u \in (P_G)_x$. Here, $\overline{\sigma} = v$, $\overline{\sigma}_1 = v_1$. Then, by the local description of the action of the vertex operator $\{\sigma\}_n$ given in Proposition 5, we have

$$(\{\sigma\}_n \sigma_1)_x = [u, \{\overline{\sigma}\}_n \overline{\sigma}_1] = [u, \{v\}_n v_1] = \{\{v\}_n v_1\}_{(-1)}.$$

This shows the commutativity of the above diagram and the map $\{\ \}_{(-1)}$ is a vertex operator algebra isomorphism.

Using the above notation and the local description of the action of $\{\sigma\}_n$ given in Proposition 5 and of $\{v\}_n$ given in Proposition 8 on $\mathcal{P}_M(\underline{V})$, we have $(\{\sigma\}_n \sigma_1)_x = [u, \{v\}_n v_1] = (\{v\}_n \sigma_1)_x$. Thus, the actions of $Y(\sigma, \zeta)$ and of $Y(v, \zeta)$ on $\mathcal{P}_M(\underline{V})$ are the same when $\sigma = \{v\}_{(-1)}$. $\square$

Elliptic genera as modules over vertex operator super algebras. The graded vector space of parallel sections in the graded vector bundle $\underline{V}$ of vertex operator super algebras over M plays a fundamental role in representation theoretic aspects of elliptic genera. The graded vector space $\mathcal{P}_M(\underline{V})$ contains geometrically interesting objects such as generalized Riemannian metric tensors, generalized Kähler forms, certain Casimir forms, etc. In the next section §5.5, we deal with the consequences of the existence of these forms on various Riemannian manifolds when Module Theorem below is applied.

Module Theorem (ELLIPTIC GENERA AS MODULES OVER VERTEX OPERATOR SUPER ALGEBRAS $\mathcal{P}_M(\underline{V})$). *Let M^{2N} be a closed Riemannian Spin manifold. Let*

$$\underline{V} = \bigoplus_{n \geq 0} \bigwedge{}^{*}_{q^{n+\frac{1}{2}}} T_{\mathbb{C}} M \left(-n - \tfrac{1}{2}\right)$$

be the graded complex vector bundle of vertex operator super algebras over M with the connection induced from the Levi-Civita connection on TM. Let $\mathcal{P}_M(\underline{V})$ be the graded vector space of parallel sections of $\underline{V}$.

(I) *The graded vector space $\mathcal{P}_M(\underline{V})$ has the structure of a vertex operator super algebra. If $G \subset \mathrm{Spin}(2N)$ is the structure group of a holonomy bundle $P_G \subset \tilde{Q}(M)$ on M, there exists a canonical isomorphism of vertex operator super algebras $\{\ \}_{(-1)} : V^G \xrightarrow{\cong} \mathcal{P}_M(\underline{V})$ which depends only on the choice of the holonomy bundle P_G.*

(II) *The elliptic genus $\Phi^*_{\mathrm{ell}}(M^{2N})$ for $* = Q, R$ is a super pair of modules over the vertex operator super algebra $\mathcal{P}_M(\underline{V})_\ell \times \mathcal{P}_M(\underline{V})_r$ each of rank N, where $\mathcal{P}_M(\underline{V})_\ell$, $\mathcal{P}_M(\underline{V})_r$ are copies of $\mathcal{P}_M(\underline{V})$. That is, there exist linear maps*

(21) $\quad Y(\ ,\zeta_1) \times Y(\ ,\zeta_2) : \mathcal{P}_M(\underline{V})_\ell \times \mathcal{P}_M(\underline{V})_r$

$$\longrightarrow \mathrm{End}\left(\Phi^Q_{\mathrm{ell}}(M^{2N})_\ell\right) [[\zeta_1, \zeta_1^{-1}]] \times \mathrm{End}\left(\Phi^Q_{\mathrm{ell}}(M^{2N})_r\right) [[\zeta_2, \zeta_2^{-1}]],$$

(22) $\quad Y(\ ,\zeta_1) \times Y(\ ,\zeta_2) : \mathcal{P}_M(\underline{V})_\ell \times \mathcal{P}_M(\underline{V})_r$

$$\longrightarrow \mathrm{End}\left(\Phi^R_{\mathrm{ell}}(M^{2N})_\ell\right) [[\zeta_1^{\frac{1}{2}}, \zeta_1^{-\frac{1}{2}}]] \times \mathrm{End}\left(\Phi^R_{\mathrm{ell}}(M^{2N})_r\right) [[\zeta_2^{\frac{1}{2}}, \zeta_2^{-\frac{1}{2}}]],$$

Here, the left and right vertex operator super algebras $\mathcal{P}_M(\underline{V})_\ell$, $\mathcal{P}_M(\underline{V})_r$ independently satisfy the Jacobi identity for modules over $\mathcal{P}_M(\underline{V})$. Namely, for any $\sigma_1, \sigma_2 \in \mathcal{P}_M(\underline{V})_\ell$ or $\in \mathcal{P}_M(\underline{V})_r$,

(23)
$$\sum_{0 \leq i \in \mathbb{Z}} (-1)^i \binom{r}{i} \left(\{\sigma_1\}_{m+r-i} \{\sigma_2\}_{n+i} - (-1)^{|\sigma_1||\sigma_2|+r} \{\sigma_2\}_{n+r-i} \{\sigma_1\}_{m+i} \right)$$

$$= \sum_{0 \leq k \in \mathbb{Z}} \binom{m}{k} \{\{\sigma_1\}_{r+k}\sigma_2\}_{m+n-k}, \qquad \textit{where}$$

(24)
$$\begin{cases} Y(\sigma_i, \zeta) = \sum_{n \in \mathbb{Z}} \{\sigma_i\}_n \zeta^{-n-1}, m, n, r \in \mathbb{Z}, \text{ on } \Phi^Q_{\mathrm{ell}}(M^{2N})_*, \\ Y(\sigma_i, \zeta) = \sum_{n \in \mathbb{Z}+\frac{1}{2}|\sigma_i|} \{\sigma_i\}_n \zeta^{-n-1}, m \in \mathbb{Z}+\frac{1}{2}|\sigma_1|, n \in \mathbb{Z}+\frac{1}{2}|\sigma_2|, \text{ on } \Phi^R_{\mathrm{ell}}(M^{2N})_*. \end{cases}$$

(III) *As a super pair of $\mathcal{P}_M(\underline{V})$-module, the elliptic genus $\Phi^*_{\mathrm{ell}}(M)$ contains the following super pairs of (possibly trivial) $\mathcal{P}_M(\underline{V})$-free modules on super pairs of*

Spin index $\hat{A}(M)$ and Witten genus $\Phi_\theta(M)$: for $* = R, Q$,

$$(25) \quad \begin{cases} \hat{A}(M)_\ell \otimes \mathcal{P}_M(\underline{V})_\ell \subset \Phi_\theta(M)_\ell \otimes \mathcal{P}_M(\underline{V})_\ell \subset \Phi_{\mathrm{ell}}^*(M)_\ell, \\ \hat{A}(M)_r \otimes \mathcal{P}_M(\underline{V})_r \subset \Phi_\theta(M)_r \otimes \mathcal{P}_M(\underline{V})_r \subset \Phi_{\mathrm{ell}}^*(M)_r. \end{cases}$$

Here $\Phi_{\mathrm{ell}}^*(M)_{\ell,r}$ are the graded vector spaces of kernels and cokernels of a graded family of twisted Dirac operators (See the diagram below in the proof). Thus, when the numerical Witten genus $\varphi_\theta(M) = \text{s-dim}_*\Phi_\theta(M)$ or the numerical $\hat{A}$-genus $\hat{A}(M) = \text{s-dim}\,\hat{A}(M)$ doesn't vanish, the diagonal action of the vertex operator algebra $\mathcal{P}_M(\underline{V})$ on the elliptic genus $\Phi_{\mathrm{ell}}^*(M)$ through the diagonal map $\mathcal{P}_M(\underline{V}) \to \mathcal{P}_M(\underline{V})_\ell \times \mathcal{P}_M(\underline{V})_r$ composed with maps in (21), (22) is faithful.

PROOF. (I) has been proved in Theorem 6 and in Theorem 9 in this section.

(II) Let $\sigma \in \mathcal{P}_M(\underline{V})$. Then, by Proposition 5, the corresponding vertex operators $\{\sigma\}_n$ for $n \in \mathbb{Z}$ commute with the covariant derivative ∇ on $\underline{V}$, i.e., $[\nabla, \{\sigma\}_n] = 0$. By Proposition 3 in §5.3, the action of the operator $\{\sigma\}_n = 1\hat{\otimes}1\hat{\otimes}\{\sigma\}_n$ on the graded vector space of sections $\Gamma(\underline{\Delta}^\pm \otimes \underline{S} \otimes \underline{V})$ commutes with the twisted graded Dirac operator $d_* = d_{\underline{S}\otimes\underline{V}}$. Hence, $\{\sigma\}_n$ induces a map on the Ker $d_{\underline{S}\otimes\underline{V}} = \Phi_{\mathrm{ell}}^Q(M)_\ell$, making the left square of the diagram below commutative.

$$\begin{array}{ccccccc}
\Phi_{\mathrm{ell}}^Q(M)_\ell \hookrightarrow \Gamma(\underline{\Delta}^+ \otimes \underline{S} \otimes \underline{V}) & \xrightarrow{d_{\underline{S}\otimes\underline{V}}} & \Gamma(\underline{\Delta}^- \otimes \underline{S} \otimes \underline{V}) \twoheadrightarrow \Phi_{\mathrm{ell}}^Q(M)_r \\
\{\sigma\}_n \downarrow \qquad \{\sigma\}_n \downarrow \{\tau\}_n & & \{\sigma\}_n \downarrow \{\tau\}_n \qquad \downarrow \{\tau\}_n \\
\Phi_{\mathrm{ell}}^Q(M)_\ell \hookrightarrow \Gamma(\underline{\Delta}^+ \otimes \underline{S} \otimes \underline{V}) & \xrightarrow{d_{\underline{S}\otimes\underline{V}}} & \Gamma(\underline{\Delta}^- \otimes \underline{S} \otimes \underline{V}) \twoheadrightarrow \Phi_{\mathrm{ell}}^Q(M)_r
\end{array}$$

This shows that the operators $\{\sigma\}_n$ preserve the graded vector space $\Phi_{\mathrm{ell}}^Q(M)_\ell$, decreasing the weight by $n + 1 - \text{wt}(\sigma)$. Since the vertex operators $\{\sigma\}_n$ act fibrewise and the Jacobi identity holds in $\underline{V}_x$ for each $x \in M$, we see that the Jacobi identity holds on the space of sections $\Gamma(\underline{\Delta}^+ \otimes \underline{S} \otimes \underline{V})$. Since $\Phi_{\mathrm{ell}}^Q(M)_\ell$ is an $\mathcal{P}_M(\underline{V})_\ell$ invariant subspace of $\Gamma(\underline{\Delta}^+ \otimes \underline{S} \otimes \underline{V})$, the Jacobi identity holds in $\Phi_{\mathrm{ell}}^Q(M)_\ell$ by restriction.

Similarly, for any $\tau \in \mathcal{P}_M(\underline{V})_r$, the action of the corresponding vertex operator $\{\tau\}_n = 1\hat{\otimes}1\hat{\otimes}\{\tau\}_n$ on $\Gamma(\underline{\Delta}^\pm \otimes \underline{S} \otimes \underline{V})$ commutes with the twisted graded Dirac operator $d_{\underline{S}\otimes\underline{V}}$. This makes the right square of the above diagram commutative, where Coker $d_{\underline{S}\otimes\underline{V}} = \Phi_{\mathrm{ell}}^Q(M)_r$.

The Jacobi identity (23) for $\mathcal{P}_M(\underline{V})_r$ acting on $\Phi_{\mathrm{ell}}^Q(M)_r$ follows by taking the quotient of $\mathcal{P}_M(\underline{V})$-module $\Gamma(\underline{\Delta}^- \otimes \underline{S} \otimes \underline{V})$, in which the Jacobi identity holds, by its $\mathcal{P}_M(\underline{V})$-submodule $d_{\underline{S}\otimes\underline{V}}\left(\Gamma(\underline{\Delta}^+ \otimes \underline{S} \otimes \underline{V})\right)$. The form of operators in (24) follow from those of vertex operators acting on V or on W described in §2.2.

(III) By the inclusion relation in Proposition 4 of §5.3, any vector $\eta \in \Phi_\theta(M)_\ell \subset \Gamma(\underline{\Delta}^+ \otimes \underline{S})$ can be regarded as a vector $\eta \otimes 1 \in \Phi_{\mathrm{ell}}(M)_\ell \subset \Gamma(\underline{\Delta}^+ \otimes \underline{S} \otimes \underline{V})$. Now consider a linear inclusion map $\Phi_\theta(M)_\ell \hat{\otimes} \mathcal{P}_M(\underline{V}) \xrightarrow{\iota} \Gamma(\underline{\Delta}^+ \otimes \underline{S} \otimes \underline{V})$ given by $\iota(\eta \otimes \sigma) = \eta \otimes \sigma$. The image of this map is actually contained in

$\Phi_{\text{ell}}^Q(M)$. To see this, we apply the twisted graded Dirac operator $d_{\underline{S} \otimes \underline{V}}$. Let $\{v_1, \ldots, v_{2N}\}$ be an orthonormal basis in $T_x M$. Then, from Definition 1' in §5.3 of the twisted Dirac operator,

$$d_{\underline{S} \otimes \underline{V}}(\eta \otimes \sigma)_x = \sum_{i=1}^{2N} v_i \cdot \nabla_{v_i}(\eta \otimes \sigma) = \sum_{i=1}^{2N} (v_i \cdot \nabla_{v_i} \eta) \otimes \sigma + \sum_{i=1}^{2N} v_i \cdot \eta \otimes \nabla_{v_i} \sigma$$
$$= d_{\underline{S}}(\eta) \otimes \sigma + \sum_{i=1}^{2N} v_i \cdot \eta \otimes \nabla_{v_i} \sigma.$$

Since $\eta \in \Phi_\theta(M)_\ell$ and $\sigma \in \mathcal{P}_M(\underline{V})_\ell$, $d_{\underline{S}}(\eta) = 0$ and $\nabla_{v_i} \sigma = 0$, by definition. Hence, $d_{\underline{S} \otimes \underline{V}}(\eta \otimes \sigma) = 0$ and consequently $\eta \otimes \sigma \in \Phi_{\text{ell}}^Q(M)$. We regard $\Phi_\theta(M)_\ell \otimes \mathcal{P}_M(\underline{V})$ as a free $\mathcal{P}_M(\underline{V})_\ell$-module by letting $\mathcal{P}_M(\underline{V})$ act on the second factor $\mathcal{P}_M(\underline{V})$ in the above tensor product. We show that the above inclusion map ι is a $\mathcal{P}_M(\underline{V})$-module map. Let $\tau \in \mathcal{P}_M(\underline{V})$. Then, by definition, for $\eta \otimes \sigma \in \Phi_\theta(M)_\ell \otimes \mathcal{P}_M(\underline{V})_\ell$, $\{\tau\}_n(\eta \otimes \sigma) = \eta \otimes \{\tau\}_n \sigma$. On the other hand, for $\iota(\eta \otimes \sigma) = \eta \otimes \sigma \in \Phi_{\text{ell}}^Q(M) \subset \Gamma(\underline{\Delta}^+ \otimes \underline{S} \otimes \underline{V})$, $\{\tau\}_n$ acts on the third factor $\underline{V}$ and we have $\{\tau\}_n(\eta \otimes \sigma) = \eta \otimes \{\tau\}_n \sigma$. Hence, the inclusion map ι is a $\mathcal{P}_M(\underline{V})$-module map. Since $\hat{\mathcal{A}}(M)_\ell \subset \Phi_\theta(M)_\ell$, the first inclusion relation in (25) follows. The case for $\Phi_{\text{ell}}^R(M)$ is similar.

Since the twisted Dirac operators are self-adjoint, we have $d_{\underline{S} \otimes \underline{V}}^* = d_{\underline{S} \otimes \underline{V}}$. See §5.3 for details. So, the cokernels can be replaced by kernels in the appropriate spaces and the above argument can be applied. From this, the second sequence of inclusion relations in (25) follows.

If $\varphi_\theta(M) \neq 0$, then, either $\Phi_\theta(M)_\ell$ or $\Phi_\theta(M)_\ell$ is a nontrivial graded vector space. Thus, $\Phi_{\text{ell}}^*(M)$ has at least one copy of a free module over $\mathcal{P}_M(\underline{V})$. Thus, if $Y(\sigma, \zeta) = 0$ as an operator on $\Phi_{\text{ell}}^*(M)$, then we must have $Y(\sigma, \zeta) = 0$ as an operator on $\Phi_\theta(M) \otimes \mathcal{P}_M(\underline{V})$, which is a free $\mathcal{P}_M(\underline{V})$-module. Hence, by the injectivity of the vertex operator map $Y(\ , \zeta) : \mathcal{P}_M(\underline{V}) \to \text{End}(\mathcal{P}_M(\underline{V}))[[\zeta, \zeta^{-1}]]$, we must have $\sigma = 0$. This shows that the diagonal action of $\mathcal{P}_M(\underline{V})$ on the elliptic genus $\Phi_{\text{ell}}^*(M)$ is faithful. Note that the nontriviality of $\hat{A}$-genus implies the nontriviality of numerical Witten genus $\varphi_\theta(M)$, because $\hat{A}(M)$ is the first term in the q-expansion of $\varphi_\theta(M)$. $\square$

REMARKS. (i) For $\sigma \in \mathcal{P}_M(\underline{V})$, the action of $\{\sigma\}_n$ on $\Gamma(\underline{\Delta}^\pm \otimes \underline{S} \otimes \underline{V})$ induces the action of $\{\sigma\}_n$ on both $\Phi_{\text{ell}}^*(M)_\ell$ and $\Phi_{\text{ell}}^*(M)_r$ at the same time. So, the natural geometric action corresponds to the diagonal action on the elliptic genus in (21) and (22) through $\mathcal{P}_M(\underline{V}) \to \mathcal{P}_M(\underline{V})_\ell \times \mathcal{P}_M(\underline{V})_r$.

(ii) Let Q_M be the principal bundle of orthonormal frames on a Riemannian manifold M. A holonomy bundle $Q_M(u)$ and its structure group G_u depends on the choice of a frame $u \in Q_M$, although different choices of u give rise to conjugate holonomy bundles and structure groups. Thus, the vector space of G_u-invariant vectors in V depends on the choice of u. On the other hand, the space of parallel sections $\mathcal{P}_M(\underline{V})$ is *independent* of the choice of $u \in Q_M$. So $\mathcal{P}_M(\underline{V})$ is geometrically more intrinsic. This is the reason why we prefer to use $\mathcal{P}_M$ instead of V^G.

By part (I) above, $\mathcal{P}_M(\underline{V})$ is isomorphic to V^G for some $G \subset \mathrm{Spin}(2N)$. Since our vertex operator super algebra V is explicitly given as a spin representation of $\hat{\mathfrak{o}}(2N)$, it can be easily checked that the vertex operators corresponding to weight 1 vectors in V generate the orthogonal affine Lie algebra $\hat{\mathfrak{o}}(2N)$. So, weight 1 vectors in $(V^G)_1 \cong \mathcal{P}_M(\underline{V})_1 = \mathfrak{g}$ also generate an affine Lie algebra $\hat{\mathfrak{g}}$, which is a subalgebra of $\hat{\mathfrak{o}}(2N)$.

For the next Corollary, recall that $\mathcal{P}_M(\underline{V})_{1/2}$ is the vector space of parallel sections in $TM(-\frac{1}{2}) \otimes \mathbb{C} = T^*_\mathbb{C}M$ and $\mathcal{P}_M(\underline{V})_1$ is the vector space of parallel sections in $\bigwedge^2 TM(-\frac{1}{2}) \otimes \mathbb{C} = \bigwedge^2 T^*_\mathbb{C}M$.

COROLLARY 10. *Let M^{2N} be a closed Riemannian Spin manifold. The vertex operator super algebra $\mathcal{P}_M(\underline{V})$ has the following subalgebras:*

(i) *(Clifford algebra) The vertex operators for weight $\frac{1}{2}$ subspace $\mathcal{P}_M(\underline{V})_{\frac{1}{2}}$ generate an infinite dimensional Clifford algebra on $\mathcal{P}_M(\underline{V})_{\frac{1}{2}}$ with respect to a symmetric bilinear pairing given by the Riemannian metric.*

(ii) *(Affine Lie algebra) The weight 1 subspace $\mathfrak{g} = \mathcal{P}_M(\underline{V})_1$ has the structure of a Lie algebra and vertex operators associated to $\mathcal{P}_M(\underline{V})_1$ generate an affine Lie algebra $\hat{\mathfrak{g}}$.*

Thus, the elliptic genus $\Phi^*_{\mathrm{ell}}(M)$ is a super pair of representations of the above algebras for $* = Q, R$.

PROOF. The part (i) is a general fact about vertex operator super algebras given in Proposition 1 of §2.1. For (ii), in a general vertex operator super algebra, one needs an extra assumption as in Proposition 4 of §2.1. But in our present context, our vertex operator super algebra $\mathcal{P}_M(\underline{V})$ is constructed from the explicitly given Spin representation V of an orthogonal affine Lie algebra, and explicit calculation shows that the above extra assumption is not necessary. $\square$

Recall that in Proposition 6 of §2.1, we constructed a Lie super algebra $\{V\}^W_0$ from a pair of a vertex operator super algebra V and its module W using the operators $\{v\}_0$ for $v \in V$. The super commutator formula implies

$$\text{(26)} \qquad \{v_1\}_0\{v_2\}_0 - (-1)^{|v_1||v_2|}\{v_2\}_0\{v_1\}_0 = \{\{v_1\}_0 v_2\}_0 .$$

We apply this construction to the present context of elliptic genera. Let

$$\mathrm{Ann}^*_\ell(M) = \{\sigma \in \mathcal{P}_M(\underline{V})_\ell \mid \{\sigma\}_0 w = 0, \text{ for all } w \in \Phi^*_{\mathrm{ell}}(M)_\ell\}, \quad \text{for } * = Q, R.$$

We define $\mathrm{Ann}^*_r(M)$ in a similar way using $\mathcal{P}_M(\underline{V})_r$ and $\Phi^*_{\mathrm{ell}}(M)_r$. We then let

$$\{\mathcal{P}_\ell\}^*_0 = \{\{\sigma\}_0 \in \mathrm{End}(\Phi^*_{\mathrm{ell}}(M)_\ell) \mid \sigma \in \mathcal{P}_M(\underline{V})_\ell\} \cong \mathcal{P}_M(\underline{V})_\ell/\mathrm{Ann}^*_\ell(M),$$

for $* = Q, R$. We also define $\{\mathcal{P}_r\}^*_0$ in a similar way as above.

COROLLARY 11. *Let M^{2N} be a closed Riemannian Spin manifold. Then, for $* = Q, R$, each of the graded vector spaces $\{\mathcal{P}_\ell\}^*_0$, $\{\mathcal{P}_r\}^*_0$ has a structure of a*

Lie super algebra, and the elliptic genus $\Phi_{\mathrm{ell}}^*(M) = [\Phi_{\mathrm{ell}}^*(M)_\ell \; ; \; \Phi_{\mathrm{ell}}^*(M)_r]$ *is a faithful representations of* $\{\mathcal{P}_\ell\}_0^* \oplus \{\mathcal{P}_r\}_0^*$.

On $\Phi_{\mathrm{ell}}^R(M)_*$, *the odd parity part of the algebra* $\mathcal{P}_M(\underline{V})$ *acts trivially and we have* $\{\mathcal{P}_*\}_0^R = \{\mathcal{P}_*^{\mathrm{eve}}\}_0^R$ *for* $* = \ell, r$.

The algebra $\{\mathcal{P}_\ell\}_0^* \oplus \{\mathcal{P}_r\}_0^*$ *acts trivially on the Spin index* $\hat{A}(M)$ *and on the Witten genus* $\Phi_\theta(M)$ *which are super subspaces of the elliptic genus* $\Phi_{\mathrm{ell}}^*(M)$ *for* $* = Q, R$.

PROOF. The first part is due to Proposition 6 in §2.1. For the second part, from (24) above, the vertex operator corresponding to $\sigma \in \mathcal{P}_M(\underline{V})_*^{\mathrm{odd}} \cong V_1^G$ for $* = \ell, r$ acting on $\Phi_{\mathrm{ell}}^R(M)_*$ is of the form $Y(\sigma, \zeta) = \sum_{n \in \mathbb{Z}+\frac{1}{2}} \{\sigma\}_n \zeta^{-n-1}$. So, $\{\sigma\}_0 = 0$. Hence, $\mathrm{Ann}_*^R(M) \supset \mathcal{P}_M(\underline{V})_*^{\mathrm{odd}}$ and operators in the algebra $\{\mathcal{P}_*^R\}_0$ comes entirely from the even part of $\mathcal{P}_M(\underline{V})_*$. This proves the second part.

For the third part, note that vectors in $\hat{A}(M)$, $\Phi_\theta(M)$ are contained in the subspace $\Gamma(\underline{\Delta}^\pm \otimes \underline{S} \otimes 1_M)$ of $\Gamma(\underline{\Delta}^\pm \otimes \underline{S} \otimes \underline{V})$. Now, Lemma 5 in §2.1 says that $\{v\}_0 \cdot 1 = 0$ for any $v \in V$, where $1 \in (V)_0$ is the "vacuum" vector. From this, we see that for any $\sigma \in \mathcal{P}_M(\underline{V})$, we have $\{\sigma\}_0 \cdot 1_M = 0$ where 1_M is the constant function with value 1 on M. Hence, the subspace $\Gamma(\underline{\Delta}^\pm \otimes \underline{S} \otimes 1_M)$ is annihilated by operators of the form $\{\sigma\}_0$ for $\sigma \in \mathcal{P}_M(\underline{V})$. Consequently, operators $\{\sigma\}_0$ annihilate $\hat{A}(M)$ and $\Phi_\theta(M)$ for any $\sigma \in \mathcal{P}_M(\underline{V})$. Hence they are annihilated by the algebras $\{\mathcal{P}_*\}_0^{**}$ for $** = R, Q$, $* = \ell, r$. $\square$

In the above, we used $\{\ \}_0$ to convert V into an associative algebra which also has the structure of a Lie super algebra with respect to a bracket product given by the super commutator. We can introduce another product structure on V making it into another associative algebra which almost has the structure of a Lie super algebra. For this, we consider operators $Y_0(v)$ for $v \in V$.

We recall the definition of the operators $Y_m(v)$ for $m \in \frac{1}{2}\mathbb{Z}$ and $v \in V$ for a vertex operator super algebra V. These operators are given by

$$Y(v, z) = \sum_{n \in \frac{1}{2}\mathbb{Z}} \{v\}_n z^{-n-1} = \sum_{m \in \frac{1}{2}\mathbb{Z}} Y_m(v) z^{-m-\mathrm{wt}(v)}.$$

Thus, we have the relation $Y_m(v) = \{v\}_{\mathrm{wt}(v)+m-1}$, or $\{v\}_k = Y_{k+1-\mathrm{wt}(v)}(v)$. Operator $Y_m(v)$ lowers the weight by m and $\{v\}_k$ lowers the weight by $k+1-\mathrm{wt}(v)$. So, $Y_0(v) = \{v\}_{\mathrm{wt}(v)-1}$ preserves the weight for any $v \in V$. From the super commutator formula for vertex operators, for any $v_1, v_2 \in V$, we have

$$(27) \quad Y_0(v_1)Y_0(v_2) - (-1)^{|v_1||v_2|}Y_0(v_2)Y_0(v_1) = \sum_{0 \leq k \in \mathbb{Z}} \binom{\mathrm{wt}(v_1)-1}{k} Y_0(\{v_1\}_k v_2).$$

Here, the summation over k is finite, $k \leq \mathrm{wt}(v_1) + \mathrm{wt}(v_2) - 1$, because the weight of the vector $\{v_1\}_k v_2$ is $\mathrm{wt}(v_1) + \mathrm{wt}(v_2) - 1 - k$. As this formula shows, operators $Y_0(v)$ don't quite form a Lie super algebra with respect to the super commutator, although these operators $Y_0(v)$ do close under super brackets. Let

$$Y_0^Q(\mathcal{P}_\ell) = \{Y_0(v) \mid v \in V\} \subset \mathrm{End}\big(\Phi_{\mathrm{ell}}^Q(M)_\ell\big).$$

Elements in this algebra satisfy the above commutation relations. We define $Y_0^Q(\mathcal{P}_r)$, $Y_0^R(\mathcal{P}_\ell)$, $Y_0^R(\mathcal{P}_r)$ in a similar way as collections of suitable operators indicated by the individual notation.

COROLLARY 12. *The elliptic genus* $\Phi_{\mathrm{ell}}^*(M) = [\Phi_{\mathrm{ell}}^*(M)_\ell \; ; \; \Phi_{\mathrm{ell}}^*(M)_r]$ *for* $* = Q, R$ *is a representation of the algebra* $Y_0^*(\mathcal{P}_\ell) \oplus Y_0^*(\mathcal{P}_r)$.

On $\Phi_{\mathrm{ell}}^Q(M)$, *the odd parity part of the algebra* $\mathcal{P}_M(\underline{V})_*$ *acts trivially through* $Y_0(\;)$ *and we have* $Y_0^Q(\mathcal{P}_*) = Y_0^Q(\mathcal{P}_*^{\mathrm{eve}})$, $* = \ell, r$. *In particular, the Spin index* $\hat{A}(M)$ *is a representation of* $Y_0^Q(\mathcal{P}_\ell^{\mathrm{eve}}) \oplus Y_0^Q(\mathcal{P}_r^{\mathrm{eve}})$, *and the Signature* $\mathcal{L}(M)$ *is a representation of* $Y_0^R(\mathcal{P}_\ell) \oplus Y_0^R(\mathcal{P}_r)$.

PROOF. The first part is a direct consequence of Module Theorem. For the second part, note that the vertex operator corresponding to $\sigma \in \mathcal{P}_M(\underline{V})_*^{\mathrm{odd}}$ for $* = \ell, r$ is given by $Y(\sigma, \zeta) = \sum_{n \in \mathbb{Z}} \{\sigma\}_n \zeta^{-n-1} = \sum_{m \in \mathbb{Z}+\frac{1}{2}} Y_m(\sigma) \zeta^{-m-\mathrm{wt}\,(\sigma)}$. Hence, $Y_0(\sigma) = 0$ if σ has odd parity. Thus, operators in the algebra $Y_0^Q(\mathcal{P}_*)$ come from even parity elements in $\mathcal{P}_M(\underline{V})_*$ for $* = \ell, r$. $\square$

We apply Module Theorem and its corollaries to oriented Riemannian manifolds and in particular to Kählerian manifolds of various types including special Kähler manifolds, hyperkähler manifolds, and quaternion-Kähler manifolds. Our interest in these manifolds stems from their holonomy groups which are contained in the groups $SO(n)$, $U(N)$, $SU(N)$, $Sp(N')$, $Sp(N') \cdot Sp(1)$, respectively, where $n = 2N$, $n = 4N'$. As we have seen before, these groups are interesting in view of the classification theory of holonomy groups of Riemannian manifolds.

Elliptic genera for Kähler manifolds are dealt with in §5.5. Here, we consider some other types of manifolds for which tangent bundles split into direct sums of subbundles. Let $TM \cong \bigoplus_{j=1}^r \xi_j^{n_j}$ be a splitting of the tangent bundle for M^n, where $\xi_j^{n_j}$ is an oriented real n_j-dimensional subbundle of TM. Let g_j be an arbitrary metric on the bundle ξ_j. Then, the direct sum metric $g = g_1 \oplus g_2 \oplus \cdots \oplus g_r$ gives a Riemannian metric on M. With respect to this metric, any two different subbundles are orthogonal to each other and parallel transports preserve each subbundles ξ_j. Hence, the holonomy group of the metric g is contained in $\prod_{j=1}^r SO(n_j)$. We consider two special cases.

First, suppose that the tangent bundle splits into a sum of real 2-dimensional plane bundles. Then, the cotangent bundle T^*M also splits into a sum of oriented plane bundles $T^*M^{2N} = \bigoplus_{j=1}^N \xi_j$, where each ξ_j is real 2-dimensional. Note that $\bigwedge^2 \xi_j$ is a trivial real line bundle. With respect to the direct sum metric $g = \bigoplus_{j=1}^N g_j$ as above, the holonomy group is contained in the maximal torus $T^N \subset SO(2N)$ and a nowhere vanishing section η_j of $\bigwedge^2 \xi_j$ of unit length is parallel for $1 \le j \le N$. So, $\mathcal{P}_M(\underline{V})_1 \supset \bigoplus_{j=1}^N \mathbb{C}\eta_j$. Here, we identify T^*M and $TM(-\frac{1}{2})$ as usual. Since this vector space corresponds to the maximal abelian subalgebra $(V)_1^{T^N} \cong \mathfrak{o}(2N)^{T^N} = \mathfrak{t}_{\mathbb{C}}^N$ of $\mathfrak{o}(2N)$, the vertex operators $Y(\eta_j, \zeta)$, $1 \le j \le N$, generate a Heisenberg algebra $\mathfrak{h}_{(N)}$.

COROLLARY 13. *Suppose the tangent bundle of a closed Spin manifold* M^{2N} *splits into a sum of oriented plane bundles. Then, the elliptic genus* $\Phi_{\mathrm{ell}}^*(M)$ *is* *a super pair of representations of a Heisenberg algebra* $\mathfrak{h}_{(N)}$.

Next, suppose that there exist r linearly independent (nowhere vanishing) vector fields $X_1, X_2, \ldots, X_r$ on M^n so that $TM \cong \xi^{n-r} \oplus \bigoplus_{j=1}^{N} \mathbb{C}X_j$. We introduce a direct sum metric corresponding to this decomposition $g = g_0 \oplus \bigoplus_{j=1}^{N} g_j$ such that $g_j(X_j, X_j) = 1$. Obviously, these vector fields X_j are parallel because the parallel translation of a unit vector must be a unit vector, and the holonomy group of M^n is contained in $SO(n-r)$. The dual statement through the Riemannian metric is that the cotangent bundle T^*M has everywhere linearly independent parallel 1-forms $\varepsilon_1, \varepsilon_2, \ldots, \varepsilon_r$ which are dual to X_j's. Thus, we have $\mathcal{P}_M(\underline{V})_{\frac{1}{2}} \supset \bigoplus_{j=1}^{r} \mathbb{C}\varepsilon_j$. In fact, the algebra $\mathcal{P}_M(\underline{V})$ contains the total exterior algebra $E_{\mathbb{C}}(\varepsilon_1, \varepsilon_2, \ldots, \varepsilon_r)$. The vertex operators $Y(\varepsilon_j, \zeta)$ generate an infinite dimensional Clifford algebra on r-dimensional vector spaces. In particular, the finite dimensional complex Clifford algebra C_r acts on the elliptic genus $\Phi_{\text{ell}}^R(M)$ preserving the weights. Here, the Clifford algebra C_r acts on the finite dimensional Spin bundle $\Delta(TM) \otimes \mathbb{C}$ which is a tensor factor of $\underline{W}$. Thus, in particular, the signature of M, $\mathcal{L}(M)$, is a super pair of modules over C_r. Recall that C_r has a unique irreducible module Δ of dimension $2^{r'}$ if r is even and $r = 2r'$, and two irreducible representations Δ^+ and Δ^- of complex dimension $2^{r'-1}$ when r is odd and $r = 2r' - 1$. Since the L-genus of M is zero unless the manifold dimension is a multiple of 4, we assume that $\dim_{\mathbb{R}} M = 4N'$. We show that the complex dimension of $\mathcal{L}(M)$ is divisible by $2^{r'}$, where $r = 2r'$ or $r = 2r' - 1$. When r is even, this is obvious, since then $\mathcal{L}(M)$ is a direct sum of Spin representations Δ of complex dimension $2^{r'}$. From the above description of Clifford modules, when $r = 2r' - 1$, the complex dimension of $\mathcal{L}(M)$ is divisible by $2^{r'-1}$. We show that it is actually divisible by $2^{r'}$. Assume $r = 2r' - 1$. The cotangent bundle of M splits as $T^*M = \xi^{4N'-r} \oplus \bigoplus_{j=1}^{2r'-1} \mathbb{C}\varepsilon_j$. Due to our choice of metric g on M, $\bigwedge^{4N'-r} \xi$ has a parallel section ε_0 of unit length. Let $s = 4N' - r$ and let $\{f_1, \ldots, f_s\}$ be an orthonormal basis in the fiber ξ_x at $x \in M$. Although we cannot define orthonormal basis globally on M, their exterior product $\varepsilon_0 = f_1 \wedge \cdots \wedge f_s$ is a globally defined object independent of the choice of orthonormal bases in tangent spaces. We consider the vertex operators $Y(\varepsilon_0, \zeta)$ and $Y(\varepsilon_j, \zeta)$ for $1 \leq j \leq r$ acting on $\underline{W}$ corresponding to $\varepsilon_0 \in (\underline{V})_{s/2}$ and $\varepsilon_j \in (\underline{V})_{1/2}$. From the definition of vertex operators on the module W, we have $Y(\varepsilon_j, \zeta) = \overline{Y}(e^{\Delta(\zeta)}\varepsilon_j, \zeta)$ for $0 \leq j \leq r$, where the operator $\Delta(\zeta)$ is quadratic and is given by

$$\Delta(\zeta) = \sum_{1 \leq i \leq N} \sum_{0 \leq r, s \in \mathbb{Z}} \frac{r-s}{2(r+s+1)} \binom{-1/2}{r}\binom{-1/2}{s} a_i(r+\tfrac{1}{2})a_i^*(s+\tfrac{1}{2})\zeta^{-r-s-1}.$$

Here, $\{a_1, \ldots, a_N\}$ and $\{a_1^*, \ldots, a_N^*\}$ are any standard bases of a decomposition of $T_x^*M \otimes \mathbb{C} = A^+ \oplus A^-$ into maximal isotropic subspaces. The above operator is independent of this choice of standard bases and so $\Delta(\zeta)$ is defined intrinsically. See (20), (22) of §2.2. Since the operator $\Delta(\zeta)$ lowers the weight at least by 1, it acts trivially on $(\varepsilon_j)_x \in ((\underline{V})_{1/2})_x$ for $1 \leq j \leq r$. So, $\Delta(\zeta)\varepsilon_j = 0$ and this implies that $e^{\Delta(\zeta)}\varepsilon_j = \varepsilon_j$ for $1 \le j \le r$. As for the action of $\Delta(\zeta)$ on ε_0, the operator $a_i(r+\frac{1}{2})a_i^*(s+\frac{1}{2})$ with $r, s \ge 1$ acts trivially on $\varepsilon = f_1(-\frac{1}{2})\cdots f_s(-\frac{1}{2})\cdot 1 \in (\underline{V})_{s/2}$ because there are no nontrivial pairings between vectors of non-complementary weights. When $r = s = 0$, the above operator can act nontrivially, but then its coefficient $C_{0,0}$ in the operator $\Delta(\zeta)$ vanishes. Hence $\Delta(\zeta)\varepsilon_0 = 0$ and so we also have $e^{\Delta(\zeta)}\varepsilon_0 = \varepsilon_0$. Thus,

$$
\begin{aligned}
Y(\varepsilon_0, \zeta) &= \overline{Y}(e^{\Delta(\zeta)}\varepsilon_0, \zeta) = \overline{Y}(\varepsilon_0, \zeta) \\
&=: \Big(\sum_{m_1 \in \mathbb{Z}} f_1(m_1)\zeta^{-m_1-\frac{1}{2}} \Big) \cdots \Big(\sum_{m_s \in \mathbb{Z}} f_s(m_s)\zeta^{-m_s-\frac{1}{2}} \Big): ,
\end{aligned}
$$

$$
Y(\varepsilon_j, \zeta) = \overline{Y}(e^{\Delta(\zeta)}\varepsilon_j, \zeta) = \overline{Y}(\varepsilon_j, \zeta) = \sum_{m \in \mathbb{Z}} \varepsilon(m)\zeta^{-m-\frac{1}{2}}, \qquad 1 \le j \le r.
$$

We recall that $Y(v, \zeta) = \sum_{m \in \frac{1}{2}\mathbb{Z}} Y_m(v)\zeta^{-m-\mathrm{wt}(v)}$. By selecting Y_0 part of these operators, we have

(28)
$$
\begin{cases}
Y_0(\varepsilon_0) = \displaystyle\sum_{m_1+\cdots+m_s=0} :f_1(m_1)\cdots f_s(m_s): \\
Y_0(\varepsilon_j) = \varepsilon_j(0).
\end{cases}
$$

Since the operators $Y_0(\varepsilon_0)$ and $Y(\varepsilon_j)$'s preserve the weight, these operators preserve the total Spin bundle $(\underline{\Delta}) \subset (\underline{W})_0 = \Delta(T^*M)$ of M.

LEMMA 14. *Suppose that a Riemannian Spin manifold $M^{4N'}$ has $r = 2r' - 1$ linearly independent vector fields on M, $r' \in \mathbb{N}$. With respect to the above notations, the vertex operators $Y_0(\varepsilon_j) : \underline{\Delta} \to \underline{\Delta}$, $0 \le j \le r$, satisfy the following relations:*

(i) $Y_0(\varepsilon_0)Y_0(\varepsilon_0) = \dfrac{(-1)^{r'}}{2^s}$,

(ii) $Y_0(\varepsilon_j)Y_0(\varepsilon_j) = \dfrac{1}{2}, \quad 1 \le j \le r.$

(iii) $Y_0(\varepsilon_0)Y_0(\varepsilon_j) = -Y_0(\varepsilon_j)Y_0(\varepsilon_0) \quad 1 \le j \le r,$

(iv) $Y_0(\varepsilon_j)Y_0(\varepsilon_k) = -Y_0(\varepsilon_k)Y_0(\varepsilon_j) \quad 1 \le j < k \le r.$

Hence the operators $\{Y_0(\varepsilon_j) \mid 0 \le j \le r\}$ generate the Clifford algebra $C_{r+1} = C_{2r'}$ acting on $\underline{\Delta}$.

PROOF. Although the above vertex operators are globally defined on M, their actions are fiberwise. So, we prove the above relations in a fibre $(\underline{W})_x$ over an arbitrary point $x \in M$. At $x \in M$, we choose a decomposition of $T^*M \otimes \mathbb{C}$ into maximal isotropic subspaces $A^+ \oplus A^-$. Let $\{a_1, \ldots, a_N\}$ and $\{a_1^*, \ldots, a_N^*\}$ be their basis vectors such that $\langle a_i, a_j^* \rangle = \delta_{ij}$, $\langle a_i, a_j \rangle = 0$, $\langle a_i^*, a_j^* \rangle = 0$. Then, as Clifford modules, over the complex Clifford algebra C_{2N} constructed from $(T_{\mathbb{C}}^*M, g_x)$, we have $(\underline{\Delta})_x \cong \bigwedge^* A^-(0)$. This isomorphism is unique up to a constant multiple. We calculate the effect of $Y_0(\varepsilon_j)$ on the Spin bundle $\underline{\Delta}$ through

$\bigwedge^* A^-(0)$ for $0 \leq j \leq r$. As before, let $\{f_1, \ldots, f_s\}$ be an orthonormal basis of a real vector space ξ_x. For any vector $v \in (\Delta)_x$, using (28), we have

$$Y_0(\varepsilon_0)v = \sum_{m_1 + \cdots + m_s = 0} : f_1(m_1) \cdots f_s(m_s) : \cdot v.$$

Since $v \in \bigwedge^* A^-(0)$, if $m_j > 0$ for some $1 \leq j \leq s$, then $f_j(m_j)$ annihilates v. So, for a nontrivial action we must have $m_j \leq 0$ for all j. But then the condition $m_1 + \cdots + m_s = 0$ implies that $m_j = 0$ for all j. Hence we have

$$Y_0(\varepsilon_0)v =: f_1(0) \cdots f_s(0) : \cdot v$$
$$= \frac{1}{s!} \Big(\sum_{\sigma \in \mathfrak{S}_s} \operatorname{sgn}(\sigma) f_{\sigma(1)}(0) \cdots f_{\sigma(s)}(0) \Big) \cdot v$$
$$= f_1(0) \cdots f_s(0) \cdot v \in (\Delta)_x.$$

The third equality above is due to the identity in the Clifford algebra

$$(29) \qquad f_i(m_i)f_j(m_j) + f_i(m_i)f_j(m_j) = 0, \qquad \text{for } i \neq j,$$

which follows from the fact that $\{f_1, \ldots, f_s\}$ is an orthonormal basis of ξ_x. A similar argument shows that the action of the vertex operator $Y_0(\varepsilon_0)$ on the vector $Y_0(\varepsilon_0)v \in (\Delta)_x$ is given by

$$Y_0(\varepsilon_0)Y_0(\varepsilon_0) \cdot v = \big(f_1(0) \cdots f_s(0)\big) \cdot \big(f_1(0) \cdots f_s(0)\big) \cdot v \in (\Delta)_x.$$

In the Clifford algebra $\operatorname{Cliff}(T^*_{\mathbb{C}}M(0), g_x)$, (29) implies

$$\big(f_1(0) \cdots f_s(0)\big) \cdot \big(f_1(0) \cdots f_s(0)\big)$$
$$= (-1)^{\frac{s(s-1)}{2}} f_1(0)f_1(0) \cdot f_2(0)f_2(0) \cdots f_s(0)f_s(0) = (-1)^{s(s-1)/2} \left(\frac{1}{2}\right)^s.$$

Since $s = 4N' - 2r' + 1$, $s(s-1)/2 = (4N' - 2r' + 1)(2N' - r') \equiv r' \mod 2$. Hence we obtain

$$Y_0(\varepsilon_0)Y_0(\varepsilon_0)v = (-1)^{r'} \left(\frac{1}{2}\right)^s \cdot v$$

for any $v \in (\Delta)_x$. This proves (i) of Lemma 14. For (ii), we note that in the Clifford algebra $\operatorname{Cliff}(T^*_{\mathbb{C}}M(0), g_x)$, we have

$$\varepsilon_j(0)\varepsilon_k(0) + \varepsilon_k(0)\varepsilon_j(0) = \langle \varepsilon_j, \varepsilon_k \rangle = \delta_{ij}, \qquad 1 \leq i, j \leq r.$$

In particular, when $j = k$ we have $\varepsilon_j(0)\varepsilon_j(0) = \frac{1}{2}$. Since $Y_0(\varepsilon_j) = \varepsilon_j(0)$ for $1 \leq j \leq r$, we have proved (ii) and also (iv). To prove (iii), we need the super commutator formula for the vertex operators. Since $Y_0(\varepsilon_j) = \{\varepsilon_j\}_{-1/2}$ as operators acting on $(W)_x$ for $1 \leq j \leq r$, the commutator formula gives

$$Y_0(\varepsilon_j)Y_0(\varepsilon_0) - (-1)^{|\varepsilon_j||\varepsilon_0|}Y_0(\varepsilon_0)Y_0(\varepsilon_j) = \sum_{0 \leq k \in \mathbb{Z}} \binom{-1/2}{k} Y_0(\{\varepsilon_j\}_k\varepsilon_0).$$

Here, $|\varepsilon_0| = 1$ since wt $(\varepsilon_0) = (4N' - r)/2 = (4N' - 2r' + 1)/2 \in \mathbb{Z} + \frac{1}{2}$. Also, $|\varepsilon_j| = 1$ for $1 \leq j \leq r$ since wt $(\varepsilon_j) = 1/2$. Hence $-(-1)^{|\varepsilon_0||\varepsilon_j|} = +1$ and we have an anticommutator on the left hand side. As for the right hand side, note that as an operator acting on $(\underline{V})_x$, $\{\varepsilon_j\}_k = \varepsilon_j(k + \frac{1}{2})$. Since $\varepsilon_0 = f_1(-\frac{1}{2})\cdots f_s(-\frac{1}{2})\cdot 1 \in (\underline{V})_x$, the action of $\varepsilon_j(k + \frac{1}{2})$ on ε_0 for $k \geq 0$ can be nontrivial only when $k = 0$. Since subbundles ξ, $\mathbb{C}\varepsilon_j$ are orthogonal in $T_{\mathbb{C}}^* M$ for $1 \leq j \leq r$, $\langle f_i, \varepsilon_j \rangle = 0$. So, $\varepsilon_j(\frac{1}{2}) = \{\varepsilon_j\}_0$ acts trivially on $\varepsilon_0 = f_1(-\frac{1}{2})\cdots f_s(-\frac{1}{2})\cdot 1$ and $Y_0(\{\varepsilon_j\}_k \varepsilon_0) = 0$ for $0 \leq k \in \mathbb{Z}$. Hence the right hand side of the above commutator formula vanishes and we have $Y_0(\varepsilon_j)Y_0(\varepsilon_0) + Y_0(\varepsilon_0)Y_0(\varepsilon_j) = 0$. This completes the proof. $\square$

Since both the even and odd vector space of $\mathcal{L}(M)$ is a representation of the Clifford algebra $C_{2r'}$ generated by $Y_0(\varepsilon_0)$, $Y_0(\varepsilon_1)$, $\ldots$, $Y_0(\varepsilon_r)$ when $r = 2r' - 1$, they are direct sums of irreducible Clifford modules $\Delta_{2r'}$ of complex dimension $2^{r'}$. Hence $L(M) = \text{s-dim}\,\mathcal{L}(M)$ is divisible by $2^{r'}$. This proves the next proposition.

PROPOSITION 15. *Suppose a closed Spin manifold M^n has everywhere linearly independent r vector fields. Then, with respect to a suitable metric, the elliptic genus $\Phi_{\text{ell}}^*(M)$ is a super-pair of representations of the infinite dimensional Clifford algebra on r dimensional vector space. In particular, the L-genus of M is divisible by $2^{r'}$, where $r' = [\frac{r+1}{2}]$.*

A proof from the G-equivariant point of view. In Proposition 2 of §5.3, we showed that the twisted Dirac operators on a manifold with a G-structure equipped with a connection commute with the bundle maps induced from G-equivariant maps between vector spaces. Applying this fact in the present context of elliptic genera, we obtain the next theorem.

THEOREM 16 (THE MODULE STRUCTURE OVER VERTEX OPERATOR SUPER ALGEBRAS). *Let $G \subset \text{Spin}(2N)$ be the structure group of a G-structure P_G with a connection on a closed Riemannian Spin manifold M. Then, the elliptic genus $\Phi(M^{2N})$, as a super pair of graded vector spaces, is a module over the direct sum of the vertex operator super algebras $(V^G)_\ell \oplus (V^G)_r$.*

PROOF. From Corollary 3 of §3.1, any G-invariant vector $v \in V^G$ gives rise to a family of G-equivariant vertex operators $\{v\}_n : V \to V$, $W \to W$ for $n \in \frac{1}{2}\mathbb{Z}$ on the graded vector spaces V, W. Here, V and W are those in (11) in §2.2. Since the holonomy group of M^{2N} is $G \subset \text{Spin}(2N)$ with respect to some frame on M, for $v \in V^G$ we have bundle maps $\{v\}_n$, $n \in \frac{1}{2}\mathbb{Z}$, defined in terms of the G-structure P_G due to Proposition 8 above.

$$\{v\}_n = 1 \otimes 1 \otimes \{v\}_n : \Delta^+ \otimes \underline{S} \otimes \underline{V} \longrightarrow \Delta^- \otimes \underline{S} \otimes \underline{V},$$
$$\{v\}_n = 1 \otimes 1 \otimes \{v\}_n : \Delta^+ \otimes \underline{S} \otimes \underline{W} \longrightarrow \Delta^- \otimes \underline{S} \otimes \underline{W},$$

induced by $\{v\}_n$ acting on the third factor. Here, $\underline{S} = \bigotimes_{\ell > 0} S^*(T_{\mathbb{C}}(-\ell))$. By Proposition 2 in §5.3, the above operators $\{v\}_n$ for $n \in \frac{1}{2}\mathbb{Z}$ preserves the kernel and cokernel of the family of twisted Dirac operator (4) in §1.1. Since we can make these operators act separately on the kernel and on the cokernel, a pair of the vertex operator super algebras $(V^G)_\ell \oplus (V^G)_r$ acts on the elliptic genus $\Phi^*_{\text{ell}}(M)$ defined in §1.1 (5) and (6).

As for Jacobi identities for modules, these identities hold for elliptic genera because operators $\{v\}_n$ act fibrewise and on each fibre these identities follow from those of the algebra V acting on V or on W. $\square$

REMARK. In Theorem 16, we assumed that the manifold is Riemannian Spin so that the twisted Dirac operators can be defined. The associated principal $\text{Spin}(2N)$ bundle $\tilde{Q}(M)$ has a canonical connection, namely the Levi-Civita connection. Any principal subbundle $P_G \subset \tilde{Q}(M)$ may not be invariant under the parallel transports with respect to the Levi-Civita connection. So, in general, the Levi-Civita connection doesn't restrict to general principal subbundles of $\tilde{Q}(M)$. It can only restrict to unions of holonomy bundles. However, we can always introduce connections on any principal bundles, with no reference to the Levi-Civita connection. These connections on P_G are the ones we use in Theorem 16. So, it applies in slightly more general context than Module Theorem.

A corollary similar to Corollary 10 can be stated as follows.

COROLLARY 17. *Let M be a closed Riemannian Spin manifold equipped with a G-structure P_G with a connection. Then, the elliptic genus $\Phi^*_{\text{ell}}(M)$ is a super pair of representations*

(1) *of an (infinite) dimensional Clifford algebra generated by vertex operators $Y(v,z)$ for $v \in (V^G)_{\frac{1}{2}}$, and*

(2) *of an affine Lie algebra generated by vertex operators $Y(v,\zeta)$ for $v \in (V^G)_1$.*

As for G, we are interested in the following groups. Here, $2N' = N$.

$$
\begin{aligned}
\text{Sp}(N') &\subset \text{SU}(N) \subset \text{U}(N)^{(2)} \subset \text{Spin}(2N) \supset \text{T}^N, \\
\text{Sp}(N') &\cdot \text{Sp}(1) \subset \text{Spin}(4N'),
\end{aligned}
$$
(30)
$$
\prod_i \text{SO}(n_i) \subset \text{SO}(2N) \leftarrow \text{Spin}(2N) \supset \text{Spin}(k)
$$

Here, $\text{U}(N)^{(2)}$ is the double covering of $\text{U}(N)$ corresponding to the double covering $\text{Spin}(2N) \twoheadrightarrow \text{SO}(2N)$, and the inclusion map $\text{U}(N) \subset \text{SO}(2N)$.

§5.5 Infinite dimensional symmetries in elliptic genera for Kähler manifolds

We apply our Module Theorem to special class of manifolds, namely, to Kähler manifolds. The types of Kähler manifolds we consider are, Kählerian, special Kähler, hyperkähler and quaternion-Kähler manifolds. Our interest in Kähler manifolds arises from the fact that these manifolds possess special differential forms, Kähler forms, associated to Riemannian metrics and integrable almost complex structures. Kähler forms are closed real differential forms of type $(1,1)$. A manifold can have several different Kähler forms. It turns out that Kähler forms are parallel and Module Theorem implies that they generate infinite families of vertex operators acting on elliptic genera. These vertex operators form an affine Lie algebra of level 1. Together with generalizations of the Riemannian metric tensor, Kähler forms and their generalizations are important elements of the vertex operator super algebra $\mathcal{P}_M(\underline{V})$ consisting of all parallel sections of a graded vector bundle $\underline{V}$ on a closed Riemannian manifold M^{2N}. Here the graded vector space V is a Spin representation of the orthogonal affine Lie algebra $\hat{o}(2N)$. This vertex operator algebra $\mathcal{P}_M(\underline{V})$ acts on the elliptic genus, which is a super pair of graded vector spaces, of a closed Riemannian Spin manifold. The vertex operator super algebra $\mathcal{P}_M(\underline{V})$ contains various affine Lie algebras, the Virasoro algebra with central charge N, and other algebras of interest. The purpose of this section is to investigate and describe these subalgebras of $\mathcal{P}_M(\underline{V})$ for various types of Kähler manifolds.

In what follows, for convenience we assume that the given Riemannian manifold M^{2N} is connected. In this case, the space of locally constant functions on M form a 1-dimensional vector space, which makes our argument a little easier. If M is not connected, the conclusions below hold for each connected component.

Elliptic genera for oriented Riemannian Spin manifolds. Let M^{2N} be a closed oriented Riemannian manifold of even dimension. Recall that a closed manifold is a compact manifold without boundary. The structure group of M is contained in $SO(2N)$. We assume that $N \geq 2$. When $N = 1$, $SO(2) = U(1)$ and we are dealing with Riemann surfaces. This case is dealt with in the section of Kähler manifolds below. The basic geometric object on a Riemannian manifold M is the Riemannian metric g which is a parallel tensor with respect to the Levi-Civita connection. Since the manifold M^{2N} is oriented, another basic object on M is the Riemannian volume form ξ which is a nowhere vanishing differential $2N$-form on M^{2N} of unit length belonging to the orientation class. In the graded vector bundle $\underline{V}$, we can consider generalized Riemannian tensors and generalized Riemannian volume forms on M^{2N}. See §5.2 for more details. In particular, a generalized Riemannian tensor $[g_{(1,0)}] \in \mathcal{P}_M(\underline{V})_2$ of weight 2 plays a basic role throughout this section.

PROPOSITION 1. *Let M^{2N} be a connected oriented Riemannian manifold whose holonomy group is precisely $SO(2N)$. Then, up to weight 2, the algebra of parallel sections $\mathcal{P}_M(\underline{V})$ in the bundle of vertex operator super algebras $\underline{V}$ is given as follows:*

$$\mathcal{P}_M(\underline{V})_0 = \mathbb{C} \cdot 1, \qquad \mathcal{P}_M(\underline{V})_{k/2} = \{0\}, \quad 1 \leq k \leq 3,$$

$$\mathcal{P}_M(\underline{V})_2 = \begin{cases} \mathbb{C}[\xi] \oplus \mathbb{C}[g_{(1,0)}], & \text{if } N = 2, \\ \mathbb{C}[g_{(1,0)}], & \text{if } N \geq 3. \end{cases}$$

Here, $\xi = \xi(-\frac{1}{2}) \in \mathcal{P}_M(\underline{V})_N$ is the parallel Riemannian volume form on M^{2N} of weight N in $\Gamma(\bigwedge^{2N} T_{\mathbb{C}} M(-\frac{1}{2}))$.

PROOF. From the general theory of parallel sections on Riemannian manifolds, we have an isomorphism Res : $V^{SO(2N)} \hookrightarrow \mathcal{P}_M(\underline{V})_x$ for any $x \in M$, as vertex operator super algebras [§4.5 Proposition 23]. From our calculation of $(V)_*^{SO(2N)}$ done in §3.12, Proposition 1 follows. □

For general Riemannian manifolds, Module Theorem implies the following.

THEOREM 2 (ELLIPTIC GENERA FOR RIEMANNIAN SPIN MANIFOLDS). *Let M^{2N} be a connected closed Riemannian Spin manifold. Then, its elliptic genus $\Phi_{ell}(M^{2N})$ is a super pair of representations of the Virasoro algebra of rank N generated by the generalized metric tensor $[g_{(1,0)}] \in (\mathcal{P}_M(\underline{V}))_2$, i.e., if we let its vertex operator be*

$$Y\left([g_{(1,0)}], z\right) = -2 \sum_{n \in \mathbb{Z}} D(n) z^{-n-2} \in \text{End}\left(\Phi_{ell}^*(M)\right)[[z, z^{-1}]],$$

for $* = Q, R$, then, the vertex operators $D(n)$, $n \in \mathbb{Z}$, satisfy

$$[D(m), D(n)] = (n - m)D(m + n) + \frac{m(m^2 - 1)}{12}\delta_{m+n,0} \cdot N \cdot \text{Id}.$$

When $N = 2$, let $\Theta = -(2\xi - [g_{(1,0)}])/4$, $\Lambda = -(2\xi + [g_{(1,0)}])/4$ and let the corresponding vertex operators be

$$Y([\Theta], z) = \sum_{n \in \mathbb{Z}} \Theta(n) z^{-n-2}, \qquad Y([\Lambda], z) = \sum_{n \in \mathbb{Z}} \Lambda(n) z^{-n-2}.$$

Then, operators $\{\Theta(n)\}_{n \in \mathbb{Z}}$, $\{\Lambda(n)\}_{n \in \mathbb{Z}}$ generate commuting Virasoro algebras with central charges 1.

$$\begin{cases} [\Theta(m), \Theta(n)] = (n - m)\Theta(m + n) + \dfrac{m(m^2 - 1)}{12}\delta_{m+n,0} \cdot \text{Id} \\[2mm] [\Lambda(m), \Lambda(n)] = (n - m)\Lambda(m + n) + \dfrac{m(m^2 - 1)}{12}\delta_{m+n,0} \cdot \text{Id} \\[2mm] [\Theta(m), \Lambda(n)] = 0, \end{cases}$$

Here, Id is the identity operator.

PROOF. First we note that the parallel tensor $[g_{(1,0)}] \in \mathcal{P}_M(\underline{V})_2$ restricts to $-2\omega_x \in (\underline{V}_x)_2$ at any $x \in M$. For notations of various elements in $\underline{V}_x$, see (17) of §5.2. In $\underline{V}_x$, the vertex operator corresponding to ω_x generates a Virasoro algebra of rank N. Since the action of the vertex operator $Y([g_{(1,0)}], \zeta)$ is fibrewise, the vertex operators $\{D(n)\}_{n \in \mathbb{Z}}$ generate a rank N Virasoro algebra acting on the graded vector bundle $\underline{V}$, and hence on the space of sections $\Gamma(\underline{V})$.

Next, we show that elements $[\Theta]$ and $[\Lambda]$ generate Virasoro algebras when $N = 2$. To see this, due to a dimensional reason, we note that

$$[h_x^2] = -2[a_1(-\tfrac{1}{2})_x a_2(-\tfrac{1}{2})_x a_1^*(-\tfrac{1}{2})_x a_2^*(-\tfrac{1}{2})_x] = -2[\xi]_x.$$

We then see that $[\Theta]_x = \frac{1}{4}(2[\omega_x] - [h_x^2])$, $[\Lambda]_x = \frac{1}{4}(2[\omega_x] + [h_x^2])$. We then apply our previous calculation of the unitary Virasoro algebra in V given in Theorem

2 in §3.6 for $N = 2$ to obtain commutation relations in each fibre V_x. We then globalize this relation using Proposition 4 in §5.4 in each connected component to obtain commutation relations stated in Theorem 2. $\square$

REMARK. (i) Appearance of unitary Virasoro algebras on Riemannian Spin manifolds (M^4, g) is not surprising since Spin$(4) = $ SU$(2) \times$ SU(2).
(ii) Although vertex operators corresponding to the Riemannian volume forms $[\xi] \in \mathcal{P}_M(\underline{V})$ are rather hard to calculate for general N, it should be an object of some interest since it is a geometric object of basic importance on Riemannian manifolds.

Elliptic genera for Kählerian manifolds. Let M^{2N} be a Kählerian manifold. It possesses a Kähler form κ which is a parallel closed real 2-form of type $(1,1)$ with respect to the almost complex structure I on M. When M^{2N} is compact, powers of the Kähler form represent nontrivial cohomology classes of M^{2N}. Since I is an isometry, we can consider two Hermitian pairings on M^{2N} (see the paragraph after Definition 3 in §5.1) which turn out to be also parallel on M^{2N}.

Our object of interest is the vertex operator super algebra $\mathcal{P}_M(\underline{V})$ of parallel sections in the bundle $\underline{V}$. In this algebra, we have generalized Hermitian tensors $\eta^\pm = \eta_I^\pm \in T_{\mathbb{C}}(-\frac{3}{2}) \otimes T_{\mathbb{C}}M(-\frac{1}{2})$ which were introduced in §5.2. Together with powers of the Kähler form κ, low weight elements in this algebra are described as follows.

PROPOSITION 3. *Let M^{2N} be a connected Kählerian manifold with* U(N) *holonomy. Vectors of weight up to 2 in the algebra $\mathcal{P}_M(\underline{V})$ are given as follows:*

$$\mathcal{P}_M(\underline{V})_0 = \mathbb{C} \cdot 1, \qquad \mathcal{P}_M(\underline{V})_{1/2} = \{0\},$$
$$\mathcal{P}_M(\underline{V})_1 = \mathbb{C} \cdot [\kappa], \qquad \mathcal{P}_M(\underline{V})_{3/2} = \{0\},$$
$$\mathcal{P}_M(\underline{V})_2 = \begin{cases} \mathbb{C}[\eta^+] \oplus \mathbb{C}[\eta^-], & N = 1, \\ \mathbb{C}[\eta^+] \oplus \mathbb{C}[\eta^-] \oplus \mathbb{C} \cdot [\kappa^2], & N \geq 2. \end{cases}$$

Here, $[\kappa]$ is the Kähler form on M^{2N}, $[\eta^\pm]$ are the generalized Hermitian tensor fields of weight 2.

PROOF. For any $x \in M$, we know that the evaluation at $x \in M$ of parallel sections on manifolds with U(N)-holonomy induces an isomorphism $\mathcal{P}_M(\underline{V}) \overset{\cong}{\longrightarrow} \mathcal{P}_M(\underline{V})_x \cong V^{U(N)}$ by Proposition 23 in §4.5. From Propositions 8, 11 in §5.2,

$$[\kappa]_x = -i[h_x], \quad [\kappa^2]_x = -[h_x^2], \quad [\eta^+]_x = -2[\omega_{+,x}], \quad [\eta^-]_x = -2[\omega_{-,x}].$$

From our previous calculation of $V^{U(N)}$ in Proposition 3 in §3.12,

$$(V)_0^{U(N)} = \mathbb{C}\Omega, \qquad (V)_{\frac{1}{2}}^{U(N)} = \{0\},$$
$$(V)_1^{U(N)} = \mathbb{C}[h], \qquad (V)_{\frac{3}{2}}^{U(N)} = \{0\},$$
$$(V)_2^{U(N)} = \begin{cases} \mathbb{C}[\omega_+] \oplus \mathbb{C}[\omega_-], & N = 1, \\ \mathbb{C}[\omega_+] \oplus \mathbb{C}[\omega_-] \oplus \mathbb{C}[h^2], & N \geq 2. \end{cases}$$

From these results, Proposition 3 follows. □

Generalized Hermitian tensors $[\eta^{\pm}]$ are closely related to generalized Riemannian tensor $[g_{(1,0)}]$ and Kähler form $[\kappa]$.

LEMMA 4. *In* $\mathcal{P}_M(\underline{V})_2$, *we have*

$$\tfrac{1}{2}([\eta^+] + [\eta^-]) = [g_{(1,0)}], \qquad \tfrac{i}{2}([\eta^+] - [\eta^-]) = -[\kappa_{(1,0)}] = D(-1)[\kappa].$$

PROOF. The first identity is obvious from the definition of Hermitian tensors fields. For the second identity, the first equality is from the definition of Hermitian tensors in Definition 10 of §5.2. For the second equality, since the Virasoro operator $D(-1)$ preserves the graded vector space $\mathcal{P}_M(\underline{V})$ of parallel sections and the evaluation map at $x \in M$ is an injection map into V_x on each connected component of M, we only have to check the agreement of these two sections at one point $x \in M$. So, we look at their local expressions at $x \in M$. For $\kappa_{(1,0)}$, it is given by

$$\kappa_{(1,0),x} = -i\sum_{j=1}^{N} a_j(-\tfrac{3}{2})_x a_j^*(-\tfrac{1}{2})_x + i\sum_{j=1}^{N} a_j^*(-\tfrac{3}{2})_x a_j(-\tfrac{1}{2})_x = i(\omega_+ - \omega_-).$$

Since $D(-1)$ acts fiberwise, and $[\kappa]_x = -i[h]$,

$$(D(-1)[\kappa])_x = D(-1)[\kappa]_x = -iD(-1)[h] = -i([\omega_+] - [\omega_-]).$$

Hence, $[\kappa_{(1,0)}]_x = -(D(-1)[\kappa])_x$ and these two parallel sections in $\mathcal{P}_M(\underline{V})$ agree at any point $x \in M$. Consequently, these two parallel sections coincide even if M is not connected. □

Applying Module Theorem to Kähler manifolds, we obtain

THEOREM 5 (ELLIPTIC GENERA FOR KÄHLER MANIFOLDS). *Let M^{2N} be a connected closed Kählerian Spin manifold. Then, the left and the right part of the elliptic genus* $\Phi_{\mathrm{ell}}^*(M^{2N}) = [\Phi_{\mathrm{ell}}^*(M^{2N})_\ell \; ; \; \Phi_{\mathrm{ell}}^*(M^{2N})_r]$ *are separately representations of the following subalgebras of the vertex operator super algebra* $\mathcal{P}_M(\underline{V})$:

(1) (Heisenberg algebra) *Let* $Y([\kappa], z) = \sum_{n \in \mathbb{Z}} K(n) z^{-n-1}$ *be the vertex operator generated by the Kähler form* $[\kappa]$. *Then, the operators* $\{K(n)\}_{n \in \mathbb{Z}}$ *are closed under the Lie brackets and generate a Heisenberg algebra with the following commutation relations:*

$$[K(m), K(n)] = nN\delta_{m+n,0} \cdot \mathrm{Id}, \qquad m, n \in \mathbb{Z}.$$

(2) (Unitary Virasoro algebra) *Let*

$$Y([\Theta], z) = \sum_{n \in \mathbb{Z}} \Theta(n) z^{-n-2}, \qquad Y([\Lambda], z) = \sum_{n \in \mathbb{Z}} \Lambda(n) z^{-n-2}.$$

be vertex operators corresponding to the following vectors in $\mathcal{P}_M(\underline{V})$:

$$[\Theta] = \frac{1}{2N}([\kappa^2] - [g_{(1,0)}]), \qquad [\Lambda] = \frac{-1}{2N}\left([\kappa^2] + (N-1)[g_{(1,0)}]\right).$$

Then, operators $\{\Theta(n)\}_{n\in\mathbb{Z}}$, $\{\Lambda(n)\}_{n\in\mathbb{Z}}$ *generate commuting Virasoro algebras with central charge 1 and* $N-1$, *respectively:*

$$\begin{cases} [\Theta(m), \Theta(n)] = (n-m)\Theta(m+n) + \dfrac{m(m^2-1)}{12}\delta_{m+n,0} \cdot \mathrm{Id}, \\[2mm] [\Lambda(m), \Lambda(n)] = (n-m)\Lambda(m+n) + \dfrac{m(m^2-1)}{12}\delta_{m+n,0}(N-1) \cdot \mathrm{Id}, \\[2mm] [\Theta(m), \Lambda(n)] = 0. \end{cases}$$

These operators are related to the canonical Virasoro operators by the formula

$$\Theta(n) + \Lambda(n) = D(n), \qquad n \in \mathbb{Z}.$$

(3) *Commutation relations between the Heisenberg algebra and the Unitary Virasoro algebra are given by*

$$[K(m), \Theta(n)] = -mK(m+n), \qquad [K(m), \Lambda(n)] = 0,$$

for any $m, n \in \mathbb{Z}$. *Thus, the Heisenberg algebra is an ideal in the Lie algebra which is a semi-direct product of the Heisenberg algebra and the unitary Virasoro algebra.*

REMARK. If we compare Theorem 2 and Theorem 5, we see that the existence of Kähler form $[\kappa] \in \mathcal{P}_M(\underline{V})_1$ and its square $[\kappa^2]$ not only give rise to the Heisenberg algebra but also splits the canonical Virasoro section $[g_{(1,0)}] \in \mathcal{P}_M(\underline{V})_2$ into two parts, generating mutually commuting Virasoro algebras. So, it would be of some interest to see the role played by higher powers of Kähler forms in the context of vertex operators. Note that the highest power $[\kappa^N]$ of the Kähler form is equal, up to a constant multiple, to the Riemannian volume form $[\xi] \in \mathcal{P}_M(\underline{V})_N$. Thus, on Kählerian manifolds, we may be able to expect some nice properties for the vertex operator $Y([\xi], \zeta)$ corresponding to the volume form $[\xi]$.

For the proof of Theorem 5, we recall some results in the vertex operator super algebra V.

LEMMA 6. *For any* $m, n \in \mathbb{Z}$, *in the vertex operator super algebra* V, *we have*

$$[h(m), h(n)] = mN\delta_{m+n,0} \cdot 1,$$
$$[h(m), D(n)] = -m \cdot h(m+n),$$
$$[h(m), \{h^2\}_{n+1}] = 2m(N-1) \cdot h(m+n).$$

PROOF. For the first formula, from commutation relations of the orthogonal affine Lie algebra $\hat{\mathfrak{o}}(2N)$ acting on $\underline{V}$, $[h(m), h(n)] = m\langle h, h\rangle \delta_{m+n,0} \cdot 1$. Since

$$\langle h, h\rangle = \sum_{j,k=1}^{N} \langle a_j a_j^*, a_k a_k^* \rangle$$

$$= \sum_{j,k=1}^{N} \left(\langle a_j, a_k^* \rangle \langle a_j^*, a_k \rangle - \langle a_j, a_k \rangle \langle a_j^*, a_k^* \rangle \right) = \sum_{j=1}^{N} 1 = N.$$

This proves the first formula. The second one also follows from the general commutation formula in V. The third one is calculated in §3.5 Proposition 11. $\square$

LEMMA 7. *Let* θ, λ *be elements in* $(\mathcal{A})_2$ *defined by*

$$\theta = \frac{1}{2N}(2\omega - h^2), \qquad \lambda = \frac{1}{2N}\{2(N-1)\omega + h^2\}.$$

Then, as operators acting on V, *the corresponding vertex operators are such that*

$$[h(m), \{\theta\}_{n+1}] = -m \cdot h(m+n), \qquad [h(m), \{\lambda\}_{n+1}] = 0.$$

PROOF. Since $\{\theta\}_{n+1} = \left(2D(n) - \{h^2\}_{n+1}\right)/(2N)$, Lemma 6 implies that

$$[h(m), \{\theta\}_{n+1}] = \frac{1}{N}[h(m), D(n)] - \frac{1}{2N}[h(m), \{h^2\}_{n+1}]$$

$$= -\frac{m}{N}h(m+n) - \frac{m(N-1)}{N}h(m+n) = -m \cdot h(m+n).$$

Similarly, since $\{\lambda\}_{n+1} = \left(2(N-1)D(n) + \{h^2\}_{n+1}\right)/(2N)$, Lemma 6 again implies that

$$[h(m), \{\lambda\}_{n+1}] = \frac{N-1}{N}[h(m), D(n)] + \frac{1}{2N}[h(m), \{h^2\}_{n+1}]$$

$$= -\frac{m(N-1)}{N}h(m+n) + \frac{m(N-1)}{N}h(m+n) = 0.$$

This completes the proof of Lemma 7. $\square$

(PROOF OF THEOREM 5). We only have to prove commutation relations of parallel vertex operators. Since the evaluation map at $x \in M$ induces an isomorphism $\mathcal{P}_M(\text{End}(\underline{V})) \xrightarrow{\cong} \text{End}^{\text{U}(N)}(V)$ on manifolds with U(N)-holonomy, we calculate commutation relations among operators locally at $x \in M$. Local expressions of these parallel sections are given by

$$[\kappa]_x = -i[h], \qquad [\kappa^2]_x = -[h^2], \qquad [g_{(1,0)}]_x = -2[\omega],$$

$$[\Theta]_x = \frac{1}{2N}(2[\omega] - [h^2]) = [\theta], \qquad [\Lambda]_x = \frac{1}{2N}\left(2(N-1)[\omega] + [h^2]\right) = [\lambda].$$

Since the action of vertex operators is local, i.e., $Y(\sigma,\zeta)_x = Y(\sigma_x,\zeta)$ for any $\sigma \in \Gamma(\underline{V})$, Lemma 6 and Lemma 7 imply that

$$[K(m), K(n)]_x = [-ih(m), -ih(n)] = -mN\delta_{m+n,0} \cdot 1_x = nN\delta_{m+n,0} \cdot 1_x,$$
$$[K(m), \Theta(n)]_x = [-ih(m), \{\theta\}_{n+1}] = im \cdot h(m+n) = -m \cdot K(m+n)_x,$$
$$[K(m), \Lambda(n)]_x = [-ih(m), \{\lambda\}_{n+1}] = 0.$$

From these, by globalization to parallel operators on M, commutation relations in (1) and (3) in Theorem 5 follow. Commutation relations in (2) follow from our local calculation of unitary Virasoro algebra in V done in §3.6. □

REMARK. In Theorem 5, we considered vertex operators corresponding to parallel sections in the graded vector bundle $\underline{V}$ of weight up to 2. We didn't discuss the vertex operator corresponding to the parallel section $[\kappa_{(1,0)}] \in \mathcal{P}_M(\underline{V})_2$. But its vertex operator $Y([\kappa_{(1,0)}],\zeta) = \sum_{n \in \mathbb{Z}} K_{(1,0)}(n)\zeta^{-n-2}$ doesn't provide new operators. Due to the general relation $\{D(-1)v\}_{n+1} = (n+1)\{v\}_n$ for $v \in V$ and for $n \in \mathbb{Z}$, and the relation $[\kappa_{(1,0)}] = -D(-1)[\kappa]$ imply that $K_{(1,0)}(n) = -(n+1)K(n)$ for $n \in \mathbb{Z}$. So, the algebra generated by $K_{(1,0)}(n)$'s forms only a subalgebra of the Heisenberg algebra.

Elliptic genera for special Kähler manifolds. A special Kähler manifold is a Kähler manifold whose Ricci tensor identically vanishes. This forces the *restricted* holonomy group of the manifold to be contained in $SU(N)$. For our present purpose, we need a stronger requirement. We require that the *entire* holonomy group be contained in $SU(N)$. In this case, there exist complex volume forms ξ^+ on M^{2N} which are nowhere vanishing parallel forms of type $(N,0)$. In fact, the parallel differential forms of type $(1,1)$ form a one dimensional vector space generated by ξ^+. Its conjugate ξ^- is also a parallel differential form and is of type $(0,N)$.

PROPOSITION 8 (PARALLEL SECTIONS ON SPECIAL KÄHLER MANIFOLDS). *Let M^{2N} be a connected Kählerian manifold whose holonomy group is precisely $SU(N)$. Then, up to weight 2 the vertex operator super algebra $\mathcal{P}_M(\underline{V})$ of parallel sections in the graded vector bundle $\underline{V}$ is given as follows:*

$$\mathcal{P}_M(\underline{V})_0 = \mathbb{C} \cdot 1, \qquad \mathcal{P}_M(\underline{V})_{1/2} = \{0\},$$

$$\mathcal{P}_M(\underline{V})_1 = \begin{cases} \mathbb{C}[\kappa], & \text{if } N \neq 2, \\ \mathbb{C}[\kappa] \oplus \mathbb{C}[\xi^+] \oplus \mathbb{C}[\xi^-], & \text{if } N = 2, \end{cases}$$

$$\mathcal{P}_M(\underline{V})_{3/2} = \begin{cases} \{0\}, & \text{if } N \neq 3, \\ \mathbb{C}[\xi^+] \oplus \mathbb{C}[\xi^-], & \text{if } N = 3, \end{cases}$$

$$\mathcal{P}_M(\underline{V})_2 = \begin{cases} \mathbb{C}[\eta^+] \oplus \mathbb{C}[\eta^-] \oplus \mathbb{C}[\kappa^2], & \text{if } N \neq 4, \\ \mathbb{C}[\eta^+] \oplus \mathbb{C}[\eta^-] \oplus \mathbb{C}[\kappa^2] \oplus \mathbb{C}[\xi^+] \oplus \mathbb{C}[\xi^-], & \text{if } N = 4. \end{cases}$$

Here, we have chosen complex volume forms $[\xi^\pm]$ in such a way that $[\xi^\pm]_x = [\rho_x^\pm]$ at any $x \in M$.

PROOF. This Proposition follows from our previous calculation of $V^{SU(N)}$ in Proposition 5 in §3.12, together with the isomorphism $\mathcal{P}_M(\underline{V}) \xrightarrow{\cong} (V)_x^{SU(N)} \xleftarrow{\cong}$

$(V)_*^{\mathrm{SU}(N)}$ in Proposition 23 of §4.5 induced by an evaluation map at $x \in M$. We only have to note that

$$[\kappa]_x = -i[h_x], \quad [\kappa^2]_x = -[h_x^2], \quad [\eta^\pm]_x = -2[\omega_{\pm,x}], \quad [\xi^\pm]_x = [\rho_x^\pm].$$

The last identity is by our choice of complex volume forms $\xi^\pm$. □

. We note that up to weight 2, the algebra $\mathcal{P}_M(\underline{V})$ for a Kählerian SU(N) manifold and that for Kählerian manifold differ only by appearance of complex volume forms $[\xi^\pm]$.

Although we are not very much concerned with weight 3 vectors in $\mathcal{P}_M(\underline{V})$, for the next Theorem, the following weight 3 vector plays a role:

$$\gamma = \kappa^3 + 3\kappa_{(1,1)} - 3\kappa \cdot g_{(1,0)} \in \mathcal{P}_M(\underline{V})_3.$$

Let the corresponding vertex operator be $Y(\gamma, \zeta) = \sum_{n \in \mathbb{Z}} \Gamma(n) \zeta^{-n-3}$.

Applying Module Theorem in §5.4 to Kählerian manifolds with SU(N) holonomy, we obtain the following Theorem.

THEOREM 9 (ELLIPTIC GENERA FOR SPECIAL KÄHLER MANIFOLDS WITH SU(N) HOLONOMY). *For a connected closed special Kählerian manifold M^{2N} whose entire holonomy group is contained in* SU(N), *each part of the elliptic genus* $\Phi_{\mathrm{ell}}^*(M) = [\Phi_{\mathrm{ell}}^*(M)_\ell ; \Phi_{\mathrm{ell}}^*(M)_r]$ *is a representation of the following subalgebras of vertex operator super algebra* $\mathcal{P}_M(\underline{V})$:

(1) (Heisenberg and unitary Virasoro algebra) *Let vertex operators corresponding to the Kähler form* $[\kappa] \in \mathcal{P}_M(\underline{V})$ *and the unitary Virasoro elements* $[\Theta], [\Lambda] \in \mathcal{P}_M(\underline{V})_2$ *be*

$$Y([\kappa], z) = \sum_{n \in \mathbb{Z}} K(n) z^{-n-1},$$

$$Y([\Theta], z) = \sum_{n \in \mathbb{Z}} \Theta(n) z^{-n-2}, \qquad Y([\Lambda], z) = \sum_{n \in \mathbb{Z}} \Lambda(n) z^{-n-2},$$

where $[\Theta] = ([\kappa^2] - [g_{(1,0)}])/(2N)$ *and* $[\Lambda] = -([\kappa^2] + (N-1)[g_{(1,0)}])/(2N)$. *Then* $\Theta(n) + \Lambda(n) = D(n)$ *for* $n \in \mathbb{Z}$, *and these operators generate a semi direct sum algebra of a Heisenberg algebra and a unitary Virasoro algebra with commutation relations given by*

$$[K(m), K(n)] = nN\delta_{m+n,0}\mathrm{Id},$$

$$[K(m), \Theta(n)] = -mK(m+n), \qquad [K(m), \Lambda(n)] = 0,$$

$$\begin{cases} [\Theta(m), \Theta(n)] = (n-m)\Theta(m+n) + \dfrac{m(m^2-1)}{12}\delta_{m+n,0} \cdot \mathrm{Id}, \\[2mm] [\Lambda(m), \Lambda(n)] = (n-m)\Lambda(m+n) + \dfrac{m(m^2-1)}{12}\delta_{m+n,0}(N-1) \cdot \mathrm{Id}, \\[2mm] [\Theta(m), \Lambda(n)] = 0, \end{cases}$$

for any $m, n \in \mathbb{Z}$.

(2) (Vertex operators generated by complex volume forms) *Let the vertex operators generated by the complex volume forms $[\xi^{\pm}] \in \mathcal{P}_M(\underline{V})_{N/2}$ be*

$$Y([\xi^+], z) = \sum_{n \in \mathbb{Z} - \frac{N}{2}} \Xi^+(n) z^{-n-\frac{N}{2}}, \qquad Y([\xi^-], z) = \sum_{n \in \mathbb{Z} - \frac{N}{2}} \Xi^-(n) z^{-n-\frac{N}{2}}.$$

Then, these operators satisfy the following commutation relations:

$$[\Xi^+(m), \Xi^+(n)]_{\pm} = 0, \qquad [\Xi^-(m), \Xi^-(n)]_{\pm} = 0,$$

for any $m, n \in \mathbb{Z} - \frac{N}{2}$, where $\pm = (-1)^{N-1}$. Namely, each family of operators $\{\Xi^+(n)\}_{n \in \mathbb{Z} - \frac{N}{2}}$, $\{\Xi^-(n)\}_{n \in \mathbb{Z} - \frac{N}{2}}$ generates an infinite dimensional abelian Lie algebra when N is even, and an infinite dimensional exterior algebra when N is odd.

(3) (Heisenberg and unitary Virasoro algebra, and complex volume operators) *Commutation relations between complex volume operators $\Xi^{\pm}(n)$ and those vertex operators in (1) are given by*

$$[K(m), \Xi^{\pm}(n)] = \mp i N \cdot \Xi^{\pm}(m+n),$$
$$[D(m), \Xi^{\pm}(n)] = \left\{ n - \left(\frac{N}{2} - 1 \right) m \right\} \Xi^{\pm}(m+n),$$
$$[\{\kappa^2\}_{m+1}, \Xi^{\pm}(n)] = 0,$$

for any $m \in \mathbb{Z}$, $n \in \mathbb{Z} - \frac{N}{2}$.

(4) (Mixed commutators) *Commutation relations between the two families of operators $\{\Xi^+(n)\}_{n \in \mathbb{Z} - \frac{N}{2}}$, $\{\Xi^-(n)\}_{n \in \mathbb{Z} - \frac{N}{2}}$ for $N \leq 4$ are given as follows.*

$$[\Xi^-(m), \Xi^+(n)]_- = iK(m+n) - m\delta_{m+n,0} \cdot \mathrm{Id}, \quad \text{if } N = 2,$$

$$[\Xi^-(m), \Xi^+(n)]_+ = N \cdot \Theta(m+n) - i\frac{(n-m)}{2} K(m+n)$$
$$- \frac{(2m+1)(2m-1)}{8}\delta_{m+n,0} \cdot \mathrm{Id}, \quad \text{if } N = 3,$$

$$[\Xi^-(m), \Xi^+(n)]_- = \frac{i}{6}\Gamma(m+n) + \left(\frac{n-m}{2} \right) N \cdot \Theta(m+n)$$
$$- \frac{i}{2}\left\{ \binom{m+1}{2} + \binom{n+1}{2} \right\} K(m+n) + \binom{m+1}{3}\delta_{m+n,0} \cdot \mathrm{Id}, \quad \text{if } N = 4.$$

Here, $m, n \in \mathbb{Z} - \frac{N}{2}$ and $\Gamma(n)$'s are vertex operators corresponding to the vector $\gamma = \kappa^3 + 3\kappa_{(1,1)} - 3\kappa \cdot g_{(1,0)} \in \mathcal{P}_M(\underline{V})_3$.

For the proof of (1) of Theorem 9 is the same as the Kählerian case. For (2) and (3), we recall some calculations in the vertex operator super algebra V involving the complex volume forms $\rho_{\pm}$ from §3.7.

PROPOSITION 10. *In the vertex operator super algebra V, we have the following (anti)commutation relations:*

$$[\{\rho_+\}_m, \{\rho_+\}_n] = 0, \qquad [\{h^2\}_{m+1}, \rho_\pm(n)] = 0, \qquad [\{\rho_-\}_m, \{\rho_-\}_n] = 0,$$
$$[D(m), \rho_\pm(n)] = (n - (\tfrac{N}{2} - 1)m)\, \rho_\pm(n+m),$$
$$[h(m), \rho_\pm(n)] = (-1)^{N-1} N \cdot \rho_\pm(n+m).$$

PROOF. The above formulae are proved in Proposition 3 and Proposition 7 of §3.7. □

(PROOF OF (2) AND (3) OF THEOREM 9). We first note that the values of parallel sections $[\kappa], [\kappa^2], [\xi^\pm]$ at $x \in M$ are given by

$$[\kappa]_x = -i[h_x], \qquad [\kappa^2]_x = -[h_x^2], \qquad [\xi^\pm]_x = [\rho_x^\pm].$$

So, in $\mathrm{End}(\underline{V})_x$, we also have the following identities among operators.

$$K(m)_x = -ih_x(m), \qquad \{\kappa^2\}_{m+1,x} = -\{h_x^2\}_{m+1}, \qquad \Xi^\pm(m)_x = \rho_x^\pm(m),$$

for any $m \in \mathbb{Z}$. Since $\mathcal{P}_M\,(\mathrm{End}(\underline{V})) \hookrightarrow \mathrm{End}(\underline{V})_x$ is an injection for any $x \in M$, we have (2) and (3) of Theorem 9 by globalizing commutation relations in Proposition 10. □

For the proof of (4) in Theorem 9, we recall our calculations of mixed commutators of vertex operators $\rho_\pm(n)$ for $N \le 4$ from Proposition 5 in §3.7.

LEMMA 11. *In the vertex operator super algebra V, the following commutation relations hold among vertex operators corresponding to parallel sections $\rho_\pm \in \mathcal{P}_M(\underline{V})_{N/2}$ on M^{2N}:*

$$[\{\rho_-\}_m, \{\rho_+\}_n]_- = h(m+n) - m \cdot \delta_{m+n,0} \cdot \mathrm{Id}, \qquad N = 2,$$

$$[\{\rho_-\}_m, \{\rho_+\}_n]_+ = -\frac{1}{2}\{h^2\}_{m+n} + D(m+n-1)$$
$$-\frac{(n-m)}{2}h(m+n-1) - \frac{m(m-1)}{2} \cdot \delta_{m+n-1,0} \cdot \mathrm{Id}, \qquad N = 3,$$

$$[\{\rho_-\}_m, \{\rho_+\}_n]_- = \frac{1}{2}\{h(-\tfrac{3}{2})\}_{m+n} - \frac{1}{6}\{h^3\}_{m+n} + \{\omega \cdot h\}_{m+n}$$
$$-\frac{(n-m)}{4}\{h^2\}_{m+n-1} + \frac{(n-m)}{2}D(m+n-2)$$
$$-\frac{1}{2}\left\{\binom{m}{2} + \binom{n}{2}\right\}h(m+n-2) + \binom{m}{3}\delta_{m+n-2,0} \cdot \mathrm{Id}, \qquad N = 4.$$

With the proof of (4) of Theorem 9 in mind, we rewrite the above relations in a form convenient for our purpose. First we note that in the above formulae, vertex operators coming from weight 2 vectors are actually vertex operators for $[\theta]$ where $\theta = (-\frac{1}{2}h^2 + \omega)/N$. Note that this θ is the same as the one we defined previously in Lemma 7. Its associated vertex operators $\{\theta\}_*$ are such that

$$N\{\theta\}_{n+1} = -\tfrac{1}{2}\{h^2\}_{n+1} + D(n),$$

for $n \in \mathbb{Z}$. Also, since $\rho_\pm(n) = \{\rho_\pm\}_{n+\frac{N}{2}-1}$, shifting subscripts in the above formulae, we obtain

PROPOSITION 12. *The following commutation relations hold in* V,

$$[\rho_-(m), \rho_+(n)]_- = h(m+n) - m \cdot \delta_{m+n,0} \cdot \mathrm{Id}, \quad \textit{if } N = 2,$$

$$[\rho_-(m), \rho_+(n)]_+ = N\{\theta\}_{m+n+1} - \frac{(n-m)}{2} h(m+n)$$
$$- \frac{m(m-1)}{2} \cdot \delta_{m+n-1,0} \cdot \mathrm{Id}, \quad \textit{if } N = 3,$$

$$[\rho_-(m), \rho_+(n)]_- = \frac{i}{6}\{\gamma\}_{m+n+2} + \left(\frac{n-m}{2}\right) N\{\theta\}_{m+n+1}$$
$$- \frac{1}{2}\left\{\binom{m+1}{2} + \binom{n+1}{2}\right\} h(m+n) + \binom{m+1}{3}\delta_{m+n,0} \cdot \mathrm{Id}, \quad \textit{if } N = 4,$$

where $\gamma = i\{h^3 - 3h(-\frac{3}{2}) - 6h \cdot \omega\}$.

REMARK. In Proposition 12, results of super commutators are described in terms of vertex operators corresponding to parallel sections of different weights. For $N = 2$ case, the commutator needs a weight 1 vector $[h]$ for its description. For $N = 3$, we need a weight 2 vector $[\theta]$ in addition to $[h]$ to describe the anticommutator. When $N = 4$, a weight 3 vector $[\gamma]$ is necessary for the description of mixed commutators of complex volume operators. It is likely that this pattern continues picking up a certain weight j vector for $N = j+1$ case, $j \geq 1$, although it is not clear what these specific weight j parallel sections mean geometrically.

(PROOF OF (4) OF THEOREM 9). We first observe that the following identities of local vertex operators hold. Here, $\gamma = \kappa^3 + 3\kappa_{(1,1)} - 3\kappa \cdot g_{(1,0)}$ is a weight 3 global section in $\underline{V}$.

$$\Gamma(m+n)_x = \{\gamma_x\}_{m+n+2}, \qquad \Theta(m+n)_x = \{\theta_x\}_{m+n+1}, \qquad \Xi^\pm(n)_x = \rho_x^\pm(n).$$

For example, for the first one

$$\Gamma(n)_x = \{\Gamma\}_{n+2,x} = \{\kappa^3\}_{n+2,x} + 3\{\kappa_{(1,1)}\}_{n+2,x} - 3\{\kappa \cdot g_{(1,0)}\}_{n+2,x}$$
$$= \{(-ih_x)^3\}_{n+2} + 3\{-ih_x(-\frac{3}{2})\}_{n+2} - 3\{(-ih_x) \cdot (-2\omega_x)\}_{n+2} = \{\gamma_x\}_{n+2}.$$

Now the part (4) of Theorem 9 follows from Proposition 12 since the evaluation map of parallel operators $\mathcal{P}_M (\mathrm{End}(\underline{V})) \rightarrow \mathrm{End}(\underline{V})_x$ is an injection due to Proposition 23 in §4.5. ☐

The statement (3) means that vector spaces

$$\mathbb{C} \cdot \mathrm{Id} \oplus \bigoplus_{n \in \mathbb{Z}} \mathbb{C}K(n) \oplus \bigoplus_{n \in \mathbb{Z}} \mathbb{C}\Theta(n) \oplus \bigoplus_{n \in \mathbb{Z}} \mathbb{C}\Lambda(n) \oplus \bigoplus_{n \in \mathbb{Z}-\frac{N}{2}} \mathbb{C}\Xi^+(n),$$

$$\mathbb{C} \cdot \mathrm{Id} \oplus \bigoplus_{n \in \mathbb{Z}} \mathbb{C}K(n) \oplus \bigoplus_{n \in \mathbb{Z}} \mathbb{C}\Theta(n) \oplus \bigoplus_{n \in \mathbb{Z}} \mathbb{C}\Lambda(n) \oplus \bigoplus_{n \in \mathbb{Z}-\frac{N}{2}} \mathbb{C}\Xi^-(n)$$

are closed under Lie brackets, hence each of the above two vector spaces have the structure of a Lie (super) algebra. Note that subspaces $\bigoplus_{n\in\mathbb{Z}}\mathbb{C}\Xi^+(n)$ and $\bigoplus_{n\in\mathbb{Z}}\mathbb{C}\Xi_-(n)$ are ideals of the above Lie algebras and the quotient Lie algebras are the same as the Heisenberg-unitary Virasoro algebra for the Kählerian manifolds case. Recall that this algebra has the Heisenberg algebra as its ideal and the quotient is the unitary Virasoro algebra of rank $(1, N-1)$.

Note that the Riemannian volume form ξ is equal to $\xi^+\xi^-$ up to a sign. It may be of some interest to study the vertex operator $Y([\xi], \zeta)$ from this point of view.

Elliptic genera for quaternion-Kähler manifolds. Let M be a closed quaternion-Kähler manifold. So, it is a real $4N'$-dimensional Riemannian manifold whose holonomy group is contained in $\mathrm{Sp}(N')\cdot\mathrm{Sp}(1)$. Here, the compact Lie groups $\mathrm{Sp}(N')$ and $\mathrm{Sp}(1)$ are commuting subgroups of $\mathrm{SO}(4N')$ whose intersection is $\{\pm 1\}$ and $\mathrm{Sp}(N')\cdot\mathrm{Sp}(1) = \mathrm{Sp}(N') \times \mathrm{Sp}(1)/\{\pm 1\}$ is the subgroup of $\mathrm{SO}(4N')$ generated by $\mathrm{Sp}(N')$ and $\mathrm{Sp}(1)$. Some properties of quaternion-Kähler manifolds relevant to our present purpose are discussed in §5.1. We recall that a quaternion-Kähler manifold $M^{4N'}$ possesses a canonical parallel closed 4-form, the quaternion-Kähler form κ_{Q-K}, such that on an open set U on which the structure bundle L trivializes,

$$\kappa_{Q-K}|_U = \kappa_I^2 + \kappa_J^2 + \kappa_K^2,$$

where I, J, K are isometric almost complex structures on U such that $L|_U = \mathbb{C}I \oplus \mathbb{C}J \oplus \mathbb{C}K$ and $I^2 = J^2 = -1$, $IJ = -JI = -K$. The differential forms κ_I, κ_J, κ_K are 2-forms defined by

$$\kappa_{\mathcal{J}}(X, Y) = g(\mathcal{J}(X), Y), \qquad X, Y \in TM,$$

for $\mathcal{J} = I, J, K$. Since almost complex structures I, J, K are not necessarily parallel on quaternion-Kähler manifolds, the differential forms κ_I, κ_J, κ_K may not be parallel nor closed. However, the sum of their squares $\kappa_I^2 + \kappa_J^2 + \kappa_K^2$ is parallel and closed on U. For details, see Proposition 18 in §5.1.

Let $[\kappa_{Q-K}]$ denote the image of $\kappa_{Q-K} \in \bigwedge^4 T^*M$ under the embedding

$$\bigwedge^4 T^*M \subset \bigwedge^* T_\mathbb{C}^*M \xrightarrow{\cong} \bigwedge^* T_\mathbb{C}M(-\tfrac{1}{2}) \subset \bigoplus_{n\geq 0} \bigwedge^* T_\mathbb{C}M(-n-\tfrac{1}{2}) = \underline{V}.$$

The low weight part of the algebra $\mathcal{P}_M(\underline{V})$ of parallel sections on a quaternion-Kähler manifold is given as follows.

PROPOSITION 13 (PARALLEL SECTIONS ON QUATERNION-KÄHLER MANIFOLDS). *Let $M^{4N'}$ is a connected quaternion-Kähler manifold whose holonomy group is precisely $\mathrm{Sp}(N')\cdot\mathrm{Sp}(1)$. Then, up to weight 2, the vertex operator super algebra $\mathcal{P}_M(\underline{V})_*$ of parallel sections in $\underline{V}$ is given as follows:*

$$\mathcal{P}_M(\underline{V})_0 = \mathbb{C}\cdot 1, \qquad \mathcal{P}_M(\underline{V})_{\frac{k}{2}} = \{0\}, \quad 1\leq k\leq 3,$$
$$\mathcal{P}_M(\underline{V})_2 = \mathbb{C}[g_{(1,0)}] \oplus \mathbb{C}[\kappa_{Q-K}].$$

PROOF. This follows from our previous calculation of $\mathrm{Sp}(N')\cdot\mathrm{Sp}(1)$-invariants in the vertex operator super algebra V, by using the fact that the evaluation of parallel sections at $x \in M$ induces an isomorphism

$$\mathcal{P}_M(\underline{V})_* \xrightarrow[\mathrm{Res}]{\cong} (\underline{V}_x)_*^{\mathrm{Sp}(N')\cdot\mathrm{Sp}(1)} \xleftarrow{\cong} (V)_*^{\mathrm{Sp}(N')\cdot\mathrm{Sp}(1)},$$

and by noting that the effect of the above evaluation map is given by $[g_{(1,0)}]_x = -2[\omega_x]$, $[\kappa_{Q-K}]_x = [\omega_{I,x}^2] + [\omega_{J,x}^2] + [\omega_{K,x}^2] = 2[\phi_{\mathfrak{sp}(N'),x}]$. $\square$

Next, we study subalgebras of the vertex operator super algebra $\mathcal{P}_M(\underline{V})$.

THEOREM 14 (ELLIPTIC GENERA FOR QUATERNION-KÄHLER MANIFOLDS).
Let M be a connected closed quaternion-Kähler Spin manifold. Then, the vertex operator super algebra $\mathcal{P}_M(\underline{V})$ contains a symplectic Virasoro algebra. Consequently, each part of the elliptic genus $\Phi_{\mathrm{ell}}^*(M^{4N'}) = [\Phi_{\mathrm{ell}}^*(M^{4N'})_\ell \; ; \; \Phi_{\mathrm{ell}}^*(M^{4N'})_r]$ is a representation of the symplectic Virasoro algebra.
More precisely, let $[\sigma], [\tau] \in \mathcal{P}_M(\underline{V})_2$ be given by

$$[\sigma] = \frac{1}{2(N+4)}\left([\kappa_{Q-K}] - 3[g_{(1,0)}]\right), \quad [\tau] = \frac{-1}{2(N+4)}\left([\kappa_{Q-K}] + (N+1)[g_{(1,0)}]\right).$$

Let the corresponding vertex operators be

$$Y([\sigma], z) = \sum_{n \in \mathbb{Z}} S(n) z^{-n-2}, \qquad Y([\tau], z) = \sum_{n \in \mathbb{Z}} T(n) z^{-n-2}.$$

Then, operators $\{S(n)\}_{n \in \mathbb{Z}}$, $\{T(n)\}_{n \in \mathbb{Z}}$ generate commuting Virasoro algebras of rank $3N/(N+4)$ and $N(N+1)/(N+4)$, respectively, such that $S(n) + T(n) = D(n)$ for $n \in \mathbb{Z}$. Namely, for any $m, n \in \mathbb{Z}$, we have

$$\begin{cases} [S(m), S(n)] = (n-m)S(m+n) + \dfrac{m(m^2-1)}{12}\left\{\dfrac{3N}{(N+4)}\right\}\delta_{m+n,0} \cdot \mathrm{Id} \\[2mm] [T(m), T(n)] = (n-m)T(m+n) + \dfrac{m(m^2-1)}{12}\left\{\dfrac{N(N+1)}{(N+4)}\right\}\delta_{m+n,0} \cdot \mathrm{Id} \\[2mm] [T(m), S(n)] = 0. \end{cases}$$

PROOF. First observe that restrictions of parallel sections $[\kappa_{Q-K}]$, $[g_{(1,0)}]$ at $x \in M$ is given by

$$[\kappa_{Q-K}]_x = [\omega_{I,x}^2] + [\omega_{J,x}^2] + [\omega_{K,x}^2] = 2[\phi_{\mathfrak{sp}(N')}], \qquad [g_{(1,0)}]_x = -2[\omega_x].$$

Thus, the parallel sections $[\sigma]$, $[\tau]$ given in Theorem 14 restrict to the following vectors in the fibre $\underline{V}_x$:

$$[\sigma]_x = \frac{1}{N+4}\left(3[\omega_x] + [\phi_{\mathfrak{sp}(N'),x}]\right), \quad [\tau]_x = \frac{1}{N+4}\left((N+1)[\omega_x] - [\phi_{\mathfrak{sp}(N'),x}]\right).$$

Commutation relations for vertex operators generated by these vectors acting on $\underline{V}_x$ are given in Theorem 2 in §4.8, or in Corollary 20 in §3.11. By Proposition 10 or by Proposition 11 in §5.4, the same commutation relations hold for parallel operators defined globally on $M^{4N'}$. $\square$

Elliptic genera for hyperkähler manifolds. Let $M^{4N'}$ be a hyperkähler manifold, that is, a Riemannian manifold whose holonomy group is contained in $\mathrm{Sp}(N')$. Since $\mathrm{Sp}(N') \subset \mathrm{SO}(4N')$ is a simply connected subgroup, we can lift this inclusion map to a map into $\mathrm{Spin}(4N')$. Thus, a hyperkähler manifold is always a Spin manifold. On a hyperkähler manifold $M^{4N'}$, there exists three integrable parallel almost complex structures I, J, K satisfying quaternion relations, $I^2 = J^2 = -1$, $IJ = -JI = -K$, and acting on the tangent bundle TM from the left as isometries. Let $\mathbb{H} = \mathbb{R} \oplus \mathbb{R}i \oplus \mathbb{R}j \oplus \mathbb{R}k$ be the skew field of quaternions. The almost complex structures I, J, K on M allow us to define a (right) $\mathbb{H}$-bundle structure on the tangent bundle TM by

$$X \cdot i = I(X), \qquad X \cdot j = J(X), \qquad X \cdot k = K(X),$$

for any $X \in TM$. The usual quaternion relations among i, j, k, $i^2 = j^2 = -1$, $ij = -ji = k$ are compatible with the above relations among I, J, K.

The existence of isometric almost complex structures satisfying the quaternionic relations implies that the manifold M carries a real two dimensional family of almost complex structures of the form $\mathcal{J} = aI + bJ + cK$ with $a^2 + b^2 + c^2 = 1$, $a, b, c \in \mathbb{R}$. Any almost complex structure $\mathcal{J}$ of this form is an integrable parallel almost complex structure acting as an isometry on TM. Thus, there is a real 2-dimensional family of compatible complex structures on the manifold. Let $\kappa_{\mathcal{J}}$ be the Kähler form associated to $\mathcal{J}$ defined by

$$\kappa_{\mathcal{J}}(X, Y) = g(\mathcal{J}(X), Y), \qquad X, Y \in TM.$$

Since $\mathcal{J}$ is parallel and the Levi-Civita connection is torsion free, this parallel form is a closed 2-form. In fact, it is a $(1, 1)$ form with respect to the complex structure $\mathcal{J}$. In §3.9, we have seen that in quaternionic vector spaces, Kähler forms $\omega_{\mathcal{J}}$ depend linearly on the complex structure $\mathcal{J}$. So, on a hyperkähler manifold $M^{4N'}$, the Kähler form $\kappa_{\mathcal{J}}$ also depends linearly on the almost complex structure $\mathcal{J}$. That is, for any $a, b, c \in \mathbb{R}$ with $a^2 + b^2 + c^2 = 1$, we have

$$\kappa_{aI+bJ+cK} = a\kappa_I + b\kappa_J + c\kappa_K \in \Gamma\left(\bigwedge^2 T^*M\right).$$

Let $TM(-n - \tfrac{1}{2})$ be a copy of TM of weight $n + \tfrac{1}{2}$. As before, we identify the tangent bundle TM with $TM(1/2)$, and the cotangent bundle T^*M with $TM(-1/2)$. Consequently, we identify the total exterior algebra $\bigwedge^* T^*M$ with $\bigwedge^* TM(-\tfrac{1}{2})$ which is a subbundle of $\underline{V}$. For any almost complex structure $\mathcal{J}$ on M, we let $[\kappa_{\mathcal{J}}]$ be the image of $\kappa_{\mathcal{J}} \in \bigwedge^2 T^*M$ under this identification. So,

$$[\kappa_{\mathcal{J}}] \in \Gamma\left(\bigwedge^2 T_{\mathbb{C}}M(-\tfrac{1}{2})\right) \subset \Gamma\left(\bigotimes_{n \geq 0}\bigwedge^* T_{\mathbb{C}}M(-n - \tfrac{1}{2})\right) = \Gamma(\underline{V}).$$

Images of all nontrivial exterior products of parallel Kähler forms $\kappa_I, \kappa_J, \kappa_K$ under the embedding of the total exterior algebra $\bigwedge^* T_{\mathbb{C}}M$ into $\underline{V}$ are also denoted with brackets, $[\kappa_I], [\kappa_J], \ldots, [\kappa_I \kappa_J], \ldots$.

Let the complementary subspace of the 1-dimensional space $\mathbb{C}[\kappa_I^2 + \kappa_K^2 + \kappa_K^2]$ in $S^2(\mathbb{C}[\kappa_I] \oplus \mathbb{C}[\kappa_K] \oplus \mathbb{C}[\kappa_K])$ be given by

$$S^2(\mathbb{C}[\kappa_I] \oplus \mathbb{C}[\kappa_K] \oplus \mathbb{C}[\kappa_K])^{\perp} = \left\{ \sum_{\substack{\mathcal{J}_1, \mathcal{J}_2 \\ = I, J, K}} a_{\mathcal{J}_1 \mathcal{J}_2}[\kappa_{\mathcal{J}_1} \kappa_{\mathcal{J}_2}] \;\middle|\; a_{II} \neq a_{JJ} \text{ or } a_{JJ} \neq a_{KK} \right\}.$$

As we have seen, the vector $\kappa_I^2 + \kappa_K^2 + \kappa_K^2$ in $\mathcal{P}_M(\underline{V})_2$ plays a special role for quaternion-Kähler manifolds. It also plays a role for hyperkähler manifolds also.

We recall that the generalized Riemannian tensor $[g_{(1,0)}]$ is a parallel tensor field of weight 2.

$$[g_{(1,0)}] \in \Gamma\left(T_\mathbb{C}M(-\tfrac{3}{2}) \otimes T_\mathbb{C}(-\tfrac{1}{2})\right)$$

We also recall that the vector bundle $\underline{V}$ is a bundle of Virasoro representations. So, the Virasoro operator $D(-1)$ acts on sections of $\underline{V}$, fibrewise.

Next Proposition describes the graded algebra of parallel sections in $\underline{V}$ when the underlying manifold is hyperkähler.

PROPOSITION 15 (PARALLEL SECTIONS ON HYPERKÄHLER MANIFOLDS). *Let* $M^{4N'}$ *be a connected hyperkähler manifold whose holonomy group is exactly* $\mathrm{Sp}(N')$. *Then, up to weight 2, the graded vector space* $\mathcal{P}_M(\underline{V})$ *of parallel sections in the graded vector bundle* $\underline{V}$ *is described as follows:*

$$\mathcal{P}_M(\underline{V})_0 = \mathbb{C} \cdot 1, \qquad\qquad \mathcal{P}_M(\underline{V})_{\frac{1}{2}} = \{0\},$$
$$\mathcal{P}_M(\underline{V})_1 = \mathbb{C}[\kappa_I] \oplus \mathbb{C}[\kappa_J] \oplus \mathbb{C}[\kappa_K], \qquad \mathcal{P}_M(\underline{V})_{\frac{3}{2}} = \{0\},$$

$$\begin{cases} \mathcal{P}_M(\underline{V})_2 = \mathbb{C}D(-1)[\kappa_I] \oplus \mathbb{C}D(-1)[\kappa_J] \oplus \mathbb{C}D(-1)[\kappa_K] \\ \qquad\qquad\qquad\qquad \oplus \mathbb{C}[g_{(1,0)}] \oplus \mathbb{C}[\kappa^2], \qquad \text{if } N' = 1, \\[2mm] \mathcal{P}_M(\underline{V})_2 = \mathbb{C}D(-1)[\kappa_I] \oplus \mathbb{C}D(-1)[\kappa_J] \oplus \mathbb{C}D(-1)[\kappa_K] \oplus \mathbb{C}[g_{(1,0)}] \\ \oplus \mathbb{C}[\kappa_I^2] \oplus \mathbb{C}[\kappa_J^2] \oplus \mathbb{C}[\kappa_K^2] \oplus \mathbb{C}[\kappa_I \kappa_J] \oplus \mathbb{C}[\kappa_J \kappa_K] \oplus [\kappa_K \kappa_I], \qquad \text{if } N' \geq 2. \end{cases}$$

When $N' = 1$, *then* $\mathrm{Sp}(1) = \mathrm{SU}(2)$ *and* $[\kappa_I^2] = [\kappa_J^2] = [\kappa_K^2] = [\kappa^2]$ *is the Riemannian volume form up to a sign. When* $N' \geq 2$, $\mathcal{P}_M(\underline{V})_2$ *can be concisely written as*

$$\mathcal{P}_M(\underline{V})_2 = \mathbb{C}[g_{(1,0)}] \oplus D(-1)\left(\mathcal{P}_M(\underline{V})_1\right) \oplus S^2\left(\mathcal{P}_M(\underline{V})_1\right).$$

PROOF. This follows from our previous calculation of $\mathrm{Sp}(N')$-invariants in the vertex operator super algebra V in Lemma 9 in §3.12. We recall from §5.2 that restrictions of the above parallel forms to a point $x \in M$ are given by

$$[\kappa_I]_x = [\omega_{I,x}], \quad [\kappa_J]_x = [\omega_{J,x}], \quad [\kappa_K]_x = [\omega_{K,x}], \quad [g_{(1,0)}]_x = -2[\omega_x].$$

We then appeal to the restriction isomorphism $\mathcal{P}_M(\underline{V}) \xrightarrow{\cong} (\underline{V}_x)_*^{\mathrm{Sp}(N')}$ to conclude that there are no other parallel sections in $\underline{V}$ up to weight 2. $\square$

For a general hyperkähler manifold, the algebra of parallel sections $\mathcal{P}_M(\underline{V})$ can be larger than the vector space described in Proposition 13 since the holonomy group can be strictly smaller than $\mathrm{Sp}(N')$, but it still contains the algebra generated by Kähler forms. This algebra of Kähler forms and the corresponding vertex operators are important as demonstrated in Theorem 16 below. We let

$$\mathcal{P}_M(\underline{V})^{\frac{1}{2}}_2 = D(-1)\,(\mathcal{P}_M(\underline{V})_1) \oplus S^2\,(\mathcal{P}_M(\underline{V})_1)^{\perp}\,.$$

Here, $S^2\,(\mathcal{P}_M(\underline{V})_1)^{\perp}$ is the part "perpendicular" to $\mathbb{C}[\kappa_{Q-K}]$ in $S^2(\mathcal{P}_M(\underline{V})_1)$. The above $\mathcal{P}_M(\underline{V})^{\frac{1}{2}}_2$ is "perpendicular" to $\mathrm{Sp}(N') \cdot \mathrm{Sp}(1)$-invariants $\mathbb{C}[g_{(1,0)}] \oplus \mathbb{C}[\kappa_{Q-K}]$ in $S^2\,(\mathcal{P}_M(\underline{V})_2)$. So,

$$\mathcal{P}_M(\underline{V})_2 = \mathbb{C}[g_{(1,0)}] \oplus \mathbb{C}[\kappa_{Q-K}] \oplus \mathcal{P}_M(\underline{V})^{\frac{1}{2}}_2\,.$$

We recall that if $M^{4N'}$ is connected, then, the vertex operator super algebra $\mathcal{P}_M(\underline{V})$ is also connected, i.e., $\mathcal{P}_M(\underline{V})_0 = \mathbb{C} \cdot 1_M$.

THEOREM 16 (ELLIPTIC GENERA FOR HYPERKÄHLER MANIFOLDS). *Let M be a connected closed hyperkähler manifold with three parallel integrable almost complex structures I, J, K. Let $\kappa_I, \kappa_J, \kappa_K$ be the corresponding Kähler forms. Then, each part of the elliptic genus $\Phi^*_{\mathrm{ell}}(M^{4N'}) = [\Phi^*_{\mathrm{ell}}(M^{4N'})_\ell \; ; \; \Phi^*_{\mathrm{ell}}(M^{4N'})_r]$ is a representation of the vertex operator super algebra $\mathcal{P}_M(\underline{V})$. The algebra $\mathcal{P}_M(\underline{V})$ contains the following (Lie) subalgebras:*

(1) (*Affine Lie algebra* $A_1^{(1)}$) *The vertex operators corresponding to Kähler forms in $\mathcal{P}_M(\underline{V})_1$ generate a representation of an affine Lie algebra $A_1^{(1)} = \widehat{\mathfrak{sl}}_2(\mathbb{C})$ of level N'. Namely, let the associated vertex operators, which we call Kähler operators, be given by*

$$Y([\kappa_{\mathcal{J}}], z) = \sum_{n \in \mathbb{Z}} K_{\mathcal{J}}(n)z^{-n-1}, \qquad \mathcal{J} = I, J, K.$$

Then, these operators satisfy the following commutation relations:

$$[K_{\mathcal{J}_1}(m), K_{\mathcal{J}_2}(n)] = [K_{\mathcal{J}_1}, K_{\mathcal{J}_2}](m+n) + m\langle K_{\mathcal{J}_1}, K_{\mathcal{J}_2}\rangle\delta_{m+n,0} \cdot \mathrm{Id},$$

where the Lie brackets $[\ ,\ ]$ and pairings $\langle\ ,\ \rangle$ are given by

$$[K_I, K_J] = 2K_K, \qquad [K_J, K_K] = 2K_I, \qquad [K_K, K_I] = 2K_J,$$
$$\langle K_I, K_I\rangle = \langle K_J, K_J\rangle = \langle K_K, K_K\rangle = -N = -2N',$$
$$\langle K_I, K_J\rangle = \langle K_J, K_K\rangle = \langle K_K, K_I\rangle = 0.$$

The standard normalized pairing $\langle\ ,\ \rangle_0$ on $\mathfrak{sl}_s(\mathbb{C})$ is such that $\langle\ ,\ \rangle = \langle\ ,\ \rangle_0$.

(2) (*Vertex operators for quadratic Kähler forms*) *Let the vertex operator corresponding to any quadratic Kähler form $\vartheta \in S^2\,(\mathbb{C}[\kappa_I] \oplus \mathbb{C}[\kappa_J] \oplus \mathbb{C}[\kappa_K])$ be*

$$Y([\vartheta], z) = \sum_{n \in \mathbb{Z}} \{\vartheta\}_{n+1}z^{-n-2}.$$

Then, the action of corresponding vertex operators $\{\vartheta\}_*$ on $[\vartheta]$ is given by

$$\{\vartheta\}_0[\vartheta] = -\tfrac{1}{2}D(-1)\left(\{\vartheta\}_1[\vartheta]\right),$$

$$\{\vartheta\}_1[\vartheta] \in S^2\left(\mathbb{C}[\kappa_I] \oplus \mathbb{C}[\kappa_J] \oplus \mathbb{C}[\kappa_K]\right) \oplus \mathbb{C}[g_{(1,0)}],$$

$$\{\vartheta\}_2[\vartheta] = 0,$$

$$\{\vartheta\}_3[\vartheta] \in \mathbb{C}\Omega,$$

$$\{\vartheta\}_k[\vartheta] = 0, \qquad k \geq 4.$$

Consequently, the vertex operators $\{\vartheta\}_*$ have the following commutation relations for any $m, n \in \mathbb{Z}$:

$$[\{\vartheta\}_{m+1}, \{\vartheta\}_{n+1}] = \frac{(m-n)}{2}\{\{\vartheta\}_1[\vartheta]\}_{m+n+1} + \binom{m+1}{3}\left(\{\vartheta\}_3[\vartheta], \Omega\right)\delta_{m+n,0}\cdot\mathrm{Id}.$$

Here, if $[\vartheta] = a[\kappa_I^2] + b[\kappa_J^2] + c[\kappa_K^2] + d[\kappa_I\kappa_J] + e[\kappa_J\kappa_K] + f[\kappa_K\kappa_I]$ for some coefficients $a, b, c, d, e, f \in \mathbb{C}$, then, by letting $2N' = N$ we have

$$\begin{aligned}
\{\vartheta\}_1[\vartheta] = -4\,\big\{&2(a+b+c)^2 + 2(N-2)(a^2+b^2+c^2) \\
&+(N-2)(d^2+e^2+f^2)\big\}\,[g_{(1,0)}]
\end{aligned}$$

$$- \left(4(N-2)a^2 + (N-4)d^2 + 4e^2 + (N-4)f^2 + 8ab - 16bc + 8ca\right)[\kappa_I^2]$$

$$- \left(4(N-2)b^2 + (N-4)d^2 + (N-4)e^2 + 4f^2 + 8ab + 8bc - 16ca\right)[\kappa_J^2]$$

$$- \left(4(N-2)c^2 + 4d^2 + (N-4)e^2 + (N-4)f^2 - 16ab + 8bc + 8ca\right)[\kappa_K^2]$$

$$- \left(2(N-8)ef + 4(N-2)ad + 24cd + 4(N-2)bd\right)[\kappa_I\kappa_J]$$

$$- \left(2(N-8)df + 4(N-2)be + 24ae + 4(N-2)ce\right)[\kappa_J\kappa_K]$$

$$- \left(2(N-8)de + 4(N-2)cf + 24bf + 4(N-2)af\right)[\kappa_K\kappa_I],$$

$$\begin{aligned}
\left(\{\vartheta\}_3[\vartheta], \Omega\right) = {}&2N(N-1)(a^2+b^2+c^2) + 4N(ab+bc+ca) \\
&+ N(N-2)(d^2+e^2+f^2).
\end{aligned}$$

(3) (Quaternionic unitary Virasoro algebras (I): Vertex operators for squares of Kähler forms) *Let $\mathcal{J} = aI + bJ + cK$, $a^2 + b^2 + c^2 = 1$ be a parallel almost complex structure on $M^{4N'}$. Let the vertex operator for the square of the corresponding Kähler form $\kappa_{\mathcal{J}} = a\kappa_I + b\kappa_J + c\kappa_K$ be given by $Y([\kappa_{\mathcal{J}}^2], z) = \sum_{n\in\mathbb{Z}}\{\kappa_{\mathcal{J}}^2\}_{n+1}z^{-n-1}$. Then, these vertex operators satisfy*

$$[\{\kappa_{\mathcal{J}}^2\}_{m+1}\{\kappa_{\mathcal{J}}^2\}_{n+1}] = 4(N-1)(n-m)D(m+n)$$

$$+ 2(N-2)(n-m)\{\kappa_{\mathcal{J}}^2\}_{m+n+1} + \frac{m(m^2-1)}{3}N(N-1)\delta_{m+n,0}\cdot\mathrm{Id},$$

(4) (Quaternionic unitary Virasoro algebras (II)) *For any parallel almost complex structure $\mathcal{J} = aI + bJ + cK$ on $M^{4N'}$ with $a^2 + b^2 + c^2 = 1$, $a, b, c \in \mathbb{R}$, let parallel sections $[\Theta_{\mathcal{J}}], [\Lambda_{\mathcal{J}}] \in \mathcal{P}_M(\underline{V})_2$ be given by*

$$[\Theta_{\mathcal{J}}] = \frac{1}{2N}\left([\kappa_{\mathcal{J}}^2] - [g_{(1,0)}]\right), \qquad [\Lambda_{\mathcal{J}}] = \frac{-1}{2N}\left([\kappa_{\mathcal{J}}^2] + (N-1)[g_{(1,0)}]\right).$$

Then, the corresponding vertex operators $Y([\Theta_{\mathcal{J}}], z) = \sum_{n \in \mathbb{Z}} \Theta_{\mathcal{J}}(n) z^{-n-2}$ *and*
$Y([\Lambda_{\mathcal{J}}], z) = \sum_{n \in \mathbb{Z}} \Lambda_{\mathcal{J}}(n) z^{-n-2}$ *generate commuting Virasoro algebras of rank*
1 and $N - 1$, *respectively, satisfying the following commutation relations:*

$$
\begin{cases}
[\Theta_{\mathcal{J}}(m), \Theta_{\mathcal{J}}(n)] = (n - m)\Theta_{\mathcal{J}}(m + n) + \dfrac{m(m^2 - 1)}{12}\delta_{m+n,0} \cdot \mathrm{Id}, \\[2mm]
[\Lambda_{\mathcal{J}}(m), \Lambda_{\mathcal{J}}(n)] = (n - m)\Lambda_{\mathcal{J}}(m + n) + \dfrac{m(m^2 - 1)}{12}\delta_{m+n,0}(N - 1) \cdot \mathrm{Id}, \\[2mm]
[\Theta_{\mathcal{J}}(m), \Lambda_{\mathcal{J}}(n)] = 0,
\end{cases}
$$

$$
[K_{\mathcal{J}}(m), \Theta_{\mathcal{J}}(n)] = -mK_{\mathcal{J}}(m + n), \qquad [K_{\mathcal{J}}(m), \Lambda_{\mathcal{J}}(n)] = 0.
$$

These operators give a decomposition of Virasoro operators $\{D(n)\}_{n \in \mathbb{Z}}$ *in the sense that* $\Theta_{\mathcal{J}}(n) + \Lambda_{\mathcal{J}}(n) = D(n)$ *for any* $n \in \mathbb{Z}$ *and for any integrable almost complex structure* $\mathcal{J}$.

(5) (Symplectic Virasoro algebra) *Let* $[\sigma], [\tau] \in \mathbb{C}[g_{(1,0)}] \oplus \mathbb{C}[\kappa_{Q-K}] \subset \mathcal{P}_M(\underline{V})_2$ *be given by*

$$
[\sigma] = \frac{1}{2(N + 4)}\left([\kappa_{Q-K}] - 3[g_{(1,0)}]\right), \quad [\tau] = \frac{-1}{2(N + 4)}\left([\kappa_{Q-K}] + (N + 1)[g_{(1,0)}]\right).
$$

Then, vertex operators $Y([\sigma], \zeta) = \sum\limits_{n \in \mathbb{Z}} S(n)\zeta^{-n-2}$, $Y([\tau], \zeta) = \sum\limits_{n \in \mathbb{Z}} T(n)\zeta^{-n-2}$
generate commuting Virasoro algebras of rank $3N/(N + 4)$, $N(N + 1)/(N + 4)$,
respectively. Namely, for any $m, n \in \mathbb{Z}$, *we have* $S(n) + T(N) = D(n)$ *and*

$$
\begin{cases}
[S(m), S(n)] = (n - m)S(m + n) + \dfrac{m(m^2 - 1)}{12}\left\{\dfrac{3N}{(N + 4)}\right\}\delta_{m+n,0} \cdot \mathrm{Id}, \\[3mm]
[T(m), T(n)] = (n - m)T(m + n) + \dfrac{m(m^2 - 1)}{12}\left\{\dfrac{N(N + 1)}{(N + 4)}\right\}\delta_{m+n,0} \cdot \mathrm{Id}, \\[3mm]
[T(m), S(n)] = 0.
\end{cases}
$$

(6) (Commutators between Kähler operators and quadratic Kähler operators)
Vertex operators corresponding to vectors from $\mathbb{C}[\kappa_I] \oplus \mathbb{C}[\kappa_J] \oplus \mathbb{C}[\kappa_K]$ *and from*
$S^2\left(\mathbb{C}[\kappa_I] \oplus \mathbb{C}[\kappa_J] \oplus \mathbb{C}[\kappa_K]\right)$ *have the following commutation relations:*

$$
\begin{aligned}
[K_I(m), \{\kappa_I^2\}_{n+1}] &= -2m(N - 1)K_I(m + n), \\
[K_I(m), \{\kappa_J^2\}_{n+1}] &= 4\{\kappa_J\kappa_K\}_{m+n+1} - 2mK_I(m + n), \\
[K_I(m), \{\kappa_K^2\}_{n+1}] &= -4\{\kappa_J\kappa_K\}_{m+n+1} - 2mK_I(m + n), \\
[K_I(m), \{\kappa_K\kappa_I\}_{n+1}] &= -2\{\kappa_I\kappa_J\}_{m+n+1} - m(N - 2)K_K(m + n), \\
[K_I(m), \{\kappa_I\kappa_J\}_{n+1}] &= 2\{\kappa_K\kappa_I\}_{m+n+1} - m(N - 2)K_J(m + n), \\
[K_I(m), \{\kappa_J\kappa_K\}_{n+1}] &= 2\{\kappa_K^2\}_{m+n+1} - 2\{\kappa_J^2\}_{m+n+1}.
\end{aligned}
$$

Commutators with K_J, K_K *can be obtained by cyclically permuting* (I, J, K) *in the above formula.*

(7) (Some relations among quadratic Kähler operators) *Some of the commutators among quadratic Kähler operators* $\{\kappa_{J_1}\kappa_{J_2}\}_*$, $J_1, J_2 \in \{I, J, K\}$ *have the following relations, where* $m, n \in \mathbb{Z}$:

$$[\{\kappa_I^2\}_{m+1}, \{\kappa_I^2\}_{n+1}] = 4(N-1)(n-m)D(m+n)$$

$$+2(N-2)(n-m)\{\kappa_I^2\}_{m+n+1} + \frac{m(m^2-1)}{3}N(N-1)\delta_{m+n,0} \cdot \text{Id},$$

$$[\{\kappa_I^2 - \kappa_J^2\}_{m+1}, \{\kappa_I^2 - \kappa_J^2\}_{n+1}] = 4[\{\kappa_I\kappa_J\}_{m+1}, \{\kappa_I\kappa_J\}_{n+1}],$$

$$[\{\kappa_I\kappa_J\}_{m+1}, \{\kappa_J^2\}_{n+1}] + [\{\kappa_J^2\}_{m+1}, \{\kappa_I\kappa_J\}_{n+1}]$$
$$= 2(N-2)(n-m)\{\kappa_I\kappa_J\}_{m+n+1},$$

$$[\{\kappa_K^2\}_{m+1}, \{\kappa_I\kappa_J\}_{n+1}] + [\{\kappa_I\kappa_J\}_{m+1}, \{\kappa_K^2\}_{n+1}]$$
$$+2[\{\kappa_J\kappa_K\}_{m+1}, \{\kappa_K\kappa_I\}_{n+1}] + 2[\{\kappa_K\kappa_I\}_{m+1}, \{\kappa_J\kappa_K\}_{n+1}]$$
$$= 2(N-2)(n-m)\{\kappa_I\kappa_J\}_{m+n+1}.$$

Commutators of similar type can be obtained by replacing (I, J, K) *by* (J, K, I), (K, I, J), $(-I, K, J)$, $(K, -J, I)$, *or* $(J, I, -K)$.

(8) (General commutators for quadratic Kähler operators) *For any quadratic Kähler form* $\vartheta_1, \vartheta_2 \in S^2(\mathbb{C}[\kappa_I] \oplus \mathbb{C}[\kappa_K] \oplus \mathbb{C}[\kappa_K])$, *the weight 3 vector* $\{\vartheta_1\}_0[\vartheta_2] \in \mathcal{P}_M(\underline{V})_3$ *belongs to the following vector subspace:*

$$\{\vartheta_1\}_0[\vartheta_2] \in D(-1)(\mathcal{P}_M(\underline{V})_2) + \sum_{J=I,J,K} K_J(-1)\left(\mathcal{P}_M(\underline{V})_{\frac{1}{2}}^{\perp}\right) \subset \mathcal{P}_M(\underline{V})_3$$

Actual commutators among quadratic Kähler operators can be calculated from the following results:

$$\begin{cases} \{\kappa_I^2\}_0[\kappa_I^2] = -2(N-1)D(-1)[g_{(1,0)}] + 2(N-2)D(-1)[\kappa_I^2], \\ \qquad = -\frac{1}{2}D(-1)\left(\{\kappa_I^2\}_1[\kappa_I^2]\right) \\ \{\kappa_I^2\}_1[\kappa_I^2] = 4(N-1)[g_{(1,0)}] - 4(N-2)[\kappa_I^2], \\ \{\kappa_I^2\}_2[\kappa_I^2] = 0, \\ \{\kappa_I^2\}_3[\kappa_I^2] = 2N(N-1)\Omega. \end{cases}$$

$$\begin{cases} \{\kappa_I^2\}_0[\kappa_J^2] = 8K_I(-1)[\kappa_J\kappa_K] + 2D(-1)\left([\kappa_I^2] - [\kappa_J^2] - [g_{(1,0)}]\right), \\ \{\kappa_I^2\}_1[\kappa_J^2] = -4\left([\kappa_I^2] + [\kappa_J^2] - 2[\kappa_K^2] - [g_{(1,0)}]\right), \\ \{\kappa_I^2\}_2[\kappa_J^2] = 0, \\ \{\kappa_I^2\}_3[\kappa_J^2] = 2N\Omega. \end{cases}$$

$$\begin{cases} \{\kappa_I^2\}_0[\kappa_I\kappa_J] = 4K_I(-1)[\kappa_K\kappa_I] - 2(N-2)K_I(-2)[\kappa_J] - 2D(-1)[\kappa_I\kappa_J], \\ \{\kappa_I^2\}_1[\kappa_I\kappa_J] = -2(N-2)[\kappa_I\kappa_J] + 2(N-2)D(-1)[\kappa_K], \\ \{\kappa_I^2\}_2[\kappa_I\kappa_J] = -4(N-2)[\kappa_K], \\ \{\kappa_I^2\}_3[\kappa_I\kappa_J] = 0. \end{cases}$$

$$\begin{cases} \{\kappa_I \kappa_J\}_0[\kappa_I^2] = -4K_I(-1)[\kappa_K \kappa_I] - 2K_I(-2)[\kappa_J] - 2(N-1)K_J(-2)[\kappa_I], \\ \{\kappa_I \kappa_J\}_1[\kappa_I^2] = -2(N-2)[\kappa_I \kappa_J] - 2(N-2)D(-1)[\kappa_K], \\ \{\kappa_I \kappa_J\}_2[\kappa_I^2] = 4(N-2)[\kappa_K], \\ \{\kappa_I \kappa_J\}_3[\kappa_I^2] = 0. \end{cases}$$

$$\begin{cases} \{\kappa_I^2\}_0[\kappa_J \kappa_K] = 4K_I(-1)\left([\kappa_K^2] - [\kappa_J^2]\right) - 2D(-1)[\kappa_J \kappa_K], \\ \{\kappa_I^2\}_1[\kappa_J \kappa_K] = -12[\kappa_J \kappa_K], \\ \{\kappa_I^2\}_2[\kappa_J \kappa_K] = 0, \\ \{\kappa_I^2\}_3[\kappa_J \kappa_K] = 0. \end{cases}$$

$$\begin{cases} \{\kappa_J \kappa_K\}_0[\kappa_I^2] = 4K_J(-1)[\kappa_I \kappa_J] - 4K_K(-1)[\kappa_K \kappa_I] + 2D(-1)[\kappa_J \kappa_K], \\ \{\kappa_J \kappa_K\}_1[\kappa_I^2] = -12[\kappa_J \kappa_K], \\ \{\kappa_J \kappa_K\}_2[\kappa_I^2] = 0, \\ \{\kappa_J \kappa_K\}_3[\kappa_I^2] = 0. \end{cases}$$

$$\begin{cases} \{\kappa_I \kappa_J\}_0[\kappa_I \kappa_J] = \frac{N-4}{2}D(-1)\left([\kappa_I^2] + [\kappa_J^2]\right) + 2[\kappa_K^2] - (N-2)[g_{(1,0)}] \\ \qquad\qquad\quad = -\frac{1}{2}D(-1)\left(\{\kappa_I \kappa_J\}_1[\kappa_I \kappa_J]\right), \\ \{\kappa_I \kappa_J\}_1[\kappa_I \kappa_J] = -(N-4)[\kappa_I^2] - (N-4)[\kappa_J^2] - 4[\kappa_K^2] + 2(N-2)[g_{(1,0)}], \\ \{\kappa_I \kappa_J\}_2[\kappa_I \kappa_J] = 0, \\ \{\kappa_I \kappa_J\}_3[\kappa_I \kappa_J] = N(N-2)\Omega. \end{cases}$$

$$\begin{cases} \{\kappa_I \kappa_J\}_0[\kappa_K \kappa_I] = 2K_I(-1)\left([\kappa_I^2] - [\kappa_K^2]\right) - 2K_J(-1)[\kappa_I \kappa_J] \\ \qquad\qquad\qquad\quad - (N-2)K_J(-2)[\kappa_K], \\ \{\kappa_I \kappa_J\}_1[\kappa_K \kappa_I] = -(N-8)[\kappa_J \kappa_K] - (N-2)D(-1)[\kappa_I], \\ \{\kappa_I \kappa_J\}_2[\kappa_K \kappa_I] = -2(N-2)[\kappa_I], \\ \{\kappa_I \kappa_J\}_3[\kappa_K \kappa_I] = 0. \end{cases}$$

All the other relations of the above type can be obtained by replacing (I, J, K) by (J, K, I), (K, I, J), $(-I, K, J)$, $(K, -J, I)$, or $(J, I, -K)$. The relevant commutator formula for arbitrary vertex operators $\{\vartheta_1\}_*$, $\{\vartheta_2\}_*$ for $[\vartheta_1], [\vartheta_2] \in \mathcal{P}_M(\underline{V})_2$ are given by

$$[\{\vartheta_1\}_{m+1}, \{\vartheta_2\}_{n+1}] = \sum_{0 \le k \in \mathbb{Z}} \binom{m+1}{k} \{\{\vartheta_1\}_k[\vartheta_2]\}_{m+n+2-k}.$$

PROOF. All the operators in Theorem 16 are parallel bundle maps arising from parallel sections of $\underline{V}$. So, we can appeal to either Proposition 4 or Proposition 5 in §5.4, which show that global identities of parallel operators can be verified by evaluating these operators at a single point $x \in M$ of the manifold.

For (1), first note that at $x \in M$ we have $[\kappa_J]_x = [\omega_{J,x}]$ for $J = I, J, K$. The corresponding vertex operator identity is given by $K_J(n)_x = \omega_{J,x}(n)$ for

$\mathcal{J} = I, J, K$. From the general theory of vertex operator super algebras, weight 1 vectors generate an affine Lie algebra with commutation relations given by

$$[x(m), y(n)] = [x, y](m + n) + m\langle x, y\rangle\delta_{m+n,0} \cdot \mathrm{Id}$$

where $x, y \in \mathrm{Cliff}(A)$ are quadratic elements of the form $: r_1 r_2 :$ for $r_1, r_2 \in A$. From Lemma 17 in §3.9, we have

$$[\omega_I, \omega_J] = 2\omega_K, \quad [\omega_J, \omega_K] = 2\omega_I, \quad [\omega_K, \omega_I] = 2\omega_J.$$

From Lemma 8 in §3.11, we have

$$\langle\omega_I, \omega_I\rangle = \langle\omega_J, \omega_J\rangle = \langle\omega_K, \omega_K\rangle = -N,$$
$$\langle\omega_I, \omega_J\rangle = \langle\omega_J, \omega_K\rangle = \langle\omega_K, \omega_I\rangle = 0.$$

These relations calculate commutation relations for $\omega_{\mathcal{J}}(n)$, $n \in \mathbb{Z}$, $\mathcal{J} = I, J, K$, and consequently they calculate commutation relations for the corresponding Kähler operators. Note that the normalized invariant pairing $\langle\ ,\ \rangle_0$ on $\mathfrak{sl}_2(\mathbb{C})$ satisfies $\langle h, h\rangle_0 = 2$, $\langle x, x^*\rangle_0 = -1$. Comparing with formulae in Lemma 8 of §3.11, we see that $\langle\ ,\ \rangle = N'\langle\ ,\ \rangle_0$. Thus the level of our representation of $\mathfrak{sl}_2(\mathbb{C})$ is N'.

All the other formulae follow from corresponding statements in the vertex operator super algebra V. (2) follows from Theorem 24 in §3.11. (3), (7) follow from Theorem 1, Proposition 4, Proposition 6 in §3.10. (4) follows from Theorem 2 in §3.6. (5) is the same as the quaternion-Kähler case. (6) follows from Proposition 26 in §3.11. The calculational result in (8) follows from Proposition 19 in §3.11. For the general statement on $\{\vartheta_1\}_0[\vartheta_2]$, the commutator $[D(-1), K_{\mathcal{J}}(-1)] = -K_{\mathcal{J}}(-2)$ implies that

$$K_{\mathcal{J}}(-2)\,(\mathcal{P}_M(\underline{V})_1) \subset D(-1)\,(\mathcal{P}_M(\underline{V})_2) + \sum_{\mathcal{J}=I,J,K} K_{\mathcal{J}}(-1)\,(D(-1)\mathcal{P}_M(\underline{V})_1).$$

By examining a computational result, we get our result. $\square$

REFERENCES

[A1]   J. F. Adams, *Lectures on Lie Groups*, Benjamin Press, New York, 1969.

[A2]   J. F. Adams, *Stable Homotopy and Generalized Homology*, Univ. Chicago Press, Chicago, Ill., 1974.

[As]   M. Aschbacher, *Homology manifolds with cell structure*, preprint (1995).

[At]   M. F. Atiyah, *Collected Works*, Oxford Univ. Press, 1988.

[AH]   M. F. Atiyah and F. Hirzebruch, *Spin-manifolds and group actions*, Essays on Topology and Related Topics (Mémoires dédiés à Georges de Rham), Springer-Verlag, New York, 1970, pp. 18-26.

[ASe]   M. F. Atiyah, G. B. Segal, *The index of elliptic operators II*, Ann. of Math. **87** (1968), 531–545.

[ASi1] M. F. Atiyah, I. M. Singer, *The index of elliptic operators I*, Ann. of Math. **87** (1968), 484–530; III **87** (1968), 546–604; IV **93** (1971), 119–138; V **93** (1971), 139–149.

[ASi2] M. F. Atiyah, I. M. Singer, *Index theory of skew adjoint Fredholm operators*, Publ. Math. I.H.E.S. **37** (1969), 305–326.

[Ba1]   A. Baker, *Hecke operators as operations in elliptic cohomology*, J. Pure and App. Algebra **63** (1990), 1–11.

[Ba2]   A. Baker, *Elliptic genera of level N and elliptic cohomology*, J. London Math. Soc. **49** (1994), 581–593.

[Ba3]   A. Baker, *Operations and cooperations in elliptic cohomology, Part I: Generalized modular forms and the cooperation algebra*, New York J. of Math. **1** (1995), 39–74.

[Be]   A. Beauville, *Variétés Kähleriennes dont la 1ère class de Chern est nulle*, J. Diff. Geom. **18** (1983), 755–782.

[BGV]   N. Berline, E. Getzler and M. Vergne, *Heat Kernels and Dirac Operators*, Grundlehren der Math. Wissenschaften 298, Springer-Verlag, New York, 1992.

[Bes]   A. L. Besse, *Einstein Manifolds*, Ergebnisse der Math., 3 Folge, Band 10, Springer-Verlag, New York, 1987.

[Bor]   L. D. Borsari, *Bordism of semi-free circle actions on Spin manifolds*, Trans. Amer. Math. Soc. **301** (1987), 479–487.

[BT]   R. Bott and C. Taubes, *On the rigidity theorems of Witten*, J. of AMS **2** (1989), 137-186.

[BD]   T. Bröcker and T. tom Dieck, *Representations of Compact Lie Groups*, Graduate Texts in Math.;98, Springer-Verlag, New York, 1985.

[Br]   J-L. Brylinski, *Representations of loop groups, Dirac operators on loop spaces and modular forms*, Topology **29** (1990), 461-480.

[C]   K. Chandrasekharan, *Elliptic Functions*, Grundlehren 281, Springer-Verlag, 1985.

[CJS]   R. L. Cohen, J. D. S. Jones, G. B. Segal, *Floer's infinite dimensional Morse theory and homotopy theory*, Proceedings of Floer Memorial Conference, 1993.

[CN]   J. H. Conway, S. P. Norton, *Monstrous moonshine*, Bull. London Math. Soc. **11** (1979), 308–339.

[DGM]   L. Dolan, P. Goddard, and P. Montague, *Conformal field theory of twisted vertex operators*, Nucl. Phys. **B338** (1990), 529–60.

[DPR]   R. Dijkgraaf, V. Pasquier, P. Roche, *Quasi quantum groups related to orbifold models*, Modern quantum field theory, (Bombay, 1990), World Scientific, 1991, pp. 375–383.

[DVVV] R. Dijkgraaf, C. Vafa, E. Verlinde, H. Verlinde, *The operator algebra of orbifold models*, Comm. Math. Phys. **123** (1989), 485–526.

[DW] R. Dijkgraaf, E. Witten, *Topological gauge theories and group cohomology*, Comm. Math. Phys. **129** (1990), 393–429.

[DM] C. Dong, G. Mason, *Vertex operator algebras and Moonshine: A survey*, Advanced Studies in Pure Math. 24, Progress in Alg. Combinatorics, 1996, pp. 101–136.

[D] V. Drinfeld, *Quantum groups*, Proc. Int'l. Cong. Math. Berkeley, American Math. Soc., 1986, pp. 798–820.

[EZ] M. Eichler and D. Zagier, *The Theory of Jacobi Forms*, Progress in Math. 55, Birkhäuser, Boston-Basel-Stuttgart, 1985.

[FF] B. L. Feigen and D. B. Fuchs, Functs. Anal. Prilozhen **16** (1982), no. 2, 47–63.

[FFR] A. J. Feingold, I. B. Frenkel, J. F. X. Ries, *Spinor Construction of Vertex Operator Algebras, Triality, and $E_8^{(1)}$*, Contemp. Math., vol. 121, Amer. Math. Soc., Providence, Rhode Island, 1991.

[F] J. Franke, *Onthe construction of elliptic cohomology*, Math. Nachr. **158** (1992), 43–65.

[Fr] I. B. Frenkel, *Two constructions of affine Lie algebra representations*, J. Funct. Anal. **44** (1981), 259–327.

[FHL] I. B. Frenkel, Y. Z. Huang, and J. Lepowsky, *On Axiomatic Approaches to Vertex Operator Algebras and Modules*, Memoirs of AMS; 494, Amer. Math. Soc., Providence, RI, 1993.

[FLM] I. B. Frenkel, J. Lepowsky, A. Meurman, *Vertex Operator Algebras and the Monster*, Pure and Applied Math., vol. 134, Academic Press, Boston, 1988.

[FZ] I. B. Frenkel, Y. Zhu, *Vertex operator algebras associated to representations of affine and Virasoro algebras*, Duke Math. J. **66** (1992), 123–168.

[Ge] R. W. Gebert, *Introduction to vertex algebras, Borcherds algebras, and the Monster Lie algebra*, hep-th/9308151, Int. J. Mod. Phys. **A8** (1993), 5441.

[Go] P. Goddard, *Meromorphic conformal field theory*, Infinite Dimensional Lie Algebras and Lie Groups, Proc. CIRM-Luminy Conf., 1988 (V. Kac, ed.), Advanced Ser. in Math. Phys., Vol 7, World Scientific, Singapore, 1989.

[GH] P. Griffith and J. Harris, *Principles of Algebraic Geometry*, John-Wiley, New York, 1978.

[H1] F. Hirzebruch, *Topological Methods in Algebraic Geometry*, Springer-Verlag, New York, 1956.

[H2] F. Hirzebruch, *Elliptic genera of level N for complex manifolds*, Diff. Geom. Methods in Theoretical Physics, Kluwer Dordrecht, 1988, pp. 37–63.

[HBJ] F. Hirzebruch, T. Berger, R. Jung, *Manifolds and Modular Forms*, Aspects of Mathematics, Vieweg, 1992.

[HS] F. Hirzebruch and P. Slodowy, *Elliptic genera, involutions and homogeneous Spin manifolds*, Geom. Dedicata **35** (1990), 309–343.

[Ho1] M. J. Hopkins, *Characters and elliptic cohomology*, London Math. Soc. Lecture Note Ser., 139, Advances in homotopy theory (Cortona, 1988), Cambridge University Press, Cambridge-New York, 1989, pp. 87–104.

[Ho2] M. Hopkins, *Topological modular forms, the Witten genus, and the theorem of the cube*, Proceedings of ICM (1994), 554–565.

[HH] M. J. Hopkins, M. A. Hovey, *Spin cobordism determines real K-theory*, Math. Z. **210** (1992), 181–196.

[HKR] M. J. Hopkins, N. J. Kuhn, D. C. Ravenel, *Morava K-theories of classifying spaces and generalized characters for finite groups*, Lecture Notes in Math., 1509, Algebraic Topology (San Feliu de Guixos, 1990), Springer, Berlin, 1992, pp. 186–209.

[Hu] J. E. Humphreys, *Introduction to Lie Algebras and Representation Theory*, Graduate Texts in Math.; 9, Springer-Verlag, New York, 1972.

[K1] V. Kac, *Infinite dimensional Lie algebras*, Third Edition, Cambridge Univ. Press, Cambridge, 1990.

[K2]   V. Kac, Proceedings ICM, Helsinki, 1978; Lecture Notes in Phys., vol. 94, 1979, pp. 441.

[KP]   V. G. Kac, D. H. Peterson, *Spin and wedge representations of infinite dimensional Lie algebras and groups*, Proc. Nat'l. Acad. Sci. USA **78** (1981), 3308–3312.

[KR]   V. Kac, A. Raina, *Highest Weight Representations of Infinite Dimensional Lie Algebras*, Advanced Studies in Math. Phys. Vol. 2, World Scientific, Singapore, 1984.

[KN]   S. Kobayashi and K. Nomizu, *Foundations of Differential Geometry I, II*, Interscience, Wiley, New York, 1963, 1969.

[KS]   M. Kreck, S. Stolz, $\mathbb{HP}^2$-*bundle and elliptic homology*, Acta Math. **171** (1993), 231–261.

[Kri]   I. Krichever, *Generalized elliptic genera and Baker-Akhiezer functions*, Math. Notes **47** (1990), 132-142.

[L1]   P. S. Landweber, *Homological properties of comodules over MU_*MU and BP_*BP*, Amer. J. Math. **98** (1976), 591–610.

[L2]   P. S. Landweber (ed.), *Elliptic Curves and Modular Forms in Algebraic Topology*, Lecture Notes in Math. (Proceedings, Princeton 1986), vol. 1326, Springer-Verlag, New York, 1988.

[LRS]   P. S. Landweber, D. C. Ravenel, R. E. Stong, *Periodic cohomology theories defined by eliptic curves*, Contemp. Math., The Čech Centennial (Boston, MA, 1993), Amer. Math. Soc., Providence, RI, 1995, pp. 317–337.

[LS]   P. S. Landweber, R. Stong, *Circle actions on Spin manifolds and characteristic numbers*, Topology **27** (1988), 145–161.

[LM]   H. B. Lawson, Jr., and M. L. Michelsohn, *Spin Geometry*, Princeton Mathematical Series, 38, Princeton Univ. Press, Princeton, 1989.

[Li]   H. Li, *Symmetric invariant bilinear forms on vertex operator algebras*, J. of Pure and Appl. Algebra **96** (1994), 279–297.

[Liz]   A. Lichnerowicz, *Spineurs harmoniques*, C. R. Acad. Sci. Paris, Sér. A-B **257** (1963), 7–9.

[Mc]   D. A. McLaughlin, *Orientation and string structures on loop space*, Pacific J. Math. **155** (1992), 143–156.

[Mi]   H. Miller, *The elliptic character and Witten genus*, Contemp. Math., vol. 96, 1989, pp. 281–289.

[MS]   J. W. Milnor and J. D. Stasheff, *Characteristic Classes*, Annals of Math. Studies; 76, Princeton Univ. Press, Princeton, 1974.

[Mo]   J. Morava, *Forms of K-theory*, Math. Z. **201** (1989), 401–428.

[Mum]   D. Mumford, *Tata Lectures on Theta I, II*, Progress in Math., vol. 28, Birkhäuser, Boston, 1982.

[O1]   S. Ochanine, *Sur les genres multiplicatifs définis par des intégrales elliptiques*, Topology **26** (1987), 143–151.

[O2]   S. Ochanine, *Elliptic genera, modular forms over KO^* and the Brown-Kervaire invariant*, Math. Z. **206** (1991), 277–291.

[PS]   A. Pressley, G. Segal, *Loop Groups*, Oxford Mathematical Monographs, Oxford Univ. Press, New York, 1988.

[R]   D. C. Ravenel, *Complex Cobordism and Stable Homotopy Groups of Spheres*, Academic Press, New York, 1986.

[S1]   G. B. Segal, *On the definition of conformal field theory*, preprint (1987).

[S2]   G. B. Segal, *Elliptic cohomology*, Seminair Bourbaki 1987-1988, Astérisque 161-162, 1988, pp. 191-202.

[S3]   G. B. Segal, *Two-dimensional conformal field theories and modular functors*, IXth International Congress on Math. Phys., Swansea, B. Simon, A. Truman, I. M. Davies (eds.), Adam-Hilger, Bristol and New York, 1988, pp. 22–37.

[S3]   G. B. Segal, *Geometric aspects of quantum field theory*, Proceedings, ICM, Kyoto (1990), 1387-1396.

[St]   S. Stolz, *A conjecture concerning positive Ricci curvature and the Witten genus*, Math. Ann. **304** (1996), 785–800.

[T1]   H. Tamanoi, *(Hyper)elliptic genera*, Thesis, The Johns Hopkins University, 141 pages, 1988.

[T2]   H. Tamanoi, *Elliptic genera and vertex operator super algebras*, Ser. A, Proc. Japan Acad. **71** (1995), 177–181.

[T3]   H. Tamanoi, *Symmetric invariant pairings in vertex operator super algebras and Gramians*, to appear in J. of Pure and Appl. Alg. (1997).

[T4]   H. Tamanoi, *Conformally Invariant Generalized Hypergeometric Differential Forms*: $\mathfrak{sl}_2$ *Conformal Field Theory on* $\mathbb{CP}^1$, in progress, 248 pages.

[T5]   H. Tamanoi, *Witten genus and vertex operator algebras*, in preparation.

[Tau]   C. H. Taubes, S^1 *Actions and elliptic genera*, Commun. Math. Phys. **122** (1989), 455–526.

[Wa]   F. W. Warner, *Foundations of Differential Manifolds and Lie Groups*, Graduate Texts in Math.;94, Springer-Verlag, New York, 1983.

[We]   R. O. Wells, *Differential Analysis on Complex Manifolds*, Graduate Texts in Math., vol. 65, Springer-Verlag, New York, 1980.

[Wey]   H. Weyl, *Symmetry*, Princeton Univ. Press, Princeton, 1952.

[WW]   E. T. Whittaker and G. N. Watson, *A Course of Modern Analysis*, fourth edition, Cambridge Univ. Press, 1927.

[W1]   E. Witten, *Supersymmetry and Morse theory*, J. Diff. Geometry **17** (1982), 661–692.

[W2]   E. Witten, *Fermion quantum numbers in Kaluza-Klein theory*, Shelter Island, II, Proc. of the 1983 Shelter Island Conference on Quantum Field Theory and the Fundamental Problems of Physics (ed. R. Jackiw, N. Khuri, S. Weinberg, E. Witten), MIT press, 1985, pp. 227–277.

[W3]   E. Witten, *The index of the Dirac operator in loop space*, Lecture Notes in Math. (Proceedings, Princeton 1986, P. S. Landweber, ed.), vol. 1326, Springer-Verlag, New York, 1988, pp. 161–181.

[W4]   E. Witten, *Topological quantum field theory*, Comm. Math. Phys. **117** (1988), 353–386.

[Y]   S. T. Yau, *On the Ricci curvature of a compact Kähler manifold and the complex Monge-Ampère equation* I, Com. Pure and Appl. Math. **31** (1978), 339–411.

[Z]   D. Zagier, *Notes on Landweber-Stong elliptic genus*, Lecture Notes in Math. (Proceedings, Princeton 1986, P. S. Landweber, ed.), vol. 1326, Springer-Verlag, New York, 1988, pp. 216–224.

[Zh]   Y. Zhu, *Modular invariance of characters of vertex operator algebras*, J. of AMS **9** (1996), 237–302.

Index of Notation

Subject Index

Lecture Notes in Mathematics

For information about Vols. 1–1510
please contact your bookseller or Springer-Verlag

Vol. 1552: J. Hilgert, K.-H. Neeb, Lie Semigroups and their Applications. XII, 315 pages. 1993.

Vol. 1553: J.-L- Colliot-Thélène, J. Kato, P. Vojta. Arithmetic Algebraic Geometry. Trento, 1991. Editor: E. Ballico. VII, 223 pages. 1993.

Vol. 1554: A. K. Lenstra, H. W. Lenstra, Jr. (Eds.), The Development of the Number Field Sieve. VIII, 131 pages. 1993.

Vol. 1555: O. Liess, Conical Refraction and Higher Microlocalization. X, 389 pages. 1993.

Vol. 1556: S. B. Kuksin, Nearly Integrable Infinite-Dimensional Hamiltonian Systems. XXVII, 101 pages. 1993.

Vol. 1557: J. Azéma, P. A. Meyer, M. Yor (Eds.), Séminaire de Probabilités XXVII. VI, 327 pages. 1993.

Vol. 1558: T. J. Bridges, J. E. Furter, Singularity Theory and Equivariant Symplectic Maps. VI, 226 pages. 1993.

Vol. 1559: V. G. Sprindžuk, Classical Diophantine Equations. XII, 228 pages. 1993.

Vol. 1560: T. Bartsch, Topological Methods for Variational Problems with Symmetries. X, 152 pages. 1993.

Vol. 1561: I. S. Molchanov, Limit Theorems for Unions of Random Closed Sets. X, 157 pages. 1993.

Vol. 1562: G. Harder, Eisensteinkohomologie und die Konstruktion gemischter Motive. XX, 184 pages. 1993.

Vol. 1563: E. Fabes, M. Fukushima, L. Gross, C. Kenig, M. Röckner, D. W. Stroock, Dirichlet Forms. Varenna, 1992. Editors: G. Dell'Antonio, U. Mosco. VII, 245 pages. 1993.

Vol. 1564: J. Jorgenson, S. Lang, Basic Analysis of Regularized Series and Products. IX, 122 pages. 1993.

Vol. 1565: L. Boutet de Monvel, C. De Concini, C. Procesi, P. Schapira, M. Vergne. D-modules, Representation Theory, and Quantum Groups. Venezia, 1992. Editors: G. Zampieri, A. D'Agnolo. VII, 217 pages. 1993.

Vol. 1566: B. Edixhoven, J.-H. Evertse (Eds.), Diophantine Approximation and Abelian Varieties. XIII, 127 pages. 1993.

Vol. 1567: R. L. Dobrushin, S. Kusuoka, Statistical Mechanics and Fractals. VII, 98 pages. 1993.

Vol. 1568: F. Weisz, Martingale Hardy Spaces and their Application in Fourier Analysis. VIII, 217 pages. 1994.

Vol. 1569: V. Totik, Weighted Approximation with Varying Weight. VI, 117 pages. 1994.

Vol. 1570: R. deLaubenfels, Existence Families, Functional Calculi and Evolution Equations. XV, 234 pages. 1994.

Vol. 1571: S. Yu. Pilyugin, The Space of Dynamical Systems with the C^0-Topology. X, 188 pages. 1994.

Vol. 1572: L. Göttsche, Hilbert Schemes of Zero-Dimensional Subschemes of Smooth Varieties. IX, 196 pages. 1994.

Vol. 1573: V. P. Havin, N. K. Nikolski (Eds.), Linear and Complex Analysis – Problem Book 3 – Part I. XXII, 489 pages. 1994.

Vol. 1574: V. P. Havin, N. K. Nikolski (Eds.), Linear and Complex Analysis – Problem Book 3 – Part II. XXII, 507 pages. 1994.

Vol. 1575: M. Mitrea, Clifford Wavelets, Singular Integrals, and Hardy Spaces. XI, 116 pages. 1994.

Vol. 1576: K. Kitahara, Spaces of Approximating Functions with Haar-Like Conditions. X, 110 pages. 1994.

Vol. 1577: N. Obata, White Noise Calculus and Fock Space. X, 183 pages. 1994.

Vol. 1578: J. Bernstein, V. Lunts, Equivariant Sheaves and Functors. V, 139 pages. 1994.

Vol. 1579: N. Kazamaki, Continuous Exponential Martingales and BMO. VII, 91 pages. 1994.

Vol. 1580: M. Milman, Extrapolation and Optimal Decompositions with Applications to Analysis. XI, 161 pages. 1994.

Vol. 1581: D. Bakry, R. D. Gill, S. A. Molchanov, Lectures on Probability Theory. Editor: P. Bernard. VIII, 420 pages. 1994.

Vol. 1582: W. Balser, From Divergent Power Series to Analytic Functions. X, 108 pages. 1994.

Vol. 1583: J. Azéma, P. A. Meyer, M. Yor (Eds.), Séminaire de Probabilités XXVIII. VI, 334 pages. 1994.

Vol. 1584: M. Brokate, N. Kenmochi, I. Müller, J. F. Rodriguez, C. Verdi, Phase Transitions and Hysteresis. Montecatini Terme, 1993. Editor: A. Visintin. VII. 291 pages. 1994.

Vol. 1585: G. Frey (Ed.), On Artin's Conjecture for Odd 2-dimensional Representations. VIII, 148 pages. 1994.

Vol. 1586: R. Nillsen, Difference Spaces and Invariant Linear Forms. XII, 186 pages. 1994.

Vol. 1587: N. Xi, Representations of Affine Hecke Algebras. VIII, 137 pages. 1994.

Vol. 1588: C. Scheiderer, Real and Étale Cohomology. XXIV, 273 pages. 1994.

Vol. 1589: J. Bellissard, M. Degli Esposti, G. Forni, S. Graffi, S. Isola, J. N. Mather, Transition to Chaos in Classical and Quantum Mechanics. Montecatini Terme, 1991. Editor: 2S. Graffi. VII, 192 pages. 1994.

Vol. 1590: P. M. Soardi, Potential Theory on Infinite Networks. VIII, 187 pages. 1994.

Vol. 1591: M. Abate, G. Patrizio, Finsler Metrics – A Global Approach. IX, 180 pages. 1994.

Vol. 1592: K. W. Breitung, Asymptotic Approximations for Probability Integrals. IX, 146 pages. 1994.

Vol. 1593: J. Jorgenson & S. Lang, D. Goldfeld, Explicit Formulas for Regularized Products and Series. VIII, 154 pages. 1994.

Vol. 1594: M. Green, J. Murre, C. Voisin, Algebraic Cycles and Hodge Theory. Torino, 1993. Editors: A. Albano, F. Bardelli. VII, 275 pages. 1994.

Vol. 1595: R.D.M. Accola, Topics in the Theory of Riemann Surfaces. IX, 105 pages. 1994.

Vol. 1596: L. Heindorf, L. B. Shapiro, Nearly Projective Boolean Algebras. X, 202 pages. 1994.

Vol. 1597: B. Herzog, Kodaira-Spencer Maps in Local Algebra. XVII, 176 pages. 1994.

Vol. 1598: J. Berndt, F. Tricerri, L. Vanhecke, Generalized Heisenberg Groups and Damek-Ricci Harmonic Spaces. VIII, 125 pages. 1995.

Vol. 1599: K. Johannson, Topology and Combinatorics of 3-Manifolds. XVIII, 446 pages. 1995.

Vol. 1600: W. Narkiewicz, Polynomial Mappings. VII, 130 pages. 1995.

Vol. 1601: A. Pott, Finite Geometry and Character Theory. VII, 181 pages. 1995.

Vol. 1602: J. Winkelmann, The Classification of Three-dimensional Homogeneous Complex Manifolds. XI, 230 pages. 1995.

Vol. 1603: V. Ene, Real Functions – Current Topics. XIII, 310 pages. 1995.

Vol. 1604: A. Huber, Mixed Motives and their Realization in Derived Categories. XV, 207 pages. 1995.

Vol. 1605: L. B. Wahlbin, Superconvergence in Galerkin Finite Element Methods. XI, 166 pages. 1995.

Vol. 1606: P.-D. Liu, M. Qian, Smooth Ergodic Theory of Random Dynamical Systems. XI, 221 pages. 1995.

Vol. 1607: G. Schwarz, Hodge Decomposition – A Method for Solving Boundary Value Problems. VII, 155 pages. 1995.

Vol. 1608: P. Biane, R. Durrett, Lectures on Probability Theory. Editor: P. Bernard. VII, 210 pages. 1995.

Vol. 1609: L. Arnold, C. Jones, K. Mischaikow, G. Raugel, Dynamical Systems. Montecatini Terme, 1994. Editor: R. Johnson. VIII, 329 pages. 1995.

Vol. 1610: A. S. Üstünel, An Introduction to Analysis on Wiener Space. X, 95 pages. 1995.

Vol. 1611: N. Knarr, Translation Planes. VI, 112 pages. 1995.

Vol. 1612: W. Kühnel, Tight Polyhedral Submanifolds and Tight Triangulations. VII, 122 pages. 1995.

Vol. 1613: J. Azéma, M. Emery, P. A. Meyer, M. Yor (Eds.), Séminaire de Probabilités XXIX. VI, 326 pages. 1995.

Vol. 1614: A. Koshelev, Regularity Problem for Quasilinear Elliptic and Parabolic Systems. XXI, 255 pages. 1995.

Vol. 1615: D. B. Massey, Le Cycles and Hypersurface Singularities. XI, 131 pages. 1995.

Vol. 1616: I. Moerdijk, Classifying Spaces and Classifying Topoi. VII, 94 pages. 1995.

Vol. 1617: V. Yurinsky, Sums and Gaussian Vectors. XI, 305 pages. 1995.

Vol. 1618: G. Pisier, Similarity Problems and Completely Bounded Maps. VII, 156 pages. 1996.

Vol. 1619: E. Landvogt, A Compactification of the Bruhat-Tits Building. VII, 152 pages. 1996.

Vol. 1620: R. Donagi, B. Dubrovin, E. Frenkel, E. Previato, Integrable Systems and Quantum Groups. Montecatini Terme, 1993. Editors:M. Francaviglia, S. Greco. VIII, 488 pages. 1996.

Vol. 1621: H. Bass, M. V. Otero-Espinar, D. N. Rockmore, C. P. L. Tresser, Cyclic Renormalization and Auto-morphism Groups of Rooted Trees. XXI, 136 pages. 1996.

Vol. 1622: E. D. Farjoun, Cellular Spaces, Null Spaces and Homotopy Localization. XIV, 199 pages. 1996.

Vol. 1623: H.P. Yap, Total Colourings of Graphs. VIII, 131 pages. 1996.

Vol. 1624: V. Brınzanescu, Holomorphic Vector Bundles over Compact Complex Surfaces. X, 170 pages. 1996.

Vol.1625: S. Lang, Topics in Cohomology of Groups. VII, 226 pages. 1996.

Vol. 1626: J. Azéma, M. Emery, M. Yor (Eds.), Séminaire de Probabilités XXX. VIII, 382 pages. 1996.

Vol. 1627: C. Graham, Th. G. Kurtz, S. Méléard, Ph. E. Protter, M. Pulvirenti, D. Talay, Probabilistic Models for Nonlinear Partial Differential Equations. Montecatini Terme, 1995. Editors: D. Talay, L. Tubaro. X, 301 pages. 1996.

Vol. 1628: P.-H. Zieschang, An Algebraic Approach to Association Schemes. XII, 189 pages. 1996.

Vol. 1629: J. D. Moore, Lectures on Seiberg-Witten Invariants. VII, 105 pages. 1996.

Vol. 1630: D. Neuenschwander, Probabilities on the Heisenberg Group: Limit Theorems and Brownian Motion. VIII, 139 pages. 1996.

Vol. 1631: K. Nishioka, Mahler Functions and Transcendence.VIII, 185 pages.1996.

Vol. 1632: A. Kushkuley, Z. Balanov, Geometric Methods in Degree Theory for Equivariant Maps. VII, 136 pages. 1996.

Vol.1633: H. Aikawa, M. Essén, Potential Theory – Selected Topics. IX, 200 pages.1996.

Vol. 1634: J. Xu, Flat Covers of Modules. IX, 161 pages. 1996.

Vol. 1635: E. Hebey, Sobolev Spaces on Riemannian Manifolds. X, 116 pages. 1996.

Vol. 1636: M. A. Marshall, Spaces of Orderings and Abstract Real Spectra. VI, 190 pages. 1996.

Vol. 1637: B. Hunt, The Geometry of some special Arithmetic Quotients. XIII, 332 pages. 1996.

Vol. 1638: P. Vanhaecke, Integrable Systems in the realm of Algebraic Geometry. VIII, 218 pages. 1996.

Vol. 1639: K. Dekimpe, Almost-Bieberbach Groups: Affine and Polynomial Structures. X, 259 pages. 1996.

Vol. 1640: G. Boillat, C. M. Dafermos, P. D. Lax, T. P. Liu, Recent Mathematical Methods in Nonlinear Wave Propagation. Montecatini Terme, 1994. Editor: T. Ruggeri. VII, 142 pages. 1996.

Vol. 1641: P. Abramenko, Twin Buildings and Applications to S-Arithmetic Groups. IX, 123 pages. 1996.

Vol. 1642: M. Puschnigg, Asymptotic Cyclic Cohomology. XXII, 138 pages. 1996.

Vol. 1643: J. Richter-Gebert, Realization Spaces of Polytopes. XI, 187 pages. 1996.

Vol. 1644: A. Adler, S. Ramanan, Moduli of Abelian Varieties. VI, 196 pages. 1996.

Vol. 1645: H. W. Broer, G. B. Huitema, M. B. Sevryuk, Quasi-Periodic Motions in Families of Dynamical Systems. XI, 195 pages. 1996.

Vol. 1646: J.-P. Demailly, T. Peternell, G. Tian, A. N. Tyurin, Transcendental Methods in Algebraic Geometry. Cetraro, 1994. Editors: F. Catanese, C. Ciliberto. VII, 257 pages. 1996.

Vol. 1647: D. Dias, P. Le Barz, Configuration Spaces over Hilbert Schemes and Applications. VII. 143 pages. 1996.

Vol. 1648: R. Dobrushin, P. Groeneboom, M. Ledoux, Lectures on Probability Theory and Statistics. Editor: P. Bernard. VIII, 300 pages. 1996.

Vol. 1649: S. Kumar, G. Laumon, U. Stuhler, Vector Bundles on Curves – New Directions. Cetraro, 1995. Editor: M. S. Narasimhan. VII, 193 pages. 1997.

Vol. 1650: J. Wildeshaus, Realizations of Polylogarithms. XI, 343 pages. 1997.

Vol. 1651: M. Drmota, R. F. Tichy, Sequences, Discrepancies and Applications. XIII, 503 pages. 1997.

Vol. 1652: S. Todorcevic, Topics in Topology. VIII, 153 pages. 1997.

4. Lecture Notes are printed by photo-offset from the master-copy delivered in camera-ready form by the authors. Springer-Verlag provides technical instructions for the preparation of manuscripts. Macro packages in T_EX, L^AT_EX2e, $L^AT_EX2.09$ are available from Springer's web-pages at http://www.springer.de/math/authors. Careful preparation of the manuscripts will help keep production time short and ensure satisfactory appearance of the finished book.

The actual production of a Lecture Notes volume takes approximately 12 weeks.

5. Authors receive a total of 50 free copies of their volume, but no royalties. They are entitled to a discount of 33.3% on the price of Springer books purchase for their personal use, if ordering directly from Springer-Verlag.

Commitment to publish is made by letter of intent rather than by signing a formal contract. Springer-Verlag secures the copyright for each volume. Authors are free to reuse material contained in their LNM volumes in later publications: A brief written (or e-mail) request for formal permission is sufficient.

Addresses:

Professor F. Takens, Mathematisch Instituut,
Rijksuniversiteit Groningen, Postbus 800,
9700 AV Groningen, The Netherlands
E-mail: F.Takens@math.rug.nl

Professor B. Teissier, DMI, École Normale Supérieure
45, rue d'Ulm,
F-7500 Paris, France
E-mail: Teissier@ens.fr

Springer-Verlag, Mathematics Editorial, Tiergartenstr. 17,
D-69121 Heidelberg, Germany,
Tel.: *49 (6221) 487-701
Fax: *49 (6221) 487-355
E-mail: C.Byrne@Springer.de